Waves and Oscillations in Plasmas

T0250242

Series in Plasma Physics

Waves and Oscillations in Plasmas, 2nd Edition

Hans L. Pécseli

High Power Microwaves, 3rd Edition

James Benford, John A. Swegle, Edl Schamiloglu

Waves and Oscillations in Plasmas

Hans L. Pécseli

Complex and Dusty Plasmas: From Laboratory to Space

Vladimir E. Fortov, Gregor E. Morfill

Plasma Kinetic Theory

Donald Gary Swanson

Aspects of Anomalous Transport in Plasmas

Radu Balescu

Magnetohydrodynamic Waves in Geospace: The Theory of ULF Waves and their Interaction with Energetic Particles in the Solar-Terrestrial Environment

A.D.M. Walker

Plasma Physics via Computer Simulation

C.K. Birdsall, A.B Langdon

Plasma Waves

Donald Gary Swanson

Waves and Oscillations in Plasmas

Second Edition

Hans L. Pécseli
University of Oslo
Norway

CRC Press
Taylor & Francis Group
Boca Raton London New York

CRC Press is an imprint of the
Taylor & Francis Group, an **informa** business

The cover photo shows the Q-machine at the Risø National Laboratory in Denmark. The device was operating from 1967 until 1992.

Second edition published 2020
by CRC Press
2 Park Square, Milton Park, Abingdon, Oxon, OX14 4RN

and by CRC Press
6000 Broken Sound Parkway NW, Suite 300, Boca Raton, FL 33487-2742

© 2020 Hans L. Pécseli
CRC Press is an imprint of Taylor & Francis Group, an Informa business

First edition published by CRC Press 2012

The right of Hans L. Pécseli to be identified as author of this work has been asserted by him in accordance with sections 77 and 78 of the Copyright, Designs and Patents Act 1988.

This book contains information obtained from authentic and highly regarded sources. Reasonable efforts have been made to publish reliable data and information, but the author and publisher cannot assume responsibility for the validity of all materials or the consequences of their use. The authors and publishers have attempted to trace the copyright holders of all material reproduced in this publication and apologize to copyright holders if permission to publish in this form has not been obtained. If any copyright material has not been acknowledged please write and let us know so we may rectify in any future reprint.

Except as permitted under U.S. Copyright Law, no part of this book may be reprinted, reproduced, transmitted, or utilized in any form by any electronic, mechanical, or other means, now known or hereafter invented, including photocopying, microfilming, and recording, or in any information storage or retrieval system, without written permission from the publishers.

For permission to photocopy or use material electronically from this work, access www.copyright.com or contact the Copyright Clearance Center, Inc. (CCC), 222 Rosewood Drive, Danvers, MA 01923, 978-750-8400. For works that are not available on CCC please contact mpkbookspermissions@tandf.co.uk

Trademark Notice: Product or corporate names may be trademarks or registered trademarks, and are used only for identification and explanation without intent to infringe.

Publisher's Note
The publisher has gone to great lengths to ensure the quality of this reprint but points out that some imperfections in the original copies may be apparent.

British Library Cataloguing-in-Publication Data
A catalogue record for this book is available from the British Library

Library of Congress Cataloging-in-Publication Data
Names: Pécseli, Hans L., 1947- author.
Title: Waves and oscillations in plasmas / Hans L. Pécseli.
Description: Second edition. | Boca Raton : CRC Press, [2020] |
Series: Series in plasma physics | Includes bibliographical references and index. |
Identifiers: LCCN 2019058642 | ISBN 9781138591295 (hardback) | ISBN 9780429489976 (ebook)
Subjects: LCSH: Plasma dynamics. | Plasma oscillations. | Plasma waves.
Classification: LCC QC718.5.D9 P43 2020 | DDC 530.4/4--dc23
LC record available at https://lccn.loc.gov/2019058642

ISBN 13: 978-1-03-223642-1 (pbk)
ISBN 13: 978-1-138-59129-5 (hbk)

DOI: 10.1201/9780429489976

Visit the eResources: https://www.crcpress.com/Waves-and-Oscillations-in-Plasmas/Pecseli/p/book/9781032236421

Contents

Preface to Second revised edition

These notes were prepared for lectures in plasma physics at the University of Tromsø and at the Department of Physics at the University of Oslo. The author would like to thank students and colleagues for their assistance. Liv Larssen from the Auroral Observatory in Tromsø helped, as so often before, in typing and editing parts of the manuscript and in particular preparing many of the figures. Her patience and help are gratefully acknowledged. Bjørn Lybekk was also very helpful with figures and solving a number of technical problems during the preparation of the text. The author is indebted to Prof. Kristian Dysthe, University of Bergen, Prof. Karl Lonngren, University of Iowa, Prof. Einar Mjølhus, University of Tromsø, Prof. Lennart Stenflo, University of Umeå and Prof. Mitsuo Kono, Chuo University, for valuable discussion on the chapters discussing nonlinear waves in fluids and plasmas. Prof. Jan Trulsen, Institute for Theoretical Astrophysics, University of Oslo, has offered help and given advice at all stages of the work. His friendly support have been indispensable. Many discussions with colleagues at the Risø National Laboratory as well as the University of Tromsø have been of great value. Comments from students over the years have helped to clarify many points in the text. I am very grateful to Prof. Wojciech Miloch, who patiently and carefully read large parts of the text and offered many useful comments and valuable criticism. Solveig Kjus read the part on drift waves and had many constructive comments. Prof. Tünde Fülöp's initiative contributed to the publication of this book, and I would like to thank her too.

The present notes do not require any particular previous knowledge in plasma sciences. Basic mathematics courses, i.e., introductions to, for instance, Fourier transforms, etc., will be a great advantage, though. The notation is standard in most places. By the letter i we usually have the standard interpretation, $i=\sqrt{-1}$. We let $\underline{1}$ denote the unit tensor. Unfortunately, standard notations can sometimes be confusing; at places we have σ denoting a cross section for a scattering process, and sometimes it is a conductivity, for instance. The correct interpretation will in all cases be evident from the context. Energy densities are as a rule denoted by W, energies by $\mathcal{W}$. Forces, for instance on a particle, are denoted $\mathbf{F}$, force densities by $\mathbf{f}$. Unit vectors are written as, e.g., $\widehat{\mathbf{x}}$. The symbols $\Re\{\}$ and $\Im\{\}$ denote the real and imaginary parts of the complex quantities appearing in the parentheses $\{\}$. In principle the space-time variables of all quantities are written explicitly, i.e., as $\phi(\mathbf{r},t)$ for the electrostatic potential, etc. At times, however, the expressions would tend to become excessively long, if this rule were to be followed strictly. In such cases $(\mathbf{r},t)$ is omitted. Care has been taken to avoid confusing notations, but even when the Greek alphabet is included, the alphabets contain too few letters to allow each one being assigned one particular physical quantity only. With a little imagination, no problems should arise, though!

The text contains a number of "Examples" and "Exercises". It is attempted to place these at, or at least in the immediate vicinity of, the discussion they are supposed to illustrate, rather than at the end of the relevant chapters. Several examples are to be seen as "Solved exercises". A number of the exercises are a little lengthy, and the solutions to most of these, marked by Ⓢ, are given separately. A few appendices cover special topics which would have taken up too much space in the text. Any textbook on plasma physics will contain very frequent references to basic electromagnetism. The present author concluded that one might as well simply include a chapter on this topic, written in a way that makes the relevant references at hand.

In many ways the book may seem to be mathematically oriented. This was not really the intention, but this is the way it turned out. It would be great if we could discuss physics in other terms, but experience (and maybe also the education we are brought up with) has demonstrated that this is a very effective way of presentation. It might be appropriate to quote Eugene Wigner here: "*The miracle of appropriateness of the language of mathematics for the laws of physics is a wonderful gift which we neither understand nor deserve. We should be grateful for it and hope that it will remain in future research and that it will extend, for better or worse, to our pleasure, even though to our bafflement, to wide branches of learning*" (Tisza 2003). He did not have plasma physics in mind, but his comments are also appropriate here. On the other hand, some words of warning may be appropriate: it is so easy to be carried away in a theoretical analysis, making some seemingly innocent approximations and assumptions, only to discover that the final results have little to do with reality. A most gloomy view was formulated by Jean Beaudrillard (1929-2007): "*Theory can be no more than this: a trap set in the hope that reality will be naïve enough to fall into it!*"[1] The author may not be quite as pessimistic, but a solid basis in experimental observations and tests of analytical results seem appropriate. Relevant sections are included in the text at appropriate chapters.

To limit the exposition, a number of topics are left out. Relativistic plasmas are not discussed, for instance. The Clemmov-Mullaly-Allis (CMA) diagram is discussed only briefly. Its main virtue is a classification of electromagnetic waves in cold plasmas. Those particularly interested in this topic should consult specialized references, where a few are provided in the appropriate chapters. In the author's personal opinion, the most significant omission concerns kinetic models for *magnetized* plasmas. Plasma media, magnetized plasmas in particular, are rich in instabilities. The book describes a few important ones, but the reader is urged to search the literature for more general expositions, such as Mikhailovkii (1974*a*), Mikhailovkii (1974*b*), Hasegawa (1975) and Cap (1976). A highly commendable, but unfortunately unpublished summary (available as a report), was written by Prof. Lehnert (1972), where the development of a classification scheme was attempted.

The present book contains material for an introductory as well as a more advanced course. The introductory course can consist of 2 or 3 lectures per week, in addition to one lecture for exercises and one for "colloquia", where one or more students can try to give a lecture on a selected, self contained, problem for the other students, not necessarily with the supervisor present (although in our experience, it works best with some guidance). Relevant suggestions for these colloquia are marked with Ⓚ. It could be argued that a detailed discussion of, for instance, the operation of a Q-machine is not necessary for the exposition as such, but on the other hand that section represents an example for a "typical" problem being suitable for a colloquia, and serves also as an introduction for the understanding of basic plasma conditions in this as well as other laboratory devices.

Plasma physics has come a long way in a relatively short time. When the present author was a student in the very end of the 1960s-beginning of the 1970s, our lecturer in plasma physics started out by telling us that a plasma physicist wrote up the Vlasov equation on a piece of paper in the morning, stared at it for a full day, and then went home late at night. After enjoying our bewilderment for a moment, he then comforted us by telling that we would be much better off than, for instance, high energy physicists, who started out with a blank piece of paper, and then stared at it for a full day before leaving for home! Today, plasma science is an integrated part of the curriculum at all major universities, with applications in basic sciences as well as industry.

A number of misprints in the First version of the book are corrected. The number of exercises is increased and the solution manual for the book significantly expanded. To keep the original length of the book, some topics and subsections have been deleted, while new ones have been added together with new illustrations. Also the list of references is significantly expanded,

The book is dedicated to my family.

[1] Conan Doyle lets Sherlock Homes express a somewhat similar view in *Scandal in Bohemia*: "It is a capital mistake to theorise before one has data".

1

Introduction

1.1 What is a plasma?

Plasma physics deals, in its most general form, with studies of the dynamics of charged particles. In principle this includes the motions of single charged particles in a priori given electric and magnetic fields, but the most interesting problems are concerned with collective interactions between many charged particles. In these cases it is the charge and current distributions resulting from all the particles which ultimately set up the electric and magnetic fields, which together with externally imposed fields determine the self-consistent motion of the plasma particles themselves.

1.2 Where do we find plasma?

When the author was a PhD student, we proudly stated that more than 99% of matter in the known universe was supposed to be in the plasma state, and therefore this particular state of matter should actually be the most interesting one! Today we have to be somewhat more modest, stating that 99% of *visible* matter is in the plasma state, but after all, this is still quite something! Even this percentage may appear surprisingly large, but one should bear in mind that, for instance, in our own solar system, the Sun represents by far the largest accumulation of matter, and because of the high temperatures there, the material in the Sun is almost completely ionized.

More recently, it has been realized that studies of the plasma state include topics which have important industrial applications (Lieberman & Lichtenberg 1994, Roth 1995). Discharges become increasingly important for ion production. To increase the reaction rate of chemical processes it is generally advantageous to increase the temperature of the mixture, and as a consequence, a significant part of the constituents can become ionized. Charged dust has been found to be a serious problem in some industrial plants, and simultaneously it was discovered that this particular state also has an important role in a number of natural phenomena as well. Charged dust particles will in many ways behave as super massive ions with positive or negative charge depending on conditions (Shukla & Mamun 2002*a*, Piel 2010), and this should be easy enough to incorporate in analytical models. By closer inspection it is, however, found that the charges are fluctuating in time, and the average charge on a dust particle will in general depend on its velocity as well as other dynamic conditions. These and related phenomena are being intensely studied, but the field is still "young" and it will not be covered in the present book.

1.3 Plasma physics – why bother?

The dynamics of hot plasmas are in many respects similar to those of neutral gases, and in many cases it might be argued that no particular new physical insight is gained by including elements from plasma physics in the analysis. Plasma sciences as an individual discipline cannot be justified by such examples. There are, however, a number of phenomena which are specific and unique for plasmas. The most notable is magneto-hydrodynamics (MHD), with Alfvén waves being an illustrative example. Also the stability properties of hot dilute plasmas as described by kinetic plasma theory are uniquely plasma related. These examples are not just of academic interest: MHD, for instance, is of central importance for our description and understanding of large scale phenomena in the magnetospheres of planets and stars, such as the Sun in our solar system.

While MHD can be considered as a sort of effective "fluid-like" model for describing the dynamics of a plasma, there is another, in a certain sense complementary, so called "kinetic" description of what is called "collisionless" plasmas. In many ways this description has very surprising implications. That analysis attempts to describe the space-time evolution of the velocity distributions of the particles constituting the plasma. It turns out that a new form of wave damping is discovered, collisionless Landau damping, which is present even though the basic dynamic equations are fully time reversible, in complete variance with experience from electrical circuits, for instance.

One major stimulus for the development of plasma sciences in the time following the mid-1950s can be identified as its importance for controlled fusion (Wilhelmsson 2000, Braams & Stott 2002). An interesting (and beautifully illustrated) account of the early U.S. fusion program is given by Bishop (1958). It seems plausible that the energy demands of mankind can be satisfied for all the foreseeable future if we succeed in harnessing fusion processes. Several concepts have been suggested, and practically all of them involve confinement of matter at temperatures of millions of degrees, a state where we can safely assume all particles to be fully ionized, and constituting what we call a plasma.

Fusion plasma physics will not have any pronounced place in the present treatise, but a few illustrative examples in the form of exercises may be in order.

Ⓢ **Exercise:** Thermonuclear fusion will not be discussed much in the present book, but it seems appropriate as a minimum to present an exercise to illustrate the importance of the topic.

In a deuterium (D) plasma at temperatures $\sim$ 100 keV and an ion density of $N = 10^{14}$ cm^{-3} we have the following dominating nuclear reactions

$$D + D \rightarrow T + p + 4.03\,\text{MeV}, \tag{1.1}$$

$$D + D \rightarrow He^3 + n + 3.27\,\text{MeV}, \tag{1.2}$$

where T stands for *tritium*, He for *helium*, p for *proton* and n for *neutron*. The reactions occur at such a rate that the number of reactions per sec are $\frac{1}{2}N^2 I$, where $I = 2.3 \times 10^{-17}$ cm^3/s for (1.1) and $I = 2.8 \times 10^{-17}$ cm^3/s for (1.2). The expressions for the reaction rates are here to be seen simply as given, but intuitively it seems reasonable that they are proportional to the densities N squared, since two atoms must be present in order to have a fusion process. (Usually we would use n for particle density, but in this particular context we want to avoid confusion with the neutron symbol.) If we had fusion of D and T, we expect the product of the respective densities N_D and N_T to enter. The I's then account for the probability of fusion processes to occur at the given temperature, provided the two atoms are present. How large is the fusion produced power density in the plasma? For how long does the process have to continue in order to produce the amount of energy we had to use for heating up the plasma from room temperature?

The processes (1.1) and (1.2) are what you have for $D-D$ reactions. If you succeed in using also the resulting tritium, T, and helium, *He*, you can at least in principle also use the processes

$$D+T \rightarrow He^4+n+17.6\,\text{MeV}, \tag{1.3}$$

$$D+He^3 \rightarrow He^4+p+14.7\,\text{MeV}. \tag{1.4}$$

How much energy can you obtain by fusion per deuterium atom, considering the processes (1.1)-(1.4)? Ⓢ

Ⓢ **Exercise:** Naturally occurring water contains approximately 0.015% deuterium. Determine the approximate number of D atoms in one liter of water. Assuming that all the deuterium is fusioned, calculate the amount of oil necessary to produce the same amount of energy, using that the energy released by burning of oil is approximately 4.6×10^4 J/g. Estimate the amount of deuterium available in the oceans. Assuming that all this is undergoing fusion, estimate the total available energy. Compare this energy with the energy received by radiation from the Sun in one year.

The energy consumption of the Earth's population can be estimated as 0.3-0.5 Q/100 y in the period 0-1850 (calculated in part from forests that disappeared in that period), 4 Q/100 y in the period 1850-1950, and 80-500 Q/100 y in the period 1950-2050 (extrapolated), in units of $\text{Q}\equiv10^{21}$ J. With the current rate of energy consumption, for how long can mankind survive on the fusion energy available from the oceans? How realistic is this scenario? The half-life of tritium is approximately 12.26 y. How does this influence the construction of a fusion reactor involving the $D+T$ process mentioned before? Ⓢ

1.4 Why study waves in plasmas?

The present book is devoted mostly to waves and oscillations in plasmas. The author will argue that this topic is important, although it definitely does not cover all of plasma sciences. We could also study equilibrium configurations in complex magnetic field geometries, etc. Waves with small or large amplitudes, linear or nonlinear, are manifestations of dynamic phenomena, and represent a well defined topic, but still we could argue that there are so many other wave phenomena in physics, so do we really learn something substantially new here? Could it be that plasmas represent just one other medium, and everything we learned from acoustics, electromagnetic wave propagation, etc. could be re-used for describing plasma dynamics with a change of some symbols here and there? No, it is not so! It *is* true that many plasma waves have strong similarities with waves in other media, and it is an advantage to identify such cases. There are at least two important cases where the analysis of wave phenomena in plasmas differs from that in other media, as discussed in the following.

1.4.1 Alfvén waves

Plasmas support wave-types that cannot be found in other media. To illustrate the idea we might first consider one of the simplest oscillating systems, a pendulum. Here the energy is kinetic at the vertical reference position while it is potential at the two turning points. Another oscillator is an ideal $L-C$ circuit without a resistor, where at one time all energy is stored in the magnetic field in the self-induction and a quarter period later in the electric field in the capacitor. A standing wave is one other form of an oscillator. For a standing electromagnetic wave, for instance, the energy is stored at one time in the electric field and later in the magnetic field, as indicated schematically in Fig. 1.1. For a standing sound wave in an organ pipe, for instance, all the energy is in the pressure, and later in the average velocity of the air.

Sound-like waves are found in plasmas and they are in many ways similar to sound waves in other media. We have electromagnetic waves: they are modified by the presence of the plasma, but remain electromagnetic all the same. We can have electrostatic waves, called Langmuir waves, where the energy forms are kinetic particle motion and electric fields. These latter waveforms are a little more exotic than the former ones, but still some you find in, for instance, solids, where they are called optical phonons. However, only in plasmas, or in plasma-like media, do you find what are called Alfvén waves, where the dynamic energy forms are kinetic particle motion and magnetic field energy; see Fig. 1.1. This waveform turns out to be very important for the *large* scale motion in the Earth's near plasma environment as well as in interstellar media in general.

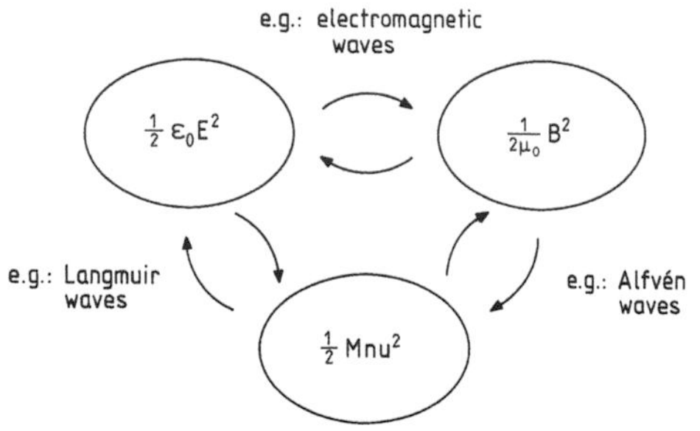

FIGURE 1.1: *Illustration of oscillation characteristics of different wave-types in plasmas.*

1.4.2 Landau damping and instability

One basic problem in statistical mechanics is to account for how fully reversible dynamics on the microscopic particle scale can give rise to irreversible macroscopic dynamics. Many dynamical models can contribute to the basic understanding of this seemingly paradoxical property of nature, but very few can be solved without making assumptions along the way. The kinetic model of plasma, based on the Vlasov equation, offers (in this author's opinion) a particularly elegant insight into how a basically time reversible description of matter can give rise to seemingly irreversible behavior. The phenomenon is called Landau damping and is an important feature of the *small* scale dynamics of plasmas.

In addition, we might find kinetic plasma instabilities. Also this result can be surprising; instabilities are well known from classical fluid dynamics, but are there always associated with spatial gradients of some sort: we can have gradients in temperature, velocity, mass density, etc. A velocity gradient can give rise to the Kelvin-Helmholtz instability, for instance. The kinetic dynamics of collisionless plasmas can be quite different; here we can find conditions where all the basic parameters in configuration space, plasma density, bulk plasma velocity, etc. are homogeneous, but for some easily satisfied conditions in phase space we can find that the plasma is unstable nonetheless.

1.4.3 Waves for diagnostics

It will be found, not surprisingly, that many wave-types depend significantly or even critically on the characteristics of the plasma where they propagate, such as density, temperature, magnetization, etc. Accurate measurements of wave propagation characteristics will therefore contains this information, albeit often concealed. With some non trivial efforts, the plasma parameters can then be deduced from such wave measurements.

2

Basics of Continuum Models

A plasma is a relatively complex medium, consisting of a mixture of free electrons and ions, susceptible to electric and magnetic forces in addition to those acting also on classical neutral gases or fluids. Although this book is dealing mainly with plasma media, it may be instructive first to summarize some basic equations and results for the simpler case dealing with neutral fluids and gases, in order to demonstrate how classical continuum mechanics forms the basis of the description also of plasma dynamics.

In the classical mechanics of discrete particles, a basic description deals with the positions of these particles, which are then functions of time, for given initial conditions and external forces. The position is a vector function of time only, $\mathbf{R} = \mathbf{R}(t)$, and a differentiation with respect to the independent variable is unambiguous. Without being misunderstood, we can use Newton's notation and write $\dot{\mathbf{R}}$ for $d\mathbf{R}/dt$, where the latter was the notation preferred by Leibniz.

Dealing with the mechanics of continuous media as gases and fluids, the problem is more complicated. The basic tools for the description are *vector fields*, where any spatial position at any time is assigned a vector quantity, be it a velocity, an electric field or something else. Simpler fields are *scalar*, and here any spatial position is at any time assigned a scalar quantity, which can be density, temperature, an electric potential or similar. These fields depend in general on time as well as position so we have several independent variables. To distinguish the differentiation it is necessary to write explicitly which one of the variables we have in mind and use *partial* differentiation.

Taking the simplest case first with a scalar field, we may discuss the space-time variation of the mass density ρ of the given medium. In general we have, as stated, $\rho = \rho(\mathbf{r}, t)$, and can ask for the time derivative of this density as observed at a fixed point. This we write as $\partial\rho/\partial t$. Similarly, we might want to differentiate with respect to the spatial variable x by $\partial\rho/\partial x$, or more generally $\nabla\rho \equiv \partial\rho/\partial x\widehat{\mathbf{x}} + \partial\rho/\partial y\widehat{\mathbf{y}} + \partial\rho/\partial z\widehat{\mathbf{z}}$, for some given fixed time t: we can assume that we have a "snapshot" of the field obtained at a given time, and differentiate with respect to either one of the components of the position variable.

The mass density of a fluid or a gas as detected at a fixed position in space can vary basically for two reasons if we disregard production or losses. Take first a simple system in one spatial dimension. We can have a spatial gradient in ρ and the gas can flow past the observation point with some constant velocity u. Let the mass density change with distance so that $\rho(x+\Delta x) - \rho(x) = \Delta\rho$. Let Δx be the length of the fluid element passing by the observer in a time unit Δt. We thus have the fluid velocity to be $u = \Delta x/\Delta t$. In this case the fixed observer will find a change of the mass density in a time Δt to be $-\Delta\rho$. Estimating the time derivative, this observer will find $-u\Delta\rho/\Delta x$ (note the negative sign!), which approximates the time derivative $-u\partial\rho/\partial x$. There is, however, one more possibility for changes in density: even if there are no gradients in ρ at the selected time, there can be gradients in the flow velocity, so that more gas or fluid is flowing *into* than out of a small volume around the observer. This case requires that the medium is *compressible*. Let the flow velocity vary so that $u(x+\Delta x) - u(x) = \Delta u$. The change in fluid density in a time Δt in a small volume with length Δx along the x-axis is then given as the difference between the amount flowing in and the amount going out, and we have $-\rho\Delta u/\Delta x$, approximating $-\rho\partial u/\partial x$, emphasizing again the negative sign. The two results can be combined into $-\partial(u\rho)/\partial x$. The general result in three dimensions, where we have gradients in both mass density and velocity, can be formulated as a partial differential equation, the *continuity equation*.

2.1 Continuity equation

In the absence of creation or loss of particles, any medium will follow a continuity equation in terms of velocity field $\mathbf{u} = \mathbf{u}(\mathbf{r},t)$ and mass density $\rho = \rho(\mathbf{r},t)$

$$\frac{\partial}{\partial t}\rho + \nabla \cdot (\mathbf{u}\rho) = 0. \tag{2.1}$$

The vector combination $\rho\mathbf{u}$ is the *flux of mass density*, or simply the material flux density. By Gauss' theorem (see Appendix D.1) we write (2.1) as

$$\int_{\mathcal{V}} \frac{\partial}{\partial t}\rho(\mathbf{r},t)dV = \frac{d}{dt}\int_{\mathcal{V}} \rho(\mathbf{r},t)dV = -\int_{\mathcal{V}} \nabla \cdot (\mathbf{u}\rho)dV = \int_{\mathcal{S}} (\mathbf{u}\rho)\cdot \widehat{\mathbf{n}}dS, \tag{2.2}$$

where $\int_{\mathcal{V}} dV$ denotes integration over an otherwise unspecified volume $\mathcal{V}$, and $\int_{\mathcal{S}} dS$ integration over a surface bounding that volume, while $\widehat{\mathbf{n}}$ is the unit vector normal to the surface, pointing outwards. When we have integrated the spatial derivative in the first term of (2.2) we write d/dt as the time derivative operator and not $\partial/\partial t$ as indicated. From (2.2) we conclude that the change of fluid mass in a volume $\mathcal{V}$ is due to a net mass flux into (or out of) the volume, through the bounding surface $\mathcal{S}$, as one could also have argued from the outset. We could as well start by postulating the integral form (2.2): from basic physical principles it merely states mass conservation. Since the result must remain correct for *any* shape of the volume $\mathcal{V}$, the integrands on the two sides of the first equality must balance, which immediately gives (2.1).

We can introduce the *total time derivative operator* or *convective time derivative*

$$\frac{D}{Dt} \equiv \frac{\partial}{\partial t} + \mathbf{u}\cdot\nabla \tag{2.3}$$

and rewrite (2.1) in the form

$$\frac{D}{Dt}\rho = -\rho\nabla\cdot\mathbf{u}. \tag{2.4}$$

A fixed observer, measuring time derivatives by $\partial/\partial t$, can experience a time variation even in incompressible fluids in case there was a density gradient from the very beginning, and this gradient is swept by the observer with velocity $\mathbf{u}$, in which case $\partial\rho/\partial t = -\mathbf{u}\cdot\nabla\rho$ is found. If the fluid is compressible, then $-\rho\nabla\cdot\mathbf{u}$ comes in addition to this. In the co-moving frame, the one moving with velocity $\mathbf{u}$, the contribution $-\mathbf{u}\cdot\nabla\rho$ vanishes. The relation (2.4) has the heuristic interpretation that in the co-moving frame, the density is enhanced if more flux vectors point into a reference volume than out of it. In the opposite case the density is depleted. Both cases imply that the medium is compressible. For *incompressible* flows, just as many flux vectors point into the volume as out of it. Well, as said, this was heuristically speaking.

The total time derivative measures the time variation following a fluid element, and the continuity equation means that in this frame, any variation in density is caused by expansion or compression of the fluid element as given by $\nabla\cdot\mathbf{u}$. *Incompressible fluids* have $\nabla\cdot\mathbf{u} = 0$. We emphasize that the velocity entering (2.3) is a function of the spatial coordinate as well as time, $\mathbf{u} = \mathbf{u}(\mathbf{r},t)$, where $\mathbf{r}$ is an *independent* variable. The total time derivative should not be confused with the time derivative in classical Newtonian mechanics, where the velocity of a particle is a function of time only. Rather, D/Dt should be seen as a shorthand for the definition (2.3) (Hazeltine & Waelbroeck 1998). In the absence of production and losses, mass conservation will be a general property of any gas or fluid, in particular also plasma media.

If matter is produced or being lost, we have to complete the continuity equation by terms accounting for these productions and losses, giving for instance

$$\frac{\partial}{\partial t}\rho + \nabla\cdot(\mathbf{u}\rho) = \alpha - \beta \tag{2.5}$$

with $\alpha = \alpha(\mathbf{r},t)$ accounting for the production of particles and $\beta = \beta(\mathbf{r},t)$ for their loss, where both can vary with position as well as time. Both α and β may depend implicitly on the density of the fluid or gas, in some cases also the velocity. If we have a mixture of gases or fluids, we have continuity equation for each of the components.

In any case, we now have a continuity equation (2.1) or (2.5) relating the density of matter to its velocity. To have a closed set of equations, we need an equation for the velocity. This we can obtain from Newton's second law.

- **Example:** Consider a charged fluid in motion, taking the density to be n and velocity to be $\mathbf{u}$. Let the charges on the individual particles be q. The charge density is then $\zeta \equiv qn$, and the current density due to the moving charges is $\mathbf{J} \equiv qn\mathbf{u}$. By (2.1) we then obtain the continuity equation for charges in the form

$$\frac{\partial}{\partial t}\zeta + \nabla \cdot \mathbf{J} = 0. \tag{2.6}$$

In order to avoid infinite electric fields we here assume that the charge density has finite spatial support.

2.2 Newton's second law

A fluid has to follow Newton's second law, giving a force equal to a mass times an acceleration. Just as in the discussions of the continuity equation, Section 2.1, we can argue that a fixed observer will experience a temporal variation of the velocity in case the velocity field has a spatial variation as it is moving past the observer with the local velocity $\mathbf{u}$, this contribution being $-\mathbf{u}\nabla \cdot \mathbf{u}$. We also have to take this into account when writing the momentum density budget for the fluid.

As in Section 2.1, we consider a volume enclosing some fluid or gas with spatially varying mass density ρ and velocity $\mathbf{u}$. The volume $\mathcal{V}$ is not enclosed by a *physical* boundary, and its surface $\mathcal{S}$ should be seen merely as an integration surface here. The net momentum of that volume is $\int_{\mathcal{V}} \rho \mathbf{u} d\mathbf{r}$. We assume that some force acting on the fluid or gas. We introduce a force density $\mathbf{f}$ so that $\mathbf{F} = \int_{\mathcal{V}} \mathbf{f} d\mathbf{r}$ is the net force acting on the volume $\mathcal{V}$. The changes in momentum are due to forces acting on the volume *and* the momentum flux through the surface of $\mathcal{V}$. We can write

$$\frac{d}{dt}\int_{\mathcal{V}} \rho \mathbf{u} d\mathbf{r} = -\oint_{\mathcal{S}} \rho \mathbf{u}\mathbf{u} \cdot \widehat{\mathbf{n}} dS + \int_{\mathcal{V}} \mathbf{f} d\mathbf{r}. \tag{2.7}$$

The first term on the right-hand side contains the velocity component normal to the integration surface, $\mathbf{u} \cdot \widehat{\mathbf{n}}$, multiplied by the momentum density $\rho\mathbf{u}$ to account for the momentum flux through the surface $\mathcal{S}$ of the reference volume. The second term on the right-hand side of (2.7) accounts for all forces acting on that volume, so (2.7) expresses Newton's second law.

Note that we ignored production and losses within the reference volume: α and β in (2.5) were included to account for changes in particle densities. To continue these ideas here we have to account for the production and loss of momentum density, which in general need not be related to α and β at all. In case a particle being lost takes with it all of its momentum, we might, however, argue for simple relations as far as losses are concerned.

Using again Gauss' theorem on each of the three components of the vector relation (2.7), we find that the left side and the first term on the right-hand side can be combined to

$$\int_{\mathcal{V}} \left(\frac{\partial}{\partial t}(\rho \mathbf{u}) + \nabla \cdot (\rho \mathbf{u}\mathbf{u}) \right) d\mathbf{r} =$$

$$\int_{\mathcal{V}} \rho \left(\frac{\partial}{\partial t}\mathbf{u} + \mathbf{u} \cdot \nabla \mathbf{u} \right) d\mathbf{r} = \int_{\mathcal{V}} \mathbf{f} d\mathbf{r}, \tag{2.8}$$

where we explicitly used the continuity equation (2.1). Requiring (2.8) to be fulfilled for *any* volume $\mathcal{V}$, we obtain

$$\rho \frac{D}{Dt}\mathbf{u} = \rho\left(\frac{\partial}{\partial t}\mathbf{u} + \mathbf{u}\cdot\nabla\mathbf{u}\right) = \mathbf{f}. \tag{2.9}$$

Any gas or fluid, in particular also plasma media, will have to follow Newton's second law, the distinctions arising when we discuss the forces represented by $\mathbf{f}$. We can have, for instance, a force density due to a gradient in pressure, $\mathbf{f} = -\nabla p$, in which case (2.9) is called *Euler's equation.* To have a closed set of equations in the present case, we then have to find a new equation for the pressure. By now we begin to get suspicious; we started out with the expectation of writing an equation for the evolution of fluid or gas density to find that it required knowledge of the velocity variation. Writing an equation for the velocity we need an equation for the pressure (and possibly other forces). It seems likely that an equation for the pressure will contain a new quantity, which requires a new equation, etc. The preferable solution is (if possible) to express the pressure in terms of quantities already known, e.g., the density. This would be an equation of state, but in general it *might* have the form of a differential equation.

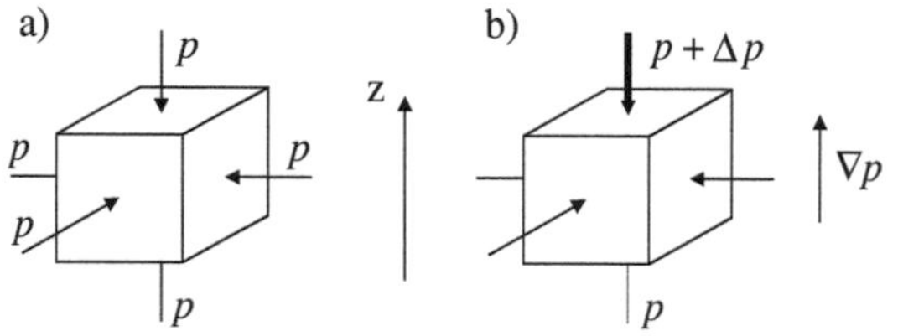

FIGURE 2.1: *Illustration of the pressure force on a volume element of a gas or a fluid. In a) we have isotropic pressures (and consequently no net force on the box) while in b) we have a net force due to a gradient in pressure.*

- **Example:** Assume that we have a gradient of pressure in a fluid or a gas. Let the pressure vary with the z-coordinate, for instance, $p = p(z)$. Assume that we have a small box of fluid or a gas immersed in the pressure gradient. In the reference case shown in Fig. 2.1a) the pressure gradient vanishes and we have isotropic pressures and consequently no net force on the box. In Fig. 2.1b) we have a finite pressure gradient: here the top of the box is exposed to a pressure $p = p(z + \Delta z)$, giving a force pointing *downwards*, while the bottom of the box feels a pressure $p = p(z)$ upwards. The horizontal components of the net force cancel at all z-positions, since the pressure is assumed to vary with the z-coordinate only. The net pressure force on the box with top and bottom surfaces having an area $d\mathcal{A}$ is then $\mathbf{F} = d\mathcal{A}(p(z) - p(z+\Delta z))\widehat{\mathbf{z}} \approx -(d\mathcal{A}\Delta z)\widehat{\mathbf{z}}dp(z)/dz$. The force density, i.e., force per volume, is then $-\widehat{\mathbf{z}}dp(z)/dz$, or more generally $-\nabla p(\mathbf{r})$ for an arbitrary direction of the pressure gradient. The negative sign indicates that the force is acting in the direction of decreasing pressure, as intuitively expected.

We can write one of the most relevant forms of (2.9) as

$$\rho\left(\frac{\partial}{\partial t}\mathbf{u} + \mathbf{u}\cdot\nabla\mathbf{u}\right) = -\nabla p + \mu\nabla^2\mathbf{u}, \tag{2.10}$$

where μ is a viscosity coefficient (sometimes denoted *dynamic viscosity*), representing a model for momentum losses in the system. We will discuss viscous effects later, in Section 2.6. For a large class of problems we can consider the medium to be incompressible, so if the mass density ρ is constant or uniform to begin with it will remain so for all later times. For such cases we might as well divide all terms by ρ and introduce the *kinematic viscosity* $\nu \equiv \mu/\rho$. The notation μ for dynamic viscosity and ν for kinematic viscosity is, however, not universally adopted.

- **Exercise:** Let a vessel containing water be placed on a scale. The reading of the scale is recorded and then a solid metal object suspended on a string is immersed in the water, so that it is completely covered. Let the weight of the object be M. What is the new reading of the scale? Does it make a difference if we cut the string or not?

2.3 Equation of state

An equation of state, relating pressure to, for instance, mass density, can be obtained from thermodynamics and statistical mechanics (Schroeder 2000). The most usual and simplest relation used is that of ideal gases

$$pV = mRT \tag{2.11}$$

where m is the number of moles of gas, $R = 8.31$ J/(mol K) is a universal constant, with K being the unit of absolute temperature, Kelvin, and V is the volume, while T is the absolute temperature. A *mole* of atoms or molecules is Avogadro's number of them, $N_A = 6.02 \times 10^{23}$. The number of molecules or atoms in the volume V is then $N = mN_A$. We can then write another, more standard version of (2.11) as

$$pV = N\kappa T \qquad \text{or} \qquad p = n\kappa T \tag{2.12}$$

with $\kappa = R/N_A = 1.281 \times 10^{-23}$ J/K being Boltzmann's constant, and $n \equiv N/V$ is the number density of the gas or fluid in question. This is a rather convenient equation, which relates pressure to the local, instantaneous, values of the density and temperature. A much more inconvenient relation would have the form of a partial differential equation; see, for instance, Braginskiĭ (1965). We used here the particle density (i.e., number of particles per unit volume), but could as well have used the *mass density*, where the two are related by $n = \rho/M$, where M is the atomic or molecular mass of the particles, assuming we have only one species present.

So far, the expression for the pressure contains *two* quantities, the density for which we have an equation already, and then the temperature T, which is for the moment unknown. We can now, for instance, *assume* isothermal processes and have pressure and density being directly proportional, $p = \kappa T n$, where the constant of proportionality is κT with T here being a given temperature and κ is Boltzmann's constant. As an alternative we can assume, depending on conditions, that the processes are *adiabatic*, i.e., $p\rho^{-\gamma} =$ const., with $\gamma \equiv C_P/C_V$ being the ratio of specific heats for constant pressure and constant volume, respectively (Feynman et al. 1963). The actual physical conditions determine which *equation of state* is applicable.

When the time evolution of a process is so slow that the thermal conductivity in the medium is large enough to maintain an essentially constant temperature throughout the medium, we can safely assume that the process is isothermal. To check this assumption we should in principle solve the equation for thermal conductivity, $\partial T/\partial t = D_T \nabla^2 T$, which has the form of a diffusion equation with D_T being the thermal diffusion coefficient, assumed to be known. In reality, it may suffice to make an order of magnitude estimate; let the characteristic scale length of the problem be $\mathcal{L}$ and a characteristic time be τ, implying a characteristic velocity $U \sim \mathcal{L}/\tau$. The diffusion equation gives a characteristic velocity U_T, which is the diffusion flux $-D_T \nabla T$ divided by T itself, i.e., as an order of magnitude $U_T \sim D_T/\mathcal{L}$. If we have $U_T \gg U$ we can assume that the process is isothermal, since the thermal conductivity will smooth out the temperature variations faster than the perturbation is able to propagate them. In the other limit, $U \gg U_T$, our best guess is an adiabatic process. Evidently, the shorter the scale lengths are, the more likely it is that the process is isothermal. For sound waves in neutral gases it turns out that thermal conductivity controls the temperature variation for wavelengths comparable to the collisional mean free path, and we are not really interested in such short waves; sound waves in neutral gases are safely assumed to be adiabatic. Note that we have not explicitly made any assumption of an ideal gas law, $p = n\kappa T \equiv \rho\kappa T/M$, but this will be implicit in many applications. The important point is that we are now able to obtain a closed dynamic set of equations (2.1) and (2.9) together with an appropriate equation of state, as long as the only relevant force originates from a pressure gradient.

2.3.1 Gas pressure on a surface

First we want to find an expression for the pressure force on a solid surface. As a first illustration we have in Fig. 2.2 a plane area element $d\mathcal{A}$ bombarded by a steady flow of particles arriving at an angle θ to the normal of the plane, all with velocity u. The number of particles arriving per time dt is $n\, d\mathcal{A}\, u\, dt \cos\theta$, where n is the particle number density. The momentum transferred by particles with mass M to the surface in this time is then $d\mathcal{A} M u^2\, dt\, n \cos\theta$, if we assume that the momentum is completely absorbed by an inelastic collision. The momentum transferred per time unit and per area unit is then $Mu^2 n\cos\theta$. This expression has to be modified if particles are reflected, but the changes depend on the momentum loss at the reflection. For the case with perfectly elastic reflections, the momentum transferred per time unit and per unit area is $2Mu^2 n\cos\theta$.

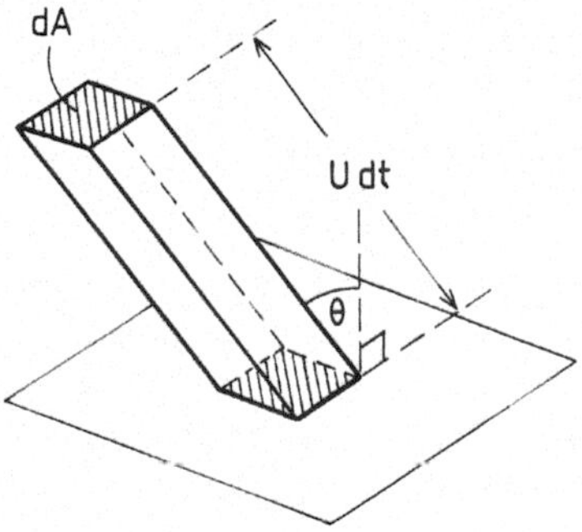

FIGURE 2.2: *Illustration for determining the flux of particles arriving at a surface element $d\mathcal{A}$. The angle between the vertical z-direction and the incoming particle flux direction is θ.*

We assume now that the particle velocity distribution is completely isotropic, so that we find any direction of propagation with equal probability. Consider a narrow velocity (or more correctly "speed") interval $u, u+du$ and let the density of particles with velocities in that interval be $f(u)du$ in units of *length*$^{-3}$. We have $n = \int_0^\infty f(u)du$. Take now a small spherical angle $d\omega$ as illustrated in Fig. 2.3. The density of particles with velocity in $u, u+du$ *and* direction in $d\omega$ is $f(u)du\,d\omega/4\pi$. The coefficient $1/4\pi$ consists of $1/2$ since we only include particles with directions *towards* the surface, and $1/2\pi$ accounting for the fraction of particles selected by the surface area $d\omega$ out of a unit radius hemisphere with area 2π. We consider again a small area element $d\mathcal{A}$ and want to find the number of particles per time unit dt hitting this surface at a spherical angle θ within $d\omega$. The relevant particles have to be in a cylinder tilted by an angle θ with bottom area $d\mathcal{A}$ and height $u\,dt\cos\theta$. Of the particles $d\mathcal{A} u f(u) du \cos\theta dt$ only the fraction $d\mathcal{A} u f(u) du \cos\theta dt\, d\omega/4\pi$ fulfills the criterion on having the correct direction. To obtain the total number of particles hitting the small area element we integrate over all velocities and angles to obtain the particle flux density Γ with dimension *length*$^{-2}\times$ *time*$^{-1}$, i.e., number of particles arriving per area element $d\mathcal{A}$, per time unit dt.

$$
\begin{aligned}
\Gamma &= \int_0^\infty u f(u) du \int_0^{\pi/2} \cos\theta \frac{d\omega}{4\pi} \\
&= n\langle|u|\rangle \int_0^{\pi/2} \cos\theta \frac{d\omega}{4\pi},
\end{aligned}
$$

where we introduced $n\langle|u|\rangle \equiv \int_0^\infty u f(u)du$ where n is the particle number density. The integrand is a function of θ only, and we can choose the angular element $d\omega$ as a narrow "belt" on a unit sphere (see Fig. 2.3) and have $d\omega = 2\pi \sin\theta d\theta$, giving

$$\Gamma = n\langle|u|\rangle \int_0^{\pi/2} \frac{1}{2} \cos\theta \sin\theta d\theta = \frac{1}{4} n\langle|u|\rangle. \tag{2.13}$$

Apart from the basic assumption of homogeneity and isotropy of the velocity distribution, we have not (yet) made any assumptions concerning $f(u)$. The assumption of isotropy is, however, rather restrictive, and most practical applications will assume in addition that $f(u)$ is a Maxwellian.

We can now calculate the pressure on the area element $\mathcal{A}$ as the momentum transferred by the bombarding particles per unit time. We here assume that the particles are reflected by a perfectly elastic collision, so that for a particle with mass M arriving at an angle θ with velocity u, the momentum transfer is $2Mu\cos\theta$. The momentum received by the surface per area element $d\mathcal{A}$, per time unit dt.

$$\begin{aligned} p &= \int_0^\infty 2Mu^2 f(u)du \int_0^{\pi/2} \cos^2\theta \frac{d\omega}{4\pi} \\ &= 2Mn\langle u^2\rangle \int_0^{\pi/2} \cos^2\theta \frac{d\omega}{4\pi} = \frac{1}{3} Mn\langle u^2\rangle. \end{aligned}$$

The result assumed the presence of one particle mass M only. Generalization to a *mixture* of masses is trivial, provided the momentum transfers from different particle types are statistically independent. Note that pressure has the physical dimension *energy*$\times$*length*$^{-3}$, i.e., energy density. This can be confusing sometimes.

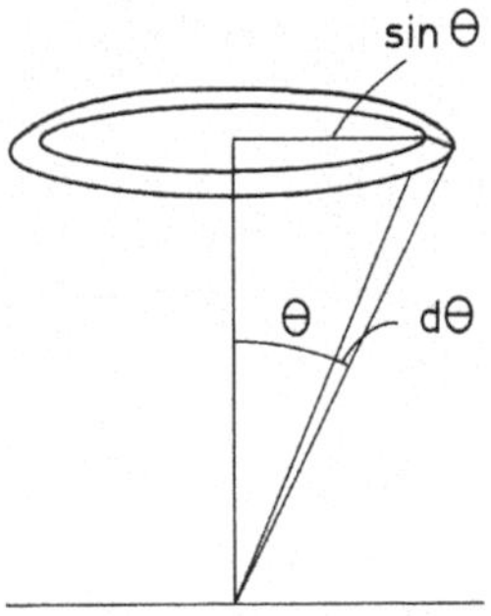

FIGURE 2.3: *Illustration of the "belt shaped" angular element $d\omega$ used for calculating the number of particles arriving at an area element $d\mathcal{A}$ in a time interval dt. We have $d\omega = 2\pi\sin\theta d\theta$.*

If we assume that the velocity distribution of the particles in three dimensional space is a Maxwellian at a temperature T, i.e.,

$$f(u_x, u_y, u_z) = n \left(\frac{M}{2\pi\kappa T}\right)^{3/2} e^{-\frac{1}{2}M(u_x^2+u_y^2+u_z^2)/\kappa T},$$

then we have

$$f(u) = 4\pi n u^2 \left(\frac{M}{2\pi\kappa T}\right)^{3/2} e^{-\frac{1}{2}Mu^2/\kappa T},$$

giving

$$\langle|u|\rangle = \sqrt{8\kappa T/\pi M} \quad \text{and} \quad \langle u^2\rangle = 3\kappa T/M.$$

Using the latter result for $\langle u^2\rangle$ in the expression for the pressure we obtained before, we have $p = n\kappa T$, which is nothing but the ideal gas law. For a Maxwellian distribution we have the simple relation $\langle|u|\rangle = \sqrt{8\langle u^2\rangle/3\pi}$, which may be useful sometimes.

2.3.2 Heat capacities

Heat capacities of ideal gases will be used or referred to quite often, and it might be worthwhile to give a short summary of the concept. By definition (Schroeder 2000), the heat capacity, C, of a substance is the amount of heat, Q, needed to raise its temperature one unit

$$C \equiv \frac{Q}{\Delta T} = \frac{\Delta U - W}{\Delta T}, \tag{2.14}$$

where we used the first law of thermodynamics, $Q = \Delta U - W$, with $W = -P\Delta V$ being the work done *on* the system, and ΔU is the change in energy of the system. Maybe a more fundamental quantity is the *specific heat capacity*, being the heat capacity per unit mass, $c \equiv C/M$. Unfortunately, the symbol c is canonized for the specific heat capacity, but we will probably never confuse it with the speed of light that is also denoted by c!

Unfortunately this definition (2.14) is ambiguous: C depends on the circumstances: in particular we might supply the heat to a volume element at constant volume, V, but variable pressure, P, or to a volume which is allowed to expand, but where the pressure is defined by the equilibrium with the surroundings. We can introduce the heat capacity at constant volume, C_V, and have $W = -P\Delta V = 0$, giving

$$C_V = \left(\frac{\Delta U}{\Delta T}\right)_V \to \left(\frac{\partial U}{\partial T}\right)_V. \tag{2.15}$$

This result has not made explicit use of any ideal gas law.

Alternatively, for the heat capacity at constant pressure we have from (2.14)

$$C_P = \left(\frac{\Delta U - (-P\Delta V)}{\Delta T}\right)_P \to \left(\frac{\partial U}{\partial T}\right)_P + P\left(\frac{\partial V}{\partial T}\right)_P, \tag{2.16}$$

with $C_P \neq C_V$, in general. For solids or liquids, the second term in (2.16) is often small and might be neglected. For gases, ideal gases in particular, this term is important, but in either case we have $(\partial U/\partial T)_V \neq (\partial U/\partial T)_P$. Note that the physical dimensions of C_V and C_P are energy divided by temperature, i.e., J/K, so that $[C_{P,V}] = [\kappa]$ with κ being Boltzmann's constant. The notation $[*]$ implies "dimension of" $*$.

Assume now that the volume of ideal gas contains N identical molecules or atoms, each with f degrees of freedom. The equipartition theorem states that $U = \frac{1}{2}Nf\kappa T$, assuming that f is independent of temperature. The assumption of *ideal* gases enters here solely by treating the molecules as independent. In this case we readily find

$$C_V = \left(\frac{\partial U}{\partial T}\right)_V = \frac{1}{2}Nf\kappa. \tag{2.17}$$

For monatomic gases, inert gases for instance, we have $f = 3$, and find $C_V = \frac{3}{2}N\kappa = \frac{3}{2}mR$, where m is the number of moles of gas, $mR = N\kappa$.

Concerning C_P we note that with the ideal gas law it does not matter whether we take $\partial U/\partial T$ for constant V or P. For such ideal gases we find

$$\left(\frac{\partial V}{\partial T}\right)_P = \frac{\partial}{\partial T}\left(\frac{N\kappa T}{P}\right) = \frac{N\kappa}{P}, \tag{2.18}$$

implying from (2.16) that

$$C_P = C_V + N\kappa = C_V + mR \tag{2.19}$$

for ideal gases, again in terms of m, being the number of moles of gas.

For an ideal monatomic gas (this is about as ideal as it can be) we find $C_P/C_V = 5/3$, which is a number we often use as an approximation in other cases.

At phase transitions (we are then no longer dealing with ideal gases), we have that the system changes as a response to the supplied heat without change in temperature: for this case the heat capacity is infinite; see (2.14). For such cases the *latent heat*, L, of the system is of particular interest, with $L \equiv Q/M$, introducing the total mass M of the substance. Also this definition is ambiguous, and one usually assumes that the pressure is kept constant.

Ⓢ **Exercise:** A calorimeter can be designed to determine the specific heat capacity for fluids, by taking a well isolated long tube, where the fluid flows past an electric heater element, completely immersed in the fluid somewhere in the middle of the tube. Assume here that the power delivered into the fluid is $P = 60$ W. The temperature of the fluid is measured at the inlet and the outlet of the tube. We ignore first heat losses to the surroundings. Assume that for a given fluid we find a temperature increase of $\Delta T = 10$ K. Determine the fluid flow in kg/s, when the specific heat of the fluid is $c = 2$ kJ/kg K.

We now allow for some heat losses to the surroundings. It so happens for the present case that ΔT remains constant when we double P and we triple the fluid flow. What are the power losses to the surroundings? Ⓢ

2.4 Dynamic properties

Basically we might be interested in two sorts of solutions for the equations (2.1), (2.9) and (2.11): steady state solutions and *dynamic* solutions. Here we will be interested mainly in the latter type. Unfortunately, the equations cannot be solved generally, and we have to be content with some approximate solutions.

2.4.1 Sound waves

Retaining (2.9) with $\mathbf{f} = -\nabla p$ we find a simple steady state solution as $\mathbf{u} = 0$ and $p = p_0 =$ const. and $\rho = \rho_0 =$ const. Making a slight perturbation of these quantities as $p = p_0 + \widetilde{p}$, $\mathbf{u} = \widetilde{\mathbf{u}}$, $\rho = \rho_0 + \widetilde{\rho}$, with $\widetilde{p} \ll p_0$ and $\widetilde{\rho} \ll \rho_0$ we obtain from the equation of continuity (2.1) that

$$\frac{\partial}{\partial t}\widetilde{\rho} + \rho_0 \nabla \cdot \widetilde{\mathbf{u}} = 0,$$

where we assumed that small terms like $\widetilde{\rho}\widetilde{\mathbf{u}}$ are negligible: we have *linearized* the equations, by ignoring small terms. Later we will find that in some cases the results from such a linearized analysis cannot be universally valid, and the assumption may eventually break down. By taking the initial perturbation sufficiently small, it seems, however, intuitively reasonable that these nonlinear terms can be made negligible, at least for a long time interval. Similarly, we have from (2.9)

$$\frac{\partial}{\partial t}\widetilde{\mathbf{u}} + \frac{1}{\rho_0}\nabla \widetilde{p} = 0.$$

Assuming that we have an equation of state given in the form $p = p(\rho)$, we can argue that for small perturbations, $\widetilde{\rho}$, we can approximate the fluctuations in pressure, $\widetilde{p}$, by

$$\widetilde{p} = \left.\frac{dp}{d\rho}\right|_{\rho_0} \widetilde{\rho}$$

and find

$$\frac{\partial}{\partial t}\widetilde{p} + \rho_0 \left.\frac{dp}{d\rho}\right|_{\rho_0} \nabla \cdot \widetilde{\mathbf{u}} = 0.$$

It is now an advantage to introduce a *velocity potential* by $\mathbf{u} \equiv \nabla\Psi$. This definition can be quite general, so we omitted the "tilde". We note that by this introduction we discard rotational (or "shear") velocity fluctuations, and have $\nabla \times \mathbf{u} = 0$, but this is of no concern for the compressional motions with a fluctuating pressure studied here.

With the velocity potential available, we find the equation for the pressure $\widetilde{p} = -\rho_0 \partial\Psi/\partial t$, giving finally the *wave equation*

$$\frac{\partial^2}{\partial t^2}\Psi - C_s^2 \nabla^2 \Psi = 0 \tag{2.20}$$

where the *sound speed* $C_s \equiv \sqrt{dp/d\rho|_{\rho_0}}$ was introduced. Note that the definition has been made without (yet) specifying the actual equation of state, $p = p(\rho)$. For sound waves in neutral gases the natural choice will be an adiabatic equation of state,

$$p = p_0 \left(\frac{\rho}{\rho_0}\right)^\gamma, \tag{2.21}$$

as mentioned before. From (2.20) we can obtain similar equations for any velocity component, for instance. To obtain the sound speed for this choice of equation of state for an ideal monatomic gas with particle masses M, we find $dp/d\rho = \gamma p_0 \rho^{\gamma-1}/\rho_0^\gamma = \gamma p_0 (\rho/\rho_0)^\gamma/\rho = \gamma p/\rho = \gamma n \kappa T/M$, giving $C_s = \sqrt{(5/3)\kappa T/M}$, where we used the value $\gamma = 5/3$ for ideal gases as an approximation.

By Fourier transform of (2.20) we obtain the dispersion relation $\omega^2 = C_s^2 k^2$, in terms of the frequency ω and the wave-number k for the waves. Note that the speed of propagation is C_s, independent of frequency: verbal communication depends critically on this feature! If you talk after filling your mouth with helium, the situation is different: here sound waves become dispersive in the audio range, which is why it sounds so funny! Explain this in simple terms!

- **Exercise:** Demonstrate that the perturbations in velocity and fluid density in a sound wave, as described by (2.20), are related by $\widetilde{u} = C_s \widetilde{\rho}/\rho_0$. By omitting the vector notation for the velocity, we have in mind here a one-dimensional situation for simplicity. Introducing the perturbation in local temperature as $\widetilde{T} = \partial T/\partial p|_0 \widetilde{p}$ and using the well known thermodynamic formula, $\partial T/\partial p|_s = (T/C_P)\,\partial V/\partial T|_{p_0}$, demonstrate that $\widetilde{T} = C_s \beta T_0 \widetilde{u}/C_P$, where $\beta = (1/V)\,\partial V/\partial T|_0$ is the coefficient of thermal expansion (Landau & Lifshitz 1987). The subscript $_0$ refers to the initial, unperturbed, state, p_0, ρ_0, and T_0.

2.5 Incompressible media

Previously we distinguished compressible and incompressible media. Strictly speaking, of course, nothing is incompressible: if we apply a large enough pressure, we can compress everything, at least a little. In many cases, however, this compressibility is of no consequence, and might be ignored. We would like to obtain criteria for the applicability of that approximation.

Consider again the equation of continuity of a fluid or a gas

$$\begin{aligned} \frac{\partial}{\partial t}\rho + \nabla\cdot(\rho\mathbf{u}) &= 0 \\ \left(\frac{\partial}{\partial t} + \mathbf{u}\cdot\nabla\right)\rho + \rho\nabla\cdot\mathbf{u} &= 0 \end{aligned} \tag{2.22}$$

or, as written before (2.4),

$$\frac{D}{Dt}\rho = -\rho\nabla\cdot\mathbf{u}.$$

We want to determine conditions under which a fluid or a gas (a plasma in particular) can be considered incompressible, i.e., $\Delta\rho/\rho_0 \ll 1$. We let τ and ℓ be the time and length scales over which the flow exhibits significant changes. From Euler's equation (see discussion of (2.9)) we have a balance between the acceleration $D\mathbf{u}/Dt$ and the pressure containing term $(1/\rho)\nabla p$. As an order of magnitude estimate we therefore have $\Delta u/\tau \sim \Delta p/\rho\ell$, or $\Delta p \sim \rho\ell\Delta u/\tau$, where Δu is the variation of u. To estimate the pressure variation we use $\Delta p/\Delta\rho \sim \partial p/\partial\rho$. Using the definition of the sound speed (see discussion of (2.20)), we have $\Delta p \sim C_s^2\Delta\rho$. Combining the two expressions for Δp we can now estimate the change in density as $\Delta\rho \sim \ell\rho\Delta u/\tau C_s^2$, giving the order-of-magnitude relation $\Delta\rho/\rho \sim \ell\Delta u/\tau C_s^2$. In order to have incompressible fluid motion, $\Delta\rho/\rho \ll 1$, we require $\ell\Delta u/\tau C_s^2 \ll 1$, or $\Delta u \ll \tau C_s^2/\ell$. Arguing that, as an estimate, we can use $\Delta u \sim \ell/\tau$, the latter inequality implies $\Delta u \ll C_s$ with $M \equiv \Delta u/C_s$ being the Mach number. We can write the same result in a different form as $\ell \ll \tau C_s$, or

$$\tau \gg \frac{\ell}{C_s},$$

implying that in order to treat a flow as incompressible, we require the time ℓ/C_s it takes a sound pulse to traverse the distance ℓ must be small compared to the characteristic time, τ, for variations in the flow. From this we can conclude that as long as we are dealing with phenomena involving motions much slower than sound speed, we can ignore compressibility of the medium, in particular also of a plasma, altogether (Batchelor 1967, Landau & Lifshitz 1987). In effect, we are setting the sound speed to infinity, a rather drastic assumption, if you think about it! Note that by introducing the estimate $\Delta u \sim \ell/\tau$, we are assuming that changes in fluid conditions are due to bulk motion of matter, and that the velocity ℓ/τ is not associated with a wave-type of motion, where it is a spatial form that propagates and not the fluid elements as such. Imagine a piece of cork floating on a water surface that supports a propagating wave: the cork performs circular motions without any average net displacement as the wave propagates with its phase velocity.

We notice that the assumption of incompressibility, $\nabla\cdot\mathbf{u} = 0$, in effect makes an additional equation of state redundant! Take, for instance, the divergence of (2.9). This gives

$$\frac{D}{Dt}\nabla\cdot\mathbf{u} = \nabla\cdot\left(\frac{\partial}{\partial t}\mathbf{u} + \mathbf{u}\cdot\nabla\mathbf{u}\right) = \nabla\cdot\left(\frac{1}{\rho}\mathbf{f}\right). \tag{2.23}$$

Using $\nabla\cdot\mathbf{u} = 0$ and the pressure gradient inserted for the force, we obtain

$$\nabla\cdot(\mathbf{u}\cdot\nabla\mathbf{u}) = -\nabla\cdot\left(\frac{1}{\rho}\nabla p\right). \tag{2.24}$$

This equation expresses the pressure by the fluid velocity and mass density. It is a rather complicated expression for sure, but an equation of state nonetheless. If we can assume that $\rho = \rho_0 = \text{const.}$, the right-hand side of the equation can be rewritten and (2.24) has the form of Poisson's equation $\nabla^2 p = -\rho_0\nabla\cdot(\mathbf{u}\cdot\nabla\mathbf{u})$, which is actually not that hard to solve for a given velocity field $\mathbf{u}(\mathbf{r},t)$. If we have $\rho = \text{const.}$ for an incompressible flow, *any* observer will experience the mass density to be constant. If we, on the other hand, have density gradients in the system, but still an incompressible flow, then in general only a co-moving observer will find the density to be constant. Also, it should be emphasized that a fluid or a gas can appear incompressible in some cases and compressible in others.

Note that in reality the assumption of incompressibility imposes restrictions on the *force density* in the momentum equation; it must not imply accelerations so large as to give large velocities. If, for instance, the motion of the fluid is initiated by a piston moving with velocity U in a tube, we require that the *Mach number*, $M \equiv U/C_s$, is much smaller than unity for the motion to be considered incompressible. If $M \ll 1$, we can assume also perturbations in air to be incompressible, in spite of air being easily compressed in other contexts, sound propagation, for instance.

- **Example:** Consider Poisson's equation for pressure on the form $\nabla^2 p(\mathbf{r}) = g(\mathbf{r})$ to be obtained for instance from (2.24) with $\rho =$ const. The function $g(\mathbf{r})$ represents the resulting spatial variation of the left-hand side of (2.24), with the time variation being immaterial here, and therefore omitted. With modest restrictions on $g(\mathbf{r})$ for $|\mathbf{r}| \to \infty$, the solution of Poisson's equation (Jackson 1975) can be written as

$$p(\mathbf{r}) = -\frac{1}{4\pi}\int \frac{g(\mathbf{r}')}{|\mathbf{r}-\mathbf{r}'|} d\mathbf{r}',$$

where the integral runs over all space. Note that the physical dimension of $d\mathbf{r} \equiv dxdydz$ is *length*^{3} in the present notation, so we have $[p] = [g] \times length^2$ with $[*]$ indicating "dimension of" $*$. For the special case where $g(\mathbf{r}) = A\delta(\mathbf{r})$ we have the well known result $p(\mathbf{r}) = (A/4\pi)/r$ (which is of little relevance here!). This latter result can be expressed in the often seen but nonetheless somewhat ambiguous notation $\nabla^2(1/|\mathbf{r}-\mathbf{r}'|) = -4\pi\delta(\mathbf{r}-\mathbf{r}')$.

In the following we denote $\nabla \cdot \mathbf{u} = 0$ as an equation of state, although it does not contain any expression for the fluid or gas pressure: incompressibility is merely a tool for obtaining the desired equation of state! The correct expression (2.24) is, however, too complicated to be written out explicitly every time (and it gets even worse when we consider magnetohydrodynamics), so we retain this slightly incorrect notation.

In summary, we can list three basic "equations of state".

1. Incompressibility
$$\nabla \cdot \mathbf{u} = 0.$$
2. Isothermal conditions
$$p = \kappa T n = p_0 \frac{n}{n_0}.$$
3. Adiabatic conditions
$$p = p_0 \left(\frac{n}{n_0}\right)^\gamma,$$
with $\gamma = C_P/C_V$.

More generally, we can have an independent dynamic equation for the space-time evolution of the pressure, where the analysis relevant for weakly collisional plasmas is summarized by, for instance, Braginskiĭ (1965).

In case the equation of state can be inverted to give density in terms of pressure, we might simplify, or at least rewrite, the equations somewhat. We can take the adiabatic equation of state, for instance, and find $\rho = \rho_0(p/p_0)^{1/\gamma}$. This, inserted into the equation of continuity (2.22), gives, after some simple manipulations

$$\frac{\partial}{\partial t}p + \mathbf{u}\cdot\nabla p = -\gamma p \nabla\cdot\mathbf{u}, \tag{2.25}$$

where the consequences of compressibility appear explicitly by the right-hand side. The equation of continuity is made redundant by (2.25). The model for incompressibility is obtained by taking $\nabla \cdot \mathbf{u} \to 0$ and letting $\gamma \to \infty$ in such a way that $\gamma p \nabla \cdot \mathbf{u}$ remains finite. This quantity will for this limit simply express the magnitude of Dp/Dt and the equation can be omitted, while the missing constraint is compensated by introducing $\nabla \cdot \mathbf{u}$ explicitly in the basic set of equations. Note that it is *not* permissible naively to let $\nabla \cdot \mathbf{u} \to 0$ in (2.25) without treating γ in a consistent way! If we do this incorrectly, we find $Dp/Dt = 0$, which is wrong as a general statement for incompressible flows.

Ⓢ **Exercise:** Introduce the *vorticity*, $\nabla \times \mathbf{u}$, into (2.10), and derive a dynamic equation for $\nabla \times \mathbf{u}$ for incompressible flows. Ⓢ

2.6 Molecular viscosity

In discussions of (2.9) we let the force on a fluid element be given by the gradient in the pressure. There are of course also other contributions, gravity for instance, which may or may not be important, depending on conditions. We expect, however, the equivalent of *friction* to be important for a wide range of conditions. The force density, $\mathbf{f}_u$, experienced by a fluid element due to the "friction" with its surroundings is usually denoted *viscous interaction*. We expect these interactions to be dependent on the local velocity, $\mathbf{u}(\mathbf{r},t)$, of the flow, and are led to an expansion of the form

$$\mathbf{f}_u = \mathbf{C}_0 + \underline{C}_1 \cdot \mathbf{u} + \underline{C}_2 \cdot \nabla \mathbf{u} + C_{3a} \nabla^2 \mathbf{u} + C_{3b} \nabla \nabla \cdot \mathbf{u} + \ldots, \tag{2.26}$$

where $\mathbf{C}_0$ is a vector, $\underline{C}_1$ is a tensor, etc. The two terms with scalar coefficients C_{3a} and C_{3b} represent the two possible combinations of the case where the ∇-operator acts twice on a vector $\mathbf{u}$. Recall that $\nabla \times \nabla \mathbf{u} = \nabla\nabla \cdot \mathbf{u} - \nabla^2 \mathbf{u}$, so there are no rotational parts missing in (2.26). Terms which contain the effects of variations in density come in addition: these will account for the modifications of the frictional force when a denser fluid element interacts with one which is slightly more rarefied.

Some of the coefficients in (2.26) can be discarded right away: it would thus be unphysical if friction were effective for a fluid at rest, and hence $\mathbf{C}_0 = 0$. Similarly, we would consider it strange if the viscous effects were to change sign just because the velocity changed sign, and consequently we expect $C_1 = 0$. Also by symmetry arguments, we similarly expect $C_2 = 0$. The first effective term will therefore be the one with coefficient C_{3a} and C_{3b}. We will first consider *incompressible* flows, and take $\nabla \cdot \mathbf{u} = 0$. For isotropic flows, we then find the simplest lowest order form to be $\mathbf{f}_u = \mu \nabla^2 \mathbf{u}$. This is the force density associated with the "friction" between a fluid element and the surrounding flow. If we divide $\mathbf{f}_u$ by the fluid mass density we obtain a quantity having the dimension of acceleration. Intuitively, we expect viscosity to give a *deceleration*, but this *need* not be so: imagine a striated velocity field where we have a local *reduction* in velocity. For this case, the friction with the surrounding flow will *accelerate* a fluid element.

Physically the viscosity is expressing the internal momentum exchange mediated by atoms or molecules in the fluid when they cross an (imagined) boundary, or "cut" $A - A'$, in the flow. Assume that the gradient is such that the horizontal velocity component *increases* with the vertical coordinate, z, in a Cartesian coordinate system. Particles entering through the "cut" $A - A'$ from *above* due to their thermal motion have on average a larger horizontal velocity component compared to those crossing $A - A'$ from below. Hence, there must be a momentum exchange across $A - A'$ in general.

The *dynamic* viscosity is a measure for the fluids resistance toward deformations. Newton's relation gives, for the shear stress τ, in the simple geometry outlined here

$$\tau = \mu \frac{d}{dz} u \tag{2.27}$$

where μ is the *dynamic* viscosity. Usually, μ depends strongly on temperature, but not so much on pressure. We can define the *kinematic* viscosity, $\nu \equiv \mu/\rho$, where again ρ is the mass density of the fluid. The dimension of τ is *force / area*, i.e., the same as *pressure*, but in this case the force acts *parallel* to the surface, so it is important to avoid confusion. The quantity μ in (2.27) has dimension *mass* / (*length* $\times$ *time*), and ν has dimension *length*^{2} / *time*, just like a diffusion coefficient.

The foregoing discussion assumes that the mass density of the fluid is constant. Even when the flow is incompressible, it *can* be "seeded" with an initial density variation. One such example is water in the oceans: the spatial variation of the salinity gives a corresponding mass density variation. If the fluid is set into motion, this density variation will give rise to an inhomogeneous mass density at later times even though the fluid dynamics as such can be assumed incompressible. The collisional momentum exchange will be modified when the mass density varies with the spatial coordinate. We will not be concerned with such effects here, but they can be important for phenomena in nature.

The expression $\mathbf{f}_u = \mu\nabla^2\mathbf{u}$ need not be sufficient when compressibility is allowed for (Aris 1962, Landau & Lifshitz 1987). In this case more general expressions are required, which will in general contain $\nabla\cdot\mathbf{u}$ explicitly. We have a useful expression in the form $\mathbf{f}_u = \mu(\nabla^2\mathbf{u} + \frac{1}{3}\nabla\nabla\cdot\mathbf{u})$.

2.6.1 An explicit calculation of the viscosity coefficient

By a simple model we are able to give a simple analytical expression for the viscosity of a fluid or a gas. We can calculate the net momentum carried from *above* the cut $A - A'$ *through* a unit area per sec. Let the density of particles be $n_0 = \text{const.}$ and the directions of the velocity vectors be uniformly distributed over the spherical angle 4π. The fraction of particles with velocities in the spherical angular element $d\Omega$ is then $n_0 d\Omega/(4\pi)$. We want to determine first the number of particles with velocity u in a narrow velocity interval du that enter through a unit area within a directional angle in the narrow interval $\{\Theta, \Theta + d\Theta\}$ within a time interval dt. These particles fill a box with base being unit area and height $u\,dt\cos\Theta$. This expression only includes the *magnitude*, u, of the velocity. Only a fraction of the particles has the correct *direction* for the velocity vector, namely, $n_0 u\,dt\cos\Theta d\Omega/(4\pi)$. We now use this expression for an angular interval of the form of two concentric conical segments with angle Θ and $\Theta + d\Theta$, respectively; see also Fig. 2.2. The corresponding angle is $2\pi\sin\Theta d\Theta$, and the corresponding number dN of particles is

$$dN = \frac{2\pi\sin\Theta d\Theta}{4\pi} n_0 u\cos\Theta\,du\,dt.$$

Integration of this expression over all angles Θ in the interval $\{0, \pi/2\}$ gives $\frac{1}{4}n_0 u\,dt$. We now integrate over all velocities, assuming a Maxwellian velocity distribution and obtain trivially $\frac{1}{4}n_0\langle|u|\rangle$ particles per sec. (Reif 1965).

We have here found the number of particles crossing the reference surface within the time dt. In order to find the *momentum* carried across the unit area of particles with mass M coming from the region above we recall that the vertical momentum component of such a particle is $Mu\cos\Theta$ and we have to integrate

$$\frac{2\pi\sin\Theta d\Theta}{4\pi} n_0 u\cos^2\Theta M u\,du$$

over all angles Θ in the interval $\{0, \pi/2\}$ to find $\frac{1}{6}n_0 M u^2 du$. We now integrate over all velocities, assuming a Maxwellian velocity distribution and obtain trivially a momentum $\frac{1}{6}n_0\langle u^2\rangle$ per sec. It is trivially demonstrated that $\langle u^2\rangle = 3\kappa T/M$ for a Maxwellian distribution, while $\langle|u|\rangle^2 = 8\langle u^2\rangle/(3\pi)$. If we are interested in the *pressure* exerted by such particles on a *solid* wall, we have to recall that the wall receives *twice* the momentum upon a completely elastic reflection, and the pressure, i.e., momentum received per area per sec., becomes $p = \frac{1}{3}n_0\langle u^2\rangle$.

We now consider the inhomogeneous velocity case where it is assumed that the average velocity is in all spatial positions much smaller than the thermal velocity. We can then, as an approximation, still use some of the previous results when estimating the net momentum carried across the unit area as the difference between the momentum received from above minus the momentum received from below. We write the vertical variation of the horizontal average velocity component as $\langle U_y(z)\rangle$. A particle crossing $A - A'$ carries the information from the last time it collided with one of the other particles in the flow. For simplicity, and without much loss of generality, we assume that the mean free collision path, ℓ_c, is identically the same for all particles.

To a good approximation we can assume, also for the present inhomogeneous flow, that the *vertical* momentum crossing $A-A'$ from above and below exactly cancels. Since, however, the *horizontal* flow component varies with z, there will be a net y-component of the momentum deposited at $A-A'$. To estimate this we write the net rate of momentum transfer across a unit area at the reference position z_0 of the cut $A-A'$ as

$$P_y(z_0,\Theta)d\Theta = \frac{2\pi\sin\Theta d\Theta}{4\pi} n_0 u \cos\Theta M \left[\langle U_y(z=z_0-\ell_c\cos\Theta)\rangle - \langle U_y(z=z_0+\ell_c\cos\Theta)\rangle\right].$$

Making a Taylor expansion of the average velocity profile as

$$\langle U(z)\rangle_y \approx \left.\frac{d\langle U_y(z)\rangle}{dz}\right|_{z=z_0}(z_0+z) \equiv \langle U_y(z)\rangle'(z_0+z)$$

with $'$ denoting differentiation, as usual, we can write, without any explicit reference to the particular z_0 position,

$$P_y(z_0,\Theta)d\Theta = -2\frac{2\pi\sin\Theta d\Theta}{4\pi} n_0 M u \ell_c \cos^2\Theta \langle U_y(z)\rangle'.$$

After integration over all Θ in the range $\{0;\pi/2\}$ and averaging over all velocities assuming a local Maxwellian, we obtain

$$P_y(z) = -\frac{1}{3}n_0 M\langle|u|\rangle\ell_c\frac{d\langle U_y(z)\rangle}{dz}, \tag{2.28}$$

where, as mentioned before, we assume that $\langle U\rangle \ll \langle|u|\rangle$. This is the net *force* per unit area exerted on the fluid *above* z_0 on the fluid *below*. The dimension of $P_y(z)$ is *force / area*. The momentum transferred is in the y-direction, so this force is then also in the y-direction. In order to obtain the net viscous drag we take a small box, or parcel of fluid, with base area A and height dz at the selected reference position, and calculate the difference in viscous forces between top and bottom, and find in terms of acceleration a_y in the y-direction that $a_y n_0 M A dz = AP_y(z-dz/2) - AP_y(z+dz/2)$. This gives

$$a_y = \frac{1}{3}\langle|u|\rangle\ell_c\frac{d^2\langle U_y(z)\rangle}{dz^2}$$

in the limit where $dz \to 0$. By direct comparison with the first non-vanishing term in (2.26) we find the kinematic viscosity of a gaseous medium to be

$$\nu = \frac{1}{3}\langle|u|\rangle\ell_c, \tag{2.29}$$

at least to the accuracy of the present calculations. The dynamic viscosity becomes $\mu = \frac{1}{3}n_0 M\langle|u|\rangle\ell_c$. This result is interesting since $\ell_c \approx 1/(n_0\sigma)$ in terms of a collisional cross section σ (see Appendix B), and μ is then, to the present approximation, independent of gas density! It varies with temperature, though, since $\langle|u|\rangle \sim \sqrt{T}$. In case we have a *linear* gradient in velocity, the box, or parcel of fluid, will be accelerated on the top by an amount equal to the deceleration at the bottom, and the net acceleration vanishes. To be effective, viscosity requires a non-vanishing second order derivative of the velocity gradient.

The foregoing analysis implicitly assumes that the density is constant, a requirement a bit stronger than the condition of incompressibility, which was made explicit. In realistic flows it is, however, rather unlikely to encounter density gradients which are comparable to the velocity gradients, at least if viscosity is to play any role at all. If required, the analysis can readily be generalized. The problem of including *compressible* flows is, however, more important, but also a more complicated problem. In this case *both* terms with C_{3a} and C_{3b} in (2.26) have to be retained, and the question is to determine these two coefficients. We will not enter these discussions here in any detail, only note the result, where (2.9) is written as

$$\rho\frac{D}{Dt}\mathbf{u} \equiv \rho\left(\frac{\partial}{\partial t}\mathbf{u} + \mathbf{u}\cdot\nabla\mathbf{u}\right) = \mathbf{f} + \mu\nabla^2\mathbf{u} + (\lambda+\mu)\nabla\nabla\cdot\mathbf{u}. \tag{2.30}$$

For incompressible flows, $\nabla \cdot \mathbf{u} = 0$, and $\mathbf{f} = -\nabla p$, (2.30) is called the *Navier-Stokes* relation. In many cases (Aris 1962) it is possible to argue for Stokes relation $\lambda + \mu = \mu/3$ in (2.30). Often, we can assume the flows to be incompressible, as already argued, but in case we want to study, for instance, damping of sound waves in gases or liquids, the last term in (2.30) becomes important. The Stokes relation is found to be good for monatomic gases, but often fails for liquids and polyatomic gases. For gases, we generally find that viscosity *increases* with temperature, but for liquids, molecular liquids in particular, it might be the other way around. For water, we thus find the viscosity to increase with decreasing temperature. Many liquids follow an approximate relation of the form $\mu \approx A + B/(\kappa T)$ for the dynamic viscosity (Kaye & Laby 1995), valid for a nontrivial temperature range with constant A and B.

A *Newtonian fluid* is one where there is a linear relationship between stress and strain (Aris 1962). This is generally given as a tensor relation, where (2.31) is a simple scalar version of this. Non-Newtonian fluids *do* exist, and have quite peculiar properties.

- **Example:** Consider a simple metal bar of length ℓ_0, and cross section $\mathcal{A}$, fixed in one end, and subject to a small tensile force F at the other end. When F is applied perpendicular to the area $\mathcal{A}$, the force gives rise to an elongation $d\ell$ of the bar, so that the *entire* length of the bar is $\ell_0 + d\ell$. *Hooke's law*, which relates the elongation to the force, has the form (Brekhovskikh & Goncharov 1985)

$$\mathcal{E}\frac{d\ell}{\ell_0} = \frac{F}{\mathcal{A}}, \tag{2.31}$$

  with $\mathcal{E}$ being a proportionality constant. If we double the length of the bar, the relative elongation becomes only half of what it was before; if we double the cross sectional area, we need twice the force to produce the same elongation. The quantity $\mathcal{E}$ is *Young's modulus*, $F/\mathcal{A}$ is the *stress*, while the relative extension $\varepsilon \equiv d\ell/\ell_0$ is the *strain*. Hooke's law (2.31) is a *linear* relation: doubling the force gives a doubling of the elongation. These definitions can be readily generalized also to cover forces in fluid or gases. More generally we can have *shear stresses*, where the force is applied parallel to the area $\mathcal{A}$ and we have here a *shear strain* and a *shear modulus*.

- **Exercise:** An elegant example due to Reif (1965) may illustrate the effect of viscosity. Consider two trains moving on parallel tracks, both of them loaded with sandbags. Let the speed of the trains be somewhat different, and assume now that workers on both trains constantly pick up sand-bags on their own train, and throw them on board the other one. There will be a net transport of momentum between the trains, and the slower one will speed up, while the faster one will be slowed down. Make a realistic model for the problem outlined here, and calculate the rate of acceleration of the two trains, assumed equal to begin with, apart from their difference in speed. Generalize the problem to *three* trains with velocities $u_1 < u_2 < u_3$, where the workers on the middle one (no. two) throw an equal number of sandbags to the right and to the left trains, the other two trains onto train no. two only. Determine the parameters which give train no. two a constant velocity.

2.7 Thermal conductivity in gases

By simplifying assumptions similar to those used in Section 2.6.1 we can obtain an expression for the thermal conductivity in gases. Consider a box like the one shown in Fig. 2.2, assuming that we have a difference in temperature from the top to the bottom of the box. We find the number of particles dN passing through the area dA per time unit to be

$$\begin{aligned} dN &= \frac{1}{4\pi}\frac{nf(u)dud\Omega dV}{dAdt} \\ &= \frac{1}{4\pi}nu\,f(u)du\sin\theta\cos\theta d\theta d\phi \end{aligned} \tag{2.32}$$

where we used the volume $dV = dA\,udt\cos\theta$ and wrote the spherical angle element as $d\Omega = \sin\theta d\theta d\phi$ in terms of the standard angles, while $f(u)$ is the velocity distribution function. An atom or molecule passing dA in the positive z-direction had on average its previous collision at a distance $h = \ell_c\cos\theta$ below the plane determined by dA, where ℓ_c is the collisional mean free path here. The temperature at this position is $T \approx T_0 - hdT/dz = T_0 - \ell_c\cos\theta dT/dz$. Each particle carries with it, on average, an energy $(D/2)\kappa T$, where D is here the number of degrees of freedom. For point-like particles we have $D = 3$, and this will be the value used mostly.

The average upward heat flux in the upward direction is

$$\begin{aligned} dJ_{hu} &= \frac{D}{2}\kappa\left(T_0 - \ell_c\cos\theta\frac{dT}{dz}\right)dN \\ &= \frac{D\kappa}{8\pi}nuf(u)du\left(T_0 - \ell_c\cos\theta\frac{dT}{dz}\right)\sin\theta\cos\theta d\theta d\phi\,. \end{aligned}$$

We can readily integrate this expression with respect to u, θ and ϕ to have

$$J_{hu} = \frac{D\kappa}{8}n\langle|u|\rangle\left(T_0 - \frac{2}{3}\ell_c\frac{dT}{dz}\right). \tag{2.33}$$

This is the heat from below: we have a similar expression for the heat flux from above the plane dA; the only difference is a change in sign in front of ℓ_c. Altogether we find

$$J_h = -\frac{D}{6}n\langle|u|\rangle\ell_c\kappa\frac{dT}{dz}\,.$$

We can identify the *heat conductivity* as $k_h \equiv (D/6)n\langle|u|\rangle\ell_c\kappa$. Using a Maxwellian distribution to calculate $\langle|u|\rangle = \sqrt{8\kappa T/\pi M}$ and using $\ell_c = 1/(\sqrt{2}n\sigma)$ we obtain

$$k_h = \frac{D}{3}\frac{\kappa}{\sigma}\sqrt{\frac{\kappa T}{\pi M}},$$

where σ is the scattering cross section, usually taken to be the geometric cross section $4\pi r_0^2$ in terms of some suitable defined radius r_0. Note that the expression for J_h has the form of Fick's law, i.e., a flux proportional to a constant multiplied by a gradient. It is interesting to note that within the present model, the thermal conductivity is independent of particle density and therefore also independent of pressure for constant temperature. Note that here D is the number of degrees of freedom: shortly the same symbol will be used also for a diffusion constant.

2.8 Diffusion in gases

Viscosity, as discussed in, e.g., Section 2.6.1, can be seen as a transport of momentum density. The thermal conductivity discussed in Section 2.7 can be understood as transport of energy density. In the present section we discuss the transport of mass density by similar arguments. Consider again a box like the one shown in Fig. 2.2; assume that we have a vertical gradient dn/dz in particle density n. An atom or molecule passing dA in the positive z-direction had on average its previous collision at a distance $h = \ell_c \cos\theta$ below the plane determined by dA, where ℓ_c is the collisional mean free path here. The following procedure is very similar to the one we used in Section 2.7; we can in reality simply replace $(D/2)\kappa T n$ by n. The mass flux along the density gradient becomes

$$J_n = -\frac{1}{3}\langle|u|\rangle\ell_c\frac{dn}{dz},$$

and has the form of Fick's law. Inserting the expression for the mean free path $\ell_c = 1/(\sqrt{2}n\sigma)$ we can define a diffusion coefficient $D \equiv \frac{1}{3}\langle|u|\rangle/(n\sigma\sqrt{2})$. The discussion can be extended to allow for a mixture of gases.

The kinematic viscosity ν and the mass diffusivity D have the same physical dimension $length^2\, time^{-1}$, so their ratio, called the Schmidt number, Sc, is a dimensionless quantity. For the present model we find $Sc \equiv \nu/D = \mu/(\rho D) = 1$. Physically Sc relates the relative thickness of a hydrodynamic layer and mass-transfer boundary layer.

Ⓢ **Exercise:** Assume that you are given a diffusion coefficient that depends on spatial position, $D = D(\mathbf{r})$. You can postulate a diffusion equation in the form $\partial n/\partial t = D(\mathbf{r})\nabla^2 n$ or alternatively $\partial n/\partial t = \nabla \cdot D(\mathbf{r})\nabla n$. Dimensionally both forms are correct. Why do you give preference to the latter version?. Ⓢ

3

Linear Wave Dynamics

The dynamic properties of media like plasmas can in most relevant cases be described by a set of differential equations, these being linear or nonlinear (Whitham 1974, Shivamoggi 1988). In this section we discuss wave propagation in general, with little or no specific reference to plasmas, also with the aim of emphasizing the similarities of wave propagation in various media.

Assume thus that a *linear* partial differential equation has been obtained and written in the form

$$\begin{aligned}\Big(a_m(x,t)\frac{\partial^m}{\partial t^m}+a_{m-1}(x,t)\frac{\partial^{m-1}}{\partial t^{m-1}}+\ldots+a_0(x,t)\\ +b_n(x,t)\frac{\partial^n}{\partial x^n}+b_{n-1}(x,t)\frac{\partial^{n-1}}{\partial x^{n-1}}+\ldots+b_0(x,t)\\ +c_p(x,t)\frac{\partial}{\partial t}\frac{\partial^p}{\partial x^p}+c_{p-1}(x,t)\frac{\partial^2}{\partial t^2}\frac{\partial^{p-1}}{\partial x^{p-1}}+\ldots\\ +c_1(x,t)\frac{\partial^p}{\partial t^p}\frac{\partial}{\partial x}+c_0(x,t)\Big)\,\phi(x,t)=0\end{aligned} \tag{3.1}$$

where ϕ denotes the relevant space-time varying quantity, potential, density or something else. Many of the coefficients a_j, b_j, c_j will be vanishing for practical cases. We wrote equation (3.1) for one spatial dimension for simplicity. Using ∇ rather than the one dimensional derivative $\partial/\partial x$, we would have to keep in mind that, for instance, ∇^2 is a scalar operator, while $\nabla^3 \equiv \nabla^2\nabla$ is a *vector* operator. For such a two or three dimensional representation all a_j, b_j and c_j should be interpreted in the sense which makes the equation vectorially consistent. Here we take ϕ to be a scalar with little loss of generality. The model equation (3.1) is clearly linear; if ϕ is a solution, then so is 2ϕ, etc.

A differential equation has to be supplemented with boundary and/or initial conditions. Considering first the *Cauchy problem*, these conditions can prescribe $\phi(\mathbf{r},0)$ and its temporal derivatives $\partial^k\phi(\mathbf{r},t)/\partial t^k\big|_{t=0}$ for $k < m-1$, assuming $a_m \neq 0$. This problem will be referred to as a pure initial condition, with the differential equation (3.1) describing the space-time evolution for all $t > 0$. This is the condition most often used, although it can be argued that there will always be boundaries somewhere in finite space, and that the conditions here must be taken into account. Formally this argument is quite correct, but, for instance, in discussions of waves in space plasmas it might be difficult to worry about boundary conditions one Earth radius away. As an alternative to the Cauchy problem, we can choose to prescribe the value of the function at a closed boundary at all times, for instance. In the investigations of, for instance, the Laplace equation, $\nabla^2\phi = 0$, such conditions have been discussed and classified in detail. Here the prescription of the full boundary value is called *a Dirichlet type problem* (Jackson 1975). As another alternative we can prescribe at all times the derivative of the function in the direction perpendicular to the boundary, $\partial\phi/\partial n$. This will be a *von Neumann type problem* (Jackson 1975). Finally, we might have prescribed "mixed" boundary conditions, where the derivative is given at a segment and the functional value given elsewhere. These types of conditions are at times called *Robin's type problems*. Such a mixed condition will be relevant if we discuss thermal conductivity in relation to a lump of material where parts of the surface are kept at a constant temperature, and the rest is thermally isolated, for instance. Basically, the moral is that the problem is not fully determined unless we have decided on the initial conditions as well as the conditions at the boundary. Even for the restrictive problem of linear equations considered here, this general problem can be immensely complicated. Fortunately, a restricted class of equations often allows a basic insight into the relevant problems. These are those where the material in question can be considered homogeneous, and all the coefficients in (3.1) are constants (in particular, some of them zero). Historically, one of the most important examples is *the wave equation*, (2.20) here rewritten for three spatial dimensions

$$\frac{\partial^2}{\partial t^2}\phi(\mathbf{r},t) - C^2\nabla^2\phi(\mathbf{r},t) = 0,$$

where C is a constant, which characterizes the velocity of propagation.

3.1 Dispersion relations

A particularly simple version of equation (3.1) is obtained when the coefficients a_j, b_j, c_j are constants for all j. The wave equation (2.20) is such an example. Although the example with constant coefficients represents a special case, it is met sufficiently often to justify a special treatment. In spite of representing a simplified limit, such an equation can be quite difficult to solve if nontrivial boundary conditions have to be taken into account. Such cases are often investigated numerically. For an initial value problem, where the variable is prescribed at $t = 0$ and the boundary conditions are "simple", an almost universally employed method of solution is based on Fourier transform in space and time. The ideas are best illustrated by an example. Take, for instance, (3.1) with constant coefficients as stated, and introduce the Fourier transform of ϕ by $\phi(\mathbf{r},t) = \int\int \phi(\mathbf{k},\omega)\exp(-i(\omega t - \mathbf{k}\cdot\mathbf{r}))d\omega d\mathbf{k}$. The virtue of this particular procedure is that all time derivatives are then replaced by a multiplication by $-i\omega$ and all spatial derivatives by a multiplier $i\mathbf{k}$. Consequently, we have that the simple one dimensional relation (3.1) is replaced by an algebraic equation

$$\begin{aligned}\int\int_{-\infty}^{\infty} &\big(a_m(-i\omega)^m + a_{m-1}(-i\omega)^{m-1} + \ldots + a_0 \\ &+ b_n(ik)^n + b_{n-1}(ik)^{n-1} + \ldots + b_0 \\ &+ c_p(-i\omega)(ik)^p + \ldots + c_1(-i\omega)^p(ik) \\ &+ c_0\big)\,\phi(k,\omega)e^{-i(\omega t - kx)}d\omega dk = 0. \end{aligned} \tag{3.2}$$

In order for the expression (3.2) to be zero for all times and at all positions, we must require

$$\begin{aligned} &a_m(-i\omega)^m + a_{m-1}(-i\omega)^{m-1} + \ldots + a_0 + b_n(ik)^n + b_{n-1}(ik)^{n-1} + \\ &\ldots + b_0 + c_p(-i\omega)(ik)^p + \ldots + c_1(-i\omega)^p(ik) + c_0 = 0. \end{aligned} \tag{3.3}$$

This is the dispersion relation for the problem associated with the differential equation (3.1). The procedure obviously presupposes that the expression used for $\phi(x,t)$ *is* a solution which satisfies the imposed boundary conditions. The formulation outlined here is easily generalized to fully three dimensional problems.

A dispersion relation is generally written as $D(\omega,\mathbf{k}) = 0$. In many, but not necessarily all cases, this relation can be written in explicit form, $\omega = \omega(\mathbf{k})$. In many cases we find that ω is real for a real wave-number, $\mathbf{k}$. In most realistic physical systems there will always be *some* damping in the system, and $\omega = \omega_1(\mathbf{k}) + i\omega_2(\mathbf{k})$ will be complex, with an imaginary part $\omega_2 < 0$. (Here we assume both ω_1 and ω_2 to be real.) We might, however, encounter amplifying or *active* media, where $\omega_2 > 0$, corresponding to unstable waves. In such cases the wave amplitude will be increasing without limit for $t \to \infty$, in the present linearized description. Physically, we cannot accept arbitrarily large wave amplitudes, and the error must be sought after in the basic equations. For a *transient* period we can, on the other hand, accept exponentially increasing wave amplitudes without reservations.

- **Example:** The dispersion relation associated with the wave equation (2.20) is $\omega = \pm Ck$.

Ⓢ **Exercise:** Consider a problem with little relevance for plasma physics, but very important for plasma physicists. Imagine one such scientists walking with a cup of coffee. Inevitably he or she spills parts of the content after walking from the coffee machine to the office. Why is this? (Hint: find the dispersion relation for shallow water waves in the literature.) Ⓢ

3.1.1 Complex notation

It is a great simplification to discuss oscillatory motion and waves in terms of a complex notation. Unfortunately, there are slight differences in various presentations in the literature, and it might be worthwhile to discuss this problem briefly. Basically, any physically observable quantity *must* be described in terms of a real expression; taking, for instance, $\phi(t) = \phi_0 \cos(t)$ to be a measurable scalar, we have in terms of a complex presentation $\widetilde{\phi}\exp(-i\omega t)$

$$\phi(t) \equiv \Re\left\{\widetilde{\phi}\exp(-i\omega t)\right\} = \frac{1}{2}\widetilde{\phi}\exp(-i\omega t) + \frac{1}{2}\widetilde{\phi}^* \exp(i\omega t) \tag{3.4}$$

with * denoting a complex conjugate as usual, and $\phi_0 \equiv \Re\{\widetilde{\phi}\}$. The representation (3.4) is exact, but unfortunately a certain sloppiness has developed, and the factor $\frac{1}{2}$ in $\widetilde{\phi}$ is sometimes omitted. This can give rise to inconsistencies, unless the actual notation is kept constantly in mind.

One virtue of the complex notation is that it makes it particularly easy to obtain time averages of, for instance, $\overline{\phi^2(t)}$, with the overline denoting the operation $(1/\mathcal{T})\int_0^{\mathcal{T}} dt$ with $\mathcal{T} \to \infty$. Then

$$\overline{\phi^2(t)} = \frac{1}{2}\left|\widetilde{\phi}\right|^2,$$

without having to actually integrate anything, since it is obvious that products like, for example, $\widetilde{\phi}\exp(-i\omega t)\widetilde{\phi}\exp(-i\omega t)$, which vary with the frequency 2ω, will vanish by averaging.

To illustrate the use of complex notation in detail we consider as an exercise the vector product

$$\begin{aligned}
\mathbf{S} &= \mathbf{E}(z,t) \times \mathbf{H}(z,t) \\
&= \frac{1}{2}\Re\{\mathbf{E}_0 e^{ikz-i\omega t}\} \times \frac{1}{2}\Re\{\mathbf{H}_0 e^{ikz-i\omega t}\} \\
&= \frac{1}{2}\left(\mathbf{E}_0 e^{ikz-i\omega t} + \mathbf{E}_0^* e^{-ikz+i\omega t}\right) \times \frac{1}{2}\left(\mathbf{H}_0 e^{ikz-i\omega t} + \mathbf{H}_0^* e^{-ikz+i\omega t}\right) \\
&= \frac{1}{4}\left(\mathbf{E}_0 \times \mathbf{H}_0^* + (\mathbf{E}_0 \times \mathbf{H}_0^*)^* + \mathbf{E}_0 \times \mathbf{H}_0 e^{2(ikz-i\omega t)} + (\mathbf{E}_0 \times \mathbf{H}_0 e^{2(ikz-i\omega t)})^*\right) \\
&= \frac{1}{2}\Re\{\mathbf{E}_0 \times \mathbf{H}_0^*\} + \frac{1}{2}\Re\{\mathbf{E}_0 \times \mathbf{H}_0 e^{2(ikz-i\omega t)}\}.
\end{aligned}$$

The average over time is then given as

$$\overline{\mathbf{S}} = \frac{1}{T}\int_0^T \mathbf{S}(z,t)dt = \frac{1}{T}\int_0^T \left(\frac{1}{2}\Re\{\mathbf{E}_0 \times \mathbf{H}_0^*\} + \frac{1}{2}\Re\{\mathbf{E}_0 \times \mathbf{H}_0 e^{2(ikz-i\omega t)}\}\right) dt$$

The second term is a propagating plane wave $e^{2(ikz-i\omega t)}$ whose time-average will be zero at any fixed position z. This gives

$$\overline{\mathbf{S}} = \frac{1}{2}\Re\{\mathbf{E}_0 \times \mathbf{H}_0^*\}.$$

The strength of the complex notations is most apparent when used in linear differential equations, where $\partial/\partial t \to -i\omega$ and $\nabla \to i\mathbf{k}$, leaving the exponential factor to appear in every term, so it subsequently cancels. The problem associated with the complex notation arises because only the real part represents a physically observable phenomenon. This means that $\exp(-i\omega t)$ and $\exp(i\omega t)$ are equally acceptable in this respect. It seems (at least for the present author) that engineers prefer the latter version with the notation $\exp(j\omega t)$, while physicists swear by the former.

3.1.2 Characteristic velocities

Given a dispersion relation, implicit or explicit, we can define a characteristic velocity, the *phase velocity*, as $u_f \equiv \omega/k$. Given a plane wave, $\cos(\omega t - kx)$, we find that any phase θ of the wave at a position x_1 given at time t_1, its local maximum or minimum, for instance, is recovered later at time t_2 at a position x_2, where $\theta = \omega t_1 - kx_1 = \omega t_2 - kx_2$ implies a constant propagation velocity $(x_2 - x_1)/(t_2 - t_1) = \omega/k$. Usually, we do not assign any *direction* for the phase velocity, only a *magnitude*. The direction will be along the wave-vector **k**.

If we are dealing *not* with a plane wave, but one which is slightly amplitude modulated, we can define another characteristic velocity, the *group velocity* $u_g \equiv d\omega(k)/dk$. When the dispersion relation is given in the explicit form $\omega = \omega(k)$ it is often easy to determine u_g. In case the dispersion relation has the implicit form $D(\omega, k) = 0$ we have

$$\frac{d}{dk}D(\omega,k) = \frac{\partial D(\omega,k)}{\partial k} + \frac{\partial D(\omega,k)}{\partial \omega}\frac{d\omega}{dk} = 0$$

giving

$$u_g = -\frac{\partial D(\omega,k)}{\partial k} \bigg/ \frac{\partial D(\omega,k)}{\partial \omega}. \tag{3.5}$$

To illustrate the physical meaning of the group velocity we write an amplitude modulated plane wave as, for instance, $U(t,z) = A_0 \cos(\omega t - kz + \theta) + \Delta A \cos\left((\omega + \Delta\omega)t - (k + \Delta k)z\right)$, where θ is some constant phase. The amplitudes of the primary plane wave and the modulation are A_0 and ΔA, respectively, while $\omega + \Delta\omega$ and $k + \Delta k$ are the frequency and wavenumber of the modulation. We have the dispersion relation $\omega = \omega(k)$ and $\omega + \Delta\omega = \omega(k + \Delta k)$. Using the trigonometric relation $\cos\alpha + \cos\beta = 2\cos\left(\frac{1}{2}(\alpha + \beta)\right)\cos(\frac{1}{2}(\alpha - \beta))$ we can write

$$\begin{aligned} U(t,z) &= (A_0 - \Delta A)\cos(\omega t - kz + \theta) \\ &\quad + 2\Delta A \cos\left(\frac{1}{2}(\Delta\omega t - \Delta k z) - \frac{1}{2}\theta\right) \cos\left((\omega + \frac{1}{2}\Delta\omega)t - (k + \frac{1}{2}\Delta k)z + \frac{1}{2}\theta\right). \end{aligned}$$

The $\Delta A \cos\left(\frac{1}{2}(\Delta\omega t - \Delta k z) - \frac{1}{2}\theta\right)$ multiplier accounts for the amplitude modulation, or the *envelope* of the *carrier wave*, which is here $\cos\left((\omega + \frac{1}{2}\Delta\omega)t - (k + \frac{1}{2}\Delta k)z + \frac{1}{2}\theta\right)$, as distinct from the primary wave $(A_0 - \Delta A)\cos(\omega t - kz + \theta)$. The speed of propagation of the envelope is $\Delta\omega/\Delta k$. In the limit $\Delta k \to 0$, we have $\Delta\omega/\Delta k \to d\omega/dk \equiv u_g$, demonstrating that a very long wavelength envelope propagates with the group velocity. In this limit also the distinction between the carrier wave and the primary wave, as defined before, vanishes apart from a phase shift. Note that since both ω and $\omega + \Delta\omega$ belong to the same dispersion relation we necessarily have $\Delta\omega \to 0$ for $\Delta k \to 0$, assuming that the dispersion relation is differentiable. If it is not, we cannot define a group velocity. Such cases are encountered in plasma physics, in particular for conditions out of equilibrium.

An expression like (3.5) refers to a spatially one dimensional model. For the fully three dimensional case we have

$$\mathbf{u}_g = -\nabla_\mathbf{k} D(\omega,k) \bigg/ \frac{\partial D(\omega,k)}{\partial \omega}, \tag{3.6}$$

with $\nabla_\mathbf{k} \equiv \{\partial/\partial k_x, \partial/\partial k_y, \partial/\partial k_z\}$. In the fully three dimensional model we can find the directions of $\mathbf{u}_g$ and $\mathbf{k}$ to be different. For one dimensional wave propagation we can have "backward waves" where the phase velocity and the group velocity has opposite directions. The group velocity characterizes also the motion of localized wave-packets. As long as the dispersion of this wave-packet can be ignored, it can be considered a "blob" of wave energy, and $\mathbf{u}_g$ is then the velocity of wave energy transport (Coulson 1955, Brillouin 1960).

The notion of phase and group velocities refers to wave-like motion (Brillouin 1960). For communication purposes, one might be interested in the velocity with which *information* is propagating, or the "signal velocity". In the early days of the special theory of relativity, this was a particularly relevant discussion, because paradoxes could be produced, indicating that information *could* be propagated with velocities exceeding that of the speed of light in vacuum. A detailed discussion of these and related problems (interesting as they might be) falls outside the present treatise, and the interested reader is referred to the literature (Brillouin 1960, Bloch 1976).

With reference to the special theory of relativity, it is often, but incorrectly, claimed that the group velocity associated with a correctly derived dispersion relation must at all frequencies and wave-numbers be less than the speed of light in vacuum. The argument should be that information is propagated by at most the speed of light, and that a wave propagates information by the group velocity, therefore $u_g \leq c$. The first part of the argument is evidently correct, but the last part is not! A modulated wave-train does not necessarily contain any information: if the waveform of the modulation is infinitely many times differentiable, we can, at least in principle, make a Taylor expansion around a given point on the basis of an observation in any finite time interval, and again in principle predict the form of the modulated wave-train at any later time (Pécseli 2000). A wave modulated in this way can therefore propagate at any speed, at least as far as the special theory of relativity is concerned. The signal velocity (that propagates information) is different from the group velocity. In particular in plasma media, there are many examples of group velocities being larger than c. Information is associated with lack of complete predictability and has to be associated with discontinuities in the signal or one of its derivatives, but such discontinuities do *not* propagate with the group velocity of the carrier wave! A wave-front, where the wave amplitude rises from zero level to some finite amplitude, could be a good candidate for carrying information, and it is the case analyzed in most detail (Brillouin 1960, Yeh & Liu 1972). Very often it is found that such a wave-front propagates with the speed of light, at least in plasmas and waveguides, to give two examples. Intuitively, this is reasonable: if we, in imagination, make a sort of local spectral analysis, the step function associated with the envelope contains frequency components extending to $\omega \to \infty$. For very high frequencies most media behave like a vacuum, and that part of the signal is therefore likely to propagate with the corresponding speed of light. *Pulses* are also obvious candidates for signal propagation (and probably most frequently used in communication).

3.1.3 Evolution of modulated waves and wave-packets Ⓚ

It was demonstrated in Section 3.1.2 that the envelope of a modulated wave or a wave-packet in a first approximation propagates with the group velocity (3.6). The result is somewhat restrictive since it explicitly assumes that the basic wavelength is much smaller than the characteristic scale length of the wave envelope, but (3.6) is used quite generally nonetheless. Higher order corrections will give a slow distortion of the wave-packet (Ikezi et al. 1978, Sugai et al. 1979). Let the wave-train be given by the analytical form

$$\psi(z,t) = \int_{-\infty}^{\infty} \Phi(\omega) \exp(ik(\omega)z - i\omega t) d\omega \tag{3.7}$$

where the dispersion relation is given as $k = k(\omega)$ rather than $\omega = \omega(k)$. We assume that the wave is excited at a boundary of the system at $z = 0$ and $\Phi(\omega)$ is the Fourier transform of $\psi(0,t)$. It is now assumed that $\Phi(\omega)$ is significant only in a narrow interval around $\omega = \omega_0$. Introducing $\delta\omega \equiv \omega - \omega_0$ we make a series expansion giving the envelope function

$$\phi(z,t) = \int_{-\infty}^{\infty} \Phi(\omega_0 + \delta\omega) \exp(i(\delta\omega/u_g - P\delta\omega^2)z - i\delta\omega t) d\delta\omega, \tag{3.8}$$

where

$$\psi(z,t) = \phi(z,t) \exp(ik_0 z - i\omega_0 t)$$

with $k_0 \equiv k(\omega_0)$. We introduced the abbreviation

$$P \equiv -\frac{1}{2}\frac{d^2k}{d\omega^2}\bigg|_{\omega=\omega_0} = \frac{1}{2u_g^3}\frac{d^2\omega}{dk^2}\bigg|_{k=k_0},$$

and used

$$k(\omega) \approx k_0 + \frac{dk}{d\omega}\delta\omega + \frac{1}{2}\frac{d^2k}{d\omega^2}\delta\omega^2 \equiv k_0 + \frac{\delta\omega}{u_g} - P\delta\omega^2.$$

By differentiation with respect to z and t it is now readily demonstrated that $\phi(z,t)$ as given by (3.8) is a solution to the partial differential equation

$$i\left(\frac{\partial\phi}{\partial z} + \frac{1}{u_g}\frac{\partial\phi}{\partial t}\right) + P\frac{\partial^2\phi}{\partial t^2} = 0, \tag{3.9}$$

having the form of a linear Schrödinger equation, albeit with variables differing from the standard form. This version is best suited for describing waves excited at a boundary. The relation (3.9) has a dispersion relation

$$K = \frac{\Omega}{u_g} \pm |P|\Omega^2, \tag{3.10}$$

that accounts for a modulation in the form of a plane wave with frequency Ω and wavenumber K.

The expression summarized so far refers to the case where a wave is excited at a boundary: after all this is the most common experimental condition. For a theoretical or numerical analysis, often an initial condition is used, and the expressions have to be modified accordingly, where it is more convenient to write the dispersion relation in the form $\omega = \omega(k)$.

Let a modulated wave be excited a $z = 0$, with a modulation index $m < 1$ so that $\psi(o,t) = (1 + m\cos\Omega_0 t)\cos\omega_0 t = \cos\omega_0 t + m(\cos(t(\omega_0 + \Omega_0)) + \cos(t(\omega_0 - \Omega_0)))/2$ giving an upper and a lower sideband, each corresponding to an envelope and a group velocity as in Section 3.1.2. For dispersive waves, the two envelopes will propagate with their respective group velocities, and there will be a "beating" between the two modes, i.e., a recurrence of the modulation. The difference in modulation wavenumbers found by use of (3.10) gives the recurrence length as $\mathcal{L} = 2\pi/(|P|\Omega_0^2)$.

For the wave modulation outlined before we can write the expression for the envelope function as $\phi(z,t) = 1 + m\cos(P\Omega_0^2 z)\cos(\Omega_0(z/u_g - t))$. The wave amplitude is spatially modulated on two scales, the long recurrence length $\mathcal{L}$ and a shorter scale $2\pi u_g/\Omega_0$. For the present case with a wave excited at a boundary, the modulation disappears at regular spatial intervals $z = (\pi/2 + n\pi)/(P\Omega_0^2)$ at all times.

By (3.9) we have $\int_{-\infty}^{\infty}|\phi(z,t)|^2 dz$ to be conserved. Using a Gaussian wavepacket as a test function in (3.9) it can be shown that the width of $|\phi(z,t)|^2$ increases linearly with z, and the peak value of the wavepacket must then decrease as z^{-1}. A physical explanation for this result can be found by taking the modulation index $m = 1$, and use a section of the modulated wave from one zero of the modulation to the next as a wavepacket. The wavepacket is generated at $z = 0$ and expands as it propagates for $z > 0$. The two sidebands propagate at their respective group velocities, so for large z the width in time $L(z)$ of the wavepacket detected at a position z increases as

$$L(z) = L(0) + \left|\frac{dk}{d\omega}\bigg|_{\omega=\omega_0-\Omega_0} - \frac{dk}{d\omega}\bigg|_{\omega=\omega_0+\Omega_0}\right| z,$$

in the present formulation. Using a series expansion for $k(\omega)$ as before we find the expansion as $|P|\Omega_0^2 z$, i.e., linearly increasing with distance z from the exciter.

- **Exercise:** Prove that $\phi(z,t)$ as given by (3.8) indeed satisfies (3.9).

- **Exercise:** The analysis in this section considers a boundary value problem, where a modulated wave train or a wave-packet is excited at a boundary. It will be instructive to the reader to go through the analysis for an initial value problem where the waves are excited over all space at some time t_0.

Ⓢ **Exercise:** Sometimes, in particular in older literature (Coulson 1955), you will find the frequency of a wave expressed not in terms of wave-number k as $\omega = \omega(k)$ but of wavelength λ as $\omega = \omega(\lambda)$. Express the group velocity and the associated differentiation in terms of the wavelength. Ⓢ

Ⓢ **Exercise:** Consider a localized wave-packet propagating according to a dispersive wave relation, i.e., a case where the group and phase velocities are different. Assume that you have obtained a "snapshot" of the wave, showing the wave-packet at a certain time. You can count a number N_G of zero-crossing on this figure. Take now instead a time record, obtained at a fixed position while the wave-packet passes by. On this figure you count a number N_P of zero crossings. Give an estimate for the ratio N_G/N_P (Sugai et al. 1979). Ⓢ

3.1.4 Doppler shifts

With the exception of electromagnetic waves in vacuum, all sorts of wave motion are supported by matter, and the wave characteristics change if we move from one frame of reference to another. The classical example is sound waves, where the detected frequency changes depending on the velocity of our reference system with respect to the sound emitter. The Austrian scientist J. C. Doppler's (1803–1853) original analysis of this phenomenon was quite amusing, involving a train, two carefully tuned trumpets with their players and an observer with absolute pitch. The essential point in the classical Doppler shift is the existence of a preferred frame of reference, namely, the one at rest with respect to the matter supporting the waves.

Note first that the phase of a plane wave is an invariant (the same in any frame of reference), since the elapsed phase of a wave is the number of wave crests passing an observer, and this involves merely a counting, not an actual measurement. Assume that we have two frames with coordinates and time $\mathbf{r}$, t and $\mathbf{r}'$, t', respectively, where the phase velocities are C and C', respectively, and the two frames are moving with a relative velocity $\mathbf{U}$. The Galilean transformation is $t = t'$, $\mathbf{r}' = \mathbf{r} - \mathbf{U}t$. Given phase-normal vectors $\mathbf{n}$ and $\mathbf{n}'$ (being distinct from the wave-vectors), the phase of the wave is $\psi = \omega(t - \mathbf{n}\cdot\mathbf{r}/C) = \omega(t' - \mathbf{n}'\cdot\mathbf{r}'/C')$, where we explicitly used the invariance of the phase ψ. We now insert the expression for t and $\mathbf{r}$ to have $\omega(t'(1 - \mathbf{n}\cdot\mathbf{U}/C) - \mathbf{n}\cdot\mathbf{r}'/C) = \omega(t' - \mathbf{n}'\cdot\mathbf{r}'/C')$. This result most hold for any time and position, so the coefficients to t' and $\mathbf{r}'$ must vanish simultaneously. This requirement results in the classical expressions for the Doppler shift

$$\mathbf{n} = \mathbf{n}', \qquad C' = C - \mathbf{n}\cdot\mathbf{U}, \qquad \text{and} \qquad \omega' = \omega(1 - \mathbf{n}\cdot\mathbf{U}/C). \tag{3.11}$$

Note that we have not yet made any decision concerning which one of the phase velocities refers to the rest frame!

The expressions for the Doppler shifts (3.11) are best understood when illustrated by some examples. Let a sound emitter with frequency $f_0 \equiv \omega_0/2\pi$ be at a fixed position with respect to the reference gas (air), which supports sound waves with constant phase velocity C_s. The frequency detected by an observer moving with velocity U is then $f_1 = f_0(C_s \pm U)/C_s$, where we use the + sign if the observer is moving *toward* the emitter, and the $-$ sign for motion in the opposite direction. Assume now that the observer is at a fixed position, and the *sound emitter* is moving with velocity U. In this case, the detected frequency is $f_1 = f_0 C_s/(C_s \mp U)$. Here we use the $-$ sign when the emitter is moving toward the observer, and vice versa. We used the frequency f measured in Hz in the examples, since this is what a detector usually measures. Since the speed of sound C_s is not so large in air at room temperature, the Doppler shift of sound is something we can easily observe, as indeed most readers would have experienced when an ambulance or similar vehicle is passing.

Ⓢ **Exercise:** A bat is flying with a velocity U along the positive direction of the x-axis, and emits sound pulses with center frequency ω_0. The sound pulse is reflected from a bug flying also along the x-axis, with a velocity V, which can have either sign. Give the analytical expressions for the frequency ω_1 detected by the bug, and ω_2 detected by the bat, when it hears the reflected sound pulse. Assume that the air can be taken to be an ideal gas of oxygen molecules O_2 at room temperature: what is the sound speed? What are the frequencies ω_1 and ω_2 if $f_0 \equiv \omega_0/2\pi = 1$ kHz and $U = 60$ km/h while $V = -40$ km/h? Ⓢ

Ⓢ **Exercise:** Assume that a wave is reflected from a moving obstacle, i.e., a wall of some suitable material, with a propagation velocity U. Let the wave be dispersive, $\omega = \omega(k)$ assuming a one dimensional problem. If the incoming wavelength is given, what is the wavelength of the reflected wave. If the incoming wave has the form of a wavepacket with some given width, what is the width of the reflected wavepacket? Express the results in terms of the propagation velocities before and after reflection, and give an implicit expression for solving the outgoing wavelengths and widths for the incoming wavelength given. You can simplify the analysis by letting the wavepacket have the form of a "box". The problem can be analyzed conveniently in terms of a space-time diagram.

Practical applications of the results from this exercise have been discussed by Yeh and Casey (1966) and Dysthe et al. (1983). The present exercise considers only "head-on" reflection. A more intricate question deals with a "side-wise" moving slab (Yeh & Casey 1966), not discussed here. Ⓢ

3.1.5 Wave polarization

Introduce a vector, the electric field vector, for instance, in the x,y-plane as

$$\mathbf{E} = \widehat{\mathbf{x}}E_x + \widehat{\mathbf{y}}E_y = \widehat{\mathbf{x}}E_{0x}e^{-i(\omega t - kz)} + \widehat{\mathbf{y}}E_{0y}e^{-i(\omega t - kz + \delta)} \equiv \left(\widehat{\mathbf{x}}E_{0x} + \widehat{\mathbf{y}}E_{0y}e^{-i\delta}\right)e^{-i(\omega t - kz)}, \tag{3.12}$$

where both E_{0x} and E_{0y} are assumed to be real without loss of generality since δ can accommodate any phase differences. The electric field is here associated with a plane wave propagating with $\mathbf{k}$ along the z-direction. We define the *polarization* of the vector as

$$\mathcal{R}_W = \frac{E_x}{E_y} = \frac{E_{0x}}{E_{0y}}e^{i\delta} \equiv \frac{E_{0x}}{E_{0y}}(\cos\delta + i\sin\delta)\,.$$

If $\delta = n\pi$ we have $\mathcal{R}_W$ being real; if $\delta = \pm\pi/2 + n\pi$ with $n = 0, 1, 2\ldots$, we find $\mathcal{R}_W$ imaginary. Real $\mathcal{R}_W$ implies that E_x and E_y are proportional for *linear polarization.* The *angle of polarization* θ_0 is given by $\sin\theta_0 = E_{0x}/E_{0y}$. For $\mathcal{R}_W$ imaginary, we have E_x being maximum when $E_y = 0$, and vice versa, so the field vector rotates in the x,y-plane, in general with the "arrow" following an ellipse with major and minor axes along the x- and y-axes, in special cases a circle, denoted elliptical or circular polarizations, respectively. If $\mathcal{R}_W$ is complex we have, in general, elliptical polarization, with the major axes of the ellipse being at an angle to the coordinate axes. Evidently, by proper choice of reference system, we can make the major axis of the ellipse coincide with an axis of the coordinate system.

Circularly polarized waves are particularly important and will be illustrated in some detail here. For this purpose we let $E_{0x} = E_{0y} \equiv E_0$ and take the relative phase to be $\delta = \pm\pi/2$. In complex notation, the wave field is then given by

$$\mathbf{E} = E_0\left(\widehat{\mathbf{x}} \pm i\widehat{\mathbf{y}}\right)e^{-i(\omega t - kz)}.$$

In a representation using real functions, the same expression becomes

$$\mathbf{E} = E_0\left(\widehat{\mathbf{x}}\cos(\omega t - kz) \mp \widehat{\mathbf{y}}\sin(\omega t - kz)\right),$$

Selecting a fixed spatial position, say $z = 0$, then for the combination $\widehat{\mathbf{x}} - i\widehat{\mathbf{y}}$ the point of field vector $\mathbf{E}(t)$ will rotate *against* the clock for when the observer is facing the oncoming wave: we define this as a *right circular* wave polarization. For the opposite case, with $\widehat{\mathbf{x}} + i\widehat{\mathbf{y}}$, the field vector rotates *with* the clock, and we call this *left circular* polarization; see also Fig. 3.1 specifying the left- and right-hand polarized waves.

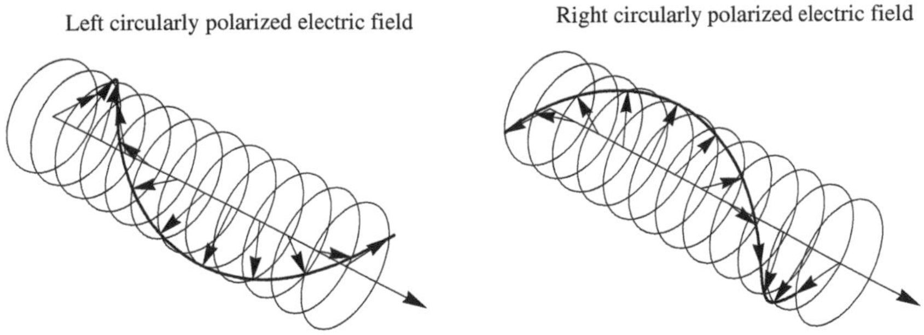

FIGURE 3.1: *Illustration of the spatial variations of left and right circular electric field polarizations at a given time. The waves are assumed to be propagating to the right in both cases. The long arrow points in the positive z-direction. The figure refers to the notation adapted in plasma physics. Be careful to distinguish circular polarization for a fixed position as a function of time, and the spatial variation for a fixed time.*

If we select a fixed time, say $t = 0$, then the spatial variation of the wave can be expressed as $\mathbf{E} = E_0(\widehat{\mathbf{x}}\cos(kz) + \widehat{\mathbf{y}}\sin(kz))$, for the left circular polarization in a representation using real functions, and similarly $\mathbf{E} = E_0(\widehat{\mathbf{x}}\cos(kz) - \widehat{\mathbf{y}}\sin(kz))$ for the right polarized wave. Introduce the complex orthogonal unit vectors

$$\widehat{\varepsilon}_\pm \equiv \frac{1}{\sqrt{2}}(\widehat{\mathbf{x}} \pm i\widehat{\mathbf{y}})$$

with properties $\widehat{\varepsilon}^*_\pm \cdot \widehat{\varepsilon}_\mp = 0$, $\widehat{\varepsilon}^*_\pm \cdot \widehat{\mathbf{z}} = 0$ and $\widehat{\varepsilon}^*_\pm \cdot \widehat{\varepsilon}_\pm = 1$. Then, as a general representation, equivalent to (3.12), we can write

$$\mathbf{E} = \widehat{\varepsilon}_+ E_+ + \widehat{\varepsilon}_- E_- = (\widehat{\varepsilon}_+ E_{0+} + \widehat{\varepsilon}_- E_{0-})\, e^{-i(\omega t - kz)}, \tag{3.13}$$

where E_{0+} and E_{0-} are in general complex amplitudes.

The notation "right" and "left" polarization originates from optics. Unfortunately, the traditions in optics and plasma sciences evolved differently, and in optics we often find the opposite definitions of wave polarizations as compared to the usage in plasma sciences. Some of the classical textbooks can therefore be confusing for a plasma physicist. A more modern usage in optics would be *positive helicity* for left circular polarization and *negative helicity* for right circular (Jackson 1975).

A circularly polarized wave can be written as a sum of two orthogonal linearly polarized waves, and a linearly polarized wave can be written as a sum of a left- and a right-hand circularly polarized wave, with the implicit assumption that both waves exist. There are media, plasma media in particular, where there are frequency ranges where only one of the circular polarizations exists, and in such cases we cannot propagate a linearly polarized transverse wave.

3.1.6 Method of stationary phase

Suppose we want to evaluate integrals of the form $\int F(x)\cos(W(x))dx$ or $\int F(x)\exp(iW(x))dx$, with real W, where the integrations run along the real x-axis. Assume that $F(x)$ is slowly varying everywhere. Then, in places where $W(x)$ varies rapidly, we have $\cos(W(x))$ or $\exp(iW(x))$ to vary rapidly, and the integral contributions from those regions are virtually vanishing (Brillouin 1960). It is therefore important to identify regions where $W(x)$ is either a local maximum or local minimum, i.e., regions where $dW(x)/dx \approx 0$. These are called points of *stationary phase*.

For illustration (Whitham 1974), we consider a function $f(x,t) = \int F(\gamma)\exp(i\chi(\gamma)t)d\gamma$ with $\chi(\gamma) = W(\gamma) - \gamma x/t$, where we take x/t to be a fixed parameter. A point of stationary phase $\gamma = k$ is defined by $\chi'(k) = W'(k) - x/t = 0$. In case this relation is *not* satisfied, the integrand oscillates rapidly and the integral becomes negligible. The functions $F(\gamma)$ and $\chi(\gamma)$ are now expanded into a Taylor series in the neighborhood of $\gamma = k$, giving $F(\gamma) \approx F(k)$ (since F was assumed to be slowly varying) and $\chi(\gamma) \approx \chi(k) + \frac{1}{2}(\gamma - k)^2\chi''(k)$, where we implicitly assume $\chi''(k) \neq 0$. With these approximations, we find for large times

$$f(x,t) \approx F(k)e^{-i\chi(k)t}\int_{-\infty}^{\infty} e^{-\frac{i}{2}(\gamma-k)^2\chi''(k)t}d\gamma.$$

By rotating the path of integration through $\pm\pi/4$, the integral is reduced to the error integral

$$\int_{-\infty}^{\infty} \exp(-\alpha z^2)dz = \sqrt{\pi/\alpha}.$$

The sign of rotation should be chosen to be the sign of $\chi''(k)$. We then have

$$f(x,t) \approx F(k)e^{-i\chi(k)t - i\frac{\pi}{4}sign(\chi''(k))}\sqrt{\frac{2\pi}{t|\chi''(k)|}}. \tag{3.14}$$

There may be several points of stationary phase, and in such cases we have a sum of terms like (3.14). The method, and its generalizations, is discussed by several authors (Brillouin 1960, Whitham 1974).

The stationary phase method is of particular value when $W(\gamma)$ represents a dispersion relation, for instance, $\omega = \omega(k)$ or the "inverted" dispersion relation $k = k(\omega)$. In the former case we have that W' is the group velocity, and the transformation giving $\chi'(k) = 0$ simply means that we prefer the reference system moving with the group velocity. In this frame the oscillatory wave field becomes $\exp(-i\chi(k)t)$ multiplied by a phase factor, and we also find that the wave amplitude is damped due to the wave dispersion, as given by $d^2\omega(k)/dk^2$. It is interesting to note that the amplitude decreases as $\sim 1/\sqrt{t}$, i.e., in the same way as we find for diffusive processes as described by the standard diffusion equation. The derivation explicitly assumed that W, and therefore $\omega(k)$, is real. Neither damping nor wave growth was included. Such cases require a special analysis (Brillouin 1960).

3.1.7 Absolute and convective instabilities

When dealing with unstable media, there is an important distinction we have to be aware of. Basically, a fixed observer can distinguish *two* different cases; either the initial perturbation is increasing in all spatial positions, or alternatively it can propagate in such a way that in any given position the wave amplitude is increasing for a while and then decreases again. In this latter case *another* observer at a *different* position will find the same overall features, but with a larger peak value for the wave pulse. The question is most easily illustrated as in Fig. 3.2.

Formally, the distinction can appear rather trivial, since it is only a question of selecting the frame of reference. However, for problems involving fixed boundaries, we have no such freedom of choice, and it is important to be able to make the distinction (Dysthe 1966). For simplicity we consider here a spatially one dimensional case (Kadomtsev 1965).

Suppose that the growth rate of the instability as expressed through $\Im\{\omega(k)\} \equiv \omega_2(k)$ has a maximum at $k = k_0$. Evidently, after a sufficiently long time, any given initial perturbation will

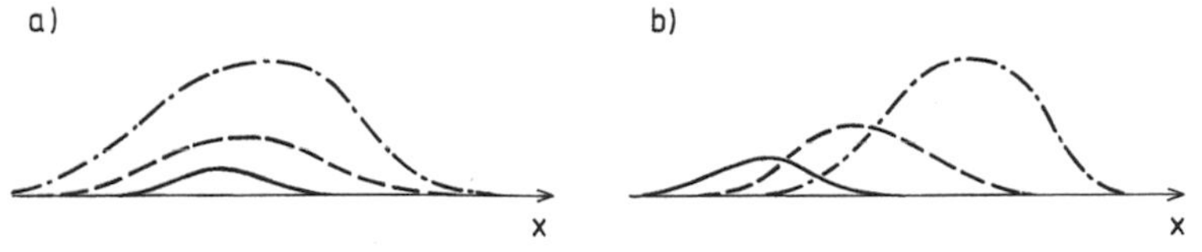

FIGURE 3.2: *Illustration of the difference between an* absolute *in a) and a* convective *instability in b). A solid line indicates an initial perturbation, dashed and dot-dashed lines the evolution at later times. The curves can show either a pulse-like perturbation, or the envelope of a localized wave-packet, centered around a wave-number k_0.*

be deformed so that it has a sharp maximum centered around k_0. It is therefore here sufficient to consider a wave-packet consisting of a superposition of plane waves with wave numbers close to k_0. We can then make a series expansion of ω_1 and ω_2 with respect to the small quantity $k - k_0$, i.e.,

$$\omega_1 \approx \omega_1(k_0) + u_g(k - k_0) + \frac{1}{2}u'_g(k - k_0)^2 \quad \text{and} \quad \omega_2 \approx \omega_2(k_0) - \alpha\frac{1}{2}(k - k_0)^2,$$

where $\alpha \equiv d^2\omega_2(k)/dk^2|_{k_0}$, $u_g \equiv d\omega_1(k)/dk|_{k_0}$ and $u'_g \equiv d^2\omega_1(k)/dk^2|_{k_0}$. Evidently we have $d\omega_2(k)/dk|_{k_0} = 0$. For simplicity, we now transform to a coordinate system moving with velocity U so that $\omega_1(k_0) - Uk_0 = 0$, making the real part of the frequency vanish. The time evolution of a wave-packet is then simply given by

$$\phi(x,t) = A(k_0)\exp(\omega_2(k_0)t - ik_0x) \int \exp\left(i(x - u_gt)(k - k_0) - \frac{1}{2}t(\alpha - iu'_g)(k - k_0)^2\right)dk, \tag{3.15}$$

where $A(k_0)$ is the initial amplitude of the perturbation. The integral can be approximated to give

$$\phi(x,t) = A(k_0)\exp\left(-ik_0x + \omega_2(k_0)t - \frac{(x - u_gt)^2}{2(\alpha^2 + u'^2_g)t}(\alpha + iu'_g)\right), \tag{3.16}$$

where the last term in the parentheses accounts for the spreading out of the wave-packet. For

$$u_g^2 < 2\omega_2(k_0)\frac{(\alpha^2 + u'^2_g)}{\alpha} \equiv U_c^2, \tag{3.17}$$

the amplitude of the wave-packet will increase in time for all positions x, while in the opposite case the increase is observed only in a suitably moving coordinate system. Physically, the term u'_g disperses a wave-packet, and (3.17) states that in case this is a fast process, the instability is absolute. We emphasize that the expression (3.15) is applicable only in the vicinity of $x = u_gt$, and the result (3.17) is therefore only approximate.

The distinction between absolute, $u_g^2 < U_c^2$, and convective, $u_g^2 > U_c^2$, instabilities is, as already mentioned, necessitated by the presence of boundaries in linearly unstable active media. If the medium is *absolutely* unstable the perturbation will be likely to increase in amplitude at all positions in the system, until the amplitudes become so large that nonlinear effects set in, and the basic equations become inapplicable. For *convectively* unstable systems, on the other hand, we might have that the amplitude of the perturbation cannot reach any appreciable value before it escapes the system, and the linear analysis can be used. Most lasers (laser: Light Amplification by Stimulated Emission of Radiation) are convectively unstable. Partially reflecting mirrors at both ends of the system are needed to reflect a significant part of the wave energy into the active medium, to have multi-pass amplification. Without this, lasers would not give an output of any appreciable intensity, with the exception of a few "super-lasing" media.

3.1.8 Pulse response

As we have seen, the concept of a dispersion relation is connected with our choice of Fourier transform of the basic dynamic equation, i.e., our choice of the ortho-normal set of functions. The description of the dynamical properties of media in terms of a dispersion relation is so well established that many authors attempt to attribute a sort of "physical reality" to plane waves, in particular also because they are so readily excited in controlled laboratory experiments. This idea *can* be misleading, and at least it deserves a brief discussion (Haskell & Case 1967, Vaĭnshteĭn 1976).

If we drop a piece of jelly placed in a rectangular box, it will be set into violent oscillations when it hits the floor. As long as the amplitudes of the oscillations are not too large, these oscillations can be *described* by a superposition of plane waves. If, however, the jelly is confined to a circular cylindrical box we can easily imagine cases where a plane wave model is not the smartest one to take, and for instance Bessel functions should be preferred. In other words, the set of complete and orthogonal basic functions can with advantage be chosen with attention to the geometry of the problem and the relevant boundary conditions.

The dynamic properties of linear media can, however, also be described in ways which accentuate the initial condition. The simplest initial condition is not a plane wave, which will be very difficult to set up in a real experiment. A pulse-like perturbation, albeit also idealized, can be more readily imposed. If we want to excite waves on the surface of a pond, our immediate choice will be to throw a pebble into it, a process described mathematically by $\delta(t)\delta(r)$. The pulse response of a medium can characterize its dynamic properties just as well as the plane wave response accounted for by the dispersion relation. The two descriptions are of course interrelated. The ideas are best explained by illustrative examples.

We can write a general medium response as

$$f(x,t) = \int \widetilde{f}(k)\, e^{-i(\omega(k)t-kx)}dk, \tag{3.18}$$

where we introduced the dispersion relation $\omega = \omega(k)$. We can require the response $f(x,t)$ to satisfy the initial condition $f(x,0) = a\delta(x)$, and find trivially that this implies $\widetilde{f} = a$. We used the formal expression for Dirac δ-function $\delta(t) = (2\pi)^{-1}\int_{-\infty}^{\infty} e^{-i\omega t}d\omega$.

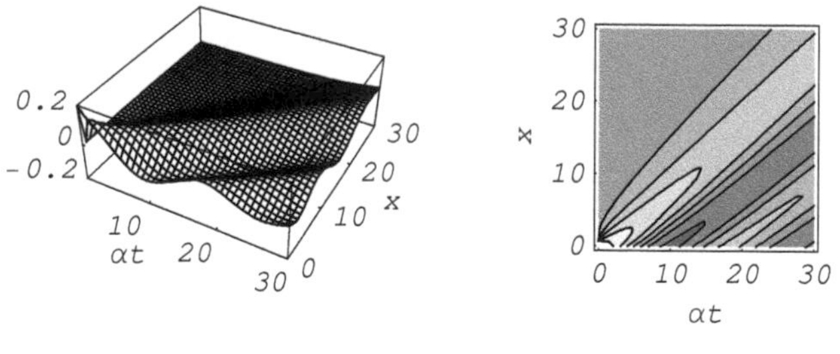

FIGURE 3.3: *Illustration of the Airy function response for t* > 0*, as given by (3.20), here shown for* $C = 1/\alpha$ *for simplicity.*

1) Consider first a spatially one dimensional model, where waves propagating along the x-axis are assumed to obey a model dispersion relation $\omega(k) = Ck + \alpha k^3$. We write the medium response as

$$f(x,t) = \int \widetilde{f}(k)\, e^{-i(Ckt+\alpha k^3 t-kx)}dk. \tag{3.19}$$

We now recall the *Airy integral*, given by

$$\mathrm{Ai}(x) \equiv \frac{1}{2\pi}\int_{-\infty}^{\infty} \exp\left(-i\left(sx+\frac{1}{3}s^3\right)\right)ds = \frac{1}{\pi}\int_0^{\infty} \cos\left(sx+\frac{1}{3}s^3\right)ds.$$

In terms of this Airy function we can express the pulse response for this particular case as

$$f(x,t) = a\frac{1}{2(3\alpha t)^{1/3}}\mathrm{Ai}\left(\frac{x-Ct}{(3\alpha t)^{1/3}}\right). \tag{3.20}$$

The result is shown in Fig. 3.3 for a selected value of C. Note that $f(x,t)$ is very small for $x > Ct$. Note that we here, as well as in the following examples, consider propagation in one direction only. Often it is so that an initial perturbation will send waves in both directions of a one dimensional model as these, but this feature is not included here.

Note that we *could* have written the general expression for the medium response (3.18) in a different way, i.e., as an integral over frequencies rather than wave-numbers

$$f(x,t) = \int \widetilde{f}(\omega)\, e^{-i(\omega t - k(\omega)x)}d\omega. \tag{3.21}$$

The two versions (3.18) and (3.21) have somewhat different mathematical properties, but *physically* they represent closely related problems. Version (3.21) is understood best by assuming that the function f is given at $x = 0$ at all times, i.e., we assume a "radiator" to be placed at $x = 0$ and determining

$$f(0,t) = \int_{-\infty}^{\infty} \widetilde{f}(\omega)\, e^{-i\omega t}d\omega, \qquad \text{with} \qquad \widetilde{f}(\omega) = \frac{1}{2\pi}\int_{-\infty}^{\infty} f(0,t)\, e^{-i\omega t}dt\,.$$

We define

$$g(x,t) = \frac{1}{2\pi}\int_{-\infty}^{\infty} e^{-i(\omega t - k(\omega)x)}d\omega\,, \tag{3.22}$$

which satisfies the initial condition $g(0,t) = \delta(t)$. We can now express $f(x,t)$ in terms of $g(x,t)$ as

$$f(x,t) = \int g(x,t-t')\, f(0,t')dt'\,, \tag{3.23}$$

where the function g is usually denoted the space-time Green's function. Causality requires $g(x,t - t') = 0$ for $t' > t$. This requirement is fulfilled for the illustrative cases in Figs. 3.4–3.6. In Fig. 3.3 this requirement is not *exactly* fulfilled, but it comes close.

2) As a general choice of dispersion relation we consider

$$k(\omega) = \frac{\omega + i\alpha}{C}\sqrt{1 + \frac{\beta^2}{(\omega + i\alpha)^2}} \qquad \text{where} \qquad \alpha \geq \beta \geq 0\,, \tag{3.24}$$

where we wrote the form $k = k(\omega)$ rather than $\omega = \omega(k)$. By C we here mean a velocity, but not necessarily the velocity of light in vacuum. It can be shown (Vaĭnshteĭn 1976) that for the case (3.24) we have

$$g(x,t) = \frac{\beta x}{C\sqrt{t^2 - (x/C)^2}} I_1\left(\beta\sqrt{t^2 - (x/C)^2}\right) e^{-\alpha t} \qquad \text{for} \qquad t > x/C\,. \tag{3.25}$$

This result contains several physically relevant special cases. In Fig. 3.4 we show (3.25) for $\beta = 1$ and $\alpha = 0.25$. The following results can be obtained as special cases of (3.25). In all cases, the analytical expression should be completed with the "free-space" contribution $\delta(t - x/C)$, which contributes along the line $x = Ct$.

3) As an example for a linearly unstable system, with dispersion relation $\omega^2 = C^2k^2 - \beta^2$ we find, by inserting $\alpha = 0$ in (3.25), the result

$$g(x,t) = \frac{\beta x}{C\sqrt{t^2 - (x/C)^2}} I_1\left(\beta\sqrt{t^2 - (x/C)^2}\right) \qquad \text{for} \qquad t > x/C\,, \tag{3.26}$$

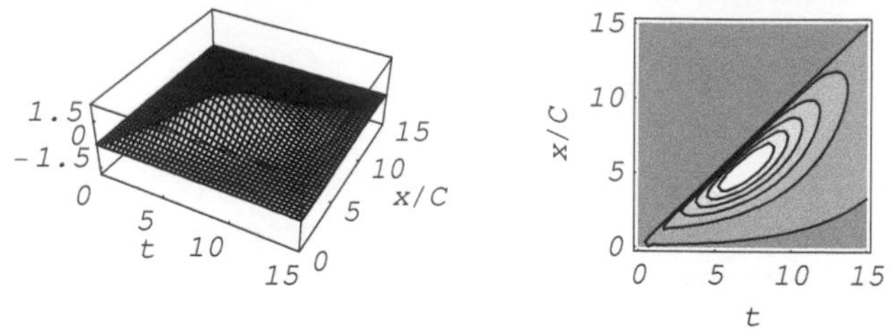

FIGURE 3.4: *Illustration of the Green function corresponding to the dispersion relation (3.24) with* $\beta = 1$ *and* $\alpha = 0.25$.

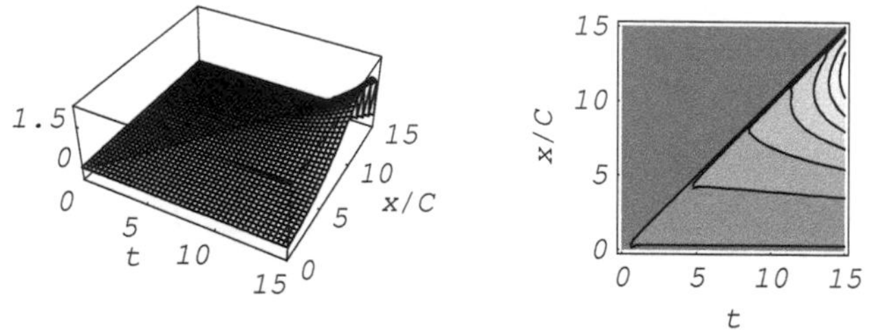

FIGURE 3.5: *Illustration of the Green function corresponding to the dispersion relation* $k(\omega) = (\omega/C)\sqrt{1+\beta^2/\omega^2}$, *which corresponds to an unstable system, here shown for* $\beta = 0.025$.

with results shown in Fig. 3.5, for $\beta = 0.025$.

4) A dispersion relation representative of several wave-types found in plasmas can be modeled as $\omega^2 = \omega_p^2 + C^2k^2$. We obtain this case by $\alpha = 0$ and $\beta = -i\omega_p$ in (3.25). For this case we then find

$$g(x,t) = -\frac{\omega_p x}{C\sqrt{t^2-(x/C)^2}} J_1\left(\omega_p\sqrt{t^2-(x/C)^2}\right) \qquad \text{for} \qquad t > x/C,$$

shown in Fig. 3.6 for $\omega_p = 1$ and also discussed by Jovanović et al. (1982*b*). Note that we have $J_1(\sqrt{x})/\sqrt{x} \approx 1/2 - x/16 + O(x^{3/2})$ for small x, implying that the response is constant along the line $x = Ct$.

3.2 Wave propagation in inhomogeneous media

Dispersion relations in their standard form assume the media supporting the wave to be homogeneous, though not necessarily isotropic. For general inhomogeneous media we would not expect this dispersion relation to be of much use (although some important special cases exist). Often we can, however, assume media to be *locally* homogeneous (and sometimes also locally isotropic), in which case the dispersion relation contains much of the information we might desire. It may easily happen, though, that the inhomogeneity of the media needs to be taken explicitly into account, and for such cases there are a few standard methods, which are good to know.

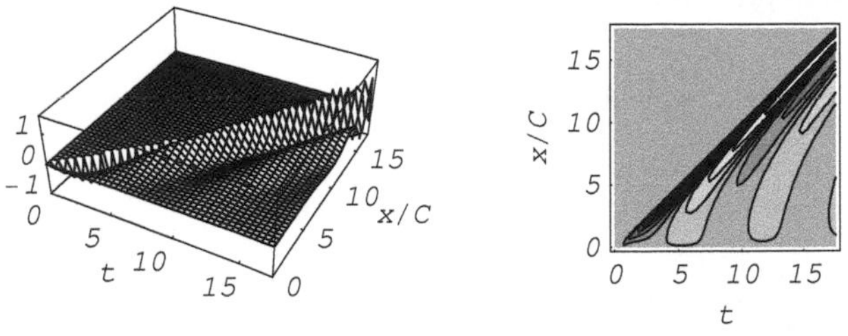

FIGURE 3.6: *Illustration of the Green function corresponding to the dispersion relation* $k(\omega) = (\omega/C)\sqrt{1-\omega_p^2/\omega^2}$, *i.e., "the plasma response".*

3.2.1 Snell's law

Diffraction of a plane wave at a sharp interface between two media at rest can be described by Snell's law expressed in terms of the index of refraction $n \equiv c/(\omega/k)$ being the ratio of the speed of light in vacuum and the phase velocity of the wave for a given frequency. Assume that two media with refractive indices n_1 and n_2 are separated by such an interface. Let the normal vector be $\widehat{\mathbf{n}}$, and let θ_1 be the angle between $\widehat{\mathbf{n}}$ and the incoming wave-vector $\mathbf{k}_1$ (i.e., the one in medium 1). The similar angle for the *outgoing* wave-vector $\mathbf{k}_2$ (i.e., the one in medium 2) is θ_2. We have

$$n_1 \sin\theta_1 = n_2 \sin\theta_2 ,$$

known as *Snell's law*. We have $n_1/n_2 = k_2/k_1$. The result applies in principle to *any* wave-type, not just electromagnetic waves. The only essential assumption is that we are dealing with a well defined wave-vector and a corresponding constant frequency, i.e., a plane wave at both sides of the interface.

3.2.2 WKB analysis

As a limiting case we envisage media with variations in characteristic properties on scale lengths much larger than the wavelength of the waves we are interested in. A wave is then represented by a localized wave-packet, with a spatial extent smaller than the scale length of the gradient of the variation of refractive index in the media. The trajectory of the "center-of-mass" of this wave-packet is then followed. We present here a somewhat simplified summary, with more detailed accounts given by Whitham (1974), Bernstein (1975) and others. WKB is an often used "initialism" of the names Wentzel-Kramers-Brillouin. It is also known as the LG or Liouville-Green method. Other often-used letter combinations include JWKB and WKBJ, where the "J" stands for Jeffreys.

Assume that we have a solution for a wave propagating in an inhomogeneous medium

$$\phi(\mathbf{r},t) = A(\mathbf{r},t)\exp(i\Psi(\mathbf{r},t)),$$

where A is the wave amplitude, and Ψ is the *phase*. In the case of a plane wave propagating in a homogeneous medium, the phase is simply $\mathbf{k}\cdot\mathbf{r} - \omega t$, giving $\nabla\Psi = \mathbf{k}$ and $\partial\Psi/\partial t = -\omega$. We use this as a starting point for the generalization $\nabla\Psi = \mathbf{k}(\mathbf{r},t)$ and $\partial\Psi/\partial t = -\omega(\mathbf{r},t)$. The media are taken to vary slowly so that $\mathbf{k}\phi \gg \nabla A$ and $\omega\phi \gg \partial\Psi/\partial t$. By differentiation of Ψ with respect to both temporal and spatial variables we readily find

$$\frac{\partial}{\partial t}\mathbf{k} + \nabla\omega = 0. \tag{3.27}$$

Assume now that frequency and wave-number can be obtained from a "local" dispersion relation $D(\omega, \mathbf{k}, \mathbf{r}, t) = 0$, where we take the medium to be slowly varying on the scale of the wavelengths in question. We can, at least in principle, assume that we have a relation $\omega = \Omega(\mathbf{k}, \mathbf{r}, t)$. By (3.27) we then have

$$\frac{\partial}{\partial t}\mathbf{k} + \nabla_r \Omega + (\nabla \mathbf{k}) \cdot \nabla_k \Omega = 0, \tag{3.28}$$

where we added a subscript on the ∇-operators to indicate the variables they act upon.

By implicit differentiation of the dispersion relation we also find

$$(\nabla_r \Omega)\frac{\partial}{\partial \omega}D + \nabla_r D = 0, \quad \text{and} \quad (\nabla_k \Omega)\frac{\partial}{\partial \omega}D + \nabla_k D = 0. \tag{3.29}$$

We have in particular $(\nabla \mathbf{k}) \cdot \nabla_k \Omega = (\nabla_k \Omega) \cdot \nabla \mathbf{k}$, with $\nabla \mathbf{k} = \nabla\nabla\Psi$ being symmetric. We can now express (3.28) as

$$\frac{\partial}{\partial t}\mathbf{k} + (\nabla_k \Omega) \cdot \nabla \mathbf{k} = \frac{\nabla_r D}{\partial D/\partial \omega}. \tag{3.30}$$

If we introduce the temporal variable t and define ray-trajectories by

$$\frac{d}{dt}\mathbf{r} = -\frac{\nabla_k D}{\partial D/\partial \omega} = \nabla_k \Omega \tag{3.31}$$

we can express (3.30) in the form

$$\frac{d}{dt}\mathbf{k} = \frac{\nabla_r D}{\partial D/\partial \omega} = -\nabla_r \Omega, \tag{3.32}$$

where we recognize the group velocity $\mathbf{u}_g = \nabla_k D/(\partial D/\partial \omega)$, so the rays are streamlines of the group velocity (Bernstein 1975). The expressions (3.31) and (3.32) have to be supplemented by the proper boundary conditions, defining the antennas, for instance. Differentiation of $\omega = \Omega(\mathbf{k}, \mathbf{r}, t)$ along the ray gives

$$\frac{d}{dt}\omega = \frac{\partial}{\partial t}\Omega.$$

The relations (3.31) and (3.32) are clearly in canonical form, with Ω playing the role of the Hamiltonian. An important special case is encountered when the medium can be considered time stationary in a properly chosen frame of reference. In that case the frequency ω is constant along the ray. In general, we like to interpret the ray as the trajectory followed by a small wave-packet, as it moves with the local group velocity.

Unfortunately, the WKB analysis is not easily generalized to dissipative systems. The problem is that, in general, different wave-vector components constituting a wave-packet are damped differently, thereby "shifting" the mass center of the wave-packet to a position which is different from that obtained by using solely the real part of the dispersion relation. An exception is the particular case where all wave numbers are damped equally much.

As already emphasized, the WKB approximation, corresponding to "geometrical optics", assumes that the medium is slowly varying on the length scales determined by the actual wavelength. In reality, it is often experienced that the analysis gives quite adequate results even when this inequality is not strictly satisfied. Unfortunately, this is not something to be counted on for *all* cases. By a *ray tracing* analysis, we can then follow the evolution of the wave-field in space and time. In most cases, this has to be done numerically.

Ⓢ **Exercise:** Some years ago the author lived a few kilometers from a railway station. Sometimes at night we could hear the trains, even the loudspeakers on the platforms, quite clearly, at other times nothing at all. It turned out that we heard the station when the wind was blowing *toward* us, while with opposite wind directions we heard nothing. How can this be? To save the reader from some trivial mistakes, we recall that the sound speed in air is typically 330 m/s, while a moderate wind velocity is some 5 m/s. Ⓢ

4

Weakly Nonlinear Waves

In real life, we will of course never encounter a wave phenomenon which can truly be called *linear* (with the exception of the classical description of electromagnetic waves in vacuum), although a description based on a linear formalism as in the foregoing section can be very good, in some cases even *extremely* good. To describe the lowest order nonlinear corrections to the basic linear wave analysis, it is an advantage to classify wave-types according to their dispersion relation. The important observation will be that many nonlinear wave phenomena can be "classified" into either Korteweg-deVries types or nonlinear Schrödinger types, as explained in detail later. These two are the most relevant ones for plasma physics, but there are also others. The basic classification is rooted in the linear dispersion relation, where we here distinguish two cases. Already from the outset, the reader should be warned that the term "simple waves" will in the following section 4.1.1 have a slightly different meaning compared to its standard usage. The terminology used in the following is standard in the literature.

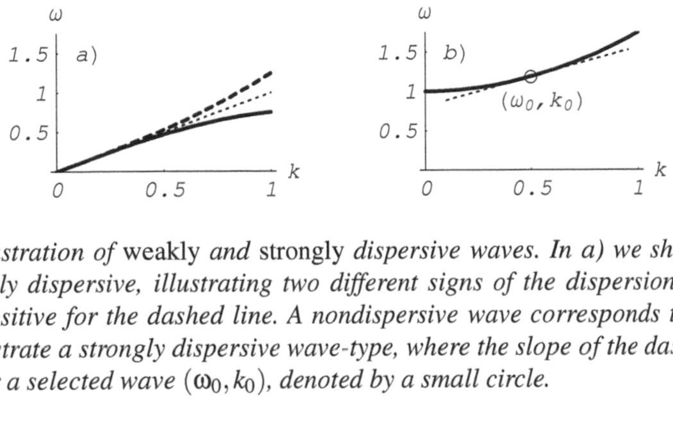

FIGURE 4.1: *Illustration of* weakly *and* strongly *dispersive waves. In a) we show the dispersion relation for weakly dispersive, illustrating two different signs of the dispersion, negative for the solid line, and positive for the dashed line. A nondispersive wave corresponds to the thin dashed line. In b) we illustrate a strongly dispersive wave-type, where the slope of the dashed line gives the group velocity for a selected wave* (ω_0, k_0)*, denoted by a small circle.*

1) We have weakly dispersive linear waves, where the phase and group velocities are similar, though not necessarily identical (one example being acoustic or sound waves). By "weakly dispersive" we understand waves where to the lowest approximation all wavelengths (i.e., wave-numbers) propagate with the velocity, phase and group velocities being identical; see Fig. 4.1a). We have a linear dispersion relation of the form $\omega^2 = C_s^2 k^2$, or equivalently $\omega = \pm C_s k$. Corrections then appear as higher order corrections to this dispersion relation. Such corrections can originate from linear as well as nonlinear effects.

2) We can have strongly dispersive linear waves, where the phase and group velocities are significantly different (one example being Langmuir waves); see Fig. 4.1b). For these waves we have a local dispersion relation $\omega = \omega_0 + u_g(k - k_0)$, with $\omega_0 \equiv \omega(k_0)$ and $u_g \equiv d\omega/dk \neq \omega_0/k_0$ being the group velocity, obtained for $k = k_0$. Corrections to this lowest order dispersion relation originate from linear as well as nonlinear effects also in this case.

4.1 Nondispersive waves

We consider first simple waves in gases, with the basic equations being the continuity equation, the momentum equation and an equation of state, which we leave unspecified to begin with

$$\frac{\partial}{\partial t}\rho + \frac{\partial}{\partial x}(\rho u) = 0 \tag{4.1}$$

$$\frac{\partial}{\partial t}u + u\frac{\partial}{\partial x}u + \frac{1}{\rho}\frac{\partial}{\partial x}p = 0 \tag{4.2}$$

$$p = p(\rho), \tag{4.3}$$

with ρ being the mass density (Whitham 1974). For simplicity, we consider one spatial dimension only, noting that this restriction is not quite trivial. We define a quantity

$$C_s^2 \equiv \frac{dp}{d\rho} \tag{4.4}$$

and will later on find that C_s is the appropriate sound speed; see also Section 2.4.1.

Consider first the linearized version of these equations. We note that we have an equilibrium solution $\rho = \rho_0 =$ const. and $u_0 = 0$. Writing $\rho = \rho_0 + \tilde{\rho}$, $u = u_0 + \tilde{u} = \tilde{u}$ and ignoring second order terms containing, for instance, $\tilde{u}\tilde{\rho}$, which are small compared to $\tilde{u}\rho_0$, etc., we find

$$\frac{\partial}{\partial t}\tilde{\rho} + \rho_0\frac{\partial}{\partial x}\tilde{u} = 0 \tag{4.5}$$

$$\frac{\partial}{\partial t}\tilde{u} + \frac{C_s^2}{\rho_0}\frac{\partial}{\partial x}\tilde{\rho} = 0 \tag{4.6}$$

giving the one dimensional *wave equation*

$$\frac{\partial^2}{\partial t^2}\tilde{\rho} - C_s^2\frac{\partial^2}{\partial x^2}\tilde{\rho} = 0 \tag{4.7}$$

implying a dispersion relation $\omega^2 = C_s^2 k^2$, or $\omega = \pm C_s k$, where we now indeed find C_s to be the sound speed obtained from (4.4), taken at $\rho = \rho_0$. The result (4.7) is independent of ρ_0.

It is easily shown that given a *one dimensional* initial condition $\rho(x,0) = F(x)$, together with the condition on the derivative $\partial\rho(x,t)/\partial t|_{t=0} = G(x)$, the solution to (4.7) is

$$\rho(x,t) = \frac{1}{2}F(x+C_s t) + \frac{1}{2}F(x-C_s t) + \frac{1}{2C_s}\int_{x-C_s t}^{x+C_s t} G(y)dy. \tag{4.8}$$

Equation (4.7) contains nondispersive sound waves, but also other wave-types as in (4.7), where C_s then is representing the proper speed of propagation. As a particular case we might have $G(x) = 0$ with $F(x) \neq 0$, but also $G(x) \neq 0$ with $F(x) = 0$. In the latter case the perturbation in ϕ will seem to grow "out of nothing," so to speak, and increase with time in a transient period. This *could*, by a naive observer, be interpreted as being indicative of an unstable medium! In reality it is merely an indication of an interference between a forward and a backward propagating wave-component.

The variations with arguments $x \pm C_s t$ in (4.8) can be appreciated by noting that in a one dimensional representation (4.7) can be written as

$$\left(\frac{\partial}{\partial t} - C_s\frac{\partial}{\partial x}\right)\left(\frac{\partial}{\partial t} + C_s\frac{\partial}{\partial x}\right)\rho(x,t) = 0, \tag{4.9}$$

which is fulfilled if either factor is zero, and each factor has a solution with a simple translation $F(x+C_s t)$ or $F(x-C_s t)$.

For equation (4.7), the solutions are thus basically represented by pulses propagating without change in shape, the shape of the pulses being determined by the initial conditions. There are many physically relevant phenomena which are described by this simple equation, but we would expect that in more general situations the pulses would be distorted and damped as they propagate. Such a situation is seemingly not accounted for by (4.7).

- **Exercise:** Prove (4.8). This result is sometimes called *d'Alembert's relation.*

- **Exercise:** Find the general solution of the inhomogeneous wave equation

$$\frac{\partial^2}{\partial t^2}\tilde{\rho} - C_s^2 \frac{\partial^2}{\partial x^2}\tilde{\rho} = S(x,t)$$

for initial condition $\rho(x,0) = F(x)$, together with $\partial\rho(x,t)/\partial t|_{t=0} = G(x)$.

Ⓢ **Exercise:** Write up the expression for spherically symmetric sound waves. In this case the problem is three dimensional, but we still only have one spatial variable, the radial position. Demonstrate how (4.8) can be used in this problem. Ⓢ

4.1.1 Simple waves

We now note that in the linearized version of the analysis in Section 4.1, we have $\tilde{\rho} = \rho_0 \tilde{u}/C_s$. *Assume* now that in the linear case we have a slightly more general relation $\rho = \rho(u)$, with $u = u(x,t)$; this is the basic assumption for *simple waves*. By this we imply that ρ is a function of the independent variables x and t, but that this functional relation is through the velocity u. For the *linear* waves this assumption is trivially fulfilled, as stated. Using

$$\frac{\partial}{\partial t}\rho = \frac{d\rho}{du}\frac{\partial}{\partial t}u \qquad \text{and} \qquad u\frac{\partial}{\partial x}\rho = u\frac{d\rho}{du}\frac{\partial}{\partial x}u,$$

we then have, by use of a full set of nonlinear equations (4.1)–(4.4), that

$$\frac{d\rho}{du}\left(\frac{\partial}{\partial t}u + u\frac{\partial}{\partial x}u\right) + \rho\frac{\partial}{\partial x}u = 0, \tag{4.10}$$

$$\frac{\partial}{\partial t}u + u\frac{\partial}{\partial x}u + \frac{C_s^2}{\rho}\frac{d\rho}{du}\frac{\partial}{\partial x}u = 0, \tag{4.11}$$

where we have taken care to distinguish partial and absolute derivatives, and used $C_s^2 = dp/d\rho$ as before. From (4.10) and (4.11) we find

$$C_s^2\left(\frac{d\rho}{du}\right)^2 = \rho^2 \qquad \text{or} \qquad \frac{d\rho}{du} = \pm\frac{\rho}{C_s},$$

which inserted into (4.10) gives

$$\frac{\partial}{\partial t}u + (u \pm C_s)\frac{\partial}{\partial x}u = 0. \tag{4.12}$$

This is the basic equation for weakly nonlinear simple waves (Kadomtsev & Karpman 1971, Blackstock 1972, Whitham 1974, Naugolnykh & Ostrovsky 1998).

One problem remains concerning C_s: it is defined as the derivative of the equation of state with respect to the mass density, but this will in general vary with the position in the wave-train because ρ is varying with position. The exception is the isothermal equation of state, where C_s is independent of ρ. If the temperature is varying, then the sound speed is varying as well. To make this dependence on spatial and temporal coordinates explicit in a useful way is not as easy as one might expect. We first introduce a new thermodynamic quantity $\lambda \equiv \int_{\rho_0}^{\rho}(C_s(\rho')/\rho')\,d\rho'$ which implies $\partial\lambda/\partial t = (d\lambda/d\rho)\partial\rho/\partial t = (C_s/\rho)\partial\rho/\partial t$, etc., recalling that in general we have $C_s = C_s(\rho(x,t))$ (Blackstock 1972). Using the definition (4.4) of C_s^2 together with (4.10) and (4.11) we find the relations

$$\begin{aligned} \frac{\partial}{\partial t}\lambda + u\frac{\partial}{\partial x}\lambda + C_s\frac{\partial}{\partial x}u &= 0 \\ \frac{\partial}{\partial t}u + u\frac{\partial}{\partial x}u + C_s\frac{\partial}{\partial x}\lambda &= 0. \end{aligned} \tag{4.13}$$

The simple wave solutions are those with $\lambda = \pm u$.

Taking explicitly the isentropic equation of state $p/p_0 = (\rho/\rho_0)^\gamma$, we find with some algebra (Blackstock 1972) the result $\lambda = 2(C_s - C_{s0})/(\gamma - 1)$, which gives

$$C_s = C_{s0} \pm (\beta - 1)u\,,$$

where $C_{s0}^2 \equiv \gamma p_0/\rho_0$ and $\beta \equiv \frac{1}{2}(\gamma + 1)$, with C_{s0} corresponding to the unperturbed mass density ρ_0. Using this result for C_s, we can write the basic equation (4.12) in the form

$$\frac{\partial}{\partial t}u + (\beta u \pm C_{s0})\frac{\partial}{\partial x}u = 0. \tag{4.14}$$

Note that for the case of isothermal motion (which will be incorrect for sound waves in neutral gases, although this seems to have been Newton's guess) we have the linear relation $p/p_0 = \rho/\rho_0$ and $\gamma = 1$. To have a finite λ we then require $C_s = C_{s0}$ and find $\beta = 1$.

- **Exercise:** Demonstrate that for simple waves we have the equation of state

$$p = p_0\left(1 \pm (\beta - 1)\frac{u}{C_{s0}}\right)^{2\gamma/(\gamma-1)},$$

with the definitions introduced before. What is the limit for small amplitude waves?

It is often an advantage (mainly cosmetic, though) to change the frame of reference to one moving with either the velocity C_{s0} or $-C_{s0}$, and obtain

$$\frac{\partial}{\partial t}u + \beta u\frac{\partial}{\partial x}u = 0. \tag{4.15}$$

The solution of (4.15) is implicitly given as

$$u(x,t) = F(x - \beta u(x,t)t)\,, \tag{4.16}$$

for an initial condition $u(x,0) = F(x)$. Introducing the inverse function $F^{-1}(x)$ so that $F^{-1}(F(x)) = x$, this solution can readily be written as $x = \beta ut + F^{-1}(u)$. Actually, the solution can most readily be explained by a construction, as shown in Fig. 4.2, illustrating the steepening and ultimate breaking of an initial condition.

The steepening of the waveform for increasing times can be understood by the following argument (Blackstock 1972). The sound wave travels with speed C_s with respect to the fluid particles, but these are themselves moving with velocity u. To a fixed observer, the net speed is $u + C_s$. This is a kinematic effect, and can be referred to as *convection*. The thermodynamic effect is due to the variation of the sound speed through the waveform for an adiabatic equation of state. Where the acoustic pressure is positive, the sound speed is enhanced, and conversely in wave troughs, where the sound speed diminishes, locally. The two effects, with entirely different physical origins, are accounted for by β in (4.15). For an isothermal gas, the sound speed would be constant, and we would have $\beta = 1$ and only the kinematic effect will remain. For any value of β we find that in the frame moving with the sound speed, regions of large positive u propagate faster in the forward direction; those with large negative u move rapidly in the negative direction. The zero-crossings, where $u = 0$, are not propagating at all. The process is readily transformed to the fixed frame of reference.

- **Exercise:** Prove (4.16).
- **Exercise:** Why did we assume $\rho = \rho(u)$, with $u = u(x,t)$, and not $u = u(\rho)$, with $\rho = \rho(x,t)$ in deriving the simple wave equation?

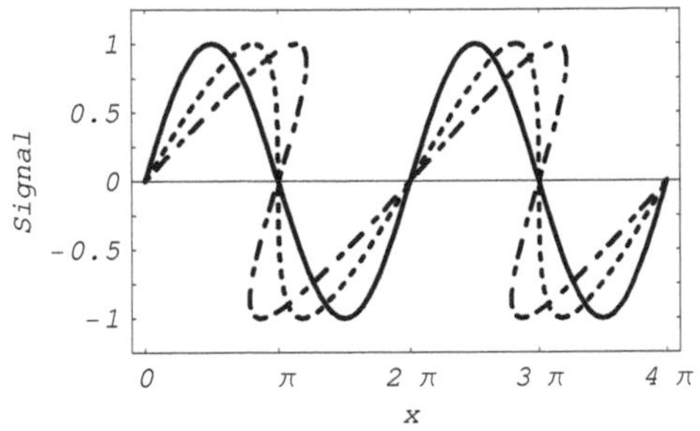

FIGURE 4.2: *Wave steepening and wave breaking for simple waves, shown in the frame of reference moving with the sound velocity. The solid line shows the initial sinusoidal waveform. These waveforms should not be confused with ocean waves breaking on a shore! In that case there is nothing unphysical in having multiple solutions for the water surface at a given position.*

The basic problem with the simple wave solutions is that the wave will break after a finite time, as readily seen by inspection of the solutions given, and also Fig. 4.2. It is not physically acceptable to have a multi-valued fluid velocity! Note that in a Fourier representation obtained by a spatial Fourier transform of the solution as a function of time, this wave breaking will appear rather dramatic! If the Fourier transform has a finite support initially (it can, for instance, be a δ-function for an initially plane wave in configuration space) the entire wave number axis will be filled within a finite time, appearing as a sort of "explosion" in wave-number space. Physically, the wave steepening is due to harmonic generation by the nonlinearity being *resonant* since all harmonics are *on* the linear dispersion relation for the waves. We start out with a wave characterized by a wavenumber k, and through the dispersion relation $\omega = \omega(k)$ also a frequency, and at later times we see wave components $(2\omega, 2k)$, $(3\omega, 3k)$, etc. develop for increasing distances.

We can also solve the simple wave equation for a boundary value problem, where $u(0,t) = G(t)$ is given. In this case it is not advisable to change the frame of reference, so we use the full equation (4.12). The solution is now $u(x,t) = G(t - x/(\beta u \pm C_s))$, or in terms of $G^{-1}(t)$ introduced as before, $t \mp x/C_{s0} = G^{-1}(u) - (x/C_{s0})\beta u/(C_{s0} \pm \beta u)$.

For a boundary value problem with a harmonic wave excitation, we have at the point of excitation a well defined frequency ω, and find higher harmonics develop at larger distances. This latter phenomenon is best illustrated by what is known as *Fubini's solution*, given by

$$u = u_0 \sum_{n=1}^{\infty} \frac{2}{n\sigma} J_n(n\sigma) \sin(n(\omega_0 t - kx)) \tag{4.17}$$

in terms of a normalized distance $\sigma \equiv x(\omega_0/C_{s0})(u_0/C_{s0})$, introducing the Bessel functions J_n (Blackstock 1962, Blackstock 1972, Naugolnykh & Ostrovsky 1998); see Fig. 4.3. The Fubini solution (also Bessel seems to deserve credit for obtaining a similar result by a different method (Blackstock 1962)) was derived for waves excited by a harmonically oscillating piston, with displacement $X(t) = (u_0/\omega_0)(1-\cos(\omega_0 t))$, with $u_0/C_{s0} \ll 1$. The result is approximate, and accounts qualitatively for a variety of waves excited at a plane boundary, also some plasma waves.

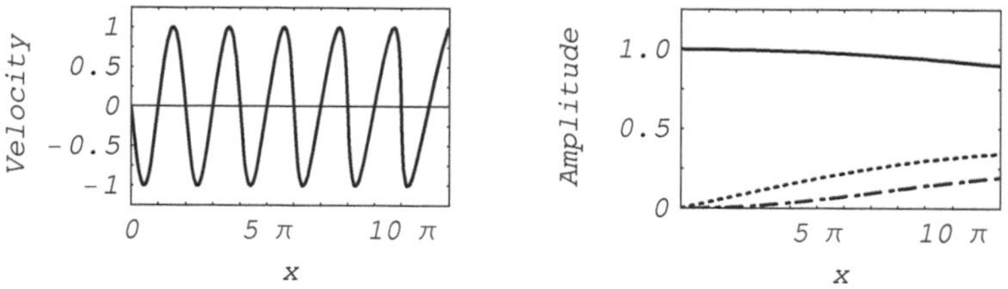

FIGURE 4.3: *The Fubini solution (4.17) for simple waves excited at a boundary at the origin, shown at a fixed time. The waves steepen as they propagate away from the exciter. The right-hand frame shows the amplitude of the fundamental wave with the solid line as given by the Bessel function J_1, while the two dashed lines give the amplitudes of the two first harmonics.*

We note that a linearized solution for simple waves can only be applied in a finite time, no matter how small the initial wave amplitude is! *Any* wave amplitude will eventually lead to breaking, i.e., physically unacceptable solutions. All we achieve by reducing the excitation amplitude is a postponement of the problem.

- **Exercise:** Derive the Fubini solution (4.17) for simple waves excited at a boundary. A relevant reference is Blackstock (1972).

There are, however, several idealizations in the foregoing analysis, and we shall address two of them here. First we note that viscosity was ignored. This is evidently an over-idealization, since *all* media are dissipative at least at standard conditions. Next, we note that the model for simple waves implies that the linear dispersion relation continues as $\omega = kC_s$ to arbitrarily small wavelengths, which is also incorrect from a physical point of view. We consider inclusion of these two corrections separately in what follows, but add one note of caution; it has *not* been demonstrated that *all* waves having a small amplitude linear dispersion relation will have breaking nonlinear solutions!

We have now understood the role of the coefficient β in (4.15), and to simplify the equations we let $\beta = 1$ in the following and ignore what was termed a thermodynamic effect. Apart from being a simplification of the notation, it is physically realizable by assuming an isothermal equation of state, as already mentioned.

4.1.2 Burgers' equation

We will include a standard viscosity term somewhat "ad hoc" by supplementing the simple wave equation (4.15) with $\beta = 1$ to obtain

Burgers' equation (Whitham 1974, Naugolnykh & Ostrovsky 1998)

$$\frac{\partial}{\partial t}u + u\frac{\partial}{\partial x}u = \mu\frac{\partial^2}{\partial x^2}u, \tag{4.18}$$

where μ is a standard kinematic viscosity coefficient (Aris 1962). It is immaterial whether we consider the reference system propagating with C_s or $-C_s$. Linearizing (4.18) we obtain the simple diffusion equation $\partial u/\partial t = \mu\partial^2 u/\partial x^2$, implying the dispersion relation $\omega = -i\mu k^2$, i.e., all wavelengths are damped, but shorter wavelengths are more damped than long wavelengths. The waves are, on the other hand, nondispersive in this model, as far as the real part of the frequency is concerned.

We now return to the nonlinear equation (4.18). By an ingenious device, known as the Cole-Hopf transformation (Whitham 1974), we can reduce this equation to a linear one. As the first step in the transformation we introduce a velocity potential by $u = \partial\psi/\partial x$ in (4.18) and integrate it once with respect to x to obtain

$$\frac{\partial}{\partial t}\psi + \frac{1}{2}\left(\frac{\partial\psi}{\partial x}\right)^2 = \mu\frac{\partial^2}{\partial x^2}\psi, \tag{4.19}$$

where we used

$$\frac{\partial\psi}{\partial x}\frac{\partial^2\psi}{\partial x^2} = \frac{1}{2}\frac{\partial}{\partial x}\left(\frac{\partial\psi}{\partial x}\right)^2. \tag{4.20}$$

As a second step, we introduce a *nonlinear* transformation by $\psi = -2\mu\ln\phi$ to obtain

$$\frac{\partial}{\partial t}\phi = \mu\frac{\partial^2}{\partial x^2}\phi, \tag{4.21}$$

and find the *linear* diffusion equation! The point is, as mentioned, that starting from the nonlinear Burgers' equation we used a nonlinear transformation to obtain this result. But it is ingenious nonetheless. The transformation is reversible, i.e., given a solution ϕ there is a unique inverse transform giving u. We can therefore conclude right away that the unphysical wave breaking is inhibited by the viscosity term for arbitrary initial conditions! The diffusion equation has only well behaved solutions (as long as $\mu > 0$), and therefore the solutions of (4.18) must be well behaved also. Physically we note that viscosity becomes increasingly important as the gradient of velocity increases, and dominates the time evolution entirely when the wave is just about to break.

The solutions to Burgers' equation can with advantage be analyzed in more detail. First we note that $A \equiv \int_{-\infty}^{\infty} u(x,t)dx = 2\mu\ln(\phi_{-\infty}/\phi_{\infty})$ is conserved; it is an *integral of motion*. We defined $\phi_{\pm\infty} \equiv \phi(x\to\pm\infty)$, where $\phi_\infty = \phi_{-\infty}\exp(-A/2\mu)$. The conservation of A is trivially demonstrated by integration of (4.18). We consider two cases: one where A is finite and one where it is *in*finite.

Case 1: Assume an initial condition $u(x,0) = F(x)$, with $F(x\to\pm\infty) = 0$. This corresponds to $\phi(x,0) = \exp\left(-\int_{-\infty}^{x} F(x')dx'/2\mu\right)$. The solution to the diffusion equation for the given conditions is

$$\phi(x,t) = \frac{1}{2\sqrt{\pi\mu t}}\int_{-\infty}^{\infty}\exp\left(-\frac{1}{2\mu}\int_{-\infty}^{\eta}F(x')dx' - \frac{(x-\eta)^2}{4\mu t}\right)d\eta, \tag{4.22}$$

or in terms of the velocity u

$$u(x,t) = \int_{-\infty}^{\infty}\frac{x-\eta}{t}\exp(-B/2\mu)d\eta\left(\int_{-\infty}^{\infty}\exp(-B/2\mu)d\eta\right)^{-1}, \tag{4.23}$$

with $B \equiv B(\eta|x,t) = \int_{-\infty}^{\eta}F(x')dx' + (x-\eta)^2/(2t)$. For $t\to\infty$, we have $u\to 0$ for any x, as long as $\mu \neq 0$. Consequently, the initial perturbation has to "spread out" in such a way that it conserves $A = \int_{-\infty}^{\infty} u dx$. See summary in Fig. 4.4.

- **Exercise:** Note that total *energy* is not conserved by (4.18): assuming a constant gas density, the kinetic energy density is $W = \frac{1}{2}\rho u^2$. Demonstrate that the total kinetic energy of the system $\int_{-\infty}^{\infty} W(x')dx'$ is dissipated by viscosity as $t\to\infty$.

Case 2: If u assumes a finite value at either $x\to\infty$ or $x\to-\infty$ we readily see that A becomes infinite, and this case requires a separate treatment. Assume for these conditions that a propagating solution exists such that $u = h(x - Ut)$, with U being a constant velocity and the solution propagates without change of shape as given by the function h. Inserting such a solution into (4.18) we find with $\xi \equiv x - Ut$

$$(h - U)\frac{dh}{d\xi} = \mu \frac{d^2 h}{d\xi^2}.$$

This equation has a one parameter family of solutions

$$h(\xi) \equiv h(x - Ut) = \frac{2U}{1 + \exp(U(x - Ut)/\mu)}, \tag{4.24}$$

known as a *shock solution*, where the shock thickness is $\Delta = \mu/U$, given the boundary condition as $U = u(\xi \to -\infty, 0)/2$, assuming that $u(\xi \to \infty, 0) = 0$. The details between an actual initial condition and the solution h will be smeared out, as in Case 1. The energy dissipated in the present Case 2 is replaced by the agency maintaining the constant velocity at $\xi \to -\infty$. See summary in Fig. 4.4.

The shock solution represents a balance between the wave steepening and the "smearing out" due to the viscosity. In classical studies of shock dynamics (Blackstock 1972, Whitham 1974) it is often described as a discontinuity separating two states of fluid, corresponding to very small $\Delta = \mu/U$ in the foregoing model. Some general relations can be obtained from basic gas dynamics, classical relations being the Rankine-Hugoniot relations (also referred to as Rankine-Hugoniot jump conditions), relating pressures, flow velocities, gas densities and temperatures on the two sides of the discontinuity.

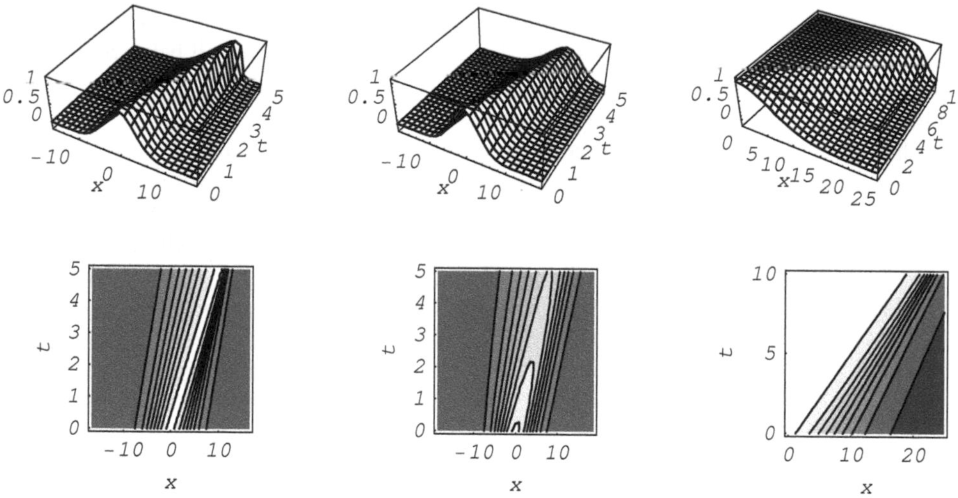

FIGURE 4.4: *Illustration of wave steepening (left) for simple waves, damping of a localized pulse as described by Burgers' equation with $\mu = 1$ (middle) and shock formation by the steepening of an initial condition with $\mu = 0.2$, where energy is supplied continuously at $x \to -\infty$ (right).*

- **Exercise:** Demonstrate that the shock width can be estimated by balancing two time scales, one for dissipation and one for nonlinear wave steepening.

- **Exercise:** Demonstrate that the analytical expression for the shock width $\Delta = \mu/U$ can be obtained (apart from a numerical factor) by dimensional reasoning.

4.2 Weakly dispersive waves

In the discussion of wave breaking of simple waves we saw that viscosity could act to arrest the breaking. However, a similar effect can be obtained by a modification which leaves the system in an energy conserving form. Introduction of *wave dispersion* has also a similar effect.

4.2.1 Korteweg-deVries equation

We will include a standard dispersive term somewhat "ad hoc" by supplementing the simple wave equation to become

$$\frac{\partial}{\partial t}u+(u+C_s)\frac{\partial}{\partial x}u+\alpha\frac{\partial^3}{\partial x^3}u=0, \tag{4.25}$$

where α is here a constant measuring the strength and sign of the dispersion. Physically, (4.25) represents the simple equation for waves propagating in the positive x-direction with velocity C_s, with this equation being modified by two small corrections, one due to nonlinearity and one originating from dispersion. It is again immaterial which of the reference systems we consider and will usually remove the term $C_s\partial u/\partial x$ by a Galilean transform. This new reference system may in other respects be a bit "unphysical" since here all fluid particles are streaming backwards with velocity C_s.

The resulting equation is the *Korteweg-deVries equation*, or simply *KdV equation*

$$\frac{\partial}{\partial t}u+u\frac{\partial}{\partial x}u+\alpha\frac{\partial^3}{\partial x^3}u=0. \tag{4.26}$$

Linearizing (4.26) we obtain the dispersion relation $\omega(k)=-\alpha k^3$, demonstrating that the phase velocity ω/k varies with wavenumber, albeit weakly for small k. When $\alpha>0$, we have $\omega/k<0$ in this frame of reference, corresponding to phase velocities below C_s in the rest frame. In this model the waves remain undamped. A general solution for waves described by this linearized version of the KdV equation is

$$u(x,t)=\int_{-\infty}^{\infty}F(k)e^{i(kx-\omega(k)t)}dk,$$

with $F(k)$ being the Fourier transform of the initial condition $u(x,0)$. For the particular case, where $u(x,0)=a\delta(x)$, we find $F(k)=a/2\pi$, and

$$\begin{aligned} u(x,t) &= \frac{a}{2\pi}\int_{-\infty}^{\infty}e^{i(kx-\omega(k)t)}dk \\ &= \frac{a}{2\pi}\int_{-\infty}^{\infty}\cos\,(kx+\alpha k^3t)dk \\ &\equiv \frac{a}{2(3\alpha t)^{1/3}}\mathrm{Ai}\left(\frac{x}{(3\alpha t)^{1/3}}\right), \end{aligned} \tag{4.27}$$

where the Airy function $\mathrm{Ai}(x)$ was introduced (Whitham 1974); see Fig. 4.5. Since the phase velocity varies with wavenumber, the various Fourier components of an initial condition will propagate with different velocities, and the entire perturbation will eventually spread out. When $\alpha>0$, the long wavelengths arrive first, and the shorter ones subsequently later on, as also evidenced by Fig. 4.5. Note, however, that during this "spreading out," the quantity $\int_{-\infty}^{\infty}u(x,t)dx$ is conserved, as readily shown by using the linearized version of (4.26).

We now return to the full nonlinear equation (4.26). It turns out that just as for Burgers' equation, we can determine stationary solutions, as shown in the following. We look for solutions traveling with some (yet undetermined) velocity U, of the form $u = v(x-Ut)$. By insertion into the Korteweg-deVries equation and introducing $\xi \equiv x - Ut$ we find

$$(v-U)\frac{dv}{d\xi} + \alpha\frac{d^3v}{d\xi^3} = 0 .$$

Using $v dv/d\xi = \frac{1}{2}dv^2/d\xi$ and integrating once we obtain

$$\frac{1}{2}v^2 - Uv + \alpha\frac{d^2}{d\xi^2}v = 0 , \tag{4.28}$$

where we set the integration constant to zero without loss of generality. Upon multiplication by $dv/d\xi$, using again the formula (4.20) and integrating once more we get

$$\frac{1}{2}\left(\frac{dv}{d\xi}\right)^2 + \frac{1}{6\alpha}v^3 - \frac{1}{2\alpha}Uv^2 = W, \tag{4.29}$$

where W is an integration constant. The relation (4.29) has an interesting physical analogy; if we in our imagination consider ξ as a temporal variable and v as a position, then

$$V(v) \equiv \frac{1}{6\alpha}v^3 - \frac{U}{2\alpha}v^2$$

can be seen as being equivalent to a potential, and (4.29) is then simply describing a "particle" with total energy W being the sum of a kinetic energy (here $\frac{1}{2}(dv/d\xi)^2$) and a potential energy (here $V(v)$), and moving in this "quasi-potential" $V(v)$. This pseudo-potential is sometimes called *Sagdeev potential.* We have $V(v) = 0$ for $v = 0$ and for $v = 3U$, and $V'(v) = 0$ for $v = 2U$. At the local minimum $V(v = 2U) = -(2/3)U^3/\alpha$.

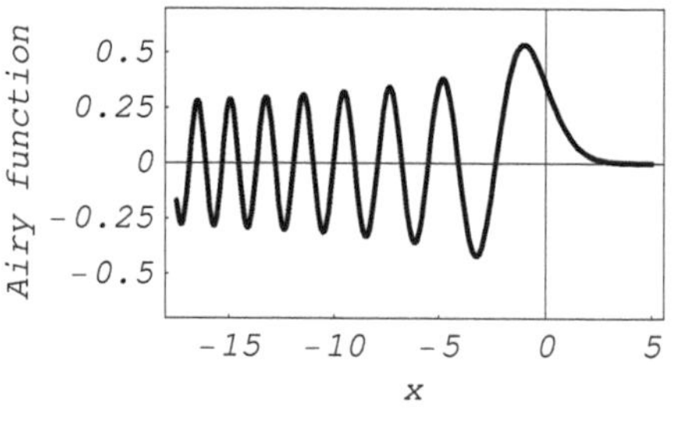

FIGURE 4.5: *The Airy function, Ai(x). Asymptotically, we have for real* $x \to \infty$ *that* $Ai(x) \to \frac{1}{2}x^{-1/4}\pi^{-1/2}\exp(-\zeta)$, *while* $Ai(-x) \to \pi^{-1/2}x^{-1/4}[\sin(\zeta + \frac{\pi}{4}) - \cos(\zeta + \frac{\pi}{4})]$, *with* $\zeta \equiv \frac{2}{3}x^{3/2}$ *everywhere. The period of the oscillations becomes shorter for x being increasingly negative.*

In Fig. 4.6a) we show the pseudo-potential $V(v)$. For $\alpha > 0$ we have nontrivial bound solutions for a proper choice of W. To be physically acceptable, we require them to be finite for all ξ and also v being finite, possibly zero, for $|\xi| \to \infty$. In a $(v, dv/d\xi)$-plane (see Fig. 4.6b)), we will have a family of curves for varying W, the physically acceptable ones being closed around the local minimum at $2U$ of V. For $-\frac{2}{3}v^3/\alpha W < 0$ we have closed orbits corresponding to oscillating solutions; for $W > 0$ the solutions are not bound and therefore physically unacceptable. For $W = 0$ we have a particularly interesting solution; here $v \geq 0$ for all ξ. It is symmetric and has one local maximum $3U$ at the origin $\xi = 0$, and goes asymptotically to zero for $|\xi| \to \infty$. By solving (4.29) we find

$$v(x,t) = 3U\,\mathrm{sech}^2\left(\frac{1}{\Delta}(x - Ut)\right), \tag{4.30}$$

where $\Delta = 2\sqrt{\alpha/U}$ with $\alpha > 0$. This is the so called *soliton* solution, illustrated in Fig. 4.7. Its peak amplitude $A \equiv 3U$ is determining its width Δ as well as its velocity $U = A/3$. *The larger the amplitude of the soliton, the more narrow it is, and the faster it moves, the velocity being in linear proportion to the amplitude!*

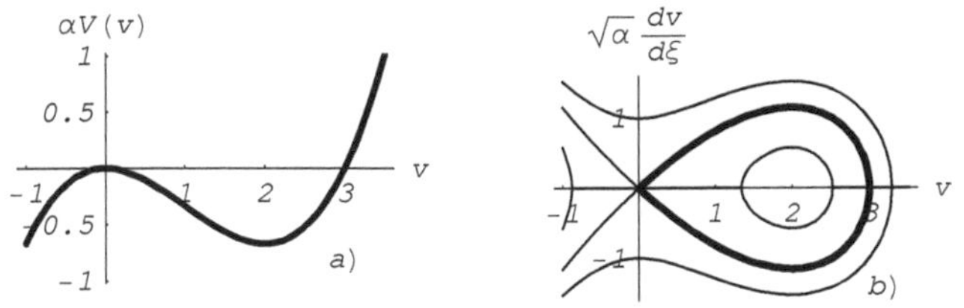

FIGURE 4.6: *The pseudo-potential $V(v)$ is shown in a) for $U = 1$, while in b) we have the plane spanned by $dv/d\xi$ and v with curves corresponding to different values of W. The heavy line corresponds to $W = 0$, giving the soliton solution (4.30). We took $\alpha > 0$ in the presentation.*

We have seen from (4.27) that the dispersion part in the *linear* equation tends to "spread out" an initial perturbation, while from the analysis of (4.15) we learned that the nonlinearity tends to create steep gradients. The soliton can be seen as a pulse shape, where dispersion is exactly balanced by nonlinearity. For the solution to exist, we require $\mathrm{sign}(\alpha) = \mathrm{sign}(U)$. The infinitely extended periodic solutions are called *cnoidal waves* (Whitham 1974).

Ⓢ **Exercise:** Demonstrate that, by a proper choice of variables, (4.25) can be brought into a normalized form

$$\frac{\partial}{\partial t}u - 6u\frac{\partial}{\partial x}u + \frac{\partial^3}{\partial x^3}u = 0 .$$

How will you change the variables to give, for instance,

$$\frac{\partial}{\partial t}u + u\frac{\partial}{\partial x}u + \frac{1}{2}\frac{\partial^3}{\partial x^3}u = 0 .$$

How about other normalizations? Ⓢ

- **Exercise:** Demonstrate that the KdV equation in the form

$$\frac{\partial}{\partial t}u - \beta u\frac{\partial}{\partial x}u + \alpha\frac{\partial^3}{\partial x^3}u = 0$$

with $\alpha > 0$ and $\beta > 0$ has a soliton solution

$$u = -A\,\mathrm{sech}^2\left((x - Ut)\sqrt{\frac{A\beta}{12\alpha}}\right),$$

where $U = A\beta/3$ with $A > 0$.

For a *linear* sound pulse with width $\mathcal{L}$, we can define a time scale, $\tau_\ell \equiv \mathcal{L}/C_s$, as the time it takes the pulse to propagate its own width. For the KdV soliton, we can in the same way define a *nonlinear* time scale as the time it takes the soliton (4.30) to propagate its own width *in the frame of reference moving with the sound speed*, i.e., $\tau_{NL} \equiv \mathcal{L}/U$, or, as expressed in terms of the soliton *amplitude*, $\tau_{NL} \equiv 6\sqrt{\alpha/A^3}$, using $\mathcal{L} = \Delta$ with $U = A/3$. Sometimes we find τ_{NL} denoted "soliton time." Evidently, $\tau_{NL} \to \infty$ as we approach the linear wave description by letting $A \to 0$.

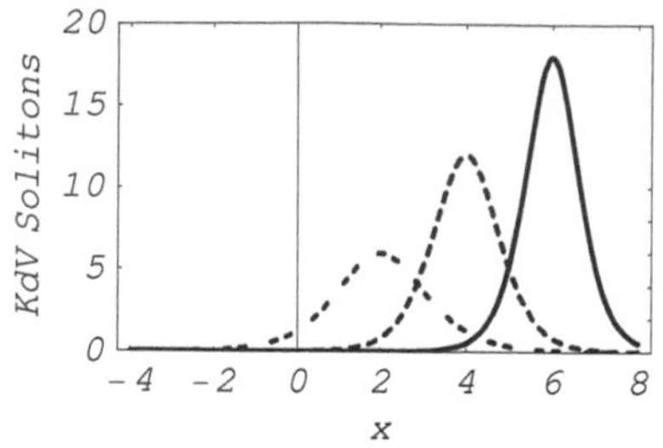

FIGURE 4.7: *Illustration of KdV solitons for different velocities,* $U = 2,4,6$, *shown with double dashed, dashed and solid lines, respectively, all at time* $t = 1$ *and* $\alpha = 1$. *A line drawn through the maxima of the three solitons will cross the origin, since the soliton velocity is proportional to the amplitude!*

At first sight, one might wonder why such a refined object as a soliton deserves so much attention! It turns out, however, that the KdV equation can be solved exactly (formally, at least) by the so called *inverse scattering transform*, and it turns out that any compact initial condition can be described asymptotically by a superposition of one or more solitons with different amplitudes and some small amplitude Airy function like oscillations (Drazin & Johnson 1989, Kono & Škorić 2010). When the solitons have different amplitudes, they also have different velocities, and asymptotically they will be separated in space. In Fig. 4.8 we show illustrative results. The left-hand case is obtained by using an initial condition that is too wide compared to its amplitude so it breaks up into three solitons moving with different velocities. The opposite polarity, also shown in Fig. 4.8, continues to form solitons at a steady rate.

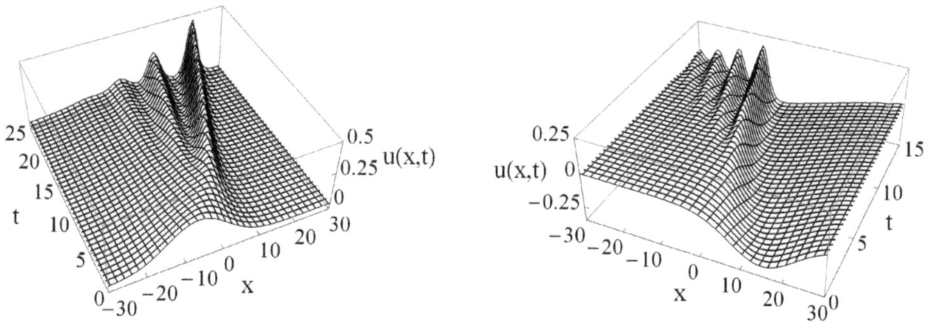

FIGURE 4.8: *Illustrative numerical solutions of the KdV equation, carried out in the frame moving with the sound speed. The equation solved has the form* $\partial u/\partial t + 6u\partial u/\partial x + \partial^3 u/\partial x^3 = 0$. *The positive initial condition breaks up into three solitons moving with different velocities. For the opposite polarity, solitons are formed in a steadily increasing number without saturation. The initial conditions are Gaussians.*

It is important to note that for the given signs in the KdV-equation considered, only initial conditions containing positive u-values (some regions *might* be negative) will produce one or more solitons. If the initial condition is *all* negative, its subsequent time evolution will be dispersing and oscillating at all times. See also the right-hand Fig. 4.8.

Two solitons with different amplitudes will collide when the largest one is overtaking the smaller; see Fig. 4.9. If you compare the initial state with the final, it appears as if the solitons were not changed at all during the collision. From Fig. 4.9 we note one other feature of the soliton-soliton interaction: if we draw a straight line through the positions of the two solitons in the $x-t$-plane before and after the collision, we note a small shift between these two lines, often called a "phase shift." Also this can be predicted by the inverse scattering transform, but we will not discuss such details here.

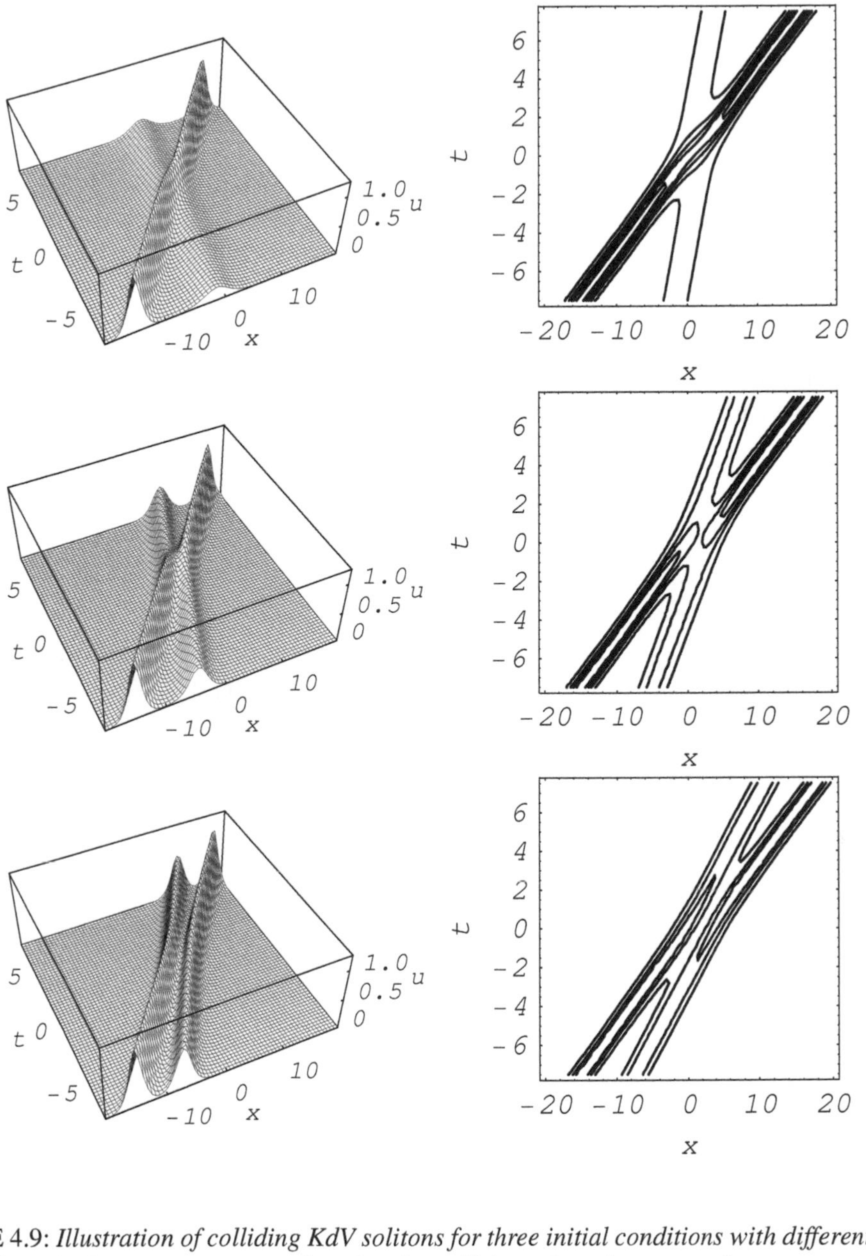

FIGURE 4.9: *Illustration of colliding KdV solitons for three initial conditions with different relative amplitudes. The figures are generated so that the collision takes place at* $t = 0$.

Returning to the laboratory frame of reference by a Galilean transformation with velocity C we observe that the solitons are all *supersonic*; the larger their amplitude, the more supersonic they are! In order to have a stationary solution we would actually expect this; if the soliton was sub-sonic, $U < C$, it would couple to the linear oscillations and radiate small amplitude waves, thus violating the time stationary assumption.

Intimately associated with the inverse scattering transform we have an infinite number of conserved quantities associated with the KdV equation. A few examples are (Drazin & Johnson 1989)

$$I_1 \equiv \int_{-\infty}^{\infty} u(x,t)dx, \tag{4.31}$$

$$I_2 \equiv \int_{-\infty}^{\infty} \frac{1}{2}u^2(x,t)dx, \tag{4.32}$$

$$I_3 \equiv \int_{-\infty}^{\infty} \left(\frac{\alpha}{3}u^3(x,t) + \frac{1}{2}\left(\frac{\partial}{\partial x}u(x,t)\right)^2 \right) dx, \tag{4.33}$$

where the proofs are left to the reader as an exercise. (In (4.32) we introduced $1/2$ for later convenience.) As already mentioned, the examples (4.31)–(4.33) are just the beginning; there are infinitely many, but more complicated, conservation laws (Drazin & Johnson 1989). One might think that the soliton solution owes its existence to this infinity of conservation laws, but it is not obviously so; there are a large number of nonlinear equations having *solitary* wave solutions, but these will in general not have the soliton property, i.e., they do *not* emerge undistorted after a collision. The corresponding equations are *not* even formally solvable by a version of an inverse scattering transform. These equations often have only a few conserved quantities!

- **Example:** The expression (4.30) represents a single soliton solution of the KdV equation. We have a *two* soliton solution (Drazin & Johnson 1989) in the form

$$v(x,t) = 2\frac{k_1^2E_1 + k_2^2E_2 + 2(k_2-k_1)^2E_1E_2 + A(k_2^2E_1 + k_1^2E_2)E_1E_2}{1 + E_1 + E_2 + AE_1E_2}$$

with $A \equiv (k_2-k_1)^2/(k_1+k_2)^2$, $E_j = \exp(\theta_j)$, where $\theta_j \equiv k_jx - k_j^3t + \alpha_j$ for $j = 1,2$. This solution was applied for constructing Fig. 4.9, using the values $(k_1,k_2) = (0.7, \sqrt{2})$, $(1.0, \sqrt{2})$ and $(1.2, \sqrt{2})$, respectively.

- **Exercise:** Demonstrate that $I_0 \equiv \int_{-\infty}^{\infty} \left(xu - \frac{1}{2}tu^2\right) dx$ is also a conserved quantity.

- **Exercise:** Show that if $u(x,t)$ is a localized solution of the KdV equation, then we have

$$\frac{\partial}{\partial t}\int_{-\infty}^{\infty} xu(x,t)dx = \text{const}.$$

Interpret this result. What is the physical meaning of the integral? In light of this, how will you interpret I_0 found before?

Traditionally, waves in shallow water channels are used as the standard example for solitons as described by the KdV equation (Whitham 1974). An amusing summary of the history of the first observations is given by, for instance, Lonngren and Scott (1978). However, several examples can be found in nature where this equation is applicable as a good approximation. In particular in plasma physics we find quite a few such cases, but we again add a note of caution; it has *not* been demonstrated that *all* weakly dispersive waves can be described by a KdV equation, not even approximately! Even though their linear dispersion relation might be well approximated by the form $\omega = C_1k + C_2k^3$ with properly chose constants C_1 and C_2, the nonlinear properties can differ from a KdV-type equation.

The KdV-equation touches upon an important problem in basic physical sciences, also historically important. Taking a linear chain, or a lattice (in two or three spatial dimensions) it can demonstrated (Symon 1960) that in a linear approximation the model equations can be decoupled into a set of independent equations specifying normal mode solutions. Exciting one mode initially, the energy will therefore remain in that mode and in this limit the system will thus not relax into a thermal equilibrium. It was anticipated that nonlinear terms left out by the linearization would "do the trick", but at the time this was merely a conjecture. When numerical computing became a possibility a group of scientists decided to simulate the one-dimensional analogue of atoms in a crystal: a long chain of linked by massless springs that obey Hooke's law (i.e., a linear interaction), but with an additional weak nonlinear term. Although the work was a significant step, it was published only as a Los Alamos report (Fermi et al. 1955). The study is now known as the Fermi-Pasta-Ulam problem, or simply the FPU-problem. (The one who did all the numerical work, Mary Tsingou, did not receive much credit for her efforts (Dauxois 2008)). The surprising observation was that in a short transient time interval, indeed a tendency towards equipartition of energy could be observed, but as time went on the initial condition re-emerged, and the process continued periodically as long as the code ran. We found recurrence in a linear model, see Section 3.1.3, but the surprising observation was to recover it also when some lowest order nonlinear terms were included. Pursuing the solution of the FPU paradox, Zabusky and Kruskal (1965) investigated the problem by a continuous KdV-model in real space rather than in Fourier space. They were able to explain the periodic behavior in terms of the dynamics of localized excitations of solitons forming after some initial time, passing through each other to reconstruct the initial condition after some later time depending on the amplitude of the initial periodic condition. Here we deal with nonlinear physical processes where interacting localized pulses do not scatter irreversibly. A summary and review of these early works is given by one of the pioneers in the field (Zabusky 2005). A particular plasma wave problem have been discussed by semi-intuitive arguments by Märk and Sato (1977), where also a nonlinear recurrence phenomenon was found.

4.2.2 Perturbations of a KdV equation

The Korteweg-deVries equation (4.26) is very appealing by its elegance, and serves to give a basic insight into many properties of weakly nonlinear, weakly dispersive waves. In most cases, however, the detailed agreement between the analytical results and some experimental observations can be rather unsatisfactory. Often it is possible to modify (4.26) by some small terms, which can improve the agreement, but this will in general be at the expense of an exact solution. Since, as stated before, the inverse scattering transform (Whitham 1974, Drazin & Johnson 1989) is in many cases to be seen as a sort of "existence theorem" anyhow, this loss need not be discouraging. We write a perturbed KdV equation as

$$\frac{\partial}{\partial t}u + u\frac{\partial}{\partial x}u + \alpha\frac{\partial^3}{\partial x^3}u = \varepsilon R[u]\,, \tag{4.34}$$

where $R[u]$ is some, for the moment unspecified, operator acting on u, and ε is a small quantity, introduced as a reminder to treat the right-hand side of (4.34) as a small perturbation. Analytical results for a systematic perturbation expansion exist (Karpman 1979*b*) for the investigations of equations like (4.34). Here we shall be content with a simpler semi-intuitive discussion, which nonetheless gives rather accurate results (Watanabe 1978, Karpman et al. 1980, Lynov 1983, Watanabe & Yajima 1984, Pécseli 1985). Although the analysis can be made slightly more general, we restrict the discussion to the case where $\int R[u]dx = 0$, implying that $I_1 \equiv \int u(x,t)dx =$ const., as for the unperturbed KdV equation. This will be the most relevant situation for practical applications. One example is a viscosity-like perturbation with $\varepsilon R[u] = \varepsilon\partial^2 u/\partial x^2$ (while a counter-example is a dissipative term like $\varepsilon R[u] = \varepsilon u$). In general we have

$$\frac{d}{dt}I_2(t) \equiv \frac{d}{dt}\int_{-\infty}^{\infty}\frac{1}{2}u^2(x,t)dx = \varepsilon\int_{-\infty}^{\infty}u(x,t)R[u(x,t)]dx \neq 0\,. \tag{4.35}$$

We now make a formal decomposition $u(x,t) \equiv u_s(x,t) + u_{ns}(x,t)$, where u_s denotes a "soliton part" and u_{ns} a "non-soliton part," respectively. Our anticipation is that for small ε the initial perturbation will develop into a soliton, apart from some small deviations which vanish with $\varepsilon \to 0$.

In case the initial perturbation produces *several* solitons, we expect these to separate in time (see Fig. 4.7), so also in this case we are free to consider one soliton at a time. At each "time step," the soliton is modified slightly by the perturbation in (4.34). Assuming the time variation to be slow on the nonlinear time scale τ_{NL} introduced before, we consider each time to be a "new" initial condition, and expect that the soliton is readjusting to this new condition in such a way that u_s remains to be described by one parameter solution, the amplitude $A(t) \equiv 3U(t)$ given in terms of the soliton velocity U. The soliton part of $u(x,t)$ is then given as $u_s(x,t) \equiv A(t)\,\mathrm{sech}^2\,[(x-A(t)t/3)/\Delta(t)]$ with $\Delta(t) \equiv 2\sqrt{3\alpha/A(t)}$. All deviations from the local soliton solution are lumped into the non-soliton part u_{ns}, which we anticipate to be of order $O(\varepsilon)$. We can consider u_{ns} to represent the material left over when u is adjusting to the instantaneous soliton parameters, while simultaneously conserving I_1. By these arguments, we can expect that the spatial overlap between the soliton and the non-soliton parts is small, i.e., that $u_{ns}u_s \approx 0$.

Illustrative results for the evolution of a soliton-like solution for a perturbed KdV equation are shown in Fig. 4.10. The non-soliton part can have either sign, as indicated by Figs. 4.10a and b. For the restricted class of perturbations conserving I_1, the proper sign of u_{ns} can be determined from

$$I_1 = \int_{-\infty}^{\infty} \left[u_{ns} + A(t)\,\mathrm{sech}^2 \left(\frac{x - A(t)t/3}{2\sqrt{3\alpha/A(t)}} \right) \right] dx = \mathrm{const}. \tag{4.36}$$

with the time evolution of $A(t)$ inserted.

We will need the expressions for the soliton parts of I_1 and I_2, with $u(x,t)$ given as a soliton. Using (4.30), we have $I_1^{(s)} = 12\sqrt{\alpha U} = 4\sqrt{3\alpha A}$ and $I_2^{(s)} = 24U\sqrt{\alpha U} = 8A\sqrt{\alpha A/3}$, where we anticipate that the velocity $U = U(t)$ and therefore also the amplitude $A(t) \equiv 3U(t)$ varies with time in case we have $\varepsilon R[u] \neq 0$. For the case where (4.36) applies, we have initially $u_{ns} = 0$, and the constant is equal to $4\sqrt{3\alpha A(0)}$. Consequently we have $\int_{-\infty}^{\infty} u_{ns} dx = 4\left(\sqrt{3\alpha A(0)} - \sqrt{3\alpha A(t)}\right)$. As an estimate we can approximate u_{ns} as a spatially constant level, say $h = h(t)$, extending from the origin to the position of the soliton at time t, this position being $X(t) = \int_0^t U(\tau)d\tau = \frac{1}{3}\int_0^t A(\tau)d\tau$. By this, we ignore the area of the oscillating "tail" at positions to the left of the origin; see, for instance, Fig. 4.10. As an estimate we therefore have $h(t) \approx 4\left(\sqrt{3\alpha A(0)} - \sqrt{3\alpha A(t)}\right)/X(t)$, which is completely determined once we have found $A = A(t)$. Since $\varepsilon R[u]$ is assumed to be a small perturbation, we might as well use the approximation $X(t) \approx 3A(0)t$ here, although the simplification we gain is meager. Note that we may very well have $h(t) < 0$. In laboratory studies of weakly dispersive plasma waves, the development of a trailing plateau has been observed (Karpman et al. 1980, Lynov 1983) quite similar to the one illustrated in Fig. 4.10.

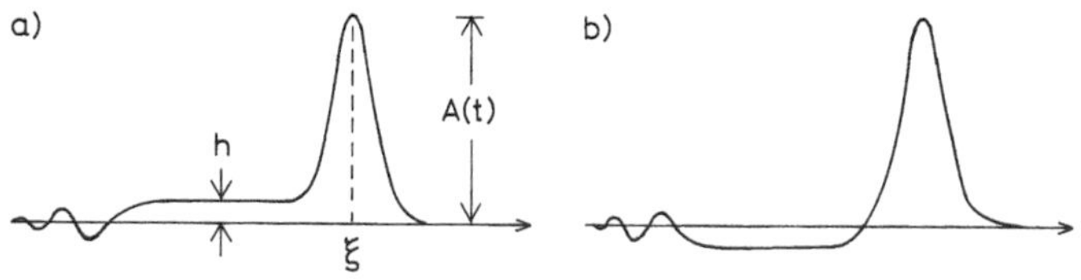

FIGURE 4.10: *Illustration of perturbed soliton evolutions for the KdV equation, including a perturbation term.*

Using (4.35) we can now obtain an approximate expression for the time evolution of the amplitude, $A(t)$. We assume that the non-soliton part $u_{ns} \sim O(\varepsilon)$, and ignore terms $O(\varepsilon^2)$ or smaller. With these assumptions and the observation $u_{ns}u_s \approx 0$ mentioned before, we find that (4.35) only contains the soliton part, and we readily have

$$\frac{d}{dt}\left(4A(t)\sqrt{\alpha A(t)/3}\right) = \varepsilon \int_{-\infty}^{\infty} A(t)\,\mathrm{sech}^2\left(\frac{x - A(t)t/3)}{2\sqrt{3\alpha/A(t)}}\right) R[u_s(x,t)]dx,$$

giving

$$\frac{d}{dt}A(t) = \frac{\varepsilon}{2\sqrt{3}}\sqrt{\frac{A(t)}{\alpha}} \int_{-\infty}^{\infty} \mathrm{sech}^2\left(\frac{x - A(t)t/3)}{2\sqrt{3\alpha/A(t)}}\right) R\left[A(t)\mathrm{sech}^2\left(\frac{x - A(t)t/3)}{2\sqrt{3\alpha/A(t)}}\right)\right] dx, \qquad (4.37)$$

which, at least formally, gives a result for $A(t)$ with given $\varepsilon R[u]$, thus solving the problem.

- **Example:** The summary of approximate solutions to the perturbed KdV equation can best be appreciated when illustrated by an example. We consider here the KdV-Burgers equation, which to the best of our knowledge cannot be solved analytically. The perturbation solution was discussed by Watanabe (1978). In normalized form, the relevant equation can be written as

$$\frac{\partial}{\partial t}u + u\frac{\partial}{\partial x}u + \frac{1}{2}\frac{\partial^3}{\partial x^3}u = \varepsilon\frac{\partial^2}{\partial x^2}u.$$

We have $dI_1/dt = 0$ (see (4.31)) and also the more interesting result

$$\frac{d}{dt}I_2 = -\varepsilon\int_{-\infty}^{\infty}\left(\frac{\partial u}{\partial x}\right)^2 dx$$

(see (4.32)). Ignoring terms $O(\varepsilon^2)$, we again find that this expression contains only the soliton part, and find by (4.37) the result

$$A(t) = \frac{A(0)}{1 + t\varepsilon A(0)8/45}, \qquad \lim_{t\to\infty} A(t) = \frac{45}{8\varepsilon t}.$$

We obtain the estimate for the tail amplitude from

$$\int_{-\infty}^{\infty} u_{ns}dx = \sqrt{24A(0)}\left(1 - \frac{1}{\sqrt{1 + t\varepsilon A(0)8/45}}\right),$$

where evidently we have $h(t) > 0$. By the estimate argued before we find

$$h(t) \approx \frac{1}{t}\sqrt{\frac{8}{3A(0)}}\left(1 - \frac{1}{\sqrt{1 + t\varepsilon A(0)8/45}}\right), \qquad \lim_{t\to\infty} h(t) = \frac{1}{t}\sqrt{\frac{8}{3A(0)}}.$$

The asymptotic limit of $h(t)$ is here independent of ε, to the given accuracy, and is reached at times $t \gg 1/\varepsilon A(0)$. Slightly inhomogeneous systems can be analyzed as well (Watanabe & Yajima 1984) by these methods.

4.2.3 Boussinesq equations

The KdV equation is explicitly derived for waves or pulses propagating in one direction, as evidenced by the operator $\partial/\partial t - C_s\partial/\partial x$ in the lowest order approximation. It is possible to obtain an equation which can account for bi-directional propagation, here given in dimensionless form (Drazin & Johnson 1989)

$$\frac{\partial^2}{\partial t^2}u - \frac{\partial^2}{\partial x^2}u - \frac{\partial^4}{\partial x^4}u + \frac{\partial^2}{\partial x^2}u^2 = 0. \qquad (4.38)$$

The two first terms correspond to (4.9), as might be expected. The third term represents a dispersion, where we note that a term like $\partial^4 u/\partial t^2 \partial x^2$ might as well have been argued. The last term represents the nonlinearity. The equation does not have any significant advantage over the KdV equation, however, at least not as long as soliton dynamics is an issue. The point is that two counter-propagating pulses overlap for only a small time, and do not manage to interact significantly. In case of *overtaking* interactions, the interaction time is much longer, and the interaction becomes significant. This limit is, however, well described by the KdV equation.

4.2.4 Generalization to three dimensions

The discussions and derivation of the KdV equation was restricted to spatially one dimensional systems. Just by looking at the equation (4.25) it is not obvious how it is to be generalized to three spatial dimensions. The problem is simply how to generalize the operator $\partial^3/\partial x^3$ to a three dimensional operator: the simplest guess $\nabla^2\nabla$ does not work, since it is a *vector* operator entering an equation containing the *scalar* operator $\partial/\partial t$. It is best to consider individual problems, as done by Zakharov and Kuznetsov (1974), who for a strongly magnetized plasma derived

$$\frac{\partial}{\partial t}u + \frac{\partial}{\partial x}\left(u^2 + \nabla_\perp^2 u\right) = 0, \tag{4.39}$$

for the scalar variable u, being the velocity component along the homogeneous magnetic field in the x-direction, while $\nabla_\perp$ is the differential operator for the coordinates perpendicular to the magnetic field. The equation (4.39) is written in normalized units.

It is interesting, in this context, that the Boussinesq equation (4.38) is quite simple to generalize into three dimensions. This problem was, for instance, studied by Kako and Yajima (1978) and later in more detail (Kako & Yajima 1982) by analyzing the equation

$$\frac{\partial^2}{\partial t^2}u - \nabla^2 u - \beta\frac{\partial^2}{\partial t^2}\nabla^2 u + \alpha\frac{\partial}{\partial t}(\nabla u)^2 = 0, \tag{4.40}$$

as obtained, for instance, for ion acoustic waves under homogeneous conditions.

4.3 Strongly dispersive waves

In contrast to the weakly dispersive waves, we can have cases as illustrated in Fig. 4.1b), where the phase velocity and the group velocity are significantly different. For strongly dispersive waves we expect, in general, that we have a frequency for the oscillations which does not change much with wavenumber, at least in a nontrivial interval, and we are dealing with a well defined characteristic frequency, as for an oscillator. Before discussing *waves* it is instructive first to discuss a simple oscillating system with well defined frequency, such as a pendulum. This analysis will reveal a basic feature, the *nonlinear frequency shift*.

4.3.1 Simple oscillators

The general dynamic equation for a simple pendulum in the form of a thin massless wire of length ℓ and a heavy point mass M acted upon by gravity is

$$\frac{d^2}{dt^2}\theta + \frac{g}{\ell}\sin\theta = 0 \tag{4.41}$$

where θ is the angular deflection of the string with respect to vertical and g is the gravitational acceleration. For small amplitudes we linearize (4.41) to obtain $d^2\theta/dt^2 + (g/\ell)\theta = 0$, and the standard oscillation with frequency $\omega_0 = \sqrt{g/\ell}$ is recovered (Symon 1960, Nelson & Olsson 1986, Lima & Arun 2006). To this order we are approximating the potential well at the minimum of the $\sin\theta$ function by a parabola. To the next order in a series expansion of $\sin\theta$ we find the nonlinear Duffing equation

$$\frac{d^2}{dt^2}\theta + \frac{g}{\ell}\theta = \frac{g}{3!\ell}\theta^3. \tag{4.42}$$

Incidentally, this equation can be solved exactly in terms of known functions, but an approximate solution is here more informative (Munakata 1952). There are many ways to solve such nonlinear equations in an approximate way (Nayfeh 1973), and they all have their respective merits. We here use the method of *strained parameters*, also called the Lindstedt-Poincaré method (Nayfeh 1973). For this purpose we measure time in units of $\sqrt{\ell/g}$ and the displacement in units of the maximum initial displacement A. We can then rewrite (4.42) as $d^2\theta/dt^2 + \theta + \varepsilon\theta^3 = 0$, where now $\varepsilon = -A^2/3!$ is a small parameter measuring the maximum displacement of the pendulum. The essence of the method is to prevent the appearance of secular terms, i.e., terms growing linearly with time. Such terms appear in case the right-hand side of (4.42) has terms containing $\exp(-i\omega_0 t)$, which will drive the system at the resonance frequency. The presence of these terms is physically unacceptable: we start with a perfectly well behaved system, a pendulum, and secular terms cannot appear, so we have to find a way to remove them. To avoid these unphysical terms we first introduce a new time-like variable by

$$t \equiv s(1 + \varepsilon\omega_1 + \varepsilon^2\omega_2 + \ldots) \tag{4.43}$$

The right-hand side of this expression can be seen as an expansion in ε with, for the moment, unknown coefficients $\omega_1, \omega_2, \ldots$. In order to account for a nonlinear dependence of the characteristic frequency we stretch the time axis with a yet unknown factor somehow depending on ε. In terms of this new time variable we obtain

$$\frac{d^2}{ds^2}\theta + (1 + \varepsilon\omega_1 + \varepsilon^2\omega_2 + \ldots)^2(\theta + \varepsilon\theta^3) = 0\,,$$

where the solution for θ will depend on ε. We write $\theta = \sum_0^\infty \varepsilon^n \theta_n(s)$ corresponding to an expansion of θ in terms of ε. The differential equation has to be fulfilled for all ε, implying that we can group terms containing equal powers of ε together, and then each of these resulting equations has to be fulfilled individually. Equating terms of like powers in ε we find to zeroth order

$$\frac{d^2}{ds^2}\theta_0 + \theta_0 = 0$$

to first order

$$\frac{d^2}{ds^2}\theta_1 + \theta_1 = -\theta_0^3 - 2\omega_1\theta_0$$

and to second order

$$\frac{d^2}{ds^2}\theta_2 + \theta_2 = -3\theta_0^2\theta_1 - 2\omega_1(\theta_1 + \theta_0^3) - (\omega_1^2 + 2\omega_2)\theta_0,$$

etc. As the general solution of the first of these equations we have $\theta_0 = a\cos(s+\phi)$, where a and ϕ are constants of integration. (The lower case a is taken here, not to be confused with A used before, as the maximum initial displacement, but the two are of course related.) To next order we have

$$\begin{aligned}\frac{d^2}{ds^2}\theta_1 + \theta_1 &= -\frac{1}{4}a^3\cos 3(s+\phi) \\ &\quad - \left(\frac{3}{4}a^2 + 2\omega_1\right)a\cos(s+\phi)\,.\end{aligned}$$

If a straightforward perturbation expansion is used, we would have secularity producing terms. With the present expansion we can use ω_1 to eliminate such terms, and find that this is achieved by the choice $\omega_1 = -3a^2/8$. We then find $\theta_1 = \frac{1}{32}a^3 \cos 3(s+\phi)$. Similarly, we have (Nayfeh 1973)

$$\begin{aligned} \frac{d^2}{ds^2}\theta_2 + \theta_2 &= \left(\frac{51}{128}a^4 - 2\omega_2\right) a\cos(s+\phi) \\ &\quad + \text{terms that do not produce secularities}. \end{aligned}$$

Secular terms are again eliminated by the choice $\omega_2 = 51a^4/256$. By straightforward insertion we then have $\theta = \theta_0 + \varepsilon\theta_2 + \ldots = a\cos(s+\phi) + \varepsilon(a^3/32)\cos 3(s+\phi) + \ldots$, giving

$$\theta = a\cos(\omega t + \phi) + \frac{\varepsilon}{32}a^3 \cos 3(\omega t + \phi) + \text{higher order terms}. \tag{4.44}$$

We now also have $s = t\omega$. Using (4.43) we find

$$\begin{aligned} \omega &= 1/(1 - 3\varepsilon a^2/8 + 51\varepsilon^2 a^4/256 + \ldots) \\ &\approx 1 + 3\varepsilon a^2/8 - 51\varepsilon^2 a^4/256 + \ldots. \end{aligned}$$

The important observation is here that the frequency of oscillation has a nonlinear contribution proportional to a^2, etc., i.e., a *nonlinear frequency shift* (Nelson & Olsson 1986, Lima & Arun 2006). (There are no terms with odd exponents of a since these would imply the frequency shift depends on the sign of the initial displacement, which would be meaningless.) In the present case of the simple pendulum, with $\varepsilon < 0$, the oscillation frequency *decreases* with amplitude. Physically, the nonlinear frequency shift means that we are correcting the parabolic potential well for the *linear* oscillation slightly, in order to accommodate the best fit to $\theta + \varepsilon\theta^3$, for a given amplitude a. A parabolic fit is very good just at the bottom of a cos-trough, but if you want to make a parabola valid for a little bit wider range, it is an advantage to make it a little "flatter" and then live with a somewhat reduced fit in the bottom! The resulting oscillation frequency diminishes with increasing a, in agreement with the analytical result.

Based on these results, we expect that one important characteristic of strongly dispersive waves will be a nonlinear frequency shift, with a corresponding effect for the phase velocity.

- **Exercise:** Apart from all this, the method of strained parameters can be useful in many cases, and it is a good exercise to go carefully through the derivations in this chapter.

Ⓢ **Exercise:** Give an expression for the period of the full nonlinear oscillator (Nelson & Olsson 1986, Lima & Arun 2006). This can be done without actually solving the full-time variation. Express the result as a series expansion including $\sin(a/2)$, where a is the initial angular displacement of the pendulum. It is thus relatively easy to find expressions for the nonlinear corrections to the oscillation period. The method of strained parameters gives also an approximation to the functional form of the nonlinearly oscillating amplitude, see (4.44), which is otherwise not always easily found. Ⓢ

4.3.2 Weakly nonlinear dispersive waves

The discussion of weakly nonlinear waves takes here the linear dispersion relation as the starting point (Lonngren 1983). We have quite generally for homogeneous media

$$D(\omega_0, \mathbf{k}_0) = 0 \qquad \text{for the wave solution} \qquad \mathbf{u}(\mathbf{r},t)\, e^{-i(\omega_0 t - \mathbf{k}_0\cdot\mathbf{r})}, \tag{4.45}$$

where we assume that the space-time variation of $\mathbf{u}(\mathbf{r},t)$ is slow compared to the exponential, i.e., the wave amplitude changes only little within a wave-period. Our aim is to first obtain a differential equation for the linear variation of the amplitude, which we subsequently extend to cover also the most important nonlinear effects. In the following we use a spatially one dimensional representation with **k** along the x-axis, noting that again here the generalization to the fully three dimensional case is not trivial.

From a solution ω_0, k_0 to (4.45) we make a series expansion (Lonngren 1983, Craik 1985)

$$D(\omega,k) \approx D(\omega_0,k_0) + \frac{\partial D}{\partial \omega_0}(\omega-\omega_0) + \frac{\partial D}{\partial k_0}(k-k_0)$$
$$+\frac{1}{2}\left(\frac{\partial^2 D}{\partial \omega_0^2}(\omega-\omega_0)^2 + \frac{\partial^2 D}{\partial k_0^2}(k-k_0)^2 + 2\frac{\partial^2 D}{\partial \omega \partial k_0}(\omega-\omega_0)(k-k_0)\right). \tag{4.46}$$

The subscripts ω_0 and k_0 serve as reminders that the derivations of $D(\omega,\mathbf{k})$ should be obtained at ω_0, k_0. To save space we introduced the notation $\partial D/\partial \omega_0$ for $\partial D/\partial \omega|_{\omega_0}$ and similarly for the wave-number derivatives. Since we assumed that (ω_0, k_0) is a natural mode of oscillation, we have obviously $D(\omega_0, k_0) = 0$.

While (ω_0, k_0) is the "carrier wave," then $(\omega-\omega_0, k-k_0)$ represents the modulation in wave-number and frequency. In order for the first few terms in the expansion to be sufficient, we implicitly assume the process to be narrow banded in frequency as well as wave-number. We now recall the observation made earlier that the dispersion relation is nothing but a different representation of the governing partial (linear) differential equation for the problem. In this spirit, we are therefore free to make the substitutions $(\omega-\omega_0) \to i\partial/\partial t$ and $(k-k_0) \to -i\partial/\partial x$ in (4.46). Since $(\omega-\omega_0)$ represents the sideband frequency, the time-differential operator derived from it will refer to the time variation of the envelope of the modulated carrier wave, and we argue similarly for the spatial operator. The resulting differential equation then describes the evolution of the slow time, long wavelength envelope of the carrier wave in a linearized model. In (4.45) we let the exponential factor account for the carrier wave, and let a scalar velocity $u = u(x,t)$ represent the slow space-time evolution of the envelope.

In the present approximation, the basic differential equation for the linear waves can then be written in the form

$$e^{-i(\omega_0 t - k_0 x)}\left[D(\omega_0,k_0)u + i\frac{\partial D}{\partial \omega_0}\frac{\partial}{\partial t}u - i\frac{\partial D}{\partial k_0}\frac{\partial}{\partial x}u\right.$$
$$\left.-\frac{1}{2}\left(\frac{\partial^2 D}{\partial \omega_0^2}\frac{\partial^2}{\partial t^2}u + \frac{\partial^2 D}{\partial k_0^2}\frac{\partial^2}{\partial x^2}u - 2\frac{\partial^2 D}{\partial \omega_0 \partial k_0}\frac{\partial^2}{\partial x \partial t}u\right)\right] = 0. \tag{4.47}$$

We now recall that

$$u_g \equiv -\frac{\partial D}{\partial k}\bigg/\frac{\partial D}{\partial \omega} = \frac{d\omega}{dk} \tag{4.48}$$

is the group velocity, while $u_g' \equiv du_g/dk$ is the lowest order dispersion term. To lowest order from (4.47) we then have $\partial u/\partial t + u_g \partial u/\partial x \approx 0$, giving, for instance, $\partial^2 u/\partial x \partial t = -u_g \partial^2 u/\partial x^2$ and $\partial^2 u/\partial t^2 = -u_g \partial^2 u/\partial t\, \partial x$. Note that with the present approximations we could as well have used $u_g^2 \partial^2/\partial x^2$ instead of $\partial^2/\partial t^2$. We have, for instance, the approximation

$$\left(\frac{\partial^2 D}{\partial \omega_0^2}\frac{\partial^2 u}{\partial t^2} + \frac{\partial^2 D}{\partial k_0^2}\frac{\partial^2 u}{\partial x^2} - 2\frac{\partial^2 D}{\partial \omega_0 \partial k_0}\frac{\partial^2 u}{\partial x \partial t}\right) \approx$$
$$\left(u_g^2\frac{\partial^2 D}{\partial \omega_0^2} + \frac{\partial^2 D}{\partial k_0^2} + 2u_g\frac{\partial^2 D}{\partial \omega_0 \partial k_0}\right)\frac{\partial^2 u}{\partial x^2}.$$

Also we have in general $u_g = u_g(\omega,k)$, so that $du_g/dk \equiv u'_g = \partial u_g/\partial k + \partial u_g/\partial\omega(d\omega/dk)$, where we need

$$\frac{\partial u_g}{\partial k} = -\frac{\partial^2 D}{\partial k^2}\bigg/\frac{\partial D}{\partial\omega} + \frac{\partial D}{\partial k}\frac{\partial^2 D}{\partial k\partial\omega}\bigg/\left(\frac{\partial D}{\partial\omega}\right)^2,$$

$$\frac{\partial u_g}{\partial\omega} = -\frac{\partial^2 D}{\partial k\partial\omega}\bigg/\frac{\partial D}{\partial\omega} + \frac{\partial D}{\partial k}\frac{\partial^2 D}{\partial\omega^2}\bigg/\left(\frac{\partial D}{\partial\omega}\right)^2.$$

With these manipulations together with (4.48), we rewrite (4.47) in a straightforward manner. We also use $D(\omega_0,k_0) = 0$. We take the resulting expression and here simply introduce a nonlinear correction $N.L.$ on the right-hand side, leaving details concerning this correction to later discussions, and obtain

$$i\left(\frac{\partial}{\partial t} + u_g\frac{\partial}{\partial x}\right)u + \frac{1}{2}u'_g\frac{\partial^2}{\partial x^2}u = N.L. \tag{4.49}$$

The equation can be simplified slightly by changing the frame of reference to one moving with the group velocity.

Linearizing the equation (4.49), we find the dispersion relation $\Omega = u_g K + \frac{1}{2}u'_g K^2$, where we recall that Ω is to be seen as the *correction* to the basic wave frequency ω in (4.45), and likewise for K. This linear dispersion relation can be interpreted as a parabolic approximation to a "true" dispersion relation $\omega = \omega(k)$ in the vicinity of a selected point (ω_0, k_0). As it stands, the equation (4.49) does not refer to strongly dispersive waves in any specific manner, although we implicitly have required that the dispersive term containing u'_g is larger than the next term in the expansion. The emphasis on the strong dispersion is introduced by specifying the nonlinearity. It will be demonstrated in Section 4.3.5 that in the absence of wave-wave interactions, the nonlinear term can be written as $N.L. = \beta|u|^2u$, giving

the *nonlinear Schrödinger equation*, or simply the *NLS equation*

$$i\left(\frac{\partial}{\partial t} + u_g\frac{\partial}{\partial x}\right)u + \frac{1}{2}u'_g\frac{\partial^2}{\partial x^2}u = \beta|u|^2u. \tag{4.50}$$

We first discuss some basic properties of this equation. Evidently, (4.50) is a complex relation, so in reality it represent *two* equations, the real and imaginary parts. We *could* write out the relation with $\Re\{u\}$ and $\Im\{u\}$, but this is not very fruitful, since these two quantities do not have a direct physical relevance. It is better to introduce $u \equiv a\exp(i\phi)$ with real a and ϕ, since we can hereby identify an amplitude modulation by $a(x,t)$ and a phase modulation by $\phi(x,t)$ of the original plane wave $\exp(-i(\omega_0 t - k_0 x))$. We then rewrite (4.50) in the form

$$\frac{\partial}{\partial t}a + u_g\frac{\partial}{\partial x}a = \frac{1}{2}u'_g\left(2\frac{\partial a}{\partial x}\frac{\partial\phi}{\partial x} + a\frac{\partial^2\phi}{\partial x^2}\right) \tag{4.51}$$

$$\frac{\partial}{\partial t}\phi + u_g\frac{\partial}{\partial x}\phi = \frac{1}{2}u'_g\left(\left(\frac{\partial\phi}{\partial x}\right)^2 - \frac{1}{a}\frac{\partial^2 a}{\partial x^2}\right) + \beta a^2. \tag{4.52}$$

It is readily seen that these equations have one particularly simple plane wave solution, namely, $a_0 =$ const. and $\phi_0 = \beta a_0^2 t$, which represents a plane wave, where βa_0^2 is a *nonlinear frequency shift*, where in particular β accounts for its sign. We recall here that a_0 is the coefficient of the plane wave, $\exp(-i(\omega t - kx))$ in a one dimensional version of (4.45). The relation (4.51) is readily rewritten as

$$\frac{\partial}{\partial t}a^2 + u_g\frac{\partial}{\partial x}a^2 = \frac{\partial}{\partial x}\left(u'_g a^2\frac{\partial}{\partial x}\phi\right), \tag{4.53}$$

implying that $\int_{-\infty}^{\infty} a^2 dx$ is conserved when $a \to 0$ for $x \to \infty$. Usually, we interpret this as an energy conservation law.

Just as the KdV equation, also the *NLS* equation possesses an infinite number of conserved quantities. We list a few here:

$$\begin{aligned} I_1 &= \int_{-\infty}^{\infty} |u|^2 dx, \\ I_2 &= \frac{u_g'}{2i} \int_{-\infty}^{\infty} \left(u^* \frac{\partial}{\partial x} u - u \frac{\partial}{\partial x} u^* \right) dx, \\ I_3 &= \int_{-\infty}^{\infty} \left(u_g' \left| \frac{\partial u}{\partial x} \right|^2 + \frac{\beta}{2} |u|^4 \right) dx, \\ &\vdots \end{aligned}$$

We can construct another conserved quantity, which might be useful sometimes, by taking

$$I_0 = \int_{-\infty}^{\infty} \left[x|u|^2 - t \frac{u_g'}{2i} \left(u^* \frac{\partial}{\partial x} u - u \frac{\partial}{\partial x} u^* \right) \right] dx.$$

where $\int_{-\infty}^{\infty} x|u|^2 dx$ can be interpreted as a center-of-mass for the waveform, just as for the KdV equation, assuming $|u|^2$ to have compact support. It is customary to discuss these conserved quantities in the frame of reference moving with the group velocity u_g.

Ⓢ **Exercise:** Find an exact solution for the *linear* part of the one dimensional NLS equation. Simplify the problem by solving it in the frame moving with the group velocity, i.e., consider

$$i \frac{\partial}{\partial t} u + \frac{1}{2} u_g' \frac{\partial^2}{\partial x^2} u = 0. \tag{4.54}$$

Seek the solution for a Gaussian wave-packet, with $u(x,0) = A \exp(-ax^2)$. Show that the wavepacket expands as $\sim t$ for large times. Ⓢ

- **Exercise:** Demonstrate by introducing variables $x - tu_g \equiv i\xi$ and $t \equiv i\tau$ in (4.54) that

$$\frac{\partial u}{\partial \tau} = \frac{1}{2} u_g' \frac{\partial^2 u}{\partial \xi^2},$$

having the form of a diffusion equation, where exact solutions are known. Find the space-time evolution of the initial condition $u(\xi, 0) = A \exp(-a\xi^2)$ and show that it expands as $\sim \sqrt{t}$ for large times. Do you see any paradox here? Compare with the foregoing exercise.

4.3.3 Modulational instability

To analyze the properties of the equilibrium plane wave solution, we now make a perturbation of $a_0 \exp(i\phi_0)$ in the form

$$\begin{aligned} a &= a_0 + \delta a \exp(-i\omega t + iKx) \\ \phi &= \phi_0 + \delta\phi \exp(-i\omega t + iKx) \end{aligned}$$

where we used capital letters Ω and K as distinguished from the lowest order frequency and wavenumber of the plane wave solution, (ω, k). The perturbation takes the form of a modulation in amplitude as well as frequency of the original plane wave solution. By insertion into the basic equations (4.51)–(4.52) and linearizing we readily obtain a dispersion relation of the form

$$\Omega - u_g K = \pm\frac{1}{2}u'_g K\sqrt{K^2 + 4\beta a_0^2/u'_g}. \tag{4.55}$$

In case the so called *Lighthill condition*

$$\beta u'_g < 0 \tag{4.56}$$

is satisfied, we have unstable solutions (i.e., Ω is complex) for sufficiently small modulational wave-numbers K. The maximum growth rate of the instability is for $K = a_0\sqrt{-2\beta/u'_g}$ and it has the value $\Im\{\Omega\} = a_0^2\beta$. Note that no matter how small, but nonzero, amplitudes we have, there will *always* be some small modulational wave-number which gives this instability. It is, however, also evident that the growth rate of the instability decreases strongly with amplitude, so that it may not be noticeable within a finite time in case a_0 is small.

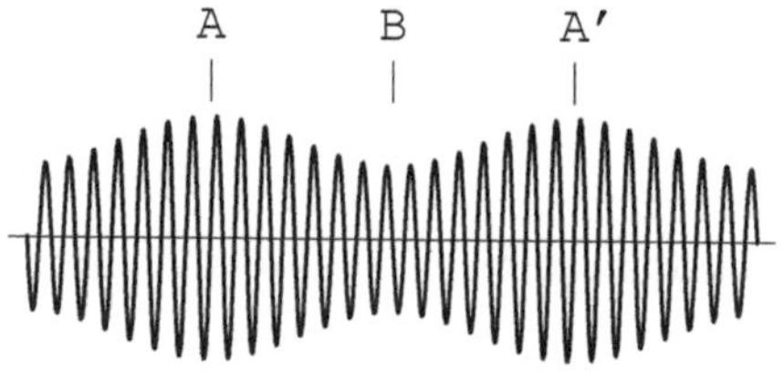

FIGURE 4.11: *Amplitude modulated wave-train for physical explanation of the Lighthill condition.*

The essence of the Lighthill condition (4.56) can be explained from first principles (Kadomtsev & Karpman 1971, Pécseli 1985). Consider, for instance, Fig. 4.11, where we show an initial condition for a modulated wave, having the zero crossings of the wave-amplitude to be equidistant by construction. Assume that we have a nonlinear dispersion relation of the form $\omega(k,a) = \omega_0(k) + \beta a^2$, where $\omega_0(k)$ is the linear dispersion relation, and βa is the correction by the nonlinear frequency shift. Assume first, for the sake of argument, that $\beta > 0$. Then the phase velocity Ω/K is maximum at positions A and A', while it is minimum at B, because of the wave amplitude dependence of the dispersion relation. At a later time the number of zero crossings of the wave, and thereby the local wave-number, will *increase* in the interval AB while it will *decrease* in BA'. If now the local group velocity is *decreasing* with increasing K, i.e., $u'_g < 0$, then the wave envelope which propagates at the local group velocity will lag in sections like AB and advance in sections like BA'. Consequently, the wave amplitude is decreased at points like B and enhanced at points like A or A'. The wave modulation will then be enhanced, and the process proceeds at an accelerated rate; the wave is modulationally unstable. The arguments can be repeated by changing the sign of β as well as u'_g, and the condition (4.56) be proved for all combinations.

It is interesting to note that the gravitational deep water waves (Stokes waves) are modulationally unstable. The dispersion relation for weakly nonlinear waves is to lowest order given by $\omega = (1 + \frac{1}{2}(ka)^2)\sqrt{gk}$, and Lighthill's criterion is fulfilled. Recall that a has the dimension *length* as appropriate for wave amplitude. The observation often made by sailors and fisherman of the "tenth wave" (in the author's childhood it was the seventh) having a particularly large amplitude may thus have a physical justification, although once discarded as superstition (Kadomtsev & Karpman 1971).

4.3.4 Soliton solutions of the NLS equation

Just as for the KdV equation, we can demonstrate the existence of soliton solutions for the nonlinear Schrödinger equation. To demonstrate this, we again use a coordinate system moving with a velocity

U, which is for the moment unspecified. We look for solutions of the form

$$u = e^{i(rx-st)}\Psi(x-Ut). \tag{4.57}$$

The physical reasoning for this choice is that in the rest frame, following such an anticipated solution, the frequency must be constant for all positions for a truly stationary case. Upon insertion into (4.50) we find

$$\frac{1}{2}u_g'\frac{d^2}{dx^2}\Psi + i\frac{1}{2}u_g'(2r-U-u_g)\frac{d}{dx}\Psi + (s-u_g r - \frac{1}{2}u_g' r^2)\Psi = \beta|\Psi|^2\Psi. \tag{4.58}$$

We now *choose* to eliminate the imaginary part by taking $r = \frac{1}{2}(U+u_g)$, thereby making Ψ real. Also, for simplicity we write $\nu \equiv 2\beta/u_g'$ and $\alpha \equiv 2(s-u_g r - \frac{1}{2}u_g' r^2)/u_g'$, and obtain the equation for Ψ in the form

$$\frac{d^2}{dx^2}\Psi - \alpha\Psi + \nu\Psi^3 = 0. \tag{4.59}$$

This equation resembles the one we already met in obtaining the KdV solitons, and it can be integrated in the same way to give

$$\left(\frac{d\Psi}{dx}\right)^2 = A + \alpha\Psi^2 - \frac{\nu}{2}\Psi^4, \tag{4.60}$$

where A is a constant. For $A = 0$, the equation can readily be solved to give a soliton solution

$$\Psi = \left(\frac{2\alpha}{\nu}\right)^{1/2} \operatorname{sech}\left(\alpha^{1/2}(x-Ut)\right), \tag{4.61}$$

with the important difference from the KdV soliton being that in the present case the soliton velocity U is independent of the amplitude; the NLS soliton belongs to a *two-parameter* family of solutions. The spatial variation of the KdV soliton (4.30) was given by sech^2, while (4.61) is only to the first power. Note, however, that by considering the wave energy density we square Ψ, and the spatial variation of the two soliton solutions becomes identical, but retains the difference in number of parameters. Considering Ψ^2 we find the same amplitude-width scaling of the NLS energy density soliton as for the KdV soliton. We can introduce an NLS soliton amplitude b and write (4.61) as

$$\Psi = b \operatorname{sech}\left((x-Ut) \Big/ \sqrt{\frac{2}{\nu b^2}}\right). \tag{4.62}$$

It is an important observation that physically acceptable soliton solutions can be found only for the case $\nu > 0$. It is readily demonstrated that this criterion is for the present case identically the same as the the Lighthill criterion for modulational instability. It might be anticipated that a soliton represents the final saturated stage of a modulational instability, and indeed it can be demonstrated that this is the case (Whitham 1974). An inverse scattering transform exists also for the NLS equation (Zakharov & Shabat 1972), and it can be demonstrated that for the modulationally unstable case, any compact initial perturbation will eventually evolve into a state which can be described by a superposition of soliton solutions of the form (4.61) or (4.62).

The modulationally *stable* situation is in comparison of minor interest. Details of the time evolution for this case are given by Whitham (1974), for instance.

4.3.5 Derivation of the nonlinear Schrödinger equation

We have already postulated the nonlinear terms $N.L.$ in (4.47) to be given as $\beta|u|^2u$. (In reality this assumption corresponds (Lonngren 1983) to the a priori assumption of a dispersion relation of the form $\omega = \omega(k, |u|^2)$ where again the wave amplitude is u.) It is timely to demonstrate that it is indeed so for strongly dispersive waves. The analysis will be carried out to third order in wave amplitude. We first note that nonlinear terms in wave equations will generate second and third harmonics, i.e., terms varying like $\exp(-2i(\omega t - kx))$ and $\exp(-3i(\omega t - kx))$ in addition to the basic wave $\exp(-i(\omega t - kx))$, which we would have in case a simple linear analysis was carried out. To simplify the notation, we introduce the phase $\theta \equiv \omega t - kx$, and write in terms of the dispersion relation D the expression

$$D\left(i\frac{\partial}{\partial t}, -i\frac{\partial}{\partial x}\right)\psi = Q_2(\psi,\psi) - \overline{Q}_2 + Q_3(\psi,\psi,\psi) \tag{4.63}$$

where the index on the Q's refer to second and third orders, respectively. We have $\psi = ae^{-i\theta} + a^*e^{i\theta} + be^{-i2\theta} + b^*e^{i2\theta}$, to third order, where $b \ll a$ since it accounts for a second order harmonic arising because of small nonlinear terms. Note that we need a term $\overline{Q}_2$ to make certain that both sides of the relation (4.63) vanish upon averaging. The term Q_2 will contain ψ^2, for instance, which vanishes trivially upon averaging, but Q_2 also contains $|\psi|^2$, which contributes with a constant if we take the time average, so we need to include $\overline{Q}_2$. The term $\overline{Q}_3$ vanishes identically, since terms such as $|\psi|^2\psi$ and ψ^3 have vanishing time averages. We ignore wave-wave interactions, and there are then no "off" frequencies. This is one restriction imposed on the derivation of the NLS equation.

We now write

$$\begin{aligned} Q_2(\psi,\psi) - \overline{Q}_2 &= A\left(a^2e^{-i2\theta} + a^{*2}e^{i2\theta}\right) \\ &\quad + B\left(ab^*e^{-i\theta} + ba^*e^{i\theta}\right) \\ &\quad + C\left(abe^{-i3\theta} + b^*a^*e^{i3\theta}\right) + O(a^2). \end{aligned}$$

We do not here retain derivatives of a and b, but in principle, Q_2 can contain differential operators acting on θ. In case a and b have a spatial and/or temporal variation, it is assumed to be so slow that its derivative does not contribute to second order terms.

Similarly, we have

$$\begin{aligned} Q_3(\psi,\psi,\psi) &= E\left(a^3e^{-i3\theta} + a^{*3}e^{i3\theta}\right) \\ &\quad + F|a|^2\left(ae^{-i\theta} + a^*e^{i\theta}\right) + O(a^2). \end{aligned}$$

We now eliminate b by use of (4.63).

$$D\left(i\frac{\partial}{\partial t}, -i\frac{\partial}{\partial x}\right)be^{-i2\theta} = Aa^2e^{-i2\theta} \quad \rightarrow \quad b = \frac{Aa^2}{D(2\omega, 2k)}.$$

It is here important to note that we require (obviously) that not only $D(2\omega, 2k) \neq 0$, but that it is of a sufficient magnitude to retain the ordering of b implied in (4.63). In other words: we require that the point $(2\omega, 2k)$ is "far away from" the dispersion relation $\omega = \omega(k)$. At first sight this seems implied in the assumption of strongly dispersive waves, but in particular in plasma media we might easily encounter cases where some *other* branches of the dispersion relation are present. This could be harmonics of a cyclotron frequency, for instance. Such cases have to be omitted from the analysis. Next we have

$$\begin{aligned} D\left(i\frac{\partial}{\partial t}, -i\frac{\partial}{\partial x}\right)ae^{-i\theta} &= e^{-i\theta}\frac{\partial D}{\partial \omega}\left[i\left(\frac{\partial}{\partial t} + u_g\frac{\partial}{\partial x}\right) + \frac{1}{2}u_g'\frac{\partial^2}{\partial x^2}\right]a \\ &\equiv (Ba^*b + Fa|a|^2)e^{-i\theta}. \end{aligned}$$

We collected terms varying as $\exp(-i\theta)$. We now obtain

$$\left[i\left(\frac{\partial}{\partial t}+u_g\frac{\partial}{\partial x}\right)+\frac{1}{2}u_g'\frac{\partial^2}{\partial x^2}\right]a = a|a|^2\left(F+\frac{AB}{D(2\omega,2k)}\right)\Big/\frac{\partial D(\omega,k)}{\partial\omega}\equiv\beta a|a|^2, \tag{4.64}$$

which is the desired NLS equation. Recall that $a^*b\sim a|a|^2$. The implied condition in the magnitude ordering of terms is that $\partial D(\omega,k)/\partial\omega$ cannot be small.

Again we emphasize that any NLS equation represents a very idealized description of the actual physics, by containing the lowest order dispersion of the waves, together with the lowest order nonlinearity. The beauty of the equation lies in its generality: it embraces *many* wave-types at the same time. Often one likes to obtain a better approximation to a given physical condition, i.e., one particular wave-type, and a *perturbed* NLS equation is obtained in the general form

$$i\left(\frac{\partial}{\partial t}+u_g\frac{\partial}{\partial x}\right)u+\frac{1}{2}u_g'\frac{\partial^2}{\partial x^2}u-\beta|u|^2u=\varepsilon R[u] \tag{4.65}$$

where the operator $R[u]$, which *can* be nonlinear, represents the corrections specific for the given problem.

- **Example:** One relevant example for plasma waves can be illustrated by $D(\omega,k)=\omega^2-\Omega^2$, independent of k. Here $\partial D(\omega,k)/\partial\omega=2\omega$. By the dispersion relation $D(\omega,k)=0$ we have $\partial D(\omega,k)/\partial\omega=2\Omega$, which is generally large, as required.

- **Example:** We can illustrate the condition for $D(2\omega,2k)$ being large by considering a case where $D(\omega,k)=\omega^2-u^2k^2-\omega_p^2$ in terms of a velocity u and a frequency ω_p relevant for the physical problem. This example is representative for certain high frequency plasma phenomena. Having a wave solution ω_0,k_0 means that $D(\omega_0,k_0)=0$. Using this we readily find that in that case $D(2\omega_0,2k_0)=3(\omega_0^2-u^2k_0^2)$, implying that in order to have $D(2\omega_0,2k_0)$ large, we have a condition on the phase velocity, i.e., $\omega_0/k_0\gg u$.

  As another case we can consider $D(\omega,k)=\omega-uk-\alpha k^3$. For this case, we have $D(2\omega_0,2k_0)=-6\alpha k_0^3$. If α is small, we find for this case that $D(2\omega_0,2k_0)$ becomes uncomfortably small, and it is unlikely that an NLS-type equation can be applicable. Fortunately, we have in many cases a KdV equation for this wave-type.

Methods for obtaining approximate soliton solutions of *perturbed* NLS equations are presented by Watanabe et al. (1979), using the conservation laws, very much in the spirit of the similar solutions for the KdV equation. A particular case relevant for plasma waves was discussed by Rypdal et al. (1982). The basic difference between the two problems is that the KdV soliton belongs to a *one*-parameter family of solutions, with, for instance, the amplitude determining width as well as velocity. The NLS solitons belong to a *two*-parameter family of solutions, as already discussed.

4.3.6 Generalization to three spatial dimensions

It is a relatively simple matter to generalize the analysis of the NLS equation to the fully two or three dimensional isotropic case, for scalar amplitudes. In three spatial dimensions we can thus use the expansion

$$\omega(\mathbf{k})\approx\omega(\mathbf{k}_0)+\mathbf{u}_g\cdot(\mathbf{k}-\mathbf{k}_0)+\frac{1}{2}\nabla_k\mathbf{u}_g:(\mathbf{k}-\mathbf{k}_0)(\mathbf{k}-\mathbf{k}_0).$$

For homogeneous and isotropic media, we have $\omega(\mathbf{k})=\omega(|\mathbf{k}|)$. For these cases, we find $\mathbf{u}_g=u_g\mathbf{k}/k$ and

$$\begin{aligned}\nabla_k\mathbf{u}_g &= \mathbf{k}\nabla_k\left(\frac{u_g}{k}\right)+\frac{u_g}{k}\nabla_k\mathbf{k}=\mathbf{k}\nabla_k\left(\frac{u_g}{k}\right)+\frac{u_g}{k}\underline{1}\\ &= \frac{\mathbf{k}\mathbf{k}}{k}\frac{d}{dk}\left(\frac{u_g}{k}\right)+\frac{u_g}{k}\underline{1}=\left(\frac{u_g'}{k^2}-\frac{u_g}{k^3}\right)\mathbf{k}\mathbf{k}+\frac{u_g}{k}\underline{1},\end{aligned}$$

where $\underline{1}$ is the unit tensor. Taking **k** to be along the x-axis, we can write

$$\omega(k) - \omega(k_0) = u_g(k_x - k_{0x}) + \frac{1}{2}u_g'(k_x - k_{0x})^2 + \frac{1}{2}\frac{u_g}{k}(\mathbf{k}_\perp - \mathbf{k}_{0\perp})^2.$$

As before, we introduce $(\omega - \omega_0) \to i\partial/\partial t$ and $(k_x - k_{0x}) \to -i\partial/\partial x$, etc. and find for this case a generalization of the NLS equation to three spatial dimensions (Kadomtsev & Karpman 1971)

$$i\left(\frac{\partial}{\partial t} + u_g\frac{\partial}{\partial x}\right)u + \frac{1}{2}u_g'\frac{\partial^2}{\partial x^2}u + \frac{1}{2}\frac{u_g}{k_0}\nabla_\perp^2 u = \beta|u|^2 u.$$

This relation describes a variety of narrow band wave processes in isotropic homogeneous media.

Most of the studies of weakly nonlinear dispersive waves assume initial value problems, which are easily realized in, for instance, numerical simulations. It is, however, perfectly feasible to assume a spatially localized plane antenna continuously emitting some wave, where the wave-field develops with distance away from the source. If the source is narrow, the primary direction for the envelope evolution will be in the direction perpendicular to the direction away from the source. For isotropic media we can then propose the approximate relation for the stationary wave envelope

$$iu_g\frac{\partial}{\partial x}u + \frac{1}{2}\frac{u_g}{k_0}\nabla_\perp^2 u = \beta|u|^2 u, \tag{4.66}$$

where now the primary coordinate x takes the role of a temporal variable along the direction of primary wave propagation. The source (i.e., the antenna) is here perpendicular to $\widehat{\mathbf{x}}$. Note that in this limit the equation does not contain u_g'. The linear evolution of the waveform in the direction $\perp \widehat{\mathbf{x}}$ is due to the spread in spectral components in this direction. The dominant nonlinear evolution will be self-focusing close to the source followed by development of soliton-like filaments fanning out from some focal point localized on the x-axis at some spatial distance from the source. The actual expression for β has to be determined for the proper physical conditions. For instance, for (4.66) it has to be obtained for steady-sate conditions.

The arguments can be generalized to anisotropic media as well, where now the primary coordinate is along the group velocity, which will in general have some angle with respect to the normal of the antenna (Dysthe et al. 1978).

The weakly nonlinear wave analysis summarized in the last part of this chapter referred to the strongly dispersive example in Fig. 4.1b). If we take the case in Fig. 4.1a) and follow the dispersion relation to shorter wavelengths or higher frequencies we can reach a parameter range where also this waveform will be classified as strongly dispersive. Ion acoustic plasma waves offer such an example. For neutral gases this limit is comparatively less interesting because of the strong viscous damping of very short wavelength sound waves.

5

Basics of Electromagnetism

The dynamics of a plasma is controlled by electric and magnetic forces in most relevant and interesting cases. The analysis of plasma dynamics must therefore be based on Maxwell's equations one way or another. We therefore start the discussion of the plasma dynamic equations with a summary of electrodynamics. The present chapter is only meant as a summary for later reference in the text: a deeper exposition of the subject is found in classic textbooks on the subject by Stratton (1941) and Jackson (1975), with more basic texts being available also (Duffin 1990, Good 1999, Griffiths 1999).

5.1 Maxwell's equations in their basic form

Maxwell's equations in terms of integrals over *closed* surfaces, $\mathcal{S}$, or along *closed* contours, $\mathcal{C}$, are given as

$$\oint_{\mathcal{C}} \mathbf{H} \cdot ds = \oint_{\mathcal{S}_C} \left(\mathbf{J} + \frac{\partial}{\partial t} \mathbf{D} \right) \cdot \widehat{\mathbf{n}} dA \equiv I + \frac{d\Phi}{dt}, \tag{5.1}$$

$$\oint_{\mathcal{C}} \mathbf{E} \cdot ds = -\oint_{\mathcal{S}_C} \frac{\partial}{\partial t} \mathbf{B} \cdot \widehat{\mathbf{n}} dA \equiv -\frac{d\Psi}{dt}, \tag{5.2}$$

$$\oint_{\mathcal{S}} \mathbf{D} \cdot \widehat{\mathbf{n}} dA = \oint_{\mathcal{V}} \zeta dV \equiv Q, \tag{5.3}$$

$$\oint_{\mathcal{S}} \mathbf{B} \cdot \widehat{\mathbf{n}} dA = 0, \tag{5.4}$$

in terms of the electric field $\mathbf{E}$, electric displacement $\mathbf{D}$, magnetic flux density $\mathbf{B}$ (alternative names for $\mathbf{B}$ are "magnetic induction" or merely "magnetic field") and magnetic field intensity, or magnetic field strength, $\mathbf{H}$. We have I being the net current through a surface $\mathcal{S}_C$ which is bounded by the contour $\mathcal{C}$, while Φ is the electric displacement flux, and Ψ the magnetic intensity flux through the surface $\mathcal{S}_C$ bounded by the contour of integration $\mathcal{C}$. At the same time, Q is the net charge contained within the relevant closed volume of integration $\mathcal{V}$ bounded by the closed surface $\mathcal{S}$, with ζ being the charge density. The unit vector normal to a surface element is denoted $\widehat{\mathbf{n}}$. The directions of the contour integrations in (5.1) and (5.2) are connected with $\widehat{\mathbf{n}}$ by the *"right-hand rule"*, i.e., if your right-hand thumb points in the direction of $\widehat{\mathbf{n}}$, then the direction of integration along $\mathcal{C}$ points in the direction of your fingers. The surfaces and contours can pass through a vacuum as well as matter. In general we will have the notation that the flux of any vector quantity $\mathbf{A}$ over a surface element with area dA is $\mathbf{A} \cdot \widehat{\mathbf{n}} dA$ and the flux through a finite surface is obtained by integration of this quantity.

The equations (5.1)–(5.4) are given as partial differential equations in the form

$$\nabla \times \mathbf{H} = \mathbf{J} + \frac{\partial \mathbf{D}}{\partial t}, \tag{5.5}$$

$$\nabla \times \mathbf{E} = -\frac{\partial \mathbf{B}}{\partial t}, \tag{5.6}$$

$$\nabla \cdot \mathbf{D} = \zeta, \tag{5.7}$$

$$\nabla \cdot \mathbf{B} = 0, \tag{5.8}$$

where $\mathbf{J}$ is the current density. The charge density ζ is often called the density of *free* or *conduction* charges. These charges and currents are different from polarization charges and polarization currents to be discussed later on. The correspondence between the two representations (5.1)–(5.4) and (5.5)–(5.8) can be demonstrated by Gauss' theorem. It might appear strange that the set of equations (5.5)–(5.8) has the position vector as a free variable, while no free spatial variable seems to appear in (5.1)–(5.4), in spite of the two sets of equations containing the same information! However, we should bear in mind that in (5.1)–(5.4) the *contours* and *surfaces* are not specified, and in a sense these contain the free spatial variables.

The integral form (5.1)–(5.4) of Maxwell's equations is used mostly to determine boundary conditions, while the differential form (5.5)–(5.8) is used mostly for studying dynamic phenomena, electromagnetic waves, in particular.

Incidentally, in their modern form we owe the Maxwell equations to Heaviside. For a time, the set of equations was even known as "Heaviside's equations" until they were renamed according to the physics they actually describe. Heaviside was a brilliant scientist, who died in deep misery, his work being recognized only later; see, e.g., www.aps.org/publications/apsnews (2010), "This month in physics history: Oliver Heaviside", *APS News* **19** (2), 2.

5.1.1 Boundary conditions

If there are sharp boundaries in the media the appropriate boundary conditions are to be included also. These boundary conditions are

$$\widehat{\mathbf{n}}\Delta \times \mathbf{H} = \mathbf{K}, \tag{5.9}$$

where $\mathbf{K}$ is the surface current density, meaning that the tangential component of $\mathbf{H}$ changes by this amount at a surface. The symbol $\widehat{\mathbf{n}}\Delta$ indicates a change by an amount Δ in the $\widehat{\mathbf{n}}$-direction. To interpret $\mathbf{K}$, we can imagine a material surface layer of some thickness δ carrying a current density $\mathbf{J}$. We then let the thickness of the layer shrink and have $\mathbf{K} \equiv \lim_{\delta\to 0}(\mathbf{J}\delta)$, implying that we formally have to let $\mathbf{J} \to \infty$ as $\delta \to 0$ so that $\mathbf{J}\delta$ remains finite. The physical dimension of $\mathbf{K}$ is *current / length*.

For the electric field we have

$$\widehat{\mathbf{n}}\Delta \times \mathbf{E} = 0, \tag{5.10}$$

stating that the tangential component of $\mathbf{E}$ is unchanged across a surface.

For the electric displacement we have

$$\widehat{\mathbf{n}}\Delta \cdot \mathbf{D} = \sigma, \tag{5.11}$$

where σ is the surface charge density, indicating that the normal component of $\mathbf{D}$ changes by this amount across a surface. As with surface current, we also here imagine a charged layer of finite thickness and have $\sigma \equiv \lim_{\delta\to 0}(\zeta\delta)$. We also here have to let $\zeta \to \infty$ as $\delta \to 0$ in such a way that the product $\zeta\delta$ remains finite. The physical dimension of σ is *charge / length*^{2}.

Finally we have

$$\widehat{\mathbf{n}}\Delta \cdot \mathbf{B} = 0, \tag{5.12}$$

with self-evident interpretation.

From (5.5) and (5.7) we have the equation for charge continuity

$$\frac{\partial}{\partial t}\zeta + \nabla \cdot \mathbf{J} = 0. \tag{5.13}$$

At a sharp boundary, this relation becomes

$$\frac{\partial}{\partial t}\sigma + \widehat{\mathbf{n}}\Delta \cdot \mathbf{J} + \nabla_s \cdot \mathbf{K} = 0 \tag{5.14}$$

with $\nabla_s \cdot \mathbf{K}$ being the surface divergence of the surface current. By the subscript $_s$, we indicate that the differential operator is restrained to the local coordinates on the plane that is tangent to the bounding surface.

5.1.2 Material relations for simple media

The equations are to be completed by material relations accounting for the material properties which relate **D** to **E**, **H** to **B** and **J** to **E**.

For simple media the relations are

$$\mathbf{D} = \varepsilon_0 \varepsilon_r \mathbf{E} \qquad \text{and} \qquad \mathbf{H} = \mathbf{B}\frac{1}{\mu_0 \mu_r} \qquad \text{and} \qquad \mathbf{J} = \gamma \mathbf{E},$$

where the constants ε_r and μ_r are the relative permittivity and permeability, respectively, and γ the conductivity of the medium. Relations in this form are often used, but they are highly idealized and will be replaced by more accurate ones later on. For a vacuum these become particularly simple, i.e., $\gamma = 0$ while

$$\mathbf{D} = \varepsilon_0 \mathbf{E} \qquad \text{and} \qquad \mathbf{H} = \mathbf{B}\frac{1}{\mu_0}.$$

If we deal with a medium following the equivalent of Ohm's law, $\mathbf{J} = \gamma\mathbf{E}$, we readily find that for finite γ this medium cannot support a surface current: for physical reasons we expect **E** to remain finite since an infinite **E** requires the sources ζ or $\partial\mathbf{B}/\partial t$ to be infinite as well. For a non-vanishing surface current $\mathbf{K} > 0$ we found that the current density **J** formally had to be infinite for finite **E**, and this is only possible if we have $\gamma \to \infty$. Finite surface currents require locally infinite conductivities.

- **Exercise:** Consider a long cylindrical solenoid of radius R, with a density of wires per unit length given by n, carrying a time varying current $I = I(t)$. Assume that $dI/dt = \text{const}$. What is the electric field (in direction and magnitude) at a position inside and outside the solenoid, i.e., at $r < R$ and $r > R$, respectively?

5.2 Discussions of Maxwell's equations

The set of Maxwell equations (5.5)–(5.8) will be discussed briefly, in reversed order:

- **MIV:** Gauss' law, i.e., the vanishing divergence of the magnetic field expresses, among other things, our belief that magnetic monopoles do not exist. For a long time it was expected that $\nabla \cdot \mathbf{B} = 0$ implied that magnetic field lines of finite length closed on themselves, those with infinite length continuing to infinity. Often it will indeed be so, but important exceptions can be found, most notably the Tokamak device, where magnetic field lines on irrational surfaces continue to wind around a torus with finite surface area, these field lines having infinite length.

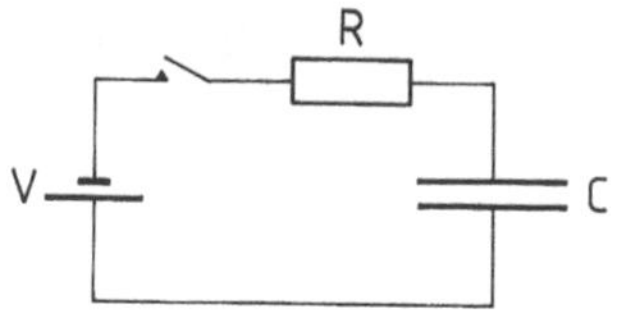

FIGURE 5.1: *A simple electrical circuit used for illustrating the role of Maxwell's displacement current.*

- **MIII:** Poisson's equation expresses that electric displacement lines can begin or end on charges. The possibility of closed **D** lines or some extending to infinity of course still exists.
- **MII:** Faraday's law of induction states that a time-varying magnetic field induces a rotation of an electric field. An arbitrary divergence can be added to this electric field without affecting the relation.
- **MI:** Without the $\partial \mathbf{D}/\partial t$ term, this is Ampere's law, relating currents to magnetic fields expressing Ørsted's discovery. Maxwell's displacement term $\partial \mathbf{D}/\partial t$ indicates that a time-varying electric displacement is just as good as a current in generating magnetic fields. Consider, for instance, the simple circuit in Fig. 5.1; prior to Maxwell this circuit would be called *open*, since it appeared intuitively evident that no current could flow through the capacitor. However, during charge-up there is obviously a current I in the wires connecting the capacitor to the battery, and this current is then apparently abruptly terminated at the capacitor plates. The charges building up on the capacitor plates generate an electric field between the plates, giving a dielectric displacement **D** if there is some material placed between the plates. Inclusion of the displacement current implies that the time-varying $\partial \mathbf{D}/\partial t$ between the capacitor plates is considered equivalent to a current, as far as generation of a rotation of the magnetic intensity, $\nabla \times \mathbf{H}$, is concerned. The circuit can then be considered as *closed.*

  Also we note that with the Maxwell displacement term we can trivially obtain the continuity equation (2.6) for charge by combining (5.5) and (5.7). *Without* the Maxwell displacement current we would have $\nabla \cdot \mathbf{J} = 0$, in disagreement with the charge continuity equation!

As they stand, Maxwell's equations are *linear*; if the magnitude of all variables is doubled (including ζ and **J**), they still remain solutions of the equations. This observation is after all rather trivial; the possibility of nonlinear phenomena enters through the interaction of fields with matter. The material relations will in general be nonlinear, although often a linearized model will prove to be fully adequate.

- **Example:** We apply Faraday's law to a linearly polarized wave-like motion of the form

$$\mathbf{E} = (E_\perp \widehat{\mathbf{e}}_\perp + E_\parallel \widehat{\mathbf{e}}_\parallel)\exp(-i(\omega t - \mathbf{k}\cdot\mathbf{r})) \qquad \text{and} \qquad \mathbf{B} = (B_\perp \widehat{\mathbf{e}}_\perp + B_\parallel \widehat{\mathbf{e}}_\parallel)\exp(-i(\omega t - \mathbf{k}\cdot\mathbf{r})),$$

  with given wave-vector $\mathbf{k} \neq 0$, and the components denoted by $\widehat{\mathbf{e}}_\perp$ and $\widehat{\mathbf{e}}_\parallel$ referring to directions perpendicular and parallel to $\mathbf{k}/k \equiv \widehat{\mathbf{e}}_\parallel$. The electric and magnetic fields of the wave are written in their general form, but we see right away that $\nabla \cdot \mathbf{B} = 0$ implies $B_\parallel = 0$. We have by **MII** that

$$\frac{|E_\perp|}{|B_\perp|} = \frac{\omega}{k} \tag{5.15}$$

  for $\omega \neq 0$, where we introduced the absolute values of the magnitudes. We can now conclude that in case the phase velocity becomes very small, $\omega/k \to 0$, this must imply $E_\perp \to 0$ for constant $B_\perp$. For this case we can *either* have $|\mathbf{E}| \to 0$, which in plasma physics (as we shall see) corresponds to an Alfvén wave, *or* that $E_\perp \to 0$ while $E_\parallel = \text{const.}$, implying that the electric field is essentially parallel to $\mathbf{k}$. This latter case will be termed "electrostatic waves" in plasma physics. In case we have $\omega/k \to \infty$, with ω finite, we will, on the other hand, expect $|\mathbf{B}| \to 0$ according to (5.15). This condition is found at wave cut-offs in plasmas and other media.

- **Exercise:** A volt meter can be used not only to detect differences in electrostatic potential, but also for measuring electromotance in general. Demonstrate, with reference to Fig. 5.2, that it can be used for measuring electric fields induced by fluctuating magnetic fields as well, although in this case one must take care to specify the way the measurement is made. Assume that a torus-shaped coil, like the one shown in Fig. 5.2a, is inducing a magnetic field through a conducting closed loop, with a certain resistance R. If the major radius of the torus is much larger than its minor radius, we can assume that the magnetic field is uniform *inside* the torus, and vanishes outside. Recall that a volt meter is in reality an ampere meter with a very large internal resistance $R_V \gg R$. Assume as an illustration that the electromotance is 12 V, and demonstrate that with the volt meter placed as in Fig. 5.2b its reading will be 3 V, while with it the position as in Fig. 5.2c will be 9 V, since the current through the volt meter is a global quantity and depends on the magnetic flux through the entire current loop.

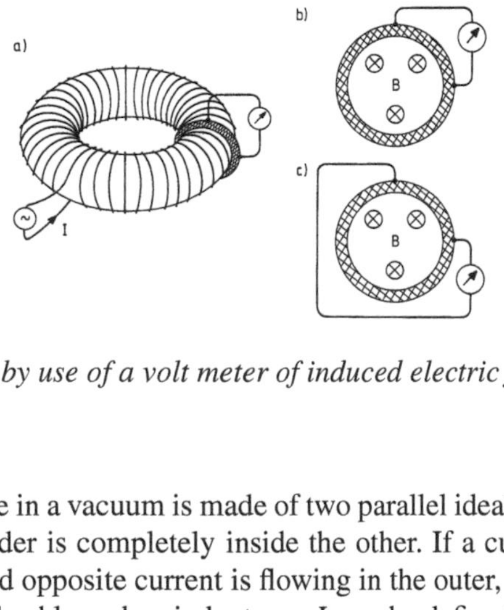

FIGURE 5.2: *Measurement by use of a volt meter of induced electric fields, i.e., electromotances.*

- **Exercise:** A coaxial cable in a vacuum is made of two parallel ideal metal cylinders of *arbitrary* cross sections; one cylinder is completely inside the other. If a current is flowing in the inner cylinder, and an equal and opposite current is flowing in the outer, then a magnetic field will be present inside the coaxial cable and an inductance L can be defined per unit length of the cable. Similarly, if we instead charge the inner and outer cylinders with charges equal in magnitude, but of opposite sign, we can view it as a capacitor, and determine its capacitance C per unit length. Show that the expression $\sqrt{LC}$ is independent of the cross section of the cylinders, and determine its value.

Ⓢ **Exercise:** Consider a uniformly charged sphere with radius a and homogeneous charge density ζ. Express the electrostatic electric field as a function of radius. What is the corresponding electrostatic potential? Assume then that a small sphere with radius b is carved out of the original sphere, as illustrated in Fig. 5.3. The center of the cavity is at a distance c from the center of the original sphere. It is assumed that $c + b < a$. Give analytical expressions for the electric field at all positions. Give in particular the magnitude and the direction of the electric field inside the cavity.

Consider the related problem where we have a long uniformly charged cylinder. In this case you can see Fig. 5.3 as representing a cut perpendicular to the cylinder. Ⓢ

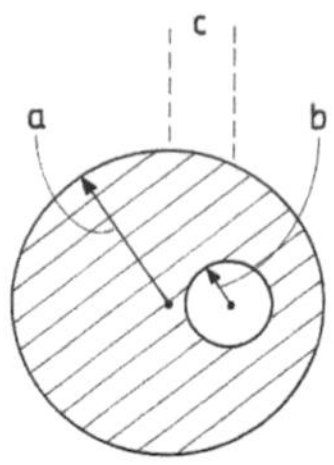

FIGURE 5.3: *Illustration of a cut through a uniformly charged sphere, having a small spherical cavity.*

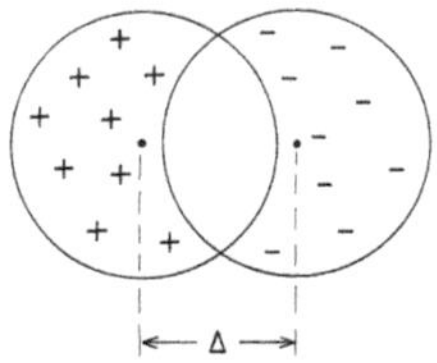

FIGURE 5.4: *Illustration of two uniformly charged spheres with opposite charges, displaced by a distance Δ. The displacement is here shown to be large, but we are mostly interested in infinitesimal Δ.*

Ⓢ **Exercise:** Consider two uniformly charged spheres, both with radius a and with opposite charges. Let the two spheres be displaced by an arbitrary distance, Δ; see Fig. 5.4. Obtain an expression for the electrostatic field inside and outside the spheres, for fixed Δ. In case the spheres are overlapping, as in Fig. 5.4, give the magnitude and direction of the electric field in the lens-shaped, charge-neutral region in the middle. Ⓢ

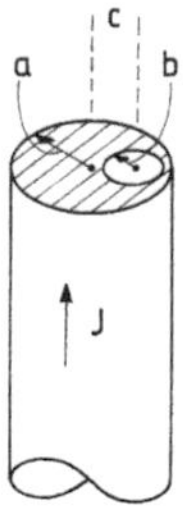

FIGURE 5.5: *Illustration of a long straight current carrying massive cylinder, or rod, with uniform current density* **J**. *The rod has a cylindrical cavity parallel to the axis of the cylinder.*

Ⓢ **Exercise:** Consider a current carrying massive cylinder, with uniform current density **J**. The radius of the cylinder is a. Express the static magnetic field as a function of radius, inside and outside the rod. Assume then that a cylindrical cavity with radius b is carved out of the original rod, as illustrated in Fig. 5.5. The center of the cavity is at a distance c from the center of the original cylinder or "rod" with axes being parallel, as shown in Fig. 5.5. It is assumed that $c+b<a$. Give analytical expressions for the magnetic field at all positions. Give in particular the magnitude and the direction of the magnetic field inside the cavity. Ⓢ

5.3 Potentials

We maintain that the forces mediated by the electric and magnetic fields represent the physical reality, but it is possible to introduce some potentials, which can facilitate the analysis considerably, in many cases. These potentials need not, in general, represent any obvious unique physical quantity.

Since the magnetic field is divergence free, $\nabla \cdot \mathbf{B} = 0$, we can introduce a vector potential $\mathbf{A}$, so that

$$\mathbf{B} = \nabla \times \mathbf{A}. \tag{5.16}$$

By this, we have assured that the divergence of the magnetic field vanishes. Evidently, the choice of $\mathbf{A}$ is not unique; we can introduce any other vector field $\widetilde{\mathbf{A}}$, provided $\mathbf{A} - \widetilde{\mathbf{A}} = \nabla\psi$, with ψ being some scalar field.

As a direct consequence of (5.16) we have the following interesting relation obtained by the Stokes theorem; see Appendix D.2

$$\iint_S \mathbf{B} \cdot \widehat{\mathbf{n}} dS = \oint_C \mathbf{A} \cdot d\ell,$$

which relates the magnetic flux through a surface S to the line integral of the vector potential around the closing contour C.

From (5.6) we find that the vector potential just introduced implies a contribution $\partial\mathbf{A}/\partial t$ to the electric field. We might, however, add any potential to this contribution and find generally

$$\mathbf{E} = -\nabla\phi - \frac{\partial \mathbf{A}}{\partial t}, \tag{5.17}$$

where ϕ is the electrostatic scalar potential, which is the *only* contribution for static problems. The relation (5.17) ensures that $\mathbf{E}$ satisfies (5.6). Again, we note that the choice of ϕ is not unique; we can add any constant to it, for instance, and the electric field is unchanged. Neither the vector potential nor the scalar potential are determined uniquely.

The freedom in choice of potential is sometimes utilized to simplify the actual analysis, by a proper *gauge transformation*. We can, for instance, choose to have $\nabla \cdot \mathbf{A} = 0$; this is the Coulomb gauge.

5.3.1 Differential equations for the potentials

We will here consider only simple media with constant ε and μ, vacuum being the most relevant example. we find

$$\mathbf{D} = -\varepsilon\left(\nabla\phi + \frac{\partial \mathbf{A}}{\partial t}\right)$$

and

$$\mathbf{H} = \frac{1}{\mu}\nabla \times \mathbf{A}.$$

Using Maxwell's equation we find

$$-\frac{1}{\varepsilon}\zeta = \nabla^2\phi + \nabla \cdot \frac{\partial \mathbf{A}}{\partial t}, \tag{5.18}$$

$$\mu\mathbf{J} = \nabla\nabla \cdot \mathbf{A} + \mu\varepsilon\nabla\frac{\partial\phi}{\partial t} - \nabla^2\mathbf{A} + \mu\varepsilon\frac{\partial^2\mathbf{A}}{\partial t^2}. \tag{5.19}$$

We find that (5.19) can be simplified considerably by imposing the *Lorenz condition* (n.b. not Lorentz condition)

$$\nabla \cdot \mathbf{A} + \mu\varepsilon\frac{\partial\phi}{\partial t} = 0. \tag{5.20}$$

In this case we find the wave equations in the form

$$\mu\varepsilon\frac{\partial^2\mathbf{A}}{\partial t^2}-\nabla^2\mathbf{A} = \mu\mathbf{J} \tag{5.21}$$

$$\mu\varepsilon\frac{\partial^2\phi}{\partial t^2}-\nabla^2\phi = \frac{\zeta}{\varepsilon}, \tag{5.22}$$

where current densities and charge densities act as generators for the potentials. The expressions can be generalized further if we assume that the medium is simple also with respect to conductivities. The choice of gauge concerns only the scalar and vector potentials. It can be demonstrated that the Coulomb gauge and the Lorenz gauge give the same observables **E** and **B** for identical sources ρ and **J**.

5.3.2 The Hertz vectors

Again consider only simple media, and impose the Lorenz condition (5.20). We readily find that this equation is fulfilled if we can find a vector field Π so that

$$\mathbf{A}=\mu\varepsilon\frac{\partial\Pi}{\partial t} \qquad \text{and} \qquad \phi=-\nabla\cdot\Pi.$$

It can be demonstrated that it is indeed possible to find such a vector field Π, but the choice is not unique, similar to the problem we encountered with the potentials **A** and φ. The vector field Π is an example of a *super potential*, and is usually called the *Hertz vector field.*

Inserting in the wave equations for **A** and φ we find the consistency equations in the form

$$\mu\varepsilon\frac{\partial}{\partial t}\left(\mu\varepsilon\frac{\partial^2\Pi}{\partial t^2}-\nabla^2\Pi\right) = \mu\mathbf{J}, \tag{5.23}$$

$$\nabla\cdot\left(\mu\varepsilon\frac{\partial^2\Pi}{\partial t^2}-\nabla^2\Pi\right) = -\frac{\zeta}{\varepsilon}. \tag{5.24}$$

It would be nice to have one equation that Π has to fulfill. To find this we note the similarity between the Lorenz condition (5.20) and the equation for charge continuity (5.13). We can argue for the existence of a current density potential **R**, so that $\mathbf{J}=\partial\mathbf{R}/\partial t$ implying $\zeta=-\nabla\cdot\mathbf{R}$. Both equations (5.23) and (5.24) will then be satisfied if the impose the inhomogeneous wave equation

$$\mu\varepsilon\frac{\partial^2\Pi}{\partial t^2}-\nabla^2\Pi=\frac{\mathbf{R}}{\varepsilon}$$

for Π. For the field vectors **B** and **E** we have in terms of the Hertz vector field

$$\mathbf{B} = \mu\varepsilon\nabla\times\frac{\partial\Pi}{\partial t} \tag{5.25}$$

$$\mathbf{E} = \nabla\nabla\cdot\Pi-\mu\varepsilon\frac{\partial^2\Pi}{\partial t^2}=\nabla\times\nabla\times\Pi-\frac{\mathbf{R}}{\varepsilon}. \tag{5.26}$$

5.4 Poynting's identity

From (5.5) and (5.6) we readily obtain

$$\mathbf{E}\cdot\nabla\times\mathbf{H}=\mathbf{E}\cdot\mathbf{J}+\mathbf{E}\cdot\frac{\partial\mathbf{D}}{\partial t}, \tag{5.27}$$

and

$$\mathbf{H}\cdot\nabla\times\mathbf{E}=-\mathbf{H}\cdot\frac{\partial\mathbf{B}}{\partial t}. \tag{5.28}$$

Using $\nabla\cdot(\mathbf{E}\times\mathbf{H}) = (\nabla\times\mathbf{E})\cdot\mathbf{H} - (\nabla\times\mathbf{H})\cdot\mathbf{E}$ we obtain Poynting's identity

$$\nabla\cdot(\mathbf{E}\times\mathbf{H}) = -\mathbf{E}\cdot\mathbf{J} - \mathbf{E}\cdot\frac{\partial\mathbf{D}}{\partial t} - \mathbf{H}\cdot\frac{\partial\mathbf{B}}{\partial t}. \tag{5.29}$$

The term $-\mathbf{E}\cdot\mathbf{J}$ accounts for the net power supplied by "external" generators, i.e., it represents the work done per unit time per unit volume by the fields. It is a conversion of electromagnetic energy into mechanical or thermal energy. Since matter is ultimately composed of charged particles, we can think of this rate of conversion as a rate of increase of energy of the charged particles per unit volume. The vector $\mathbf{S} = \mathbf{E}\times\mathbf{H}$ is known as Poynting's vector and accounts for the energy flux carried by the electromagnetic field, i.e., the rate of arrival of wave energy (Coulson 1955, Jackson 1975).

We can now integrate (5.29) over an arbitrary volume and use Gauss' divergence theorem to obtain

$$\oint_S \mathbf{n}\cdot(\mathbf{E}\times\mathbf{H})dS = -\int_V \left(\mathbf{E}\cdot\mathbf{J} + \mathbf{E}\cdot\frac{\partial\mathbf{D}}{\partial t} + \mathbf{H}\cdot\frac{\partial\mathbf{B}}{\partial t}\right)dV. \tag{5.30}$$

Since the propagation velocity of electromagnetic waves is finite (equal to c in vacuum), we know that the fields have finite support, unless they are "turned on" at an infinite past. We assume that this is not the case, and by letting the volume be all space, we find

$$-\int_V \mathbf{E}\cdot\mathbf{J}dV = \int_V \left(\mathbf{E}\cdot\frac{\partial\mathbf{D}}{\partial t} + \mathbf{H}\cdot\frac{\partial\mathbf{B}}{\partial t}\right)dV. \tag{5.31}$$

The left side of (5.31) is the net power delivered by the "external" generators to all space. To maintain energy conservation, it is then necessary to consider

$$\frac{\partial}{\partial t}\mathcal{W} = \int_V \left(\mathbf{E}\cdot\frac{\partial\mathbf{D}}{\partial t} + \mathbf{H}\cdot\frac{\partial\mathbf{B}}{\partial t}\right)dV \tag{5.32}$$

as the rate of change in the total energy $\mathcal{W}$ of the electromagnetic field. Note that since **D, E, H** and **B** appear explicitly, we have not yet made any assumptions on the material relations between these quantities. The identification of an electric, $\mathcal{W}_e$, and a magnetic, $\mathcal{W}_m$, energy is self-evident. It is tempting, but not logically self-evident to consider $\mathbf{E}\cdot\partial\mathbf{D}/\partial t + \mathbf{H}\cdot\partial\mathbf{B}/\partial t$ as the rate of change in the energy *density* associated with the electromagnetic fields. The uncertainty in this interpretation arises from the formal possibility of adding a scalar field which is spatially varying, but vanishing when integrated over all space.

5.4.1 Poynting's identity for a vacuum

In a vacuum, it is particularly simple to interpret Poynting's identity. In this case, with $\mathbf{D} = \varepsilon_0\mathbf{E}$, $\mathbf{B} = \mu_0\mathbf{H}$ and $\mathbf{J} = 0$ we have

$$\begin{aligned}
\frac{1}{\mu_0}\nabla\cdot(\mathbf{E}\times\mathbf{B}) &= -\mathbf{E}\cdot\frac{\partial\mathbf{D}}{\partial t} - \mathbf{H}\cdot\frac{\partial\mathbf{B}}{\partial t} \\
&= -\frac{\partial}{\partial t}\left(\frac{1}{2}\varepsilon_0 E^2 + \frac{1}{2\mu_0}B^2\right) \\
&\equiv -\frac{\partial}{\partial t}(W_e + W_m),
\end{aligned} \tag{5.33}$$

which can readily be integrated with respect to time in order to give the energy density W. We find that (5.33) has the form of a *continuity equation* for the energy density $W \equiv W_e + W_m$ if we associate $\mathbf{S} \equiv \mathbf{E}\times\mathbf{B}/\mu_0$ with an energy density flux, i.e., we have $\partial W/\partial t + \nabla\cdot(\mathbf{E}\times\mathbf{B}/\mu_0) = 0$, having precisely the same structure as, for instance, the continuity equation (2.1), thus substantiating the interpretation of the Poynting vector **S** as an energy density flux.

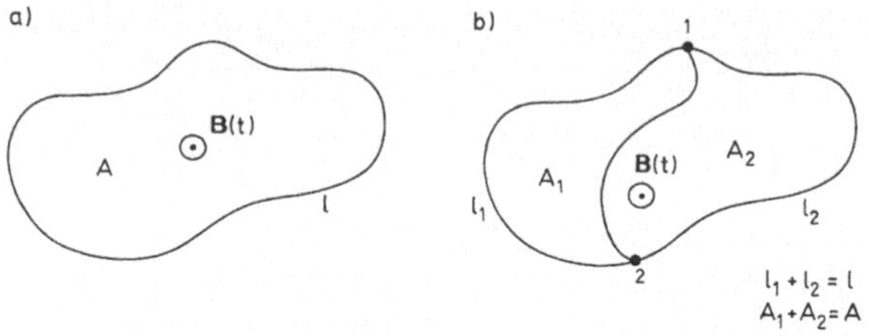

FIGURE 5.6: *a) A plane closed loop with a time varying magnetic field* $\mathbf{B}(t)$ *perpendicular to the plane of the wire. Assume the wire to be thin. In b) two points are connected by a wire, also in the same plane as the rest.*

Ⓢ **Exercise:** Consider a plane closed wire of length l, as shown in Fig. 5.6. The Ohmic resistance of the wire is r per unit length. The enclosed area is A. A time varying magnetic field $\mathbf{B}(t)$ is at all times perpendicular to the plane of the wire; the positive direction is out of the plane, as shown. Let $\mathbf{B} = \mathbf{B}_0 + \widehat{\mathbf{b}}\alpha t$. Determine the induced current and its direction. Will this current depend on $\mathbf{B}_0$? As a numerical example use the Earth's surface magnetic field at the pole, and let the wire represent the ground loop of an electric power supply, taking it to be approximated by a circle with radius 25 km. Take the wire diameter to be 10 mm. Use the resistivity of copper. Calculate the current induced in the loop by a strong magnetic storm that changes intensity but not the direction of the Earth's surface magnetic field by 10% in 2 s.

We now connect the positions 1 and 2 on the original wire, as shown in Fig. 5.6b. We have in this case two areas A_1 and A_2, and segments with lengths l_1 and l_2, as shown on the figure. Determine the area A_1 and the length l_1 necessary to give vanishing current in the new wire that connects the two points 1 and 2. How does the result depend on the length of the new wire segment connecting the points 1 and 2? Ⓢ

- **Exercise:** From Maxwell's equations, derive the wave equation

$$\frac{\partial^2}{\partial t^2}\mathbf{E} - \frac{1}{\varepsilon_0\mu_0}\nabla^2\mathbf{E} = 0 \tag{5.34}$$

for electromagnetic waves in a vacuum, and a similar equation for the magnetic field. Give the solutions for linearly, as well as elliptically polarized waves. Demonstrate that $|E|/|B| = c = \sqrt{1/\varepsilon_0\mu_0}$, being the speed of light in a vacuum.

Ⓢ **Exercise:** Consider a long coaxial arrangement like the one shown in Fig. 5.7. A coaxial cable consists of an inner conductor with radius r_1 and a thin cylindrical shell with radius r_2. The cable has a resistive load of R in one end and a battery with voltage V in the other. The outer shell is grounded. The current is assumed to be uniformly distributed on the surface of the inner conductor and the return current uniformly on the outer thin shell. Calculate the electric and magnetic fields, as well as the Poynting flux for this case with $r_1 > 0$. What is the result in the limit of $r_1 \to 0$? Does it make any difference if the current in the inner conductor is distributed on its *surface* rather than uniformly? Ⓢ

- **Exercise:** Consider an ideal magnetic dipole with magnetic dipole moment $\mathcal{M}$. Place upon this dipole a charge q. Calculate the magnetic and electric fields for this case and give an expression for the magnitude and direction of the resulting Poynting flux.

- **Exercise:** Consider a long straight wire with radius a and length $\mathcal{L} \gg a$. The resistance of the wire is R, and it carries a current I. Compute the magnetic and electric fields at the surface of the wire and give an expression for the magnitude and direction of the resulting Poynting flux there. Show that the rate of energy flow into the wire is I^2R/L per unit length.

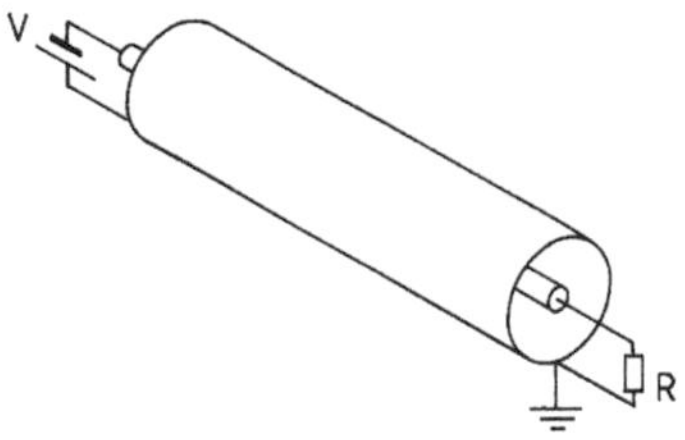

FIGURE 5.7: *Schematic diagram showing a coaxial arrangement consisting of an inner conductor with radius r_1 and a thin cylindrical shell with radius r_2. The coaxial cable has a resistive load of R in one end and a battery with voltage V in the other. The outer shell is grounded.*

5.5 Electromagnetic forces

Electric and magnetic fields are detected by the forces they give rise to. In general, matter is distributed somehow, but some idealized limits using point-charges and thin current carrying wires are most often used to bring out the basic features (Duffin 1990).

5.5.1 Electromagnetic forces on particles and currents

The electric force between two point charges q_1 and q_2 is given by Coulomb's law

$$\mathbf{F}_{q_1,q_2} = \frac{q_1 q_2}{4\pi\varepsilon_0 r^2}\widehat{\mathbf{r}} \equiv \frac{q_1 q_2}{4\pi\varepsilon_0 r^3}\mathbf{r} \tag{5.35}$$

where r is their separation and $\widehat{\mathbf{r}}$ a unit vector in the $\mathbf{r}$-direction.

If we know the magnetic field $\mathbf{B}$ due to the surrounding currents we have the simple relation for the force on a vectorial current element $d\mathbf{l}$ carrying a current I

$$d\mathbf{F}_B = Id\mathbf{l} \times \mathbf{B}. \tag{5.36}$$

A current carrying wire produces a magnetic field where we have the B-field contribution from a current element $d\mathbf{l}$ to be

$$d\mathbf{B} = \frac{\mu_0 Id\mathbf{l} \times \mathbf{r}}{4\pi r^3}, \tag{5.37}$$

known as the Biot-Savart law.

If we have two thin current carrying wires with currents I_1 and I_2, the force between two current elements $d\mathbf{l}_1$ and $d\mathbf{l}_2$ separated by r is

$$d^2\mathbf{F}_{I_1,I_2} = \frac{\mu_0 I_1 I_2 d\mathbf{l}_2 \times (d\mathbf{l}_1 \times \mathbf{r})}{4\pi r^3}, \tag{5.38}$$

as obtained by (5.37).

Ⓢ **Exercise:** Consider two small metal spheres, each with mass M and a charge Q uniformly distributed on the surfaces. Both spheres are hanging in a string of length ℓ and negligible mass, both tied to the same point at one end. Gravity is pointing downwards, with acceleration g. Calculate the angle α between the two strings, assuming that this angle is small. The radius of the spheres is much smaller than ℓ. What is the horizontal distance between the two spheres? Ⓢ

5.5.2 Electromagnetic forces on matter

Currents and charges give rise to electric and magnetic fields as described by Maxwell's equations. The electric and magnetic fields then act back upon the charged and current carrying matter by the Lorentz force, with $\mathbf{J}$ and ζ being the current and charge densities, respectively. The Lorentz force is expressed by the force density

$$\mathbf{f} = \zeta\mathbf{E} + \mathbf{J}\times\mathbf{B}. \tag{5.39}$$

Note that this is not necessarily the *total* force density. Charges and currents are associated with matter, which is influenced also by other forces, gravity, for instance.

Introducing ζ and $\mathbf{J}$ by the Maxwell equations the force density $\mathbf{f}$ is rewritten as

$$\begin{aligned}
\mathbf{f} &= (\nabla\cdot\mathbf{D})\mathbf{E} + (\nabla\times\mathbf{H})\times\mathbf{B} + \mathbf{B}\times\frac{\partial}{\partial t}\mathbf{D} \\
&= \nabla\cdot(\mathbf{DE}) - \mathbf{D}\cdot\nabla\mathbf{E} + (\nabla\cdot\mathbf{B})\mathbf{H} + \mathbf{B}\cdot\nabla\mathbf{H} \\
&\qquad -(\nabla\mathbf{H})\cdot\mathbf{B} + \mathbf{B}\times\frac{\partial}{\partial t}\mathbf{D} \\
&= \nabla\cdot(\mathbf{DE}) - (\nabla\mathbf{E})\cdot\mathbf{D} + \mathbf{D}\times\nabla\times\mathbf{E} + \nabla\cdot(\mathbf{BH}) \\
&\qquad -(\nabla\mathbf{H})\cdot\mathbf{B} + \mathbf{B}\times\frac{\partial}{\partial t}\mathbf{D} \\
&= \nabla\cdot(\mathbf{DE}) - (\nabla\mathbf{E})\cdot\mathbf{D} - \mathbf{D}\times\frac{\partial}{\partial t}\mathbf{B} + \nabla\cdot(\mathbf{BH}) \\
&\qquad -(\nabla\mathbf{H})\cdot\mathbf{B} + \mathbf{B}\times\frac{\partial}{\partial t}\mathbf{D} \\
&= \nabla\cdot(\mathbf{DE}+\mathbf{BH}) - (\nabla\mathbf{E})\cdot\mathbf{D} - (\nabla\mathbf{H})\cdot\mathbf{B} - \frac{\partial}{\partial t}(\mathbf{D}\times\mathbf{B}),
\end{aligned} \tag{5.40}$$

where we inserted $(\nabla\cdot\mathbf{B})\mathbf{H}$ (this we can do since the magnetic fields are divergence free) to obtain the term $\nabla\cdot(\mathbf{BH})$.

Not much can be done with (5.40) without additional assumptions. Here we assume that the media are simple, linear and homogeneous, implying a general tensor notation $\mathbf{D} = \varepsilon_0\underline{\varepsilon}_r\cdot\mathbf{E}$ and $\mathbf{B} = \mu_0\underline{\mu}_r\cdot\mathbf{H}$, where $\underline{\varepsilon}_r$ and $\underline{\mu}_r$ are here the (constant) relative permittivity and permeability *tensors* as distinct from their *scalar* counterparts. The tensor notation adds slightly to the generality of the following results. We have

$$\begin{aligned}
\nabla\cdot(\mathbf{D}\cdot\mathbf{E}\underline{1}) &= \nabla(\mathbf{D}\cdot\mathbf{E}) = (\nabla\mathbf{D})\cdot\mathbf{E} + (\nabla\mathbf{E})\cdot\mathbf{D} \\
&= (\nabla(\mathbf{E}\cdot\varepsilon_0\underline{\varepsilon}_r))\cdot\mathbf{E} + (\nabla\mathbf{E})\cdot\mathbf{D} \\
&= (\nabla\mathbf{E})\cdot\varepsilon_0\underline{\varepsilon}_r\cdot\mathbf{E} + (\nabla\mathbf{E})\cdot\mathbf{D} \\
&= 2(\nabla\mathbf{E})\cdot\mathbf{D}
\end{aligned}$$

and similarly

$$\nabla\cdot(\mathbf{B}\cdot\mathbf{H}\underline{1}) = 2(\nabla\mathbf{H})\cdot\mathbf{B}.$$

All in all, we thus find

$$\mathbf{f} = \nabla\cdot\underline{T} - \frac{\partial}{\partial t}(\mathbf{D}\times\mathbf{B}) \tag{5.41}$$

with the tensor

$$\underline{T} \equiv \mathbf{DE} - \frac{1}{2}\mathbf{D}\cdot\mathbf{E}\underline{1} + \mathbf{BH} - \frac{1}{2}\mathbf{B}\cdot\mathbf{H}\underline{1} \tag{5.42}$$

where $\underline{1}$ is the unit tensor, with all diagonal elements being unity, and all off-diagonal elements being zero. The tensor $\underline{T}$ is usually called Maxwell's stress tensor.

For time stationary conditions in particular, we can write

$$\mathbf{F} = \int_V \mathbf{f}\,dV = \int_V \nabla\cdot\underline{T}\,dV = \oint_S \widehat{\mathbf{n}}\cdot\underline{T}\,dS. \tag{5.43}$$

This relation implies that the force acting on the volume V can be interpreted as, or considered equivalent with, a distribution of surface forces with density $\widehat{\mathbf{n}}\cdot\underline{T}$. This is the Faraday-Maxwell interpretation of the force density associated with an electromagnetic field; the force acting on charges and currents in a given volume can be interpreted as being the action of the surroundings through the surface of the volume by a force density $\widehat{\mathbf{n}}\cdot\underline{T}$. For a time the tensor $\underline{T}$ was interpreted as the elastic stress tensor associated with the "ether".

For a time varying electromagnetic field we find

$$\mathbf{F} = \int_V \mathbf{f}\,dV = \oint_S \widehat{\mathbf{n}}\cdot\underline{T}\,dS - \int_V \frac{\partial}{\partial t}(\mathbf{D}\times\mathbf{B})\,dV. \tag{5.44}$$

For a *finite system*, taking the integration volume large enough, we can make the surface integral vanish, and find

$$\mathbf{F} = \int_V \mathbf{f}\,dV = -\int_V \frac{\partial}{\partial t}(\mathbf{D}\times\mathbf{B})\,dV = -\frac{d}{dt}\int_V (\mathbf{D}\times\mathbf{B})\,dV. \tag{5.45}$$

A force can be considered as the time derivative of a mechanical momentum, and the result (5.45) invites the interpretation of $\int_V \mathbf{D}\times\mathbf{B}\,dV$ as the momentum associated with electromagnetic fields in a vacuum. Correspondingly, $\mathbf{D}\times\mathbf{B}$ can be interpreted as a momentum density.

In a vacuum, the expression (5.45) is particularly simple. For this case we find $\mathbf{D}\times\mathbf{B} = \varepsilon_0\mathbf{E}\times\mathbf{B} = \varepsilon_0\mu_0\mathbf{E}\times\mathbf{H} = \mathbf{S}/c^2$, relating the expression to the Poynting flux. Both E and B are fluctuating with the electromagnetic wave frequency. These frequencies are often very high (for instance, for visible light), so for practical applications, often only the time average is relevant.

The time average of $\mathbf{S}$ is

$$\overline{\mathbf{S}}(z) = \frac{1}{T}\int_0^T \mathbf{S}(z,t)dt = \frac{1}{T}\int_0^T \left(\frac{1}{2}\Re\{\mathbf{E}_0\times\mathbf{H}_0^*\} + \frac{1}{2}\Re\left\{\mathbf{E}_0\times\mathbf{H}_0 e^{2(ikz-i\omega t)}\right\}\right)dt.$$

See also Section 3.1.1. The second term is a propagating plane wave $e^{2(ikz-i\omega t)}$ whose time average will be zero at any fixed position z. This gives

$$\overline{\mathbf{S}} = \frac{1}{2}\Re\{\mathbf{E}_0\times\mathbf{H}_0^*\}.$$

The time averaged Poynting vector is an important quantity and is given a special name, *irradiance*, to distinguish it from the rapidly fluctuating Poynting vector. For electromagnetic waves propagating in a vacuum $\mathbf{E}\perp\mathbf{H}$ and we have the simple relations $B_0 = E_0/c$ from Faraday's law, so that

$$|\mathbf{S}| = \Re\left\{\frac{1}{c\mu_0}E_0^2 e^{2(ikz-i\omega t)}\right\} = \Re\left\{c\varepsilon_0 E_0^2 e^{2(ikz-i\omega t)}\right\}$$

and $\overline{\mathbf{S}} = c\varepsilon_0 E_0^2/2$.

A radiation pressure can be introduced by noting the analogy to thermal pressure, which is the momentum transferred per unit surface area per unit time. In the same spirit we can define the rate of momentum transfer due to electromagnetic waves. If we take $\mathbf{D} \times \mathbf{B}$ to be the momentum density associated with the electromagnetic field, then the momentum flux is this quantity multiplied by the speed of light in the medium. For a vacuum in particular this will be $c\mathbf{D} \times \mathbf{B} = c\varepsilon_0 \mathbf{E} \times \mathbf{B}$. For normal incidence on a surface in a vacuum we find the radiation pressure by time-averaging $c\mathbf{D} \times \mathbf{B}$ and obtain

$$P_{rad} = \frac{\overline{EB}}{\mu_0 c} = \frac{\overline{\mathbf{S}}}{c} \tag{5.46}$$

assuming that the wave energy is totally absorbed. Again for a vacuum we have $B = E/c$, so that the radiation pressure can be expressed in terms of the electromagnetic field energy density as $P_{rad} = \varepsilon_0 \overline{E^2}$, recalling that for this case the electric and magnetic field energy densities are the same. If we have total reflection, the result in (5.46) should be multiplied by 2 because the momentum changes direction instead of simply being absorbed. The energy density at the surface is doubled and so is the pressure. Physically, the radiation pressure can be explained as an integrated $\overline{\mathbf{J} \times \mathbf{B}}$-force density, where the fluctuating electric fields drive the currents on the surface or inside the medium which in turn interact with the fluctuating magnetic fields there. Electromagnetic wave reflections from perfectly conducting surfaces serve as a good illustrative example.

- **Example:** Consider a plane electromagnetic wave propagating in a vacuum. Assume that we now introduce a plate of thickness $\mathcal{L}$, consisting of a completely transparent medium placed so that the wave is normally incident. The material does thus in no way disturb the wave propagation. According to the previous arguments, you could argue for a radiation pressure on the surface of the medium where the wave enters! The example can be made consistent by noting that at the surface where the wave *exits* there is a recoil on the exiting surface, in such a way that the net radiation pressure on the slab of imagined material is vanishing.

Ⓢ **Exercise:** Estimate the radiation pressure originating from sunlight at the Earth's equator, above the atmosphere. Ⓢ

Ⓢ **Exercise:** Consider an idealized solar system, where we have a central sun very much like our own but assume that we have only small dust particles in its environment. These particles are subject to the solar gravity, and to the radiation pressure from the sun, while the effects of the solar wind are ignored. Assume that you can use the numerical values for the mass and the radiation from our Sun, and take all dust particles to be spherical with radius r_0 and mass density ρ, all particles being identical. Let the particles be perfectly absorbing. Can the gravitational force and the time averaged radiation pressure balance each other? If yes, at what distance from the sun? Discuss the result (Duffin 1990). Ⓢ

5.6 Waves in simple conducting media Ⓚ

A simple dielectric medium was defined as one where there is a linear relation between electric fields **E** and electric displacements **D**. We define a simple conducting medium as one where there is a linear relation between electric fields and currents, corresponding to Ohm's law for electronics. The relation can be tensorial, but the outline here will consider only scalar relations, for simplicity. Such a conducting medium can be used as a model also for plasmas in certain limits, so this particular case deserves some attention. Thus, for simple conducting media we here write

$$\mathbf{J} = \gamma \mathbf{E} \tag{5.47}$$

$$\mathbf{D} = \varepsilon_r \varepsilon_0 \mathbf{E} \tag{5.48}$$

$$\mathbf{B} = \mu_r \mu_0 \mathbf{H} \tag{5.49}$$

where γ is a conductivity, ε_r is a relative permittivity (or dielectric constant) while μ_r is the relative permeability. Note that **J** follows **E** instantly: there is no inertia in this simple model for conductors. For this medium it is now an easy matter to obtain Maxwell's equation in the form

$$\nabla \times \mathbf{B} = \mu_0 \mu_r \gamma \mathbf{E} + \frac{\varepsilon_r \mu_r}{c^2} \frac{\partial \mathbf{E}}{\partial t}, \tag{5.50}$$

$$\nabla \times \mathbf{E} = -\frac{\partial \mathbf{B}}{\partial t}, \tag{5.51}$$

$$\varepsilon_0 \nabla \cdot \mathbf{E} = \zeta, \tag{5.52}$$

$$\nabla \cdot \mathbf{B} = 0 \tag{5.53}$$

If we assume that we have no free charges, i.e., $\zeta = 0$, we find that only (5.50) is affected by the presence of the medium. From (5.50)–(5.53) we readily find

$$\nabla^2 \mathbf{E} = \mu_0 \mu_r \gamma \frac{\partial \mathbf{E}}{\partial t} + \frac{\varepsilon_r \mu_r}{c^2} \frac{\partial^2 \mathbf{E}}{\partial t^2}. \tag{5.54}$$

A similar equation can be found for **B**.

We find that (5.54) has propagating transverse wave solutions in the form $\mathbf{E} = \widehat{\mathbf{e}}_x E_0 \exp(-i\omega t + ikz) \equiv \widehat{\mathbf{e}}_x E_0 \exp(-i\omega t + ik_r z) \exp(-k_i z)$ for $z > 0$, where the complex wave-number is

$$k \equiv \frac{\omega}{c'} \sqrt{1 + i \frac{\gamma}{\omega \varepsilon_r \varepsilon_0}} \equiv k_r + ik_i \tag{5.55}$$

where $c' \equiv c/\sqrt{\varepsilon_r \mu_r}$ is the speed of wave propagation in the medium when $\gamma = 0$. We took ω to be real, corresponding to a wave excited somewhere in the medium at $z = 0$ by an antenna. The wave is damped exponentially according to the factor $\exp(-k_i z)$ due to the dissipation of energy by the Ohmic losses.

The limit of a very good conductor $\gamma \gg \omega \varepsilon_r \varepsilon_0$ becomes particularly simple. Here we find $k = (1+i)\sqrt{\omega \mu_0 \mu_r \gamma / 2}$, with $\sqrt{i} = (1+i)/\sqrt{2}$. We define a *skin depth* as $\delta \equiv \sqrt{2/(\omega \mu_0 \mu_r \gamma)}$, and may write $\mathbf{E} = \widehat{\mathbf{e}}_x E_0 \exp(-i\omega t + iz/\delta) \exp(-z/\delta)$. For large γ we have δ to be small, demonstrating that it is difficult for an electromagnetic field to propagate into a good conductor. This is known as the *skin effect*. Later we will encounter another type of skin effect, the collisonless skin effect in plasmas that is related to inertia effects, so it might be better to specify the present one as the resistive skin effect. For the case where $\gamma = 0$ we have the wavelength to be $\lambda_0 = 2\pi c'/\omega$. With $\gamma \gg \omega \varepsilon_r \varepsilon_0$ we find $\lambda = 2\pi\delta$, indicating that the wave speed is strongly reduced in the conductor to be $c_g = \omega \delta \ll c'$. Note that the waves become dispersive due to the conductivity: $\delta \sim 1/\sqrt{\omega}$. Note that (5.54) reduces to

$$\nabla^2 \mathbf{E} = \mu_0 \mu_r \gamma \frac{\partial \mathbf{E}}{\partial t} \tag{5.56}$$

for $\gamma \to \infty$. In this limit the only sources of magnetic fields are Ampere's law, entirely dominating the contribution from Maxwell's displacement term.

The analysis outlined in this section can be made compact by introducing an effective complex permittivity or dielectric constant $\varepsilon_{ef} \equiv \varepsilon_r + i\gamma/(\omega \varepsilon_0)$ to replace ε_r in Maxwell's equations. By this we obtain the formal simplification that all media properties are accounted for by two parameters μ_r and ε_{ef}. Alternatively we *could* introduce a complex effective conductivity $\gamma_{ef} \equiv \gamma - i\omega\varepsilon_0(\varepsilon_r - 1)$, but not much is gained by this.

- **Exercise**: Let a plane electromagnetic wave with electric field amplitude $\mathbf{E}_1$ be normally incident from a vacuum into a semi-infinite conductor with some finite conductivity γ. Express the incoming, reflected and transmitted Poynting fluxes. Demonstrate explicitly that the difference between the incoming and the reflected Poynting fluxes equals the rate of energy dissipation in the conductor (Coulson 1955).

Ⓢ **Exercise**: Generalize the results of this section to the limit of $\gamma \ll \omega\varepsilon_r\varepsilon_0$, i.e., for poor conductors. What is, in particular, the phase velocity on this limit? Ⓢ

Ⓢ **Exercise**: Write the full expression for the real and imaginary parts of (5.55) and discuss the limiting cases for $\omega \to 0$, $\omega \to \infty$, very good conductors and poor conductors. Ⓢ

Ⓢ **Exercise:** For $\gamma = 0$ the electric and magnetic fields of the electromagnetic wave are in phase. Demonstrate that this is no longer true when $\gamma \neq 0$. Obtain the space-time varying magnetic field by Faraday's law and determine the phase difference for $\gamma \gg \omega\varepsilon_r\varepsilon_0$. Ⓢ

Ⓢ **Exercise:** Determine the ratio of electric and magnetic field energy densities for $\gamma = 0$ and for $\gamma \gg \omega\varepsilon_r\varepsilon_0$. Demonstrate that for a good conductor a large fraction of the wave energy is carried by the magnetic field. What is the Poynting flux for large γ? Ⓢ

Ⓢ **Exercise:** Consider an electromagnetic wave with normal incidence on a conducting medium. Obtain the transmission and reflection coefficients. Consider in particular the ideally conducting limit where $\gamma \to \infty$. Discuss the surface current for finite and for infinite conductivities. Ⓢ

Ⓢ **Exercise:** Consider a plane electromagnetic wave with normal incidence on a plane surface of a good conductor. Let the wave propagate in the z-direction and assume that the conducting material fills the half space $z > 0$. Calculate the $\mathbf{J} \times \mathbf{B}$ force density in the medium and demonstrate that the net force obtained by integration over all $z > 0$ is consistent with the electromagnetic radiation pressure at the surface. Ⓢ

5.7 Polarization description

The distinction between **D** and **E** fields, as well as **H** and **B** fields is historically motivated. When Maxwell's equations were formulated around 1870, the structure of matter was not well understood; recall that the electron was discovered by J.J. Thomson by the end of the 19th century. With the present day understanding, the distinction between "free" and "bound" charges is somewhat arbitrary; after all they are electrons and protons all the same, no matter which context they occur in. Maxwell's equations can be reformulated in a form without such an artificial distinction. The reason that these equations still appear in the form (5.5)–(5.8) is due to the experience that it is often quite convenient to describe matter by some material parameters, without going into detail in the deeper physical meaning of these parameters.

Basically, we can operate with the vacuum electric and magnetic fields **E** and **B**, and consider *all* charges as embedded in this vacuum, there acting as sources for the fields. As a first step, the equations (5.5)–(5.8) are readily simplified for electric and magnetic fields in a vacuum. Introducing the electric and magnetic polarizations, the equations can then be brought into a form where the similarity between "true" charges and currents and those due to polarizations of materials is made evident.

$$\frac{1}{\mu_0}\nabla \times \mathbf{B} = \mathbf{J} + \mathbf{J}' + \varepsilon_0 \frac{\partial \mathbf{E}}{\partial t}, \tag{5.57}$$

$$\nabla \times \mathbf{E} = -\frac{\partial \mathbf{B}}{\partial t}, \tag{5.58}$$

$$\varepsilon_0 \nabla \cdot \mathbf{E} = \zeta + \zeta', \tag{5.59}$$

$$\nabla \cdot \mathbf{B} = 0, \tag{5.60}$$

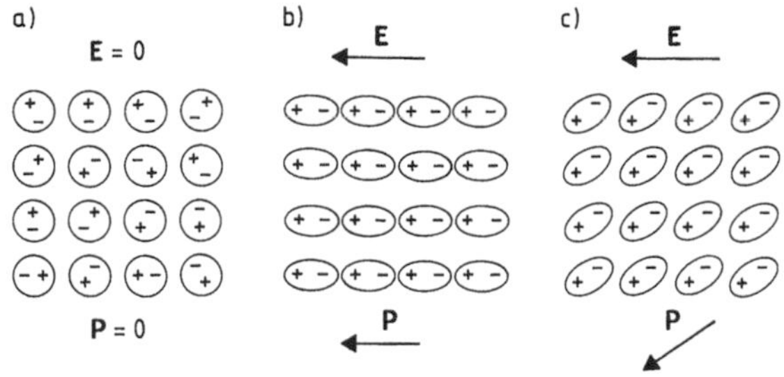

FIGURE 5.8: *Polarization of a medium by an electric field. In c) we have that the directions of* **E** *and* **P** *are different.*

where $\mathbf{J}' = \mathbf{J}'(\mathbf{r},t)$ and $\rho' = \rho'(\mathbf{r},t)$ are the polarization current density and polarization charge density, respectively. They are derived from new quantities **M** and **P** defined as

$$\mathbf{J}' = \nabla \times \mathbf{M} + \frac{\partial \mathbf{P}}{\partial t}, \tag{5.61}$$

and

$$\zeta' = -\nabla \cdot \mathbf{P}. \tag{5.62}$$

Physically, we can interpret **P** as being the *electric dipole density* in the medium, as illustrated in Fig. 5.8. Recall that two point charges $+q$ and $-q$, separated by a vector **d** are associated with an electric dipole moment $\mathbf{p} = q\mathbf{d}$, where **d** points from $-q$ to $+q$.

We have trivially the continuity equation similar to (5.13), but here for the polarization charges and currents

$$\frac{\partial}{\partial t}\zeta' + \nabla \cdot \mathbf{J}' = 0. \tag{5.63}$$

The electric polarization **P** is related to **E** and **D** as

$$\mathbf{P} = \mathbf{D} - \varepsilon_0 \mathbf{E}. \tag{5.64}$$

For simple media we have $\mathbf{P} = (\varepsilon_r - 1)\varepsilon_0 \mathbf{E}$. In the most general case we need not have **E** and **P** be in the same direction; see also Fig. 5.8c. The magnetic polarization **M** is defined similarly by

$$\mathbf{M} = \frac{1}{\mu_0}\mathbf{B} - \mathbf{H}. \tag{5.65}$$

The relations (5.57)–(5.60) are trivially proved by insertion of **M** and **P** by their definition.

It is important to note that for static conditions we have $\nabla \times \mathbf{E} = 0$, but $\nabla \times \mathbf{D} \neq 0$ for the general inhomogeneous problem. We thus have $\nabla \times \mathbf{D} = \nabla \times \mathbf{P}$ for static problems.

In analogy with (5.9)–(5.12), the boundary conditions appropriate for the polarization description become

$$\frac{1}{\mu_0}\widehat{\mathbf{n}}\Delta \times \mathbf{B} = \mathbf{K} + \mathbf{K}', \tag{5.66}$$

where $\mathbf{K}' = \widehat{\mathbf{n}}\Delta \times \mathbf{M}$ is the polarization contribution to the surface current. For the electric field

$$\widehat{\mathbf{n}}\Delta \times \mathbf{E} = 0, \tag{5.67}$$

as before. Instead of the boundary condition for the electric displacement we find

$$\varepsilon_0 \widehat{\mathbf{n}}\Delta \cdot \mathbf{E} = \sigma + \sigma', \tag{5.68}$$

where $\sigma' = -\widehat{\mathbf{n}}\Delta \cdot \mathbf{P}$ is the polarization contribution to the surface charge density. Finally we have

$$\widehat{\mathbf{n}}\Delta \cdot \mathbf{B} = 0, \tag{5.69}$$

as before. In the present formulation, all fields are vacuum fields and all charges enter Maxwell's equations on the same level, after all an intuitively rather appealing way to look at fields in matter!

We often find the notation "free" and "bound" charges for ζ and ζ', the latter referring to the charges bound in dielectrics, which can be displaced by polarization. In most cases the distinction is self-evident, but there are cases where one should be careful!

The magnetic field (or rather its rotation) has two sources: time-varying electric fields and currents (including polarization currents). One cannot distinguish the source by considering the magnetic field alone. The electric field can be generated by time-varying magnetic fields or charges, including polarization charges. The resulting electric fields *can* be distinguished: those originating from charges only are rotation free, those from time-varying magnetic fields only are divergence free.

In the polarization description it might be argued that (5.60) is somewhat trivial; the divergence of (5.58) gives $\partial\nabla \cdot \mathbf{B}/\partial t = 0$, stating that all time-varying magnetic fields are divergence free. Then (5.60) says that this holds also for time-stationary magnetic fields.

5.7.1 Method of images

There is an elegant method, the method of images, for determining the fields in cases where regions in space are divided by sharp boundaries between materials with different material relations. The basis of the method is the uniqueness theorem, stating that given a charge distribution in space and given boundary conditions, the problem has one and only one solution (Stratton 1941, Jackson 1975). If we have *a* solution, we have *the* solution! The method is usually applied only for time-stationary problems.

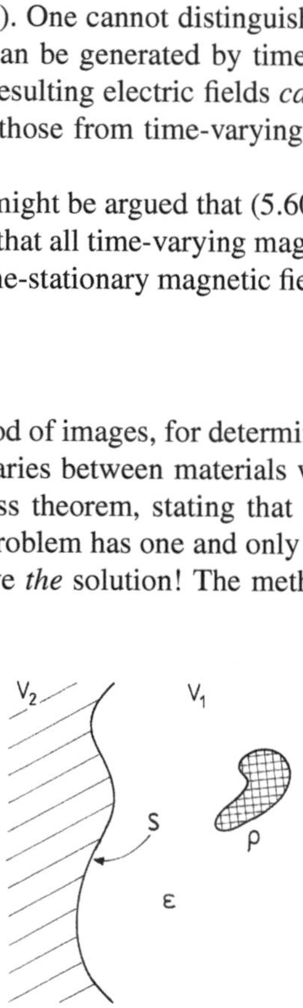

FIGURE 5.9: *Illustration for the method of images. The left-hand side volume V_2 contains a metal, and V_1 a dielectric with coefficient ε, while ζ indicates a distributed charge. S is a separating boundary.*

Basically, the method of images can be summarized by the following steps, referring to Fig. 5.9 for an electrostatic problem:

1. Remove the conductor in V_2.
2. Fill the volume V_2 with a material having the same dielectric constant as V_1.

3. Introduce charges in V_2 in such a way that the boundary conditions on S as well as at infinity are fulfilled. The charges in V_1 must not be changed, nor any new charges introduced there!

By these steps we have ensured that the solution for the electric field satisfies the boundary condition on S and is consistent with the presence of charges and charge densities in V_1. By the uniqueness theorem this is then the correct solution of the problem. In V_2, the solution obtained has of course no relation to our problem, but we are not interested in this region.

The by far simplest, and standard, illustration of the method of images considers a point charge q in a vacuum, facing a grounded infinite plane conductor at a distance ℓ. The question is to determine the electric field outside the conductor in the half-space containing the charge. Physically, the problem is actually quite complicated, if we go into detail. The electric field inside the conductor vanishes, and this is achieved by the accumulation of surface charges on the conductor, the charge being of opposite sign of q. The plane is assumed to be conducting, and we have the surface to be equipotential, here at the ground potential. (If the boundary material is insulating, the problem is different, and actually much more complicated.) All these discussions are made redundant by the method of images; remove the metal plate, place a charge $-q$ at $-\ell$, and the boundary condition $\phi = 0$ at the original plane surface is satisfied at all positions. Determining the electric field from the two point charges is then close to trivial. If attempting to determine the field energy, remember to have the original geometry in mind and not count the energy in the half-space occupied by the conductor. For the present particularly simple case this is close to trivial, but for more complicated geometries, it can be a bit tricky.

- **Exercise:** Let an ideally conducting sphere with radius r be placed in empty space. Now add a point-charge Q at a distance $R > r$ from the center of the sphere. Give the analytical expression for the resulting spatial variation of the electrostatic potential.

- **Exercise:** Consider two disconnected metallic non-deformable objects. Moving the amount of charge Q from one to the other, we find a resulting potential difference, say $\Delta\phi$. The capacitance C is defined as the ratio $C \equiv Q/\Delta\phi$. Let the two metallic objects be two parallel wires with length ℓ and radius a, separated by a distance b, where $\ell \gg a$ and $b \gg a$. Calculate the capacitance of the two wires. Use the method of images to calculate the capacitance of a long wire parallel to a conducting plane. Repeat the analysis for two metallic spheres with radii a, separated by a distance $b \gg a$.

5.8 Lorentz transformations

In the special theory of relativity it is required that the laws of physics shall be the same in any inertial system of reference. By this requirement we are led to the Lorentz transformations. It turned out that to this end a number of modifications were required in what is known as the classical mechanical physics. Electrodynamics as described by Maxwell's equations could, on the other hand, be made Lorentz invariant. By this we mean that it is possible to define relations between the field vectors in two arbitrary inertial systems in such a way that the dynamic equations have the same form in both systems of reference. For details we refer to Stratton (1941), Jackson (1975) or the more popular account by Born (1965), and give here only the expressions relating the field vectors, currents and charges as observed in two different frames of reference, with a relative translational motion with a constant velocity **U**.

$$\mathbf{B}' = \beta\mathbf{B} + \frac{1-\beta}{U^2}\mathbf{U}\cdot\mathbf{B}\mathbf{U} - \frac{\beta}{c^2}\mathbf{U}\times\mathbf{E} \quad (5.70)$$

$$\mathbf{E}' = \beta\mathbf{E} + \frac{1-\beta}{U^2}\mathbf{U}\cdot\mathbf{E}\mathbf{U} + \beta\mathbf{U}\times\mathbf{B} \quad (5.71)$$

$$\mathbf{H}' = \beta\mathbf{H} + \frac{1-\beta}{U^2}\mathbf{U}\cdot\mathbf{H}\mathbf{U} - \beta\mathbf{U}\times\mathbf{D} \quad (5.72)$$

$$\mathbf{D}' = \beta\mathbf{D} + \frac{1-\beta}{U^2}\mathbf{U}\cdot\mathbf{D}\mathbf{U} + \frac{\beta}{c^2}\mathbf{U}\times\mathbf{H} \quad (5.73)$$

and

$$\mathbf{J}' = \beta\mathbf{J} - \frac{1-\beta}{U^2}\mathbf{U}\cdot\mathbf{J}\mathbf{U} - \beta\zeta\mathbf{U} \quad (5.74)$$

$$\zeta' = \beta\zeta - \frac{\beta}{c^2}\mathbf{J}\cdot\mathbf{U} \quad (5.75)$$

$$\mathbf{M}' = \beta\mathbf{M} + \frac{1-\beta}{U^2}\mathbf{U}\cdot\mathbf{M}\mathbf{U} + \beta\mathbf{U}\times\mathbf{P} \quad (5.76)$$

$$\mathbf{P}' = \beta\mathbf{P} + \frac{1-\beta}{U^2}\mathbf{U}\cdot\mathbf{P}\mathbf{U} - \frac{\beta}{c^2}\mathbf{U}\times\mathbf{M}, \quad (5.77)$$

where primed and unprimed quantities refer to the two frames of reference and

$$\beta \equiv \frac{1}{\sqrt{1-(U/c)^2}}.$$

The forms (5.70)–(5.77) are written in a compact vector notation. The meaning of the relations can be more readily understood by noting that, for instance, (5.70) implies that the component of the magnetic field parallel to the velocity vector **U** transforms as $B'_\parallel = B_\parallel$, while for the perpendicular (vector) component we have $\mathbf{B}'_\perp = \beta(\mathbf{B}_\perp - \mathbf{U}\times\mathbf{E}/c^2)$. Similar interpretations can be made for the other relations.

For students of the history of physics it may be interesting to note that the transformations named after Lorentz were discovered independently by several scientists. Thus Larmor (1897) wrote that "*... individual electrons describe corresponding parts of their orbits in times shorter for the [rest] system in the ratio* $(1-U^2/c^2)^{1/2}$", although he later ended up in opposition to the special and general theory of relativity as formulated by Einstein. An interesting short historical summary of early discussions relating to the special theory of relativity is given by Born (1965) mentioning contributions by Voigt as early as 1887 (although this study was still founded on the elastic theory of light), and later by Larmor (around 1900) and Poincaré (around 1905). Contributions from Heaviside (1888), Thomson (1889) and Searle (1896) can also be mentioned.

We shall not dwell on *all* the implications of the set of equations (5.70)–(5.77), but note that, for instance, (5.74) means that currents, **J**, cannot be distinguished from charges in motion, $\zeta\mathbf{U}$. Charges are invariant; an electron will not disappear just because we see it from a moving frame of reference, but *charge densities* will vary because of the Lorentz contraction. If we define the charge density ζ in a rest frame as the number of electrons in a box divided by the volume $\mathcal{L}^3$ of the box, this quantity will be different when seen from a moving frame of reference, since there the length is found to be $\mathcal{L}' = \mathcal{L}/\beta$, due to the Lorentz contraction. The volume becomes $\mathcal{L}^3/\beta$, since only the side of the box parallel to the direction of motion will be contracted.

Since the speed of light is the same in all reference systems, one might naively think that electromagnetic waves should have no Doppler shift. This is not so, but for a somewhat subtle reason: the wavelength of an electromagnetic wave is also subject to the Lorentz contraction and depends on the frame of reference where it is measured. The dispersion relation is $\omega = \pm ck$ in any reference frame, with c being the (same) speed of light, but when the wavelength depends on the frame, so will the corresponding wave-number and hence also the frequency. After some simple calculations (Jackson 1975), we find the Doppler shifted frequency f_1 for an electromagnetic wave propagating in a vacuum with frequency f_0 in the frame of the source, assuming that the relative velocity is U measured along the direction of wave propagation

$$f_1 = f_0\sqrt{\frac{c-U}{c+U}} \equiv f_0\sqrt{\frac{(c-U)^2}{c^2-U^2}} \approx f_0\,(c-U)/c.$$

The approximate expression given (valid for $U \ll c$), coincides with the classical expression we found for instance for sound waves in Section 3.1.4. If we have the electromagnetic wave directed *perpendicular* to the vector **U**, we find a relation

$$f_1 = f_0\sqrt{1-U^2/c^2}$$

where the correction to f_0 is a purely relativistic effect, being $O(U^2/c^2)$ for $U^2 \ll c^2$.

- **Example:** As an illustration consider a point charge Q moving with a constant velocity **U** in empty space. In the rest frame of the charge we find the electric field

  $$\mathbf{E}' = \frac{1}{4\pi\varepsilon_0}\frac{Q}{|\mathbf{r}'|^3}\mathbf{r}'$$

  and $\mathbf{B} = 0$. Assuming **U** to be in the $\widehat{x}'$-direction, we have $x' = \beta(x-Ut)$, $y' = y$ and $z' = z$. We find

  $$|\mathbf{r}'|^2 \equiv x'^2 + y'^2 + z'^2 = R^2\left(1 + \frac{U^2}{c^2-U^2}\cos^2\Theta\right)$$

  where $\mathbf{R} \equiv \{x-Ut, y, z\}$, and Θ is the angle between the x-axis and the vector $\{x,y,z\}$. We then have

  $$\mathbf{E} = \frac{\beta}{4\pi\varepsilon_0}\frac{Q}{|\mathbf{r}'|^3}\mathbf{R}$$

  which can be rewritten as

  $$\mathbf{E} = \frac{Q}{4\pi\varepsilon_0}\frac{1-(U/c)^2}{(1-(U/c)^2\sin^2\Theta)^{3/2}}\frac{\mathbf{R}}{R^3}.$$

  Finally, we can determine also the magnetic field as

  $$\begin{aligned}\mathbf{B} &= \frac{\beta}{c^2}\mathbf{U}\times\mathbf{E}' = \frac{1}{c^2}\mathbf{U}\times\mathbf{E}\\ &= \frac{Q\mu_0}{4\pi}\frac{1-(U/c)^2}{(1-(U/c)^2\sin^2\Theta)^{3/2}}\frac{\mathbf{U}\times\mathbf{R}}{R^3},\end{aligned}$$

  completing the analysis.

- **Exercise:** Demonstrate that $\mathbf{E}\cdot\mathbf{B}$ is an invariant with respect to relativistic transformations.

For most cases we will be content with the classical limit $U^2 \ll c^2$, and have $\beta \approx 1$. In this limit equations (5.70), (5.73) (5.75) and (5.77) give rather trivial identities. On the other hand (5.71), (5.72) and (5.76) imply that even in the classical limit the fields can be different in different frames of reference. Consider, for instance, (5.71) in the limit $U^2 \ll c^2$. We then have

$$\mathbf{B}' = \mathbf{B}, \tag{5.78}$$

while in the same limit

$$\mathbf{E}' = \mathbf{E} + \mathbf{U} \times \mathbf{B}. \tag{5.79}$$

At first sight this result can appear strange! For steady state, the only sources for electric fields are charges, so if we consider a case where we have in one frame of reference a magnetic field **B** produced by currents in conductors and no electric fields, i.e., $\mathbf{E} = 0$, then how come we find $\mathbf{E}' \neq 0$ in a moving frame? Seemingly there are charges in one frame of reference and not in another! To resolve this seeming paradox we follow Feynman et al. (1963) or Sears and Brehme (1968), and consider a case where the magnetic field is generated by currents along an infinitely long straight wire. To be specific, we can assume in the rest frame a charged hollow plastic rod with a uniform charge density $\zeta^- = -\zeta$, and inside it there is a plastic rod with uniform charge density $\zeta^+ = \zeta$ moving with velocity U, both charge densities referring to the fixed laboratory frame. By this construction we ensure that all charges on the individual rods can be assumed to have the same velocity, at least ideally. We let the area of a cross section be $\mathcal{A}$ of both rods for simplicity. An observer standing at a distance r from the symmetry axis of the rods (i.e., not *inside* them) in the fixed frame of reference will therefore not find any net charge density, but a net current of magnitude $I = U\zeta\mathcal{A}$ and a magnetic field

$$B_\theta = \frac{\mu_0 \zeta U \mathcal{A}}{2\pi r} \tag{5.80}$$

in the azimuthal direction. The other magnetic field components are vanishing. If we place an electron at rest at any position (ignoring gravity), the net Lorentz force $\mathbf{F} = -e(\mathbf{E} + \mathbf{u} \times \mathbf{B})$ will be vanishing, since there is no electric field, and the electron velocity $\mathbf{u} = 0$. If we now change the frame of reference to one moving, for instance, with velocity U, we have a current and a magnetic field which in the classical limit are the same as before, but in this moving frame the selected electron is moving with velocity $-U$. We have now a contribution to the force on the electron $-e\mathbf{U} \times \mathbf{B}$, just by the change of frame. Since a simple translation cannot produce a net force, we *must* have another force contribution to balance it. (A different argument states that the electron is seen as moving with constant velocity vector, so there cannot be any acceleration and thus no force.) This force must be of electro- or magneto-static origin, so we write it as $e\mathbf{E}'$. The existence of this force is exactly what the transformation expressions guarantee, but we still need to account for the sources of these electric fields. The essential point in the arguments is the observation that the charge density depends on the frame of reference (Sears & Brehme 1968). If in a rest frame we have a charge density $\zeta \equiv Q/(\mathcal{A}\mathcal{L})$, we observe $\zeta' = \zeta\beta$ in a moving frame. When the negatively charged rod is seen moving we thus observe its charge density to be $\zeta^{-\prime} = -\zeta\beta$ when seen from the new rest frame. On the other hand, the rod with positive charge density ζ which was moving in the original frame of reference is now at rest and has a charge density $\zeta^{+\prime} = \zeta/\beta$. The net charge density on the two charged rods in this new frame of reference is therefore

$$\zeta' \equiv \zeta^{+\prime} + \zeta^{-\prime} = -\zeta \frac{U^2/c^2}{\sqrt{1 - U^2/c^2}}. \tag{5.81}$$

This is the exact result obtained by the Lorentz transformations. Since we used the classical limit in (5.79), we retain only the lowest order term in an expansion in U^2/c^2 and find $\zeta' \approx -\zeta U^2/c^2$. The composed system of the two charged rods appears to have a net charge density in this new frame of reference, as stated. This charge density produces a radial electric field

$$E_r = -\frac{\zeta'\mathcal{A}}{2\pi\varepsilon_0 r} \approx -\frac{\zeta\mathcal{A}U^2/c^2}{2\pi\varepsilon_0 r} = -\frac{\mu_0 \zeta \mathcal{A} U^2}{2\pi r} \tag{5.82}$$

with $c^2 = 1/\mu_0\varepsilon_0$. This result is identical to the electric field $\mathbf{E} = -\mathbf{U} \times \mathbf{B}$ obtained by (5.80), and it gives a force $-eE_r$, which exactly balances eUB_θ, as expected. The number of charges is not changed by the transformation, but because of the Lorentz contraction, the length of the box confining the charges will change, and for the observer consequently the charge *density* will change.

We have found that the observed charge density of a system will in general depend on the frame of reference, and it does not really make sense to distinguish the electric and magnetic origins of a force either! One obvious question arises; in which frame of reference do we find the electric field to be vanishing, provided such a frame exists at all? This question does not have any general answer (Good 1999), but in the simple example discussed before we can argue by considering the entire history of the arrangement. Assume that the solid inner and hollow outer rods are charged while at rest in such a way that the charges per unit length are the same in magnitude, but of opposite sign on the two parts. The overall system has no current and is charge neutral, no matter which frame of reference we use. To retain the net charge neutrality we can then start the current by letting the inner rod move with velocity $U/2$ to the left, and the outer hollow part with velocity $-U/2$ to the right. This produces the required current, and by symmetry arguments it retains the overall charge neutrality, resulting in a vanishing electric field in this particular frame of reference. The symmetry is broken when we go to a moving frame of reference, and we will find a net electric field there. Note that the system we discussed before does *not* correspond to this model, since in that case the two rods would have different charge densities if brought to rest in the same frame of reference!

- **Exercise:** Use (5.81) and give a derivation of the results (5.70) and (5.71), without making the expansion to lowest order in U^2/c^2.

The foregoing discussion was explicitly restricted to the transforms of special relativity, i.e., to coordinate systems moving with constant velocity. We are interested in handling *rotating* coordinate systems as well. This problem addresses non-inertial frames of reference, and this is much more difficult. We can state here that transforms like (5.78) and (5.79) remain applicable as long as the local velocity $|U| \ll c$. There are certain intriguing aspects of electrodynamics in rotating systems, but these fall outside the present treatise. Discussions of some of these problems are given by, for instance, Schiff (1939), Trocheris (1949), Irvine (1964), Corum (1980), with a summary by Parks (2004).

5.9 Dielectric properties

The relations between the electric field **E** and the electric displacement **D** in matter are determined by the physical properties of the media in question. This relation is usually expressed in terms of a dielectric constant, ε. More generally the dielectric properties may depend on frequency ω and wave-number **k** of the electric field, implying $\varepsilon = \varepsilon(\omega, \mathbf{k})$.

5.9.1 Simple media

For simple media we have in terms of constant permittivities and permeabilities

$$\mathbf{P} = (\varepsilon - \varepsilon_0)\mathbf{E} \equiv (\varepsilon_r - 1)\varepsilon_0\mathbf{E},$$

and

$$\mathbf{M} = (\mu - \mu_0)\mathbf{H}/\mu_0 \equiv (\mu_r - 1)\mathbf{H},$$

where ε and μ are assumed to be constants, independent of frequency, in particular. More generally ε and μ may be constant tensors. In terms of the relative permittivity ε_r and relative permeability μ_r introduced before we have $\varepsilon = \varepsilon_r\varepsilon_0$ and $\mu = \mu_r\mu_0$. With the assumed scalar relations we can also write $\varepsilon_r = 1 + P/(\varepsilon_0 E)$, which is a useful relation. It is sometimes convenient to use the susceptibility parameter $\chi = \varepsilon/\varepsilon_0 - 1$, and similarly for the magnetic permeability. We have, for instance,

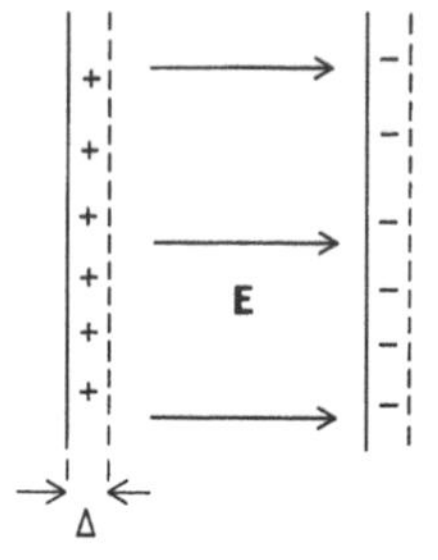

FIGURE 5.10: *A homogeneous plasma slab, where the electrons have been displaced a distance* Δ *in the direction perpendicular to the slab surface. The resulting surface charges are indicated with* $+$ *and* $-$ *signs.*

$\mathbf{P} = \varepsilon_0 \chi \mathbf{E}$, i.e., the susceptibility χ is a measure for the "polarizability" of the medium. For a vacuum evidently $\chi = 0$. For simple media, we might have $\varepsilon = \varepsilon(\omega)$ and $\mu = \mu(\omega)$ depending on frequency, but not on wave-number.

- **Example:** We derive the dielectric function associated with a cold electron gas, in a background of stationary heavy ions. This simple model serves as good, illustrative, example for many cases. In Fig. 5.10 we show a slab of plasma with uniform density n_0 inside the slab. The electron component is displaced by a distance Δ with respect to the singly charged ions in the direction perpendicular to the surface of the slab. The displacement is taken to be in the $\widehat{\mathbf{x}}$-direction.

  The electric dipole moment density is $\mathbf{P} = -en_0\Delta\widehat{\mathbf{x}}$, but since the plasma density is uniform, there is no net charge separation *inside* the slab, i.e., $\nabla \cdot \mathbf{P} = 0$ there. We have, on the other hand, surface charges $\sigma' = \pm en_0\Delta$, which give rise to electric fields, with $E = \sigma'/\varepsilon_0 = \Delta en_0/\varepsilon_0$; see Fig. 5.10.

  The equation of motion for the slab is $n_0 m d^2\Delta/dt^2 = -en_0 E$, with m being the electron mass, the heavy ions assumed immobile. We ignore pressure forces by assuming that the electron temperature vanishes, $T_e = 0$. Assuming harmonic oscillations $\Delta \sim \exp(-i\omega t)$, we find $E = \omega^2(m/e)\Delta$. Using the definition $\varepsilon_r \equiv 1 + P/(\varepsilon_0 E)$ we find the dielectric function for the plasma to be $\varepsilon_r = 1 - (\omega_{pe}/\omega)^2$, with $\omega_{pe} \equiv \sqrt{e^2 n_0/(\varepsilon_0 m)}$ being the electron plasma frequency. In the absence of external charges we have $\mathbf{D} = \mathbf{P} + \varepsilon_0\mathbf{E} = 0$ or $\varepsilon_0\varepsilon_r\mathbf{E} = 0$. Requiring $E \neq 0$ for a nontrivial solution we must have $\varepsilon_r = 0$, i.e., $\omega^2 = \omega_{pe}^2$. In this case there are no external forces to maintain the surface charges, and the system is "freely oscillating", with the natural frequency of oscillation $\omega = \pm\omega_{pe}$.

5.9.2 Material relations

Generally, we consider here a Volterra type, nonlocal, material relation which relates the electric displacement $\mathbf{D}(\mathbf{r},t)$ and the electric field $\mathbf{E}(\mathbf{r},t)$ where also the possibility of nonlocal spatial dependence is allowed

$$\mathbf{D}(\mathbf{r},t) = p_0\mathbf{E}(\mathbf{r},t) + \iiint_{-\infty}^{\infty}\int_{-\infty}^{t} p(\mathbf{r}-\xi, t-\tau)\mathbf{E}(\xi,\tau)\, d\tau\, d^3\xi, \tag{5.83}$$

with ξ being the integration vector. The first term in (5.83) accounts for the instantaneous response, with p_0 =const. More generally, p can be taken to be a tensor in the case of anisotropic media. The electric displacement at a position $\mathbf{r}$ at a time t depends generally not only on the electric field at that position, but also on $\mathbf{E}$ in neighboring positions with a spatially varying weight given through p. In addition, the actual value of $\mathbf{D}$ depends in general also on the "history" of the electric field in the medium. The upper integration limit t ensures that $\mathbf{D}$ depends only on the past and not future values of $\mathbf{E}$, i.e., the response is *causal*. The special case of a medium with temporal, but no spatial, dispersion corresponds to $p(\mathbf{r}-\xi, t-\tau) = \delta(\mathbf{r}-\xi)p(t-\tau)$. In this simple case the response is local in space, reducing (5.83) to the simpler form

$$\mathbf{D}(t) = p_0\mathbf{E}(t) + \int_0^\infty p(\tau)\mathbf{E}(t-\tau)d\tau, \tag{5.84}$$

where now the spatial variable is redundant, and therefore omitted. Note the change in the temporal integration variable.

Fourier transformation with respect to the spatial variable of the general formulation (5.83) gives

$$\mathbf{D}(\mathbf{k},t) = p_0\mathbf{E}(\mathbf{k},t) + \int_{-\infty}^{t} p(\mathbf{k},t-\tau)\mathbf{E}(\mathbf{k},\tau)d\tau. \tag{5.85}$$

Causality of the response is also here expressed by the upper limit t of integration in (5.85). (Some authors prefer to let the integration start at some finite initial time.) In other words, $\mathbf{D}(\mathbf{k},t)$ depends only on the past, i.e., on $\mathbf{E}(\mathbf{k},t)$ for times prior to the observation time t. The *spatial* variable does not give rise to similar considerations, so $\mathbf{k}$ enters only as a label in the following. For simple media, without spatial dispersion, the wave-number dependence is trivial and can be omitted altogether. Fourier transformation with respect to also the temporal variable in (5.84) or (5.85) gives

$$\begin{aligned}\mathbf{D}(\omega,\mathbf{k}) &= \mathbf{E}(\omega,\mathbf{k})\left(p_0 + \int_0^\infty p(\tau,\mathbf{k})e^{i\omega\tau}d\tau\right) \\ &\equiv \mathbf{E}(\omega,\mathbf{k})\varepsilon_0\left(1+\chi(\omega,\mathbf{k})\right),\end{aligned} \tag{5.86}$$

where now the frequency dependent susceptibility

$$\chi(\omega,\mathbf{k}) = \int_0^\infty p(\tau,\mathbf{k})\exp(i\omega\tau)d\tau$$

of the medium is introduced, where for most relevant cases we have $\chi(\omega\to\infty,\mathbf{k})\to 0$. It may be instructive to see the analysis in a slightly different way: assume that we have a harmonically varying "cause", here the electric field $\mathbf{E}(\mathbf{r},t) = \Re\{\mathbf{E}_0\exp(-i\omega t + i\mathbf{k}\cdot\mathbf{r})\}$, and a harmonically varying linear response, here $\mathbf{D}(\mathbf{r},t) = \Re\{\mathbf{D}_0\exp(-i\omega t + i\mathbf{k}\cdot\mathbf{r})\}$, with complex $\mathbf{E}_0$ and $\mathbf{D}_0$. With the given simplifications stated from the outset we then have a complex scalar response function defined as $\varepsilon(\omega,\mathbf{k}) \equiv D_0/E_0$. Inserting the assumed harmonic variations of $\mathbf{E}(\mathbf{r},t)$ and $\mathbf{D}(\mathbf{r},t)$ in (5.84) or (5.85) gives (5.86). More generally we have a tensor relation $\mathbf{D}(\mathbf{k},\omega) = \varepsilon_0\underline{\varepsilon}_r(\mathbf{k},\omega)\cdot\mathbf{E}(\mathbf{k},\omega)$ where now $\underline{\varepsilon}_r$ is a tensor. Note the change in result if we assume another convention for the plane wave, as, for instance, $\mathbf{E}(\mathbf{r},t) = \Re\{\mathbf{E}_0\exp(i\omega t - i\mathbf{k}\cdot\mathbf{r})\}$, which has a real part that is indistinguishable from the previous definition, so it is not *physically* distinguishable!

The relative dielectric function is defined as $\varepsilon_r(\omega,\mathbf{k}) \equiv 1+\chi(\mathbf{k},\omega)$ and is in general complex. We will use the notation $\varepsilon_1(\omega,\mathbf{k}) = \Re\{\varepsilon_r(\omega,\mathbf{k})\}$ and $\varepsilon_2(\omega,\mathbf{k}) = \Im\{\varepsilon_r(\omega,\mathbf{k})\}$ for the real and imaginary parts, respectively.

Physically, we argue that all media become transparent, or "vacuum-like", for infinite frequencies (X-rays or gamma-rays go unaffected through almost everything), and therefore the constant ε_0 is determined by requiring that $\chi(\omega,\mathbf{k})\to 0$ for $\omega\to\infty$. In particular, the constant ε_0 will be the response function for a vacuum, where $\mathbf{D} = \varepsilon_0\mathbf{E}$. As an alternative notation one might see ε_0 included in the real part $\varepsilon_1(\omega,\mathbf{k})$. Here we retain the separation between ε_0 and ε_r. A relation of the form (5.83) can be written for any spatially and temporally varying "cause" (here $\mathbf{E}$) and corresponding linear "response" (here $\mathbf{D}$) with $\varepsilon(\omega,\mathbf{k})$ being the Fourier transform of the transfer function.

The real and imaginary parts $\varepsilon_1(\omega,\mathbf{k})$ and $\varepsilon_2(\omega,\mathbf{k})$ are related by the Kronig-Kramers relations, here written in terms of the real, $\chi_1 \equiv \Re\{\chi\}$, and imaginary parts, $\chi_2 \equiv \Im\{\chi\}$, of the susceptibility

$$\chi_2(\omega,\mathbf{k}) = -\frac{1}{\pi}P\int_{-\infty}^{\infty}\frac{\chi_1(\omega_1,\mathbf{k})}{\omega_1-\omega}d\omega_1,$$

$$\chi_1(\omega,\mathbf{k}) = \frac{1}{\pi}P\int_{-\infty}^{\infty}\frac{\chi_2(\omega_1,\mathbf{k})}{\omega_1-\omega}d\omega_1,$$

where $P\int$ denotes the principal value of the integral. The relations are valid for any choice of $\mathbf{k}$. These relations are a consequence of causality, or if you like, the properties of complex functions. They imply that we cannot just take any two functions, one real and one imaginary, add them together and call it a dielectric function. Think about this for a moment: it is something really amazing that a theorem from mathematical analysis of a complex function implies a constraint on the physical properties of continuous media! The integral transform involved in the Kronig-Kramers relations is called the Hilbert transform. The implications of the Kronig-Kramers relations have been elaborated in some detail, for instance, also by the present author (Pécseli 2000). Details can be found in several textbooks (Champeney 1973, Jackson 1975) and will not be discussed further here. The Kronig-Kramers relations can be obtained under very general assumptions regarding transfer functions and they have found applications for many physical systems, for instance, also for servo-mechanisms (Bode 1956) and for electric circuits.

- **Exercise:** Start with a general causal relation between polarization and electric field

$$P_i = \varepsilon_0\int_{-\infty}^{\infty}\int_0^t \chi_{ij}(\mathbf{r},\mathbf{r}',t,\tau)E_j(\mathbf{r}',\tau)d\tau d\mathbf{r}',$$

  written in tensor notation for generality, implying summation over repeated indices. Demonstrate that you obtain (5.83) by assuming the medium to be temporary stationary and spatially homogeneous (although not necessarily isotropic) (Yeh & Liu 1972).

- **Exercise:** Demonstrate the consequences of the reality conditions (i.e., real $\mathbf{E}(\mathbf{k},t)$ and $\mathbf{D}(\mathbf{k},t)$) on $\varepsilon(\omega,\mathbf{k})$ and $\chi(\omega,\mathbf{k})$. Rewrite the Kronig-Kramers relations when these conditions are imposed.

5.9.3 Definition of the dielectric function

Fourier transforming Poisson's equation (5.59) we can use it for defining the dielectric function. We take here an electric field derived from a scalar potential, $\mathbf{E} = -\nabla\phi$. By this we restrict the result to electrostatic fields, which have particular interest for plasma wave phenomena. From (5.59) we have

$$i\varepsilon_0\mathbf{k}\cdot\mathbf{E} = \varepsilon_0 k^2\phi(\omega,\mathbf{k}) = \zeta(\omega,\mathbf{k}) + \zeta'(\omega,\mathbf{k}), \tag{5.87}$$

where we again distinguish the density of free charges ζ, and polarization charges ζ', just as in Section 5.7.

We here introduce as a definition of the $(\omega,\mathbf{k})$-depending scalar dielectric function $\varepsilon(\omega,\mathbf{k})$

$$k^2\phi(\omega,\mathbf{k}) = \frac{\zeta(\omega,\mathbf{k})}{\varepsilon(\mathbf{k},\omega)}. \tag{5.88}$$

This relation corresponds to $\nabla\cdot\mathbf{D} = \nabla\cdot\varepsilon\mathbf{E} = \zeta$, with $\mathbf{E} = -\nabla\phi$. Solving for $\varepsilon(\omega,\mathbf{k})$, we eliminate ζ from (5.87) and (5.88) to find

$$\varepsilon(\omega,\mathbf{k}) = \varepsilon_0 - \frac{\zeta'(\omega,\mathbf{k})}{k^2\phi(\omega,\mathbf{k})}, \tag{5.89}$$

which can be taken as the definition of the (longitudinal) dielectric function $\varepsilon(\omega,\mathbf{k}) \equiv \varepsilon_r(\omega,\mathbf{k})\varepsilon_0$ expressed in terms of the electrostatic potential and the charges ζ' it induces.

It is formally possible to define a dielectric function separately for electrons and for ions by

$$\varepsilon_{e,i}(\omega,\mathbf{k}) = \varepsilon_0 - \frac{\zeta'_{e,i}(\omega,\mathbf{k})}{k^2\phi(\omega,\mathbf{k})},$$

where $\zeta'_{e,i}$ are the induced electron and ion charges, indicated by the proper subscript. Since $\zeta' = \zeta'_e + \zeta'_i$ in (5.89), it is trivial to obtain the *total* dielectric function as

$$\varepsilon(\omega,\mathbf{k}) = \varepsilon_0 + (\varepsilon_e(\mathbf{k},\omega) - \varepsilon_0) + (\varepsilon_i(\omega,\mathbf{k}) - \varepsilon_0),$$

or more generally for the contribution of N different species, electrons and different ions, for instance, as

$$\varepsilon(\omega,\mathbf{k}) = \varepsilon_0 + \sum_j^N (\varepsilon_j(\omega,\mathbf{k}) - \varepsilon_0). \tag{5.90}$$

The relation (5.90) turns out to be most useful in studies of solids and other media. Recall that a vacuum contributes with $\varepsilon = \varepsilon_0$.

- **Example:** Assume singly charged, massive immobile ions with density n_0, and isothermally Boltzmann distributed electrons, where $n_e = n_0 \exp(e\phi(r)/\kappa T_e) \approx n_0(1 + e\phi/\kappa T_e)$. The charge distribution associated with the ions is then en_0, while for the electrons it is $-en_e$. The linear potential response to an externally introduced fixed point charge q_0 is then $k^2\phi = q_0/\varepsilon_0\varepsilon_r$, with

$$\begin{aligned}\varepsilon_r &= 1 - \frac{e(n_0 - n_e)}{\varepsilon_0 k^2\phi} \\ &= 1 + \frac{en_0 e\phi/\kappa T_e}{\varepsilon_0 k^2 \phi} \\ &= 1 + \frac{1}{(k\lambda_D)^2},\end{aligned} \tag{5.91}$$

where $\lambda_D \equiv \sqrt{\varepsilon_0 \kappa T_e / e^2 n_0}$ is the Debye length for the electrons. By an inverse Fourier transform (left to the reader as an exercise) we find the result (5.91) corresponds to an exponential shielding of the charge q_0.

The discussion outlined before explicitly referred to a scalar dielectric function. For a tensor form $\underline{\varepsilon}(\omega,\mathbf{k})$ we have

$$\mathbf{k}\cdot\underline{\varepsilon}\cdot\mathbf{k}\phi = \zeta$$

replacing (5.88). By use of this expression we find

$$\varepsilon_0 - \mathbf{k}\cdot\underline{\varepsilon}(\omega,\mathbf{k})\cdot\mathbf{k}\frac{1}{k^2} = \frac{\zeta'(\omega,\mathbf{k})}{k^2\phi(\omega,\mathbf{k})}, \tag{5.92}$$

which replaces (5.89). The relations given here are trivially rewritten in terms of currents, since $-i\omega\zeta' + i\mathbf{k}\cdot\mathbf{J}' = 0$ by the continuity equation for charges. This gives

$$\varepsilon_0 - \mathbf{k}\cdot\underline{\varepsilon}(\omega,\mathbf{k})\cdot\mathbf{k}\frac{1}{k^2} = \frac{\mathbf{k}\cdot\mathbf{J}'(\omega,\mathbf{k})}{k^2\omega\phi(\omega,\mathbf{k})},$$

which can be useful sometimes.

Being particularly interested in the electrostatic properties of media, we seek an expression relating the dielectric function to charges and potentials in the media. By definition we consider fluctuations to be electrostatic if the electric field can be derived from a potential, $\mathbf{E} = -\nabla\phi$, and that no magnetic component is associated with the fluctuations. This implies that, for this case, $\mathbf{J} + \partial\mathbf{D}/\partial t = 0$ to give $\mathbf{H} = 0$ in (5.5).

5.10 Energy density in dielectrics

When electric and magnetic fields are applied to dielectric media, usually work is done to induce the electric and magnetic polarizations. The energy density with matter present will therefore differ from what we would have with the same fields in a vacuum (Landau et al. 1984).

5.10.1 Simple media

For simple media, and a vacuum in particular, the energy density of the electric field is easily obtained from Poynting's identity as $W_e = \frac{1}{2}\varepsilon_0\varepsilon_r E^2(t)$ while for the magnetic field it is $W_m = \frac{1}{2}B^2(t)/\mu_0\mu_r$; see (5.32) and (5.33). For a monochromatic wave as $E(t) = E_0\cos(\omega t)$ with E_0 real, the energy density averaged over one period of oscillations is $\overline{W_e} = \frac{1}{4}\varepsilon_0\varepsilon_r E_0^2$ for the electric field, and similarly for the magnetic component. The generalization of the energy density in electric and magnetic fields is not trivial if we consider a harmonically varying electric field in a temporally dispersive medium, where $\varepsilon_r = \varepsilon_r(\omega)$, as discussed in Section 5.10.2.

5.10.2 Real dielectric functions – dispersive media

Consider first the case where $\varepsilon = \varepsilon_0\varepsilon_r(\omega)$ is a real function, at least in the frequency range of interest. We first attempt to use the expression (5.32) in order to determine the energy density for a strictly harmonically varying electric field. Using the notation of Section 3.1.1, we write $\Re\{\mathbf{E}_0 e^{-i\omega_0 t}\}$, with $\Re\{\}$ again denoting the real part of the expression in the brackets $\{\}$. There is here no ambiguity in defining the frequency, and using $\mathbf{D} = \varepsilon_0\varepsilon_r(\omega_0)\mathbf{E}$ we expect to have the time derivative of the electric field energy density to be given as $\partial W_e/\partial t = \mathbf{E}\cdot\partial\mathbf{D}/\partial t = \frac{1}{2}\varepsilon_0\varepsilon_r(\omega_0)\,\partial E^2/\partial t$, giving $W_e = \frac{1}{2}\varepsilon_0\varepsilon_r(\omega_0)E^2 + \text{const}$. The arbitrary integration constant cannot here be determined by setting $E = 0$ at $t \to -\infty$ since a strictly harmonic time variation was assumed at all times also at $t \to -\infty$. Consequently, the energy density is not well defined by this simple procedure. The problem can be remedied by using the form $\mathbf{E}(t) = \frac{1}{2}\mathbf{E}_0(t)e^{-i\omega_0 t} + \text{c.c.}$ for the electric field, where $\mathbf{E}_0(t)$ is slowly varying and $\mathbf{E}_0(t \to -\infty) \to 0$, while c.c. denotes "complex conjugate". In effect, we assume that the electric field is being slowly "turned on". In this case, however, the electric field is no longer strictly harmonically varying, and we cannot use $\varepsilon_r(\omega_0)$ right away in the expression for the energy density in the electric field, since the frequency ω_0 is no longer uniquely defined. In order to take into account the frequency broadening we write out the details in the expression for the electric displacement as

$$\mathbf{D}(t) = \varepsilon_0\int_{-\infty}^{\infty}\varepsilon_r(\omega)\mathbf{E}(\omega)e^{-i\omega t}d\omega, \tag{5.93}$$

giving

$$\begin{aligned}\frac{\partial\mathbf{D}}{\partial t} &= -i\varepsilon_0\int_{-\infty}^{\infty}\omega\varepsilon_r(\omega)\mathbf{E}(\omega)e^{-i\omega t}d\omega\\ &\approx -i\varepsilon_0\int_{-\infty}^{\infty}\left(\omega_0\,\varepsilon_r(\omega_0) + \left.\frac{d\omega\varepsilon_r(\omega)}{d\omega}\right|_{\omega=\omega_0}(\omega-\omega_0)\right)\mathbf{E}(\omega)e^{-i\omega t}d\omega,\end{aligned} \tag{5.94}$$

where the latter approximation is justified for $\mathbf{E}(\omega)$ being sharply peaked around $\omega = \omega_0$, which is the case when $\mathbf{E}_0(t)$ is slowly varying. We assume $d\omega\varepsilon_r(\omega)/d\omega$ to be small in the vicinity of ω_0. The spatial variable is understood, and need not be made explicit in the present context. As a sort of reminder, we retain the partial derivative $\partial/\partial t$ rather than d/dt, noting that in general also the energy density will be spatially varying. We find

$$\begin{aligned}\frac{\partial \mathbf{D}}{\partial t} &\approx -i\varepsilon_0\omega_0\varepsilon_r(\omega_0)\mathbf{E}(t) - i\varepsilon_0 \left.\frac{d\omega\varepsilon_r(\omega)}{d\omega}\right|_{\omega=\omega_0} \int_{-\infty}^{\infty} (\omega-\omega_0)\mathbf{E}(\omega)e^{-i\omega t}d\omega \\ &= -i\varepsilon_0\omega_0\varepsilon_r(\omega_0)\mathbf{E}(t) + \varepsilon_0 \left.\frac{d\omega\varepsilon_r(\omega)}{d\omega}\right|_{\omega=\omega_0} \left(i\omega_0\mathbf{E}(t) + \frac{\partial \mathbf{E}(t)}{\partial t}\right). \end{aligned} \tag{5.95}$$

With $\partial \mathbf{E}/\partial t = -i\omega_0\mathbf{E}(t) + \exp(-i\omega_0 t)\partial \mathbf{E}_0/\partial t$, we finally have

$$\frac{\partial \mathbf{D}}{\partial t} = -i\varepsilon_0\omega_0\varepsilon_r(\omega_0)\mathbf{E}_0(t)e^{-i\omega_0 t} + \varepsilon_0 \left.\frac{d\omega\varepsilon_r(\omega)}{d\omega}\right|_{\omega=\omega_0} e^{-i\omega_0 t}\frac{\partial \mathbf{E}_0(t)}{\partial t}. \tag{5.96}$$

Using this expression, $\mathbf{E}\cdot\partial\mathbf{D}/\partial t$ is readily obtained. It is readily seen that the result will be fluctuating in time and only its time-average over a period of oscillation, indicated by an overline, is really interesting. Using

$$\Re\{\mathbf{E}(t)\} = \frac{1}{2}\mathbf{E}_0(t)e^{-i\omega_0 t} + \text{c.c.} \tag{5.97}$$

where c.c. again denotes "complex conjugate", we have

$$\begin{aligned}\overline{\Re\{\mathbf{E}\}\cdot\frac{\partial\Re\{\mathbf{D}\}}{\partial t}} &= \frac{1}{4}\overline{(\mathbf{E}+\mathbf{E}^*)\cdot\left(\frac{\partial \mathbf{D}}{\partial t}+\frac{\partial \mathbf{D}^*}{\partial t}\right)} \\ &= \frac{1}{4}\left(\mathbf{E}\cdot\frac{\partial \mathbf{D}^*}{\partial t} + \mathbf{E}^*\cdot\frac{\partial \mathbf{D}}{\partial t}\right) \\ &= \frac{1}{4}\varepsilon_0\left.\frac{d\omega\varepsilon_r(\omega)}{d\omega}\right|_{\omega=\omega_0}\frac{\partial|\mathbf{E}_0(t)|^2}{\partial t}, \end{aligned} \tag{5.98}$$

where it was used that the averages $\overline{\mathbf{E}\cdot\partial\mathbf{D}/\partial t}$ and $\overline{\mathbf{E}^*\cdot\partial\mathbf{D}^*/\partial t}$ vanish. The result (5.98) inserted into (5.32) finally gives the energy density for the electric field

$$\overline{W_e} = \frac{1}{4}\varepsilon_0\left.\frac{d\omega\varepsilon_r(\omega)}{d\omega}\right|_{\omega=\omega_0}|\mathbf{E}_0(t)|^2, \tag{5.99}$$

recalling that $\mathbf{E}_0$ can be complex, in general. Entirely similar arguments can be applied for the energy density in the magnetic field, if relevant. In (5.99) there is no need to have an overline on $\mathbf{E}_0(t)$, since it is slowly varying within a period of oscillation, $2\pi/\omega_0$. If we take the particularly simple case where $\varepsilon_r(\omega)$ is a constant, independent of ω, we find as expected the standard result $\frac{1}{4}\varepsilon_0\varepsilon_r|\mathbf{E}_0(t)|^2$, with a similar term for the magnetic component. Note that the numerical factor $\frac{1}{4}$ appears here, rather than the $\frac{1}{2}$ often seen, because of the present definition of the complex electric field amplitude, $\mathbf{E}_0$. (If we define $\Re\{\mathbf{E}(t)\} = \mathbf{E}_0(t)e^{-i\omega_0 t} + \text{c.c.}$ without the factor $\frac{1}{2}$, we obtain the energy density $\frac{1}{2}\varepsilon_0\varepsilon_r|\mathbf{E}_0(t)|^2$.)

- **Exercise:** For the foregoing analysis we had no need to discuss the relation between the electric displacement $\mathbf{D}$ and $\mathbf{E}$ for the case where $\mathbf{E} = \mathbf{E}_0(t)\exp(-i\omega_0 t)$, with $\mathbf{E}_0(t)$ slowly varying. It is left to the reader as an exercise to discuss this relation.

5.10.2.1 Electrostatic waves

Physically, the case examined in this section refers to the situation where we have a dielectric placed between, e.g., two plates of a capacitor and an oscillating electric field is applied externally. This is, however, not really the case which is interesting in a plasma. In the present context we would like to discuss freely propagating plasma waves, i.e., some which satisfy Poisson's equation $\nabla\cdot\mathbf{D} = \zeta(t)$

without an external charge ζ to maintain the fluctuations. We can assume such charges to be present for a while, driving up the wave amplitude, but when a predetermined amplitude **E** has been achieved these external generators are supposed to be switched off to let the waves propagate freely. Fourier transforming equation (5.7) with respect to time this implies $\varepsilon_0\varepsilon_r(\omega)\nabla\cdot\mathbf{E}=0$. For electrostatic waves where $\nabla\cdot\mathbf{E}\neq 0$, this relation has nontrivial solutions only when $\varepsilon_r(\omega)=0$, which determines ω_0. Using (5.99) for this case we find the energy density to be

$$\overline{W_e}=\frac{1}{4}\varepsilon_0\,\omega_0\,\frac{d\varepsilon_r(\omega)}{d\omega}\bigg|_{\omega=\omega_0}|\mathbf{E}_0(t)|^2. \tag{5.100}$$

- **Example:** For the particularly simple case with $\varepsilon_r(\omega)=1-\omega_{pe}^2/\omega^2$, the foregoing arguments give the standard cold plasma oscillations with $\omega^2=\omega_{pe}^2$. Calculating the electrostatic energy density associated with these waves we find $\overline{W_e}=\frac{1}{2}\varepsilon_0|\mathbf{E}|^2$, i.e., twice the energy density we would have with the same electric field in a vacuum. This is fair enough; the wave energy includes the sloshing motion of the electrons in the electric field, and in this simple model with temperature effects ignored, the electrons have, on average, exactly the same kinetic energy density as the density of the electric field energy. Hence the resulting expression for the energy density associated with the wave is twice the energy density of the electric field itself. The time-averaged kinetic energy density $\overline{W}_{kin}$ of the particles in the motion associated with the wave is to second order in the perturbed quantities $\overline{W}_{kin}=\frac{1}{2}mn_0u^2$, which is easily calculated to be $\frac{1}{4}\varepsilon_0|\mathbf{E}|^2$, with $\mathbf{u}=-i\mathbf{E}(e/m\omega)$ and $\omega^2=\omega_{pe}^2\equiv e^2n_0/\varepsilon_0 m$. It is interesting that this kinetic energy is independent of the wave-number **k** and hence the phase velocity of the wave, at least to the present approximation.

  There are no magnetic fields, and therefore no magnetic field energy density associated with electrostatic waves, since the currents originating from the moving electrons are exactly canceled by the displacement currents. It will be found that the wave energy density obtained for cold electrons can be used as an approximation also for finite electron temperatures, in the limit of small wave-numbers, $|\mathbf{k}|\ll 1/\lambda_D$.

5.10.3 Inclusion of spatial dispersion

Assume now that the medium, for instance a plasma, is also *spatially* dispersive, meaning that the dielectric function depends explicitly also on the wavenumber **k**, i.e., $\varepsilon=\varepsilon_0\varepsilon_r(\mathbf{k},\omega)$, assuming negligible dielectric losses at least for the set of $(\mathbf{k},\omega)$ being considered. In this case Poisson's equation becomes $\varepsilon_0\varepsilon_r(\omega,\mathbf{k})\mathbf{k}\cdot\mathbf{E}(\omega,\mathbf{k})=0$. The existence of free electrostatic waves implies again that $\varepsilon_r(\omega,\mathbf{k})=0$, which is tantamount to a dispersion relation $\omega=\omega(\mathbf{k})$. A broadening in the frequency domain consequently implies a broadening in the wave-number domain also. The expression (5.94) is consequently generalized (Yeh & Liu 1972) as

$$\begin{aligned}\frac{\partial\mathbf{D}}{\partial t} &= -i\varepsilon_0\int_{-\infty}^{\infty}\omega\varepsilon_r(\omega,\mathbf{k})\mathbf{E}(\omega,\mathbf{k})e^{-i(\omega t-\mathbf{k}\cdot\mathbf{r})}d\omega d\mathbf{k}\\ &\approx -i\varepsilon_0\int_{-\infty}^{\infty}\Bigg(\omega_0\varepsilon_r(\omega_0,\mathbf{k}_0)+\frac{d\omega\varepsilon_r(\omega,\mathbf{k})}{d\omega}\bigg|_{\omega=\omega_0,\mathbf{k}=\mathbf{k}_0}(\omega-\omega_0)\\ &+\omega\nabla_{\mathbf{k}}\varepsilon_r(\omega,\mathbf{k})|_{\omega=\omega_0,\mathbf{k}=\mathbf{k}_0}\cdot(\mathbf{k}-\mathbf{k}_0)\Bigg)\mathbf{E}(\omega)e^{-i(\omega t-\mathbf{k}\cdot\mathbf{r})}d\omega d\mathbf{k},\end{aligned} \tag{5.101}$$

where no relation between ω and **k** has yet been imposed. We now assume $\varepsilon_r(\omega,\mathbf{k})=0$, which determines a dispersion relation $\omega=\omega(\mathbf{k})$.

Using the approximation (5.101) in $\mathbf{E}\cdot\partial\mathbf{D}/\partial t$ we find the time-averaged expression

$$\frac{\partial\overline{W_e}}{\partial t} - \nabla\cdot\overline{\mathfrak{S}}_e, \tag{5.102}$$

with

$$\frac{\partial \overline{W_e}}{\partial t} = \frac{1}{4}\varepsilon_0\omega \frac{\partial \varepsilon_r(\omega,\mathbf{k})}{\partial \omega}\bigg|_{\omega=\omega_0,\mathbf{k}=\mathbf{k}_0} \frac{\partial |\mathbf{E}_0(t)|^2}{\partial t}, \tag{5.103}$$

as before, and

$$\begin{aligned}\overline{\mathbf{S}}_e &= -\frac{1}{4}\varepsilon_0\,\omega\nabla_{\mathbf{k}}\varepsilon_r(\omega,\mathbf{k})\bigg|_{\omega=\omega_0,\mathbf{k}=\mathbf{k}_0} |\mathbf{E}_0(t)|^2 \\ &= \frac{1}{4}\varepsilon_0\,\omega \frac{\partial \varepsilon_r(\omega,\mathbf{k})}{\partial \omega}\bigg|_{\omega=\omega_0,\mathbf{k}=\mathbf{k}_0} \mathbf{u}_g(\mathbf{k})\,|\mathbf{E}_0(t)|^2,\end{aligned} \tag{5.104}$$

where we used implicit differentiation on $\varepsilon_r(\omega,\mathbf{k}) = 0$ to determine the group velocity

$$\mathbf{u}_g(\mathbf{k}) \equiv \nabla_{\mathbf{k}}\omega = -\nabla_{\mathbf{k}}\varepsilon_r(\omega,\mathbf{k}) \bigg/ \frac{\partial \varepsilon_r(\omega,\mathbf{k})}{\partial \omega}$$

(see also (3.5)).

A consistent interpretation of the continuity relation (5.102) is that, for spatially dispersive media, the wave energy density is given by W_e, and that wave energy is convected by the group velocity; i.e., $\mathbf{S}_e$ is the energy flux. The Poynting flux vanishes for electrostatic waves.

By their derivation, the results in this section are restricted to cases with vanishing dielectric losses; i.e., ε_r is real.

5.10.4 Negative energy waves

In the cold plasma dispersion relation $\omega = \pm\omega_{pe}$ the wave propagation is independent of $\mathbf{k}$, and the group velocity vanishes. If we assume that the entire medium has a translational velocity $\mathbf{U}$, the frequency in the laboratory frame has to be corrected with the Doppler shift $\mathbf{U}\cdot\mathbf{k}$ and the dielectric function becomes $\varepsilon_r(\omega) = 1 - \omega_{pe}^2/(\omega - \mathbf{U}\cdot\mathbf{k})^2$. The dispersion relation in the rest frame then becomes $\omega = \pm\omega_{pe} + \mathbf{U}\cdot\mathbf{k}$. We can again calculate the energy density and obtain

$$\overline{W_e} = \frac{1}{4}\varepsilon_0 \frac{\mathbf{U}\cdot\mathbf{k} \pm \omega_{pe}}{\pm\omega_{pe}} |E_0|^2. \tag{5.105}$$

Choosing the minus sign, i.e., the wave with $\omega = \mathbf{U}\cdot\mathbf{k} - \omega_{pe}$, we find $\overline{W_e} < 0$; the wave has negative energy for $\mathbf{U}\cdot\mathbf{k} > \omega_{pe}$! This result simply states that the medium is in a state of lower energy *with* the wave than *without*. This might be explained by a simple example. Consider the linearized one dimensional continuity equation $\partial n/\partial t + n_0\partial u/\partial x + U\partial n/\partial x = 0$ appropriate for these conditions. After a Fourier transform we obtain

$$\frac{n}{n_0}\left(\frac{\omega}{k} - U\right) = u\,.$$

For $\omega/k > U$, the density n and the velocity u are *in* phase; for $\omega/k < U$ they are in *counter* phase. In the latter case the kinetic energy density $\frac{1}{2}(n+n_0)(u+U)^2$ is *smaller* than $\frac{1}{2}n_0U^2$ everywhere, because in regions where the local density $n+n_0$ is larger than n_0 it has to be multiplied by a velocity $u+U$ which is smaller than U, and the deficit is not compensated in the regions of large $u+U$ because $n+n_0$ is smaller here. Consequently, the energy of the system *with* the wave is lower than *without*, as stated.

These arguments assume a priori that there exist branches of the dispersion relation where $\omega/k < U$. We have seen that this is possible. Calculating the energy flux as in (5.104) in this simple case with $\varepsilon_r(\omega) = 1 - \omega_{pe}^2/(\omega - \mathbf{U}\cdot\mathbf{k})^2$, we find that the energy propagates with the velocity $\mathbf{U}$, which was to be expected.

5.10.5 Complex dielectric functions – dispersive media

Consider now the case where $\varepsilon_r(\omega) = \varepsilon_1(\omega) + i\varepsilon_2(\omega)$ is a complex function, with ε_1 and ε_2 being real. In this case it turns out that it is no longer possible to give a unique definition of the electrostatic energy density. This can be demonstrated by the following arguments. The expression (5.96) remains valid also for complex ε and can be inserted into the first term in (5.32) or (5.33) to give

$$\begin{aligned} \frac{\partial \overline{W_e(t)}}{\partial t} &= \frac{1}{4}\varepsilon_0 \frac{d\omega\varepsilon_1(\omega)}{d\omega}\bigg|_{\omega=\omega_0} \frac{\partial |\mathbf{E}_0(t)|^2}{\partial t} + \frac{1}{2}\omega_0\,\varepsilon_0\,\varepsilon_2(\omega_0)\,|\mathbf{E}_0(t)|^2 \\ &+ \frac{1}{4}\varepsilon_0 \frac{d\omega\varepsilon_2(\omega)}{d\omega}\bigg|_{\omega=\omega_0} \left(\mathbf{E}_{02}(t)\cdot\frac{\partial \mathbf{E}_{01}(t)}{\partial t} - \mathbf{E}_{01}(t)\cdot\frac{\partial \mathbf{E}_{02}(t)}{\partial t}\right), \end{aligned} \tag{5.106}$$

where $\mathbf{E}_0(t) \equiv \mathbf{E}_{01}(t) + i\mathbf{E}_{02}(t)$. When integrating $\overline{W_e}(t)$ in (5.106), we get from the first term the results which were obtained before. The contribution from the second term can unambiguously be attributed to the dielectric losses, but the third term is troublesome because the result will depend on the "history" of the electric field. Consequently, $\overline{W_e(t)}$ will depend not only on $\mathbf{E}_0$ at time t but also on the specific way this electric field was reached. With such restrictions we cannot use the resulting $\overline{W_e(t)}$ as an expression for energy density in a thermodynamic sense. This will be possible only if $d\omega\varepsilon_2(\omega)/d\omega\,|_{\omega=\omega_0}$ is negligible. Fortunately, this will often be the case, at least in certain frequency ranges.

5.10.6 Damping by dielectric losses – electrostatic waves Ⓚ

The dielectric losses will give rise to damping of waves. The damping rate is, however, determined by the real as well as the imaginary parts of $\varepsilon(\omega, \mathbf{k})$. Here we consider for simplicity only electrostatic waves. Assume as before that the plasma (or any other medium for that matter) supports weakly damped electrostatic waves with dispersion relation $\omega(\mathbf{k}) = \omega_1(\mathbf{k}) + i\omega_2(\mathbf{k})$ with $\omega_1(\mathbf{k}) \gg \omega_2(\mathbf{k})$. These waves have to be associated with a zero of the dielectric function $\varepsilon(\omega(\mathbf{k}), \mathbf{k})$, as discussed previously in Section 5.10.2.1. For weakly damped waves we look for zeroes close to the real ω-axis, and approximate the relative dielectric function as

$$\begin{aligned} \varepsilon_r(\omega, \mathbf{k}) &\equiv \varepsilon_1(\omega_1 + i\omega_2, \mathbf{k}) + i\varepsilon_2(\omega_1 + i\omega_2, \mathbf{k}) \\ &\approx \varepsilon_1(\omega_1, \mathbf{k}) + i\varepsilon_2(\omega_1, \mathbf{k}) - \omega_2\frac{\partial\varepsilon_2}{\partial\omega_1} + i\omega_2\frac{\partial\varepsilon_1}{\partial\omega_1}. \end{aligned} \tag{5.107}$$

We assume that the third term is small, because both ω_2 and ε_2 were assumed small. The condition $\varepsilon_r(\omega, \mathbf{k}) = 0$ then implies $\varepsilon_1(\omega_1, \mathbf{k}) = 0$, which determines the real part of the dispersion relation for the waves, $\omega_1 = \omega_1(\mathbf{k})$. Taking the imaginary part of (5.107) gives

$$\omega_2(\mathbf{k}) = -\varepsilon_2(\omega_1(\mathbf{k}), \mathbf{k}) \bigg/ \frac{\partial\varepsilon_1(\omega, \mathbf{k})}{\partial\omega}\bigg|_{\omega=\omega_1(\mathbf{k})}. \tag{5.108}$$

The numerator is given by the dielectric losses, while the denominator is determined by the same term which enters the expression for the energy density of electrostatic waves.

The foregoing discussions were implicitly obtained for *temporally* damped waves; we imagined an initial condition where a plane wave was released and we subsequently followed its damping in time. Physically, the alternative situation is more interesting; a harmonic wave is excited at a certain spatial position by an antenna, and we follow the damping of the wave as it propagates away from the exciter. This problem can be analyzed in much the same way as the previous one. Assume now that ω is real, while $\mathbf{k} = \mathbf{k}_1 + i\mathbf{k}_2$ with $k_2 \ll k_1$. We then have the expansion

$$\varepsilon_r(\omega, \mathbf{k}) \approx \varepsilon_1(\omega, \mathbf{k}_1) + i\varepsilon_2(\omega, \mathbf{k}_1) - \mathbf{k}_2\cdot\nabla_{\mathbf{k}}\varepsilon_2(\omega, \mathbf{k}) + i\mathbf{k}_2\cdot\nabla_{\mathbf{k}}\varepsilon_1(\omega, \mathbf{k}). \tag{5.109}$$

As before we obtain $\varepsilon_1(\omega, \mathbf{k}_1) = 0$ determining $\omega = \omega(\mathbf{k}_1)$, which is essentially the same result as before, but now the damping is determined by

$$\mathbf{k}_2 \cdot \nabla_{\mathbf{k}} \varepsilon_1(\omega, \mathbf{k}_1) = -\varepsilon_2(\omega(\mathbf{k}_1), \mathbf{k}_1), \tag{5.110}$$

where in general $\nabla_{\mathbf{k}} \varepsilon_1$ need not be parallel to $\mathbf{k}_2$. In particular we find the relation between the temporal and the spatial dampings by combining (5.108) and (5.110) to give

$$\omega_2 = -\mathbf{k}_2 \cdot \nabla_{\mathbf{k}} \omega_1, \tag{5.111}$$

using again

$$\mathbf{u}_g \equiv \nabla_{\mathbf{k}} \omega_1 = -\nabla_{\mathbf{k}} \varepsilon_1(\omega_1(\mathbf{k}), \mathbf{k}) \Bigg/ \left. \frac{\partial \varepsilon_1(\omega, \mathbf{k})}{\partial \omega} \right|_{\omega = \omega_1(\mathbf{k})}$$

for the group velocity, $\mathbf{u}_g$; see also (3.5). The relation (5.111) is perhaps not quite as useful as one might think; it implicitly assumes that only one dispersion relation enters the problem. Unfortunately, this is not always so, in particular not in the kinetic theory for ion acoustic waves. The problem can be outlined as follows: assume for the sake of argument that we have two relevant branches of a dispersion relation, one with a large group velocity u_g and a large value of the damping rate $\omega_2(\mathbf{k})$, and another branch with small u_g and small $\omega_2(\mathbf{k})$. For the *initial value problem* we should be concerned with the latter branch, and might tend to forget completely about the other one. However, if we should then calculate a *spatial* damping for a given fixed real frequency ω, the result would be in error because waves with the least *spatial* damping (small $\mathbf{k}_2$) might be associated with the other branch, all depending on the actual values of u_g and $\omega_2(\mathbf{k})$.

5.11 Force on a fluid or a gas

If we place a sample of a dielectric medium in an electric field, the sample will in general be attracted into the field region. The reason is that the medium becomes polarized, and the polarization charges are influenced by the electric field. You can say that the bound charges tend to accumulate near the free charges of opposite sign, assuming that the electric field is due to distributions of such free charges. The calculation of these forces is, however, no simple matter. To illustrate the problem, and its subsequent solution, we first consider a simple plane parallel capacitor with quadratic plates of length L and plate separation d; see Fig. 5.11. Usually, in a discussion of such simple capacitors, it is always assumed that the electric field between the plates can be taken to be uniform and perpendicular to the plates *inside* the capacitor, with the field vanishing outside. This can of course not be exact, as easily demonstrated by integrating $\mathbf{E}$ around a closed loop with one part inside and another part outside the capacitor. Let the line of integration be parallel to $\mathbf{E}$ inside the capacitor (i.e., the line is perpendicular to the capacitor plates), and close outside at a large distance from the edges of the plates. For the assumed (oversimplified) configuration, the integral thus gives a result different from zero, in variance with (5.2) for this time stationary situation! The error lies in ignoring the *fringing field*; see Fig. 5.11. The fringing field contributes to the line integral to cancel the part originating from inside the capacitor. Usually this error is of little consequence, because the volume associated with the edges is always assumed much smaller than the volume of the interior of the capacitor. As far as the force on the dielectric is concerned, the situation is quite different, since it is basically these nonuniform fields that account for the force on a dielectric. Unfortunately, fringing fields are difficult to calculate accurately, and we seem at a loss when facing the present problem. Fortunately, an energy principle can solve our dilemma.

Let $\mathcal{W}$ be the energy of the system consisting of capacitor plates with charges $\pm Q$, with a slab of dielectric partially inserted, as in Fig. 5.11. The value of $\mathcal{W}$ of course depends on how much of

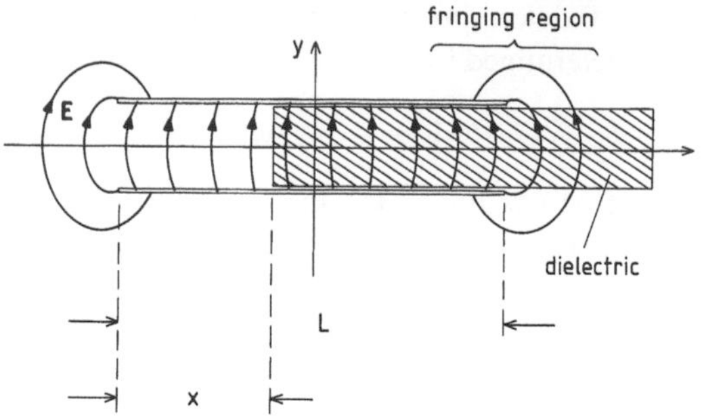

FIGURE 5.11: *Illustration of the fringing field around capacitor plates, where a sample of a dielectric medium is being inserted.*

the dielectric is in between the capacitor plates. If we now pull out the dielectric by a tiny distance dx, the energy is changed by an amount equal to the work done, i.e.,

$$d\mathcal{W} = F_{ex}dx$$

where F_{ex} is the force we must exert to counteract the force F on the dielectric, i.e., $F_{ex} = -F$. Since the force lines inside the capacitor are, to a good approximation, perpendicular to the plates, and thus perpendicular to the displacement of the dielectric, they do not contribute to the work: the important part of F is basically originating from the fringing fields.

We have also

$$F = -\frac{d}{dx}\mathcal{W}. \tag{5.112}$$

As is well known (Duffin 1990), the energy stored in the capacitor is

$$\mathcal{W} = \frac{1}{2}CV^2$$

where V is the voltage across the capacitor, and its capacitance is $C = \varepsilon_0 L(\varepsilon_r(L-x)+x)/d$, as easily demonstrated by standard arguments with ε_r being the relative dielectric constant and d being the plate separation, while x is defined in Fig. 5.11. Note that for this calculation we *ignored* the fringing fields, but as already mentioned the error in doing so can be made arbitrarily small by reducing d and augmenting L. We now assume that the charges $\pm Q$ on the capacitor plates are kept constant when displacing the dielectric plate in Fig. 5.11. Note that V will change when we do this. We have

$$\mathcal{W} = \frac{1}{2}Q^2/C.$$

From (5.112) we then have

$$F = -\frac{d}{dx}\mathcal{W} = \frac{1}{2}\frac{Q^2}{C^2}\frac{d}{dx}C = \frac{1}{2}V^2\frac{d}{dx}C.$$

On the other hand,

$$\frac{d}{dx}C = -\frac{\varepsilon_0 \chi L}{d}$$

where we for simplicity introduced the susceptibility $\chi \equiv \varepsilon_r - 1$ of the dielectric. Finally we have

$$F = -\frac{\varepsilon_0 \chi L}{2d}V^2 \tag{5.113}$$

The minus sign indicates that the force is in the negative x-direction; the dielectric is pulled *into* the capacitor. Note how smoothly we avoided all intricate questions regarding the detailed nature of the fringing fields!

- **Exercise:** The foregoing analysis assumed that we keep Q fixed. It is evident that we could as well keep the potential V fixed by some suitably connected batteries. Discuss this case. Will (5.113) remain valid in this case?

Using the experience from the foregoing simple case, we consider now more generally an electric field arising from charges located on the surfaces of conductors (antennas) embedded in an isotropic fluid or gas-like dielectric, a plasma for instance. The electric fields may be stationary, but, as argued before, it is possible to have so-called electrostatic fields also for time-varying problems (Stratton 1941). We consider the time-stationary problem, assuming simple media, where $\mathbf{D} = \varepsilon_r\varepsilon_0\mathbf{E}$, where $\varepsilon_r = \varepsilon_r(\mathbf{r})$ can be a continuous function of position. For this case we have the total electrostatic energy $\mathcal{W} = \frac{1}{2}\varepsilon_0\int \varepsilon_r E^2 dV$, where the volume integration is over all space. We now introduce an infinitesimal but otherwise arbitrary displacement $\mathbf{s}(\mathbf{r})$ of any point in the dielectric, with $\mathbf{s}(\mathbf{r})$ being a continuous vector function. Sometimes these are called *virtual displacements*. The conductors are assumed to remain fixed, and in the near vicinity of these the displacement of the dielectric is therefore tangential to the conductor surfaces. The distribution of the dielectric is now changed, and consequently also the energy density is changed by an amount which we write as $\delta\mathcal{W} = -\frac{1}{2}\varepsilon_0\int E^2\delta\varepsilon_r dV$. At the same time the distribution of charges on the conductor surfaces will be changed also. Since these charges are necessarily in electrostatic equilibrium before the displacement is introduced, their static equilibrium is minimum with respect to small displacements of the charge distribution. Since the conductors as such are kept fixed, we can therefore safely assume that the variation in energy due to the redistribution of the surface charges is negligible as compared to $\delta\mathcal{W}$.

We now make the additional simplifying assumption that $\varepsilon_r = \varepsilon_r(n(\mathbf{r}),\mathbf{r})$ depends on position $\mathbf{r}$ and the density of matter, $n = n(\mathbf{r})$. Note that we do *not* assume $\varepsilon_r = \varepsilon_r(n(\mathbf{r}))$. We then have

$$\delta\varepsilon_r = -\mathbf{s}\cdot\nabla\varepsilon_r + \frac{\partial\varepsilon_r}{\partial n}\delta n. \tag{5.114}$$

We simplify (5.114) by noting that an initial volume element dU_1 becomes $dU_2 = (1+\nabla\cdot\mathbf{s})dU_1$ after the distortion induced by the infinitesimal displacement. The mass of the volume element is conserved, and we therefore have $n_1 dU_1 = n_2(1+\nabla\cdot\mathbf{s})dU_1$. The change in density becomes $\delta n = -n\nabla\cdot\mathbf{s}$, and we can rewrite (5.114) as

$$\delta\varepsilon_r = -\mathbf{s}\cdot\nabla\varepsilon_r - n\frac{\partial\varepsilon_r}{\partial n}\nabla\cdot\mathbf{s}. \tag{5.115}$$

The change in electrostatic energy can now be written as

$$\delta\mathcal{W} = \frac{1}{2}\varepsilon_0\int\left(\mathbf{s}\cdot\nabla\varepsilon_r + n\frac{\partial\varepsilon_r}{\partial n}\nabla\cdot\mathbf{s}\right)E^2 dV \tag{5.116}$$

which we can rewrite after a little algebra (Stratton 1941) as

$$\begin{aligned}\delta\mathcal{W} = {} & \frac{1}{2}\varepsilon_0\int\left(E^2\nabla\varepsilon_r - \nabla\left(E^2 n\frac{\partial\varepsilon_r}{\partial n}\right)\right)\cdot\mathbf{s}\,dV \\ & + \frac{1}{2}\varepsilon_0\int\nabla\cdot\left(E^2 n\frac{\partial\varepsilon_r}{\partial n}\mathbf{s}\right)dV.\end{aligned} \tag{5.117}$$

We recall that the electric field vanishes inside conductors and that by assumption the normal component of $\mathbf{s}$ at the conductor surfaces was vanishing. Using the divergence theorem on the last integral in (5.117) we find that this integral vanishes, and consequently

$$\delta \mathcal{W} = \frac{1}{2}\varepsilon_0 \int \left(E^2 \nabla \varepsilon_r - \nabla \left(E^2 n \frac{\partial \varepsilon_r}{\partial n} \right) \right) \cdot \mathbf{s} dV. \tag{5.118}$$

We now want to associate this change in energy to a force, and recall that the work done by a force density $\mathbf{f}$ is $\mathbf{f} \cdot \mathbf{s}$ per unit volume. This work must be equivalent to a change in energy density. Consequently we can write $\delta \mathcal{W} = -\int \mathbf{f} \cdot \mathbf{s} dV$ and have

$$\int \mathbf{f} \cdot \mathbf{s} dV = -\frac{1}{2}\varepsilon_0 \int \left(E^2 \nabla \varepsilon_r - \nabla \left(E^2 n \frac{\partial \varepsilon_r}{\partial n} \right) \right) \cdot \mathbf{s} dV. \tag{5.119}$$

Since the displacement $\mathbf{s}$ was assumed to be arbitrary, the relation (5.119) can therefore only be generally fulfilled if

$$\mathbf{f} = -\frac{1}{2}\varepsilon_0 \left(E^2 \nabla \varepsilon_r - \nabla \left(E^2 n \frac{\partial \varepsilon_r}{\partial n} \right) \right). \tag{5.120}$$

This expression gives the force density resulting from inhomogeneous electric fields and nonuniform dielectrics, and is at times called the "ponderomotive force". By entirely similar methods we may consider the case with a spatially varying magnetic susceptibility. The results are readily obtained by replacing $\mathbf{E}$ and $\mathbf{D}$ by $\mathbf{H}$ and $\mathbf{B}$, respectively, and changing ε to μ (Kentwell & Jones 1987).

The foregoing analysis implicitly assumes media without strain. In solids, for instance, we can have the electric forces balanced by elastic forces, and the deformation of solids need not be accompanied by any change in density. The derivation of forces on solids is in general much more complicated, and we refer to, for instance, Stratton (1941) for further details. We ignored also pressure forces, arising from density inhomogeneities, but these are easily taken into account.

The analysis explicitly assumed time-stationary conditions. The analogous time-varying problem, dealing with a spatially modulated wave, for instance, is generally more complicated because the electric field has a source $\partial \mathbf{B}/\partial t$, in addition to the charged conductors. Similarly, we have new sources for the magnetic field, by Maxwell's displacement current. We shall not consider this more general problem here, but only note (once more) that we *can* encounter cases where the magnetic fields are negligible even for time-varying fields: the so-called electrostatic waves, where the currents are exactly, or almost exactly, canceled by Maxwell's displacement current. For such cases, the analysis of this section will apply with some modifications, where elements from Section 5.10 will enter into the full expression for the energy density (Kentwell & Jones 1987).

6

Plasmas Found in Nature

The plasma state, a fully or at least significantly ionized gas, can occur in many environments. One example is a gas in thermal equilibrium at a high temperature. This may be our intuitive first guess, but it turns out that we seldom meet gases in true thermal equilibrium at relevant temperatures. The problem is interesting nonetheless. The closest we come to thermal equilibrium is probably *coronal equilibrium*, which is relevant for parts of the solar atmosphere, for instance. The most abundant natural occurrence of plasmas near the Earth is the ionosphere, where the primary source of plasma is ionization by ultra-violet (UV) radiation from the Sun. In this chapter we discuss different models for plasma production, focusing mainly on those relevant for the Earth's ionosphere and magnetosphere. Important topics, such as technological applications of plasmas, are left out.

6.1 Saha's equation

We should like to have an expression for the degree of ionization of a gas. If we (optimistically, as it turns out) assume that the gas is in thermal equilibrium, at a temperature T, it is actually possible to derive this result theoretically from first principles. The basis for the derivation is the result from statistical mechanics that the probability P_i for finding a certain state is

$$P_i = g_i \exp(-\mathcal{W}_i/\kappa T), \tag{6.1}$$

where $\mathcal{W}_i$ is the energy of that state and g_i is its statistical weight, sometimes called the *degeneracy factor*. Note how powerful statistical mechanics is; we do not need to know anything about the detailed mechanics of the system, nor its actual microscopic states! The density ratio between two states a and b, both at the same temperature, is consequently

$$r \equiv \frac{n_a}{n_b} = \frac{g_a}{g_b} \exp(-\Delta\mathcal{W}/\kappa T) \tag{6.2}$$

with $\Delta\mathcal{W} \equiv \mathcal{W}_a - \mathcal{W}_b$. The ratio of states a as a fraction of *all* states is

$$\alpha \equiv \frac{n_a}{n_a + n_b} = \frac{r}{1+r} \tag{6.3}$$

assuming the system only has the two states in question. In the simplest case with $g_a = g_b$ we see that r increases with temperature to reach the asymptotic value $1/2$. If a and b were to denote an ionized and a neutral state, we could only reach a degree of ionization of 50% at infinite temperatures! However, in general the relevant ratio is $g_a/g_b \gg 1$ for plasmas, and we can obtain the temperature $T_{1/2}$ where $\alpha = 1/2$ or $r = 1$ as $\kappa T_{1/2} = \Delta\mathcal{W}/\ln(g_a/g_b)$. By plotting α as function of T it is easily realized that for large g_a/g_b the variation of α looks almost like a step function, with the transition from state a to state b appearing almost like a phase transition.

We shall now follow a semi-heuristic derivation of the ratio g_a/g_b for the case where a refers to the ionized and b to the neutral state, respectively (Thompson 1962). We assume that the system is enclosed in a box with volume L^3. A free particle can have any momentum p and any energy $p^2/2M$. The total number of particles inside the box having momentum components within a narrow region p_x and $p_x + dp_x$, p_y and $p_y + dp_y$, etc. will be

$$dN = \frac{L^3\, dp_x dp_y dp_z}{h^3} \tag{6.4}$$

where we made use of the uncertainty principle, and introduced Planck's constant h. The uncertainty principle states essentially that the number of distinguishable states in a phase space element $dp_x dp_y dp_z dxdydz$ is given as $dn = dp_x dp_y dp_z dxdydz/h$. Alternatively, the wave function of a free particle can be written as $\exp(i\mathbf{k}\cdot\mathbf{r})$ where $\mathbf{k} = \mathbf{p}/h$, with $\mathbf{k} = (2\pi/L)(n_1, n_2, n_3)$, for integer n_i's, where each set of distinct triads corresponds to a single wave function.

Using (6.1) with $\mathcal{W} = p^2/2M$, we obtain the probability for finding a free particle state dP with momentum p in the interval $dp_x dp_y dp_z$

$$dP = \frac{L^3}{h^3} \exp\left(-\frac{p^2}{2M\kappa T}\right) dp_x dp_y dp_z\,. \tag{6.5}$$

The total probability is then obtained by integration as

$$P = \frac{L^3}{h^3} \iiint_{-\infty}^{\infty} \exp\left(-\frac{p^2}{2M\kappa T}\right) dp_x dp_y dp_z = L^3 \left(\frac{2\pi M\kappa T}{h^2}\right)^{3/2}, \tag{6.6}$$

with the integrals being over all of momentum space.

Taking the energy difference between the neutral state and the ion + electron to be the ionization energy eV_i, where V_i is the ionization potential, we have for a box of unit side length, $L = 1$,

$$\frac{n_+ n_-}{n_0} = \frac{\left(2\pi M\kappa T/h^2\right)^{3/2} \left(2\pi m\kappa T/h^2\right)^{3/2} \exp(-eV_i/\kappa T)}{\left(2\pi M_0 \kappa T/h^2\right)^{3/2}}, \tag{6.7}$$

with m and M being the electron and ion masses, respectively. Since $m \ll M$ we can take the mass of the neutral state to be $M_0 \approx M$, and find

$$\frac{n_+ n_-}{n_0^2} = \frac{1}{n_0} \left(\frac{2\pi m\kappa T}{h^2}\right)^{3/2} \exp\left(-\frac{eV_i}{\kappa T}\right). \tag{6.8}$$

or

$$r \equiv \frac{n_+}{n_0} = \frac{1}{n_-} \left(\frac{2\pi m\kappa T}{h^2}\right)^{3/2} \exp\left(-\frac{eV_i}{\kappa T}\right). \tag{6.9}$$

This is the simplest expression for degree of ionization in thermal equilibrium, with the more general result known as Saha's equation. For our purpose (6.8) will suffice. The approximation consists essentially of considering only a mixture of bound states and free ions and electrons, ignoring excited states, internal degrees of freedom, etc. These can easily be included at the expense of more complicated results. If we insert densities in particles per cubic meter we find

$$\begin{aligned} \frac{1}{n}\left(\frac{2\pi m\kappa T}{h^2}\right)^{3/2} &= 2.405 \cdot 10^{21} \frac{T^{3/2}}{n} \qquad \text{with } T \text{ in K} \\ &= 3.00 \cdot 10^{27} \frac{T^{3/2}}{n} \qquad \text{with } T \text{ in eV} \end{aligned} \tag{6.10}$$

For hydrogen plasmas we have $n_+ = n_-$ exactly, for other plasmas at least approximately, unless the temperatures are very high. For small densities, the variation of α resembles a step function, as we would have for a phase transition. To some extent it is therefore justified to call plasma a *fourth state of matter.*

In case we consider, for instance, dissociation of molecules by heating a neutral gas, we can obtain results similar to (6.8) with U_i now being the dissociation energy and the mass M being the reduced mass of atoms formed in the dissociation process. If the molecule dissociates into a pair of oppositely charged particles, the resulting mixture constitutes a classical plasma. If the density of such a plasma is increased, the electric forces acting between the charged components tend to reduce the energy of the system so that dissociation proceeds more easily than in a rarefied plasma. The degree of dissociation (ionization) will in such a non-ideal plasma be larger than the result obtained from the simple version of Saha's equation (6.8).

Ⓢ **Exercise:** Illustrate the variation of α for varying temperatures. Try different ionization potentials and net plasma densities n and show how r approaches a step-function for small n. Ⓢ

6.2 Coronal equilibrium Ⓚ

Most plasmas, and certainly almost any laboratory plasma, cannot properly be assumed to be in thermal equilibrium, the basic problem being that they are too small to trap radiation. In such cases, we can still have a sort of local thermodynamic equilibrium, provided the most important mechanism for recombination is three-body recombination in preference to radiative recombination, assuming also that radiative ionization can be ignored. However, for three-body recombination to be more important than radiative recombination, it is found that the plasma density must exceed a critical density, which turns out to be of the order of 10^{22} m^{-3} for temperatures in the region of a few eV, this critical density increasing with increasing temperature.

At lower densities, where radiative recombination is dominating, we can have what is called *coronal equilibrium*. The name refers to the fact that the corona of the Sun is in such a state. Note that in the cases discussed here, the systems are continuously losing energy by the escaping radiation, and if a steady state should be maintained, the energy must be supplied by some external source.

In order to obtain an expression for the degree of ionization in a state of coronal equilibrium, we first consider the source of electrons. This is primarily due to ionization of neutrals by fast particles, since we have implicitly assumed that radiation is escaping from the plasma. Take first a neutral atom to be ionized, place in our imagination a small box around it and consider the flux of particles that can be the cause of such an "impact ionization." If we assume that the average ion and electron temperatures are at least approximately equal, we have $\frac{1}{2}m\langle u_e^2\rangle \approx \frac{1}{2}M\langle u_i^2\rangle$ for the ions and electrons. Neutrals can be treated in the same way as ions in this estimate. A typical electron velocity is much larger (about $\sqrt{M/m}$ as large) than typical ion or neutral velocities. The number of electrons entering the reference box per second is therefore much larger than the number of ions or neutrals. It is then most likely that it will be an electron which ionizes the neutral atom by a collision. The number of ion-electron pairs being created per second per unit volume by such collisions is

$$S_c = n_e n_n \sigma_{ion} U_e \,, \tag{6.11}$$

where we have n_e being the average electron density (which for singly charged ions is equal to the ion density, due to the assumed overall charge neutrality) while n_n is the average neutral density. The electron flux density is $n_e U_e$, where U_e is a typical electron velocity, which will be of the same order of magnitude as the average electron thermal velocity. The cross section for electron impact ionization is denoted σ_{ion}. For illustrating the basic ideas, we first let the cross sections be constant, independent of particle velocities.

In statistical equilibrium there will also be ions and electrons recombining to form neutrals. By the definition of coronal equilibrium, we assume that the dominant recombination is radiative recombination, i.e., an ion and electron recombining by emission of a photon. The rate of neutrals produced by this process is

$$S_r = n_e n_i \sigma_{rec} U_e \,, \tag{6.12}$$

where by the same arguments as before, the particle flux is controlled by the electrons. Strictly speaking, the velocity entering the expression (6.12) is not exactly the same as that entering (6.11), but for the present order of magnitude we retain the simple versions of the two equations.

The degree of ionization in a homogeneous plasma in coronal equilibrium is given by balancing the source of electrons created by collisional ionization against the sink or loss of electrons due to radiative recombination. For coronal equilibrium we thus require $S_r = S_c$, implying

$$\frac{n_i}{n_n} \approx \frac{\sigma_{ion}}{\sigma_{rec}} \,. \tag{6.13}$$

Since we assume the plasma to be macroscopically neutral, $n_e \approx n_i = n$ is the plasma density. The result (6.13) is interesting by implying that the degree of ionization in coronal equilibrium is solely determined by the cross sections, and independent of densities. The scaling implied in (6.13) is intuitively reasonable: if the cross section for recombination is large, then the plasma density is low and vice versa. An important shortcoming of the present discussion is the omission of recombination at the surface of the confining vessel, which will be important for laboratory applications.

To do a somewhat better job in deriving S_r and S_c we should note from the outset that the collisional cross sections are in general velocity dependent. We can then write expressions like (6.11) and (6.12) for each given velocity and then average over velocities. We introduce an average over a spherically symmetric distribution, $F(u)$, as

$$\langle \sigma(u) u \rangle = \frac{1}{\int_0^\infty F(u) u^2 du} \int_0^\infty u F(u) \sigma(u) u^2 du \,. \tag{6.14}$$

We used spherical geometry, as appropriate for the problem, with the volume element being $du_x du_y du_z = u^2 \sin\Theta \, du d\Theta d\Phi$ and integrated over Φ and Θ. In the present case the velocity distribution is isotropic, for instance, a Maxwellian, $F(u) = n(m/2\pi\kappa T)^{3/2} \exp(-\frac{1}{2} m u^2/\kappa T)$ in three dimensions, $u^2 = u_x^2 + u_y^2 + u_z^2$. In a detailed analysis, we have an expression of the form (6.14) entering S_c and S_r, with the proper cross section inserted. The result (6.14) can of course always be written as a "typical" cross section, multiplied by an average velocity, but it is also clear that this average velocity will in general not be exactly the same in (6.11) and (6.12). *Qualitatively*, the results are correct as presented. Approximate analytical expressions for electron impact ionization cross sections, averaged over Maxwellian velocity distributions, are given, for instance, by Goldston and Rutherford (1995), and similarly for the cross section for radiative recombination.

Considering, for instance, hydrogen, we find by using the proper cross sections that an electron temperature corresponding to the ionization potential (i.e., 13.6 eV) gives $n_i/n_n \approx 10^5$ so the plasma is almost fully ionized (Goldston & Rutherford 1995). A significant neutral component is found only for small electron temperatures like 1.5 eV, where the ionization is approximately 50%.

6.3 Chapman ionosphere Ⓚ

To illustrate the presence of plasma in the Earth's near environment, and to demonstrate calculations relevant for realistic conditions of plasma production, we consider the Chapman model for the ionosphere. In this model, the ionospheric plasma is created by ultraviolet (UV) radiation from the Sun, and we want to analyze this process in some detail.

Consider first a horizontally stratified neutral atmosphere, made up by some gas which can be ionized by radiation. The parameters of this atmosphere are thus assumed to depend on the altitude z only. The incoming radiation is assumed to be unidirectional, with a given directional angle χ. Following Appendix B we find a decrease in radiation dI due to absorption in a slab of thickness dz to be

$$dI = -I\sigma n_n(z)ds = I\sigma n_n(z)\sec\chi dz \tag{6.15}$$

where I is the intensity of the radiation, n_n is the altitude dependent neutral gas density, σ is the ionization cross section of the neutrals and $\sec\chi \equiv 1/\cos\chi$; see Fig. 6.1. Note the change in sign in (6.15) by defining $dz = -ds\cos\chi$; this will be convenient later on. Introducing the definition

$$\tau(z) \equiv -\int_{\infty}^{z} \sigma n_n(z')dz' , \tag{6.16}$$

where we note that the lower integration limit is $+\infty$, we find trivially

$$I = I_\infty e^{-\tau(z)\sec\chi} , \tag{6.17}$$

where now I_∞ is the intensity of the incoming radiation at $z \to \infty$. It is here we have the convenience of the choice of sign mentioned before. Often, we see that $\tau(z)$ is called the "optical depth" of the atmosphere at altitude z. Note that with the sign convention chosen here we have $\tau(z) > 0$, with $\tau(z \to +\infty) \to 0$.

The loss of intensity per unit altitude is now easily obtained from (6.15) with the result

$$\frac{d}{dz}I = I_\infty\sigma n_n(z)\sec\chi e^{-\tau(z)\sec\chi} . \tag{6.18}$$

In the following we shall be concerned with the *volume* production of ion-electron pairs by the absorbed radiation. What we need is not the energy absorbed per unit altitude, but the energy loss per volume. Taking the bottom area of the cylinder segment in Fig. 6.1 to be $\mathcal{A}$, we have the corresponding volume to be $\mathcal{A}ds = \mathcal{A}dz\sec\chi$. In other words, if we take a unit area for $\mathcal{A}$ and a unit length for ds, the volume with the base area and height dz will be $\sec\chi$ smaller. We consequently find the volume intensity loss as

$$\frac{1}{\sec\chi}\frac{d}{dz}I = I_\infty\sigma n_n(z)e^{-\tau(z)\sec\chi} . \tag{6.19}$$

It does not matter which of the two cylinder segments in Fig. 6.1 we consider. For a given dz, the volume traversed by the radiation varies with χ. This is accounted for by the factor $1/\sec\chi$ in (6.19).

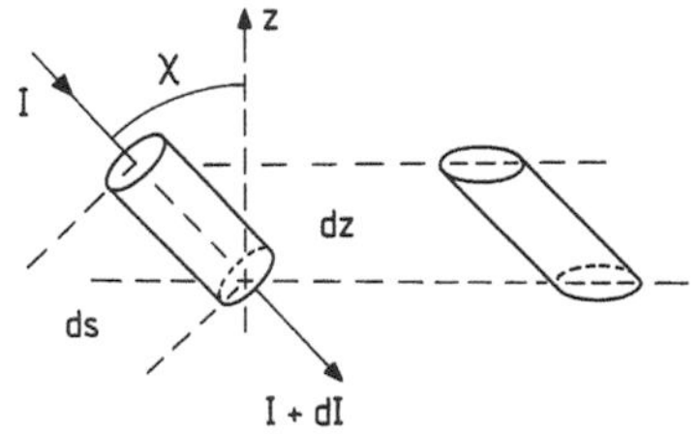

FIGURE 6.1: *Illustration defining the angle χ and the notation used in deriving the loss in the incoming intensity of radiation, as a function of altitude. The two cylinder segments to the left and right have the same volume.*

Assume now that we have η electron-ion pairs produced by an absorbed unit of energy. We denote the absorbed energy in a cylinder with area A and length ds by $W_{in} - W_{out} = IA - (I + \Delta I)A$.

The number of particle pairs produced is then

$$q = \eta \frac{IA - (I + \Delta I)A}{\Delta s A} \to \eta \frac{dI}{ds}.$$

Given a neutral density $n_n(z)$ at altitude z and given χ, we have that the production of charged particle pairs is $q(\chi, z) = \eta \sigma n_n(z) I$, or

$$q(\chi, z) = \eta I_\infty \sigma n_n(z) e^{-\tau(z) \sec \chi}. \tag{6.20}$$

Note that the result (6.20) is quite general and can be used for any given $n_n(z)$, to be inserted also in the definition (6.16) of $\tau(z)$. The relation can be used, for instance, for an empirically determined density variation obtained by measurements. The constant η can also be obtained by measurements.

We now specify the altitude variation of the neutral density by assuming a stationary isothermal atmosphere. The assumption of stationarity implies that at all times we have a balance between the gravitational force on a volume element and the force arising from the difference in pressure $p(z)$ at the top and the bottom of this volume element. The gravitational force acts downwards; the pressure on the bottom surface acts upwards; the pressure on the top surface acts downwards, see also Fig. 2.1. We need not take the pressure on the side-walls into account since the pressure on one side is exactly compensated by the pressure on the opposite side, for the assumed horizontal stratification. For such a horizontally stratified atmosphere we take the volume of the element to be Adz, where dz is its vertical height and A is the area of the horizontal part, and find $g\rho Adz = Ap(z + dz) - Ap(z)$. With $dp \approx p(z) - p(z + dz)$ this gives

$$dp = -\rho g dz. \tag{6.21}$$

See also the discussion in Section 2.2. For the equation of state we take the ideal gas law $p = n_n \kappa T$ or $p = \rho \kappa T/M$, where T is the temperature, while $\rho \equiv M n_n$ and M is the mass of a neutral atom. Inserting this into (6.21) we readily obtain $n_n = n_0 \exp(-zMg/\kappa T)$. A *scale height* for the isothermal atmosphere can be defined as $H_n \equiv \kappa T/Mg$. This is the height a particle with mass M would reach in a constant gravitational potential g if it were released with a vertical velocity $\sqrt{2\kappa T/M}$.

For the particular case of an atmosphere in isothermal equilibrium we can readily integrate (6.17) to give $\tau(z) = \sigma n_n(z) H_n$, or $\tau(z) = \sigma H_n n_0 \exp(-z/H_n)$. The result for the production rate (6.20) can now be expressed as

$$q(\chi, z) = \eta I_\infty \frac{\tau(z)}{H_n} e^{-\tau(z) \sec \chi}. \tag{6.22}$$

The maximum value of q for a given χ is then

$$q_m(\chi, z_m) = \frac{\eta I_\infty}{H_n \exp(1) \sec \chi}, \tag{6.23}$$

where we wrote $\exp(1)$ rather than e to avoid any possible confusion with the electron charge. The altitude value where we have the maximum ionization rate is given implicitly by $\tau(z_m) = 1/\sec \chi$. For the present isothermal atmosphere we find

$$z_m = H_n \ln(\sigma H_n \sec \chi \, n_0). \tag{6.24}$$

For varying χ, the production rate obtains its maximum value for $\chi = 0$, corresponding to perpendicular incidence, i.e., $q_{m0}(0, z_{m0}) = \eta I_\infty/(H_n \exp(1))$. We denote the corresponding altitude value z_{m0}. We can evidently write (6.24) as

$$z_m = z_{m0} + H_n \ln(\sec \chi) \tag{6.25}$$

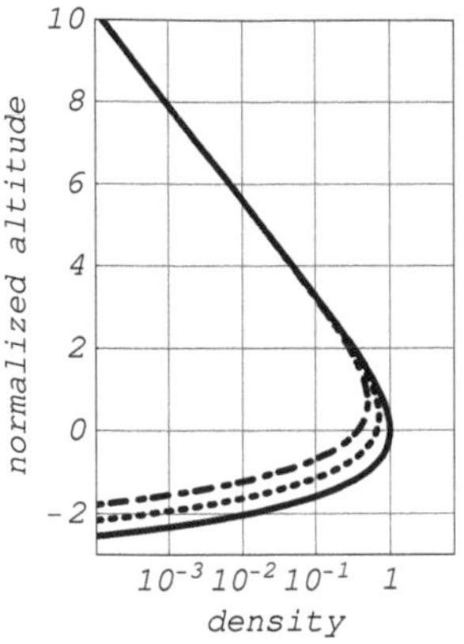

FIGURE 6.2: *Illustration of the Chapman ionosphere for three angles* $\chi = 0$, $\pi/4$ *and* $\pi/3$, *with solid, dotted and dash-dotted lines, respectively. The normalized density is shown on a logarithmic scale as a function of the normalized altitude.*

Physically, we have a high intensity of the incoming radiation at large altitudes z, but little material to ionize, so the ionization rate is small. At low altitudes we have neutrals in abundance, but now the radiation intensity has been depleted by the absorption by the neutral in the intermediate altitudes. Consequently, the ionization at low altitudes is also small.

The ionization as described by $q(\chi, z)$ is balanced by a recombination of ions and electrons to form neutrals. A dynamic equation for ion production can be written as

$$\frac{d}{dt} n_i = q - \alpha n_i n_e \,, \tag{6.26}$$

where α is a recombination coefficient. The first term in (6.26) is the ionization discussed before, while the second term accounts for the recombination. It is taken proportional to the product $n_i n_e$, since this is a simplest expression which vanishes in case we have no electrons to recombine with, and also vanishes if there are no ions to recombine. We take α to be a constant here, noting that in more general models it will depend on altitude through parameters like temperature, etc.

For a coarse-grained description, we cannot expect any significant deviation between n_i and n_e. If such an imbalance should occur, large electric fields would set up and these would rapidly restore a quasi neutral plasma. Only in case we look at the plasma on very small scales can we expect to find significant deviations between the electron and ion densities. With this hindsight we can rewrite (6.26) to the desired accuracy as

$$\frac{d}{dt} n_i = q - \alpha n_i^2 \,. \tag{6.27}$$

A steady state solution of this equation is $n_i = \sqrt{q/\alpha}$. By use of (6.22) we then have

$$n_i(z) = e^{-\frac{1}{2}\tau(z)\sec\chi} \sqrt{\eta I_\infty \frac{\tau(z)}{H_n \alpha}} \,. \tag{6.28}$$

By this result, the problem is basically solved, since $\tau(z)$ is a known function, and we have $n_i \approx n_e$ at all altitudes under steady state conditions. Introducing normalized variables $\gamma \equiv (z - z_{m0})/H_n$ we find

$$n_i(\gamma) = n_{im} e^{\frac{1}{2}[1 - \gamma - \exp(-\gamma)\sec\chi]} \,, \tag{6.29}$$

where $n_{im} = \sqrt{q_{m0}(0, z_{m0})/\alpha}$. Examples of a Chapman ionosphere are shown in Fig. 6.2.

- **Exercise:** Express the optical depth $\tau(z)$ for an isothermal atmosphere in terms of the number of particles in a cylinder with cross section σ and height H_n, extending from $z = 0$.

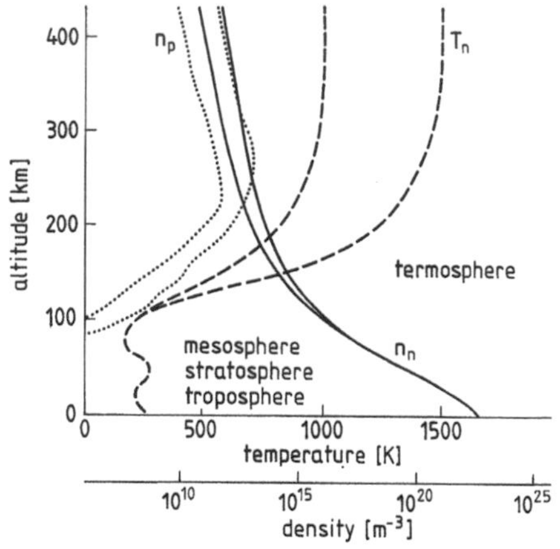

FIGURE 6.3: *Illustration for the variation with altitude of the densities n_p and n_n of the plasma and the neutral atmosphere, respectively, together with the variation of neutral temperature, T_n. Curves are shown for night as well as day time conditions, where the plasma density is largest for the latter case. Densities are given on a logarithmic scale for both plasma and neutral components.*

The present *Chapman model* for the ionosphere assumes that the neutral density is a priori given, i.e., it is unaffected by the ionization. Implicitly this requires that only a tiny fraction of the neutral atmosphere is ionized, so that it is unnecessary to distinguish n_n and $n_n - n_i$. This restriction is of minor consequence, maybe with the exception of the very top layer of the ionosphere, where the density is small anyhow. Also, only one neutral species was assumed (one η value) but this is easily remedied. The assumption of the isothermal neutral atmosphere is much more questionable. Note, however, that we are in a position to deal with *any* a priori given neutral background, so if this problem becomes essential, we might solve (6.20) with the given $n_n(z)$, as already mentioned. The horizontal stratification is a natural assumption, and it is difficult to imagine that a steady state ionosphere can be described without making this assumption.

We have here considered only the plasma density variation with altitude and have not accounted for the electron and ion temperatures. As far as the ions are concerned we can assume that they to a good approximation take the temperature of the neutral background (see Fig. 6.3), since the energy exchange per collision is large. The problem is more complicated for the electrons and we refer the reader to the literature (Schunk & Nagy 1978).

The overall properties of the Earth's ionosphere are reasonably well accounted for by the model discussed here, but by closer scrutiny it is found that it is the ionospheric E-region at 95–120 km altitude, which is following the predictions of the model most faithfully; see Fig. 6.3. A Chapman ionosphere is *not* in coronal equilibrium, because ionization by particle impact is negligible. As far as recombination is concerned, the two models are similar, of course. Radiation is the basic ionization mechanism in the Chapman ionosphere simply because of the abundance of energetic radiation coming from the Sun.

There are some evident shortcomings in the Chapman model: according to it, no ionosphere should be found on the night-side of the Earth, in variance with observations. Albeit with strongly reduced plasma density, there *is* a night-side ionosphere (see Fig. 6.3), which is due to convective transport from the morning and evening sides of the illuminated ionosphere, and also to other forms of radiation which can be ignored on the day side as compared to UV.

7

Single Particle Motion

When the particle distribution is very dilute, i.e., the particle density is low, we have as a limiting case that the electric and magnetic fields induced by the presence of the charged particles are negligible in comparison with the externally imposed fields. In such cases it suffices to consider the motion of the particles individually, and the problem is essentially reduced to investigating the motion of a charged particle with given initial position and velocity in a priori given electromagnetic force fields.

In this chapter we summarize some of the basic dynamics of charged particles in electric and magnetic fields. The equations of motion are basically Newton's second law in terms of the Lorenz force for the particle velocity $\mathbf{U}$ and particle position $\mathbf{r}$, both quantities being functions of time

$$M\frac{d}{dt}\mathbf{U}(t) = q\mathbf{E}(\mathbf{r}(t),t) + q\mathbf{U}(t)\times\mathbf{B}(\mathbf{r}(t),t), \tag{7.1}$$

$$\frac{d}{dt}\mathbf{r}(t) = \mathbf{U}(t), \tag{7.2}$$

for particles with charge q and mass M. In the present case we assume the electric and magnetic fields to be a priori given. Evidently, the fields have to be obtained at the actual position of the particle at time t, i.e., along the particle orbit. In this chapter we use capital letters for velocity and position to emphasize that we are here dealing with quantities referring to individual particles, rather than fluid volume elements.

There is an interesting similarity between the motion of a charged particle in a magnetic field and the motion of a sphere rolling on a rotating disc (Eriksen & Vøyenli 1991). Readers are recommended to become familiar with this analogy, but it will not be pursued further here.

7.1 Single particle orbits

The general problem of determining the orbits of charged particles in arbitrary electric and magnetic fields is quite complicated, even for the case where these fields are time stationary. To get an overview of the sort of trajectories one can expect, it is an advantage to consider some individual and relatively simple cases. Although we rather seldom experience fields being constant in space and time, we consider also this somewhat over idealized case.

7.1.1 E || B

Assume that the electric as well as the magnetic fields are homogeneous, i.e., $\mathbf{B}$ and $\mathbf{E}$ are constant vectors, independent of time and of the spatial coordinate. If the electric field has a component $E_\parallel$ along the magnetic field, we have a steady acceleration in this direction, implying

$$U_\parallel = \frac{q}{M}E_\parallel t + U_{0\parallel} \qquad \text{and} \qquad r_\parallel = \frac{q}{2M}E_\parallel t^2 + U_{0\parallel}t + r_{0\parallel}, \tag{7.3}$$

unaffected by the magnetic field.

If we first assume that there is no electric field component in the direction $\perp \mathbf{B}$, we find that a particle will gyrate in a circular orbit with a radius determined by the initial velocity. The equation of motion is

$$\frac{d}{dt}\mathbf{U}_\perp = \frac{q}{M}\mathbf{U}_\perp \times \mathbf{B}. \tag{7.4}$$

The acceleration is $\perp \mathbf{U}_\perp$, implying that $|\mathbf{U}_\perp|$ is constant. A magnetic field cannot impart energy to a charged particle, since the force and the displacement are perpendicular. The particle energy and thus the magnitude of the velocity are then constants. The momentum can change, though, by changing the *direction* of the velocity vector.

For circular motions with radius R of a point-like mass M with constant velocity $\mathbf{U}_\perp$, we have generally the acceleration toward the center as $a = U_\perp^2/R \equiv \Omega^2 R$. For the charged particle considered here, the magnitude of the acceleration is $|qU_\perp B|/M$, with a direction perpendicular to the velocity vector. For a locally homogeneous magnetic field, the charged particle orbit is a circle with radius r_L, the Larmor radius, and an angular frequency Ω_c, the cyclotron frequency, where

$$r_L = \frac{MU_\perp}{qB} \qquad \text{and} \qquad \Omega_c = \frac{qB}{M} \tag{7.5}$$

A special case is a particle at rest, where $r_L = 0$. The *Larmor frequency* is defined as $\frac{1}{2}\Omega_c$, although the definition is at times given incorrectly without the factor $1/2$.

If you have problems remembering the *direction* of motion in the circular gyro-orbit, you can try to imagine the particle motion as a tiny loop-current: the current (being of the same direction of motion as for a positively charged particle) is in the direction which tries to *reduce* the external magnetic field: this is sometimes taken to be a special case of Lenz's law.

Usually, we consider only the magnitudes of r_L and Ω_c, but there are times where it is important to remember that their sign follows the signs of the charge entering the definition, where $\pm\Omega_c$ consequently denotes the two directions of gyration around **B**, referring to a coordinate system fixed with respect to the magnetic field.

Ⓢ **Exercise:** Demonstrate that (7.4) can be written in the form $d^2\mathbf{U}_\perp/dt^2 = -(qB/M)^2\mathbf{U}_\perp$ and write up the solution. Ⓢ

- **Exercise:** Discuss the motion of a magnetized charged particle in a frame rotating with the angular frequency equaling the gyro-frequency of this particle. The rotation axis is parallel to **B** and passes through the gyro-center of the particle. In this frame the particle velocity is zero so there is no magnetic force. Do we have a paradox here? Why not?

Ⓢ **Exercise:** There is a theorem ("Larmor's theorem") regarding charged particles in weak magnetic fields (Symon 1960), stating that for a system of charged particles, all having the same ratio of charge to mass, moving in a central field of force, the motion in a uniform magnetic induction B is, to first order in B, the same as a possible motion in the absence of B except for the superposition of a common precession of angular frequency equal to the Larmor frequency, $\frac{1}{2}\Omega_c$. Set up the transformation to a rotating frame of reference (Symon 1960, Pedlosky 1987) and demonstrate this theorem and explain the meaning of weak magnetic fields: weak compared to what? Ⓢ

- **Exercise:** Estimate the magnetic field in the Earth's polar regions (use the library or literature available otherwise) and calculate the ion cyclotron frequencies for singly charged oxygen atoms and oxygen molecules. Compare with the corresponding numbers for hydrogen and nitrogen.

Restricting the analysis to a plane perpendicular to the magnetic field, we can introduce the average position of the particle, $\mathbf{R}_c$, which is called the guiding center. By "average" we here mean average over a gyro-period, $T_c \equiv 2\pi/\Omega_c$,

$$\overline{\mathbf{r}(t)} = \mathbf{R}_c(t) = \frac{\Omega_c}{2\pi}\int_t^{t+2\pi/\Omega_c} \mathbf{r}(\tau)d\tau .$$

Similarly, we have the average velocity

$$\overline{\mathbf{U}(t)} \equiv \frac{d}{dt}\mathbf{R}_c(t) = \frac{\Omega_c}{2\pi}\int_t^{t+2\pi/\Omega_c} \mathbf{U}(\tau)d\tau = \frac{\Omega_c}{2\pi}\int_t^{t+2\pi/\Omega_c} \frac{d\mathbf{r}(\tau)}{d\tau}d\tau ,$$

which is then the velocity of the guiding center. For a constant magnetic field and vanishing electric fields, we find $\overline{\mathbf{r}} = \mathbf{R}_c$ being a constant, and $\overline{\mathbf{U}} = 0$.

The averaged quantities $\mathbf{R}(t)$ and $\overline{\mathbf{U}(\mathbf{R}(t))}$ can be introduced also in cases where the magnetic field is slowly varying in space and time, in such a way that Ω_c in the integration limits changes only little during a cyclotron period; otherwise we cannot define the meaning of "averages over a gyro-period". In terms of the variables $\mathbf{R}_c = \overline{\mathbf{r}(t)}$ and $\overline{\mathbf{U}} = \overline{\mathbf{U}(t)}$ we can make a "coarse grained" description of the motion of charged particles in magnetic fields, by letting t be the time scale associated with the slow motion. Mathematically, the description can be formulated by using two time scales, one for the rapid, $\sim 1/\Omega_c$, and one for the slow variation (Nayfeh 1973). The guiding center motion of charged particles can be described by a more general formalism (Littlejohn 1983), which will not be discussed here.

A particle moving in a circular orbit is subject to a constant acceleration toward the guiding center. A magnetized charged particle moving in such an orbit will constantly be radiating energy, *cyclotron radiation*, and slowly be decelerated to have diminishing radius in the cyclotron motion. Cyclotron radiation has interesting implications for diagnostics of magnetic fields in distant astronomical objects. A phenomenological model for this radiation loss is given in Section 7.3.

Ⓢ **Exercise:** Consider an ideal pendulum, with a point-like mass M on a string with length ℓ, subject to a gravitational force and the Coriolis force $\mathbf{F}_c = -2M\Omega \times \mathbf{U}$, where the vector Ω gives the angular rotation of the system and $\mathbf{U}$ is the velocity of the pendulum. This is, in essence, Foucault's pendulum. Let the radius of the Earth be $R \gg \ell$ and ρ be the length of the projection of the position of the pendulum at the Earth's surface on the rotation axis Ω of the Earth. Analyze the motion of the pendulum in terms of suitably defined deflection vectors and rotation angles. Note the mathematical similarity to the description of the motion of charged particles in magnetic fields. Ⓢ

7.1.2 $\mathbf{E} \perp \mathbf{B}$

Consider now the case where we have a constant electric field in the direction $\perp \mathbf{B}$ (see Fig. 7.1), the magnetic field still being constant in space and time. We know that by a suitable change of reference this electric field can be made to vanish; see Section 5.8. The appropriate frame is the one moving with velocity $\mathbf{E}_\perp \times \mathbf{B}/B^2$. In the rest frame the orbit is therefore a combined rotation and drift, with a trajectory known as a "cycloid". The same result can be substantiated by a simple calculation also. We can introduce a new velocity $\mathbf{U}_* \equiv \mathbf{U} - \mathbf{E}_\perp \times \mathbf{B}/B^2$, where $\mathbf{U}$ is the "true" particle velocity. By simple insertion into the equation of motion

$$M\frac{d}{dt}\mathbf{U}(t) - q\mathbf{U}(t) \times \mathbf{B} = q\mathbf{E}_\perp , \tag{7.6}$$

it is readily demonstrated that the velocity $\mathbf{U}_*(t)$ follows the equation

$$M\frac{d}{dt}\mathbf{U}_*(t) - q\mathbf{U}_*(t) \times \mathbf{B} = 0 .$$

Consequently, we have the usual circular gyro-orbit solution for $\mathbf{U}_*(t)$. The actual orbit is therefore obtained by changing the reference system back to the laboratory frame by adding the velocity

$$\overline{\mathbf{U}} = \frac{\mathbf{E}_\perp \times \mathbf{B}}{B^2}. \tag{7.7}$$

This is an average velocity associated with the gyro-center. In addition comes the circular motion.

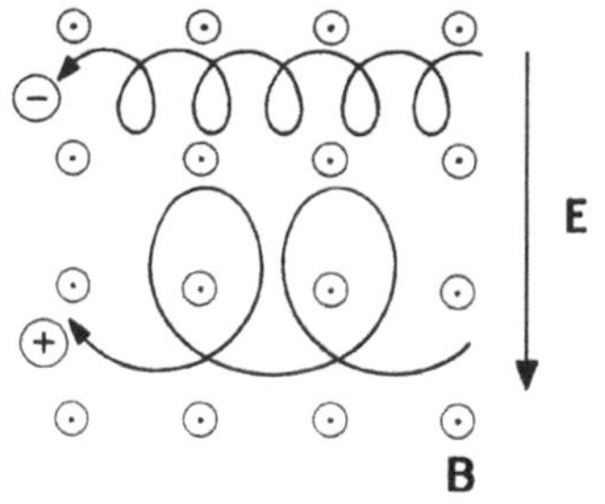

FIGURE 7.1: *Schematic illustration of particles moving in crossed electric and magnetic fields, with the magnetic field being constant in direction as well as intensity.*

Taking the time average of (7.6) we have $M\overline{d\mathbf{U}(t)/dt} = q\overline{\mathbf{E}}_\perp + q\overline{\mathbf{U}(t)} \times \mathbf{B}$. With $\overline{d\mathbf{U}(t)/dt} = 0$ we recover (7.7) by taking the cross product with $\mathbf{B}$. This derivation seems simple, but presupposes $\overline{\mathbf{U}}$ to be constant in time, and we have no a priori guarantee for this although it seems a natural assumption.

Physically, the origin of the $\mathbf{E} \times \mathbf{B}$-drift is easy to explain; since a stationary magnetic field can not change the energy of a charged particle, the changes in velocity in the particle orbit are solely caused by the electric field, which accelerates ions in the positive field direction, and electrons in the negative field direction. The radius of the curvature of the particle trajectory is small where its velocity is large, i.e., at the top on Fig. 7.1 for ions and at the bottom for electrons. The average drift, including its direction, is then readily explained by inspection of Fig. 7.1.

7.1.3 $\mathbf{F}_c \perp \mathbf{B}$

If the electric field is replaced by another force field, say gravity with acceleration $\mathbf{g}$, we can replace the electric force $q\mathbf{E}_\perp$ by the gravitational force $M\mathbf{g}$, and find that the particle has drift $M\mathbf{g} \times \mathbf{B}/(qB^2)$ in the rest frame, in addition to its gyrating motion. Generally, we have for an arbitrary constant force, $\mathbf{F}_c$ a drift velocity

$$\overline{\mathbf{U}}_D = \frac{\mathbf{F}_c \times \mathbf{B}}{qB^2}. \tag{7.8}$$

This is again an *averaged* velocity, which gives the displacement of the gyro-center. The result (7.8) contains the $\mathbf{E} \times \mathbf{B}$-drift as a special case.

In Fig. 7.2 we show four examples of cycloid motion. A trivial example is a straight line (case D), found when the initial velocity of the particle is exactly $\overline{\mathbf{U}}_D$, as given in (7.8), that is, the particle is at rest in the frame moving with the drift velocity, corresponding to the case with $r_L = 0$. The important difference between (7.8) and the simple $\mathbf{E} \times \mathbf{B}/B^2$ drift is that (7.8) depends on the sign of the charge: electrons drift in one direction, ions in the other, when F_c is independent of q.

A special case of the gravitational drift is found when a charged particle moves along curved field lines with velocity $U_\parallel$, where we here let the $\mathbf{B}$-perpendicular velocity be negligible, corresponding to negligible Larmor radius. If we let R be the radius of curvature of the field line, we have that the particle experiences a centrifugal force $MU_\parallel^2/R$ in the direction pointing away from the center of the curvature. Seen from the particles frame of reference this centrifugal force can not be distinguished from a gravitational force, and it gives rise to a drift as in (7.8). Consequently, the particle experiences a curvature drift $U_{DC} \equiv MU_\parallel^2/(qBR)$ in the direction perpendicular to the magnetic field lines. The centrifugal force, correct in magnitude and direction, is

$$\mathbf{F}_{Cf} = MU_\parallel^2 \frac{\mathbf{R}}{R^2} \tag{7.9}$$

with $\mathbf{R}$ being the particle position. By the assumption of strictly $\mathbf{B}$-parallel particle motion, there is no need to distinguish the gyro-center and particle positions here. Using (7.8) we arrive at the averaged gyro-center drift velocity

$$\overline{\mathbf{U}}_{DC} = \frac{MU_\parallel^2}{q} \frac{\mathbf{R} \times \mathbf{B}}{B^2 R^2}, \tag{7.10}$$

where electrons and positive ions drift in opposite directions $\perp \mathbf{B}$.

If we take a thermal velocity u_{The} to be representative for $U_\parallel$ we find curvature drifts $Mu_{The}^2/(qBR)$. If the electron and ion temperatures are equal or at least comparable, we have $Mu_{Thi}^2 \approx mu_{The}^2$, and find comparable curvature drifts for electrons and ions. This reminds us to be a bit cautious: it is so tempting to argue that because of their small masses, the electrons are unaffected by gravitational forces, at least in comparison with forces on ions. For this particular case, the argument will be completely in error for reasons obvious when analyzed in detail!

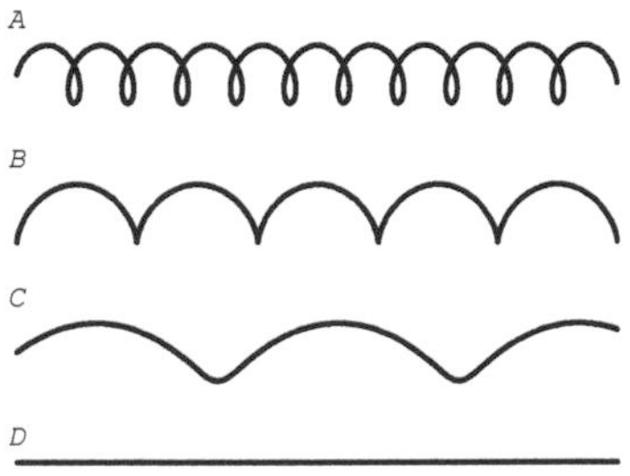

FIGURE 7.2: *Four examples for cycloid motion, for a magnetic field perpendicular to the plane of the paper, and a constant force in the vertical direction.*

The result (7.10) assumes implicitly that the particle velocity vector is parallel to the magnetic field lines, i.e., that $r_L = 0$. It is tempting to assume that a finite Larmor radius is of little consequence, but it is not necessarily so (Chen 2016). If we bend magnetic field lines to have some finite curvature, we necessarily introduce a magnetic field variation with $\nabla B \perp \mathbf{B}$. To demonstrate this we express the radius of curvature R in terms of the magnetic field by the following arguments (Chen 2016): for a current free region and steady state magnetic fields we have $\nabla \times \mathbf{B} = 0$ together with the general result $\nabla \cdot \mathbf{B} = 0$. Assuming that the magnetic field lines are bent, so they are locally well approximated by a segment of a circle, we can assume $\mathbf{B}$ to have a θ-component only. This model then implies that ∇B has an r-component only. In imagination, you can take a long solenoid and bend it into a torus or doughnut, as the one illustrated in Fig. 5.2a. Then we have a magnetic field variation with only a θ-component. For positions inside the torus, we have $B_\theta \sim 1/r$, where r is measured from the symmetry center. We can here use such a toroidal magnetic field as a local approximation for curved magnetic field lines.

We can also use

$$\nabla \times \mathbf{B} = \frac{1}{r}\frac{\partial}{\partial r}(rB_\theta) = 0,$$

i.e., $B_\theta \sim 1/r$. We can take a locally curved magnetic field with $B \sim 1/R$ or more generally $\nabla|B|/|B| = -\mathbf{R}/R^2$, with $\mathbf{R}$ being a vector in the direction *from* the center of curvature. This will necessarily imply $\partial B_\theta/\partial r \approx B_\theta/r$. A curvature of a magnetic field will necessarily imply a decrease of the magnetic field in the direction of the radius of curvature pointing outwards. The motion of particles in inhomogeneous magnetic fields with $\nabla B \perp \mathbf{B}$ will be discussed separately. Particles with finite Larmor radius $r_L \neq 0$, moving along curved magnetic field lines, will sample an inhomogeneous magnetic field.

Ⓢ **Exercise:** Determine initial conditions that give the four orbits in Fig. 7.2. Determine the analytical expressions for the trajectories of the particle orbits illustrated in a velocity plane $\{U_x, U_y\}$ perpendicular to the magnetic field $\mathbf{B}$. Ⓢ

Comment on field line curvature: You might need to determine the curvature of field lines. For this purpose we present a short summary of the appropriate analysis, assuming a vector curve given in a parametric representation as $\mathbf{r} = \mathbf{r}(\gamma)$. We assume that $\mathbf{r}(\gamma)$ is at least twice differentiable, using the notation $d\mathbf{r}(\gamma)/d\gamma \equiv \mathbf{r}'(\gamma)$, etc. We consider first the simplest case with plane curves.

1. Plane curves: in terms of the "plane-product" $[\mathbf{r}'(\gamma)\mathbf{r}''(\gamma)]$, the curvature is given as

$$\rho(\gamma) \equiv \frac{1}{R(\gamma)} = \frac{|[\mathbf{r}'(\gamma)\mathbf{r}''(\gamma)]|}{|\mathbf{r}'(\gamma)|^3},$$

where we recall that the radius of curvature R is the inverse of the curvature ρ. The plane-product is defined as the scalar

$$[\mathbf{r}'(\gamma)\mathbf{r}''(\gamma)] \equiv \begin{vmatrix} r_1' & r_1'' \\ r_2' & r_2'' \end{vmatrix} = \begin{vmatrix} r_1' & r_2' \\ r_1'' & r_2'' \end{vmatrix}$$

where r_1', etc. are vector components in a suitably defined coordinate system.

Note that you may sometimes see curvature defined with a sign: here it is taken positive always.

For the special case where we have in Cartesian coordinates $\mathbf{r}(\gamma) = x(\gamma)\widehat{\mathbf{x}} + y(\gamma)\widehat{\mathbf{y}}$ we find

$$\rho(\gamma) \equiv \frac{1}{R(\gamma)} = \frac{\begin{vmatrix} x' & y' \\ x'' & y'' \end{vmatrix}}{|(x')^2 + (y')^2|^{3/2}}.$$

For the particular case where $y = F(x)$ we can let x itself be the parameter and find

$$\rho \equiv \frac{1}{R} = \frac{|F''|}{|1 + (F')^2|^{3/2}}.$$

Consider the case where the curve is expressed in polar coordinates $\mathbf{r}(\gamma) = r(\gamma)\widehat{\mathbf{r}}(\gamma)$ where the unit vector $\widehat{\mathbf{r}}(\gamma)$ is at an angle $\Phi(\gamma)$ with some reference line. We have a unit vector $\widehat{\Phi}(\gamma) \perp \widehat{\mathbf{r}}(\gamma)$. We have then $\mathbf{r}' = r'\widehat{\mathbf{r}} + r\Phi'\widehat{\Phi}$. Since $\widehat{\mathbf{r}}' = \Phi'\widehat{\Phi}$ and $\widehat{\Phi}' = -\Phi'\widehat{\mathbf{r}}$ we have also $\mathbf{r}'' = (r'' - r(\Phi')^2)\widehat{\mathbf{r}} + (r\Phi'' + 2r'\Phi')\widehat{\Phi}$. With these expressions we can write

$$\rho \equiv \frac{1}{R} = \frac{\begin{vmatrix} r' & r\Phi' \\ r'' - r(\Phi')^2 & r\Phi'' + 2r'\Phi' \end{vmatrix}}{|(r')^2 + (r\Phi')^2|^{3/2}}.$$

For the case where we have $r = F(\Phi)$ we can again take Φ to be the parameter and find

$$\rho \equiv \frac{1}{R} = \frac{|2(dr/d\Phi)^2 - r\,d^2r/d\Phi^2 + r^2|}{(r^2 + (dr/d\Phi)^2)^{3/2}},$$

2. For curves, magnetic field lines in particular, that are fully three dimensional we can generalize the two-dimensional results as

$$\rho(\gamma) \equiv \frac{1}{R(\gamma)} = \frac{|\mathbf{r}'(\gamma) \times \mathbf{r}''(\gamma)|}{|\mathbf{r}'(\gamma)|^3}.$$

For the Cartesian case we find here

$$\rho = \frac{\sqrt{\begin{vmatrix} x' & y' \\ x'' & y'' \end{vmatrix}^2 + \begin{vmatrix} y' & z' \\ y'' & z'' \end{vmatrix}^2 + \begin{vmatrix} z' & x' \\ z'' & x'' \end{vmatrix}^2}}{|(x')^2 + (y')^2 + (z')^2|^{3/2}}.$$

- **Exercise:** Try out the expressions for the curvature on the simple case where $r = r_0 \sin\Phi$. What is the radius of curvature here? (A figure might be a good idea!)

Ⓢ **Exercise:** Approximate the Earth's magnetic field by a magnetic dipole. Find analytical expressions for the magnetic field lines associated with such a magnetic dipole. These are lines where the local field vector is tangent in any position.

Calculate the radius of curvature of a magnetic field line as a function of θ measuring the angle from the magnetic pole, so that $\theta = \pi/2$ at the magnetic equator. Take a field line starting at some distance r_0 from the center, as measured in the magnetic equatorial plane. Ⓢ

Ⓢ **Exercise:** Consider a simple magnetized torus.

1. Give the analytical expression for the magnetic field lines. What is the radius of curvature for the magnetic field lines? Give an intuitive as well as an analytical argument.
2. Add a constant homogeneous magnetic field $\mathbf{B}_V$ perpendicular to the the plane of the torus. Give the analytical expression for the resulting magnetic field and field lines also for this case. What is the radius of curvature for the magnetic field lines? Ⓢ

7.1.4 Finite Larmor radius corrections for inhomogeneous electric fields

Let a time stationary electric field component in the direction $\perp \mathbf{B}$ depend on position. Even though this Eulerian electric field is stationary, the electric field as sampled along the orbit of the gyrating particle will in general vary with time. We assume, however, also here that the characteristic time for the variations in electric field is much larger than the cyclotron period. It is evident that for almost all relevant problems it will be the modifications in the $\mathbf{E} \times \mathbf{B}/B^2$-drift of the *ion* component which is relevant; the electron Larmor radius will usually be negligible in comparison to that of the ions, and the guiding-center model is sufficient for describing the electron motion perpendicular to a magnetic field in low frequency electric fields.

Consider a particle with charge q gyro-orbiting in a homogeneous magnetic field, but inhomogeneous electric fields (Knorr et al. 1988, Hansen et al. 1989). Let the instantaneous particle position be $\mathbf{r}(t)$; its guiding center is at $\mathbf{R}(t)$ so that $\mathbf{r}(t) = \mathbf{R}(t) + \mathbf{r}_L$, with the constant Larmor radius being $r_L \equiv |\mathbf{r}_L|$.

As long as the particle experiences an electric field which in the rest frame is quasi stationary, the *effective* electric field relevant for its *finite Larmor radius corrected* orbit will be the average of the electric field experienced along this orbit. This average is indicated by an overline in the following.

At any given position of the gyro-center $\mathbf{R}$ we average the electric field experienced by the particle along the orbit with radius r_L. This is expressed as

$$\overline{E_x(\mathbf{R})} \equiv \frac{1}{2\pi r_L}\int_{-\infty}^{\infty} E_x(\mathbf{r})\delta(|\mathbf{R}-\mathbf{r}|-r_L)dxdy, \tag{7.11}$$

where r_L is the Larmor radius of the particle.

We represent the electric field by its Fourier series, i.e., write $\mathbf{E}(\mathbf{R}) = \sum_{\mathbf{k}} \mathbf{E}_{\mathbf{k}} \exp(i\mathbf{k}\cdot\mathbf{R})$ and have

$$\overline{E_x(\mathbf{R})} = \sum_{\mathbf{k}} E_{x\mathbf{k}} \frac{1}{2\pi r_L}\int_{-\infty}^{\infty} \exp(i\mathbf{k}\cdot\mathbf{R})\delta(|\mathbf{R}-\mathbf{r}|-r_L)dxdy, \tag{7.12}$$

and similarly for the E_y-component.

We now use (Champeney 1973, Hansen et al. 1989) the relation for the Bessel function of the first kind, J_0, of zeroth order

$$J_0(kr_L) = \frac{1}{2\pi r_L}\int_0^{\infty}\int_0^{2\pi} s\,\delta(s-r_L)\exp(iks\cos\phi)dsd\phi$$

to obtain from (7.12) the result

$$\overline{\mathbf{E}(\mathbf{R})} = \sum_{\mathbf{k}} J_0(kr_L)\mathbf{E}_{\mathbf{k}}\exp(i\mathbf{k}\cdot\mathbf{R}). \tag{7.13}$$

From this result we can obtain the finite Larmor radius (FLR)-corrected $\overline{\mathbf{E}}\times\mathbf{B}/B^2$-drift for the particles guiding center. It is important to distinguish this drift velocity from a *fluid* velocity! In the latter case we take the average velocity of particles passing through a small area-element, and this will in general give a result quite different from the guiding center drift.

Applications of the result (7.13) are feasible numerically (Knorr et al. 1988) for a code based on fast Fourier transforms (FFT). For analytical use, we might prefer a simpler expression. To obtain this, we restrict the analysis to electric fields varying only gently on a scale given by r_L. We then recall the one-to-one correspondence between multiplication by wave-number $\mathbf{k}$ in the Fourier domain, and differentiation with respect to position in configuration space. We can make a series expansion of the Bessel function J_0 and then "translate" the result into a differential operator by the replacement $i\mathbf{k}\to\nabla$. We have $J_0(\xi) = \sum_{n=0}^{\infty}(-1)^n(\xi/2)^{2n}/(n!)^2 \approx 1-\xi^2/4$, for small ξ, giving

$$\overline{\mathbf{E}(\mathbf{R})} \approx \sum_{\mathbf{k}}\left(1-\frac{1}{4}(kr_L)^2\right)\mathbf{E}_{\mathbf{k}}\exp(i\mathbf{k}\cdot\mathbf{R}) \tag{7.14}$$

implying that to lowest order in r_L^2, the FLR-corrected $\mathbf{E}\times\mathbf{B}/B^2$-drift of the guiding center is given by the simple expression (Chen 2016)

$$\overline{\mathbf{U}}_{\mathbf{E}\times\mathbf{B}} \approx \frac{1}{B^2}\left(1+\frac{1}{4}r_L^2\nabla^2\right)\mathbf{E}(\mathbf{R})\times\mathbf{B}, \tag{7.15}$$

where it is implicit in the notation that $\nabla^2\mathbf{E}(\mathbf{R})$ is to be obtained at the gyro-center position.

Physically, the reason for the FLR-corrections is clear: considering a plane electrostatic wave, for instance, with wave-number $\mathbf{k}\perp\mathbf{B}$, we find that a particle with its gyro-center at maximum electric field will actually spend a significant part of the time in regions of weaker fields, as it is gyrating with a finite Larmor radius. The average guiding center drift is consequently smaller than what would be obtained by using the electric field value at the guiding center (Chen 2016).

For numerical use, the relation (7.15) has an unpleasant effect by enhancing very short wavelength small amplitude noise components, and therefore one might prefer the more robust result (7.13).

In this section we only discussed the FLR-corrections to the single particle drift velocity. In case the particle contribution to the bulk plasma density is needed, then for consistency, also the particles have to be "smeared-out" along their Larmor radius. One particle with vanishing Larmor radius contributes to the density as $n(\mathbf{r}) = \delta(\mathbf{R} - \mathbf{r})$. For a particle with finite r_L we can express the averaging as $n(\mathbf{r}) = (1/2\pi r_L)\delta(|\mathbf{R} - \mathbf{r}| - r_L)$. For a distribution of particles with distributed guiding centers with density $n(\mathbf{R})$ we have the particle density given by

$$\overline{n(\mathbf{r})} \equiv \frac{1}{2\pi r_L} \int_{-\infty}^{\infty} n(\mathbf{R})\delta(|\mathbf{R} - \mathbf{r}| - r_L) dX dY\,, \tag{7.16}$$

with the ensuing calculations being identical to those already carried out here (Knorr et al. 1988). Recall that here the dimension of $\delta(\mathbf{R} - \mathbf{r})$ is $1/length^2$ for the present two-dimensional problem, while $\delta(|\mathbf{R} - \mathbf{r}| - r_L)$ has the dimension $1/length$.

Ⓢ **Exercise:** Consider a particle with charge q gyrating with a Larmor radius r_L in a stationary potential variation $\phi(\mathbf{r})$, where $\mathbf{r} \perp \mathbf{B}$. Determine the gyro-averaged electrostatic potential experienced by the particle. Ⓢ

7.1.5 Polarization drifts, $dE/dt \neq 0$

Consider a plasma in a homogeneous stationary magnetic field, and a time-varying electric field perpendicular to **B**. Generally, the electric field can vary in both intensity and direction. For simplicity we here assume the direction to remain constant, but let $dE/dt \neq 0$. It is implicit in the following that the variation of $\mathbf{E}(t)$ is slow in the sense that it changes little during the time it takes the particle to move its gyro radius, i.e. $r_L/(E/B) \gg \Omega_c^{-1}$, implying $u_{The} \gg E/B$.

We consider the motion of the guiding center of a charged particle in the frame of reference moving with the instantaneous $\mathbf{E}(t) \times \mathbf{B}/B^2$-drift. For an electric field which is slowly varying, the velocity varies so this is not an inertial frame. The charged particle will experience an acceleration which a co-moving observer can be interpreted as an effective gravitational acceleration, **g**. In the given frame, the electric field will be negligible, and $M\mathbf{g}$ will be the only force experienced by the particle. This effective, or virtual, gravity gives rise to an $M\mathbf{g} \times \mathbf{B}/(qB^2)$-velocity, varying with time; see (7.8). The acceleration experienced by an observer moving with the particle is $\mathbf{g} = -(1/B^2)d(\mathbf{E} \times \mathbf{B})/dt$. Since **B** was assumed constant we have $\mathbf{F} = M\mathbf{g} = -(M/B^2)(d\mathbf{E}/dt \times \mathbf{B})$, which can be directly inserted into (7.8). The result can be seen as an expansion to first order in a small quantity ω/Ω_c, where ω here denotes a characteristic frequency for the time variation of **E**. Note that the polarization drift has opposite sign for electrons and positively charged ions. Inserting $\mathbf{F} = M\mathbf{g}$ in (7.8), we find the polarization drift

$$\overline{\mathbf{U}}_p = \frac{M}{qB^2} \frac{d}{dt} \mathbf{E}\,, \tag{7.17}$$

which appears as a first order correction to the $\mathbf{E} \times \mathbf{B}/B^2$-velocity. The analysis treats the gyro-center as a physical particle so the time derivative should be interpreted in this sense: here we might as well have used Newton's notation $\dot{\mathbf{E}} \equiv d\mathbf{E}/dt$.

For the special case where the electric field is varying harmonically with time, $\mathbf{E} = \mathbf{E}_0 \exp(-i\omega t)$, we find the simpler result

$$\overline{\mathbf{U}}_p = -i \frac{\omega}{\Omega_c} \frac{\mathbf{E}_0 \exp(-i\omega t)}{B}\,, \tag{7.18}$$

from (7.17). The result (7.18) clarifies the expansion in ω/Ω_c mentioned before.

The velocity (7.17) should be added to $\overline{\mathbf{U}} = \mathbf{E}_0 \times \mathbf{B}/B^2$. The net result is an elliptical motion of the gyro-center in the plane perpendicular to **B**. The ratio of the minor and major axis of this ellipse is ω/Ω_c. The polarization drift of the electrons is negligible in comparison to that of the ions due to the smallness of ω/ω_{ce} compared to ω/Ω_{ci}.

We can discuss the present problem from a different point of view. Consider a charged particle moving in a plane perpendicular to a homogeneous magnetic field and now an inhomogeneous time stationary electric field, $\mathbf{E} = -\nabla\phi$. The simplest model we can imagine assumes that the gyro-center of the particle at any time moves with the local $\mathbf{E}(\mathbf{r}) \times \mathbf{B}/B^2$-velocity. We will find that this model is basically inconsistent, and violates energy conservation. The remedy of the shortcomings will be the polarization drift. The important point here is that as it moves through an inhomogeneous electric field, the particle experiences $\partial\mathbf{E}/\partial t \neq 0$, so the electric field in the moving particle frame varies in time, although in the fixed laboratory frame the field was time stationary, but varying in space.

Let the total energy of the particle be given as $\mathcal{W} \equiv \frac{1}{2}MU^2 + q\phi$. We decompose the velocity into the local $\overline{\mathbf{U}} = \mathbf{E}(\mathbf{R}(t)) \times \mathbf{B}/B^2$-velocity, obtained at the particle gyro-center position $\mathbf{R}(t)$, and the velocity of the cyclotron motion $\mathbf{U}_s$ where $U_{sx} = U_s \cos(\omega_c t)$ and $u_{sy} = u_s \sin(\omega_c t)$ and have $\mathcal{W} = \frac{1}{2}M\overline{U}^2 + \frac{1}{2}Mu_s^2 + q\phi$. Later on, in Section 7.1.8, we will recognize $\frac{1}{2}Mu_s^2/B$ as the magnetic moment of the particle, being a constant of motion (Chandrasekhar 1960). We will here simply assume that $\frac{1}{2}Mu_s^2 = \text{const.}$ but note that as a special illustrative case we are free to consider a particle released with zero velocity, in which case $u_s = 0$ (and then also finite Larmor radius effects are vanishing).

We now differentiate the particle energy to find $d\mathcal{W}/dt = M\overline{\mathbf{U}} \cdot d\overline{\mathbf{U}}/dt + qd\phi/dt$. We have $\overline{\mathbf{U}}$ being the gyro-center velocity which depends on time after release only, while $\phi(\mathbf{R}(t))$ is the electrostatic potential at the particle gyro-center position, which by assumption varies with time. With the definition of $\overline{\mathbf{U}}$, we have $d\mathcal{W}/dt = (M/B^2)\mathbf{E} \cdot d\mathbf{E}/dt + qd\phi/dt$. With $\phi(\mathbf{r})$ being constant in time, we have $d\phi/dt = \partial\phi/\partial t + \mathbf{U}_p \cdot \nabla\phi = -\mathbf{U}_p \cdot \mathbf{E}$, where the velocity $\mathbf{U}_p$ is for the moment unspecified. For a conservative system the total energy (kinetic + potential) of the charged particle must be conserved. This requires $d\mathcal{W}/dt = (M/B^2)\mathbf{E} \cdot d\mathbf{E}/dt - q\mathbf{U}_p \cdot \mathbf{E} - (M/B^2)\mathbf{E} \cdot \left(d\mathbf{E}/dt - (qB^2/M)\mathbf{U}_p\right) = 0$, or $\mathbf{U}_p = (M/qB^2)\,d\mathbf{E}/dt$, which is the polarization drift (7.18). We thus find that in order to have energy conservation, it is necessary to introduce a drift velocity $\mathbf{U}_p \neq 0$. We have then a velocity component parallel to $\mathbf{E}$, which gives rise to a change in kinetic energy, but this is exactly compensated by the change in potential energy. *If* the gyro-center velocity is taken to be strictly the local $\mathbf{E}(\mathbf{r}) \times \mathbf{B}/B^2$-velocity, then the scalar product of this velocity with $\mathbf{E}$ is vanishing, and the force can not do any work. Since the particle is moving in a spatially inhomogeneous potential, its potential energy *has* to change, and it must have a velocity component parallel to $\mathbf{E}$.

Finite Larmor radius (FLR) effects and polarization drifts appear as higher order corrections to the basic $\mathbf{E}(\mathbf{r}) \times \mathbf{B}/B^2$-velocity. The respective ordering of the two corrections is not always evident. In one limit, with time-varying spatially constant electric fields there is no FLR-correction and only the polarization drift appears. As an alternative limit, we can have, for instance, $\mathbf{E} = \{0, E_y(y), 0\}$ with $E_y(y)$ constant in time and $\mathbf{B} = \{0, 0, B\}$. For this case we have only the FLR-corrections to a temporally constant $\mathbf{E} \times \mathbf{B}/B^2$-drift in the x-direction, to be derived at the y-position of the gyro-center.

7.1.6 $\nabla B \perp \mathbf{B}$

Assume now that the electric field is vanishing, $\mathbf{E} = 0$, but the magnetic field, albeit constant in time, has a gradient in intensity in the direction $\perp \mathbf{B}$. Magnetic fields alone cannot do work on a particle, so we know from the outset that the magnetic field can only change the direction of the velocity vector, $\mathbf{U}$, not its magnitude. As a specific example we consider $\mathbf{B} \equiv \{B_x, B_y, B_z\} = \{0, 0, B_0(y)\}$, giving $\nabla|B| \perp \mathbf{B}$; see Fig. 7.3. Note that $\nabla \cdot \mathbf{B} = 0$ for the model field, so it is physically acceptable but $\nabla \times \mathbf{B} \neq 0$, so there needs to be a spatially distributed current density to support this magnetic field. The radius of curvature for a particle trajectory is stronger in a part of the gyrating orbit, where the particle is located at the stronger magnetic field, as compared to its value half a gyro-period later. As a result, we find the particles to drift in a direction $\perp \mathbf{B}$ as well as $\perp \nabla B$, as illustrated in Fig. 7.3. To derive this drift velocity we simplify the problem by considering only gentle variations in the magnetic field, and assume that we can approximate the magnetic field locally by $\mathbf{B} \approx \mathbf{B}_0(1 + y/\ell)$ where ℓ is a characteristic scale length for the variation and $\mathbf{B}_0 \parallel \widehat{\mathbf{z}}$. Mathematically, we make a Taylor expansion of the magnetic field's y variation around the position of the particles gyro-center, and have $\ell = B/|\nabla B|$. By assuming that the variation of this field is small on a scale length determined by the particles Larmor radius, r_L, we retain only the first term in this expansion. Since the magnetic field has no dependence on the coordinate z for the present case, there is no reason to be concerned with the z-component of velocities. With the Lorentz force $\mathbf{F} \equiv q\mathbf{U} \times \mathbf{B}$ we have

$$M\frac{d}{dt}\mathbf{U} = q\mathbf{U} \times \mathbf{B} \approx q\mathbf{U} \times \mathbf{B}_0 + q\mathbf{U} \times \mathbf{B}_0\frac{y_0(t)}{\ell} \tag{7.19}$$

where we have to use the expression for the magnetic field entering the Lorentz force at the actual particle position $y = y_0(t)$. We assume ℓ to be large, and the second term in the expression for the Lorentz force in (7.19) is therefore small. It is convenient to rewrite (7.19) as

$$M\frac{d}{dt}\mathbf{U} - q\mathbf{U} \times \mathbf{B}_0 = q\mathbf{U} \times \mathbf{B}_0\frac{y_0(t)}{\ell}. \tag{7.20}$$

We expect the y-component of the particle position to oscillate (at least approximately) between $\pm r_L$, and the right-hand side of (7.20) is therefore in magnitude of order of (r_L/ℓ) compared to the second term on the left side. We express this by the notation $O(r_L/\ell)$, where O means "of the order of". We can always write the solution of (7.20) in the form $\mathbf{U}(t) \equiv \mathbf{U}_0(t) + \widetilde{\mathbf{U}}(t)$, where $\mathbf{U}_0$ is the solution of (7.20) with the right-hand side omitted. By integration of $\mathbf{U}$ we obtain the y-component of the particle position as $y_p(t) \equiv y_0(t) + \widetilde{y}(t)$. Since the right-hand side of (7.20) is small, we expect $\widetilde{\mathbf{U}}(t)$ and $\widetilde{y}(t)$ to be small corrections to $\mathbf{U}_0$ and $y_0(t)$, the corrections being $O(r_L/\ell)$. We have y_0/ℓ itself being $O(r_L/\ell)$ by construction. Then we have products like $\widetilde{\mathbf{U}}(t)y_0(t)/\ell$, and similar, to be $O(r_L^2/\ell^2)$, which we assume to be so small that they can be ignored. This will be the basic assumption in the ensuing analysis. We expect that $\widetilde{y}$ is $O^2(r_L/\ell)$ since it is a small correction to $y_0(t)$, so $\widetilde{y}$ will not appear in the analysis.

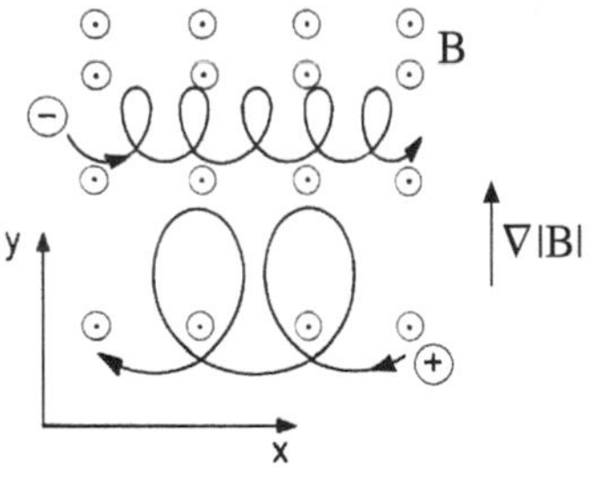

FIGURE 7.3: *Schematic illustration of a charged particle moving in an inhomogeneous magnetic field, where the gradient is perpendicular to* **B**.

1) As the first step, we ignore the right-hand side of (7.20), and solve $Md\mathbf{U}_0/dt - q\mathbf{U}_0 \times \mathbf{B}_0 = 0$. Take a positive charge, $q > 0$, and the magnetic field in the positive z-direction. To this first approximation we find for the particle positions $\{x_0; y_0\} = \{r_L \sin(\Omega_c t); r_L \cos(\Omega_c t)\}$, and velocity components $\{U_{0x}; U_{0y}\} = \{r_L\Omega_c \cos(\Omega_c t); -r_L\Omega_c \sin(\Omega_c t)\}$, where $U_\perp = r_L\Omega_c \equiv \sqrt{U_{0x}^2 + U_{0y}^2}$ see (7.5). We have $\overline{\mathbf{U}_0} \equiv (\Omega_c/2\pi)\int_0^{2\pi/\Omega_c} \mathbf{U}_0\, dt = 0$.

Ignoring small terms, as outlined before, we find the approximate expression

$$M\frac{d}{dt}\widetilde{\mathbf{U}} - q\widetilde{\mathbf{U}} \times \mathbf{B}_0 = q\mathbf{U}_0 \times \mathbf{B}_0 \frac{y_0(t)}{\ell}, \tag{7.21}$$

where the right-hand side is a force term that we will analyze in more detail.

2) We will now include the effects of the right-hand side in (7.21) by an iteration using the first order expressions. The term on the right-hand side of (7.20) gives a correction of order r_L/ℓ to the zero order solutions (x_0, y_0) and U_{0x}, U_{0y}. In an expansion in the parameter r_L/ℓ, we find to lowest order the x-component of the additional force $\mathbf{F}$ in (7.21) as

$$\begin{aligned} F_x &\approx qU_{0y}(t)B_0\frac{y_0(t)}{\ell} \\ &= -qU_\perp \sin(\Omega_c t)B_0\frac{r_L\cos(\Omega_c t)}{\ell}, \end{aligned}$$

where we here retain only the terms containing $\mathbf{U}_0$, considering those with $\widetilde{\mathbf{U}}(t)$ later. We inserted the velocity expression for a positive charge, assuming the magnetic field pointing out of the paper as in Fig. 7.3.

We are not interested here in the fine details of the particle orbit, only in the average. To obtain this, we average in time over one cyclotron period, $2\pi/\Omega_c$, the averaging here again indicted by an "overline". Upon this averaging over time, we find $\overline{F_x} \equiv (\Omega_c/2\pi)\int_0^{2\pi/\Omega_c} F_x dt = 0$, so that this term is uninteresting.

We also have

$$\begin{aligned} F_y &\approx -qU_{0x}(t)B_0\frac{y_0(t)}{\ell} \\ &= -qU_\perp \cos(\Omega_c t)B_0\frac{r_L\cos(\Omega_c t)}{\ell}. \end{aligned}$$

Averaging over time again, we now find

$$\overline{F_y} \equiv \frac{\Omega_c}{2\pi}\int_0^{2\pi/\Omega_c} F_y dt = -\frac{1}{2}qU_\perp B_0\frac{r_L}{\ell}. \tag{7.22}$$

We consequently have an average force on the particle acting in the y-direction. We now have the time-averaged expression for the velocity correction $\widetilde{\mathbf{U}}(t)$ as

$$M\overline{\frac{d\widetilde{\mathbf{U}}}{dt}} - q\overline{\widetilde{\mathbf{U}}} \times \mathbf{B}_0 = \overline{F_y}\widehat{\mathbf{y}}.$$

This is an expression of a form we have met already; see (7.6). According to (7.8) for $q > 0$ the force $\overline{F_y}$ gives rise to a steady particle drift in the direction of $-\widehat{\mathbf{y}} \times \mathbf{B}_0$, which is also the $\mathbf{B}_0 \times \nabla B$ direction. The drift velocity has a magnitude $\frac{1}{2}|U_\perp r_L/\ell|$ and $\overline{d\widetilde{\mathbf{U}}/dt} = 0$. The analysis is readily repeated for $q < 0$. The resulting drift velocity vector is given by

$$\overline{\mathbf{U}}_{\nabla B} = \pm\frac{1}{2}U_\perp r_L\frac{\mathbf{B} \times \nabla B}{B^2} \tag{7.23}$$

where we used the definition of $1/\ell = |\nabla B|/B$ given before, and noted the proper direction of the average force obtained, i.e., the $\mathbf{B} \times \nabla B$-direction. We need not distinguish the small difference between B_0 and B in the denominator of (7.23). Note that the drift (7.23) is opposite in direction for electrons and ions, as indicated by the $\pm$ sign. We can insert the definition of $r_L = MU_\perp/(qB)$ from (7.5) and obtain a different version of (7.23) as

$$\overline{\mathbf{U}}_{\nabla B} = \frac{MU_\perp^2}{2qB}\frac{\mathbf{B} \times \nabla B}{B^2} \tag{7.24}$$

which need not have the $\pm$ sign in front since the particle charge now appears explicitly. Note the similarity between (7.24) and (7.10), but note also the difference in factor $\frac{1}{2}$.

Physically, the origin of the average drift $\overline{\mathbf{U}}_{\nabla B}$-drift is readily explained by inspection of Fig. 7.3. Typically, electron Larmor radii are small, and one might wonder why the electron nonetheless has an averaged drift, when it hardly samples any variation in the magnetic field. The point here is that although the magnetic field variation is indeed small over the electron Larmor radius, the electrons cyclotron frequency is usually much larger so it receives many "kicks" as compared to an ion at the same kinetic energy and the net result is an average electron drift.

As mentioned, electrons and positive ions are drifting in opposite directions due to the $\nabla B \perp \mathbf{B}$ drift. It is, however, important to emphasize that this does not necessarily imply that we have a current density everywhere in the plasma. Try to draw a small vertical cut $A - A'$ on Fig. 7.3, and let the cut represent a small area element perpendicular to the plane of the figure. The length of $A - A'$ is supposed to be much smaller than the local Larmor radius. The local current is then given by the number of charges crossing this area element per sec. If the plasma is uniformly distributed so that the number density of particles and their temperatures are independent of position, at least in the vicinity of $A - A'$, we will find that the number of, say, negative particles per sec. coming from above $A - A'$ and propagating to the right on Fig. 7.3 will be exactly canceled by the negative particles passing through $A - A'$ from below, again because the densities are the same by assumption and the thermal velocities cannot change with position, since a stationary magnetic field alone is unable to change the energy of a charged particle. The same arguments will, of course, hold also for the ion component. Any current density induced by the $\nabla B \perp \mathbf{B}$-drift is therefore restricted to regions of inhomogeneities in plasma density or temperature.

- **Exercise:** Make a qualitative argument for (7.22) by using a small square made of conducting current carrying wire instead of a gyro orbiting particle. The length of one side in the square is $2r_L$. Let the square be placed perpendicular to $\mathbf{B}$ with two sides parallel to ∇B. Is this latter assumption essential for the result?

Ⓢ **Exercise:** Using a simple dipole approximation for the Earth's magnetic field, consider first the motion of electrons and ions in an altitude around 110 km (ionospheric E-region) over the magnetic pole. Assume that some external mechanism is imposing a horizontal electric field of intensity $E = 50$ mV/m in the east-west direction in the region above the magnetic pole. What are the direction and magnitude of the $\mathbf{E} \times \mathbf{B}$-drift velocity of the electrons and for hydrogen ions? Ignore possible collisions with neutrals.

Consider now charged particles at the magnetic Equator, again at altitude 110 km. Let a particle have vanishing $\mathbf{B}$-parallel velocity. What are the direction and magnitude of the $\nabla B \times \mathbf{B}$-drift velocity of electrons and hydrogen ions if their energies are 0.5 eV? Consider now 0.5 eV particles with directed velocities *along* the magnetic field lines. What is the direction and magnitude of the curvature-drift velocity of electrons and hydrogen ions? Ignore now magnetic field inhomogeneities, but include the effect of the Earth's gravitational field on atomic oxygen ions and on electrons. What are the magnitudes and directions of the drift velocities for these two species? Ⓢ

- **Exercise:** In their circular guiding center orbital motion by the $\nabla B \times \mathbf{B}$-drift in the Earth's magnetic field ions and electrons are subject to a centrifugal force, which also gives rise to a drift velocity. Give its direction and magnitude, and compare it to the corresponding drift caused by the Earth's gravity.

Ⓢ **Exercise:** Consider a superposition of an electric and a magnetic field, where the electric field of constant intensity E_0 is in the y-direction and a spatially varying magnetic field points in the

z-direction. We prescribe the magnetic field variation as

$$B(x) = \begin{cases} B_1 & \text{for} \quad x < -a \\ \frac{1}{2}(B_1 + B_2) + \frac{1}{2}(B_2 - B_1)\sin(\pi x/2a) & \text{for} \quad -a < x < a \\ B_2 & \text{for} \quad x > a \end{cases}$$

with $B_2 > B_1 > 0$.

1) Make a sketch of the electric and the magnetic field vectors, giving direction and strength.
2) Give an expression for a spatially distributed current, which gives rise to the prescribed magnetic field.

A particle with mass M and charge q is released at a position $(x,y,z) = (x_0, 0, 0)$ with a velocity in the x-direction u_0, and at $x_0 \ll -a$. The particle velocity in the z-direction vanishes, and this coordinate is irrelevant for the problem. The Larmor radius is assumed to be small, $r_L \ll a$.

3) Determine the variation of the particle energy, $\overline{W} \equiv \frac{1}{2}M\overline{u^2}$, averaged over a gyro-period, as the particle moves, subject to the electric and magnetic forces.
4) Determine the particle trajectory in the three spatial regions, $x < -a$, $-a < x < a$ and $x > a$.

Assume that a plane source of charged particles is placed at $x \to \infty$, where $s/2$ positive and $s/2$ negative particles are emitted per unit area each second. The current due to these particles is considered so small that it does not give any appreciable contribution to the magnetic fields: the plasma β is negligibly small.

5) Express the continuity equation for the emitted particles under the given steady state conditions.
6) Determine the current due to the emitted particles for all x, y-positions. Ⓢ

- **Exercise:** A point particle with mass M and dipole moment $\mathbf{m}$ moves in a circular obit with radius R around a fixed magnetic dipole with moment $\mathbf{m}_0$, the point particle being subject to the magnetic force from $\mathbf{m}_0$. The two moment vectors $\mathbf{m}$ and $\mathbf{m}_0$ are anti-parallel and perpendicular to the plane of the orbit. Calculate the velocity of the orbiting particle in terms of the given parameters. Is the orbit stable against small perturbations?

7.1.7 $\nabla B \parallel \mathbf{B}$

We here consider a case without electric fields, so the force of a particle taken in the laboratory frame is $\mathbf{F} = q\mathbf{U} \times \mathbf{B}$ and assume $\nabla B \parallel \mathbf{B}$; this is typically the feature associated with a *magnetic mirror*. Assume for simplicity that the magnetic field is axi-symmetric, with $B_\Theta = 0$ and $\partial B/\partial\Theta = 0$, as illustrated in Fig. 7.4, with Θ being in the "azimuthal" direction around the z-axis, being the symmetry axis of the magnetic field. Since we have assumed the magnetic field lines to converge or diverge, there will be a radial component, B_r of this magnetic field. Writing $\nabla \cdot \mathbf{B} = 0$ in cylindrical coordinates, we have

$$\frac{1}{r}\frac{\partial}{\partial r}(rB_r) + \frac{\partial}{\partial z}B_z = 0. \tag{7.25}$$

As an approximation to this result we find locally near the symmetry axis

$$B_r \approx -\frac{1}{2}r\frac{\partial}{\partial z}B_z, \tag{7.26}$$

where we assumed $\partial B_z/\partial z$ to be almost constant, i.e., ignore its local variation with r. We can argue in more detail that for symmetry reasons, $\partial B_z/\partial z$ must vary as an even power of r near the symmetry axis, e.g., as a constant plus a term proportional to r^2, where the latter correction is negligible for small r. On the other hand, we found that B_r varies proportional to r. The result (7.26) is most easily checked by direct insertion into (7.25). We assume in the following that the **B**-parallel velocity, $U_\parallel$, of the particle is much smaller than the perpendicular velocity, $U_\perp$, so that the particle gyrates many times before moving an appreciable distance along **B**.

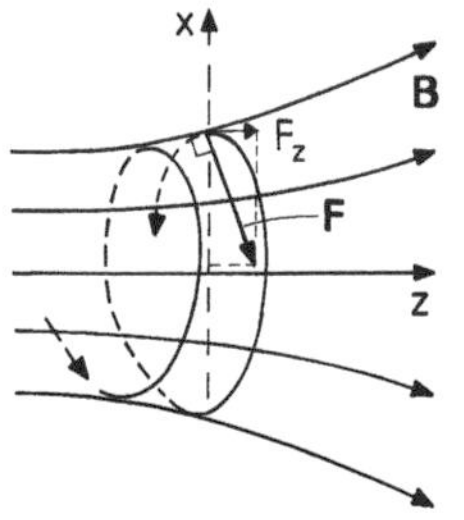

FIGURE 7.4: *Schematic illustration of a particle moving in an inhomogeneous magnetic field, where the gradient is parallel to* **B**. *The z-axis is here chosen as the symmetry axis. Note the z-component,* F_z, *of the Lorentz force.*

The x and y components of the force **F** acting on the charged particle vanish upon averaging; see also Fig. 7.4. These force components are consequently of little interest here, where we are concerned only with the motion of the guiding center determined by the z-component F_z.

The z-component of the Lorentz force (i.e., the component along the symmetry axis of **B**) on a particle with charge q and azimuthal velocity $U_\perp$ gives $F_z = qU_\perp B_r$. Using (7.26) we find $F_z = -\left(\frac{1}{2}qU_\perp r\right)\partial B_z/\partial z$, where $r = r(t)$ is the radial orbital particle position. Here the notation $\perp$ refers to the direction perpendicular to the symmetry axis; see Fig. 7.4. We average the force component F_z over one period of gyration. This averaging is easy, since $|r| = |r_L| \equiv |U_\perp/\Omega_c|$ is constant with the assumed smallness of $U_\parallel$. It takes *many* cyclotron orbits to change $U_\perp$ appreciably. We find

$$\overline{F}_z = -\frac{1}{2}qU_\perp r_L \frac{\partial}{\partial z}B_z = -\frac{1}{2}\frac{MU_\perp^2}{B}\frac{\partial}{\partial z}B_z \tag{7.27}$$

independent of the sign of the charge q, where we explicitly used that the direction of the velocity $U_\perp$ depends on the sign of q. Strictly speaking, we should divide by B_z in (7.27), but to the given accuracy, we here write B.

Charged particles always experience a force (7.27) pointing away from a region with an axially strong magnetic field, independent of the sign of their charge. This can be understood by a simple geometrical construction; see again Fig. 7.4. $\overline{F}_z$ in (7.27) is often called the *mirror* force, referring to its association with a magnetic mirror.

It might be appropriate to add one comment on momentum conservation: if we release a charged particle with a finite magnetic moment and vanishing axial initial velocity, it will be accelerated, and one might wonder whether momentum is conserved in this process. The point is that the magnetic field accelerating the particle is produced by a coil, or a system of coils. The magnetic moment of the particle can be considered as a tiny current loop, as already mentioned, and its associated current acts back on the macroscopic coil system. Since we can assume these macroscopic coils to be infinitely massive as compared to the particle, the coils do not receive any energy, but they do take up momentum, in the very same way as a massive wall takes up momentum but not energy when a tennis ball bounces off.

Sections 7.1.6 and 7.1.7 treat two limiting cases for spatial variations of the magnetic field. More generally, we can have combinations of the two, and might be able to describe the particle motion as a superposition of two drifts, one along **B** and one in the perpendicular direction.

7.1.8 Magnetic moment

Introducing the magnetic moment

$$\mu \equiv \frac{1}{2}\frac{MU_\perp^2}{B} = \pi r_L^2 \left| q\frac{\Omega_c}{2\pi} \right| \tag{7.28}$$

we can write

$$\overline{F}_z = -\mu \frac{\partial}{\partial z} B_z . \tag{7.29}$$

The physical dimension of $q\Omega_c/2\pi$ is *charge* per *time*, i.e., current, so the dimension of the magnetic moment is *current* × *area*. We can also write the magnetic moment as $\mu = \pi r_L^2 B(q^2/2\pi M)$, which for small r_L can be interpreted as the magnetic flux enclosed by the Larmor orbit multiplied by the constant $q^2/(2\pi M)$.

As easily realized, the origin of this magnetic *mirror force* is an axial component of the $q\mathbf{U}\times\mathbf{B}$-force on the charged particles as they gyrate in an axially inhomogeneous magnetic field. Since this force points in the direction of the gyro-center irrespective of the sign of the charge, the mirror force is repelling particles from regions with high magnetic field in either case. At thermal equilibrium, the magnetic moment is the same for electrons and for ions, the Larmor radius of the electrons is small, but their velocity is large, and vice versa for the ions. The force (7.29) acts equally on both components in this case, without giving rise to a charge separation. Consequently, we have no ambipolar electric fields induced by (7.29) for this case. If $T_e \gg T_i$, which is often the case, we can anticipate charge separations due to the force given by (7.29).

The magnetic moment can be interpreted as the product of the area of the circular orbit traced out by the particle, πr_L^2, and the current associated with its motion, $I = q|\Omega_c|/(2\pi)$, where with this definition I is the charge per sec. crossing a small area element perpendicular to the cyclotron orbit of the charged particle.

- **Example:** In the analysis of the force on a small ring current placed in an inhomogeneous magnetic field, we associate a magnetic dipole **m** with the ring, so that $|\mathbf{m}| = I\mathcal{A}$ with I being the current and $\mathcal{A} = \pi R^2$ is the area of the ring with radius R. The *direction* of **m** is perpendicular to the area bounded by the ring. The force on the ring current can then be expressed as

$$\mathbf{F} = \mathbf{m}\cdot\nabla\mathbf{B} .$$

It is important to note that this expression coincides with (7.27) in the proper special case. The basic difference between the two problems is that at any time we are free to change the orientation of the small loop current (i.e., the associated magnetic dipole, **m**), while the orbit of the charged particle for given $\mathbf{B}(\mathbf{r})$ is completely determined by the initial conditions.

As the particle moves along the z-axis, subject to the force F_z, it enters regions with changing magnetic fields. We find, however, that the **B**-perpendicular velocity is changing in such a way that the magnetic moment remains constant. We demonstrate this by considering the z-component of Newton's law

$$M\frac{d}{dt}U_z(t) = \overline{F}_z = -\mu \frac{\partial}{\partial z} B_z \bigg|_{z=Z(t)} \tag{7.30}$$

where we have indicated that the z-derivative of B_z has to be obtained at the guiding center position of the particle, $Z(t)$, which is close to being also the z-position of the particle itself, since we assumed

B_z to be slowly varying with position. Note that $\partial B_z/\partial z|_{z=Z(t)}$ is a function of time and not position, just like U_z. There is no need to have a line over U_z since it is the same with or without averaging over a gyro-period, again to the given accuracy of the analysis.

We multiply both sides of equation (7.30) by U_z, and use $U_z(t) = dZ/dt$, with $Z = Z(t)$ being the z-coordinate of the guiding center position. Using $dB_z(Z(t))/dt = (dB_z(Z)/dZ)(dZ/dt)$, we find

$$\frac{1}{2}M\frac{d}{dt}U_z^2 = -\mu\frac{d}{dt}B_z(Z(t)). \tag{7.31}$$

A stationary magnetic field cannot do work on a gyrating particle, so the kinetic energy must be conserved, i.e.,

$$\frac{1}{2}M\left(U_z^2 + U_\perp^2\right) = \frac{1}{2}MU_z^2 + \mu B_z = \text{const.} \tag{7.32}$$

We used the definition (7.28) of μ in expressing $U_\perp$ in (7.32), and assumed again that the dominant part of the magnetic field is the B_z-component. We now differentiate (7.32) with respect to time. Since by (7.31) we also have $\mu dB_z/dt = -\frac{1}{2}M\,dU_z^2/dt$ we then find

$$\frac{d}{dt}\left(\frac{1}{2}MU_z^2 + \mu B_z\right) = \frac{1}{2}M\frac{d}{dt}U_z^2 + \mu\frac{d}{dt}B_z + B_z\frac{d}{dt}\mu = 0 \quad \text{giving} \quad \frac{d}{dt}\mu = 0, \tag{7.33}$$

demonstrating the conservation of μ. The magnetic moment is conserved to the accuracy B_z/B, i.e., to the accuracy of μ determined solely in terms of B_z in (7.32). In the case where (7.32) is exact, we have homogeneous magnetic fields, i.e., $B = B_z$, and the magnetic moment is trivially constant. A constant magnetic moment in a time-stationary magnetic field implies that the particle encloses the same magnetic flux tube as its gyro-center moves along $\mathbf{B}$, i.e., the magnetic flux enclosed by the Larmor orbit is a constant of motion.

Note that for a co-moving observer, one that follows the gyro-center along the z-axis in Fig. 7.4, the particle energy *is* changing! This observer will, however, not see a time-stationary magnetic field, and in his/her frame of reference $\partial\mathbf{B}/\partial t$ will imply an electric field by Faraday's law (5.6); this electric field *can* change the particle energy, in that frame of reference, as stated before.

Knowing now that μ is a constant, we can by (7.29) interpret B_z as an effective potential for the mirror force, but care should be taken to differentiate it only with respect to z, rather than using the ∇-operator. For this reason, this interpretation is rarely used.

Conservation of magnetic moment invites a simple means of heating a plasma (Chandrasekhar 1960): If we slowly increase a homogeneous magnetic field we will find that also $\frac{1}{2}M\,dU_\perp^2$ is increasing in such a way that $\mu \equiv \frac{1}{2}MU_\perp^2/B$ remains constant. Faraday's law implies that the time-varying magnetic field induces a rotation of an electric field, and this electric field is subsequently mediating the acceleration of the particles. The increase in magnetic field is to be slow, so that B changes only little in one ion cyclotron period: we often meet the term "adiabatic heating". This heating mechanism was expected to be promising in the early days of fusion research, but, alas, these ideas proved to be too optimistic! The suggested heating was to be achieved in so-called pinch devices that (unfortunately: it was a good idea) proved to be inherently unstable (Chandrasekhar 1960).

Summary of the averaged single particle drifts in electric and magnetic fields:

- Motion in a plane $\perp$ $\mathbf{B}$ with stationary fields:

  a) $\mathbf{E} \parallel \mathbf{B}$:

  $$U_\parallel = \frac{q}{M}E_\parallel t + U_{0\parallel} \qquad \text{and} \qquad R_\parallel = \frac{q}{2M}E_\parallel t^2 + U_{0\parallel}t + R_{0\parallel}$$

  b) $\mathbf{E} \perp \mathbf{B}$:

  $$\overline{\mathbf{U}}_{\mathbf{E}\times\mathbf{B}} = \frac{1}{B^2}\left(1 + \frac{1}{4}r_L^2\nabla^2\right)\mathbf{E}(\mathbf{r}) \times \mathbf{B}$$

  where the lowest order correction for FLR-corrections were included. We have $E =$ const. as a special case. It is implicit that $\nabla^2\mathbf{E}(\mathbf{r})$ is obtained at the gyro-center position.

c) More generally we have for $\mathbf{F} \perp \mathbf{B}$:

$$\overline{\mathbf{U}}_D = \frac{\mathbf{F} \times \mathbf{B}}{qB^2}$$

where $\mathbf{F}$ is an arbitrary constant force, gravity for instance, in which case $\mathbf{F} = M\mathbf{g}$.

d) $\nabla B \perp \mathbf{B}$:

$$\overline{\mathbf{U}}_{\nabla B} = \pm \frac{1}{2} U_\perp r_L \frac{\mathbf{B} \times \nabla B}{B^2} = \frac{MU_\perp^2}{2qB} \frac{\mathbf{B} \times \nabla B}{B^2}$$

- Motion in a plane $\perp \mathbf{B}$ with time-varying electric fields:

$$\overline{\mathbf{U}}_p = \frac{M}{eB^2} \frac{d}{dt} \mathbf{E}$$

- Three-dimensional motion with $\nabla B \parallel \mathbf{B}$:

$$\overline{F}_z = -\mu \frac{\partial}{\partial z} B_z$$

is the *mirror force*, with $\mu \equiv \frac{1}{2} MU_\perp^2 / B$ being the magnetic moment.

In all cases the overline indicates averaging over the cyclotron period.

- **Example:** The Earth's magnetic field offers a relevant example for discussing the single particle drifts. Assume the magnetic field to be dipolar. Starting at the Earth's surface and continuing radially in the magnetic equatorial direction with $r > R_E$, the Earth's radius, we compare the gravitational drift and the magnetic curvature and ∇B-drifts of an ion with charge q and mass M and kinetic energy corresponding to the average for a population at temperature T. For generality we can let q have either sign, i.e. allow for negative ions also.

  In empty space we can to a good approximation take the Earth's magnetic field to be a simple dipole field, here written in spherical coordinates

$$B_\phi = 0, \qquad B_\theta = \mu_0 \mathcal{M}_E \frac{\sin\theta}{4\pi r^3}, \qquad B_r = \mu_0 \mathcal{M}_E \frac{2\cos\theta}{4\pi r^3}. \tag{7.34}$$

  At times you find the expressions given in terms of the angle λ measured from the equator, i.e. $\lambda \equiv 90° - \theta$, where θ used here is measured from the magnetic pole. We introduced the Earth's magnetic dipole moment $\mathcal{M}_E$ which can formally to be interpreted as describing the ring current, I, generating the magnetic field. For a tiny circular wire with radius R we have the current multiplied with the area of the circle, i.e. $\mathcal{M}_E = I\pi R^2$. For the Earth we have $\mathcal{M}_E \approx 8 \times 10^{22}$ A m^2. We need not specify the current nor the circle radius, all we need is their product.

  At a position r at the magnetic equator we have the radius of curvature for the magnetic field line at this position to be $R = r/3$. A particle with velocity $U_\parallel$ along the magnetic field lines experiences a centrifugal force $\mathbf{r}3MU_\parallel^2/r^2$ resulting in a curvature drift of

$$\mathbf{U}_C = 3\frac{MU_\parallel^2}{q} \frac{\mathbf{r} \times \mathbf{B}}{B^2 r^2} = 3\frac{MU_\parallel^2}{q} \frac{\mathbf{r} \times \widehat{\mathbf{b}}}{Br^2} = 12\pi \frac{MU_\parallel^2}{q\mu_0 \mathcal{M}_E} r\mathbf{r} \times \widehat{\mathbf{b}},$$

  with $\widehat{\mathbf{b}}$ being a unit vector in the direction of $\mathbf{B}$.

  At $\theta = \pi/2$ we have also $\nabla B = \partial B/\partial r = -3B/r$. The ∇B-drift is

$$\mathbf{U}_{\nabla B} = -3\frac{MU_\perp^2}{2qB} \frac{\widehat{\mathbf{b}} \times \mathbf{r}}{r^2} = 6\pi \frac{MU_\perp^2}{q\mu_0 \mathcal{M}_E} r\mathbf{r} \times \widehat{\mathbf{b}},$$

which adds to $\mathbf{U}_C$. With $\langle U_\perp^2 + U_\parallel^2 \rangle = \langle U_x^2 + U_y^2 + U_z^2 \rangle = 3\langle U_x^2 \rangle = 3\langle U_y^2 \rangle = 3\langle U_z^2 \rangle = 2\kappa T/M$ we have the net result

$$\mathbf{U}_{\nabla B} + \mathbf{U}_C = \frac{16\pi\kappa T}{q\mu_0 \mathcal{M}_E} r\mathbf{r} \times \widehat{\mathbf{b}}.$$

For a fixed particle energy, the effects of the gradient in magnetic field increases as r^2: the weaker the magnetic field, the larger is the particle's Larmor radius, and then it samples a larger variation in magnetic field, thereby increasing the effect of ∇B on the average drift velocity.

The gravitational force on the ion is $-\mathbf{r}GMM_E/r^3$ with M_E being the mass of the Earth. The drift velocity is

$$\mathbf{U}_G = -\frac{4\pi GMM_E}{q\mu_0 \mathcal{M}_E} \mathbf{r} \times \widehat{\mathbf{b}}.$$

The magnitude of $\mathbf{U}_G$ is proportional to r, i.e. it increases with distance to the Earth. For a dipole, the magnetic field decreases faster than the gravitational force. The direction of the two drift velocities, $\mathbf{U}_G$ and $\mathbf{U}_{\nabla B} + \mathbf{U}_C$ are opposite, irrespective of the sign of q. The ratio of their magnitudes is

$$\frac{|U_G|}{|U_{\nabla B} + U_C|} = \frac{GMM_E}{4\kappa T} \frac{1}{r}.$$

We see that a natural length scale for the present problem is $L \equiv GMM_E/\kappa T$ which depends on the particle energy. In order to ignore the gravitational drift we require $L/r \ll 1$, i.e. large radial positions.

To estimate the magnitude of L we insert the appropriate numerical values of the parameters entering and have $L = 6.7 \times 10^{-11} \times 1.66 \times 10^{-22} \times 6 \times 10^{24}/1.4 \times 10^{-23} \times 10^7 \approx 5 \times 10^8$ m, using the atomic mass unit for M and $T = 10^7$ K, corresponding to approximately 1 keV particles. Note that the magnetic moment $\mathcal{M}_E$ of the Earth does not enter. For comparison we have the Earth's radius $R_E = 6.4 \times 10^6$ m. For these parameters we find an equilibrium $|U_G| \approx |U_{\nabla B} + U_C|$ at a distance of $r \approx 25R_E$. For distances smaller than this we find that particles at these energies have their steady state motions dominated by gravitational drifts. (We keep in mind that the magnetic dipole approximation is applicable only for $r < 10R_e$ for the Earth.) If we take a higher energy such as 1 MeV, corresponding to $T = 10^{10}$ K as appropriate for radiation belt particles, this equilibrium is reached at $r \approx 0.025R_E$, i.e., inside the Earth, and in this case the effects of the gravitational force on the steady state particle motion is negligible for all relevant radial positions.

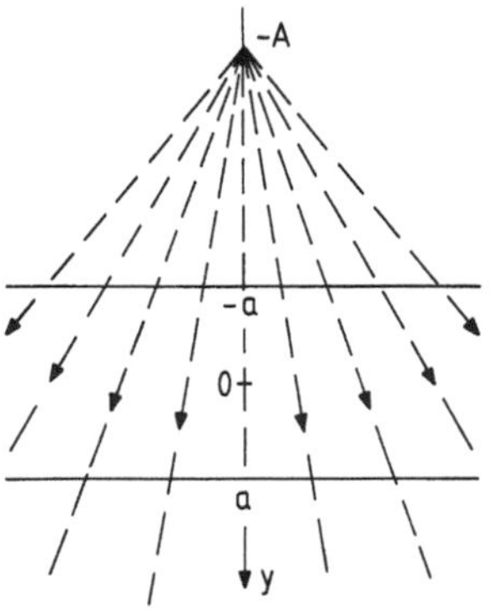

FIGURE 7.5: *Illustration of the direction of the model magnetic field, and positions of conducting plates. The directions of the magnetic field lines are shown by dashed lines and arrows. The y-axis for the cylindrical symmetry is here vertical, and the r-axis horizontal.*

- **Exercise:** Express the average drift velocity (7.23) in terms of the magnetic moment, μ.

Ⓢ **Exercise:** An interesting relation for the magnetic field generated by variation in the radiation belts of magnetized planets have been presented by Dessler and Parker. The basic results were restrictive in the sense that they dealt only with particles confined to the magnetic equatorial plane. The results were later generalized by Sckopke (1966).

To discuss this problem, determine the $\nabla B \times \mathbf{B}$ drift velocities of particles with given kinetic energy placed at a radius r at the magnetic equator plane in the Earth's magnetic field. Given the assumptions, what is the magnetic field line curvature drift velocity?

The motion of charged particles, ions and electrons gives rise to currents, which should be calculated. These currents give rise to magnetic fields, which can be detected at the Earth's surface. It is important that the gyrating particles give current contributions for two reasons. First, a charged particle gyrating around a magnetic field line can be viewed as a small current loop, with an associated magnetic dipole moment, which gives rise to a magnetic field. This will be the case also for homogeneous magnetic fields. In addition to this we have for the charged particles moving with the $\nabla B \times \mathbf{B}$-drift an additional current contribution. A simplified description can be obtained by considering a plastic disc with a small circular groove close to the edge, at some distance from the center of the disc. If we let a small charged particle move in a circular orbit in the groove, it gives a current. If we in addition let the disc rotate we get an additional current contribution. Calculate the resulting current density when the plasma density is n. Estimate the magnetic field at the surface of the Earth due to a population of particles with kinetic energy $\mathcal{W}$ and density n. Ⓢ

Ⓢ **Exercise:** Consider a cylindrical coordinate system, (r, y), where the y-axis is the symmetry axis. At the positions $y = -a$ and $y = a$ we have two parallel metal plates. The system is embedded in an inhomogeneous magnetic field, where we can assume, as an approximation, that all magnetic field lines emerge from a point $(r, y) = (0, -A)$, where $A \gg a$; see also Fig. 7.5. Note that by this assumption, we are *not* implying that there is spherical symmetry: the intensity of the magnetic field lines varies with direction, as explained in the following.

In the plane perpendicular to the y-axis at the position $y = 0$ we assume we have the y-component of the magnetic field B_0 independent of the radial position r. Make a sketch of the magnetic field vectors along the line $y = 0$. (Why is the proposed magnetic field only possible as an approximation?) Note that we do *not* assume the system to be in a vacuum: there *is* a distributed current system which maintains the magnetic field.

1): Demonstrate that with the given assumptions (see in particular Fig. 7.5), we have the magnetic field component B_y independent of r *everywhere.*

2): Determine B_y, $\partial B_y / \partial y$, and the radial magnetic field component B_r as a function of r as well as y between the plates. Give the exact expression, and present also a series expansion, accurate to first order in y/A.

A charged particle with mass M, charge q and velocity $\mathbf{U}_0 \perp \widehat{\mathbf{y}}$ is gyrating around the magnetic field lines at the origin, $(r, y) = (0, 0)$.

3): Determine the particle gyro-radius r_L and its magnetic moment.

4): Because of the magnetic field inhomogeneity, the particle is subject to a force. Determine this force in magnitude as well as direction, assuming r_L to be small.

This force is now neutralized at $y = 0$ by charging the metal plates with positive and negative surface charges, $\pm\sigma$, respectively.

5): How strong do you need the electric field to be, and what is the surface charge σ needed to achieve this field?

In a short time interval, the particle is now given a small velocity component $U_{\parallel} \ll U_0$ along the magnetic field.

6): How is the motion of the particle's gyro-center, if we can assume that $\partial B_y/\partial y = \text{const.}$?

7): Describe the motion of the gyro-center when for $\partial B_y/\partial y$, you use the series expansion from question 1). It is assumed that $U_{\parallel}$ is so small that the particle does not reach any of the conducting plates within time scales of interest.

8): Determine the spatially distributed current system, which gives rise to the postulated magnetic field. Ⓢ

7.1.9 Magnetic mirror confinement

A charged particle can be reflected by the mirror force (7.27), and in principle we can envisage that charged particles can be trapped indefinitely between two regions of enhanced magnetic fields. The trapping of a charged particle between two stationary magnetic mirrors is, however, not perfect: a particle moving parallel to, or nearly parallel to, the magnetic field lines will pass through such a mirror, since the magnetic moment is negligible, and correspondingly also the mirror force is negligible.

By (7.32) we have generally

$$\mathcal{W}_{kin} = \frac{1}{2}MU_{\parallel}^2 + \frac{1}{2}MU_{\perp}^2 = \frac{1}{2}MU_{\parallel}^2 + \mu B,$$

or

$$\frac{1}{2}MU_{\parallel}^2 = \mathcal{W}_{kin} - \mu B. \tag{7.35}$$

The left-hand side of this expression is always ≥ 0, imposing restrictions on the right-hand side, i.e., limiting the values of magnetic field being accessible. Particle reflection will occur at positions where $U_{\parallel} = 0$, i.e., where $\mathcal{W}_{kin} = \mu B$. Since $\mathcal{W}_{kin}$ is constant for particles in time-stationary magnetic fields, and also μ is a constant, at least to the relevant accuracy, reflection occurs at positions where the magnetic field is $B = \mathcal{W}_{kin}/\mu$.

In order to estimate the critical ratio between parallel and perpendicular velocity components, which determine the criteria for reflection, we recall the invariance of μ, and also that a steady state magnetic field cannot change the energy of a particle. We thus have two constants of motion, $\frac{1}{2}MU_{\perp}^2/B$ and $\frac{1}{2}MU^2$. The problem is to make proper use of these two constants for a given magnetic field configuration.

We can write the appropriate value of μ at two positions with different magnetic fields, B_0 and B_1

$$\mu = \frac{1}{2}MU_{0\perp}^2\frac{1}{B_0} = \frac{1}{2}MU_{1\perp}^2\frac{1}{B_1},$$

where it is implied that the magnetic field intensities refer to the z-direction. This expression will be correct at any two accessible particle positions labeled $_0$ and $_1$.

At any position, we can define a local "pitch angle" for the particle as the angle between the velocity vector and the magnetic field, $\Theta \equiv \text{ArcSin}(U_{\perp}/U_T)$, where $U_T \equiv |\mathbf{U}| \equiv \sqrt{U_{\perp}^2 + U_{\parallel}^2}$; see Fig. 7.6. For particles propagating strictly *along* the magnetic field we have $\Theta = 0$.

We now seek a relation which, for a given magnetic configuration and given pitch angle, can determine whether particles are reflected by points of local maxima in the magnetic field intensity, and relate this result to the pitch angle. At the turning point we have by definition $U_{\parallel} = 0$. We assume that the particle has a **B**-parallel velocity component $U_{0\parallel}$ at the position where the magnetic field is B_0. Conservation of energy implies $U_{1\perp}^2 = U_{0\parallel}^2 + U_{0\perp}^2 \equiv U_T^2$, since, by assumption, all energy is in the motion perpendicular to **B** at the position of reflection. We therefore have

$$\frac{B_0}{B_1} = \frac{U_{0\perp}^2}{U_{1\perp}^2} = \frac{U_{0\perp}^2}{U_T^2} \equiv \sin^2\Theta . \tag{7.36}$$

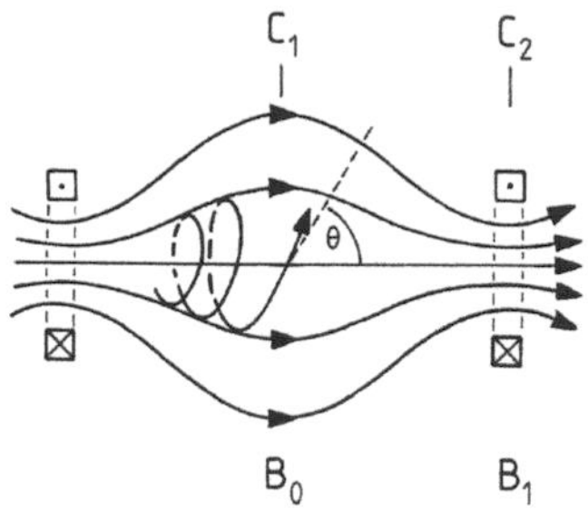

FIGURE 7.6: *Schematic illustration of a particle moving between two magnetic mirrors. The magnetic field is minimum at C_1 with a value B_0 and maximum at C_2 with a value B_1.*

The physical implication of this result is that the guiding center of a particle which has a pitch angle Θ at the position where the magnetic field is B_0 will move along the magnetic field until the particle is reflected at a position characterized by a magnetic field B_1, at which point *all* the kinetic energy of the particle is in the **B**-perpendicular motion, and $U_\parallel = 0$. The particle will propagate without reflection, in case such a magnetic field value is nowhere to be found,

Referring to a specific magnetic mirror made up by two (or more) coil systems, we now let B_1 be the maximum value of the magnetic field at the "throat" of the mirror, at position C_2 in Fig. 7.6 and B_0 be the minimum value, located between the two coil systems, at position C_2 in Fig. 7.6. The ratio B_0/B_1 is then called the *mirror ratio.* (Sometimes this quantity is defined as the ratio B_1/B_0 so be careful!) We find that a particle starting out at the position with B_0 with a pitch angle larger than Θ_0 will be trapped between the mirrors. Particles with pitch angle smaller than Θ_0 will escape. This will in general be an irreversible process; a particle once lost is unlikely to be "recaptured". Since microscopic processes, collisions, fluctuations, etc., will give rise to scattering of the particles, there will always be some which are scattered into the *loss cone*, and consequently any magnetic mirror will eventually be depleted (the magnetic mirror is "leaking"), unless its particle content is maintained somehow. Since the magnetic field coils take up some space, it might in reality be difficult to obtain mirror ratios smaller than 1/5 in a laboratory (or, as you may prefer, larger than 5), implying that particles with pitch angles in the range $\{0°; 25°\}$ will be lost right away. This is a fairly large fraction. It can be demonstrated that the strongly anisotropic "loss cone" velocity distribution, which results from some of the particles being lost, while others are confined, is *unstable* in a kinetic description.

When we have two magnetic mirrors, i.e., two regions with strong axial magnetic field enhancements, a charged particle can be trapped between them. Some fusion device schemes rely on such configurations, and magnetic mirrors are also important for the motion of particles in geophysical and astrophysical plasmas. In particular the Earth's magnetic field is strongest at the magnetic poles and weaker at the equator. These two magnetic mirrors are able to trap even energetic particles. These regions of the Earth's magnetosphere are called the *radiation belts.*

- **Example:** For a magnetic mirror as illustrated in Fig. 7.6 we can approximate the local variation of the magnetic field intensity by $B = B_0(1 + x^2/L^2)$, where x is along the horizontal symmetry axis on the figure, with the position $x = 0$ being at the midpoint between the two coils. Take a charged particle with kinetic energy $W = \frac{1}{2}M(U_\perp^2 + U_\parallel^2)$ at $x = 0$. Determine the pitch angle for the particle if it is to be confined between $x = -L$ and $x = L$, and the same question now with the two positions being $x = \pm L/2$. Set up the equation of motion for the gyrocenter of the particle, and demonstrate that the solution is a harmonic oscillation and give its characteristic frequency.

Ⓢ **Exercise:** Take a dipole model for the Earth's magnetic field. What is the mirror ratio and loss cone angles for magnetic flux tubes that start over the magnetic pole in the top ionosphere at $\sim 10^3$ km altitude and continue to a distance of a) $\sim 5\ R_E$ over the equator, and b) $\sim 10\ R_E$ over the magnetic equator. Ⓢ

7.2 Adiabatic invariants

It is well known in classical mechanics that for a system having a periodic motion, in terms of generalized momentum and position p and q, the action integral $\oint p dq$ taken over a period is a constant of the motion (Goldstein 1980). The theorem is understood best when illustrated by some examples.

1) The generic periodic motion for magnetized plasmas is the gyration of charged particles with angular frequency Ω_c. We take here $p = MU_\perp r$, the angular momentum, and $q = \theta$ being the angle of the coordinate vector with respect to a reference line. The angular momentum is here independent of θ so the q-integral is simply 2π. Since $r = r_L = MU_\perp/qB$, the action integral becomes $4\pi\mu M/|q|$, where we introduced the magnetic moment $\mu \equiv MU_\perp^2/2B$. Thus μ is a constant of motion as long as $M/|q|$ remains constant, as already demonstrated in (7.33). This will not hold, for instance, for charged dust grains moving in a magnetic field, since their charges will in general vary with time.

2) A second invariant can be associated with a particle trapped between two magnetic mirrors at position a and b. In this case the particle bounces back between the two mirrors, and we can define a "bounce frequency". A constant of motion is then $\frac{1}{2}MU_\parallel ds$, where ds is an element of the path length of the guiding center motion along a magnetic field line. We introduce a *longitudinal invariant* J as

$$J = \int_a^b U_\parallel ds \tag{7.37}$$

where the two "turning points" are denoted a and b, where these positions will in general vary slowly with time, i.e., $a = a(t)$ and $b = b(t)$.

Consider, for instance, a charged particle trapped between two approaching magnetic mirrors, as illustrated in Fig. 7.7. Assume that we consider the problem in a frame of reference where the mirror to the left is at rest and the mirror at the right-hand side is approaching with a velocity U. By reflection from the moving mirror the charged particle gains a velocity increment of $2U$ along $\mathbf{B}$, as the reader may verify. In order to have adiabatic (i.e., slow) variations of $U_\parallel$ during a transit, we require $U_\parallel \gg U$. Although the changes in velocity are, by nature, occurring at discrete times, we apply a "coarse grained" description, where we assume $U_\parallel$ to be continuously varying with time. With this approximation, we have that the frequency with which the particle encounters the moving mirror varies in such a way that the rate of velocity change becomes

$$\frac{\Delta U_\parallel}{\Delta t} \approx \frac{d}{dt}U_\parallel = \frac{U_\parallel}{2L}2U = -\frac{U_\parallel}{L}\frac{dL}{dt} \tag{7.38}$$

at an instant where the two mirrors are separated by a distance $L = L(t)$. We used that the velocity change (positive or negative) upon reflection is $\Delta U_\parallel = 2U$, while the the transit time

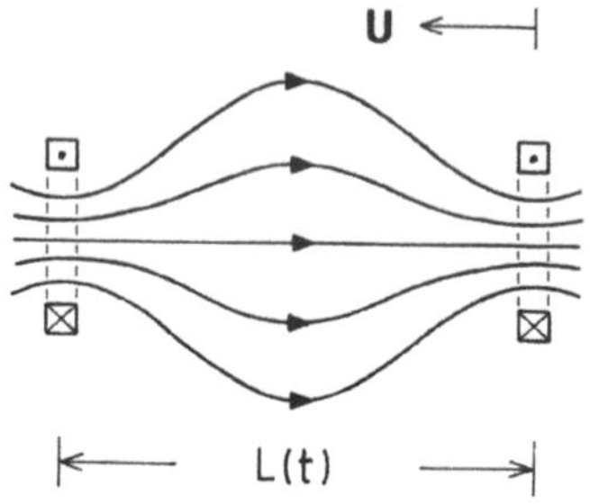

FIGURE 7.7: *Schematic illustration of a particle trapped between two magnetic mirrors, one stationary (the one to the left) and one moving. With the right magnetic mirror approaching the left one we have the separation being a function of time, $L = L(t)$.*

is $\Delta t = 2L/U_\parallel$. We also used $U = -dL/dt$. The term "coarse grained" description here means that we have many reflections by the moving mirror in the time span we consider. Using (7.38) we find $d(U_\parallel L)/dt = 0$, i.e., the desired result $U_\parallel L =$ const. implying that the particle will gain energy if L is decreasing, corresponding to approaching reflectors or mirrors. The result applies equally well for *receding* mirrors, in which case a particle is *losing* energy by multiple reflections. It is readily seen that the result (7.38) is consistent with (7.37).

Fermi (1949) suggested that parts of the cosmic radiation could be explained by large current systems moving in space, where, given time, charged particles reflected by the mirror forces could in principle be accelerated to arbitrary large velocities, according to (7.38), for instance. He distinguished two mechanisms, type I and type II, where the former considers the acceleration as being due to one large impact by a rapidly moving magnetic mirror or similar, while in type II acceleration the increase in energy is due to the accumulated effect of many small impacts. The explanation may not (at least today) be considered as adequate for the generation of the cosmic radiation, but Fermi's ideas formed the basis of a vast amount of physical and mathematical studies of acceleration processes in general (Lichtenberg & Lieberman 1983, Pécseli 2000).

- **Example:** The simple pendulum is often considered as an example for illustrating adiabatic invariants (Clemmow & Dougherty 1969). For a constant length ℓ of the string and with small angular deflections θ, the energy is $\mathcal{W} = \frac{1}{2}m\ell^2(d\theta/dt)^2 + \frac{1}{2}mg\ell\theta^2$, being the sum of kinetic and potential energies. With constant ℓ we have $\mathcal{W}$ being a constant. The tension of the string is readily found as $T = mg + m\ell(d\theta/dt)^2 - \frac{1}{2}mg\theta^2$, recalling that the centrifugal force for a particle with velocity U in local circular motion is mU^2/ℓ. The projection of the gravitational force on the direction of the string is $mg\cos\theta \approx mg(1 - \frac{1}{2}\theta^2)$. The average over one period (illustrated by an overline) is $\overline{T} = mg + \mathcal{W}/(2\ell)$, the period being $2\pi\sqrt{\ell/g} \equiv 2\pi/\omega$.

  When ℓ is made to vary while the pendulum oscillates, then work is done on the system by the tension in the string. With the increment $\Delta\ell$ of the string being small during one period, we have the work to be $-\int T dl \approx -m(g + \mathcal{W}/(2\ell))\Delta\ell$. The first term represents the change in gravitational potential energy; the second is the energy in the oscillations. Thus $\Delta\mathcal{W} = \frac{1}{2}(\mathcal{W}/\ell)\Delta\ell = \mathcal{W}\Delta\omega/\omega$, with $\Delta\omega$ being the change in frequency of oscillation for the pendulum. On the other hand, we have $\Delta(\mathcal{W}/\omega) = \Delta\mathcal{W}/\omega - \mathcal{W}\Delta\omega/\omega^2$, so the previous result demonstrates that $\Delta(\mathcal{W}/\omega) = 0$ and thus $\mathcal{W}/\omega$ is an adiabatic invariant. Energy divided by frequency has the dimension of "action" just as Planck's constant, so the result obtained here was considered important in the discussions of the early days of quantum mechanics.

Ⓢ **Exercise:** Consider two magnetic poles, one "north" and one "south", made of some magnetic material, as illustrated in Fig. 7.8. In the central plane we have the magnetic field as $\mathbf{B}(\mathbf{r}) = B_0(r)\widehat{\mathbf{z}}$, where r is in the horizontal direction $\perp$ $\mathbf{z}$, and $\widehat{\mathbf{z}}$ is a unit vector. Assume that we placed one free electron (charge e and mass m) at the position $(r_0, 0)$ in the mid-plane. Let the electron velocity be U_0.

a): Demonstrate that the electron velocity (or better: speed) is U_0 at all later times. Give the magnitude and direction of the electron velocity we need to have the electron performing a circular orbit with radius r_0 in the x,y-plane; see Fig. 7.8b). Express the magnetic moment of the charged particle in terms of the magnetic field as averaged over the area of the circle encompassed by the particle. Is the orbit stable with respect to vertical perturbations? Is this the case for any cylindrically symmetric configuration of the magnetic poles in Fig. 7.8a)?

We now replace the electron path with a thin circular conducting wire, with resistivity R. We ignore a possible self-inductance. The wire is placed precisely so it follows the circular orbit discussed before. We now let the magnetic field be time varying, $\mathbf{B}(\mathbf{r},t) = B_0(r)\sin(\omega t)\widehat{\mathbf{z}}$, where ω is a constant, assumed to be given.

b): Assume that the electric field in the x,y-plane can be written as $\mathbf{E}(\mathbf{r},t) = E(r,t)\widehat{\phi}$, where $\widehat{\phi}$ is a unit vector in the direction given by $\mathbf{z} \times \mathbf{r}$. Determine $E(r,t)$.

c): Give an expression for the induced current $I(t)$ in the wire. Give the magnitude and direction of the current at a time $t = \pi/(4\omega)$.

We now retain the time-varying magnetic field given before, but remove the wire, and re-introduce the electron. Assume that the initial position is $(r_0, 0)$ as before, but its initial velocity is now vanishing.

d): Determine the condition on the magnetic field variation $B_0(r)$ which makes the electron follow the circular orbit discussed before. Note that we require only that it follows the *orbit*, without requirements on the velocity, which will be time varying! Give an expression for the electron momentum, $p(t)$.

e): Demonstrate that the results from question d) are correct even if the electron moves with relativistic velocities. The latter question is relevant for the so-called "Betatron". Ⓢ

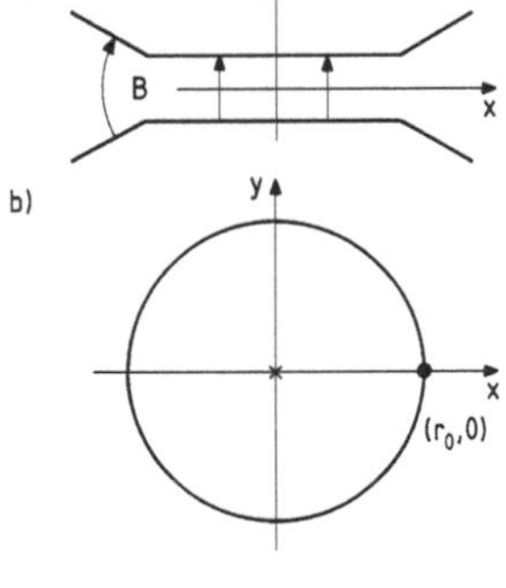

FIGURE 7.8: *Simple illustration of a set-up with magnetic north and south poles in a.) The geometry is cylindrically symmetric with respect to the z-axis, and symmetric with respect to the horizontal mid-plane as well.*

7.3 Radiation losses

Charged particles radiate electromagnetic waves when they are accelerated or decelerated (generation of electromagnetic waves with antennas relies on this). Energy conservation implies that the radiated electromagnetic wave energy must be taken from the energy of the particle, and will appear as an effective friction. A non-relativistic phenomenological model equation accounting for this loss can be taken as

$$M\frac{d}{dt}\mathbf{U}(t) = \mathbf{F} + M\tau\frac{d^2}{dt^2}\mathbf{U}(t), \tag{7.39}$$

where τ is an "effective friction coefficient" and $\mathbf{F}$ is an unspecified external force. For an electron with mass m and charge $-e$ it can be demonstrated (Clemmow & Dougherty 1969, Jackson 1975) that

$$\tau = \frac{e^2}{6\pi\varepsilon_0 mc^3}.$$

Evidently, τ is related to the classical electron radius $r_e = e^2/(4\pi\varepsilon_0 mc^2)$ m.

The strict accuracy of the simple model (7.39) can easily be criticized for various reasons (Clemmow & Dougherty 1969, Pécseli 2000). Since, however, τ is a rather small quantity for electrons ($\tau \approx 6.3\times 10^{-24}$ s), the expression can be used for a wide range of parameters to a reasonable accuracy. Consider, for instance, a simple model problem, where an electron oscillates in one spatial dimension with frequency ω_0 around a reference position, here taken as the origin. The model equation for this case is

$$\frac{d^2}{dt^2}X(t) + \omega_0^2 X(t) = \tau\frac{d^3}{dt^3}X(t).$$

A general solution for the frequency has $\omega^2 = \omega_0^2 + i\tau\omega^3$. For $\omega_0 \ll 1/\tau$, we easily find $X(t) = \exp(-i\omega_0 t - \frac{1}{2}\tau\omega_0^2 t)$, corresponding to a damped oscillation. The model discussed here can be valuable in, for instance, discussing cyclotron radiation from magnetized charged particles. This phenomenon is of great importance as a diagnostic of stellar atmospheres, for instance. The analysis can become very complicated (Trulsen & Fejer 1970, Trulsen 1971), in particular if relativistic effects are to be included (Bekefi 1966, Clemmow & Dougherty 1969).

Note that in this section we discussed only the energy losses due to radiation for accelerated/decelerated particles in a simple model, and not the radiation *patterns* of the resulting radiation. The quantum corrections in the limit of very small gyro radii are outside the scope of the present treatise.

Ⓢ **Exercise:** Apart from a numerical constant, derive the classical electron radius

$$r_e \equiv \frac{1}{4\pi\varepsilon_0}\frac{e^2}{mc^2}$$

by dimensional reasoning. The numerical value is $r_e = 2.8179403227\times 10^{-15}$ m. Find the physical interpretation of this quantity in the literature. Ⓢ

- **Exercise:** Consider a numerical example by taking an electron in the Earth's magnetic field at one of the geomagnetic poles. Calculate approximately the energy lost due to cyclotron radiation in one period of the cyclotron motion.

8

Basic Plasma Parameters

If the density of the charged particles is sufficiently large, we have to account for their interaction by electric and magnetic forces. Basically, this interaction can take place in two different forms. First it can be by a collisional process which essentially involves two particles only. Alternatively we can imagine situations where a large number of particles *simultaneously* influence the trajectory of a selected reference particle. In this limit we talk about "collective interaction" of charged particles. Properly speaking this is what we also call the *plasma limit*, as discussed later.

A plasma has a number of characteristic parameters with dimensions time and length. When classifying plasma conditions it is a great advantage to use these as basic units.

8.1 Plasma frequencies

A characteristic time scale can be obtained as the inverse of the electron plasma frequency

$$\omega_{pe} \equiv \sqrt{\frac{e^2 n}{\varepsilon_0 m}} \tag{8.1}$$

where n is the number density of the electrons, and m is the electron mass. The plasma period is then $\tau_p \equiv 2\pi/\omega_{pe}$. (We have already, in Chapter 7, encountered the cyclotron frequency ω_{ce}, but this is a frequency referring to the motion of a *single particle* and is of limited use for characterizing a plasma which is an *ensemble* of particles. The cyclotron frequency applies only for magnetized plasmas, while the plasma frequency is relevant for any macroscopic ensemble of particles that can be assigned a particle density.)

Evidently, we can associate different plasma frequencies with different species, ions and electrons in particular, by introducing the respective densities, charges and masses. Usually, $-e$ is the electron charge with $e > 0$, but in case multi-charged ions are considered it is of course the appropriate ion charge and mass with the corresponding densities that should be inserted. The physical meaning of the plasma frequency was briefly addressed in Section 5.9.1, and will be discussed again in Section 12.1.

- **Exercise:** Derive the plasma frequency, ω_{pe}, by dimensional arguments; see Appendix A.

Ⓢ **Exercise:** A plasma with electron density 10^{11} cm^{-3} is placed in a homogeneous magnetic field B. Determine the value of B that gives the electron plasma frequency ω_{pe}, equal to the electron gyro-frequency ω_{ce}. Ⓢ

8.2 The Debye length

From dimensional considerations it can be demonstrated that a quantity having dimension length can be constructed from density n, temperature T, Boltzmann's constant κ, electric charge e and the vacuum permeability ε_0. These have (written in MKSA units) dimensions $[n] = length^{-3}$, $[\kappa T] = mass \times length^2 \times time^{-2}$, $[e] = coulomb = ampere \times time$, $[\varepsilon_0] = ampere^2 \times time^4 \times mass^{-1} \times length^{-3}$, respectively, with angular brackets $[a]$ denoting the dimensions of a quantity a. It is readily demonstrated (see Appendix A) that the only nontrivial dimensionally correct combination of the quantities having dimension *length* is

$$\lambda_D = \sqrt{\frac{\varepsilon_0 \kappa T}{e^2 n}}, \tag{8.2}$$

called the *Debye length*, as the subscript D refers to this scientist's name. (A trivial length scale would be $n^{-1/3}$.) We can obtain a Debye length characterizing a particular species by inserting the appropriate temperature and density. We can, for instance, have a mixture of different ions with different densities, requiring only net charge neutrality of the plasma. We also anticipate that different species can have different temperatures, in particular electrons and ions, so that the time for thermal relaxation between electrons and ions differs from the time it takes the electrons to reach an "internal" thermal equilibrium.

The thermal velocity of a plasma particle is here given as $u_{The} \equiv \sqrt{\kappa T/M} = \omega_{pe}\lambda_D$. The Debye length can be interpreted as the distance traveled by a plasma particle within one appropriate plasma period, where admittedly we have been somewhat sloppy with a numerical coefficient in defining the thermal velocity.

8.3 Debye shielding

Physically, the Debye length is characterizing a shielding distance as demonstrated in the following. The dimensional arguments given before were based solely on *electron* parameters, but a shielding can also be obtained by a simple re-definition of λ_D when the ions are included in the analysis. The two cases (mobile and immobile ions) will be considered separately. It is important to emphasize that we here consider the re-adjustment of the plasma density when a point-like charge perturbation is introduced. We do *not* here consider losses of charged particles interacting with a solid surface: these effects are important when the body immersed in the plasma has a finite size. In two or three spatial dimensions we see no problems with this: if the external charge is located at a point (in 3D) or a line (in 2D), we can ignore the probability for plasma electrons and ions to reach physical contact with the reference charge. We will, however, consider also a one-dimensional model, which physically corresponds to a charged plane. Admittedly, in this particular case it is a bit artificial to ignore physical contact between the external charge and the plasma particles. For such a model plasma particle losses are important. This and related cases will be considered separately in more detail in relation to plasma probes.

8.3.1 Immobile ions

Assume that an immobile point charge q is placed at the origin of the coordinate system and surrounded by a plasma with electron and ion densities n_e and n_i, respectively. Poisson's equation is written as

$$\nabla^2\phi(\mathbf{r}) = \frac{e}{\varepsilon_0}\left(n_e(\mathbf{r}) - n_i(\mathbf{r}) - \frac{q}{e}\delta(\mathbf{r})\right), \tag{8.3}$$

in terms of the electrostatic potential ϕ. The electron charge is with the present notation $-e$. Assume first that the ions are immobile and $n_i = n_0$. The problem is time stationary and it is physically plausible to assume the electron distribution being of the form

$$f_e(\mathbf{u},\mathbf{r}) = n_0\left(\frac{m}{2\pi\kappa T_e}\right)^{3/2}\exp\left(-\frac{\frac{1}{2}mu^2 - e\phi(\mathbf{r})}{\kappa T_e}\right), \tag{8.4}$$

where T_e is a constant electron temperature. Integration of (8.4) with respect to velocity yields the isothermal Boltzmann distribution

$$\int_{-\infty}^{\infty} f_e(\mathbf{u},\mathbf{r})d\mathbf{u} \equiv n_e(\mathbf{r}) = n_0\exp\left(\frac{e\phi(\mathbf{r})}{\kappa T_e}\right), \tag{8.5}$$

for the electrons in the potential $\phi(\mathbf{r})$.

Unfortunately, we are not in general able to solve (8.3) for the electrostatic potential, $\phi(\mathbf{r})$. Expecting relevant magnitudes of the potential to be small, we use $e^x \approx 1 + x + \frac{1}{2}x^2 + \ldots$ and linearize (8.5) to give

$$n_e(\mathbf{r}) = n_0\left(1 + \frac{e\phi(\mathbf{r})}{\kappa T_e}\right), \tag{8.6}$$

and insert this approximation in (8.3). With immobile ions, $n_i = n_0$, we then have a simple closed equation for ϕ in the form

$$\nabla^2\phi(\mathbf{r}) = \frac{e}{\varepsilon_0}\left(n_0\frac{e\phi(\mathbf{r})}{\kappa T_e} - \frac{q}{e}\delta(\mathbf{r})\right).$$

8.3.1.1 Shielding in three spatial dimensions with spherical symmetry

We can write Poisson's equation in spherical coordinates in terms of the corresponding ∇-operator, see Appendix D.8.2. This will be appropriate when the disturbance of the plasma is caused by a point charge or a multipole. With the given spatial symmetry we find the potential varying with the radial variable only and we have

$$\nabla^2 \to \frac{1}{r^2}\frac{\partial}{\partial r}r^2\frac{\partial}{\partial r}.$$

Because of the given symmetry, n_e is also a function of the radial coordinate only and Poisson's equation is here readily solved to give

$$\phi(r) = \frac{q}{4\pi\varepsilon_0}\frac{\exp(-r/\lambda_D)}{r}, \tag{8.7}$$

where the Debye length was introduced. We assumed the reference potential $\phi(r\to\infty)\to 0$. Close to the origin, i.e., close to the charge q, the solution has the form of the free-space solution $(q/4\pi\varepsilon_0)/r$, and a shielded potential for larger distances, with a shielding distance given by λ_D; see Fig. 8.1. For small distances the linearization used in (8.6) breaks down, but this is of little consequence since the plasma shielding is negligible there anyhow. (For large values of the charge q, the linearization may give rise to nontrivial modifications of the exact result.) Three-dimensional nonlinear shielding models have been discussed by, for instance, Lin and Zhang (2003).

- **Exercise:** Demonstrate that the electric field surrounding the *unshielded* charge is in all spatial positions smaller than the field for the *shielded* charge, although the two are very close for $r \ll \lambda_D$.

Ⓢ **Exercise:** Determine the electron Debye length, λ_{De}, for high grade copper at room temperature. Find the specific weight in the literature, and assume one free electron per atom (the others do not contribute to the current). Compare this λ_{De} with an estimated distance between the atoms. What is the corresponding plasma parameter, $N_p = n\lambda_D^3$? Ⓢ

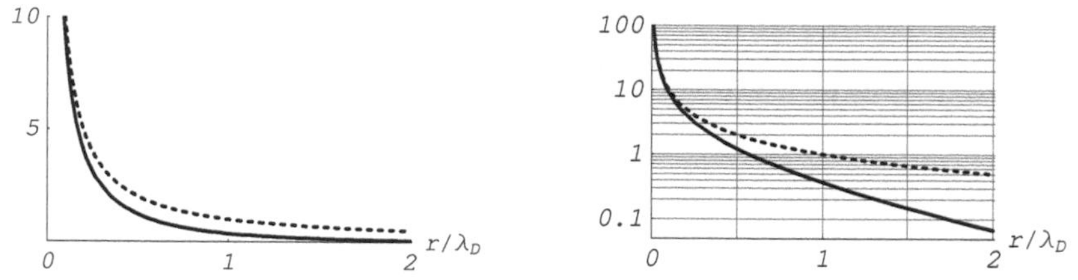

FIGURE 8.1: *Shielded potential in three spatial dimensions resulting from an externally introduced point-charge q. Distance is normalized by λ_D. Dashed line gives the unshielded potential. The result is shown in linear as well as semi-logarithmic scale.*

From the foregoing results it is now possible to calculate the net amount of charge within the Debye sphere. Using (8.6), we first write the density change induced by the charge q as

$$n_e(r) - n_0 = \frac{q}{e}\frac{1}{4\pi\lambda_D^3}\frac{\exp(-r/\lambda_D)}{r/\lambda_D}, \tag{8.8}$$

obtained by inserting ϕ from (8.7) into (8.6). Integration of the charge density over all space gives

$$-e\int_0^{2\pi}\int_0^{\pi}\int_0^{\infty}(n_e(r) - n_0)r^2\sin\theta dr\, d\theta d\xi = -q\int_0^{\infty}\gamma\exp(-\gamma)d\gamma = -q. \tag{8.9}$$

The net induced charge in the Debye sphere is thus exactly what is needed to compensate the charge q which was introduced, a result which could have been guessed right away. Assume that $q = e$. Then the net surplus of electrons in the Debye sphere is 1, as compared to the case without the perturbation in density induced by q.

Take a sphere with radius λ_D placed at a randomly selected position in the plasma. The number N of electrons in such spheres is a random variable and similarly for the ions. Assume the electrons to be placed independently of each other. This is not strictly correct (van Kampen & Felderhof 1967), but the error in the assumption is small when the plasma parameter is large. With the given assumption the number N will follow a Poisson distribution and the fluctuations in the number of electrons in a given volume are then $\sqrt{\langle(N - \langle N\rangle)^2\rangle} = \sqrt{\langle N^2\rangle - \langle N\rangle^2} = \sqrt{\langle N\rangle}$, where $\langle N\rangle$ is the average number of electrons in that volume. Since we took the volume in question to be the Debye sphere, we have $\langle N\rangle = N_p \gg 1$. In terms of the plasma parameter $N_p = n\lambda_D^3$ (see also Section 8.6) the fluctuations $\sqrt{N_p}$ in the number of electrons will generally be much larger than 1, and the Debye sphere will have a somewhat hazy appearance. If you could see individual atoms you would hardly notice the electron cloud surrounding a test charge, unless $q \gg e$.

8.3.1.2 Shielding with cylindrical symmetry

We write Poisson's equation in cylindrical coordinates in terms of the appropriate ∇-operator, see Appendix D.8.3. This limit will be appropriate for describing the perturbation caused for instance by a long thin charged wire. For the presently assumed cylindrical symmetry we have

$$\frac{1}{r}\frac{d}{dr}\left(r\frac{d}{dr}\phi\right) = \frac{e}{\varepsilon_0}\left(n_e - n_0 - \frac{q}{e}\delta(\mathbf{r})\right)$$

or, after linearization by use of (8.5),

$$\frac{1}{r}\frac{d}{dr}\left(r\frac{d}{dr}\phi\right)-\frac{1}{\lambda_D^2}\phi=-\frac{q}{\varepsilon_0}\delta(\mathbf{r}),$$

where we can use λ_D as a normalizing quantity for "length."

We find the solution $\phi(r)=aK_0(r/\lambda_D)$, where K_0 as also I_0 are modified Bessel functions of order "zero," and the coefficient a is determined by $q\lambda_D^2/\varepsilon_0$. The approximation for the Bessel function $K_0(r/\lambda_D)\approx-\ln(\frac{1}{2}r/\lambda_D)I_0(r/\lambda_D)$ for small $r\ll\lambda_D$ corresponds to the near field limit, with negligible shielding, while for large $r\gg\lambda_D$ we have exponential Debye shielding by $K_0(r/\lambda_D)\approx\exp(-r/\lambda_D)\sqrt{\pi\lambda_D/2r}$. It is important to note here that the dimension of $\delta(\mathbf{r})=\delta(x)\delta(y)$ is *length*$^{-2}$, so that the physical dimension of q is here *charge*×*length*$^{-1}$, i.e., a *line charge*. In Fig. 8.2 we show the potential associated with shielded as well as unshielded line charges.

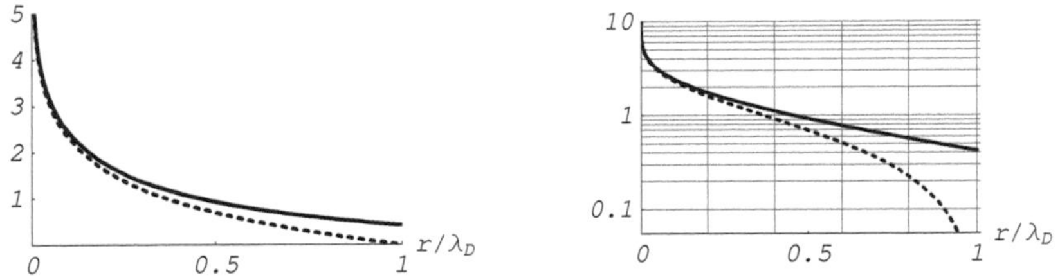

FIGURE 8.2: *Shielding in two spatial dimensions. The figure shows the shielding by K_0 of a line charge (solid line) and its unshielded logarithmic counterpart (dashed line).*

The two-dimensional problem may seem somewhat peculiar by giving a shielded potential being *larger* in amplitude than the unshielded one. We note, however, that the shielded electric field is $E_r\sim dK_0(r)/dr=-K_1(r)$, as compared to the unshielded field $E_r\sim-1/r$, where $K_1(r)<1/r$ for $r>0$, so that also in this case we have the shielded electric field being the smallest.

8.3.1.3 Shielding in one spatial dimension

In one spatial dimension we have Poisson's equation in the form

$$\frac{d^2}{dx^2}\phi(x)=\frac{e}{\varepsilon_0}\left(n_e(x)-n_0-\frac{q}{e}\delta(x)\right). \tag{8.10}$$

In the present case $q\delta(x)$ corresponds to a "slab" of charge. Again with the assumption of Boltzmann distributed electrons, (8.5), we linearize (8.10). We then readily find the solution $\phi(x)=a\exp(-|x|/\lambda_D)$, where a is an integration constant. At the origin, $x=0$, the second derivative is singular for this exponential solution, and the δ-function of (8.10) is thus recovered at this position.

In the absence of plasma, the potential will vary linearly with position on either side of the charged slab, consistent with constant electric fields having opposite directions on the two sides of the slab.

8.3.2 Mobile ions

The assumption of immobile ions can be trivially relaxed, as long as it is possible to assume an isothermal Boltzmann distribution for the ions also, with temperature T_i. Assuming singly charged ions we have

$$n_i(\mathbf{r})=n_0\exp\left(-\frac{e\phi(\mathbf{r})}{\kappa T_i}\right)\approx n_0\left(1-\frac{e\phi(\mathbf{r})}{\kappa T_i}\right) \tag{8.11}$$

with the assumption of singly charged ions. We linearized $\exp(-e\phi(\mathbf{r})/\kappa T_i)$ also here. Poisson's equation is again readily solved, this time to give

$$\phi(r) = \frac{q}{4\pi\varepsilon_0}\frac{e^{-r/\lambda_{ef}}}{r}, \tag{8.12}$$

in terms of an effective Debye length λ_{ef} given by

$$\frac{1}{\lambda_{ef}^2} = \frac{1}{\lambda_{De}^2} + \frac{1}{\lambda_{Di}^2},$$

where λ_{Di} is defined as λ_{De} with T_i replacing T_e. At first sight it seems obvious that the assumption of a steady state solution necessarily implies that the ions *must* be considered mobile. However, it can be argued that the time it takes the electrons to reach a Boltzmann equilibrium is of the order of the Debye length divided by a characteristic velocity such as the thermal velocity $u_{The} \equiv \sqrt{\kappa T_e/m}$, i.e., the time is $\sim \lambda_D/u_{The} = 1/\omega_{pe}$ in terms of the electron plasma frequency $\omega_{pe} = \sqrt{e^2 n_0/\varepsilon_0 m}$. The corresponding time for the ions to reach equilibrium is $\sim 1/\Omega_{pi}$, where $\Omega_{pi} = \sqrt{e^2 n_0/\varepsilon_0 M}$ is the *ion* plasma frequency. Although these estimates are rather crude, the relative error is the same for both ions and electrons, as long as the velocity distributions are close to Maxwellians. Since $\omega_{pe} \gg \Omega_{pi}$, it is possible to argue for the existence of a time interval where the ions are immobile but the electrons have reached an isothermal Boltzmann quasi-equilibrium.

With mobile ions, it is again possible to calculate the net amount of charge within the Debye sphere. The modification of the electron density induced by the charge q as contribution is obtained as before by inserting ϕ from (8.12) into (8.6). Integration over all space gives

$$-e\int_0^{2\pi}\int_0^{\pi}\int_0^{\infty}(n_e(r) - n_0)r^2 \sin\theta\, dr\, d\theta\, d\xi = -q\frac{T_e}{T_e + T_i}. \tag{8.13}$$

The modification in ion density is obtained the same way as

$$e\int_0^{2\pi}\int_0^{\pi}\int_0^{\infty}(n_i(r) - n_0)r^2 \sin\theta\, dr\, d\theta\, d\xi = -q\frac{T_i}{T_e + T_i}, \tag{8.14}$$

and the two contributions to the net charge add up to $-q$ as before. Unless the ion and electron temperatures are dramatically different, we can argue that the fluctuations in the densities of the ions as well as the electrons are generally much larger than the excess number of ions and electrons in a screening cloud of charge q, recalling that for, e.g., a positive charge q the local electron density *in*creases slightly, while it *de*creases for the ions.

The present analysis assumed explicitly that the test charge was at rest with respect to the plasma. Only then can be assume Boltzmann distributed ions as well as electrons. As far as the latter component is concerned this can be a safe assumption as long as a velocity is well below u_{The}, the electron thermal velocity, but the ions are so much slower in comparison, so with finite ion mobility is included we have to be careful (Wang et al. 1981). We will come back to this question in Section 21.

Finally a word of warning: for gravitational systems we have a force varying with distance as for charged particles, and by dimensional arguments it is possible also here to define an equivalent of a Debye length. Since we only have attractive forces for gravity, we have no shielding and for gravitational systems the physical meaning of the Debye length in the sense discussed before is lost.

- **Exercise:** Let the plate discussed before in the one dimensional model have a constant surface charge density σ. Express Φ_0 and the electric field E_0 just outside the plate in terms of σ.

Ⓢ **Exercise:** For a three-dimensional analysis it was necessary to linearize the expression for the isothermal Boltzmann distribution. It turns out that for the one-dimensional problem we are in a position to solve the problem without such a linearization (Sivukhin 1966). Assume that the plasma is terminated by a solid wall at a prescribed potential, Φ_0, with respect to the zero level at infinity. Ⓢ

- **Exercise:** The previous exercise considers a semi-infinite plasma, bounded at one end with a (metal?) plate at a prescribed potential. Re-analyze the problem where the plasma is bounded by *two* plates, separated by a distance $\mathcal{L}$.

8.4 Interaction energy Ⓚ

The induced potential, ϕ_{ind}, is defined as the difference between the actual potential in the plasma and eigen-potential, i.e., the $\sim 1/r$ potential variation associated with the charge q in free space. The induced potential is in effect a measure of the plasma's ability to shield the inserted charge q. Assuming for simplicity again the ions to be immobile, we find

$$\phi_{ind}(r) = \frac{q}{4\pi\varepsilon_0}\frac{\exp(-r/\lambda_D) - 1}{r}. \tag{8.15}$$

The interaction energy, i.e., the potential energy of the charge q in the potential induced in the plasma is then

$$q\phi_{ind}(r \to 0) = -\frac{q^2}{4\pi\varepsilon_0}\frac{1}{\lambda_D}, \tag{8.16}$$

since the charge will not have any potential energy in its eigen-potential. In obtaining (8.16) we used the series expansion $e^{-x} = 1 - x + \frac{1}{2}x^2 \ldots$

Introducing the test charge q, we change the plasma density distribution slightly, and thereby also the thermal energy distribution, $n\kappa T_e$, in the plasma. We can compare the perturbation in the thermal energy in a Debye sphere with the potential energy (8.16). Using (8.8), we find the change in thermal energy to be

$$\int_0^{2\pi}\int_0^{\pi}\int_0^{\infty} \kappa T_e(n_e(r) - n_0)r^2 \sin\theta \, dr \, d\theta \, d\xi = \frac{q}{e}\kappa T_e. \tag{8.17}$$

The ratio between the two energies (8.16) and (8.17) is

$$\begin{aligned} \left|\frac{q\phi_{ind}(r=0)}{\kappa T_e q/e}\right| &= \frac{q^2 e}{4\pi\varepsilon_0\lambda_D q\kappa T_e} \\ &= \left|\frac{q}{e}\right|\frac{1}{4\pi n\lambda_D^3} \equiv \left|\frac{q}{e}\right|\frac{1}{4\pi N_p}. \end{aligned} \tag{8.18}$$

We can now argue that for plasmas of interest, where $N_p = n\lambda_D^3 \gg 1$, the interaction potential energy is much less than the change in thermal energy in a Debye sphere, unless the charge q is very large. For instance, for charged dust particles we can have very large values of q, and in these cases the interaction potential energy can be important.

- **Example:** To illustrate the smallness of interaction energy between two charges as compared to their average thermal energy we can make an order of magnitude argument by considering two electrons separated by a distance $r \ll \lambda_D$. In that case we can let $\exp(-r/\lambda_D) \approx 1$. The potential energy of one charge in the field of the other one is then $\mathcal{W}_p = e^2/(4\pi\varepsilon_0 r^2)$. Taking $r \sim \lambda_D/4\pi$, which is certainly less than λ_D, we find after a little algebra that $\mathcal{W}_p \approx \kappa T_e/N_p$. Now, this result refers to electron separations $\ell_{ad} \ll r \ll \lambda_D$, where ℓ_{ad} is an average distance between particles. If r becomes very small, $r \ll \ell_{ad}$, the interaction potential can become very large. The probability of finding such close electrons is, however, very small, since this probability scales with the volume of the sphere having radius r. For plasmas relevant here, with $N_p \gg 1$, we therefore again find that the interaction potential energy is, on average, much smaller than the thermal energy of particles. We note that the discussion in the present section provides a basis for assuming electrons and ions to be uniformly distributed in space, at least to a first approximation, and thus justifying the use of a Poisson distribution in Section 8.3, when estimating the fluctuations in the number of particles in a Debye sphere.

An important conclusion can be obtained from (8.18) by considering the interaction energy of a selected plasma particle in plasmas with $N_p \gg 1$. We can argue that in order to have a large deflection of such a particle, its potential energy in the induced potential must be at least comparable to its kinetic energy. We learn from (8.18) that this is rarely so when $N_p \gg 1$, and expect that in this case collisions are of minor importance. More detailed investigations, summarized in the following, substantiate this argument.

8.5 Evacuation of a Debye Sphere Ⓚ

Assume that we want to evacuate all electrons from a small sphere with radius R. Considering again the ions as an immobile background with density n_0 we argue that removing the first electron leaves one net positive charge behind, the next one two, etc., and each time we remove an additional electron we have to use an energy equivalent of what is needed to overcome the attraction due to these ions. To calculate the potential energy of the selected electron due to the ion charges we use Gauss' law with the radial component of the electric field $E_r = -d\phi/dr$

$$4\pi r^2 \frac{d\phi}{dr} = -\frac{4\pi}{3} r^3 \frac{e(n_0 - n_e)}{\varepsilon_0}, \tag{8.19}$$

for $r < R$, giving

$$\phi(r) = -\frac{e\,(n_0 - n_e(r))}{6\varepsilon_0} r^2 + \text{const.} \tag{8.20}$$

To calculate the energy needed to remove the last electron we take the difference in potential energy from the center of the sphere at $r = 0$ to its edge at $r = R$ when $n_e(r \le R) = 0$. We have then

$$e\Delta\phi = -\frac{e^2 n_0}{6\varepsilon_0} R^2 .$$

If we set $e\Delta\phi$ equal to the thermal energy $\frac{3}{2}\kappa T_e$ of an electron, we can determine the radius R of that sphere which the electrons, at least in principle, can evacuate by their thermal energy. The result is $R = 3\lambda_D$. Thus, if for some reason all electrons in a localized region suddenly had their velocity vectors in the thermal motion point away from a fixed point, they would evacuate a small sphere with radius $\sim \lambda_D$. Of course without a Maxwell's demon this cannot happen in practice, but *energetically* it would be feasible. It might, however, safely be concluded that on scales comparable to or smaller than the Debye length, we can expect large fluctuations in electron density.

The results of this subsection could have been argued from (8.18), which simply states that the ratio of the interaction energy to the thermal energy for one electron is $1/4\pi N_p$. Since the Debye sphere contains approximately N_p electrons, the accumulated interaction energy of *all* the electrons is κT_e, apart from a numerical constant.

The arguments presented before took the case $R = \lambda_D$. If we take $R \gg \lambda_D$ it is, on the other hand, evident that by thermal motion the electrons will not be able to make any significant deviations from charge neutrality. On *large* scales we expect the plasma to be "quasi neutral," $n_e \approx n_i$, at least as long as we are dealing with phenomena in quasi-equilibrium.

8.6 The plasma parameter

From the Debye length and the plasma density a dimensionless number can be constructed as

$$N_p = n\lambda_D^3, \tag{8.21}$$

which apart from a number of order unity is the number of particles in a sphere with radius λ_D. This number, called *the plasma parameter*, turns out to be important for classifying plasma conditions of interest. The definition is not unambiguous; sometimes the inverse of this number is called the plasma parameter, but this will be evident from the context. For plasmas of interest we expect $N_p \gg 1$. Note that N_p actually *decreases* for increasing plasma density with constant temperature, because the Debye length decreases as $\sim 1/\sqrt{n}$ for increasing density. Plasmas of interest (those with large N_p) will be *hot* and *dilute*.

We can also interpret the plasma parameter as being a measure of the ratio of two length scales, the Debye length λ_D and an inter-particle separation $n^{-1/3}$. Large N_p implies that the average separation between particles, here ions and electrons, is much smaller than the Debye length.

From these purely dimensional arguments the physical implications of the Debye length and the plasma parameter are not clear. This is illustrated by examples in the following, in particular in Section 8.8. Note that the definition of the plasma parameter only makes physical sense in three spatial dimensions. Formally, of course, we can consider, for instance, a one-dimensional plasma, and introduce the number of particles along a line with a length of one λ_D, but the procedure will lead to little of substance when applications are concerned.

8.7 Collisions between charged particles

Consider first a central, or "head-on," collision between two charged particles, each with charge e and initial kinetic energies $\frac{1}{2}mu^2$ each, as the particles start out at infinity. As these particles approach, they will slow down, losing kinetic energy and gaining potential energy in their electrostatic interaction. At the time when the particles are reflected, i.e., at their distance of closest approach p_0, their entire initial kinetic energy is converted to potential energy, i.e., $mu^2 = e^2/(4\pi\varepsilon_0 p_0)$. If we consider the similar collision of a light particle with a massive one at rest and having the same charge, the result would for this case be $mu^2 = e^2/(2\pi\varepsilon_0 p_0)$. Taking $mu^2 \approx \kappa T$ as a representative value for relevant kinetic energies, we obtain

$$p_0 = \frac{e^2}{4\pi\varepsilon_0 \kappa T} = \frac{\lambda_D}{4\pi N_p} \ll \lambda_D \tag{8.22}$$

for plasma conditions of interest. This is a simple, but nevertheless an important observation; the relation (8.22) ensures that for $N_p \gg 1$, the Debye sphere surrounding one ion, for instance, is almost uniformly filled out by electrons. In the less interesting case (at least here) with $p_0 \sim \lambda_D$ found for $N_p \approx 1$, we would find that a "typical" incoming electron will never get closer to our reference electron than λ_D, and we would, on average, experience electrons placed in small "voids" with radius λ_D.

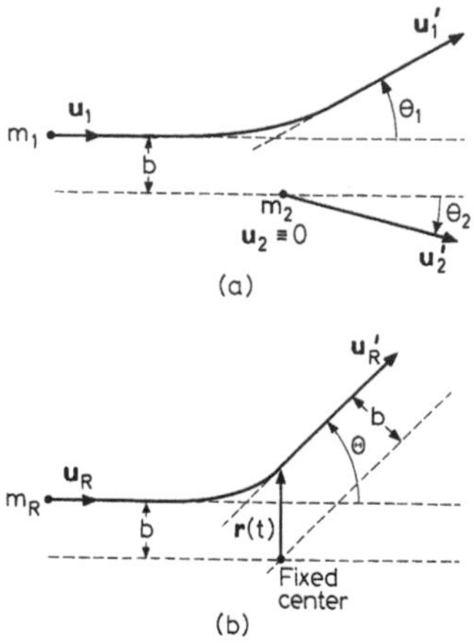

FIGURE 8.3: *Collision between two charged particles with charge e in fixed frame (a) and center-of-mass frame (b) of reference. The line with the arrow indicates the trajectory of the particle. The figures are shown for repulsive particle interactions, for instance, electron-electron collisions, and are readily generalized also to attractive interactions, as for electron-ion collisions. We can take* $\mathbf{U}_2 = 0$ *in the laboratory frame, with only little loss of generality. In particular, (a) defines the angles* θ_1 *and* θ_2.

Consider now the somewhat more interesting problem of a glancing collision between the two charged point particles, as illustrated in Figs. 8.3 and 8.4. In particular, the collision process is illustrated in the fixed frame (or laboratory system) as well as the center-of-mass frame of reference in Fig. 8.3. The center-of-mass coordinate and the relative coordinates for two point-particles with masses m_1 and m_2 are defined by

$$\mathbf{R} \equiv \frac{m_1\mathbf{r}_1 + m_2\mathbf{r}_2}{m_1 + m_2} \quad \text{and} \quad \mathbf{r} \equiv \mathbf{r}_1 - \mathbf{r}_2 .$$

The center-of-mass and the relative velocities are then

$$\mathbf{U} \equiv \frac{m_1\mathbf{U}_1 + m_2\mathbf{U}_2}{m_1 + m_2} \quad \text{and} \quad \mathbf{U}_r \equiv \mathbf{U}_1 - \mathbf{U}_2 .$$

For the simplest case, where one of the masses is infinite, we have the center-of-mass being at the position of that particle. For equal masses, $m_1 = m_2$, we have $\mathbf{U} = \frac{1}{2}(\mathbf{U}_1 + \mathbf{U}_2)$.

We know from basic mechanics courses that, in the center-of-mass frame, the angle between the incoming and the outgoing asymptotes, θ, for collisions between two charged particles with charges e is given by

$$\cot\frac{\theta}{2} = b\frac{4\pi\varepsilon_0 m u^2}{e^2}, \tag{8.23}$$

with b being the impact distance or impact parameter; see Fig. 8.3 or 8.4. In the laboratory frame of reference, we have to obtain, for instance, θ_1 from θ. In case we are dealing with, say, two electrons both having velocity u at $r \to \infty$, we can use symmetry arguments to find $\theta_1 = \theta/2$ in Fig. 8.3.

Consider now the case where an electron is scattered off a fixed scatterer with charge e. Again taking, as an order of magnitude estimate, $mu^2 \approx \kappa T$ for this case, we can determine the impact parameter $p_{90} = e^2/(4\pi\varepsilon_0 \kappa T)$, which gives a deflection $\theta \geq 90°$. Defining a collision as an interaction which gives a deflection larger than 90°, we can calculate a cross section for these collisions as $\sigma_{90} = \pi p_{90}^2$. The corresponding mean free path for collisions in a plasma with density n becomes

$$\ell_c = \frac{1}{n\sigma_{90}} = 16\pi\lambda_D N_p\,, \tag{8.24}$$

which is much larger than λ_D. Again we note that the plasma parameter is entering. One could argue that the definition of a cross section in terms of a 90° scattering is somewhat arbitrary. By reference to Fig. B.1 in Appendix B, it is a rather self-evident definition; if we send in a beam on one side of the foil with thickness dz, we are able to collect all particles (e.g., electrons) on the other side, if the scattering angles are less than 90°, otherwise not.

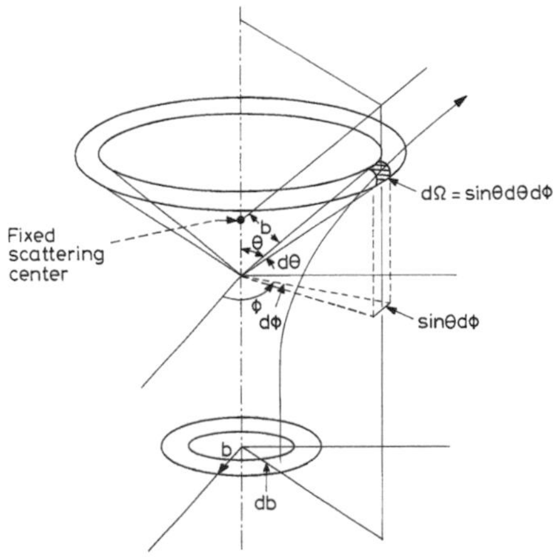

FIGURE 8.4: *Geometry for the collision between two charged particles with charge e, shown in the center-of-mass frame of reference for the repulsive forces. The scattering center will correspond to a particle position in the case where one of the particles is much heavier than the other.*

8.7.1 Simple arguments for collisional cross sections

We can present a semi-heuristic argument (Chen 2016) for the foregoing results by making an order of magnitude analysis; the Coulomb force $F = -e^2/(4\pi\varepsilon_0 r^2)$ is experienced by an incoming electron, with charge $-e$ and velocity u, within the time τ where the electron is close to the stationary scatterer, i.e., an ion, with charge e. We estimate $\tau \approx b/u$, where b is again the impact parameter. A change in electron momentum can consequently be estimated as $\Delta(mu) \approx \tau F = e^2/(4\pi\varepsilon_0 bu)$, taking $r \sim b$ in the Coulomb force. To estimate the impact distance $b = p_{90}$ for a 90° deflection we expect that the change in momentum is of the order of the incoming momentum itself, i.e., $\Delta(mu) \approx mu \approx e^2/(4\pi\varepsilon_0 p_{90} u)$, which implies that $p_{90} \approx e^2/(4\pi\varepsilon_0 mu^2)$. (For a head-on reflection, $\theta = 180°$, we have $\Delta(mu) = 2mu$, so half of this seems reasonable for a deflection of half the angle!) The cross section for that particular collision is then $\sigma_{90} \equiv \pi p_{90}^2 \approx e^4/(16\pi\varepsilon_0^2 m^2 u^4)$. The argument summarized here works equally well for two particles having charges of the same or opposite signs.

For an estimate we use $mu^2 \approx \kappa T$ to find

$$\sigma_{90} \approx \frac{e^4}{16\pi(\varepsilon_0 \kappa T)^2} = \frac{1}{16\pi}\left(\frac{\lambda_D}{N_p}\right)^2, \tag{8.25}$$

being a "representative" cross section for the given plasma, as characterized by a temperature T. The result obtained here is consistent with the expression used in (8.24), although the arguments were somewhat different.

A more careful and detailed analysis will reveal that many small collisions can add up to a nontrivial contribution. These effects can be included as a correction term (the Spitzer logarithm (Spitzer 1956)) as $\ell_c = 1/(n\sigma_{90} \ln \Lambda)$ with $\Lambda \approx 12\pi N_p$. For relevant plasma conditions we will find $\ln \Lambda$ in the range 1–10. (Plasma physicists often excel in blunt statements like "the logarithm of something is a number between 1 and 10"!) The collisional mean free path, as it is defined here, remains much larger than the Debye length for hot dilute plasmas. Ignoring the Spitzer correction and some numerical factors including π, we readily find $\ell_c \sim \lambda_D N_p$.

A typical time between collisions of an electron can be obtained by an order of magnitude estimate as

$$\tau_c \sim \ell_c/u_{The} \sim \lambda_D N_p \Big/ \sqrt{\kappa T_e/m} = N_p/\omega_{pe}. \tag{8.26}$$

As long as we are interested in moderate scale lengths and time spans (as measured in units of λ_D and $1/\omega_{pe}$, respectively), we are free to ignore collision altogether; we have a *collisionless plasma*.

An obvious, but important, inequality can be stated for plasmas with large plasma parameters as $\ell_{ad} \ll \lambda_D \ll \ell_c$, where we introduced the average distance between particles as $\ell_{ad} \sim 1/n^{1/3}$ in terms of the appropriate particle density n. We have $\ell_{ad} \sim \lambda_D/N_p^{1/3}$.

Consider again the case where the collisional mean free path is much longer than the Debye length. As we have seen, a selected electron (or ion) will experience the electric field from all charged particles within a distance of λ_D. For larger distances, the potential will be shielded, and these distant particles are therefore immaterial. We can now distinguish basically *two* sorts of interactions, or two limiting cases: one where the field from just one, or a few, nearby particles dominates and one where many particles contribute to the local field. This will be true for $N_p \approx 1$, as well as $N_p \gg 1$. In principle we can distinguish the two cases by taking all the particles within the Debye sphere and moving them randomly around, just a little bit. In the latter of the two cases mentioned we will not expect to feel any significant change in the electric field. In the former case, however, just a tiny change in the position of a close particle can make a substantial difference because of the $1/r^2$ variation of the potential. We now realize that this case is just what we called "Coulomb collisional interaction," namely, the case where the interaction takes place between two clearly identifiable charged particles. The other limit, where the electric fields are caused by larger scale charge imbalances of many particles, is what we might call "collective interaction." When there are many particles in a Debye sphere, $N_p \gg 1$, we expect the collective interactions to be dominating, and might ignore collisional interactions altogether!

By explicitly introducing a plasma temperature, the analysis summarized in the present section assumes implicitly velocity distributions corresponding to classical thermal equilibrium. In classical mechanics such equilibria are mediated by collisional interactions. In the limit of large plasma parameters N_p we have argued that collisions are rare, and it takes a long time to establish an equilibrium distribution. It is therefore worthwhile to consider Coulomb collisional interactions with other velocity distributions. So called kappa-distributions (or simply κ-distributions) with superthermal velocity "tails" have been suggested as relevant (Marsch & Livi 1985, Livadiotis 2017).

8.7.2 Center-of-mass dynamics

The foregoing sections were based on some order of magnitude estimates. It is interesting to carry out the calculations in more detail for at least one specific example (Trubnikov 1965, Lieberman & Lichtenberg 1994). Such calculations are most easily made in the center-of-mass coordinate $\mathbf{R}$ and the separation coordinate $\mathbf{r}$ as given before (Lieberman & Lichtenberg 1994). The accompanying center-of-mass velocity is, as already mentioned, $\mathbf{U} = (m_1\mathbf{U}_1 + m_2\mathbf{U}_2)/(m_1 + m_2)$, while the relative velocity is $\mathbf{U}_r = \mathbf{U}_1 - \mathbf{U}_2$. The equations of force for the particle interactions are $m_1 d\mathbf{U}_1/dt = \mathbf{F}_{1,2}(r)$ and $m_2 d\mathbf{U}_2/dt = \mathbf{F}_{2,1}(r) = -\mathbf{F}_{1,2}(r)$, where we explicitly assumed that the direction of the force between the two particles is along the line joining the two particles, and depends on the distance alone. By addition of these two latter equations we find $d\mathbf{U}/dt = 0$ for the center-of-mass motion. By dividing the first force equation by m_1, the second by m_2 and using the definition of $\mathbf{U}_r$, we find $m_R d\mathbf{U}_r/dt = \mathbf{F}_{1,2}(r)$, in terms of the *reduced mass* $m_R \equiv m_1 m_2/(m_1 + m_2)$. We can consider this equation as an equation of motion for a fictitious particle with mass m_R at position $r(t)$, subject to a force $\mathbf{F}_{1,2}(r)$. Figure 8.3b is shown for this frame of reference, defining also the scattering angle θ. In the rest frame (where $\mathbf{U}_2 = 0$), we find the scattering angles θ_1 and θ_2 (see Fig. 8.3a), and we have

$$\tan\theta_1 = \frac{\sin\theta}{(m_1/m_2)/(U_{Rb}/U_{Ra}) + \cos\theta} \quad \text{and} \quad \tan\theta_2 = \frac{\sin\theta}{(U_{Rb}/U_{Ra}) - \cos\theta},$$

where U_{Rb} and U_{Ra} are the velocities in the center-of-mass coordinates before and after the collision. These relations can be proven to be a consequence of momentum conservation for the special case shown in Fig. 8.3.

Note that up to now we have not specified $\mathbf{F}_{1,2}(r)$, only assumed that it depends on the magnitude of the vector $|\mathbf{r}|$ connecting particle 1 and particle 2. For an elastic collision, the interaction force can be written as a gradient of a potential, i.e., $\mathbf{F}_{1,2}(r) = -\nabla\Psi(r)$, where $\Psi(r \to \infty) \to 0$. Then the kinetic energy of the fictitious particle is conserved in the center-of-mass system, implying $U_{Rb} = U_{Ra}$ and

$$\tan\theta_1 = \frac{\sin\theta}{(m_1/m_2) + \cos\theta}.$$

Using the double angle formula for the tangent, we find $\theta_2 = \frac{1}{2}(\pi - \theta)$. In case we are dealing with electrons colliding with ions or neutrals, we have $m \ll M$, and we find $m_R \approx m$, giving $\theta_1 \approx \theta$. For electron-electron collisions, on the other hand, we have $m_1 = m_2$, giving $m_R = m_1/2$ and $\theta_2 = \theta/2$.

Consider now the specific case where the interaction potential is given for Coulomb interaction, $\Psi(r) = q_1 q_2/(4\pi\varepsilon_0 r)$. In polar coordinates (see Figure 8.3b with $\mathbf{r} = r(t)\widehat{\mathbf{r}}$), we have in the plane spanned by the unit vectors $\widehat{\theta}$ and $\widehat{\mathbf{r}}$ that

$$\frac{d}{dt}\mathbf{r} = \frac{dr(t)}{dt}\widehat{\mathbf{r}} + r\frac{d\Theta(t)}{dt}\widehat{\theta},$$

and

$$\frac{d^2}{dt^2}\mathbf{r} = \left[\frac{d^2 r(t)}{dt^2} - r\left(\frac{d\Theta(t)}{dt}\right)^2\right]\widehat{\mathbf{r}} + \left[2\frac{dr(t)}{dt}\frac{d\Theta(t)}{dt} + r\frac{d^2\Theta(t)}{dt^2}\right]\widehat{\theta},$$

where we used $d\widehat{\theta}/dt = -\widehat{\mathbf{r}}\, d\Theta(t)/dt$ and $d\widehat{\mathbf{r}}/dt = \widehat{\theta}\, d\Theta/dt$, which can be obtained by noting (Symon 1960) that $\widehat{\mathbf{r}} = \widehat{\mathbf{x}}\cos\Theta + \widehat{\mathbf{y}}\sin\Theta$ and $\widehat{\theta} = -\widehat{\mathbf{x}}\sin\Theta + \widehat{\mathbf{y}}\cos\Theta$. We use here $\Theta = \Theta(t)$ to denote the *variable* angle, to be distinguished from the fixed angle θ entering, for instance, Figs. 8.3 and 8.4. We then have for the scattering of a charge e on a charge q

$$\frac{d^2 r(t)}{dt^2} - r\frac{d^2\Theta(t)}{dt^2} = \frac{eq}{4\pi\varepsilon_0 m_R r^2} \equiv \frac{a_1}{r^2} \tag{8.27}$$

$$r^2\frac{d\Theta(t)}{dt} = \text{const.} = bu. \tag{8.28}$$

The relation (8.28) expresses the conservation of angular momentum. The solution for $r(t)$ can be written as

$$r(t) = \frac{1}{A\cos(\Theta(t) + \delta) - a_1/(bu)^2},$$

where A and δ are to be determined from the initial conditions. We have from the initial conditions $(\Theta = 0, r \to \infty) \to A\cos(\delta) = a_1/(ub)^2$ and $(\Theta = 0, dr/dt = -u) \to A\sin(\delta) = -1/b$ giving $\tan(\delta) = -u^2 b/a_1$. We used that initially, at $t \to -\infty$, the velocity and the position vectors point in opposite directions. The final conditions, on the other hand, give $(\Theta = \pi - \theta, r \to \infty) \to A\cos(\pi - \theta + \delta) = a_1/(ub)^2$, implying $\theta = \pi + 2\delta$. Finally we obtain $\cot(\theta/2) = bu^2/a_1$, for scattering of incoming electrons on a fixed scatterer, in agreement with (8.23).

8.8 Plasma resistivity by electron-ion collisions

The collisions relevant for the resistivity are those between electrons and ions (leaving out a neutral component for the moment). Electron-electron collisions are irrelevant. This can be readily understood by considering two electrons that are about to collide. Before the collision their effective contribution to the current is given by the motion of their center-of-mass. Since the electron-electron Coulomb collisions are elastic, the motion of this center-of-mass is the same before and after the collision, so electron-electron collisions, as modeled by some collision frequency $\nu_{e,e}$, are irrelevant for the resistivity. Ion-ion collisions are irrelevant by the same arguments. Momentum losses relevant for the resistivity of a fully ionized plasma are solely due to electron-ion collisions in fully ionized plasmas.

For the stationary unmagnetized case (or **B**-parallel electric fields for magnetized plasma) we have the relation $0 = -e\mathbf{E} - m\nu_{e,i}\mathbf{u}$ for an electron with $\nu_{e,i}$ representing a collisional drag on electrons due to collisions with ions, i.e., $m\nu_{e,i}\mathbf{u}$ is the electron momentum loss per time unit. The model simply states that the momentum increase due to the acceleration by the electric field is balanced by the momentum loss due to collisions. Taking a plasma density n, we have the electron current density $\mathbf{J} = -en\mathbf{u}$, or $\mathbf{J} = e^2 n\mathbf{E}/m\nu_{e,i}$, and we readily find the plasma conductivity $\sigma \approx e^2 n/(m\nu_{e,i}) = \omega_{pe}^2 \varepsilon_0/\nu_{e,i}$, see also Appendix B. We take this as a phenomenological relation between conductivity and collision frequency and introduce explicitly the electron-ion collision frequency $\nu_{e,i} = \sqrt{\kappa T_e/m}/\ell_c \approx \omega_{pe}/N_p$ to get the plasma conductivity

$$\sigma \approx N_p \omega_{pe} \varepsilon_0 . \tag{8.29}$$

It is easily demonstrated that σ is independent of the plasma density n. The notation σ is standard for conductivity, and should not be confused with "cross section," which is, unfortunately, also designated by the same symbol. Ion motion has been ignored here because ion velocities are usually too slow to give a significant contribution to the currents. For magnetized plasmas the situation will usually be different when we consider the conductivity in the direction perpendicular to the magnetic field.

A fully ionized plasma with a large plasma parameter, N_p, is thus a good conductor, as also intuitively expected. The temperature variation is given explicitly by N_p. A more detailed treatment (Spitzer 1956), being careful also with numerical coefficients, π's, etc. gives the expression for the resistivity of a fully ionized plasma

$$\xi \equiv \frac{1}{\sigma} = \frac{m^{1/2} Z e^2 \ln\Lambda}{16\pi\varepsilon_0^2 (\kappa T)^{3/2}} \tag{8.30}$$

where eZ is the ion charge, assuming all ions to be in the same charged state. As before, we denote by $\ln\Lambda$ the Spitzer logarithm, with $\Lambda \approx 12\pi N_p$ (Spitzer 1956). Note that ξ depends on the plasma density *only* through N_p in Λ. This is a rather weak dependence. The numerical correction due to the Spitzer logarithm in (8.30) is not trivial and can contribute with an order of magnitude, since we have $N_p \gg 1$ for plasmas of interest. On the other hand, it is also seen that increasing N_p from 10^3 to 10^4 will change the Spitzer logarithm by only a factor 4/3. The units for resistivity are *Ohm m.*

The expression for the resistivity obtained here assumed implicitly that the plasma was *unmagnetized.* If we have a homogeneous time-stationary magnetic field imposed on the plasma, the results apply for an electric field parallel to **B**. If, on the other hand, a homogeneous externally applied electric field has a component $\perp$ **B**, this component can always be removed by the proper change of reference, by moving with a velocity $\mathbf{U} = \mathbf{E}_\perp \times \mathbf{B}/B^2$; see also Section 5.8. Electron-ion collisions will not give rise to any conductivity $\perp$ **B** for this case. This result is consistent with the discussion in Section 8.8.1, where it is found that electrons and ions upon collisions "jump," on average, the same distance in the direction $\perp$ **B**. Note, though, that by a simple change of reference frame, only a homogeneous electric field can be removed over all space. Obviously, no change of reference can remove an electric field component $\parallel$ **B** for non-relativistic velocities.

The resistivity of a plasma as originating from collisions between neutrals and charged particles will be discussed separately in Section 8.9. In the very top regions of the ionosphere, the collisions between charged particles can become just as important as collisions with the neutral component. Collisions between electrons and other electrons do not contribute to the resistivity, since momentum lost by one electron is gained by another one, so the net current remains unaffected. Ion-ion collisions will not contribute to a resistivity either, by the very same arguments. Collisions between electrons and ions give to a resistivity as given by (8.30).

Summary of basic plasma parameters, expressed in terms of the appropriate charges q, masses M, densities n and temperatures T

$$\omega_p = \sqrt{\frac{q^2 n}{\varepsilon_0 M}} \qquad \lambda_D = \sqrt{\frac{\varepsilon_0 \kappa T}{q^2 n}} \qquad \omega_p \lambda_D = \sqrt{\frac{\kappa T}{M}} = u_{The}$$

$$N_p \equiv n\lambda_D^3$$

$$\ell_c \sim \lambda_D N_p \qquad \tau_c \sim N_p/\omega_p \qquad \nu_c \sim \omega_p/N_p\,.$$

Approximate expression for the resistivity ξ (with conductivity σ) of a fully ionized plasma

$$\xi \equiv \frac{1}{\sigma} = \frac{1}{N_p\,\omega_{pe}\varepsilon_0}\,.$$

The present summary does not include the corrections due to the Spitzer logarithm, and also some numerical factors are omitted.

The analysis summarized here implicitly assumes weak electric fields, in the sense that the energy gain of a particle from one collision to the next is small compared to the thermal energy. In this limit the representative particle energy will be the thermal energy. When the electric fields are strong, the energy gain between collisions is increased to an extent that the approximation $mu^2 \approx \kappa T$ in (8.25) is violated. We can find conditions where the collisional cross section is steadily decreasing from one collision to the next for increasing particle velocities and the particles can be accelerated without limit: we have *runaway electrons* (Spitzer 1956).

Ⓢ **Exercise:** Check the difference between the simple expression $\xi = 1/(N_p\,\omega_{pe}\varepsilon_0)$ for the plasma resistivity and the more accurate Spitzer expression (8.30). Estimate the error when using the simple expression rather than the Spitzer result for the plasma resistivity at the ionospheric F-region maximum, for a Q-machine plasma and for the JET-Tokamak experiment. Ⓢ

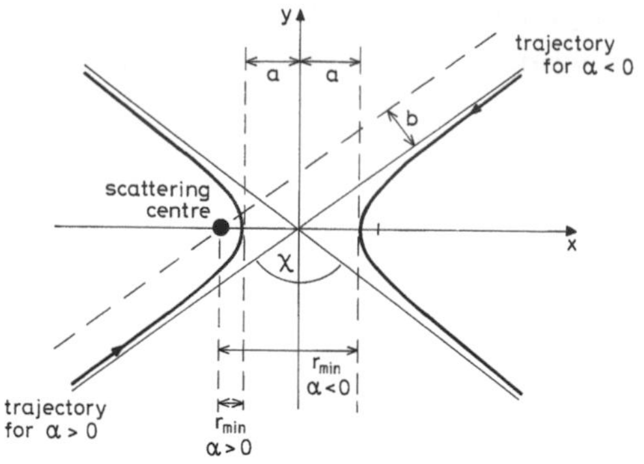

FIGURE 8.5: *Geometry for discussions of the Rutherford scattering formula; see also Fig. 8.4. The figure shows attractive as well as repulsive forces between the scattering and the scattered particles.*

Ⓢ **Exercise:** For the Earth's near ionosphere (i.e., the E and F regions), the plasma density varies with altitude in the range $10^{10} - 10^{15}$ m^{-3}, and the electron temperature increases with altitude from ~ 500 K in the lower parts of the E region to ~ 2500 K at high altitudes (Schunk & Nagy 1978). Make a table or a figure for the corresponding variation of the electron plasma frequency both in rad/s and in Hz. Make a similar table or plot for the Debye length and also for the plasma parameter. Ⓢ

Summary of results for particle collisions

We summarize some basic results for collisional interactions (Woan 2000). Scattering from a Coulomb potential is historically called "Rutherford scattering" in memory of that scientist's pioneering experimental studies. The summary is directly related to the discussion of collisions between charged particles, as summarized before.

Rutherford scattering: Scattering potential energy at a distance r is written as:

$$\mathcal{W}_{pot} = -\frac{\alpha}{r},$$

with $\alpha < 0$ for repulsion and $\alpha > 0$ for attraction.

- Scattering angle:

$$\tan\frac{\chi}{2} = \frac{|\alpha|}{2\mathcal{W}_{tot} b}.$$

The total energy, W_{tot}, is also the particle kinetic energy when the particle is at infinite distance from the scattering center.

- Closest approach:

$$r_{min} = \frac{|\alpha|}{2\mathcal{W}_{tot}}\left(\csc\frac{\chi}{2} - \frac{\alpha}{|\alpha|}\right) \equiv a(e \pm 1),$$

where we introduce the hyperbola semi-axis $a = |\alpha|/(2\mathcal{W}_{tot})$ and the eccentricity $e = \sqrt{1 + 4\mathcal{W}_{kin}^2 b^2/\alpha^2} = \csc\chi/2$. See Fig. 8.5 for definition of symbols.

- Motion trajectory:

$$\frac{4\mathcal{W}_{tot}^2}{\alpha^2}x^2 - \frac{y^2}{b^2} = 1,$$

where x, y are positions with respect to the hyperbola center.

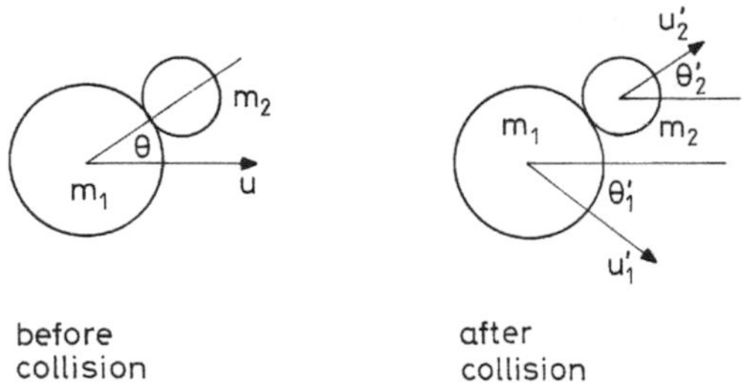

FIGURE 8.6: *Geometry for elastic collisions. Particle 2 is assumed at rest prior to the collision. This can always be obtained by a proper change in the frame of reference.*

- Scattering center, being also the focal point of the trajectory:

$$x = \pm\sqrt{b^2 + \frac{\alpha^2}{4\mathcal{W}_{tot}^2}}\,.$$

- Rutherford scattering formula for the differential cross section:

$$\frac{d\sigma}{d\Omega} = \frac{1}{n}\frac{dN}{d\Omega} = \left(\frac{\alpha}{4\mathcal{W}_{tot}}\right)^2 \csc^4\frac{\chi}{2},$$

where n is here the number of particles per second passing through a unit area, perpendicular to the beam, and dN is the number of particles scattered into a spherical angle $d\Omega$.

Perfectly elastic hard sphere collisions: As a reference, and for completeness, we summarize here some results also for perfectly elastic hard sphere collisions. See Fig. 8.6 for definition of symbols.

- Directions of motion:

$$\tan\theta_1' = \frac{m_2 \sin 2\theta}{m_1 - m_2\cos 2\theta} \quad \text{and} \quad \theta_2' = \theta,$$

with

$$\theta_1' + \theta_2' \begin{cases} > \pi/2 & \text{for} \quad m_1 < m_2 \\ = \pi/2 & \text{for} \quad m_1 = m_2 \\ < \pi/2 & \text{for} \quad m_1 > m_2. \end{cases}$$

- Final velocities:

$$u_1' = u\,\frac{\sqrt{m_1^2 + m_2^2 - 2m_1 m_2 \cos 2\theta}}{m_1 + m_2} \quad \text{and} \quad u_2' = \frac{2m_1 u}{m_1 + m_2}\cos\theta\,.$$

The analysis of Coulomb collisions presented previously assumed perfectly elastic collisions. We might anticipate models for inelastic collisions to be relevant as well, in particular when dealing with partially ionized atoms or molecules, or other cases where the scatterers have internal degrees of motion.

8.8.1 Charged particle collisions in magnetic fields

The foregoing analysis assumed implicitly that particles propagated along straight lines of orbit between collisions. In the following we give a brief discussion of the case where the charged particles are confined by a homogeneous magnetic field. As discussed in Section 7.1, the basic orbit of a charged particle in a constant magnetic field is a circle, with the Larmor radius r_L, and an orbital period given by the cyclotron frequency as $2\pi/\Omega_c$. When a light charged particle collides with a heavy neutral, we find that its guiding center is displaced by a distance which is at most r_L. If energy exchange of the charged particle and the neutral is important, the guiding center displacement can be modified somewhat, but r_L will remain as a useful order of magnitude for the particle displacement in the direction $\perp$ **B**.

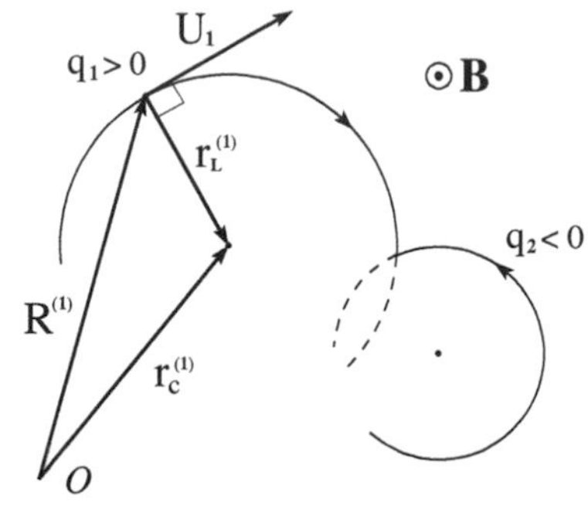

FIGURE 8.7: *Explanation of position-vectors used in (8.31). The vector* $\mathbf{r}_L$ *has the length of the Larmor radius for given velocity U, and points from the particle to the gyrocenter. We have* $\mathbf{r}_L \perp \mathbf{U}$. *The origin is marked by O. Electric fields are assumed to be vanishing in this frame of reference.*

The case where two charged magnetized particles collide is somewhat more interesting. In order to describe this process, we first write their gyro-center positions as

$$\mathbf{r}_c^{(1)} = \mathbf{R}^{(1)} + \frac{M_1}{q_1 B^2}\mathbf{U}^{(1)} \times \mathbf{B} \quad \text{and} \quad \mathbf{r}_c^{(2)} = \mathbf{R}^{(2)} + \frac{M_2}{q_2 B^2}\mathbf{U}^{(2)} \times \mathbf{B}, \tag{8.31}$$

where $\mathbf{R}^{(j)}$ is the position of particle j at some given time, and $\mathbf{U}^{(j)}$ its velocity in the orbital motion, with a corresponding notation for particle masses M_j and charges q_j. The particles are assumed to collide at time t_0. Assume that the particles have slightly different velocity components along the magnetic field, so that we need to consider only one encounter. See Fig. 8.7 for an illustration of the geometry of the problem.

We assume now that the time duration of the collision is short. This means that the particle positions slightly before and slightly after the collision are virtually the same and $\mathbf{R}^{(1)} \approx \mathbf{R}^{(2)}$, so only the directions and magnitudes of the velocity vectors have changed. This implies that

$$\Delta\mathbf{r}_c^{(1)} = \frac{M_1}{q_1 B^2}\Delta\mathbf{U}^{(1)} \times \mathbf{B} \quad \text{and} \quad \Delta\mathbf{r}_c^{(2)} = \frac{M_2}{q_2 B^2}\Delta\mathbf{U}^{(2)} \times \mathbf{B}, \tag{8.32}$$

where $\Delta\mathbf{U}^{(j)}$ is the change in velocity vector for particle j. The particles do not change position during the collisions, but the guiding centers move due to the change in direction of the velocity vectors. Conservation of momentum implies $M_1\Delta\mathbf{U}^{(1)} + M_2\Delta\mathbf{U}^{(2)} = 0$. With (8.32) this gives

$$q_1\Delta\mathbf{r}_c^{(1)} + q_2\Delta\mathbf{r}_c^{(2)} = 0, \tag{8.33}$$

irrespective of M_1 and M_2. If $q_1 = q_2$ we have $\Delta\mathbf{r}_c^{(1)} + \Delta\mathbf{r}_c^{(2)} = 0$, i.e., the position of the average gyro-center for two particles is not changed by collisions. If, however, $q_1 = -q_2$ we have $\Delta\mathbf{r}_c^{(1)} = \Delta\mathbf{r}_c^{(2)}$. In this latter case the gyro-center of the two particles moves in the same direction, and the magnitude of *both* "jumps" is the same. If the two particles in question are a singly charged ion and an electron, we can again argue that the electron gyro-center displacement can be at most an electron Larmor radius, r_{Le}. This implies that also the *ion* gyro-center displacement can be at most r_{Le}. Also we note that in a coarse grained description, with a spatial resolution larger than the Larmor radii, we cannot have any charge separation as a consequence of collisions between like or differently charged particles. Energy is of course also conserved in the collision, which is assumed to be perfectly elastic, but we do not need the corresponding expression for the arguments here.

- **Exercise:** Figure 8.7 illustrates the problem in 2 spatial dimensions. Will the analysis be different if we allow the velocity vectors to have 3 significant spatial components? If so, explain when the velocity component along the magnetic field will be relevant.

- **Exercise:** Discuss the case where $|q_1| \neq |q_2|$ in (8.33). Consider both the case where the charges are of the same sign, as well as the case where they have opposite sign. Discuss also the change in the center-of-mass position.

Ⓢ **Exercise:** The foregoing analysis of particle collisions assumed that the magnetic field is homogeneous and that there are no electric fields. Introducing a constant electric field is trivial since it merely corresponds to a change in frame of reference. The analysis of spatially varying electric fields $\mathbf{E}_\perp(\mathbf{r}_\perp)$ is more difficult. Analyze this problem and demonstrate that an electric field with an intensity varying linearly in the direction $\perp \mathbf{B}$, e.g., $\mathbf{E} = \{E_x, E_y\} = \{0, E_0 y/\ell\}$, will not give rise to any net particle transport for collisions between like particles. If the variation of the electric field contains a quadratic component, e.g., $\mathbf{E} = \{E_x, E_y\} = \{E_0\left(y/\ell_1 + y^2/\ell_2^2\right), 0\}$, there *will* however be a net particle transport $\perp \mathbf{B}$ along the electric field upon collisions between like particles (Pécseli 2016). This phenomenon has importance for currents across magnetic field lines, where this current will mostly be due to ions. Warning: a detailed analysis of this problem can become rather lengthy. Ⓢ

8.9 Plasma resistivity by neutral collisions Ⓚ

Collisions between two point-like charged particles is a relatively simple problem, in the sense that the basic equations for the problem are quite well defined. For collisions between a charged particle and a neutral, the problem can become much more complicated. In particular, if we consider collisions between a charged particle and a molecule, the neutral component can be polarized by the moving charge, and the velocity variation of the cross section for the interaction can be complicated. The best thing is to look it up in a table; Appendix B contains some relevant references. Of course, as long as we treat the problem as a sort of "billiard ball" or hard sphere collision with a constant cross section, we can describe the process relatively easily, and for many purposes the results can be quite adequate, at least as far as orders of magnitude are concerned. A more precise analysis will depend on many parameters, and a complete description falls far outside the scope of this book. In Appendix B we attempt a classification of various relevant collision processes.

In this section, we will discuss the dc-conductivities and resistivities of a magnetized plasma, in the case where the dominant collision process is between a neutral component and charged particles. The basic phenomenological model is expressed by the equation

$$M\frac{d}{dt}\mathbf{U} = q\mathbf{E} + q\mathbf{U}\times\mathbf{B} - M\nu\mathbf{U}, \tag{8.34}$$

where $\mathbf{U}(t)$ is the velocity of the selected charged particle moving in prescribed electric and magnetic fields, where ν is here a collision frequency, see Appendix B.3. The particle position is $\mathbf{R}(t)$, with $\mathbf{U}(t) = d\mathbf{R}/dt$. In (8.34) we can let q and M represent the electron as well as the ion charges and masses, respectively. In the present phenomenological model we can take ν to represent a momentum loss of particles per time unit, without going into detail in discussing the underlying physical mechanisms. A great simplifications can be achieved here by assuming that ν is a constant, irrespective of the relative particle velocities (Pécseli 2000). This implies that the collisional mean free path $\ell_c \sim u/\nu$ is assumed to decrease for decreasing velocity, giving the implicit assumption that the collisional cross section σ decreases with increasing velocity, since $\ell_c \sim 1/(n\sigma)$. This assumption is not unphysical, although admittedly somewhat restrictive.

For time-stationary conditions, we can obviously set the time derivative to zero in (8.34). It is readily seen that as an approximation we can also ignore $d\mathbf{U}/dt$ in case the time variations of the electric field are very slow on the time scale given by $1/\nu$. Consequently, we solve first the simpler equation

$$q\mathbf{E} + q\mathbf{U}\times\mathbf{B} - M\nu\mathbf{U} = 0. \tag{8.35}$$

First we take the cross product $\times\mathbf{B}$ "from the right," use $(\mathbf{U}\times\mathbf{B})\times\mathbf{B} = -\mathbf{U}_\perp B^2$ and find $q\mathbf{E}\times\mathbf{B} - q\mathbf{U}_\perp B^2 - M\nu\mathbf{U}\times\mathbf{B} = 0$. Insertion of $\mathbf{U}\times\mathbf{B}$ from (8.35) gives

$$q\mathbf{E}\times\mathbf{B} - q\mathbf{U}_\perp B^2 + \nu\frac{M}{q}(q\mathbf{E}_\perp - M\nu\mathbf{U}_\perp) = 0,$$

and we can express the velocity component in the direction $\perp\mathbf{B}$. The component *along* $\mathbf{B}$ is easily obtained as $U_\parallel = E_\parallel q/(M\nu)$. Note that we require $\nu \neq 0$ for this solution to exist with $E_\parallel \neq 0$. *If* we have $\nu = 0$, we would have a steady acceleration of the particle along $\mathbf{B}$, and the steadily increasing particle velocity would be violating the basic assumption of time stationarity.

We can write the $\mathbf{B}$-perpendicular component of the particle velocity as

$$\mathbf{U}_\perp = \frac{q\mathbf{E}\times\mathbf{B} + M\nu\mathbf{E}_\perp}{qB^2 + M^2\nu^2/q} = \left(\frac{q}{M}\right)^2 \frac{\mathbf{E}\times\mathbf{B} + (M/q)\nu\mathbf{E}_\perp}{\Omega_c^2 + \nu^2}, \tag{8.36}$$

or by introducing the definition

$$\mathbf{U}_\perp \equiv \beta\frac{\mathbf{E}\times\mathbf{B}}{B^2} + \alpha\frac{\mathbf{E}_\perp}{B}, \tag{8.37}$$

with

$$\alpha \equiv \frac{\nu M}{qB + M^2\nu^2/(qB)} \equiv \frac{\nu/\Omega_c}{1+\nu^2/\Omega_c^2} \quad \text{and} \quad \beta \equiv \frac{1}{1+\nu^2/\Omega_c^2}, \tag{8.38}$$

in terms of the appropriate cyclotron frequency $\Omega_c \equiv qB/M$. Note that α has the sign of qB. Since the electric field has a component $\parallel\mathbf{B}$, we wrote the $\mathbf{B}$-perpendicular component explicitly as $\mathbf{E}_\perp$. We have $0 \leq |\alpha| < 1/2$ and $0 < \beta < 1$, respectively.

The expressions can be used for any particle species specified by the appropriate mass M and charge q, and cyclotron frequencies Ω_c, noting, though, that we have to interpret these quantities with the correct sign of the charges and magnetic fields entering!

One could argue that the electric field component $\mathbf{E}_\perp$ could be made to vanish by a "smart" change of reference, moving with velocity $\mathbf{U}_* = -\mathbf{E}_\perp\times\mathbf{B}/B^2$. In this case, however, the neutral component is moving with velocity $-\mathbf{U}_*$ in this new frame of reference, and the last term in (8.35) has to be modified to $M\nu(\mathbf{U}-\mathbf{U}_*)$, so that nothing is gained (as is to be expected!).

To derive the current $\mathbf{J}$ induced in the plasma by the imposed electric field $\mathbf{E}$, we note that in general the ions as well as the electrons contribute, as given by the appropriate index. *Along* the magnetic field we expect the electron contribution to the current to be dominant, because of their small mass. In the direction $\perp\mathbf{B}$ we expect that the ratios ν/Ω_c can vary significantly, and for some neutral densities the ion, and for other cases the electron contribution can dominate the current. Generally, we have the plasma current given as $\mathbf{J} \equiv e(n_i\mathbf{U}_i - n_e\mathbf{U}_e)$. We ignore the possibility of deviations from local charge neutrality and set the densities of the two components equal, i.e., $n_e = n_i \equiv n$, and have

$$\mathbf{J}_\perp \equiv \sigma_P \mathbf{E}_\perp + \sigma_H \mathbf{E} \times \widehat{\mathbf{b}} \tag{8.39}$$

$$J_\parallel \equiv \sigma_\parallel E_\parallel , \tag{8.40}$$

where $\widehat{\mathbf{b}}$ is a unit vector along $\mathbf{B}$, and

$$\sigma_P \equiv \frac{en}{B}(\alpha_i - \alpha_e) \qquad \text{and} \qquad \sigma_H \equiv \frac{en}{B}(\beta_i - \beta_e)$$

are the "Pedersen" and "Hall" conductivities, respectively. Sometimes we see a presentation with a change in sign in σ_H; this is because some authors consider it preferable to have the cross product between the electric field and and the unit vector $\widehat{\mathbf{b}}$ the other way around! Note that α_e and α_i have opposite signs, ensuring here that $\sigma_P \geq 0$ for all plasma parameters. With $M \gg m$ we have $\beta_i - \beta_e < 0$, generally.

For the motion *along* the magnetic field we introduce $\sigma_\parallel \equiv e^2 n[1/(M\nu_i) + 1/(m\nu_e)] \approx e^2 n/(m\nu_e)$ to give $J_\parallel = \sigma_\parallel E_\parallel$. The dynamics are here unaffected by the magnetic field intensity.

To understand the altitude variations of σ_P and σ_H in the ionosphere, we first note that both these quantities are proportional to the plasma density $n_e \approx n_i \equiv n$, which can be assumed to follow a Chapman model (see Section 6.3) at least as an approximation. The neutral density is, on the other hand, to an approximation, following an exponential profile, implying that ν_e as well as ν_i have the same functional altitude variation (but of course with different numerical values).

As a heuristic argument we can assume that the cross sections, σ_c, for ion-neutral and electron-neutral collisions are approximately the same, so the mean free paths for collisions, $\ell_c = 1/(n_n\sigma_c)$, are also approximately the same for both species, with n_n being the neutral gas density. (Note that *this* σ_c is *not* a conductivity!) The collision frequency is then $\nu \approx u_{The}/\ell_c$ with $u_{The} = \sqrt{\kappa T/m}$ being the appropriate thermal velocity for the plasma particles. Consequently, $\Omega_{ci}/\nu_i \approx (eB/M)\ell_c\sqrt{M/\kappa T_i}$ and $\omega_{ce}/\nu_e \approx (eB/m)\ell_c\sqrt{m/\kappa T_e}$.

Assume that at some altitude we have $\Omega_{ci} \approx \nu_i$. Provided $T_e \sim T_i$, we then find $\omega_{ce}/\nu_e \approx \sqrt{M/m}\sqrt{T_i/T_e} \gg 1$. The foregoing discussion ignored the variations of cross sections with velocity. These effects will in general strengthen the arguments, since cross sections usually decrease at large velocities. Consequently, it is possible to have $\omega_{ce}/\nu_e \gg 1$ and $\Omega_{ci}/\nu_i \ll 1$ also in cases where $T_e \gg T_i$. The electrons are strongly magnetized while the ions are hardly influenced by the magnetic field in this case!

For all practical purposes, we can assume the magnetic field to be constant for relevant altitudes, and find that there is an altitude where $\omega_{ce} \approx \nu_e$, and a different, somewhat higher, altitude where $\Omega_{ci} \approx \nu_i$. The corresponding α_e and α_i have their maxima at these two altitudes, respectively, while β_e and β_i are steadily decreasing with altitude. For high altitudes above, say, 150 km, both $\beta_e \approx 1$ and $\beta_i \approx 1$, implying that $\sigma_H \approx 0$. Similarly, for altitudes below 60 km or so, we have $\beta_e \approx 0$ and $\beta_i \approx 0$, implying that $\sigma_H \approx 0$ also here.

As a numerical example we have in the day time ionospheric E-region at an altitude of 110 km over northern Scandinavia: $\nu_i \approx 600\ \mathrm{s}^{-1}$ and $\nu_e \approx 1.5 \times 10^4\ \mathrm{s}^{-1}$. The temperatures are $T_i \approx 200$ K and $T_e \approx 400$ K. The magnetic fields are $B \approx 50\ \mu$T, giving $\Omega_{ci} \approx 180$ rad/s, corresponding to an average ion mass of 31 AMU, while $\omega_{ce} = 8.8 \times 10^6$ rad/s, giving $\Omega_{ci}/\nu_i \approx 0.3 < 1$ and $\omega_{ce}/\nu_e \approx 600 \gg 1$. At around 70 km altitude we have $\omega_{ce} \approx \nu_e$ and then $\Omega_{ci} \ll \nu_i$. At such high collision frequencies, the ions are not magnetized at all, since they never manage to make a full turn in the cyclotron orbit, and as far as they are concerned, the magnetic field could as well have been absent.

In Figure 8.8 we show the altitude variation of the various components of the conductivity tensor for night time conditions. For large altitudes, where the neutral density is small, we have a large mobility for the charged particles in the direction $\perp \mathbf{B}$, but since this mobility is virtually the

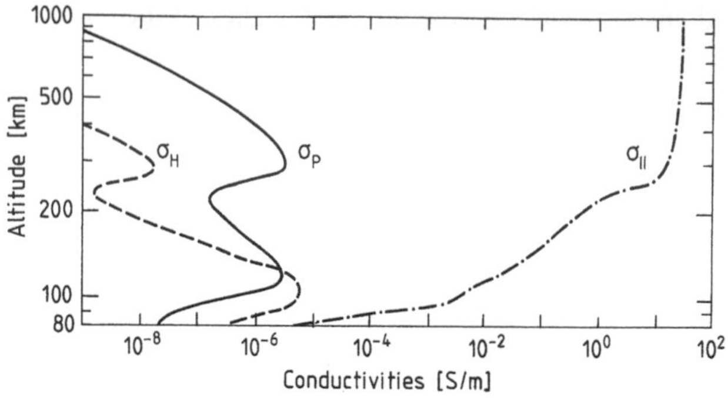

FIGURE 8.8: *Illustrative example of the altitude variation of the night time ionospheric plasma conductivity. In the altitude range shown, the most important contribution to the resistivity comes from collisions between the charged particles (ions and electrons) and the neutral component there. The figure is adapted from Boström (1964) with permission of the American Geophysical Union. Copyright 1964 American Geophysical Union.*

same for both species, the net current will be small in that direction. If we let the magnetic field be vertical, as appropriate for the polar regions, the altitude variation of σ_H and σ_P will imply that the direction of the current density vector $\mathbf{J}$ with respect to the imposed electric field is varying with altitude; large σ_H and small σ_P gives a current predominantly in the $\mathbf{E} \times \mathbf{B}$-direction. Detailed discussions for night and day time conditions are given by Boström (1964), for instance.

The $\sigma_\parallel$-conductivity is, in general, much larger than both σ_H and σ_P. If we, quite formally, consider the externally imposed electric field as being due to a generator with finite inner resistivity, we can easily imagine cases where the ionospheric plasma in effect "short circuits" this generator along the magnetic field lines, and therefore it will only be the $\perp \mathbf{B}$ component of the electric field which in such cases will reach a significant magnitude in the plasma.

- **Exercise:** Demonstrate that the total current can be expressed in tensor notation as

$$\mathbf{J} = \underline{\sigma} \cdot \mathbf{E}$$

with the conductivity tensor $\underline{\sigma}$ given as

$$\underline{\sigma} \equiv \begin{pmatrix} \sigma_P & \sigma_H & 0 \\ -\sigma_H & \sigma_P & 0 \\ 0 & 0 & \sigma_\parallel \end{pmatrix} \tag{8.41}$$

This formulation is particularly useful in case a transformation of the frame of reference is needed.

- **Exercise:** Demonstrate that at large altitudes we have $\sigma_H/\sigma_P \approx \nu_i/\Omega_{ci} \ll 1$. Give an expression for the angle θ between $\mathbf{E}_\perp$ and $\mathbf{J}_\perp$ in terms of σ_H and σ_P, and demonstrate that for high altitudes, the current is almost entirely a Pedersen current.

8.9.1 Time-varying electric fields

The foregoing discussions were restricted to dc-electric fields. It is, however, evident that the analysis can be generalized to time-varying, but spatially constant, electric fields quite easily. We can thus consider any externally applied forcing, $e\mathbf{E}(t)$, and make a Fourier expansion. Taking one Fourier component $\exp(-i\omega t)$, characterized by an angular frequency ω, we can then write (8.34) as

$$-i\omega M\mathbf{U} = q\mathbf{E} + q\mathbf{U}\times\mathbf{B} - M\nu\mathbf{U},$$

which can be written as

$$q\mathbf{E} + q\mathbf{U}\times\mathbf{B} - M(\nu - i\omega)\mathbf{U} = 0. \tag{8.42}$$

Obviously, we can make the trivial replacement $\nu \to (\nu - i\omega)$ in the previous results (8.38) and obtain, for instance,

$$\alpha = \frac{\nu\Omega_c - i\omega\Omega_c}{\Omega_c^2 + (\nu - i\omega)^2}$$

and

$$\beta = \frac{\Omega_c^2}{\Omega_c^2 + (\nu - i\omega)^2}.$$

From these quantities we can trivially obtain the elements of the frequency varying conductivity tensor. For finite frequencies, we will in general have that the velocity and the external electric fields are out of phase.

The foregoing results apply also to the collisionless case, and in the limit $\nu \to 0$ we find

$$\alpha = -i\frac{\omega\Omega_c}{\Omega_c^2 - \omega^2} \qquad \text{and} \qquad \beta = \frac{\Omega_c^2}{\Omega_c^2 - \omega^2}.$$

This limit applies also to the case where $\omega \gg \nu$. Note the resonance as $\omega \to \Omega_c$. The sign of Ω_c follows the sign of the particle charge q also here.

For the motion along magnetic field lines we can ignore $\mathbf{B}$ altogether, and in this case we have the simple result $U_\parallel = E_\parallel q/(m(\nu - i\omega))$.

In the limit where the relevant frequencies are small compared to the relevant collision frequencies, $\omega \ll \nu$, we can ignore the frequency variation of α and β and have a constant mobility tensor σ as for the steady state case treated in Section 8.9. In this low frequency limit we can consider the plasma as a conductor, as discussed in Section 5.6; see also Appendix B.3.2.

In Appendix B.3 we present a statistical model for collisions between charged particles and neutrals, and also present an appropriate model equation which gives drift velocities consistent with the statistical averages.

- **Exercise:** Consider a model ionosphere consisting of a mixture of neutral atoms with density 10^{18} m^{-3} and an ionized component (a plasma) with ion density $n_i = 10^{13}$ m^{-3}. The plasma is charge neutral. We have a homogeneous magnetic field of 50×10^3 nT. All temperatures are taken to be 1000 K. How much is this in eV? Assume that the cross section, σ, for electrons colliding with neutral atoms is 10^{-16} cm^2 (i.e. 1 Å^2) for electrons with the thermal velocity, and use this cross section as representative also for the ion neutral collisions. What is the free mean path for electron-neutral collisions?

  Assume that the effective ion mass is that of nitrogen atoms. What is the electric conductivity in the direction parallel to **B**, for electrons and for ions? What is the contribution to the electric conductivity in the direction perpendicular to **B** for electrons and for ions? Discuss the limiting case where you let $\sigma \to 0$. Estimate the ratio σ_P/σ_H in this limit.

8.10 Plasma as a dielectric ⓚ

The foregoing discussion described a plasma as a resistive medium, and for stationary conditions, this may indeed be most fruitful. For time-varying cases, a description in terms of a dielectric model can be more useful. We will consider two simple cases, but first it might be appropriate to give some general remarks.

The most straightforward procedure for deriving a dielectric function is based on the polarization model discussed in Section 5.7. From some basic equations we derive the polarization charges and currents of the medium (in our case the plasma) when subject to electric and magnetic fields. Next, we determine the polarization and magnetization. From these results we find a relation between **E** and **D** to give the dielectric function, usually as a function of frequency and wave-number in the most general case. In particular in a plasma we presume that by accounting, at least formally, for the motion of all charged particles, then at least in principle we can express all currents in the plasma as polarization currents, $\partial \mathbf{P}/\partial t$, implying that $\nabla \times \mathbf{M}$ is vanishing, so we set $\mathbf{M} = 0$. Consequently we have $\mathbf{B} = \mu_0 \mathbf{H}$, just as in a vacuum, as far as the description of plasmas is concerned. This is a great simplification.

8.10.1 Plasma as a dielectric at high frequencies

Dielectric plasma models depend critically on parameters and frequency ranges. We consider here unmagnetized plasmas and high frequencies. The frequency dependent polarization of a medium can be obtained by calculating the average displacement of the electrons subject to an externally applied electric field $\mathbf{E}e^{-i\omega t}$. Due to their large mass, the ions are assumed to be essentially immobile, forming a stationary background of positive charge density. As the direction of the electric field is unambiguous, the vector notation will be suppressed in the following. The displacement, z, of an electron in this electric field satisfies the differential equation

$$\frac{d^2 z}{dt^2} + \nu \frac{dz}{dt} = -\frac{e}{m} E \exp(-i\omega t). \tag{8.43}$$

The electron charge is $-e$. The collisions give rise to an effective friction, which is characterized by the coefficient ν; see also Section 8.9. Note, however, that ν^{-1} is here a relaxation time which is introduced without reference to the actual underlying physical process. It is simply stated by the equation (8.43) that the "memory" of z decays with the time-constant ν^{-1}. It can happen after one collision but could as well be due to many collisions which gave partial contributions adding up to a decay time ν^{-1}.

In complex notation for this simple model, the polarization P of the plasma is given by

$$P = -enz(t) = \frac{ne^2 E}{i\omega m} \frac{\exp(-i\omega t)}{i\omega - \nu}, \tag{8.44}$$

giving the electric displacement $D = \varepsilon_0 E + P \equiv \varepsilon_0 \varepsilon_r E$, or $\varepsilon_r - 1 = P/(\varepsilon_0 E)$. For the present case the complex relative dielectric function is easily obtained as $\varepsilon_r(\omega) \equiv \varepsilon_1(\omega) + i\varepsilon_2(\omega) = 1 + i\omega_{pe}^2/[\omega(i\omega - \nu)]$. The real and imaginary parts of the dielectric function are then

$$\boxed{\varepsilon_1(\omega) = 1 - \frac{\omega_{pe}^2}{\nu^2 + \omega^2} \qquad \text{and} \qquad \varepsilon_2(\omega) = \frac{\nu\omega_{pe}^2}{\omega(\nu^2 + \omega^2)}} \tag{8.45}$$

with the plasma frequency $\omega_{pe} = (e^2 n/\varepsilon_0 m)^{1/2}$ introduced. Note that $\varepsilon_2 \sim 1/\omega$ for $\omega \ll \nu$. In the collisionless limit, $\nu \to 0$, we find the standard results $\varepsilon_2 = 0$ and $\varepsilon_1 = 1 - \omega_{pe}^2/\omega^2$ for $\omega \neq 0$. In this case the polarization $P \to \infty$ for $\omega \to 0$. For a collisionless plasma and $\omega \neq 0$ the finite polarization is solely due to finite electron inertia.

The expression ε_r as given by (8.45) is local in space in the sense that the expression does not depend on $\mathbf{k}$ nor ω. Generally we can have $\varepsilon_r = \varepsilon_r(\mathbf{r})$ through the dependence of ω_{pe} on n.

- **Exercise:** Derive the characteristic frequency and the damping rate of the oscillations of a cold, weakly collisional, electron plasma by using the analysis of Section 5.10.6. Obtain the results accurate to order ν^2.

8.10.2 A magnetized plasma as a dielectric at low frequencies

To illustrate that also a magnetized plasma can be considered as a dielectric, we assume that the plasma is confined by a homogeneous strong magnetic field, and restrict the analysis to concern its polarization in the direction perpendicular to the externally imposed magnetic field. In the first approximation, ions and electrons propagate with the $\mathbf{U}_{E\times B} \equiv \mathbf{E}\times\mathbf{B}/B^2$-velocity. We now follow the particles in this moving frame of reference. For an electric field which is varying on a low frequency, this is not an inertial frame, and the ions as well as electrons experience a time-varying gravitational acceleration, $\mathbf{g}$, originating from the variation in the velocity $\mathbf{U}_{E\times B}$. This effective, or virtual, gravity gives rise to an $m_{i,e}\mathbf{g}\times\mathbf{B}/(q_{i,e}B^2)$-velocity, varying with time. The fictive acceleration experienced by a moving observer is $\mathbf{g} = -(1/B^2)\partial(\mathbf{E}\times\mathbf{B})/\partial t$. The resulting drift velocity

$$\mathbf{U}_p \equiv \frac{m_{i,e}}{q_{i,e}B^2}\frac{\partial}{\partial t}\mathbf{E}, \tag{8.46}$$

is called the polarization drift, because it gives rise to a polarization of the plasma since it is different for the ions and electrons; see also Section 7.1.5. Due to the smallness of the electron mass, we can ignore the polarization drift on the electrons for many applications. The polarization drift has opposite sign for electrons and positive ions.

Considering a harmonic oscillation, we find for the ion displacement, $\partial\mathbf{r}_p/\partial t = \mathbf{U}_p$, the result

$$\mathbf{r}_p = \frac{M}{eB^2}\mathbf{E} \equiv \frac{1}{\Omega_c B}\mathbf{E}, \tag{8.47}$$

implying that the polarization is $P = enr_p$, and the relative dielectric constant of the plasma $\varepsilon_r = 1 + P/(\varepsilon_0 E)$ becomes

$$\boxed{\varepsilon_r = 1 + \frac{\rho c^2}{B^2/\mu_0} = 1 + \left(\frac{\Omega_{pi}}{\Omega_{ci}}\right)^2 = 1 + \left(\frac{r_L}{\lambda_D}\right)^2,} \tag{8.48}$$

where we introduced the ion Larmor radius, r_L, with $\rho \equiv Mn$. Weakly magnetized dense plasmas will have large ε_r, irrespective of their temperature.

Note that ε_r as given by (8.48) is local in time as well as space. We can have $\varepsilon_r = \varepsilon_r(\mathbf{r},t)$ through the dependence on n and B. (Nonlocal material relations were discussed in Section 5.9.2.) This implies that as long as we consider low frequencies and polarizations in the direction perpendicular to the externally imposed magnetic field we can here write Poisson's equation (5.7) in the form $\nabla\cdot\mathbf{D} = \nabla\cdot\varepsilon_r(\mathbf{r},t)\varepsilon_0\mathbf{E}(\mathbf{r},t) = \zeta(\mathbf{r},t)$, with $\zeta(\mathbf{r},t)$ being the space-time varying density of free or conduction charges.

- **Example:** It may be illustrative to consider a different derivation of the relative dielectric constant for a magnetized plasma. We first assume that we have one charged particle gyrating in constant electric and magnetic fields, where $\mathbf{E}\perp\mathbf{B}$. The particle is moving with a constant $|\mathbf{U}_b| = |\mathbf{E}\times\mathbf{B}|/B^2$-velocity and the position of its gyro-center is in the co-moving frame given by

$$\mathbf{r}_b = \mathbf{R} + M\frac{\mathbf{U}_b\times\mathbf{B}}{qB^2}, \tag{8.49}$$

where $\mathbf{R}$ is the position of the particle, and $\mathbf{U}_b$ its velocity. The index b means "before," with the understanding before a change in the electric field. Now, we let the electric field change with a small increment $\Delta\mathbf{E}$ within a negligibly short time interval. The new drift velocity is then $\mathbf{U}_a = (\mathbf{E}+\Delta\mathbf{E})\times\mathbf{B}/B^2$ with index a for "after," and the difference between the two velocities is $\Delta\mathbf{U}_D = \Delta\mathbf{E}\times\mathbf{B}/B^2$. Since we assumed the change in electric field to be almost instantaneous, the particle velocity has no time to change because of the finite inertia, so we can assume that the particle velocity is the same right before and right after the change in $\mathbf{E}$, i.e., $\mathbf{U}_a = \mathbf{U}_b$, i.e., it takes some time for the electric field to do work on the particle to change its energy. The *position of the gyro-center* changes instantaneously, however. This is physically acceptable, because the gyro-center has no associated inertia. To find this position we change into the new frame in which the particle will move with its new (constant) velocity, $\mathbf{U}_a - \Delta\mathbf{U} = \mathbf{U}_b - \Delta\mathbf{U}$. In this frame we have

$$\mathbf{r}_a = \mathbf{R} + M\frac{(\mathbf{U}_b - \Delta\mathbf{U}_D)\times\mathbf{B}}{qB^2}, \tag{8.50}$$

and find

$$\Delta\mathbf{r} = -M\frac{\Delta\mathbf{U}_D\times\mathbf{B}}{qB^2} = \frac{M\Delta\mathbf{E}}{qB^2}. \tag{8.51}$$

The polarization of the plasma subject to this varying electric field is $P = qn\Delta r$. If $\mathbf{E}$ varies continuously with time we have $\Delta\mathbf{E}/\Delta t \to d\mathbf{E}/dt$ and can obtain the polarization current density contribution from the appropriate species

$$\mathbf{J} = \frac{\rho}{B^2}\frac{d}{dt}\mathbf{E},$$

where we used $J \equiv qn\Delta r/\Delta t \approx qndr/dt = qnU$, and $\rho = Mn$ is the mass density also here. Note that the ion charge q in (8.51) cancels. The ions contribute most to the polarization current, since $\rho_i \gg \rho_e$, at plasma quasi-neutrality. At least compared to the ions, the electron contribution to the dielectric constant is close to that of a vacuum, $\varepsilon_r \approx 1$; see (5.90). Using the definition $\varepsilon_r \equiv 1 + P/(\varepsilon_0 E)$ we readily reproduce (8.48).

- **Exercise:** Consider a circularly cylindrical system with an inner conductor and an outer conductor with an imposed potential difference V. The gap between the two cylinders is filled with plasma, where we can let the temperatures be zero. The system is embedded in a uniform magnetic field, B_0, with direction parallel to the axis of the system. Calculate the angular $\mathbf{E}\times\mathbf{B}_0/B_0^2$-rotation velocity of the plasma as a function of radius. Taking into account the centrifugal force, show that this velocity is slightly different for electrons and ions, and obtain the net current in magnitude as well as direction. Give the energy density in the vacuum electric field as well as kinetic energy density to lowest order, and write up the sum, W_{tot}, of these. We have $W_{tot} \equiv \frac{1}{2}\varepsilon_r\varepsilon_0 E^2$. Demonstrate that this definition is consistent with (8.48).

Ⓢ **Exercise:** Consider a hydrogen plasma with density $n \approx 10^{14}$ cm^{-3} embedded in a magnetic field of $B \approx 1$ T. Determine ε_r for this case. Ⓢ

It seems that at least in some cases it is fruitful to consider a plasma as a dielectric to be described by a relative permittivity or dielectric constant (which can in some cases be frequency and wavenumber dependent so it is not really a constant, but it is at least independent of the electric field). It would be tempting to see if it is possible to describe a plasma by a similarly constant relative permeability μ_r accounting for the magnetic dipole density (Chen 2016). Considering a magnetized case we see that indeed the plasma can be seen as composed by many small circular currents (the gyro-orbits). These small ring currents depend, however, on the magnetic field since the dipole moment $dm = IA$ associated with a charged particle is here the product of the current $I = q\Omega_c$ and the area $\pi r_L^2 = \pi(MU_\perp)^2/(qB)^2$, giving $dm = \pi MU_\perp^2/B \sim 1/B$. This result implies that such a relative permeability will depend on the magnetic field intensity, and thus not be a constant. This was not a good idea.

9

Experimental Devices

One of the most versatile experimental set-ups for studying the dynamics of plasma media has been the *Q-machine* (Motley 1975), a device for producing a relatively low temperature alkali plasma, which can be studied by means of *Langmuir probes*. Historically, it was one of the first devices to produce a steady state plasma with a modest noise level, and to celebrate this, it was given the name Q-machine, with Q being an abbreviation for *quiet*. Many plasma phenomena were observed in these devices for the first time, and in order to understand, for instance, the results concerning observations of low frequency electrostatic drift waves, it is an advantage to know at least a little concerning the basic features of Q-machines. The insight gained here will be useful in a more general sense for understanding laboratory experiments carried out in magnetized plasma columns. In hindsight it can be argued that Q-machines are expensive and not easy to operate. At the time of writing, there are only few in the world. Nonetheless, in the author's opinion, these devices have contributed significantly to our understanding of basic plasma phenomena, waves and instabilities in particular.

It is of interest also to investigate the full three-dimensional evolution of disturbances in *unmagnetized* plasma as well. (For one spatial dimension, the Q-machine is fine). For this purpose a *double plasma device* was constructed (Taylor et al. 1972). This experimental set-up also had the advantage of producing plasmas with a large electron/ion temperature ratio, which is interesting for investigations of low frequency ion acoustic waves, for instance, since the damping of these waves turns out to depend critically on the electron-ion temperature ratio.

The two devices mentioned here are merely examples. Many more have been constructed over time. It has been argued that major advances in the development of the science of plasmas have frequently been triggered by the invention of a new plasma source (Chen 1995). This is particularly true when we think of fusion plasma experiments. A summary of experimental devices is given also by Piel (2010) and Chen (2016). Some beautiful illustrations of early fusion related plasma experiments are found in the book by Bishop (1958).

9.1 The Q-machine Ⓚ

The Q-machine, in its standard version, see Fig. 9.1, consists of a vacuum vessel at low background pressure, $p < 10^{-4}$ Pa, where the plasma is produced by surface or contact ionization on a hot metal plate, a cathode. The plasma is confined by an axial magnetic field, produced by external coils. The figure shows a Q-machine in single ended operation. These devices can be in *double ended* operation as well, with a hot cathode in both ends (Motley 1975).

Contact ionization was discovered by Langmuir and Kingdon (1925), who found that an electron could be "stripped" from an atom upon contact with a hot metallic surface. The efficiency of the process approached 100% when the work function of the metal exceeded the ionization potential of the atom. The physical explanation of this process need not concern us here. Work functions of metals are usually of the order of a few electron volts (eV), while the ionization energy of, for instance, hydrogen is 13.6 eV. It is evident that we have to look for materials with particularly low ionization energies. Alkali metals are such candidates; the energies are 5.14 eV for Na, 4.34 eV for K, and 3.89 eV for Cs, for instance (Michaelson 1977, Kaye & Laby 1995). If we are prepared to live with the inconveniences of alkali metals it is no problem to achieve an ionization of the appropriate metal vapor. Also other materials can be used, Uranium (ionization energy 6.19 eV) for instance (Hashmi & van Oordt 1971), and it might be possible to use Q-machines for isotope separation.

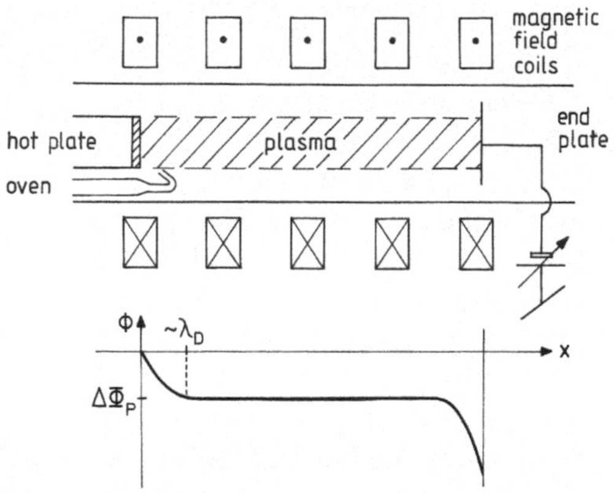

FIGURE 9.1: *A Q-machine shown schematically. An "oven" indicates the source of the neutral gas (usually alkali metals) ionized by contact with the hot cathode. The cathode is heated to temperatures in excess of 2000°C by electron bombardment from a hot filament (not shown) placed on the back side of the hot plate. Typical electron energies there are 2 keV, with 0.5 A currents. The hot plate is fairly thick, 5 mm or so, while the supporting cylinder is thin, in order to reduce the heat conduction away from the hot plate. In order to avoid the exposure of this support to the damaging impact from the energetic electrons inside the cylinder, a certain minimum magnetic field is necessary for the standard operation of most Q-machines. The lower figure shows the axial variation of the dc-electrostatic plasma potential. The curvature of the potential near the hot surface is positive, since the local space charge density is negative, due to the surplus of electrons emitted.*

To have a macroscopically charge neutral plasma, the metal plate must be able to supply also the electrons. This can be achieved by heating it to temperatures high enough to give an appreciable Richardson emission of electrons. As we shall see, this requires temperatures in excess of 1000°C, which in effect restricts the choice of metals to tungsten or tantalum, with melting points of 3380°C and 3000°C, respectively. In the following we assume that the confining axial magnetic field is sufficiently strong to justify a description in one spatial dimension.

The basic physics of the standard single ended operation of a Q-machine can be outlined as follows: Electrons are emitted by the Richardson effect from the hot surface, while ions are emitted by surface ionization. The plasma losses are determined by conditions at the two ends, in particular by the potential of a terminating metal plate at the end of the plasma column. The plasma potential adjusts itself in such a way that the plasma becomes neutral in the central part of the column (Kuhn 1979). Deviations from local charge neutrality occur at the sheaths at the two ends of the column.

9.1.1 Electron emission

The absolute value of the emitted electron current density (measured in A m^{-2}) is

$$I_R = AT^2 \exp(-eW/\kappa T) \tag{9.1}$$

for a cathode temperature T in K, while $A \approx 120$ A cm^{-2} K^{-2} is a constant and W the work function of the cathode material, i.e., the energy we have to supply an electron at rest inside the metal in order to make it escape through the surface. The constants A as well as W are given with some uncertainty even when assuming perfectly clean surfaces. It depends on the crystalline state of the cathode, etc.

Assume the normalized velocity distribution of the emitted electrons to be a Maxwellian. We integrate over the velocity components perpendicular to the Q-machine axis to obtain a one-dimensional distribution. Just outside the cathode surface we have

$$f(u) = \sqrt{\frac{m}{2\pi\kappa T}} e^{-\frac{1}{2}mu^2/\kappa T},$$

where all electron velocities are directed away from the cathode giving a current I_R flowing in the direction *into* the cathode. The absolute value of the electron current emitted from the hot surface is

$$I_R = en_0 \int_0^\infty uf(u)du = en_0\sqrt{\frac{\kappa T}{2\pi m}},$$

where n_0 is a reference electron density and $-e$, m are the electron charge and mass, respectively, with $e > 0$. The reference density n_0 should not be confused with the emitted electron density, since here only positive velocities contribute in $f(u)$ and the normalization of the distribution is obtained by the interval $-\infty < u < \infty$.

We can now anticipate two basically different scenarios: one where the hot plate is very efficient in emitting electrons, and one where the ions are emitted in abundance. The two cases are usually denoted *electron rich* and *ion rich* conditions, respectively. Q-machines are usually operated in the former condition, so we assume that there is a surplus of electrons being emitted. This implies that there will be a negative space charge in front of the hot plate, which will "push" some of the electrons back into the plate, and the macroscopically charge neutral plasma column will assume a negative potential, $-\Delta\Phi_p$, with respect to the hot plate, assumed to be at ground potential. The region with a net charge density will have a spatial extent of the order of a few Debye lengths. Only electrons with a velocity large enough to surmount this potential barrier will reach into the plasma. Assume that this minimum electron velocity is u_{em}, i.e., $\frac{1}{2}mu_{em}^2 = e\Delta\Phi_p$. The absolute value of the current associated with the electrons entering the plasma column is

$$I_1 \equiv en_0 \int_{u_{em}}^\infty uf(u)du = n_0 e\sqrt{\frac{\kappa T}{2\pi m}} \exp(-e\Delta\Phi_p/\kappa T) = I_R \exp(-e\Delta\Phi_p/\kappa T), \tag{9.2}$$

as readily shown. One might argue simply that $W + \Delta\Phi_p$ is an "effective" work function for the cathode plus the surplus charge region. Physically, the interpretation of $e\Delta\Phi_p$ is that of the potential energy (with respect to cathode ground potential) of a plasma electron with zero velocity inside the plasma column.

We assume now that the metal plate terminating the plasma column (see Fig. 9.1) is sufficiently negative to reflect *all* plasma electrons. Literally speaking, this would of course require an infinitely large negative potential, but we assume that the error by using a finite potential is negligible. The absolute value of the current contribution from the reflected electrons is then

$$I_2 \equiv en_p \int_0^\infty uf(u)du = n_p e\sqrt{\frac{\kappa T}{2\pi m}}, \tag{9.3}$$

where now n_p is the electron density in the plasma column, $n_p \neq n_0$. Some emitted electrons are reflected already near the hot plate by the charged sheath; the rest are reflected by the negative end plate. The temperature of the electrons at the hot plate surface and in the plasma is the same since they are in thermal equilibrium with the same hot plate, so there is no need to distinguish $f(u)$ from the electron velocity distribution in the plasma column. Since we assumed *all* electrons to be reflected, we cannot have any net electron current through the system, and require $I_1 = I_2$, which determines the plasma potential in terms of the plasma density n_p in the device

$$\Delta\Phi_p = \frac{\kappa T}{e} \ln\left(\frac{AT^2}{en_p\sqrt{\kappa T/(2\pi m)}}\right) - W \tag{9.4}$$

Since the plasma in the main column (outside the surface charge area) must be macroscopically charge neutral, we have n_p being equal to the electron as well as the ion plasma density there. The plasma potential will depend on the plasma density, which is reasonable after all: if we have many ions available, they can compensate more of the electrons, and only a small $\Delta\Phi_p$ is necessary to reflect the surplus electrons.

For small variations in the plasma density, $n_p = n_{p0} + \widetilde{n}_p$, with corresponding small variations in plasma potential $\Delta\Phi_p = \Delta\Phi_{p0} + \widetilde{\Phi}_p$, the relation (9.4) implies $e\widetilde{\Phi}_p/\kappa T \approx -\widetilde{n}_p/n_{p0}$. Note the minus sign! A *decrease* in plasma density, $\widetilde{n}_p < 0$, implies that the plasma potential becomes more negative, $\widetilde{\Phi}_p > 0$ with the present sign convention, as argued before.

- **Example:** For relevant parameters as $T = 2500$ K, $n_p = 10^{16}$ m^{-3} and $W = 4.1$ eV for a tantalum cathode, we find $\Delta\Phi_p = 1.6$ V, the plasma potential being negative with respect to the cathode ground potential.

9.1.2 Ion emission

Ions originating from surface ionization are accelerated through the space charge region, and are *lost* at the end plate. In order to obtain an expression for the ion distribution function in the plasma we assume that for a given neutral beam illumination of the hot plate (see Fig. 9.1) we have a current density associated with ions in the velocity interval $(v, v+dv)$ just at the cathode surface

$$dI_{i1} = n_{i0} e v \sqrt{\frac{M}{2\pi\kappa T}}\, e^{-\frac{1}{2}Mv^2/\kappa T} dv$$

where we take the temperature to be that of the cathode, since ions must be emitted in thermal equilibrium with the cathode material. The local ion density is n_{i0}. After acceleration through the space charge, an ion has an energy $\frac{1}{2}Mu^2 = \frac{1}{2}Mv^2 + e\Delta\Phi_p$, for $u > 0$. In the plasma, the same ions give a current contribution

$$dI_{i2} = n_{i0} e u \sqrt{\frac{M}{2\pi\kappa T}}\, e^{-\frac{1}{2}Mu^2/\kappa T}\, e^{e\Delta\Phi_p/\kappa T} du$$

since $udu = vdv$. Assume that the ion velocity distribution *in* the plasma is $g(u)$, normalized to unity, $\int_{-\infty}^{\infty} g(u)du = 1$, with the ion density being $n_i = n_p$, as already mentioned. For steady state conditions, the ion current is the same through any cross section of the plasma column. We can therefore write $dI_{i2} = n_p e u g(u) du$, and find

$$n_{i0} \sqrt{\frac{M}{2\pi\kappa T}}\, e^{-\frac{1}{2}Mu^2/\kappa T}\, e^{e\Delta\Phi_p/\kappa T} = n_p g(u) . \tag{9.5}$$

It would be tempting here to argue that $n_p = n_{i0}\exp(e\Delta\Phi_p/\kappa T)$, but this would be in error! The problem is that only ions with velocities $u > \sqrt{2e\Delta\Phi_p/M} \equiv u_{iM}$ are present in the plasma column, outside the space charge region. We have required $g(u)$ to be normalized, but have

$$\sqrt{\frac{M}{2\pi\kappa T}} \int_{u_{iM}}^{\infty} e^{-\frac{1}{2}Mu^2/\kappa T} du \neq 1 ,$$

so the normalization of $g(u)$ must be ensured in a more subtle manner. The correct normalization is obtained by

$$g(u) = \frac{1}{1 - \mathrm{erf}\sqrt{e\Delta\Phi_p/\kappa T}} \sqrt{\frac{2M}{\pi\kappa T}}\, e^{-\frac{1}{2}Mu^2/\kappa T} , \qquad \text{for} \qquad u > \sqrt{\frac{2e\Delta\Phi_p}{M}} \tag{9.6}$$

and $g(u) = 0$ for $u < \sqrt{2e\Delta\Phi_p/M}$. We introduced the error function as

$$\text{erf}(x) \equiv \frac{2}{\sqrt{\pi}} \int_0^x e^{-y^2} dy \qquad \text{or} \qquad \frac{1}{2}\text{erf}\left(x/\sqrt{2}\right) = \frac{1}{\sqrt{2\pi}} \int_0^x e^{-y^2/2} dy.$$

With $\text{erfc}(x) \equiv \frac{2}{\sqrt{\pi}} \int_x^\infty e^{-y^2} dy$ being the complementary error function and $\text{erf}(x) \equiv 1 - \text{erfc}(x)$, we can write

$$\frac{1}{2}\text{erfc}\left(x/\sqrt{2}\right) = \frac{1}{2}\left(1 - \text{erf}(x/\sqrt{2})\right) = \frac{1}{\sqrt{2\pi}} \int_x^\infty e^{-y^2/2} dy.$$

In Fig. 9.2 we show the variation of the ion distribution function for varying normalized plasma potentials, $e\Delta\Phi_p/\kappa T = 0, 0.5, 1.5$ and 2.5. The electrons are in thermal equilibrium with the hot plate; the ions are not. Consequently, the ion velocity distribution is not Maxwellian in the present case, only a part of one. The ion density n_p is obtained by integration of (9.5) with respect to u, using the present normalizations.

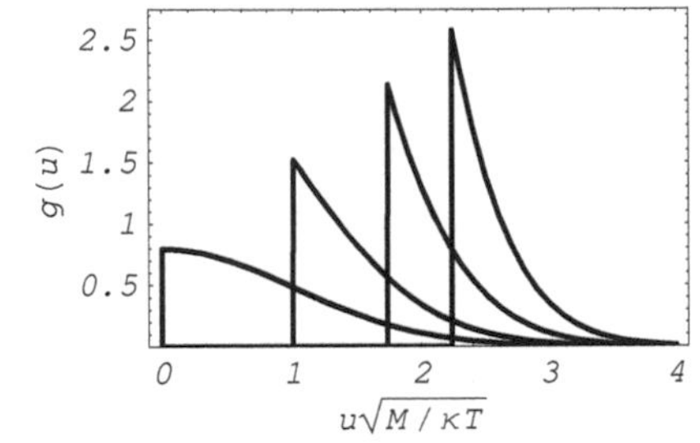

FIGURE 9.2: *Variation of the ion distribution function with normalized plasma potential. It is visually apparent how the width of the distribution decreases with increasing potential drop; from left to right we have* $e\Delta\Phi_p/\kappa T = 0, 0.5, 1.5$ *and* 2.5. *All curves are normalized.*

We can obtain an "effective temperature" for the ion distribution as $T_{eff} \equiv (M/\kappa)\langle(u - \langle u\rangle)^2\rangle = \int (u - \langle u\rangle)^2 g(u) du$. Also this result has an analytical expression, but it is rather lengthy, and is not reproduced here. The result is shown graphically in Fig. 9.3. The effective "cooling" due to the bulk ion acceleration over the potential drop $\Delta\Phi_p$ is sometimes called "adiabatic cooling," and is generic to similar phenomena. It is important to note that the cooling only affects the **B**-parallel part of the velocity distribution function. The **B**-perpendicular part has the same Maxwellian as at the hot plate surface, and the ion velocity distribution is therefore strongly anisotropic. For a very negative end-plate bias the electron component can be assumed to be in thermal equilibrium with the hot plate, and therefore assume its temperature, but the ions are not in equilibrium.

In Fig. 9.3 we note an effect on the effective temperature for $e\Delta\Phi_p/\kappa T = 0$. This is merely a consequence of the ion distribution being a drifting "half Maxwellian". The net current through the Q-machine is relatively small in single ended electron rich operation with negative end-plate bias, because the ion velocities are rather small, in spite of the acceleration, and plasma densities in Q-machines seldom exceed 10^{16} m^{-3}.

The sheaths at the two ends of the plasma column, as mentioned before, will be important for low frequency wave propagation in the Q-machine plasma (Chen 1965*a*, Chen 1979). The highly conducting warm sheath will act as short circuit at low frequencies, while the cold sheath will have a high resistance. In this particular respect, the Q-machine plasma will act as sort of "organ pipe" for low frequency, long wavelength waves, i.e., a pipe closed at one end (at the warm sheath) and open at the other (at the cold sheath). The spatial widths of the sheaths will be of the order of the electron Debye length λ_D, but the more detailed nonlinear analysis is appropriate for large potential drops, see also the solved exercise on page 143.

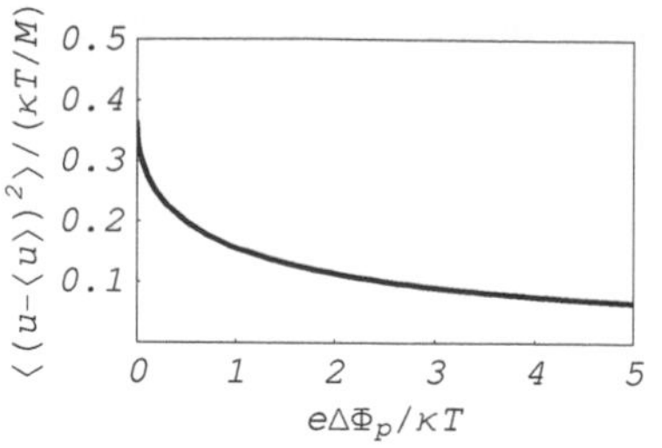

FIGURE 9.3: *Variation of the normalized effective temperature of the ion distribution with normalized plasma potential.*

The analysis outlined before can be repeated for the *ion rich* conditions, usually obtained by operating the hot plate at low temperatures. *Some* heating will always be necessary to maintain a clean cathode surface, but the contact ionization is effective also at temperatures below $\sim 1000°$C, where the Richardson emission is rather ineffective. In this case the plasma potential will become positive, in order to reflect surplus ions back to the cathode. This mode of operation is found to be more noisy than the electron rich operation, and is seldom used.

9.1.3 Discussion

If the temperature of the hot plate was ideally constant, and the illumination by the neutral gas were ideally uniform, the plasma density should be constant within the cross section of the plasma column as terminated by the hot plate. Actually, even this is a bit optimistic, since the surface ionization depends on the crystal surface orientation, and after long time operation at high temperatures, the cathode material will often be crystallized to some degree. This can easily be observed visually. In reality all these idealized conditions are far from being realized, and the plasma column will have radial density gradients. These should not be confused with the strong gradient at the edge of the column, i.e., at the *scrape off layer*. The plasma conditions in this region are quite complicated, in particular because of the strong radial electric fields there (Kent et al. 1969, Sugai et al. 1983).

Although cesium is a rather attractive element for the ion production, it suffers from one drawback, only seldom mentioned: cesium has only one electron in the outer shell, and the cross section for electrons colliding with neutral cesium is rather large (Itikawa 1973). Even with effective cooling of the confining vessel, the region close to the hot plate will unavoidably have a large partial pressure due to un-ionized cesium atoms. Near the hot cathode, the electrons can therefore experience collisional interactions sufficient to invalidate a completely collisionless model. Ions will experience charge exchange collisions in the same region: these types of collisions are resonant, and have a large cross section.

Ⓢ **Exercise:** Express the plasma density n_p in the central part of the Q-machine plasma column (where the plasma potential is constant) in terms of n_{i0}, $\Delta\Phi_p$ and other parameters, such as hot-plate temperature. Ⓢ

Ⓢ **Exercise:** The ions are accelerated at the potential drop at the hot plate surface in the Q-machine. Express the average ion velocity in the central part of the plasma column in terms of n_{i0}, $\Delta\Phi_p$ and other parameters, such as hot-plate temperature. Plot this velocity as a function of a normalized variable containing $\Delta\Phi_p$. Ⓢ

Ⓢ **Exercise:** Give the analytical expression for the average ion flux $\langle n_p u\rangle$. What is the asymptotic limit for the average ion velocity and the average ion flux when $\Delta\Phi_p$ increases? Ⓢ

9.2 Double plasma devices

The simplest way of producing a plasma is by making a discharge in a neutral gas, i.e., accelerating electrons to energies sufficient for ionizing the neutrals. This process can be quite noisy, in the sense of introducing enhanced levels of fluctuations in the plasma. One reason for this enhanced noise is that a simple discharge is rather ineffective: many of the accelerated electrons do not contribute to the ionization, but are simply lost to the walls in the device. By a careful construction, it is, however, possible to make an effective use of the accelerated electrons in a large device, and in this case an acceptable noise level can be achieved. The double plasma device (Taylor et al. 1972, Chen 1995) is constructed to provide a large volume of essentially unmagnetized plasma, usually in two chambers, separated by a negatively biased grid. In the standard operation of the device, plasma is produced in both chambers. One chamber is usually at ground potential; the other one can be biased. By this we can inject ions from one chamber through the separating grid into the other chamber, and thereby create controlled disturbances. The confinement of the charged particles can be significantly improved by an arrangement of permanent magnets around the walls (Limpaecher & MacKenzie 1973), for instance in a "checker-board" arrangement of alternating north and south poles is pointing into the camber. The local magnetic fields give a magnetic picket fence (Knorr & Merlino 1984), while the magnetic field deep inside the device remains small or negligible.

The basic principle of plasma production is simple: electrons emitted from hot filaments are accelerated to energies sufficient for ionizing a neutral background gas. Typical energies can be 100 eV with a popular neutral gas being argon, having a minimum ionization energy of 15.76 eV. Also DP devices have an inherent plasma potential just like Q-machines, but for a different reason. Since the electron energies are much larger than the ion energies, negatively charged particles try to leave the plasma, to be lost at the walls. The plasma potential therefore becomes positive, in general, in order to retain electrons, and to "expel" surplus ions. Since the collisional ionization gives most of the surplus energy to the light electron, we expect $T_e \gg T_i$, i.e., large electron to ion temperature ratios. This would indeed be so, but since the plasma potential is never quite constant either in space or time, ions created with low velocity at one plasma potential can appear as fairly fast ions at another more negative potential part of the device. This effect deteriorates the electron-ion temperature ratio, but we can still easily obtain $T_e > 10T_i$.

One problem with the operation of DP devices can be that multi-charged ions are present, and it is not easy to determine the density ratio of the various charged states. If this is a problem, one might consider using hydrogen as a filling gas, or alternatively helium where multiple charging is less likely for realistic discharge voltages.

9.3 Langmuir probes

Fluctuations in plasma density and to some extent also plasma potential can be detected by a Langmuir probe. Basically, the Langmuir probe consists of a metal wire with a small exposed tip immersed in the plasma (Lochte-Holtgreven 1968, Hutchinson 1987, Piel 2010). The probe can be biased by a variable battery to collect ions or electrons from the plasma, where the collected current can be obtained and visualized as a function of the applied probe voltage. Usually, the geometry of the probe is simple, such as a sphere or a long cylinder. In some cases it might be advantageous to use other shapes (Weber et al. 1979). Probes represent, when they can be applied, some of the most versatile and simple diagnostic methods for studies of plasma phenomena, but they are of course not the only ones (Heald & Wharton 1965, Hutchinson 1987). One of their limitations is met in the very hot plasmas of fusion experiments, where the probes might evaporate rapidly. Other diagnostic methods are fortunately available, where, for instance, the effects of a plasma on the propagation of electromagnetic waves can be used as a remote diagnostic (Huddlestone & Leonard 1965).

9.3.1 A simple example

In this section we discuss in detail a simple realistic physical model for an electric circuit that is sufficiently simple to allow a transparent demonstration of how moving charges induce currents into an electric circuit. A large capacitor is here made up of two conducting plates with area $\mathcal{A}$ and a small separation $\mathcal{L}$. Assume one pair of charged particles, one electron and one ion, to be placed between the plates. Assuming the two terminals of the capacitor to be short circuited, there will be no charge accumulations on the plates and the electron and the ion will for a time interval propagate along straight orbits, until they hit one of the plates. During such time intervals any of the charged particles will induce a current in the circuit by inducing charges q on the capacitor plates. This current can in principle be recorded by standard methods. The electron-ion pair here represents the plasma and the current the diagnostic.

The charge q induced by one particle can be determined by the following argument beginning with the open circuit (without short-circuiting): the amount of work, U, done to induce a charge q on a plate, when the potential across the capacitor is V_c, must equal the energy W gained (or lost) by a charged particle moving through a certain potential difference V. While $U = qV_c$ we have $W = -V e = -V_c e z/\mathcal{L}$, i.e., $q = -ez/\mathcal{L}$ with $-e < 0$ being the electron charge and z a position. The current is obtained by $i = dq/dt$, giving the intuitively reasonable result

$$i = -ew/\mathcal{L}, \tag{9.7}$$

with w being the velocity component perpendicular to the plates, taken to be the $\widehat{z}$-direction. For the *open* circuit this current is exactly canceled by the displacement current, i.e., the $\partial E/\partial t$ term in Maxwell's equations. When the capacitor plates are short-circuited and $E = 0$, while the current i is induced in the wire connecting the two plates. Note that the electron does not contribute particularly much to the current at the instant where it hits a capacitor plate; at that time it only neutralizes the induced charge. The variation of the current is not found correctly by simply counting electrons crossing the surface. The ion contribution to (9.7) is ignored here, since they in general have low velocities. The role of ions will here be to maintain overall charge neutrality. The current is steady state until the electron hits the plate surface. The underlying process is not time stationary since the electron position varies with time.

We now assume that we have N electron-ion pairs present, and let again the ions be at rest, and all electrons have velocity w. The net current is then $I = Ni = -eNw/\mathcal{L} = -ewn\mathcal{A}$, where $n \equiv N/(\mathcal{L}\mathcal{A})$ is the electron density. We let n be sufficiently large to allow the particle discreteness to be ignored and treat the system as a continuum. The stationary current density is then a constant $-ewn$, and the net current to the capacitor plates has precisely the value we would expect by taking the number of charges crossing the plate surface per time unit. The derivation presented here gives insight into the detailed mechanism of how the current is generated. We emphasize that the example considers time-stationary conditions only, and one electron velocity, but the basic ideas are readily generalized (Pécseli 2000).

Assume now that the plates are shielded by some ideal material with free-space permittivity so it always reflects particles. Now the particles will bounce back and forth between the plates. It will in this case always be possible to find two orbits (one approaching and one receding) that give canceling current distributions, and in this limit the net current will become negligible. If we now make a small hole in one of the plates covering the capacitor plates, some electrons will manage to escape. These orbits will not have any canceling, or compensating, counterparts and we will find a net current, which under stationary conditions is correctly accounted for by counting the number of charges arriving per time unit.

The argument can be applied also with an externally imposed magnetic field parallel to the plates: in this case the particles can be confined for long times without hitting one of the plates. If only one electron is orbiting in the magnetic field it will induce a harmonically varying current in the wire short-circuiting the capacitor plates. If we replace the particle by a rotating uniformly charged ring, the current contribution from any charge will at any time be compensated by another charge on the opposite side of the ring. Under such homogeneous conditions we again find a vanishing net current in the system. The induced current is caused by discrete particles.

Temporally nonstationary conditions can become more complicated. One observation based on the present simple model is that, for instance, numerical simulations of plasma probe performances of nonstationary inhomogeneous conditions should account also for the induced charges.

- **Exercise:** Consider an electron performing a circular motion in the plane perpendicular to the capacitor plates discussed in Section 9.3.1. Let the radius of the orbit be much smaller than $\mathcal{L}$. Obtain the analytical expression for the current generated by the electron into a wire short-circuiting the plates. Does this result depend on the average position of the electron?

- **Exercise:** Consider a Langmuir probe with a spherical probe-tip with radius R. The center is placed at a distance $L \gg R$ from an infinitely extended ground plane. The sphere and the plane are connected by a conducting wire. A particle with charge q and a velocity vector $\mathbf{U}_0$ parallel to the ground plane at a distance $L-(R+\Delta)$ passes the sphere so that $\Delta \ll R$. Give an estimate (not an exact calculation) of the time varying current $I(t)$ in the wire to the probe. Make a sketch of $I(t)$.

9.3.2 Plane probes Ⓚ

A simple, one-dimensional model serves to illustrate the basic principles of the Langmuir probe. We can, for instance, consider a strongly magnetized plasma, and let the probe be a disc with a diameter much larger than the ion Larmor radius, which is again much larger than the *electron* Larmor radius, for normal conditions. The plane of the probe is perpendicular to the magnetic field. Assume that the plasma potential is Φ_p, and that the probe is biased to a certain potential $\Phi \leq \Phi_p$. The probe potential will reflect all electrons with velocities *below* $u_{em} \equiv \sqrt{2|e(\Phi-\Phi_p)|/m}$. Electrons with larger velocities will reach the probe and be absorbed. Even when $\Phi = \Phi_p$, only particles (ions or electrons) with a velocity directed toward the probe will be absorbed, i.e., only one half of the distribution, assuming that only the front plane of the probe is exposed to plasma; the back side is usually covered by some non-conductive material. For large positive probe potentials, Φ, all electrons are absorbed, and all ion reflected, and vice versa for very negative potentials. The physical meaning of the plasma potential is, as already mentioned, that a charged particle with zero velocity has the potential energy $q\Phi_p$ with respect to a suitably defined ground.

The electron current collected by the probe under steady state conditions will be

$$I_{ep} = -en\mathcal{A}\int_{u_{em}}^{\infty} u f_e(u)du,$$

in terms of the electron velocity distribution function, $f_e(u)$, while $\mathcal{A}$ is the probe collector area and the electron charge $-e < 0$. In addition to I_{ep} we have an ion current, $I_{ip} = en\mathcal{A}\int_0^{\infty} u f_i(u)du$, which is a constant as long as $\Phi \leq \Phi_p$. This simple expression ignores acceleration of ions through the probe sheath: this assumption is in error, as we will find later, but the error becomes significant only when the ions are cold compared to the electrons, i.e., $T_i \ll T_e$. Obviously, we have $I_{ip} \ll I_{ep}$, unless $\Phi \ll \Phi_p$. It is only for very large negative probe potentials we are able to detect the ion current, and I_{ip} is often ignored in discussions of Langmuir probe performance.

With $\Phi \leq \Phi_p$ and remembering the $u_{em} \sim \sqrt{|\Phi - \Phi_p|}$ dependence, it is now a simple matter to differentiate I_{ep} with respect to Φ, and we find

$$\begin{aligned} \frac{d}{d\Phi} I_{ep} &= \frac{d I_{ep}}{d\sqrt{2|e(\Phi - \Phi_p)|/m}} \frac{d}{d\Phi}\sqrt{2|e(\Phi - \Phi_p)|/m} \\ &= \frac{1}{2} en\mathcal{A}\frac{2e}{m} f_e\left(\sqrt{2|e(\Phi - \Phi_p)|/m}\right) \propto f_e\left(\sqrt{2|e(\Phi - \Phi_p)|/m}\right). \end{aligned} \tag{9.8}$$

Apart from a constant coefficient, it is thus possible to obtain the electron energy distribution function by a relatively simple differentiation of the current to a Langmuir probe, at least in the present idealized one-dimensional operation.

- **Exercise:** Assume that the electron energy distribution is a Maxwellian. Demonstrate that you can determine the electron temperature by plotting the logarithm of the probe current as a function of the probe potential, Φ. Demonstrate that the electron saturation current in the present simple model can be expressed as $I_{ep} = -\frac{1}{4}\langle|u|_e\rangle ne\mathcal{A}$, where by the symbol $\langle\rangle$ understand averaging over the entire three-dimensional Maxwellian, giving $\langle|u|_e\rangle = \sqrt{8\kappa T_e/\pi m}$. The factor $1/4$ in I_{ep} is obtained by the requirement that the particles collected must have their velocity vector directed toward the collecting surface; see Section 2.3.1.

Assuming the velocity distributions of the electrons and ions to be Maxwellians (although not necessarily at the same temperatures), we can give analytical expressions for the probe current contributions, I_{ep} and I_{ip}. Using $\int_{\sqrt{a}}^{\infty} x\exp(-\frac{1}{2}x^2)dx = \exp(-a/2)$ we find (see also the foregoing exercise) the probe current contributions as a function of probe potential Φ measured with respect to the plasma potential Φ_p

$$\begin{aligned} I_{ep}(\psi) &= -\frac{1}{4}\langle|u|_e\rangle ne\mathcal{A}e^{-e|\Phi-\Phi_p|/\kappa T_e} \\ &\equiv -I_{0e}\, e^{-e|\Phi-\Phi_p|/\kappa T_e} \quad \text{for} \quad \Phi < \Phi_p \\ &\qquad \text{and} \quad -I_{0e} \quad \text{for} \quad \Phi \geq \Phi_p. \end{aligned} \tag{9.9}$$

Similarly we have for the ions

$$\begin{aligned} I_{ip}(\psi) &= \frac{1}{4}\langle|u|_i\rangle ne\mathcal{A}e^{-e|\Phi-\Phi_p|/\kappa T_i} \\ &\equiv I_{0i}\, e^{-e|\Phi-\Phi_p|/\kappa T_i} \quad \text{for} \quad \Phi > \Phi_p \\ &\qquad \text{and} \quad I_{0i} \quad \text{for} \quad \Phi \leq \Phi_p, \end{aligned} \tag{9.10}$$

with $I_{0e,i} \equiv \langle|u|_{e,i}\rangle ne\mathcal{A}/4$. The generalization to the case where we have multiply charged ions or a mixture of different ion components is trivial. The ion current is small and can be difficult to observe undisturbed. The total probe current $I_p = I_{ep} + I_{ip}$ in this simple model is shown in Fig. 9.4. The plasma potential Φ_p is taken to be negative in this case. We note another characteristic potential, the "floating potential," Φ_{fl}, where the net probe current vanishes (Kennedy & Allen 2002).

Under steady state conditions (i.e., constant currents) the net current to a probe is given as the difference in the number of electrons and ions hitting the probe surface per sec. In general the thermal electron flux is much larger than the corresponding ion flux. It is only for quite unrealistic electron-ion temperature ratios that the electron and ion current contributions are equal for a probe biased at the plasma potential Φ_p. To obtain a vanishing current, we have to make the probe bias negative in order to reflect a significant fraction of the electrons before they can arrive at the surface of the probe, while on the other hand all ions are collected. To simplify the notation we let Φ_{fl} be the probe potential with respect to the plasma potential, and find for Maxwellian velocity distributions for ions and electrons

$$\frac{1}{4}\langle|u|_e\rangle ne\mathcal{A}e^{-e|\Phi_{fl}|/\kappa T_e} - \frac{1}{4}\langle|u|_i\rangle ne\mathcal{A},$$

i.e.,

$$|\Phi_{fl}| = \frac{\kappa T_e}{e} \ln\left(\frac{\langle |u|_e \rangle}{\langle |u|_i \rangle}\right).$$

For Maxwellian plasmas we readily find

$$|\Phi_{fl}| = \frac{\kappa T_e}{e} \ln\left(\sqrt{\frac{T_e M}{T_i m}}\right),$$

with $T_e M \gg T_i m$, usually. Obviously we find $\Phi_{fl} \to -\infty$ for $M/m \to \infty$, since the ion saturation current vanishes in this limit and cannot compensate any finite electron probe current. For $T_e \approx T_i$ we find $|\Phi_{fl}| \approx 3.75\,\kappa T_e/e$ for a hydrogen plasma where $M/m = 1823$. For slow variations in the plasma we can assume the sheath conditions to be in local equilibrium with isothermal electrons; if this assumption is correct we can assume the difference between the plasma and floating potentials to be constant, and use variations in Φ_{fl} as indicators for variations in Φ_p.

If a probe, or any other object for that matter, is left floating in the plasma, it assumes the floating potential, and not, as one might naively have expected, the plasma potential. In particular, we expect a spacecraft, rocket or satellite to be at floating potential with respect to the surrounding plasma, as any other object floating in space, be it a dust grain or similar. Evidently, there will be no net current to or from an isolated object embedded in a plasma in space, and the potential it has to assume in order to achieve this state is the floating potential, just like a probe at vanishing net current. The floating potential of an isolated object is an important characteristic of "dusty plasmas" where a significant fraction of the charged plasma particles is absorbed on tiny dust particles embedded in the plasma (Shukla & Mamun 2002*a*, Kennedy & Allen 2002, Piel 2010).

As seen in Fig. 9.4, the plasma potential is located at the position where the derivative of the probe characteristic is maximum. Often the plasma potential is identified by the position where $dI_p/d\Phi$ is maximum also for realistic, and not nearly as nice, probe characteristics, without much argument.

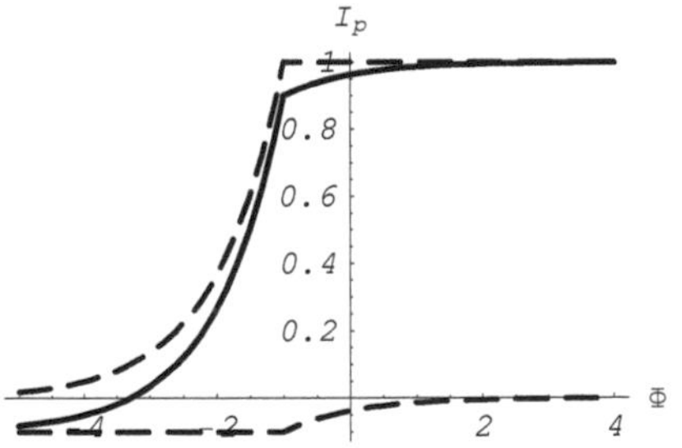

FIGURE 9.4: *Illustration of an idealized probe characteristic for a one-dimensional model. The solid line shows the (idealized) probe current, while the dashed lines show the electron and ion current contributions. For the sake of illustration we took an ion to electron mass ratio of* $M/m =$ 100. *Otherwise, the ion current would hardly be noticeable. We choose here to show the electron current positive: there are no established conventions for this.*

The foregoing analysis assumed steady state conditions, but the arguments do not change much if there are low frequency long wavelength waves propagating in the plasma. In such cases the probe current will be modulated by the fluctuations in plasma potential, plasma density and temperature. The probe can be used to detect waves. The criterion for applying the results of the simple analysis is that the wavelengths are much longer than the probe diameter plus the probe sheaths, and that the wave frequencies are so low that ions as well as electrons have time to propagate through the sheaths with only negligible perturbations. In practice this latter condition implies that the wave frequency ω is much smaller than $\Omega_{pi} \ll \omega_{pe}$.

- **Exercise:** Discuss the relation (9.10) for the case where we have a mixture of different ion species and different charged states.

Ⓢ **Exercise:** Make a figure like Fig. 9.4 for the case where we have two electron components with densities n_{e1} and n_{e2} so that $n_{e1} + n_{e2} = n_0$, and two temperatures T_{e1} and T_{e2}. Consider $n_{e1} < n_{e2}$ as well as $n_{e1} > n_{e2}$. In nature and many laboratory experiments we often find the warm component to have the smallest density. Ⓢ

- **Exercise:** Determine the floating potential for the temperature ratio $T_e/T_i = 2$ and 10, considering different ion species such as singly charged argon (Ar), helium (He), often used in plasma discharges, and cesium (Cs) and sodium (Na), often used in Q-machines.

- **Exercise:** With a potential V_A applied to the Langmuir probe, we draw a current I_A, and can associate a resistance to the probe by Ohm's law. Express the plasma resistance experienced by the ideal plane Langmuir probe as a function of the applied potential.

Because of the absorbed ions, the probe surfaces will eventually be coated (although pump oil, etc., will usually pose a greater problem in this respect, at least in laboratory experiments). The work function of a probe depends often critically on the surface conditions. Usually, probes are heated by some means to retain a clean surface.

When discussing plane probes, we usually ignore the edge effects. The electric field can be very inhomogeneous there, unless special precautions are taken. One simple device is a "guard-ring": let the probe consist of a disc surrounded by a ring with a width comparable to the radius in the disc. The separation "gap" between the disc and the ring has to be very narrow. By arranging the biasing wires properly, we can let the potential of the disc and ring be the same, but when the current is collected, only the part to the central disc is taken into account. In practice, this can be done by placing a resistor in the piece of wire connected to the disc: measuring the voltage across this resistor, we find the current in that part of the circuit alone. Since both disc and ring are on the same potential, there are no particularly large electric fields at the edges of the disc. The edge effects are of course still present at the outer edges of the ring, but we need not be concerned with these, since the ring does not contribute to the detected probe current.

Implicit in most of the foregoing arguments is an assumption of Maxwellian velocity distributions. This is of course a reasonable starting point, but generally we will find that many interesting plasma processes occur far from thermal equilibrium and we can encounter pronounced deviations from equilibrium distributions. We can, for instance, have an energetic electron beam penetrating the plasma: in this case a negative probe potential can reflect all plasma electrons, but the beam electrons can reach the probe and contribute to the probe current. As a consequence, the floating potential can deviate significantly from the one expected for the background plasma alone. This effect can be significant even when the beam electron density is low, provided the beam velocity is large. Under such conditions, the floating potential is a poor representative for the plasma potential.

The probes discussed so far are passive: also active probes are in use, in particular some emitting current due to direct or indirect heating of the probe tip (Schrittwieser et al. 2002). If the probe is emitting electrons, the resulting current-voltage probe characteristic will appear as if the ion current was strongly enhanced, since an electron leaving the probe will give a current contribution as for an ion arriving with the same velocity.

9.3.3 Double probes Ⓚ

A Langmuir probe at its standard operation will draw a relatively large electron current from the plasma when biased to the plasma potential. In some cases this can cause unacceptable disturbances. These can be reduced by using a double probe, where the potential is applied not to one probe with respect to ground, but as a difference potential between two probes (Johnson & Malter 1950). In

Fig. 9.5 we have a sketch of the arrangement. We usually assume that the two probes are identical, although this is not a trivial restriction. The entire system is ideally floating so the net current vanishes. Depending on the voltage to the variable battery giving the potential difference V_D, both probes will draw current, but in such a way that the two currents cancel. It can be instructive to consider Fig. 9.4 for constructing the situation where one probe tip draws a positive current and the other a negative one of the same numerical magnitude. If V_D is small we can approximate the probe characteristic with its tangent, and then one probe will be slightly above the floating potential while the other one will be slightly below. For large potential differences, the negative probe has to reach a very negative potential and the other one a potential between floating and plasma potentials. The positive probe can never be near or above the plasma potential: if this was the case, the current to negative probe would not be able to compensate the current, at least not for the two probe tips being of comparable sizes. The relevant probe potentials will thus both be negative with respect to the plasma potential Φ_p.

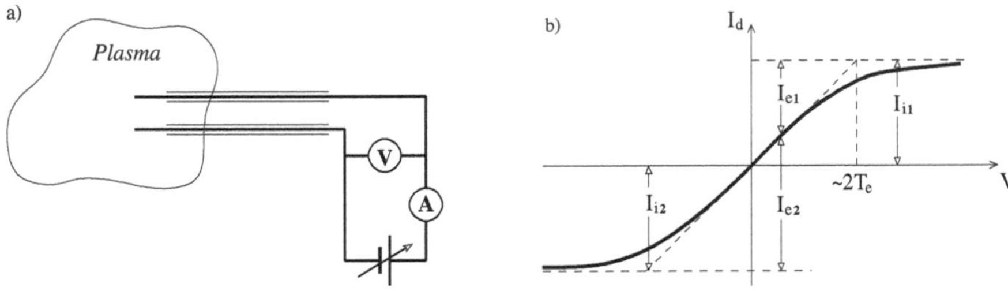

FIGURE 9.5: *Diagram showing the set-up for a double probe in a plasma. An insulation on the probe wires is illustrated by thin lines. The volt-meter shown in a) is assumed to have infinite inner resistivity so the ampere-meter gives the correct current through the system. An idealized probe characteristic is shown in b).*

Denoting the ion contributions to the two probes I_{i1} and I_{i2}, respectively (see Fig. 9.5b)), and similarly for the electron currents $I_{e1,2}$, we have $I_S \equiv I_{i1} + I_{i2} = I_{e1} + I_{e2}$, introducing the notation I_S for the sum of ion currents. As far as the electron currents are concerned, we have

$$I_{e1} = I_{01} \exp(e\Phi_1/\kappa T_e) \quad \text{and} \quad I_{e2} = I_{02} \exp(e\Phi_2/\kappa T_e),$$

with $e > 0$ and $\Phi_{1,2} < 0$ being the potential of the two probe tips with respect to the plasma potential; see (9.9). We have the voltage difference of the two probe tips to be $V_D = \Phi_2 - \Phi_1$ as measured by the voltmeter in Fig. 9.5a). The current measured by the ampere-meter in Fig. 9.5a) is given as $I_D = I_{i1} - I_{e1} = I_{e2} - I_{i2}$. We assume that the two probes constituting the double probe are identical, i.e., $I_{01} = I_{02}$ and $I_{i1} = I_{i2}$ so that $I_S = 2I_{i2}$. With a little algebra we obtain

$$\ln\left(\frac{I_S}{I_{e2}} - 1\right) = -\frac{eV_D}{\kappa T_e}. \tag{9.11}$$

This expression will allow us, in principle, to obtain the electron temperature. To reach a more manageable form we make some simplifications. Assume that the following approximation (see Fig. 9.5b)) is valid

$$I_{e2} \approx I_{i2} + \left.\frac{dI_D}{dV_D}\right|_{V_D=0} V_D,$$

which with $dI_D/dV_D|_{V_D=0} \equiv 1/R_0$ gives

$$\frac{eV_D}{\kappa T_e} \approx -\ln\left(\frac{2}{1 + V_D/(R_0 I_{i2})} - 1\right).$$

For small $V_D/(R_0 I_{i2})$ we obtain

$$\frac{eV_D}{\kappa T_e} \approx -\ln\left(2\left(1-\frac{V_D}{R_0 I_{i2}}\right)-1\right),$$

i.e., $\kappa T_e \approx \frac{1}{2} e R_0 I_{i2}$ which contains easily measurable quantities to determine T_e. In Fig. 9.5b) we show the position on the probe characteristic that estimates $2T_e$ measured in eV.

- **Exercise:** Prove the relation (9.11).
- **Exercise:** Discuss a double probe with two different probe areas. Consider in particular the limit where one probe is much larger than the other one.

9.3.3.1 Plasma sheaths

So far we have mostly discussed the influence of the plasma on the probe, the probe current in particular. The probe will also have an effect on the surrounding plasma due to the particles reflected before they reach the surface (Piel 2010, Chen 2016). This will be true for any surface exposed to the plasma, walls of confining vessels, etc., not just probe surfaces.

If the potential of the surface differs from the plasma potential, some plasma particles (ions or electrons, depending on the potential) will be reflected and contribute to the local space charge in the plasma. Assuming that the net current to the surface vanishes, this being the most common case, the potential will be at the floating potential, Φ_{fl}, negative with respect to the plasma potential. All electrons with kinetic energy less than $|e(\Phi_{fl} - \Phi_p)|$ will be reflected at some distance from the surface. At this position the velocity component normal to the surface will be vanishing and those electrons contribute particularly much to the local space charge. This electron charge imbalance constitutes a space charge that reflects the surplus of electrons and on the other hand accelerates ions toward the surface to give the ion current. Under time-stationary conditions this ion acceleration will not change their contribution to the current, which is given by the product of the ion charge density and ion velocity, i.e., the ion flux, which does not change by the acceleration. This non-neutral region close to a solid surface is called the plasma sheath. The curvature of the local potential variation $\phi(x)$ depends on the local charge density, and the sign of the potential with respect to the plasma potential is so that surplus charges are reflected. The thickness of the sheath is typically a few Debye lengths since the main plasma is assumed to be locally charge neutral, and charge imbalances can usually be maintained over distances of a few Debye lengths only. The sheath at the hot plate of the Q-machine serves as a good example; see Section 9.1. The sheath conditions change if we change the applied potential of the surface with respect to the plasma potential.

9.3.4 Orbit theory for thin cylindrical probes Ⓚ

It is possible to obtain an analytical expression for the ion contribution to the probe current in cases where the geometry is more complicated than the simple plane-probe case. Consider, for instance, a cylindrical probe with a radius $r_p \ll \lambda_D$, where the probe is so long that we do not have to worry about end effects nor the particle motion parallel to the probe axis (Allen 1992). Assume that it is possible to assign a well defined radius r_s for the boundary between the plasma and the sheath surrounding the probe. We expect the sheath thickness $r_s - r_p$ to be of the order of λ_D, or more. Since the probe has a finite radius, it will absorb some of the plasma particles. When discussing shielding of point-like test charges in Section 8.3 we did not have to take this into account.

The following steady state analysis is based on conservation of energy and angular momentum of an ion with mass M as it moves in the stationary probe sheath, as illustrated in Fig. 9.6. With U, V and W being three orthogonal velocity components, we thus have

$$Energy = \frac{1}{2}M(U^2+V^2+W^2)+q\Phi(r),$$

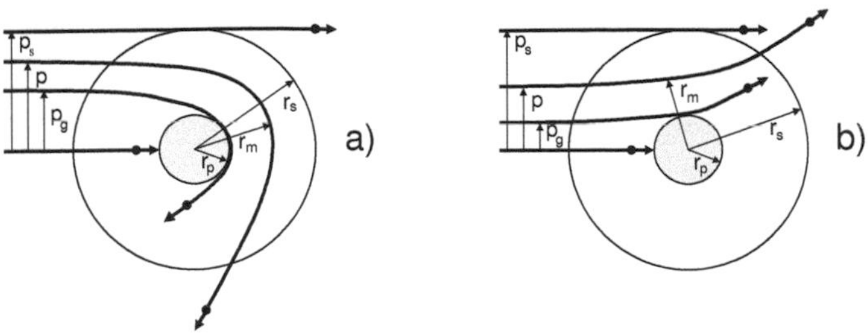

FIGURE 9.6: *Schematic representation of orbits around a cylindrical Langmuir probe. In a) we have an attracting potential; in b) it is reflecting. The probe sheath boundary is given by the thin circle: in reality, this boundary is not so well defined.*

where Φ is the electrostatic potential, with $q > 0$ being the ion charge (we just *might* have multi-charged ions), while

$$Angular\,Momentum \equiv MVp\,,$$

with V being the ion velocity component at the probe sheath, taking $U = 0$ there, while p is the impact parameter of the particle; see Fig. 9.6. Evidently we require $r > r_p$ for the argument of the potential function Φ.

To proceed it is an advantage to specify a geometry: we here assume that the probe is a long cylinder with radius r_p, and let W be the ion velocity component along the probe axis. For symmetry reasons, there are thus no forces having a component along the probe axis. This way the W-component becomes immaterial, and we can for realistic velocity distributions integrate with respect to W. Consequently we take in cylindrical geometry

$$\frac{1}{2}MV^2 = \frac{1}{2}M\left(\left(\frac{dR}{dt}\right)^2 + R^2\left(\frac{d\Theta}{dt}\right)^2\right) + q\Phi(r)\,, \tag{9.12}$$

and have for the angular momentum

$$MVp = MR^2\frac{d\Theta}{dt}\,, \tag{9.13}$$

in terms of the particle position $R = R(t)$ and the given particle velocity V. Eliminating $d\Theta/dt$ from (9.12) and (9.13), we find for a particle entering the sheath with a given velocity V

$$\frac{dR}{dt} = \frac{V}{R}\sqrt{R^2\left(1 - 2\frac{q\Phi(R)}{MV^2}\right) - p^2}\,.$$

Note that $q\Phi > 0$ for retarding potentials, and $q\Phi < 0$ for accelerating potentials; see Fig. 9.6. It is required that $R^2(1 - 2q\Phi(R)/MV^2) > 0$ for $r_p < R < r_s$ to ensure that the particle full orbit exists (i.e., is real valued for all R).

The minimum distance, r_m, from the particle to the probe is reached when $dR/dt = 0$, which corresponds to a certain impact parameter where $p^2 \equiv r_m^2(1 - 2q\Phi(r_m)/MV^2)$, where the velocity V is determined by the conditions at $r = r_s$. The condition is here that $r_m < r_s$. For the orbit that "grazes" the probe surface, as indicated on Fig. 9.6, we then have

$$p_g^2 \equiv r_p^2\left(1 - 2\frac{q\Phi_0}{MV^2}\right),$$

with Φ_0 being the probe potential. We now have basically two cases:

- Retarding fields $q\Phi > 0$: for this case all particles with velocities $V > \sqrt{2q\Phi_0/M}$ will reach the probe surface.

- Accelerating fields $q\Phi < 0$: here we have two possibilities as determined by a critical velocity, V_c. This is determined by the criterion $r_s = p_s$, giving

$$V_c = \sqrt{-2\frac{q\Phi_0}{M}\frac{r_p^2}{r_s^2 - r_p^2}}\,.$$

  All particles entering the sheath with velocity $V \le V_c$ will be absorbed, i.e., any particle with impact parameter less than r_s will also reach the probe surface ("sheath limitation"). In case $V > V_c$ we have all particles with impact parameter $p \le p_g < r_s$ reaching the probe surface, while particles with $p > p_g$ leave the sheath again without reaching the probe surface ("orbit limitation").

We are now in a position to determine the probe current. The contribution per unit probe length to a long cylindrical probe originating from charged particles in a velocity interval $\{V; V + dV\}$ is $dI = p_g q n_0 V F(V) dV$, with $F(V)$ being the velocity distribution normalized to unity $\int F(V)dV = 1$ with n_0 being the plasma density. The "cross section" as seen by these particles is p_g, as is evident from Fig. 9.6: particles with impact parameter $p > p_g$ will miss the probe; particles with $p < p_g$ are collected for collisionless plasmas. If we were to discuss *spherical probes*, the corresponding cross section would be πp_g^2. We see that p_g depends on the particle velocity V.

1) Apart from a constant factor accounting also for the complete probe surface, we have the expression for the ion contribution to the current for the case of retarding potentials

$$I(\Phi_0) = n_0 q r_p \int_{\sqrt{2q\Phi_0/M}}^{\infty} \left(1 - 2\frac{q\Phi_0}{MV^2}\right) V F(V) dV\,. \tag{9.14}$$

   For retarding potentials, the cross section of the probe is given by its geometrical cross section, and the sheath thickness does not enter at all.

2) For the accelerating potential we have to separate the integration into orbit limited and sheath limited contributions.

$$I(\Phi_0) = n_0 q \left(r_s \int_0^{V_c} V F(V) dV + r_p \int_{V_c}^{\infty} \left(1 - 2\frac{q\Phi_0}{MV^2}\right) V F(V) dV \right), \tag{9.15}$$

where the first integral gives the sheath limited contribution, and the second one is for orbit limited particles.

- **Exercise:** Discuss the thin-sheath limit, $r_s \approx r_p$, and demonstrate that the current for the accelerating potential is saturated and approximately independent of Φ_0. Consider the thick sheath, $r_p \ll r_s$, and demonstrate that we, again for accelerating potentials, have approximately the probe current $I \approx C_1 - C_2\Phi_0$, for arbitrary velocity distribution functions, with C_1 and C_2 being constants.

We consider the ion collection of a negatively biased cylindrical probe. The most reasonable assumption is to take the ion distribution, $F(V)$, to be the appropriate two-dimensional Maxwellian outside the probe sheath,

$$F(V) = \frac{M}{\kappa T_i 2\pi} \exp\left(-\frac{1}{2} M V^2/\kappa T_i\right) 2\pi V\,,$$

where the last $2\pi V$-factor originates from the cylindrical volume element. For retarding probe potentials the result is relatively simple, giving the ion contribution to the current $I_i =$

$-I_0 \exp(-q\Phi_0/\kappa T_i)$, with $I_0 = 2\pi r_p \ell n_0 q\sqrt{\kappa T_i/(2\pi M)}$, with $\ell \gg r_p$ being the probe length, so that $2\pi r_p \ell$ is the surface of the probe. Note the sign convention used here for the ion current: it is the one also used in Fig. 9.4. We recognize $\langle |u| \rangle = \sqrt{8\kappa T_i/\pi M}$ as a part of I_0. (Do not confuse I_0 with the Bessel function having the same abbreviation!) We here consider only the contribution from the ions; the electron part can be analyzed along quite the same lines.

We are now in a position to carry out the integration (9.15) for accelerating probes. With $q\Phi_0 < 0$, the result is

$$I_i = -I_0 \frac{r_s}{r_p}\left(1 - \operatorname{erfc}\left(\sqrt{\frac{-r_p^2 q\Phi_0/\kappa T_i}{r_s^2 - r_p^2}}\right)\right) - I_0 \exp\left(\frac{-q\Phi_0}{\kappa T_i}\right) \operatorname{erfc}\left(\sqrt{\frac{-r_s^2 q\Phi_0/\kappa T_i}{r_s^2 - r_p^2}}\right) \tag{9.16}$$

where $\operatorname{erfc}(x) = (2/\sqrt{\pi}) \int_x^\infty \exp(-y^2)dy$ and I_0 was defined before. Consistently with our convention in Fig. 9.7, we took the ion current to be negative for accelerating potentials. The functional form $\Phi(r)$ does not enter the results because we treat the system as conservative. To ease the notation, we now introduce $\xi \equiv q\Phi_0/\kappa T_i$, and assuming $r_s \gg r_p$, we can simplify (9.16) as

$$I_i = -I_0 \frac{r_s}{r_p}\left(1 - \operatorname{erfc}\left(\frac{r_p}{r_s}\sqrt{-\xi}\right)\right) - I_0 \exp(-\xi) \operatorname{erfc}\left(\sqrt{-\xi}\right). \tag{9.17}$$

It turns out that for $\xi < -1$, we can approximate I_i by

$$I_i \approx -I_0\sqrt{1-\xi}. \tag{9.18}$$

Illustrative results are shown in Fig. 9.7 for different values of r_s/r_p. It may be illustrative to compare with the ion current sketched on Fig. 9.4. The effect of the varying probe current with potential for fixed r_s/r_p illustrates also that the collecting area of the probe is different from the geometrical surface area. The present summary contains only the basic elements of the orbit theory for plasma probes. A critical review of the results, including comparisons with experiments, is given by, for instance, Allen (1992).

The present summary emphasizes some approximate, limiting cases. The full solutions will usually require a numerical approach where some results can be found in the early work by Chen (1965*b*), see also Chen (1965*d*).

As the reader might imagine, there has been a wealth of literature published on Langmuir probes (Lochte-Holtgreven 1968), and also more recently (Allen 1992, Hutchinson 2003). In particular for magnetized plasmas with velocity distributions deviating significantly from Maxwellians, the discussions concerning the interpretations of the current characteristics are far from settled. Langmuir probes will remain one of the most versatile methods for plasma diagnostics, in particular for space plasma explorations. They are inexpensive, robust, and easy to operate. In most fusion plasma experiments, however, the temperatures have now reached a level where solid probes are of little use.

- **Example:** Spherical probes can be analyzed along the same lines as the long cylindrical probe. It is found that an approximate expression for the ion probe current in this case becomes $I_i \approx -I_0(1-\xi)$ with $I_0 = 4\pi r_p^2 n_0\sqrt{\kappa T_i/(2\pi M)}$. The requirement of $r_p \ll \lambda_D$ can be difficult to obtain in a laboratory experiment, when taking into account also the thickness of the wire leading to the probe. In space experiments we will, however, often have Debye lengths of 1 m or more, and for such cases the results from orbit theory can be readily applied.

- **Exercise:** An ideal sphere with given radius can be difficult to manufacture. A short cylinder is easier. Let such a cylinder have a length L and radius R. As we know, a sphere is the geometrical form that encloses the largest volume for a given surface. Find the optimum choice for L and R to enclose as much volume possible for a given cylinder surface. How close can you make the volume-surface ratio to the one you have for a sphere with the same surface as your cylinder?

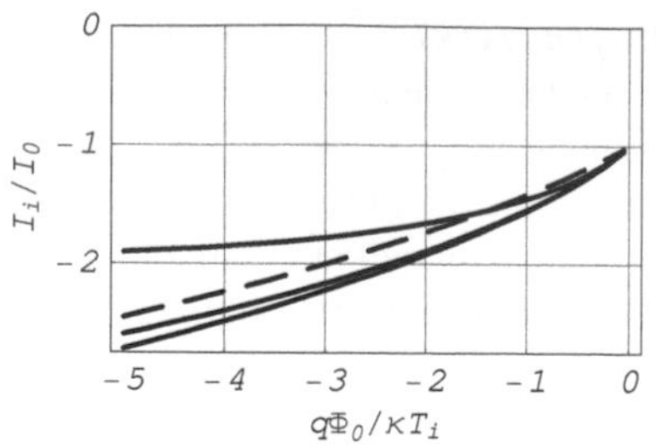

FIGURE 9.7: *Illustration of the probe characteristic shown with the solid line as given by (9.16) for varying probe potential Φ_0. The lowest one of the full curves is for $r_s/r_p = 2$ and increasing $|I_p/I_0|$ are found for $r_s/r_p = 5$ and 10. The dashed line gives the approximation (9.18). The probe potential is given with respect to plasma potential.*

- **Exercise:** As one possible means for re-fueling a future fusion reactor it has been considered to inject a small solid pellet of the proper material, Hydrogen, Deuterium and Tritium, into the hot plasma (Chang 1991). Such a pellet injection can also have diagnostic purposes in experimental conditions. Take a small Tokamak of major radius $R = 1$ m, minor plasma radius $r = 0.1$ m and plasma density $n = 10^{18}$ m^{-3}. How large a Hydrogen pellet can you have before the number of atoms in the pellet exceeds the number of plasma particles?

9.3.5 The Bohm condition Ⓚ

The foregoing summary assumed that electrons and ions arrive independently to the collecting surface of the probe. If the electron and ion temperatures are the same, this turns out to be a reasonable assumption. If, however, the electron temperature is much higher than the ion temperature, the electrons build up a charged sheath around a negatively biased probe, and the ions are accelerated through this sheath. As a result, ions arrive with velocities comparable to the sound speed, rather than an ion thermal velocity. This is the *Bohm sheath condition*, so called in acknowledgment of the scientist first pointing out the relation. For simplicity, the condition can be analyzed in detail by considering a one-dimensional situation, where the plasma fills the space to the left side, and ions are accelerated in the right-hand direction.

Assume that the electron to ion temperature ratio is so large that, in effect, the ions can be assumed cold, i.e., all having the same velocity. For steady state conditions we have ion energy conservation expressed as in (9.12), but this relation now accounts for *all* singly charged and accelerated ions, and we have

$$u = \sqrt{u_0^2 - \frac{2e\Phi(x)}{M}}, \tag{9.19}$$

where u_0 is some, for the moment unspecified, initial ion velocity, $\Phi(x)$ is the potential variation in the sheath with stationary conditions assumed and e is the ion charge. Let the edge of the sheath be at $x = 0$, and the kinetic ion energy density be $\frac{1}{2}Mn_0u_0^2$ at the sheath edge, where $\Phi(x = 0) = 0$. The ion collecting biased plate is at a position $x = x_s > 0$, where $\Phi(x_s) = \Phi_0 < 0$. Again for stationary conditions, the ion continuity equation can readily be integrated to give the ion density as

$$n_0u_0 = n_i(x)u(x) \qquad \text{giving} \qquad n_i(x) = \frac{n_0u_0}{\sqrt{u_0^2 - 2e\Phi(x)/M}}. \tag{9.20}$$

As far as the electrons are concerned we assume that they have an isothermal Boltzmann distribution for stationary conditions; this is quite reasonable. Therefore $n_e = n_0\exp(e\Phi(x)/\kappa T_e)$. From

Poisson's equation we then find

$$\varepsilon_0 \frac{d^2}{dx^2}\Phi(x) = en_0 \left(\exp\left(e\Phi(x)/\kappa T_e\right) - \frac{u_0}{\sqrt{u_0^2 - 2e\Phi(x)/M}} \right). \tag{9.21}$$

Multiplying both sides by $d\Phi/dx$ and integrating with respect to the spatial variable we find

$$\begin{aligned} \frac{\varepsilon_0}{en_0} \int_0^x \frac{d\Phi(\gamma)}{d\gamma} \frac{d^2\Phi(\gamma)}{d\gamma^2} d\gamma = \\ \int_0^x \frac{d\Phi(\gamma)}{d\gamma} \exp(e\Phi(\gamma)/\kappa T_e) d\gamma - \int_0^x \frac{d\Phi(\gamma)}{d\gamma} \frac{u_0 d\gamma}{\sqrt{u_0^2 - 2e\Phi(\gamma)/M}}. \end{aligned}$$

Since by assumption we have $\Phi = 0$ and $n = n_0$ for $x = 0$, at the edge of the sheath, the integration is readily carried out and we find

$$\begin{aligned} \frac{\varepsilon_0}{2en_0} \left(\left(\frac{d\Phi(x)}{dx} \right)^2 - \left(\frac{d\Phi(x)}{dx} \right)^2 \Bigg|_{x=0} \right) = \\ + \frac{\kappa T_e}{e} \exp\left(\frac{e\Phi(x)}{\kappa T_e} \right) - \frac{\kappa T_e}{e} + \frac{u_0^2 M}{e} \left(\sqrt{1 - \frac{2e\Phi(x)}{Mu_0^2}} - 1 \right). \end{aligned}$$

To ease the notation, we introduce the abbreviations $\chi \equiv -e\Phi/\kappa T_e$ and $\mathcal{M} \equiv u_0/\sqrt{\kappa T_e/M}$, with the normalized spatial variable $\xi \equiv x/\lambda_D = x\sqrt{n_0 e^2/\varepsilon_0 \kappa T_e}$, and have

$$\frac{1}{2} \left(\left(\frac{d\chi(\xi)}{d\xi} \right)^2 - \left(\frac{d\chi(\xi)}{d\xi} \right)^2 \Bigg|_{\xi=0} \right) = \mathcal{M}^2 \left(\sqrt{1 + \frac{2\chi}{\mathcal{M}^2}} - 1 \right) + e^{-\chi} - 1.$$

This equation is not easily solved, but if we assume moderate values of Φ, we can make a series expansion of the right-hand side to obtain

$$\begin{aligned} \frac{1}{2} \left(\left(\frac{d\chi(\xi)}{d\xi} \right)^2 - \left(\frac{d\chi(\xi)}{d\xi} \right)^2 \Bigg|_{\xi=0} \right) &= \mathcal{M}^2 \left(1 + \frac{\chi}{\mathcal{M}^2} - \frac{\chi^2}{2\mathcal{M}^4} + \ldots - 1 \right) \\ &\quad + 1 - \chi + \frac{1}{2}\chi^2 + \ldots - 1 \\ &\approx \frac{1}{2}\chi^2 \left(1 - \frac{1}{\mathcal{M}^2} \right). \end{aligned} \tag{9.22}$$

If we now argue that deep inside the plasma we have vanishing electric fields, $E \approx 0$ for $x < 0$, we can set $d\Phi/dx = 0$ at $x = 0$. Even if the sheath edge is not sufficiently well defined to let us claim these equalities to be exact (Godyak & Sternberg 2002), it can still be argued that there is a spatial region where $d\Phi/dx$ is very small. This implies that the left side of (9.22) is positive, and consequently we must have $\mathcal{M}^2 > 1$, or $u_0 > \sqrt{\kappa T_e/M} = C_s$ to get the right-hand side positive as well. This is the *Bohm condition*, and it states that the ions have to arrive at the sheath edge with a velocity exceeding the speed of sound for a steady state sheath to exist (Chen 2016, Riemann 1991, Godyak & Sternberg 2002, Sternberg & Godyak 2007). We recognize $\mathcal{M}$ as the *Mach number* for the problem. Note that we consistently use the cold ion expression for the sound speed, as appropriate for the present basic assumptions. An early mathematical study of collisionless plasma sheaths is given by Caruso and Cavaliere (1962). Particular attention is given there to the transition from the region where the ionization is dominant to the region where the space charge is important, for a collisionless model of a low-pressure arc discharge. A kinetic model for the ion dynamics with finite ion temperatures was used by Riemann (2003) in describing the steady state plasma sheath.

How the ions are accelerated to this velocity outside the sheath is a different question, and has from time to time been vividly discussed in the literature. Often a *pre-sheath* has been invoked (see Fig. 9.8) but also this explanation has been questioned. The presence of a quasi-neutral pre-sheath accelerating the ions has one important consequence; we took the reference position $x = 0$ to be at the sheath edge, taking $n(0) = n_0$ there. If the ions are accelerated prior to arriving at this position, then their density must be larger than n_0 for $x < 0$. To estimate the plasma density variation across a pre-sheath we note that to accelerate slow ions to the sound speed we need $\frac{1}{2}MC_s^2 = e\Phi_{pr}$. With $C_s^2 = \kappa T_e/M$ this gives $\Phi_{pr} \approx \frac{1}{2}\kappa T_e/e$. With Boltzmann distributed electrons we have $n = n_0 \exp(e\Phi_{pr}/\kappa T_e)$ with $|\Phi_{pr}|$ being the pre-sheath potential drop. From $x = 0$ the potential decreases toward the probe surface at $x > 0$, and increases as $x \to -\infty$. We find $n(x = -\infty) \approx n_0 \exp(\frac{1}{2}) \approx 1.65\, n_0$. With the pre-sheath model we thus find that a plane probe surface at some negative potential will give a perturbation of the plasma density with a long spatial range.

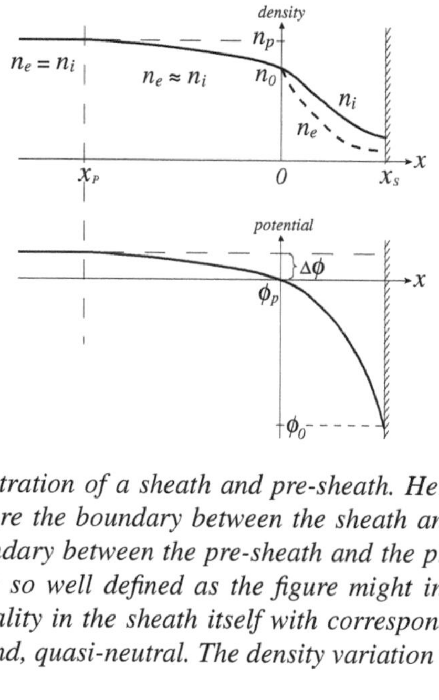

FIGURE 9.8: *Schematic illustration of a sheath and pre-sheath. Here the pre-sheath is localized in the range $x \in \{x_P; 0\}$, where the boundary between the sheath and the pre-sheath is at $x = 0$, while the corresponding boundary between the pre-sheath and the plasma is at $x = x_P$: these two boundaries are in reality not so well defined as the figure might indicate. There are significant deviations from charge neutrality in the sheath itself with corresponding large electric fields; the pre-sheath is, on the other hand, quasi-neutral. The density variation with x is not to scale!*

The outline given here represents what can be called the "conventional wisdom" concerning the Bohm sheath condition. The present author may have reached a slightly different opinion by considering an initial value problem, where a probe is biased to some negative potential at some initial time $t = 0$. The problem can be analyzed even in one spatial dimension. A sheath is formed near the surface, and the perturbation propagates away from the surface at approximately the sound speed. In this model the position x_P in Fig. 9.8 is not constant and the separation between the fixed sheath position (here placed at the origin) and x_P increases steadily with time. If the plasma is infinite, a steady state is never achieved, while it takes a time $\sim \mathcal{L}/C_s$ for a finite system of size $\mathcal{L}$. In this latter case, however, a significant part of the potential drop associated with a pre-sheath will be found at the boundary of the system.

Ⓢ **Exercise:** Discuss the Bohm condition for a single ended Q-machine in electron rich conditions. Is the Bohm condition fulfilled? Discuss the case with varying end-plate bias, but still so that the bias is more negative than the plasma potential. Ⓢ

Ⓢ **Exercise:** Let the ion component of a plasma have mass M_1 and density n_1. Introducing a minority ion mass M_2 and density $n_2 \ll n_1$, determine the velocity of minority ions at the sheath edge. Ⓢ

9.4 Ion energy analyzers

The Langmuir probe characteristic can be used to determine the entire electron energy distribution with an error given by the small ion current contribution. The corresponding ion distribution is more difficult to obtain because the ion current is small and even a tiny electron contribution is disturbing when we attempt to analyze the part of the probe current where some of the slow ions are reflected from the sheath around the probe at a potential larger than the plasma potential.

The standard construction of an ion energy analyzer consists of one or more grids in front of a collecting metal plate. The purpose of the grid(s) is to reflect electrons, so that only the ion current is detected. In this way, the ion energy analyzer acts as a Langmuir probe, which detects only the ion current. The analyzer performance is then trivially described by the results in Section 9.3, and we find, for instance, from (9.8) the result

$$\begin{aligned}\frac{d}{d\Phi}I_{ip} &= \frac{dI_{ip}}{d\sqrt{2|e(\Phi-\Phi_p)|/M}}\frac{d}{d\Phi}\sqrt{2|e(\Phi-\Phi_p)|/M)} \\ &= -\frac{1}{2}en\mathcal{A}\frac{2e}{m}f_i\left(\sqrt{2|e(\Phi-\Phi_p)|/M}\right) \propto f_i\left(\sqrt{2|e(\Phi-\Phi_p)|/M}\right). \qquad (9.23)\end{aligned}$$

In other words, if we differentiate the steady state analyzer current with respect to the collector bias Φ, we recover the ion velocity distribution function of the particles entering the collector. The result is independent of the actual form of the distribution. The differentiation can readily be carried out electronically, by imposing slight variations in the collector bias (Andersen et al. 1971*a*). The ion analyzer current is essentially the ion current contribution to the Langmuir probe, as shown in Fig, 9.4, there for a Maxwellian distribution.

For the low temperature plasmas in a Q-machine, a single grid for reflecting electrons will usually suffice. If the particles energies are larger, there may be problems with secondary emission when an energetic electron is absorbed, and then it may be necessary to introduce more grids.

In Q-machines, we encountered the problem of determining the plasma potential, i.e., the energy with respect to ground of an ion with zero velocity. As demonstrated by Andersen et al. (1971*a*), it is feasible to inject a low density neutral cloud at a low temperature, say around 270 K, into the plasma chamber. Choosing the material properly (the most obvious choice is, of course, to inlet cesium into a cesium plasma experiment, etc.) one can achieve that a small fraction of the hot ions in the plasma undergoes charge exchange collisions with the neutral component. These will be seen as a small peak in the ion energy analyzer signal. Since these particles have low velocities, their energy is basically the potential energy with respect to the ground potential of the laboratory; they will act as markers for the plasma potential. Without such an independent marker of the zero velocity component, the practical use of the electrostatic ion energy analyzer is limited: you cannot easily guess where the zero point is! The method has only been used for single ended operation, where the energy analyzer terminates the plasma column, although in principle it should be possible to build small ion energy analyzers which can be inserted from the side in a double ended operation.

In some cases (Weber et al. 1979) it can be possible to use the electron saturation current part of a Langmuir probe signal for detection of ion beams, for instance. This is a simple method and requires a flat probe. The method works best for high electron to ion temperature ratios. Recently more advanced methods for obtaining the space-time varying ion energy distribution have been developed, using optical methods (Skiff et al. 2001).

9.4.1 Space charge limited currents ⓚ

One particular problem in the construction of ion energy analyzers deserves particular attention: the separation between the grid and the collector must be made sufficiently small to avoid space charge limitations of the collector current. The problem offers also an illustration of applied plasma physics, and is discussed here in part for that reason.

The problem with space charge limitation arises when the most energetic part of the ion distribution is analyzed by the ion energy analyzer. Ions passing through the grid (or system of grids) which is reflecting the electrons are assumed to have an average velocity u_0 just inside the grid at position $x = 0$. Their current density contribution in the direction toward the collecting metal plate is $I_+ = en_0u_0\mathcal{A} =$ const., implying $I_+ = en(x)u(x)\mathcal{A}$ at any spatial position x. Part of these ions is absorbed at the collector, and gives rise to a current proportional to the number of ions having sufficiently large energy to overcome the reflection from the potential, Φ_a, applied to the collector. That part of the ions which does *not* have sufficient energy is reflected, giving a current contribution, I_-, in the direction *away* from the collector. Assuming that only the few most energetic ions are absorbed, we have $I_+ \approx I_-$. We make this assumption in the following, and for the present estimate we assume that it is sufficient to attribute one characteristic velocity u to all ions, so that their kinetic energy density is $\frac{1}{2}Mnu^2$, with $u = u(x)$. In particular $u(x=0) = u_0$. The grid potential is $\Phi(x=0) = \Phi_g$. The position of the collector is $x = d$, giving the potential $\Phi(x=d) = \Phi_a$. It is assumed that the grid is efficient in reflecting all electrons, so the only charged particles in the spatial interval $\{0;d\}$ are ions. The distribution of ion charge in the spatial interval $0 < x < d$ serves as a "barrier" for the ion current. We want to estimate the consequences of this space charge distribution.

The total ion energy is $W_i = \frac{1}{2}Mu^2(x) + e\Phi(x)$, being a constant since we assume that there are no losses in the system. We can select a reference potential so that $W_i = 0$, giving $\frac{1}{2}Mu^2(x) = -e\Phi(x)$, so that $\Phi(x) < 0$. With Poisson's equation this gives

$$\frac{d^2}{dx^2}\Phi = -\frac{I_+ + I_-}{\varepsilon_0 u(x)\mathcal{A}} \approx -2\frac{I_+}{\varepsilon_0 u(x)\mathcal{A}} = -2\frac{I_+}{\varepsilon_0\mathcal{A}}\sqrt{\frac{M}{-2e\Phi(x)}}\,.$$

We multiply by $2d\Phi/dx$, and find after some elementary manipulations and an integration

$$\left(\frac{d\Phi}{dx}\right)^2 = 8\frac{I_+}{\varepsilon_0\mathcal{A}}\sqrt{\frac{M}{2e}}\sqrt{-\Phi(x)}\,,$$

where an integration constant was taken to zero. Taking the square root and integrating with respect to x we find

$$-\frac{4}{3}(-\Phi(x))^{3/4} = 2(x-d)\sqrt{2\frac{I_+}{\varepsilon_0\mathcal{A}}}\left(\frac{M}{2e}\right)^{1/4}.$$

The maximum current we can have is then

$$\max\{I_+\} = \frac{2}{9}(\Phi_a - \Phi_g)^{3/2}\frac{\varepsilon_0\mathcal{A}}{d^2}\sqrt{\frac{2e}{M}} \tag{9.24}$$

recalling that I_+ is the current density *toward* the collector in the direction *from* the grid. If we try to exceed this current limit, any excess ion will be reflected by the space charge of ions in the region between the grind and collector. Note that $\max\{I_+\}$ scales as $1/d^2$. To avoid the space charge limitation, the distance between grid and collector in an electrostatic ion energy analyzer should be the absolute minimum, usually a fraction of a millimeter. There are no paradoxes in (9.24) for $\Phi_a = \Phi_g$; since we implicitly assume that essentially all ions had to be reflected from the positively biased collector, we require $\Phi_a > \Phi_g$.

10

Magneto-Hydrodynamics by Brute Force

In this section, we present a relatively simple, yet very useful, model for the dynamics of fully ionized plasmas, as they are found in, for instance, the Earth's upper ionosphere, in the magnetosphere or in the more distant regions of space. The model is based on some simple assumptions, by arguing that a plasma can be considered as a medium which follows the same basic laws as any other continuum, i.e., the equation of continuity and Newton's second law. The electromagnetic parts of the fields that are entering the total force have to be consistent with Maxwell's equations, subject to some simplifications, which are valid within certain parameter regions. The model ignores the fact that a plasma is composed by a mixture of two interacting "gases", an electron and an ion "gas". Instead the plasma is described as one medium, characterized essentially by a large conductivity, but we do not here address the question of *how* this conductivity is achieved!

10.1 Ideal magneto-hydrodynamics

In the absence of creation or loss of particles, any medium will follow a continuity equation in terms of velocity $\mathbf{u}$ and mass density ρ

$$\frac{\partial}{\partial t}\rho + \nabla \cdot (\mathbf{u}\rho) = 0, \tag{10.1}$$

and this equation is therefore nothing special in the present context.

Attempting to formulate a simple dynamic model for describing a plasma, we may apply a sort of "brute force" argument (van Kampen & Felderhof 1967) to produce a simple version of what is usually known as the Magneto-HydroDynamic (MHD) equations (Alfvén 1950, Lighthill 1960, Goedbloed & Poedts 2004). It can thus be argued that a plasma is a gas which happens to be a good conductor, but a gas nonetheless. Then it is reasonable to assume that it can be adequately described by the usual Navier-Stokes equation

$$\rho\left(\frac{\partial}{\partial t}\mathbf{u} + \mathbf{u}\cdot\nabla\mathbf{u}\right) = -\nabla p + \mathbf{J}\times\mathbf{B} + \rho\mathbf{g}, \tag{10.2}$$

where we, in addition to the force from a pressure gradient, also included a $\mathbf{J}\times\mathbf{B}$ force density. With the expectation of the plasma being a good conductor, we presume that the space time varying currents flowing in it, and the magnetic fields generated by them, will be important for the dynamics. For generality we included a gravitational acceleration $\mathbf{g}$. We have, on the other hand, *not* included any forces due to charge distributions and electric fields. This is because we expect that in a good conductor even tiny deviations from overall charge neutrality will give rise to large currents, and that the $\mathbf{J}\times\mathbf{B}$ forces due to these will dominate any $\zeta\mathbf{E}$-force densities, with ζ being the charge density. Assuming charge separations to be short-circuited instantaneously, as we do here, implies that we shall be concerned with large time scales (slow phenomena) where the inertia effects of the electrons can be neglected. If we are to study the dynamics of, for instance, the solar plasma, the contributions of gravitational forces are of central importance, as included in (10.2). To obtain a closed set of equations, we now have to account for the space time variations of the pressure p, the magnetic fields $\mathbf{B}$ and the currents $\mathbf{J}$.

In principle, in its basic form Magneto-Hydro-Dynamics applies not only for plasmas but liquid metallic conductors (like Mercury) as well. In practice the resistivity of the metals is found to be a problem and the equivalent of the plasma parameter is small, but early experiments on the basic properties of the MHD model were performed in such liquid metals, studies of wave propagation in particular. By now, many of these works have mostly historical interest, but may be worthwhile reading nonetheless (Lundquist 1949, Lehnert 1954, Lehnert & Bullard 1955, Dattner et al. 1958, Lehnert & Sjögren 1960).

10.1.1 Faraday's law

Faraday's law (5.58), given here in differential form $\nabla \times \mathbf{E} = -\partial \mathbf{B}/\partial t$, relates magnetic fields to electric fields. To facilitate the discussion it is convenient to use a test function in the form of a plane wave, with the usual $\exp\big(-i(\omega t - \mathbf{k}\cdot\mathbf{r})\big)$ variation in space and time. We split the electric and magnetic fields into components $\mathbf{E}_\parallel$ and $\mathbf{E}_\perp$ parallel and perpendicular to $\mathbf{k}$. Since $\nabla \times \mathbf{E} \to i\mathbf{k} \times \mathbf{E}$ is a vector in the direction $\perp \mathbf{k}$, we have from Maxwell's equation (5.58) that the direction of the fluctuating magnetic field $\mathbf{B}$ has to be $\perp \mathbf{k}$ also, consistent with $\nabla \cdot \mathbf{B} = 0$. We also have $\partial \mathbf{B}/\partial t \to -i\omega \mathbf{B}$. This in turn implies that $|E_\perp|/|B| = \omega/k$, which turns out to be a most convenient and useful relation. The electric field *can* have a component $\parallel \mathbf{k}$, but this is not yet determined. On the other hand, we note that *if* the rotation $\nabla \times \mathbf{E}$ vanishes *simultaneously* with the fluctuating magnetic field, then the ratio $|E_\perp/B|$ is undetermined, and a relation for the characteristic phase velocity of the waves must be found by a different procedure. This will be the case for purely electrostatic waves, such as Langmuir waves, but these do not enter the magneto-hydrodynamic description.

The present discussion used a plane electromagnetic wave as a reference case, since it is here easy to distinguish $\mathbf{E}_\parallel$ and $\mathbf{E}_\perp$. More generally we can define the transverse component as the one where $\nabla \times \mathbf{E}_\perp \neq 0$ and the longitudinal part as the one where $\nabla \cdot \mathbf{E}_\parallel \neq 0$. Introducing the vector and scalar potentials (see Section 5.3) we have $\mathbf{E} = -\nabla\phi - \partial \mathbf{A}/\partial t$ by (5.17) where again the distinction between longitudinal and transverse components becomes evident.

10.1.2 Ampere's law

We are now in a position to make an order of magnitude estimate for the various terms in the general law $\nabla \times \mathbf{B} = \mu_0 \mathbf{J} + \mu_0 \varepsilon_0 \partial \mathbf{E}/\partial t$ from (5.57), where $\mathbf{J}$ includes all current densities in the plasma. Assume that we have a characteristic time scale $\mathcal{T}$ (for instance, the period of a harmonic oscillation), a characteristic length scale $\mathcal{L}$ (for instance, the wavelength of a plane wave) and characteristic amplitudes for the fluctuations in electric and magnetic fields, $\widetilde{\mathcal{E}}_\perp$ and $\widetilde{\mathcal{B}}$. We use these quantities for normalizing time and length and field quantities, and find

$$\nabla' \times \mathbf{B}' = \mu_0 \mathbf{J} \frac{\mathcal{L}}{\widetilde{\mathcal{B}}} + \left(\frac{1}{c^2} \frac{\widetilde{\mathcal{E}}_\perp}{\widetilde{\mathcal{B}}} \frac{\mathcal{L}}{\mathcal{T}} \right) \frac{\partial}{\partial t'} \mathbf{E}', \tag{10.3}$$

where the prime denotes dimensionless variables and quantities. Now, a characteristic velocity is $\mathcal{L}/\mathcal{T}$ and so is $\widetilde{\mathcal{E}}_\perp/\widetilde{\mathcal{B}}$ according to the foregoing arguments of Faraday's law. Provided this characteristic velocity is much smaller than the speed of light, we find that the last term in (10.3) is negligible. We can see this estimate as a consistency requirement for ignoring Maxwell's displacement current. We then end up with Ampere's law in its original form

$$\nabla \times \mathbf{B} = \mu_0 \mathbf{J} \,. \tag{10.4}$$

With this relation we can eliminate the current density $\mathbf{J}$ from (10.2). We have equation (10.4), to connect current densities directly with magnetic fields, which is a great simplification. If we, for the moment, ignore the pressure and set the density to be constant, we see that (10.2), with (10.4) inserted, essentially relates magnetic fields to velocities for this simple limit. See also a related discussion for electromagnetic waves in conducting media in Section 5.6. Implicit in the argument is the assumption that $\widetilde{\mathcal{E}}_\perp$ is an acceptable normalization factor for the electric field. This implies that $|\mathbf{E}_\parallel|$ cannot be larger than $|\mathbf{E}_\perp|$ for the MHD model to apply. In many ways it can be argued that the omission of Maxwell's displacement current is the basic assumption in MHD.

One often finds the argument that $\partial\mathbf{E}/\partial t$ can be ignored for slow variations, but the statement is meaningful only when it is stated "slow compared to what"! In the present case, the time variation of the electric field should be compared to a time variation of an electromagnetic perturbation having the same scale size $\mathcal{L}$, propagating in free space.

Since we have ignored the force from electric fields acting on charges in (10.2) it is consistent to ignore also Maxwell's equation (5.59), the Poisson equation. Now, this has a price; for consistency we then have to ignore the charge density also in the continuity equation for currents and charge densities $\partial\zeta/\partial t + \nabla\cdot\mathbf{J} = 0$, see (5.13). When we assume that ζ is in effect negligible, we implicitly also assume $\nabla\cdot\mathbf{J} \approx 0$ for the current densities in the plasma.

10.1.3 Ohm's law

We thus have four essential dependent variables, **J**, **E**, **B** and **u**, and three equations, (5.58), (10.2) and (10.4). It seems intuitively evident to look for one more equation relating some or all of these variables. Such a relation is Ohm's law. It is, however, important to recall that the proper frame of reference for this equation needs consideration. We have most generally

$$\mathbf{J} = \sigma(\mathbf{E} + \mathbf{u}\times\mathbf{B}), \tag{10.5}$$

where σ is the plasma conductivity, here assumed to be a scalar. Since we have ignored charge separations in obtaining (10.2), we omit a current contribution originating from a moving net charge in (10.5). In (10.5) we have contributions from the electric field, **E**, in the absolute frame of reference, which we would have under all circumstances. In addition we note that in the frame of reference following the plasma, there is one more electric field contribution, namely, the one induced by the motion with velocity **u** across magnetic field lines, this being the second term in (10.5); see the discussion in Section 5.8, in particular (5.79). This velocity is assumed to be nonrelativistic, which can hardly be seen as an additional restriction, the accuracy of all other approximations considered.

As a limiting case, we can assume the plasma to be an ideal conductor, and let $\sigma \to \infty$. In order to keep the currents finite we then have to require

$$\mathbf{E} = -\mathbf{u}\times\mathbf{B}, \tag{10.6}$$

for this ideal MHD limit. This implies that all electric fields are induced by motion of plasma across magnetic field lines, and the electric fields can be eliminated by inserting (10.6) into Faraday's law (5.58). If electrostatic fields are to be present, they have to be imposed "externally" by free charges, for instance. The relation (10.6) implies that we have no electric field components along **B** within the framework of ideal MHD. This implication on the electric field direction from (10.6) is quite important by stating that, within ideal MHD, electric fields cause no particle acceleration along magnetic field lines. We *can* have velocity components parallel to **B**, but these are not caused by electric fields; see, for instance, (10.2).

Since all fluctuating electric fields are in the direction perpendicular to the local magnetic field, we can argue that in case we follow a small volume element of the plasma, its local velocity component in the direction $\perp \mathbf{B}(\mathbf{r},t)$ will be determined by the $\mathbf{E}\times\mathbf{B}/B^2$-velocity; see also Section 5.8. In the rest frame of the plasma we have $\mathbf{E} = 0$ at the position of that volume element. At a neighboring position, the velocity $\mathbf{u}(\mathbf{r},t)$ will be different as a consequence of a spatial variation of $\mathbf{E}(\mathbf{r},t)$. By going to a "smart" frame of reference we can, in general, only transform away the electric field in one point at a given time. In general, with inhomogeneous electric fields, we therefore can have $\nabla \times \mathbf{E} \neq 0$ in any frame, implying that $\partial \mathbf{B}/\partial t \neq 0$, from (5.58).

10.1.4 Equation of state

Referring to the arguments in Section 2.5, we assume that the motion, at least to relevant accuracy, can be considered as incompressible, $\nabla \cdot \mathbf{u} = 0$, and reduce the continuity equation to

$$\frac{\partial}{\partial t}\rho + \mathbf{u} \cdot \nabla \rho = 0 \tag{10.7}$$

which is equivalent to $D\rho/Dt = 0$. Note that incompressibility does not necessarily imply that the density is constant; if it is initially inhomogeneous, it will remain so, and a stationary observer will experience a time-varying density according to (10.7), due to gradients in density being swept with the velocity $\mathbf{u}$ past the observer.

Taking the divergence of (10.2), ignoring gravitational forces and using the incompressibility condition, we obtain

$$\nabla \cdot \left(\frac{1}{\rho} \nabla p \right) = -\nabla \cdot (\mathbf{u} \cdot \nabla \mathbf{u}) + \nabla \cdot (\mathbf{J} \times \mathbf{B}) . \tag{10.8}$$

This relation implies that the pressure can, at least in principle, be obtained once the velocity field and the current distributions are known. The relation (10.8) can be seen as an equation of state, or a *constitutive* equation in the sense discussed in Section 2.5, and no additional relation establishing the connection between pressure and, for instance, density is needed for incompressible flows. The basic requirement for an equation of state is that the pressure is completely determined by, say, the actual density and temperature. The pressure shall *not* depend on the "history" of the system, i.e., on the path in phase space along which the system arrived at the actual state. If we allow for compressibility here, $\nabla \cdot \mathbf{u} \neq 0$, (10.8) would contain an explicit time derivative, and in this case the equation would relate pressure to the earlier values of the velocity. Such an equation would therefore not be acceptable as an equation of state, and that relation has to be imposed by other means.

- **Exercise:** Take the density ρ to be only incompressible, but also constant in space. Write the result (10.8) with this simplification. Derive an equivalent form of (10.8), where $\nabla^2 p$ appears explicitly, and use (10.4) to simplify the $\mathbf{J} \times \mathbf{B}$-term. Demonstrate that this term in (10.4) can be written as the divergence of a tensor containing the magnetic field, and obtain an analytical expression for this tensor. By this exercise, a Poisson equation is obtained for the plasma pressure. This equation has a rather complicated right-hand side and may not be readily solved, but at least we know that its solution *exists*, and quite formally this solution can be written down. Do this!

Summary:

The basic equations of simple incompressible, ideal MHD, can be summarized as follows, noting that $\nabla \cdot \mathbf{B} = 0$ enters as an auxiliary relation. We find that $\partial \nabla \cdot \mathbf{B}/\partial t = 0$ follows from (10.9), but this condition does not exclude stationary solutions with $\nabla \cdot \mathbf{B} \neq 0$, so the requirement $\nabla \cdot \mathbf{B} = 0$ comes separately in Maxwell's equations.

$$\frac{\partial}{\partial t}\mathbf{B} = \nabla \times (\mathbf{u} \times \mathbf{B}), \tag{10.9}$$

$$\frac{\partial}{\partial t}\rho + \mathbf{u} \cdot \nabla \rho = 0, \tag{10.10}$$

$$\nabla \cdot \mathbf{u} = 0, \tag{10.11}$$

$$\rho \left(\frac{\partial}{\partial t}\mathbf{u} + \mathbf{u} \cdot \nabla \mathbf{u} \right) = -\nabla p + \frac{1}{\mu_0} (\nabla \times \mathbf{B}) \times \mathbf{B}. \tag{10.12}$$

Magneto-hydrodynamics, even in its ideal form, describes highly complicated dynamics! It is not correct simply to argue that electric fields are due to motion of plasma across magnetic field lines; magnetic forces, together with the plasma pressure, determine the plasma velocity. Electric fields are derived from Faraday's law (5.58), but the $\perp \mathbf{B}(\mathbf{r},t)$-velocity component has to be *consistent* with the $\mathbf{E}\times\mathbf{B}/B^2$-velocity, and it cannot be argued that one is *caused* by the other, but rather that the ideal Ohm's law (10.6) acts as a *constraint*; see, for instance, an illuminating discussion by Vasyliūnas (2001). Note, finally, that the equations of ideal MHD are *time reversible*, as expected for a system without losses.

Ⓢ **Exercise:** Normalize the magnetic field in such a way that it acquires the dimension "velocity" by letting $\mathbf{B}\to\mathbf{B}/\sqrt{\rho\mu_0}$, with ρ constant, and also let $p\to p/\rho$. Introduce the *Elsasser variables* $\Psi\equiv\mathbf{B}-\mathbf{u}$ and $\Phi\equiv\mathbf{B}+\mathbf{u}$. Rewrite the basic equations for incompressible ideal magneto-hydrodynamics in terms of these new variables (Yoshizawa et al. 2003). Demonstrate that for ideal MHD, we have $\Psi^2+\Phi^2=2(B^2+u^2)$ and $\Psi^2-\Phi^2=2\mathbf{B}\cdot\mathbf{u}$, where the latter quantity is called "MHD cross helicity". Prove that $u^2+B^2\geq 2|\mathbf{B}\cdot\mathbf{u}|$. When does the "equal" sign apply here? Ⓢ

Ⓢ **Exercise:** Include a gravitational force into the momentum equation of the Magneto-HydroDynamic (MHD) equations, assuming that the (constant) vertical gravitational acceleration $\mathbf{g}$ is in the direction perpendicular to the magnetic field lines in the horizontal direction. The plasma density and temperature are taken homogeneous.

1) Write-up the MHD-equations for steady state conditions.
2) Demonstrate that in the given equilibrium steady state conditions, there must be a net current density, $\mathbf{J}$, in the plasma. Find the magnitude and direction of $\mathbf{J}$. Explain the physical origin of this current. Ⓢ

- **Exercise:** Prove the relations

$$\frac{\partial}{\partial t}\mathbf{B}=\mathbf{B}\cdot\nabla\mathbf{u}-\mathbf{u}\cdot\nabla\mathbf{B}-\mathbf{B}(\nabla\cdot\mathbf{u}), \qquad \text{and} \qquad \frac{\partial}{\partial t}\mathbf{B}=\nabla\cdot(\mathbf{B}\mathbf{u}-\mathbf{u}\mathbf{B}),$$

where the latter expression can be seen as a conservation form of Faraday's law. How are these relations simplified if we assume incompressible flows?

10.2 Compressible MHD

The restrictive assumption of incompressible flows is easily removed in the basic ideal or non-resistive MHD equations, and we find the basic equations for compressible MHD in the form

$$\frac{\partial}{\partial t}\mathbf{B} = \nabla\times(\mathbf{u}\times\mathbf{B}), \tag{10.13}$$

$$\frac{\partial}{\partial t}\rho+\mathbf{u}\cdot\nabla\rho = -\rho\nabla\cdot\mathbf{u}, \tag{10.14}$$

$$\rho\left(\frac{\partial}{\partial t}\mathbf{u}+\mathbf{u}\cdot\nabla\mathbf{u}\right) = -\nabla p+\frac{1}{\mu_0}(\nabla\times\mathbf{B})\times\mathbf{B}. \tag{10.15}$$

$$p = P(\rho) \tag{10.16}$$

where now an equation of state was introduced; see also Section 2.3. We might here take the adiabatic ideal gas law and assume

$$p=C\rho^{\gamma}, \tag{10.17}$$

where $\gamma \equiv C_P/C_V$, the ratio of specific heats, and C is a constant. Alternatively, we have

$$T\rho^{1-\gamma} = C_1 \qquad \text{or} \qquad T\,p^{(1-\gamma)/\gamma} = C_2,$$

where T is the temperature of the plasma. We used $p = \kappa T\rho/M$. Again we note that the basic requirement for an equation of state is that the pressure is completely determined by, say, the actual density and temperature. The pressure will not depend on the "history" of the system, i.e., on the path in phase space along which the system arrived at the actual state.

In static or quasi-static conditions one will alternatively assume an isothermal condition and find

$$p = \left(\frac{\kappa T}{M}\right)\rho. \tag{10.18}$$

The choice of equation of state to be used depends often on the problem at hand. The choice of the ideal gas law can also appear somewhat arbitrary; after all, a plasma can hardly be expected to behave *exactly* as an ideal gas, but in view of all the other approximations which were made, the restrictions implied by this choice are probably immaterial.

In case the equation of state can be inverted to give density in terms of pressure, we might simplify, or at least rewrite, the equations somewhat. We can take the adiabatic equation of state, for instance, and find $\rho = \rho_0(p/p_0)^{1/\gamma}$. This, inserted into the equation of continuity, gives, after some simple manipulations,

$$\frac{\partial}{\partial t}p + \mathbf{u}\cdot\nabla p = -\gamma p\nabla\cdot\mathbf{u}, \tag{10.19}$$

where the consequences of compressibility appear explicitly by the right-hand side; see also the discussion of (10.19). The equation of continuity is made redundant by (10.19). Incompressibility results by taking $\nabla\cdot\mathbf{u}\rightarrow 0$ and letting $\gamma\rightarrow\infty$ in such a way that $\gamma p\nabla\cdot\mathbf{u}$ remains finite. This quantity will for this limit simply express the magnitude of Dp/Dt and the equation can be omitted, while the missing constraint is compensated by introducing $\nabla\cdot\mathbf{u}$ explicitly in the basic set of equations.

More generally, we can write the equation of state as $p = f(s)\rho^\gamma$, where s is the entropy per unit mass. The adiabatic process is obtained with

$$\frac{\partial}{\partial t}s + \mathbf{u}\cdot\nabla s = 0.$$

We have

$$\frac{\partial p/\partial t + \mathbf{u}\cdot\nabla p}{\partial\rho/\partial t + \mathbf{u}\cdot\nabla\rho} = f(s)\gamma\rho^{\gamma-1} = \frac{\gamma p}{\rho} \equiv C_s^2 \tag{10.20}$$

introducing again the speed of sound C_s. This definition is consistent with (4.4). The function f is here explicitly given as $f(s) = A\exp(s/C_V)$, with A being a constant.

In principle we could add a classical viscosity term to the force equation in all of the MHD equations. By this we make the equation time *ir*reversible in the ideal MHD limit. It is, however, questionable to what extent such a viscosity term is physically relevant. The MHD equations as discussed here can describe slow and large scale phenomena only. In that case classical viscosity has a negligible effect, so its inclusion is purely cosmetic (although it might give computational advantages for numerical solutions).

- **Exercise:** Demonstrate the validity of (10.19). How will the result appear if you assume isothermal conditions rather than adiabatic?

Ⓢ **Exercise:** We are now in a position to make different, but equivalent, representations of the ideal compressible MHD equations. Although they describe the same physics, some of them may have particular advantages in, for instance, numerical studies of the basic nonlinear equations. Write the basic equations of simple compressible, ideal MHD, expressed in terms of i) $\mathbf{u}$, $\mathbf{B}$, w and ρ, ii) $\mathbf{u}$, $\mathbf{B}$, p and s, iii) $\mathbf{u}$, $\mathbf{D}$, ρ and s In all cases $\nabla\cdot\mathbf{B} = 0$ enters as an auxiliary relation. Ⓢ

10.3 Dissipative MHD

Collisional momentum exchange does not affect the plasma momentum equation for MHD applied for a fully ionized plasma since momentum lost/gained from electrons is a gain/loss for the ions: the net momentum of the *entire* flow is conserved. The momentum exchange gives rise, however, to a finite electrical conductivity of the plasma. This conductivity is usually large, but there are significant differences between the MHD model with infinite and finite conductivity.

The MHD equations become somewhat more complicated when we retain a finite plasma conductivity σ. Using $\nabla \times (\nabla \times \mathbf{B}) = \nabla(\nabla \cdot \mathbf{B}) - \nabla^2 \mathbf{B} = -\nabla^2 \mathbf{B}$, together with (10.5) to eliminate the current from (10.4), we find

$$\frac{\partial}{\partial t}\mathbf{B} = \nabla \times (\mathbf{u} \times \mathbf{B}) + \frac{1}{\mu_0 \sigma}\nabla^2 \mathbf{B}. \tag{10.21}$$

This relation can be interpreted as a combination of a diffusion equation for the vectorial magnetic field and a sort of "advection" term. In analogy to a simple diffusion equation, we can define a dimensionally correct diffusion velocity for the magnetic field as $-\nabla B/(B\mu_0\sigma)$, which can be compared to the local flow velocity $\mathbf{u}$. The reason for the diffusion velocity only being a dimensionally, and not physically correct quantity is that we are here dealing with a non-traditional diffusion equation, in comparison with the standard diffusion problem, which concerns dispersion of a *scalar* density.

To estimate the relative magnitude of the two terms on the right-hand side of (10.21), we introduce a characteristic length scale $\mathcal{L}$ and a characteristic velocity $\mathcal{U}$ and rewrite (10.21) in terms of normalized quantities $u' \equiv u/\mathcal{U}$, $\partial/\partial t' \equiv (\mathcal{L}/\mathcal{U})\partial/\partial t$ and $\nabla' \equiv \mathcal{L}\nabla$, giving

$$\frac{\partial}{\partial t'}\mathbf{B} = \nabla' \times (\mathbf{u}' \times \mathbf{B}) + \frac{1}{R_L}\nabla'^2 \mathbf{B}, \tag{10.22}$$

where we introduced a characteristic quantity

$$R_L = \mu_0 \sigma \mathcal{L} \mathcal{U}, \tag{10.23}$$

which can be interpreted as a dimensionless number estimating the relative importance of the two terms on the right-hand side of (10.21). We have implicitly assumed that the typical scale length $\mathcal{L}$ is the same for the x, y and z variations, although this may not necessarily always be correct. Note that (10.21) is a *vector* equation, and that a scaling parameter R_L applies to each of its components: strictly speaking we should have introduced characteristic lengths and velocities for each component.

If the assumption of incompressibility is retained, the only significant modification of the ideal MHD is contained in the diffusive part of (10.21). To generalize the equations somewhat, we allow the plasma dynamics to be compressible and then have to retain the full continuity equation. To complete the set of equations we now also have to find an equation of state. One generally assumes that such a relation can be achieved by assuming the pressure to be a function of the density ρ alone, i.e., $p = P(\rho)$. Of course this density varies with time and with the spatial position, $\rho(\mathbf{r},t)$, so p is implicitly a function of space and time as well. See the discussion in Section 10.2.

As mentioned, a viscous term $\mathbf{f}_u$ can be introduced "ad hoc" in the momentum equation (10.12). We *might* take $\mathbf{f}_u = \mu\nabla^2\mathbf{u}$ with μ being the dynamic (not kinematic) viscosity, but strictly speaking this expression need not be sufficient when plasma compressibility is allowed for; see Section 2.6. A more general expression would be $\mathbf{f}_u = \mu(\nabla^2\mathbf{u} + \frac{1}{3}\nabla\nabla\cdot\mathbf{u} + 2\mathbf{S}\cdot\nabla\ln\rho)$, where $S_{ij} \equiv \frac{1}{2}(u_{ij} + u_{ji}) - \frac{1}{3}\delta_{ij}\nabla\cdot\mathbf{u}$ is the traceless "rate of strain" tensor. Admittedly, a certain "sloppiness" has developed, where a phenomenological viscosity is considered to be sufficient, and the simple form $\mathbf{f}_u = \mu\nabla^2\mathbf{u}$ is used without further ado. In numerical simulations often a *super-viscosity* or "hyper-viscosity" of the form $\mathbf{f}_u = \mu \nabla^{2m}\mathbf{u}$ with integer $m > 1$ is introduced for practical reasons. As mentioned in the discussion of compressible MHD, inclusion of molecular viscosity effects in the MHD equations is somewhat questionable from a physical point of view. Inclusion of finite resistivity, on the other hand, is necessary for most realistic studies.

Ⓢ **Exercise:** Consider the postulate that the Earth's magnetic field is "fossil", i.e., being a remnant of a magnetization trapped in the Earth when it was created, and remaining there even in the absence of any dynamo motions. Use as an estimate the conductivity for copper, and find other relevant quantities in the literature. Obtain a qualitative result by dimensional reasoning. Ⓢ

10.3.1 Frozen-in field lines

Magnetic field lines can at any given time be defined by the equation

$$\frac{dx}{B_x} = \frac{dy}{B_y} = \frac{dz}{B_z}. \tag{10.24}$$

These lines have the magnetic field vectors as *tangents* in any point. In particular for representing spatially two-dimensional phenomena, they can be most useful. For the general case we might prefer to have field lines described through a parameter s varying along the field line. For this problem we extend (10.24) as

$$\frac{dx}{B_x} = \frac{dy}{B_y} = \frac{dz}{B_z} = \frac{ds}{\sqrt{B_x^2 + B_y^2 + B_z^2}}. \tag{10.25}$$

If we know $\mathbf{B}(\mathbf{r})$, or more precisely the direction of $\mathbf{B}$ as a function of $\mathbf{r}$, we can integrate $d\mathbf{r}/ds = \mathbf{B}(\mathbf{r}(s))/|\mathbf{B}|$.

- **Example:** You might of course have a representation of field lines in spherical geometry as well, in which case (10.24) becomes

$$\frac{dr}{B_r} = \frac{rd\theta}{B_\theta} = \frac{r\sin\phi\, d\phi}{B_\phi}.$$

Actually, this form is most suitable, for instance, for representing magnetic dipoles.

It is a well known fact of basic electrodynamics that electric and magnetic field lines are entities visualizing instantaneous fields, and that it does not make sense to follow the motion of such field lines with time. It turns out, however, that in the limit of ideal MHD this can be considered formally possible. Properly speaking, it can be argued that in the limit of $\sigma \to \infty$ we can identify magnetic flux tubes with particles, and in this sense we are not making any logical error in claiming that we *can* follow the motion of magnetic field lines in space, as illustrated in Fig. 10.1. *If* one agrees to let the magnetic field lines be carried along by the flow of particles, we have the correct behavior of $\mathbf{B}$ as a function of time (at least as long as the MHD model applies!). Some advantage can be obtained by this model, although also a number of paradoxes have been produced, which have their origin in a superficial application of this concept of "frozen-in field lines". The analysis in this section allows for plasma compressibility, i.e., we do not assume $\nabla \cdot \mathbf{u} = 0$. We will in reality be using only the expression (10.13).

First we introduce a notation $\Phi(\mathcal{S}_{t+\Delta t}, B_t)$ for a magnetic flux as

$$\Phi(\mathcal{S}_{t+\Delta t}, B_t) \equiv \int_{\mathcal{S}_{t+\Delta t}} \mathbf{B}_t \cdot \mathbf{ds}$$

where we introduced a vector $\mathbf{ds}$ normal to the surface S. Note that the notation allows us to consider a magnetic flux through a surface $\mathcal{S}$ defined at a time $t + \Delta t$, while the magnetic vector field is taken

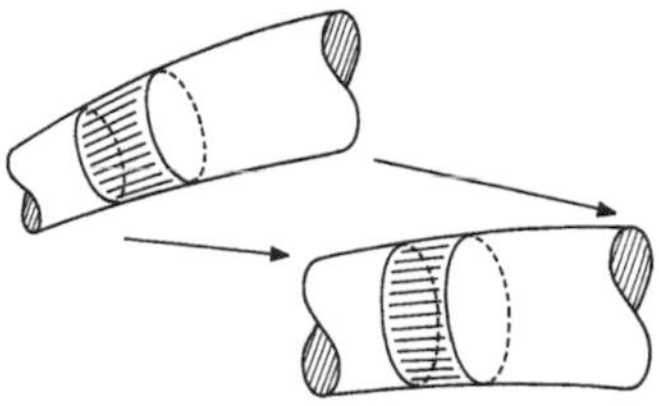

FIGURE 10.1: *Simple illustration of "frozen-in" field lines. Displaced and distorted magnetic flux tubes can be identified by the particles they enclose.*

at a *different* time B_t! This is a perfectly legal (although seemingly strange), and it turns out to be also a *smart* thing to do (Sturrock 1994). Time, t, is here placed as a subscript to indicate that we consider a fixed time, and do not have time as a variable.

We then write the change of magnetic flux through a surface S, which is following the motion of the fluid; see, for instance, Fig. 10.2. Introducing a vector $\mathbf{d}s$ normal to the surface S, we have

$$\begin{aligned}\Phi(\mathcal{S}_{t+\Delta t}, B_{t+\Delta t}) &= \int_{\mathcal{S}_{t+\Delta t}} \mathbf{B}_{t+\Delta t} \cdot \mathbf{d}s \\ &\approx \Delta t \int_{\mathcal{S}_{t+\Delta t}} \frac{\partial}{\partial t}\mathbf{B}_t \cdot \mathbf{d}s + \int_{\mathcal{S}_{t+\Delta t}} \mathbf{B}_t \cdot \mathbf{d}s \\ &\approx \Delta t \int_{\mathcal{S}_t} \frac{\partial}{\partial t}\mathbf{B}_t \cdot \mathbf{d}s + \int_{\mathcal{S}_{t+\Delta t}} \mathbf{B}_t \cdot \mathbf{d}s,\end{aligned} \tag{10.26}$$

where we made a series expansion of the magnetic field $\mathbf{B}$ around the time t, and kept only the first two terms. Assuming Δt to be small, we have that the difference between $\int_{\mathcal{S}_{t+\Delta t}} \mathbf{B}_t \cdot \mathbf{d}s$ and $\int_{\mathcal{S}_t} \mathbf{B}_t \cdot \mathbf{d}s$ is of the order Δt. In the first integral on the right-hand side we subsequently replaced the index $t+\Delta t$ with t; this term remains correct to first order in Δt. This can be seen best by considering the integration contour of $\int_{\mathcal{S}_{t+\Delta t}} \partial\mathbf{B}_t/\partial t \cdot \mathbf{d}s$ as a function of $t+\Delta t$, and making a series expansion giving

$$\int_{\mathcal{S}_{t+\Delta t}} \frac{\partial}{\partial t}\mathbf{B}_t \cdot \mathbf{d}s \approx \int_{\mathcal{S}_t} \frac{\partial}{\partial t}\mathbf{B}_t \cdot \mathbf{d}s + O(\Delta t).$$

We retained the first term of this expansion in (10.26), while ignoring the second, since it would give rise to a term of order Δt^2, as already mentioned. From (10.26) we then obtain

$$\Phi(\mathcal{S}_{t+\Delta t}, B_t) \equiv \int_{\mathcal{S}_{t+\Delta t}} \mathbf{B}_t \cdot \mathbf{d}s = \Phi(\mathcal{S}_{t+\Delta t}, B_{t+\Delta t}) - \Delta t \int_{\mathcal{S}_t} \frac{\partial}{\partial t}\mathbf{B}_t \cdot \mathbf{d}s, \tag{10.27}$$

which we need later on.

Since $\nabla \cdot \mathbf{B} = 0$, the net magnetic flux through the entire closed volume defined by the two contours $\mathcal{S}_t$ and $\mathcal{S}_{t+\Delta t}$ in Fig. 10.2 must vanish. The box has the surfaces $\mathcal{S}_t$ and $\mathcal{S}_{t+\Delta t}$ as "top" and "bottom" and the sidewall $\mathcal{S}_{sw}$ indicated in Fig. 10.2. Consequently, we have

$$\Phi(\mathcal{S}_t, B_t) = \Phi(\mathcal{S}_{t+\Delta t}, B_t) + \Phi(\mathcal{S}_{sw}, B_t),$$

where all magnetic fields are (evidently!) taken at time t. Inserting (10.27) for the first term on the right-hand side we find

$$\Phi(\mathcal{S}_t, B_t) = \Phi(\mathcal{S}_{t+\Delta t}, B_{t+\Delta t}) - \Delta t \int_{\mathcal{S}_t} \frac{\partial}{\partial t}\mathbf{B}_t \cdot \mathbf{d}s + \Phi(\mathcal{S}_{sw}, B_t).$$

Note that if the vector $\mathbf{d}s$ on $\mathcal{S}_{t+\Delta t}$ points *into* the volume, then $\mathbf{d}s$ on $\mathcal{S}_t$ points *out* of the volume, and vice versa; see Fig. 10.2. We now need to find $\Phi(\mathcal{S}_{sw}, B_t)$. Integrate the magnetic flux over the area of the sidewall of the tube defined in Fig. 10.2 with $|\mathbf{u}\Delta t|$ being the local distance between the two contours $\mathcal{S}_{t+\Delta t}$ and $\mathcal{S}_t$. Note that the vector $\mathbf{d}l \times \mathbf{u}\Delta t$ is normal to the sidewall $\mathcal{S}_{sw}$, and find the magnetic flux out through $\mathcal{S}_{sw}$ at the time t to be

$$\Phi(\mathcal{S}_{sw}, B_t) = \oint \mathbf{B}_t \cdot (\mathbf{d}l \times \mathbf{u}\Delta t) = \Delta t \oint (\mathbf{u} \times \mathbf{B}_t) \cdot \mathbf{d}l = \Delta t \int_{\mathcal{S}_t} \nabla \times (\mathbf{u} \times \mathbf{B}_t) \cdot \mathbf{d}s, \tag{10.28}$$

where the contour integral denoted by $\oint$ is performed along the boundary of $\mathcal{S}_t$. We used the standard vector relations like $\mathbf{a} \cdot (\mathbf{b} \times \mathbf{c}) = \mathbf{c} \cdot (\mathbf{a} \times \mathbf{b}) =$ etc. Physically, we have traced the "footpoint" of the vector $\mathbf{u}\Delta t$ in Fig. 10.2 around the closed contour containing $\mathcal{S}_t$. Note that this is the place where we explicitly use that the contour follows the flow! By taking the cross-product, $\mathbf{d}l \times \mathbf{u}\Delta t$, we ensure that the area of the element on $\mathcal{S}_{sw}$ is correctly represented. The length, $\mathbf{u}\Delta t$, of the sidewall of the tube $\mathcal{S}_{sw}$ is infinitesimal, i.e., of lowest order in Δt. The second relation in (10.28) is obtained by use of Stokes' theorem; see, for instance, Appendix D. The vector $\mathbf{d}s$ is normal to the surface bounded by $\mathcal{S}_t$. By (10.28) we can associate the magnetic flux through the sidewall $\mathcal{S}_{sw}$ to the flux of $\nabla \times (\mathbf{u} \times \mathbf{B})$ through the surface $\mathcal{S}_t$.

The result (10.28) is now used together with (10.26). Upon division by Δt, taking the limit $\Delta t \to 0$, we find

$$\frac{d}{dt}\Phi(\mathcal{S}, B) \equiv \frac{d}{dt}\int_{\mathcal{S}} \mathbf{B} \cdot \mathbf{d}s = \int_{\mathcal{S}} \left[\frac{\partial}{\partial t}\mathbf{B} - \nabla \times (\mathbf{u} \times \mathbf{B})\right] \cdot \mathbf{d}s, \tag{10.29}$$

where the magnetic field $\mathbf{B} = \mathbf{B}(t)$ and also the integration surface $\mathcal{S}$ depends on time, since it is carried along with the plasma.

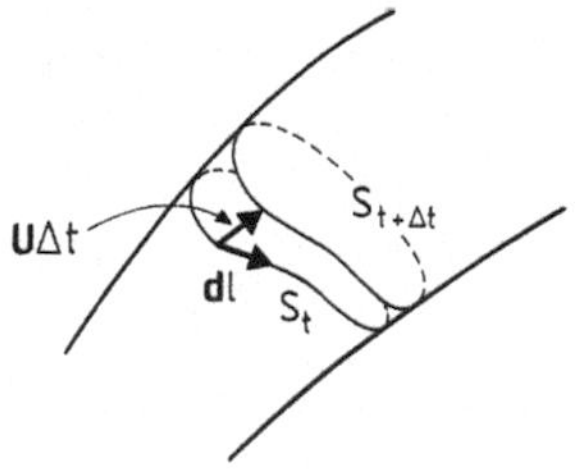

FIGURE 10.2: *Simple illustration of a surface enclosed by the contour, S_t, being displaced by the flow with velocity $\mathbf{u} = \mathbf{u}(\mathbf{r}, t)$. The contours need not be perpendicular to the magnetic field lines. The vectors* $\mathbf{dl}$ *and* $\mathbf{u}\Delta t$ *need not be perpendicular either. The area of a small parallelogram is given by the length of the vector* $\Delta t\, \mathbf{u} \times \mathbf{dl}$*, which is normal to the surface of the parallelogram.*

For ideal MHD we have $\partial \mathbf{B}/\partial t = \nabla \times (\mathbf{u} \times \mathbf{B})$. The relation (10.29) implies that, within ideal MHD, the magnetic field $\mathbf{B}$ at each point varies in such a way that its flux through any material surface S following the fluid remains constant, i.e.,

$$\frac{d}{dt}\Phi(\mathcal{S}_t, B_t) \equiv \frac{d}{dt}\int_{\mathcal{S}_t} \mathbf{B}(\mathbf{r}, t) \cdot \mathbf{d}s = 0. \tag{10.30}$$

One more argument can be presented for frozen-in magnetic field lines, this time based directly on the expression for infinite conductivity (10.6). We can *define* a velocity $\mathbf{u}_M$ of a magnetic field line as one where the electric field is vanishing (Sturrock 1994). If it were not so, the moving field lines would induce electric fields. This assumption gives $\mathbf{E} + \mathbf{u}_M \times \mathbf{B} = 0$. Only a velocity component across magnetic field lines can be defined this way, but, on the other hand, a motion along the field lines does not give rise to any motion of **B**-field lines. We therefore have $\mathbf{B}\cdot\mathbf{u}_M = 0$. For ideally conducting plasmas as given by (10.6) we have $\mathbf{E}+\mathbf{u}\times\mathbf{B} = 0$, implying $\mathbf{u}_M = \mathbf{u}_\perp$ being the **B**-perpendicular component of the bulk plasma velocity. By this we can argue (again) that magnetic field lines move with the plasma.

If we agree to let the magnetic field lines be carried along by the flow of particles, *then* the equation (10.30) is satisfied at all times, so that we have the correct behavior of **B** as a function of time. For nonideal, resistive MHD we have $\partial\mathbf{B}/\partial t \neq \nabla\times(\mathbf{u}\times\mathbf{B})$ and magnetic field lines are no longer frozen in.

It must again be emphasized that the concept of frozen-in field lines is an artifact: Maxwell's equations describe a force field, i.e., the space-time evolutions of vector fields $\mathbf{E}(\mathbf{r},t)$ and $\mathbf{B}(\mathbf{r},t)$. The field lines are something *we* as observers draw, as an aid to the eye and, as already mentioned, it does not make sense to follow their motion in time. Neither magnetic nor electric field lines can be assigned a physical existence or reality. You cannot paint a magnetic field line so you can recognize it at a later time! Within ideal MHD, the charged particles act as identifiers, and thanks to these we do not make any *error* in pretending that we can follow magnetic field lines, at least as long as we remain within the validity of the ideal MHD model. We note, in particular, that if a volume element of plasma is initially void of magnetic field, it will remain so for all later times within a description based on ideal MHD.

In order to identify a magnetic field line with a gyrating particle, we require any characteristic time for changes in the plasma conditions to be much greater than $1/\Omega_{ci}$; otherwise the particles cannot be assumed to be completely magnetized: the MHD equations describe slow phenomena, as already stated. Similarly, we cannot localize a magnetic field line better than the Larmor radius of the gyrating particle. Any characteristic length scale for plasma variations must be $\gg r_L$: the MHD equations describe large scale phenomena.

Some aspects of the frozen-in magnetic field lines can be understood by the conservation of the magnetic moment; see Section 7.1.8. Consider a long cylindrical plasma column with radius R, confined by a homogeneous magnetic field. If we increase the magnetic field intensity from B_1 to $B_2 > B_1$ magnetic flux density increases so that $\pi R_1^2 B_1 = \pi R_2^2 B_2$. The radius of the plasma column consequently decreases from R_1 to R_2. The plasma column is compressed and if it is very long, its contact with the surroundings can be ignored. According to simple thermodynamic principles, the plasma temperature increases due to the compression by adiabatic heating. From the derivation of the MHD equations outlined before, it is not self-evident by which basic physical mechanisms this heating is achieved. By the conservation of magnetic moment, $\mu \equiv \frac{1}{2}MU_\perp^2/B$ (see Section 7.1.8) it becomes obvious how the particle kinetic energy increases for increasing magnetic fields.

The reader should be warned that the concept of frozen-in magnetic field lines can also be deceiving. Rather meaningless paradoxes can be created, and several examples can be found in the literature. These are usually associated with model constructions where ideal MHD applies to a part of space, and not to others. Joining the mathematical models at the interface between the two regions can lead to strange conclusions. Other aspects of the frozen-in field line model have been discussed in the literature (Stern 1973*a*).

The arguments in this section do not apply to super-conductors; in this case we have the *Meissner effect*, which implies that the magnetic field in a super-conductor is vanishing *always*, independent of the strength of the magnetic field in the material before it was cooled down below the critical temperature, T_c. If it were not like this, we could not consider the super-conducting state as a distinct thermodynamic state (Kuper 1968).

10.3.2 Magnetic pressure

It is at times convenient to have a different representation of (10.2). Using the relation $\mathbf{B}\times(\nabla\times\mathbf{B}) = (\nabla\mathbf{B})\cdot\mathbf{B} - (\mathbf{B}\cdot\nabla)\mathbf{B} = \frac{1}{2}\nabla(\mathbf{B}\cdot\mathbf{B}) - (\mathbf{B}\cdot\nabla)\mathbf{B} \equiv \frac{1}{2}\nabla B^2 - (\mathbf{B}\cdot\nabla)\mathbf{B}$ to re-express the $(\nabla\times\mathbf{B})\times\mathbf{B}$-force, we rewrite the plasma momentum equation (10.2) in the form

$$\rho\left(\frac{\partial}{\partial t}\mathbf{u}+\mathbf{u}\cdot\nabla\mathbf{u}\right)=-\nabla\left(p+\frac{1}{2\mu_0}B^2\right)+\frac{1}{\mu_0}(\mathbf{B}\cdot\nabla)\mathbf{B}, \tag{10.31}$$

possibly completed with other forces, originating from gravity, for instance. The following analysis addresses only the momentum equation and is therefore independent of any assumptions of equations of state, incompressibility in particular.

We readily see that the magnetic field enters (10.31) in a way similar to the plasma pressure, at least partly. To emphasize this feature in more detail we can write the force

$$\mathbf{J}\times\mathbf{B}=\frac{1}{\mu_0}(\nabla\times\mathbf{B})\times\mathbf{B}=\nabla\cdot\underline{\Pi}^M, \tag{10.32}$$

in terms of a magnetic tensor

$$\underline{\Pi}^M=\frac{1}{\mu_0}\left\{\begin{array}{ccc} B_x^2-\frac{1}{2}B^2 & B_xB_y & B_xB_z \\ B_yB_x & B_y^2-\frac{1}{2}B^2 & B_yB_z \\ B_zB_x & B_zB_y & B_z^2-\frac{1}{2}B^2 \end{array}\right\}. \tag{10.33}$$

Since also the pressure can be expressed as a divergence of a tensor, $-\nabla\cdot\underline{\mathrm{P}}$ with $\underline{\mathrm{P}}\equiv p\underline{1}$, we can readily combine the expressions for the two forces. We encountered this magnetic tensor already in the discussion of the virial theorem; see the relation (10.42) in Section 10.3.5.

For the particularly simple case where the field lines are straight and parallel, and the intensity of the magnetic field varies only in the direction $\perp\mathbf{B}$, the term $(\mathbf{B}\cdot\nabla)\mathbf{B}$ is zero. In this case the total effect of the magnetic field is contained in the extra pressure-like term B^2/μ_0, justifying the notion of *magnetic pressure*. We can more generally transform $\underline{\Pi}^M$ to the local principal axes, choosing the z-axis parallel to $\mathbf{B}$. Here, it can be done simply by setting $B_x=B_y=0$ in (10.33) to find

$$\underline{\Pi}^M=\frac{1}{2\mu_0}\left\{\begin{array}{ccc} -B^2 & 0 & 0 \\ 0 & -B^2 & 0 \\ 0 & 0 & B^2 \end{array}\right\}, \tag{10.34}$$

and find that there appears to be a local pressure $B^2/2\mu_0$ in the direction perpendicular to the field lines, and a *tension* $B^2/2\mu_0$ along the field lines. In the case of a homogeneous field, this tension is immaterial, since it does not contribute to $\nabla\cdot\underline{\Pi}^M$. One can also say that we have an isotropic pressure $B^2/2\mu_0$ in addition to a tension B^2/μ_0 along magnetic field lines. A short basic introduction to the concept of magnetic stress is given by Jensen (1995).

The concept of magnetic pressure can be most convenient, in a discussion of plasma stability, for instance.

- **Exercise:** Prove the relation (10.32) with (10.33).

Ⓢ **Exercise:** Particular attention has been given to the so-called *force free magnetic fields*, which have the property

$$\nabla\times\mathbf{B}=\alpha\mathbf{B}. \tag{10.35}$$

These fields are called force free because $\mathbf{J}\times\mathbf{B}=(\nabla\times\mathbf{B})\times\mathbf{B}/\mu_0=0$. The condition (10.35) only implies that the $\mathbf{J}\times\mathbf{B}$-force vanishes: there might be other forces acting on the plasma. In particular we should bear in mind that by arguing for the ideal MHD model, we assumed that the $\mathbf{J}\times\mathbf{B}$-term dominated the electric forces; this argument needs revision in the case in which we have force free magnetic fields.

The relation (10.35) has the form of an eigenvalue relation, where $\nabla\times$ is an operator, $\mathbf{B}$ is the eigenfunction and α the eigenvalue (Yoshida & Giga 1990, Knorr et al. 1990). Sometimes α is called the *abnormality* of the field (Aris 1962)

Find the solution for (10.35) for the special case with α constant. Demonstrate (van Kampen & Felderhof 1967) that a force free field cannot vanish sufficiently fast at infinity to make the total magnetic field energy finite. Ⓢ

- **Exercise:** Consider a cylindrically symmetric case where $\alpha = \alpha(|\mathbf{r}|)$. Write (10.35) for this condition, and find a general solution independent of the variables ϕ and z.

- **Exercise:** Consider the case with $\mathbf{u} = 0$ in (10.21) giving

$$\frac{\partial}{\partial t}\mathbf{B} = \frac{1}{\mu_0 \sigma}\nabla^2 \mathbf{B}.$$

Let the magnetic field be force free, $\nabla \times \mathbf{B} = \alpha \mathbf{B}$, and demonstrate that a solution is $\mathbf{B} = \mathbf{B}_0 \exp(-t/\tau)$ with $\nabla^2 \mathbf{B}_0 + \alpha^2 \mathbf{B}_0 = 0$ and $\tau = \mu_0 \sigma / \alpha$, where α is a constant.

10.3.3 Plasma β

As a simple stationary solution with $\mathbf{u} = 0$, we have from (10.31)

$$\nabla\left(p + \frac{1}{2\mu_0}B^2\right) = \frac{1}{\mu_0}(\mathbf{B}\cdot\nabla)\mathbf{B}. \tag{10.36}$$

We consider (10.36) in plane geometry, where the pressure depends on a Cartesian coordinate, $p = p(x)$ and the magnetic field is in the z-direction, with $\mathbf{B} = \{0, 0, B_z(x)\}$. For this case, the right-hand side in (10.36) is vanishing, and we find

$$\frac{d}{dx}\left(p + \frac{1}{2\mu_0}B_z^2\right) = 0, \qquad \text{or} \qquad p + \frac{1}{2\mu_0}B_z^2 = \text{const.} \tag{10.37}$$

The integration constant can be determined by some "external" boundary condition, typically given by the magnetic field in a vacuum, i.e., as $B_0^2/(2\mu_0)$. Note that we have not specified any equation of state so far!

We readily find that a plasma is *diamagnetic*, i.e., a high plasma pressure reduces the magnetic pressure; see also the discussion in Section 11.1.1.

> As a measure of the relative importance of the magnetic and thermal pressures it is convenient to introduce the ratio of plasma pressure and magnetic pressure as β, with
>
> $$\beta \equiv \frac{n\kappa T}{B^2/2\mu_0}. \tag{10.38}$$

We again used the ideal gas law for the pressure p. The "plasma β" defined here turns out to be an important parameter for classifying plasmas, as we will find, in particular, when discussing waves in plasmas. In the solar wind we often have β in the range 0.5–1; in some laboratory plasmas, such as in a Q-machine (Motley 1975), we find $\beta \sim 10^{-8}$.

10.3.4 Plasma pinches

Broadly speaking, "pinches" denote experiments where large currents are passed through plasmas so that the magnetic fields generated by these currents contribute significantly, or even dominate, to the total magnetic field. The basic features of these devices are well described by MHD. They are also interesting by being some of the first experiments attempting to reach thermonuclear fusion in laboratory plasmas. Unfortunately the study of plasma instabilities was not well developed at the time, and the dream of a fusion plasma generator was not fulfilled, but we learned a lot of physics.

If we take a simple cylindrical case without radial magnetic field components, where p and $\mathbf{B}$ do not depend on the variable z and θ, i.e., $\mathbf{B} = B_\theta(r)\widehat{\theta} + B_z(r)\widehat{\mathbf{z}}$, we can relate the current density to the magnetic field by Ampere's law to give

$$\mathbf{J} \equiv \{J_r, J_\theta, J_z\} = \frac{1}{\mu_0}\left\{0, -\frac{dB_z}{dr}, \frac{1}{r}\frac{d(rB_\theta)}{dr}\right\}. \tag{10.39}$$

Both vectors $\mathbf{J}$ and $\mathbf{B}$ lie on surfaces with constant r (Boyd & Sanderson 2003). Obtaining again the force density $\mathbf{J}\times\mathbf{B}$, we have the relation for steady state conditions

$$\frac{d}{dr}\left(p + \frac{1}{2\mu_0}\left(B_\theta^2 + B_z^2\right)\right) = -\frac{1}{\mu_0 r}B_\theta^2. \tag{10.40}$$

This result could of course be obtained from (10.36) as well. We have, for the present case, $(\mathbf{B}\cdot\nabla)\mathbf{B} = (B_\theta/r)\partial(B_\theta\widehat{\theta})/\partial\theta$; see Appendix D.7. It is important to note that, for the present cylindrical geometry, an increment in the θ-direction turns the unit vector $\widehat{\theta}$ slightly, in such a way that the "correction" to the original vector becomes a vector in the $-\widehat{\mathbf{r}}$-direction, and (10.40) is obtained.

10.3.4.1 θ-Pinch

We can generally classify pinches according to some simple basic geometries. For the special case with $B_\theta = 0$, which can be realized by purely azimuthal currents, we have $\mathbf{B} = \{0, 0, B_z(r)\}$. In this case we find from (10.31) or (10.40)

$$\frac{d}{dr}\left(p + \frac{1}{2\mu_0}B_z^2\right) = 0,$$

which is readily integrated to give (10.37) again. Since the magnetic field lines do not "bend", we have no tension in the magnetic field, and the right-hand side of (10.36) vanishes.

When $B_\theta = 0$ in cylindrically symmetric plasma configurations, the current has to be flowing in the θ-direction, and this configuration is consequently termed a "θ-pinch". Such azimuthal currents are induced by some external discharge, with very fine illustrations given by Bishop (1958).

10.3.4.2 Z-Pinch

For the case with $B_z = 0$, the currents are flowing in the z-direction in cylindrically symmetric configurations, and they are therefore called Z-pinches. The magnetic field is then in the $\widehat{\theta}$-direction. In may ways this is the simplest of the pinches. The basic idea is that parallel currents attract and thereby compress and heat the plasma. Given that this could happen before energy was lost from the plasma, a sufficiently large current could in principle induce fusion reactions. For the present geometry we have from (10.40)

$$\frac{d}{dr}\left(p + \frac{1}{2\mu_0}B_\theta^2\right) = -\frac{1}{\mu_0 r}B_\theta^2.$$

Ⓢ **Exercise:** Prove the *Bennett relation* (Boyd & Sanderson 2003) for a cylindrically symmetric Z-pinch of radius R and constant electron and ion temperatures T_e and T_i, respectively

$$I^2 = \frac{8\pi}{\mu_0}\kappa(T_e + T_i/Z)N_{p},$$

where the integrated current is $I \equiv \int_0^R 2\pi r J(r)dr$, and $N_p \equiv \int_0^R 2\pi r n(r)dr$ is the number of electrons per unit length (plasma line density), while Z is here the ion charge state, so that $Zn_i(r) = n_e(r)$ with the assumption of local charge neutrality. Bennett's relation determines the total current needed to contain a plasma of a specified temperature and line density. ⓢ

10.3.4.3 Screw-pinch

As a sort of combination of a Z-pinch and a θ-pinch, we have "screw-pinches", where the magnetic field lines wind on cylindrical surfaces in helical paths. Such configurations can be used as local approximations for long, thin toroidal Tokamak plasmas.

10.3.4.4 Pinch instabilities

The concept of magnetic pressure allows us to obtain a simple qualitative understanding of some plasma instabilities, where we here address Z-pinches. We will discuss two simple types of MHD instabilities, sausage and kink instabilities, as illustrated in Fig. 10.3a) and b), respectively.

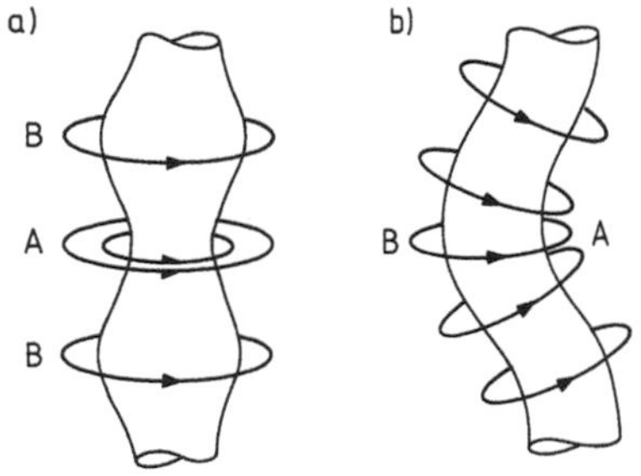

FIGURE 10.3: *Illustration of sausage in a) and and kink perturbations of a pinch in b).*

We consider a plasma column with an axial current, I. It is not necessary for the arguments that the cross section is circular, but it is usually drawn that way. For such a current carrying cylindrical rod, we have an azimuthal magnetic field which increases linearly with radius inside the cylinder, $B_\theta = \mu_0 Ir/(2\pi R^2)$ for $r < R$, and then decreases again as $B_\theta = \mu_0 I/(2\pi r)$ for $r > R$, with R being the radius of the plasma column. The $\mathbf{J} \times \mathbf{B}$-force is in the inward radial direction, compressing the plasma. For this reason, the configuration is usually called a *pinch* (Bishop 1958, Chandrasekhar 1960, Braams & Stott 2002). It turns out that the configuration described here is unstable; if we perturb it slightly, then this perturbation will increase in time. Usually two types of perturbations are distinguished; see Fig. 10.3a) and b). With the current through the plasma column assumed constant, it is now intuitively clear that the magnetic field intensity is enhanced by the perturbation at the positions A in the sausage (Fig. 10.3a) as well as for the kink (Fig. 10.3b) perturbations, while it is diminished at positions B in both cases. Consequently, the magnetic pressure is enhanced at positions A and diminished at positions B, and the result is an enhancement of the initial perturbation, and the difference in magnetic fields at A and B will be further enhanced; the system is unstable since the force enhances the perturbation. The kink instability can also have the form of a spiral. To determine the actual growth rate for the instabilities, we will have to analyze the the problem in terms of the full set of MHD equations (Chandrasekhar 1960). The purpose of the present summary was, however, limited to give a physical understanding of the instability in terms of magnetic pressure. An interesting review of the early studies of pinches is given by Cole (1959). Some early observations of pinch instabilities studied in liquid metals (Dattner et al. 1958, Lehnert & Sjögren 1960) are interesting and illustrative, see Fig. 10.4. Experiments carried out in conducting fluids offer simpler means of diagnostics in comparison to those in hot plasmas.

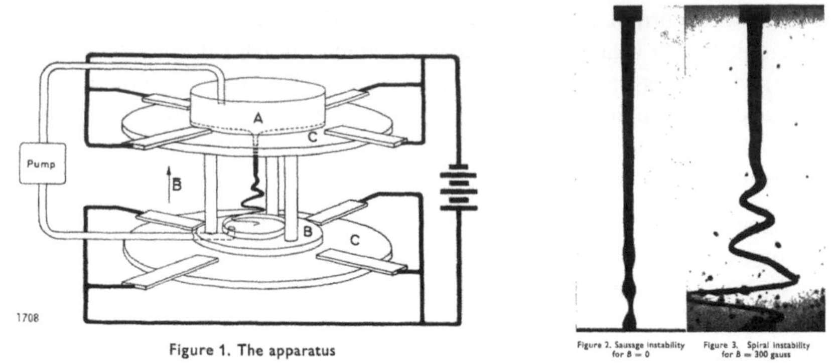

FIGURE 10.4: *Illustration of pinch instabilities studied in liquid metals. The left figure shows the experimental set-up schematically. The current carrying liquid is flowing down in the vertical direction, so that an instability that is aperiodic in the liquid rest frame here appears as a propagating instability. The two right-hand figures show observations of the sausage instability and the spiral instability. For the latter case a vertical homogeneous weak magnetic field is applied externally. The figures are reproduced with permission from Elsevier from a conference contribution by Dattner et al. (1958).*

Ⓢ **Exercise:** Consider a long linear plasma discharge (a *pinch*) having a circular cross section with radius a, supporting a homogeneous current density J. The magnetic permeability of the plasma is μ_0. For the following problem, we assume that the discharge retains its shape, and ignore possible instabilities. Calculate the pressure inside the pinch at any radial position from the z-axis. Assume that the pressure at the surface is given as p_0. Taking $a = 10^{-2}$ m, and $J = 10$ kA/cm^2, calculate the pressure at the center of the pinch, with p_0 corresponding to atmospheric pressure. Ⓢ

Ⓢ **Exercise:** Consider a long solenoid with radius R and N closely wound turns per meter. The current in the wire is I, see also Fig. 10.5.

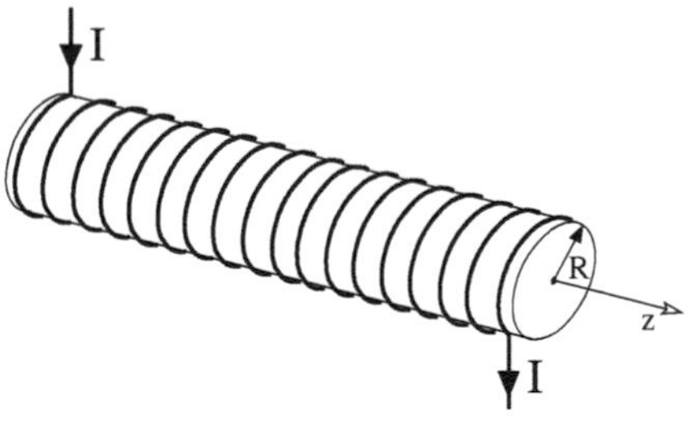

FIGURE 10.5: *Illustration of a long solenoid with current I. An ideal solenoid assumes the coils to be strictly perpendicular to the axis, while the windings have to be tilted slightly for the realizable case in order to cover the entire length of the solenoid. To model a* physical *solenoid you have to take an ideal one and add a linear wire carrying a current I parallel to the axis. For experimental conditions you have to compensate for this.*

1) What is the magnetic field inside the solenoid?

Assume now that beginning at $t = 0$, the current is "ramped up" to be linearly increasing from I to $2I$ within a time t_1. The solenoid current remains constant $2I$ for $t > t_1$.

2) Determine the magnitude and direction of the electric field in the system as a function of time. Make a small sketch of a plane perpendicular to the solenoid axis, illustrating your result.
3) Give the direction and magnitude of the $\mathbf{E}\times\mathbf{B}/B^2$-velocity as a function of time. Illustrate the direction of this velocity on your figure from question 2).

Let the solenoid be partially filled with a hot ideally conducting cylindrical plasma column. Initially, the plasma is at a spatially uniform temperature T_0 and has a constant density n_0 within a radius $r_0 < R$ and vanishes outside r_0. Let the solenoid be so long that plasma losses at the ends can be ignored. Assume an adiabatic process and use the ideal gas law for the plasma for $t > 0$. Use a one-fluid plasma model.

4) What is the outer radius r_1 of the plasma column at time t_1? Explain the result in terms of "frozen-in" magnetic field lines.
5) Demonstrate explicitly that for the velocity field derived in this exercise, an initially spatially constant density will remain spatially constant at all later times, only varying with time. The flow is compressible. What happens if the initial density is not constant but varies with radius?
6) Give the equation of state for an adiabatic process. How is it distinguished from the equation of state for an isothermal process?
7) Give the plasma temperature T_1 and density n_1 for the plasma at $t = t_1$. You can assume that the plasma density n_1 is uniform for $r < r_1$. Argue why it is so. Ⓢ

Ⓢ **Exercise:** It was pointed out by Alfvén (1939) that there are limits to how much current you can pass through a plasma column. When the so-called Alfvén limit (Davies 2006) is exceeded, the magnetic field generated by the current itself becomes so strong that the current carriers (usually the electrons) get magnetized. Discuss this effect qualitatively by discussing the current of a long plasma column with diameter $2R$, where a uniformly distributed current is flowing axially. What is the role of the collisional mean free path? Ⓢ

10.3.4.5 Kink instability of a long thin pinch

A limiting case of a pinch discharge can be discussed without introducing the plasma pressure. As an illustration, we consider a long thin pinch; see Fig. 10.6. The net pinch current is **I**. The equilibrium position is shown with a thin vertical dashed line. Selecting a position $\mathbf{r}_0$ on the pinch, we calculate the magnetic field contribution from the rest of the pinch at $\mathbf{r}_0$ and then calculate the $\mathbf{I}\times\mathbf{B}$-force.

Biot-Savart's law gives a magnetic field contribution $d\mathbf{B} = \mu_0 I d\mathbf{s}\times\mathbf{r}/(4\pi r^3)$ from an element $d\mathbf{s}$ on the pinch at a distance $\mathbf{r}$ from the reference position $\mathbf{r}_0$; see also Fig. 10.6c). The net force at the line element $d\mathbf{s}'$ at position $\mathbf{r}_0$ is then

$$d\mathbf{F} \equiv \mu_0 I^2 d\mathbf{s}' \times \int_0^L \frac{d\mathbf{s}\times\mathbf{r}}{4\pi r^3} \tag{10.41}$$

where the $d\mathbf{s}$-integration is taken along the pinch. It turns out that the integration in (10.41) becomes singular (Hama & Nutant 1961) if we retain a model with a line-pinch with vanishing cross section, so in practice we will have to allow for a small but finite pinch diameter. The divergence is due to the magnetic field becoming infinite at a line current as $r \to 0$. For a qualitative discussion we need not be concerned with this divergence.

We assume that the pinch shown with the thick line in Fig. 10.6a) is in the plane of the paper. The magnetic field lines as obtained by Biot-Savart's law are then crossing that plane in the perpendicular direction. Consequently, the force $d\mathbf{F}$ is *in* the plane of the pinch in Fig. 10.6. Taking a position $\mathbf{r}_0$ we take the contribution to the integral in (10.41) from the lower limit $z = 0$ to a point a small distance ε before $\mathbf{r}_0$ and then add the integral contribution starting at a position at a distance ε after the reference position $\mathbf{r}_0$ to the end point at $z = L$. (This trick with $\varepsilon \neq 0$ is one way to avoid the singularity mentioned.) When the curvature of the perturbed pinch is constant as in Fig. 10.6a) we find that the magnetic field at $\mathbf{r}_0$ is out of the plane with the given current direction and consequently the force is in the direction away from the equilibrium position given by the dashed vertical line in Fig. 10.6a) so the perturbation is unstable since the force enhances the perturbation. The same result would be obtained with the opposite direction for the current. Due to the $|\mathbf{r}/r^3| = 1/r^2$ variation of the integrand, we find that by far the largest contribution to the integral comes from parts of the pinch close to the reference position $\mathbf{r}_0$, so the precise shape of the perturbation at large distances is of little consequence. Therefore also the perturbation in Fig. 10.6b) is unstable, although parts of the integral path contribute to reduce the magnetic field in the vicinity of position ②. At, for instance, the central part ② in Fig. 10.6b) the contribution from regions ① and ③ are immaterial. The only significant criterion for the validity of the argument is that the length scale of the perturbation should be much larger than the diameter of the pinch. This criterion is satisfied in Fig. 10.6.

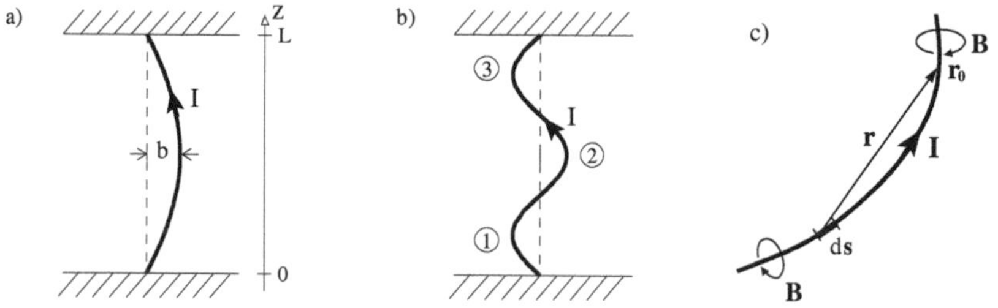

FIGURE 10.6: *A long thin pinch, slightly displaced from the reference position as shown in a) and b) with two different perturbations. In c) we enlarge a part of the figure in order to specify the vectors and line elements entering (10.41).*

Ⓢ **Exercise:** In order to illustrate the singularity of the integral in (10.41) we consider a parabolic displacement of the pinch in Fig. 10.6a). In this case it is possible to solve the integral in several relevant limits (Hama & Nutant 1961). Discuss this plane parabolic perturbation and demonstrate that the singularity is logarithmic for small ε when the integral is written in two parts for $\{0; r_0 - \varepsilon\}$ and $\{r_0 + \varepsilon; L_s\}$ with $\varepsilon \to 0$. We have the limits of the s-integration as $\{0; L\}$ and $\{0; L_s\}$. Logarithmic singularities are usually not considered so serious! Ⓢ

Ⓢ **Exercise:** Calculate the average electron drift velocity in a a copper wire of 2 mm diameter carrying a current of 1 A. Assume the wire to be at room temperature: what is the ratio between the average electron drift velocity and the electron thermal velocity? Is a cold electron model acceptable here?

Calculate the average electron drift velocity in a an idealized MHD-pinch 20 mm diameter with plasma density $n = 10^{16}$ m^{-3}, carrying a current of 1 kA. Assume the pinch to be at a temperature of 1 keV: what is the ratio between the average electron drift velocity and the electron thermal velocity? Is a cold plasma model acceptable here? The pinch need not be assumed to be in steady state, the question is meaningful anyway. Ⓢ

10.3.5 Virial theorem Ⓚ

We express the force density $\mathbf{f} \equiv -\nabla p + \mathbf{J} \times \mathbf{B} = \nabla \cdot \underline{\mathsf{T}}$ in the standard MHD model as a divergence of a stress tensor, T_{ij}, by using $\mathbf{J} = \nabla \times \mathbf{B}/\mu_0$ from Ampere's law (10.4). The general idea is very similar to the one used in Section 5.5. We write the i-th component of $\mathbf{f}$ as $f_i = \partial \mathsf{T}_{ij}/\partial r_j$, where r_j is the j-th component of the position vector $\mathbf{r}$. In this section the usual Einstein summation convention will be used extensively. The convention means that any set of repeated indices are to be summed, i.e., A_{jj} should be interpreted as $\sum_j A_{jj}$, etc. The simplification in the notation by doing so will be evident. We have

$$\mathsf{T}_{ij} = -p\delta_{ij} + \frac{1}{\mu_0}B_iB_j - \frac{1}{2\mu_0}B^2\delta_{ij}, \tag{10.42}$$

where δ_{ij} is the Kronecker delta, which is unity for $i = j$ and zero otherwise.

As the first application of the form (10.42), we first discuss the possibility of finding a self-consistent plasma configuration where $\mathbf{f} = 0$ everywhere. The analysis is restrictive in the sense that a finite spatial support for the plasma is assumed: for $|\mathbf{r}| \to \infty$ we have no plasma, only the electric and magnetic fields generated by sources inside the plasma. The so-called "force free fields" require a separate analysis, which is used as an exercise, see page 196.

We form the quantity $M \equiv \int r_j f_j d\mathbf{r}$ with $\int d\mathbf{r} \equiv \int dr^3 \equiv \int dxdydz$, with the integration volume unspecified for the moment. Inserting $f_i = \partial \mathsf{T}_{ij}/\partial r_j$ and using (10.42) we find after an integration by parts that

$$M = \int_S \widehat{n}_j r_i \mathsf{T}_{ij} dS - \int_V \mathsf{T}_{jj} d\mathbf{r}, \tag{10.43}$$

where the first integral is over the surface S bounding the volume V, with $\widehat{\mathbf{n}}$ being the unit vector normal to this surface, again with summation over repeated indices.

We now consider only a plasma occupying a finite part of space, and let the surface S expand to infinity. Since the magnetic field must vanish at least as $1/r^3$ at infinity, corresponding to a dipole contribution in the multipole expansion, we readily find that the surface contribution is vanishing in this limit. Consequently we have

$$M = \int_V \left(3p + \frac{1}{2\mu_0}B^2\right) d\mathbf{r}. \tag{10.44}$$

If we had the force density $\mathbf{f} = 0$ everywhere, then $M = 0$ by its definition. From (10.44) we find, on the other hand, that $M > 0$ for nontrivial cases with $p \neq 0$ and $B \neq 0$, which means that we can *not* find a solution with $\mathbf{f} = 0$ everywhere. A self-consistent force-free situation cannot be supported by ideal MHD, or in other words, we can never have pressure gradients and $\mathbf{J} \times \mathbf{B}$-forces balancing each other. This is the simplest formulation of the *virial theorem* (Sturrock 1994).

An extension of the virial theorem can be found by writing the equation of motion as

$$\frac{\partial}{\partial t}(\rho u_i) = \frac{\partial}{\partial r_j}(-\rho u_i u_j + \mathsf{T}_{ij}), \tag{10.45}$$

written again by implicit use of the summation convention. We used the equation of continuity, (10.1), to obtain (10.45). We now consider the inertia tensor given as $\mathsf{I}_{ij} \equiv \int \rho r_i r_j dr^3$, obtained for $\rho = \rho(\mathbf{r}, t)$. We have

$$\frac{d}{dt}\mathsf{I}_{ij} = \int r_i r_j \frac{\partial}{\partial t}\rho\, d\mathbf{r} = -\int r_i r_j \frac{\partial}{\partial r_k}(\rho u_k)\, d\mathbf{r},$$

where we also used the equation of continuity for the plasma density, allowing for compressibility, in general. Integration by parts gives $d\mathsf{I}_{ij}/dt = \int \rho\,(r_i u_j + r_j u_i)\, d\mathbf{r}$, where we ignored a surface contribution, with the experienced gained from discussing (10.44). Differentiation with respect to time once more gives

$$\frac{d^2}{dt^2}\mathsf{I}_{ij} = \int \left(r_i \frac{\partial}{\partial t}(\rho u_j) + r_j \frac{\partial}{\partial t}(\rho u_i)\right) d\mathbf{r}.$$

Note that when differentiating under the integral sign, we introduce the *partial* differentiation. The reasons are rather self-evident. Using (10.45) we obtain

$$\frac{d^2}{dt^2}\mathsf{I}_{ij} = \int \left(r_i \frac{\partial}{\partial r_s}\mathcal{T}_{js} + r_j \frac{\partial}{\partial r_s}\mathcal{T}_{is}\right) d\mathbf{r}, \tag{10.46}$$

where we used the abbreviation $\mathcal{T}_{ij} \equiv -\rho u_i u_j + \mathsf{T}_{ij}$ for simplicity, making the notation more compact. Integrating by parts, and again assuming a plasma with compact support, we have

$$\frac{1}{2}\frac{d^2}{dt^2}\mathsf{I}_{ij} = -\int \mathcal{T}_{ij} d\mathbf{r}, \tag{10.47}$$

since the tensor $\mathcal{T}_{ij}$ is symmetric. Contracting the tensors in (10.47) we find, by using the definitions of $\mathcal{T}_{ij}$ and T_{ij}, that

$$\frac{1}{2}\frac{d^2}{dt^2}I = \int \left(\rho u^2 + 3p + \frac{1}{2\mu_0}B^2\right) d\mathbf{r}, \tag{10.48}$$

with $I \equiv \mathsf{I}_{jj}$. Since all terms in the integral on the right-hand side are positive, we have that $d^2I/dt^2 > 0$. This necessarily implies that a plasma will expand, as long as it is subject to self-consistent forces alone. Note that changing the arrow of time by the replacement $t \to -t$ does not make any difference. The basic equations are time reversible, as no dissipation mechanism is included. Inclusion of gravity contributes to (10.48) with a term having a negative sign (Sturrock 1994). A star in a fully ionized state can remain confined to a finite part of space, as we would expect.

Ⓢ **Exercise:** Express the result (10.48) in terms of the total magnetic, kinetic and thermal energies of the plasma. (This is almost trivial!) Ⓢ

Ⓢ **Exercise:** Introduce a gravitational force density as $F_j^g = -\rho\partial\Psi/\partial r_j$. The scalar potential satisfies the Poisson equation $\nabla^2\Psi = 4\pi G\rho$. Introduce the self-consistent gravitational force in the virial theorem, and demonstrate that it contributes with a term $-\frac{1}{2}\rho\Psi d\mathbf{r}$ on the right-hand side of (10.48), giving the possibility for a stable configuration with $I =$ const. (Sturrock 1994). Ⓢ

Virial theorems offer a convenient basis for a stability analysis of equilibrium situations with $I =$ const. We emphasize that the present derivation of the virial theorem is based on simple MHD, allowing for compressibility. The result is therefore limited in applicability, and restricted to low frequency, large scale phenomena in magnetized plasmas with high conductivity. Virial theorems in classical mechanics are discussed, for instance, by Goldstein (1980).

- **Exercise:** To demonstrate the usefulness of the summation convention, show that the condition for the three vectors **a**, **b**, and **c** to be co-planar can be written as $\varepsilon_{ijk}a_i b_j c_k = 0$ with the summation convention.

10.4 Applications of MHD to the Earth's magnetosphere

We will apply some of the results from the foregoing discussions of MHD to understand some basic features of the interactions between the solar wind and the Earth's magnetosphere.

10.4.1 The solar wind Ⓚ

In a stationary isothermal neutral atmosphere above a flat Earth (implying a constant gravitational acceleration, **g**, at all altitudes) we can have a balance between the thermal pressure force $-\nabla p$ and the gravitational force $\rho\mathbf{g}$ acting on a small volume element, provided we have an exponential distribution of the density with altitude, $n = n_0 \exp(-z\overline{M}g/\kappa T)$. This approximation is applicable only when the scale height $\kappa T/\overline{M}g$ is much smaller than the radius of curvature of the surface (i.e., when the Earth indeed looks flat!). We let $\overline{M}$ be an average mass, so that the gas density is

$\rho = \overline{M}n$; see also Section 15.5. Observations are, however, at variance with such a static model. The pressure at large interstellar distances from the Sun becomes negligible, and the problem has to be re-analyzed (Parker 1965, Dessler 1967), with a detailed monograph presented by Hundhausen (1972). Basically, the problem is to derive a coherent formalism that includes an energy source somewhere at or near the solar surface and include the basic forces, here gravity and pressure, in a model consistent with the negligible pressure at infinite distances. An important result of the *Parker model* is that it predicts the existence of a *solar wind.* Observations support the model for this solar wind having significant, supersonic, streaming velocities at distances of ~ 1 AU. The aim of the following discussion is to make the presence of this solar wind plausible. For the static atmosphere, the force from the pressure gradient is exactly balanced by gravity, to give a vanishing net acceleration and a vanishing net flow. It turns out that relevant solutions of the basic equations exist, where such a balance is not achieved, and the plasma will be accelerated away from the Sun.

Since we are interested in time-stationary conditions, the basic equations can be simplified to give $\nabla \cdot (\rho \mathbf{u}) = 0$. To make the problem even simpler we assume, with only a little loss of generality, that the problem can be taken to be spherically symmetric. By these assumptions we have

$$4\pi r^2 \rho U = \text{const.}, \tag{10.49}$$

with U being a radial velocity, and the surface of a sphere surrounding the origin at radius r outside the Sun is $4\pi r^2$. We use the capital letter for velocity to indicate this particular steady state case. The relation (10.49) simply states that the net mass flux through any spherical surface is constant; it *can* be zero for $U = 0$. In the same way, the momentum equation (Newton's second law) becomes

$$-\frac{d}{dr}p - \rho\frac{M_\odot G}{r^2} = \rho U \frac{d}{dr}U, \tag{10.50}$$

with $M_\odot$ being the solar mass, and $G = 6.67 \times 10^{-4}$ N m^2 kg^{-2} is the gravitational constant. For the solar wind, the plasma pressure is relatively large as compared to the magnetic pressure, $\beta \sim 0.5 - 1$, and we have for *simplicity* ignored the magnetic forces in the momentum equation. Although correct for regions where the dominant magnetic field component is radial, this omission can be criticized, but has minor consequences for the basic physical arguments presented in the following.

The left side of (10.50) is the sum of the forces acting on a volume element, i.e., gradient in pressure and solar gravity, while the right-hand side is the acceleration a volume element experiences due to its radial motion with velocity U. Again, we have as a possibility $U = 0$, in which case we have an exact balance between gravity and pressure gradient. This is, however, not the relevant solution in this case, as we shall see from the following. Since the plasma expands as it propagates in the radial direction, we do not here assume incompressibility, but take $\nabla \cdot \mathbf{u} \neq 0$.

To the same accuracy as for (10.49) and (10.50) the expression for the radial energy conservation equation has the form (Cranmer 2004)

$$\frac{1}{r^2}\frac{d}{dr}\left(r^2\left[F_H + F_C + \rho U\left(\frac{U^2}{2} + \frac{5}{2}\frac{p}{\rho} - \frac{GM_\odot}{r}\right)\right]\right) = -\rho^2\Phi(T). \tag{10.51}$$

where F_H is the input energy density, while $\Phi(T)$ is the radiative loss function, being a function of temperature. Some of the energy deposited into the corona is transported downward to the lower atmosphere via heat conduction (the radial component of the conductive flux F_C is negative), and some of it is converted back and forth between kinetic energy, thermal energy, and gravitational potential energy. Some of this energy is also lost in the form of radiation through Φ in (10.51).

We make the assumption that the pressure is a function of the spatial variables by being a function of the spatially varying mass density, $p(r) = p(\rho(r))$; see, for instance, Section 2.4.1. This is true, for instance, for both isothermal or adiabatic expansions. We introduce, as in, e.g., (4.4), the sound speed by $C_s^2 \equiv dp/d\rho$. We then easily eliminate $dp/dr = dp/d\rho(d\rho/dr) = C_s^2 d\rho/dr$ from

(10.50). Using (10.49) we can then also eliminate ρ to obtain

$$\frac{1}{U(r)}\left(\frac{U^2(r)}{C_s^2(r)}-1\right)\frac{d}{dr}U(r)=\frac{1}{r}\left(2-\frac{M_\odot G}{rC_s^2(r)}\right). \tag{10.52}$$

Since each term in (10.50) contains ρ, either explicitly or implicitly through p, we find that the integration constant from the right-hand side of (10.49) cancels. This is fortunate.

The sound speed can in general be a function of the radial position, as explicitly indicated through the spatial variation of the mass density. We arrived at (10.52) without use of (10.51). The results derived by (10.52) can consequently be only partial, and indeed we will find that they give an adequate expression for the solar wind velocity, but limited information on the mass and energy loss of the Sun.

Qualitative results for $U(r)$ can be found for the case where C_s varies monotonically with r. To arrive at a solution of (10.52) we here make a further simplification and consider the particular case where we can take the solar wind is isothermal and the sound speed is constant, $C_s^2 = \kappa T/\overline{M}$, or at least slowly varying with r. It is readily seen that (10.52) has solutions for U which increase with r, eventually to become supersonic, $U > C_s$ at a position $r > r_c$, the critical radius. A constant sign of dU/dr requires $U = C_s$ at $r_c = M_\odot G/(2C_s^2)$, which determines the critical radius (Dessler 1967). In this case the two bracket-terms in (10.52) change sign at the same radial position. This is the solution appropriate for the Sun and similar stars. We have not demonstrated that it is the actual evolution for a given boundary condition. The important point here is to demonstrate that analytical solutions *exist*, with properties consistent with observations. Precisely what the boundary condition is and precisely where it should be imposed near the solar surface are questions not addressed here. To a large extent, these problems are still under debate (Mahajan et al. 2002). Also the solar wind is highly intermittent both in time and space, and not nearly as nice and steady state as assumed here. In spite of the uncertainties and limitations mentioned here, the Parker model is accepted as the best coherent model that includes the basic features of the problem.

In Fig. 10.7 we show a schematic illustration of solutions to (10.52). We do not need the actual radial variation of U here, but in principle it can be obtained from (10.52). With $U = U(r)$ given, we can determine $\rho = \rho(r)$ from (10.49) once the boundary conditions at the solar surface are known, determining the integration constant in (10.49). Unfortunately, it is not so simple to account for this integration constant. The question is addressed in an exercise.

For a constant sound speed, we can integrate (10.52) to give the equivalent analytic form (Parks 2004). The velocity $U(r_c) = C_s$ at the critical position, r_c, enters as an integration constant. We find

$$\frac{U^2}{C_s^2}-2\ln\frac{U}{C_s}=4\ln\frac{r}{r_c}+4\frac{r_c}{r}-3. \tag{10.53}$$

We note that $U \to 2C_s\sqrt{\ln(r/r_c)}$ for $r \to \infty$, and $\rho \to \frac{1}{2}\rho_0(r_c/r)^2/\sqrt{\ln r/r_c}$ in the same limit, with the reference mass density being $\rho(r = r_c) \equiv \rho_0$. We found a result where $U \to \infty$ for $r \to \infty$, but in reality, the solar wind will eventually encounter a dilute cold interstellar plasma, and (presumably) a shock is formed there (Hundhausen 1972).

The result (10.53) is one of four possible different solutions of (10.52). The other three are here excluded by the experimental observations.

For the Sun we have $r_c \approx 3.5R_\odot$, taking a temperature 10^6 K, and a sound speed of $C_s \approx 1.7 \times 10^5$ m/s. Solar gravity is trying to retain the plasma at the surface, while the pressure gradient is "pushing" the plasma away from the hot Sun. Both the solar gravity and the pressure gradient are decreasing with distance, but as it turns out, the gravity is decreasing faster, so the pressure gradient is "winning". As a consequence we have the radial plasma flow called "the solar wind" extending to infinity. Basically, it is a part of the thermal distribution at the surface which "boils-off". The interaction between the highly conducting solar wind and the magnetic field of the Earth is important in shaping the Earth's magnetosphere. The model outlined so far is not affected by the solar rotation.

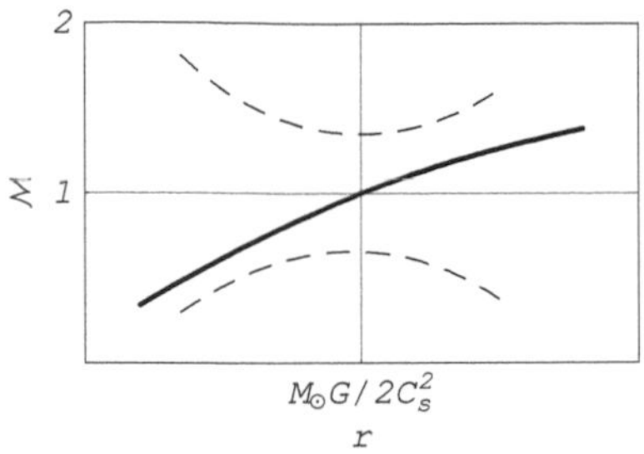

FIGURE 10.7: *Schematic illustration of the variation of the Mach number $\mathcal{M} \equiv U/C_s$ as given by (10.52), with the solid line indicating the solution with subsonic to supersonic transition at r_c. Dashed lines show other possible solutions, which are, however, not relevant here. The derivative of the solid line is positive everywhere, in particular also at $r = r_c$, while the sign of the derivatives of the dashed lines change sign at $r = r_c$.*

The Parker model for the solar wind as outlined in the present section has similarities to the description of a *deLaval nozzle* (Dessler 1967). A similar principle is found in the jet engine after-burner, or "re-heater" in British usage.

The model discussed so far ignored the magnetic field of the rotating Sun. In reality, there is a weak magnetic field that has little effect on the flow as such since the plasma β-value is high. The solar wind is in the radial direction in the rotating frame. At the solar equator, the rotation period is 24.47 days. This is called the "sidereal rotation period", and should not be confused with the synodic rotation period of 26.24 days, which is the time for a fixed feature on the Sun to rotate to the same apparent position as viewed from Earth. (The Sun does not rotate as a solid body: The Sun's rotation rate decreases with increasing latitude, so that the rotation rate is slowest near its poles. At its poles the synodic period of the Sun is 36 days). This solar rotation was first detected by observing the motion of sunspots. Because of the high conductivity of the solar wind plasma, the magnetic field lines become frozen in to the solar wind flow, that is, the magnetic field behaves a passive tracer, like drops of ink in a flow of water. A possible outdoor demonstration of this effect can be performed with a persistent, rotating source of water flow, like a lawn sprinkler. Despite the fact that all of the water droplets are flowing radially away from the center, a snapshot at any time shows them arranged in a spiral "streakline" (Aris 1962). This field is analogous to the *Parker spiral*, i.e. the magnetic field pattern in the solar wind, which carries the imprint of the Sun's rotation, but still channels the particle flow to be radial (Cranmer 2004). Another useful visualization can be the pick-up of a gramophone: it moves in a radial direction, while the groves of the record on the rotating turntable here represent the spiraling magnetic field lines in the plane perpendicular to the rotation axis of the Sun.

Ⓢ **Exercise:** Analyze the case of a stationary solar atmosphere, i.e., the case where we set $U = 0$ in (10.50). Why is this solution unrealistic from a physical point of view? Discuss also the case with adiabatic plasma expansion. Ⓢ

Ⓢ **Exercise:** Introduce Lambert's W-function (or the "ProductLog" function) that often appears in physical problems (Cranmer 2004). The function $W(z)$ returns the principal solution for ζ in $z = \zeta \exp(\zeta)$, where $W(z)$ is real for $z > -\exp(-1)$. In Fig. 10.8 we give some basic details, specifying in particular the two real branches W_0 and W_{-1}. Assuming a constant sound speed, write (10.52) in the form

$$\left(U - \frac{C_s^2}{U}\right)\frac{dU}{dr} = \frac{2C_s^2}{r} - \frac{GM_\odot}{r^2}.$$

Demonstrate how to solve this equation by using Lambert's W-function. Ⓢ

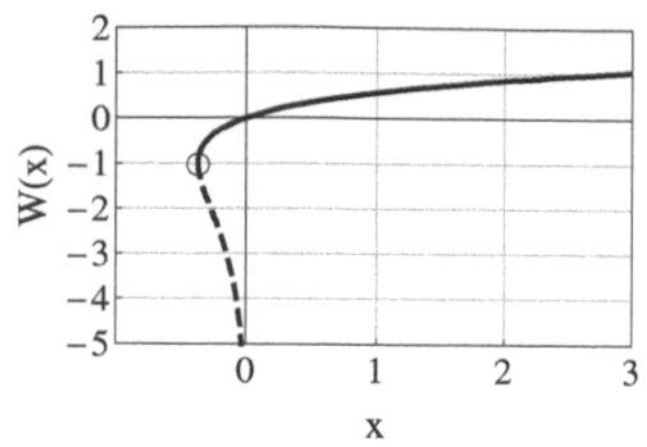

FIGURE 10.8: *Illustration of the two real branches of Lambert's W-function. Full line gives* $W_0(x)$, *dashed line* $W_{-1}(x)$. *There is a branch-cut at* $x = -1/\exp(1)$, *marked by* ○.

Ⓢ **Exercise:** The solution of the solar wind velocity $U(r)$ contains only limited information for the mass loss of the Sun. Relaxing the assumption of strictly isothermal conditions, simplify (10.51) to give a result that is tractable in terms of the Lambert's W-function, and find a result for the conductive flux. You can use an empirical form for the radiative loss function $\rho\Psi = n^2 A T^{-1/2}$ where $A = 51.9 \times 10^{-32}$ Wm^{3}K$^{1/2}$. Use the Spitzer expression for the the classical conductive flux $F_c = -\mathcal{K} T^{5/2} dT/dr$, where $\mathcal{K}$ is the heat conduction constant in an ionized plasma, with a value 8.8×10^{-12} Wm^{-1} K$^{-7/2}$ for the range of densities and temperatures of the corona (Cranmer 2004). Ⓢ

Ⓢ **Exercise:** Consider a compressible flow, streaming along the x-axis, through a Laval nozzle with varying cross section, $A(x)$. Derive the conditions for a supersonic streaming, assuming that the inclination of the walls of the nozzle with the x-axis is everywhere small, so we can restrict the analysis to one independent spatial variable only. Ⓢ

10.4.2 The Earth's magnetosphere

As an example of a realistic stationary MHD plasma state, we discuss the magnetosphere of a magnetized planet, such as the Earth, being exposed to a solar wind in the plasma state streaming out from a central star, such as our Sun.

In empty space we approximate the Earth's magnetic field by a simple dipole (Duffin 1990, Parks 2004), here written in spherical coordinates

$$B_\phi = 0, \qquad B_\theta = \mu_0 \mathcal{M} \frac{\sin\theta}{4\pi r^3}, \qquad B_r = \mu_0 \mathcal{M} \frac{2\cos\theta}{4\pi r^3}. \tag{10.54}$$

At times you find the expressions given in terms of the angle λ measured from the equator, i.e., $\lambda \equiv 90° - \theta$. We introduced the magnetic dipole moment $\mathcal{M}$, which can formally to be interpreted as describing the ring current, I, generating the magnetic field. For a tiny circular wire we have the current multiplied by the area of the circle, i.e., $\mathcal{M} = I\pi R^2$. For the Earth, $\mathcal{M} \approx 8 \times 10^{22}$ A m^2. We have an exact analogy between electric and magnetic dipoles as far as the spatial field variations are concerned. The axis of the effective magnetic dipole of the Earth is tilted approximately 11° with respect to the rotation axis. An even better approximation to the actual magnetic field is obtained by displacing the magnetic dipole axis approximately 400 km with respect to the Earth's rotation axis, so that the magnetic south pole is located in northern Canada. In the vicinity of the Earth this approximation is adequate for many purposes. At larger distances, we find that the interaction between the solar wind and the Earth's magnetic field gives rise to current distributions in space which give significant modifications of the simple dipolar model.

Ⓢ **Exercise:** Introduce a magnetic moment vector **M** with magnitude $\mathcal{M}$ and direction along the magnetic axis of the dipole, where the components in spherical coordinates are given by (10.54).

Demonstrate that the magnetic field from the dipole can be written as

$$\mathbf{B}(\mathbf{r}) = \mu_0 \frac{3\widehat{\mathbf{r}}(\widehat{\mathbf{r}}\cdot\mathbf{M}) - \mathbf{M}}{4\pi r^3},$$

where $\widehat{\mathbf{r}}$ is a unit vector in the direction of $\mathbf{r}$, as before. In current free regions and static fields we have from Ampére's law $\nabla \times \mathbf{B} = 0$, and we can introduce a magnetic potential Φ_M, so that $\mathbf{B} = -\nabla\Phi_M$. Demonstrate that

$$\Phi_M(\mathbf{r}) = \mu_0 \frac{\mathbf{M}\cdot\widehat{\mathbf{r}}}{4\pi r^2}.$$

Find an analytic expression for the equi-potential contours of Φ_M.

Show that the absolute value of the dipole magnetic field is given by

$$|B(\mathbf{r})| = \frac{\mu_0 \mathcal{M}}{4\pi r^3}\sqrt{1 + 3\cos^2\theta}.$$

Discuss the (rather obvious) consequences of this expression for the magnetic field intensity at the magnetic poles and the equator.

Determine the magnetic field intensity at an altitude of 110 km (i.e., in the ionospheric E-region), and make a plot of the electron cyclotron frequency ω_{ce} as you move from magnetic Equator to the magnetic pole at this altitude. Ⓢ

Ⓢ **Exercise:** In two spatial dimensions you can construct something equivalent to a dipole by assuming that two close parallel wires perpendicular to the plane carry a current I and $-I$, respectively. Demonstrate that for this case we have $|B|$ independent of θ. Give the full expression for $|B| = |B(\mathbf{r})|$. This dipole model can be useful for simplified two-dimensional numerical simulations of the interaction between a solar wind and a magnetized planet. Ⓢ

- **Exercise:** Find an expression for the interaction force between two magnetic dipoles placed at a relative distance a, and with arbitrary relative orientations of the dipole moments. Compare with the corresponding expression for *electric* dipoles.

Ⓢ **Exercise:** Demonstrate that the magnetic dipole field (10.54) can be derived from a magnetic scalar potential. Find the corresponding scalar magnetic potential for a constant homogeneous magnetic field. Ⓢ

Assume as a first approximation that the solar wind can be considered as a "wall" of ideally conducting unmagnetized material leaving the Sun at a certain instant, and approaching the Earth with some not necessarily constant velocity, as discussed in Section 10.4.1. According to the previous results, we can assume that this ideally conducting medium retains its original magnetic field, which at a large distance from the Earth can be assumed to be vanishing (we are for the time being ignoring the magnetic field of the Sun itself). As a consequence, we will at any instant have a magnetic field free solar wind approaching the Earth. To maintain this field free condition we must have surface currents induced in the solar wind, in such a way the the Earth's dipolar field together with the magnetic fields originating from the surface currents exactly cancel inside the model solar wind. This situation is illustrated in Fig. 10.9. For a stationary observer it will appear as if the magnetic field lines near the Earth are "compressed". The magnetic field outside the ideal solar wind can be determined by the method of images, where an image magnetic dipole is placed inside the solar wind, as indicated by the arrow to the left in Fig. 10.9. We will not need this exact solution here, but will be content with the overall variation. Note the two cusp points labeled Q on the figure; these are important by allowing material to flow freely from the solar wind to the Earth. The magnetic field intensity is vanishing at Q. Using the mirror image and the basic expressions for a magnetic dipole, the construction of Fig. 10.9 is straightforward.

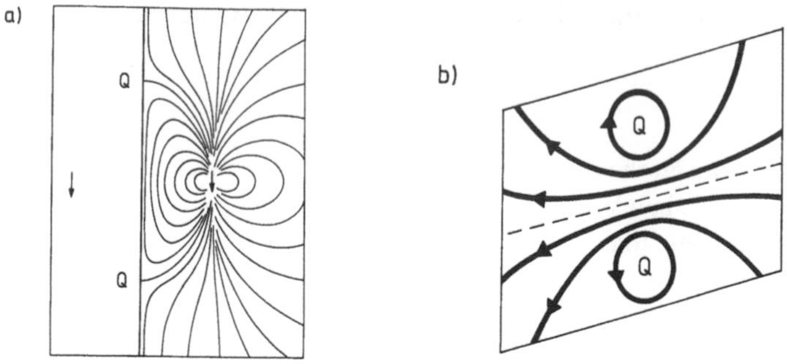

FIGURE 10.9: *Simple illustrative model for the magnetosphere, obtained by considering the solar wind as an ideally conducting wall (Chapman & Bartels 1940, Alfvén 1950). Figure a) shows the magnetic field lines, and b) selected surface current paths at the interface, being representative for a continuous distribution. The magnetic field vanishes at the cusp points labeled Q.*

The force between the Earth's magnetic dipole and the surface current on the solar wind is repulsive. The distance between the Earth and the solar wind surface is determined as the balance between this repulsion and the force with which the solar wind is pressed toward the Earth.

- **Example:** The mirror image of a magnetic dipole can be understood by taking first an ideally conducting wall and a single charge. The mirror image of this charge is one of opposite sign, placed symmetrically with respect to the wall; see Section 5.7.1. We now let the reference charge perform a small circular orbit, say clockwise, in the plane perpendicular to the surface of the conductor. The mirror charge will then perform a circular orbit in the opposite sense, i.e., counterclockwise for this case. The two small orbits can be seen as small ring currents, which give rise to *parallel* magnetic dipoles, since the two charges have opposite sign; see Fig. 10.9. To make the overall system charge neutral, we can introduce two stationary charges of appropriate sign in the center of the rings; their presence will not affect the argument.

Ⓢ **Exercise:** Consider the simple model illustrated in Fig. 10.9, with a plane conducting wall and a magnetic dipole axis parallel to the wall. Calculate the angle between the vector pointing from the magnetic dipole toward the cusp points Q and the line from the dipole normal to the wall. Does this angle depend on the distance between the dipole and the wall? What is the intensity of the magnetic field at the surface of the conductor at the position facing the dipole halfway between the two cusps? Ⓢ

The foregoing simple solar wind model can be criticized for several reasons; first of all, the solar wind can by no means be adequately described as a "wall;" on the contrary, it is in a gaseous state and its surface will be deformed as it approaches the Earth's magnetic field. Second, the solar wind is a good conductor, but it is not an *ideal* conductor, and according to (10.21), the magnetic field will eventually diffuse into it. The first one of these objections is not so severe; near the Earth, the repulsion between the Earth's magnetic field and the surface current is the strongest and a relatively large distance between the Earth and the solar wind surface can be maintained. At larger distances (top and bottom in Fig. 10.9) this force is smaller, since the magnetic field as well as the surface current density reduce with distance. The pressure in the solar wind is then able to push the surface more toward the Earth, and the surface is "folded" around the Earth. This deformation is illustrated in Fig. 10.10, with illustrative surface current paths indicated. The difference between the Figs. 10.9 and 10.10 is thus only qualitative; topologically they are similar. Note in particular that the cusp points Q remain. A number of classical papers on this subject can be consulted for details, for instance, works by Hurley (1961*a*, *b*), Dungey (1961), Mead (1964) and Mead and Beard (1964)

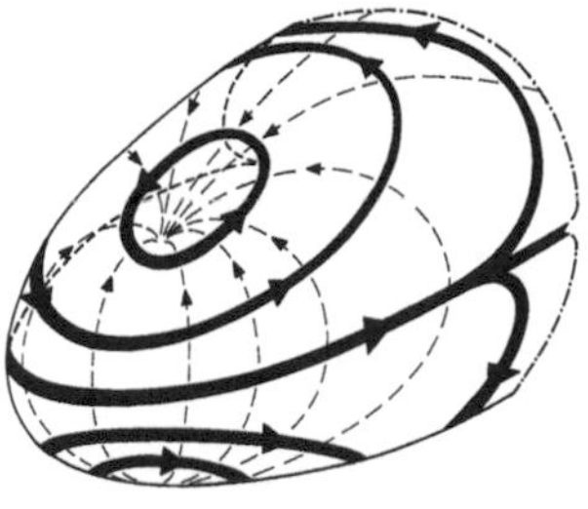

FIGURE 10.10: *Deformation of surface current paths in Fig. 10.9b when the gaseous phase of the solar wind is taken into account.*

The second objection concerning the conductivity of the solar wind is more serious. According to (10.21) we might expect that in the stationary limit, the magnetic field will have diffused into any plasma with finite conductivity, no matter how large this conductivity is. Formally, this is indeed correct, but as the following heuristic argument will show, the magnetic field inside the plasma can be so small that it is in practice immaterial for the dynamics. We thus estimate the quantity R_L from (10.23) for the solar wind plasma by first determining its conductivity to be $\sigma \approx 5 \times 10^6$ S/m. As we find in a different context, the solar wind velocity is quite high, $U \approx 10^5$ m/s or larger, at the Earth's orbit, at one "astronomical unit (AU)", i.e., 150×10^6 m, from the Sun. So, if we restrict ourselves to consider only length scales of, say, $\mathcal{L} \approx 10^3$ km or more, we find $R_L \approx 6 \times 10^8 \gg 1$. Even though several of our estimates are rather uncertain and can easily vary orders of magnitude one way or the other, we see that there is ample margin for these uncertainties when we claim $R_L \gg 1$ for the present problem. Consequently, the diffusive term in (10.21) is negligible for the present case. Physically this means that as the magnetic field of the Earth is diffusing into the solar wind, there will be new unmagnetized solar wind plasma flowing toward the Earth at a high velocity. The Earth's magnetic field will therefore hardly be felt at a distance of a few Earth radii ($R_E \approx 6.4 \times 10^3$ km) in the direction toward the Sun, and we will not make any significant error in ignoring the resistivity of the solar wind, at least not in the present context.

An effective surface current in the magnetosheath can be estimated by the observed value $B \sim$ 20 nT of the Earth's magnetic field in the vicinity of the stagnation point between the Earth and the Sun. A change in magnetic field from zero to 20 nT implies a surface current $K = \Delta B/\mu_0 \approx$ 16 mA m^{-1}; see (5.9) or (5.66). This may sound like a very small current, but the enormous distances involved should be kept in mind. The current passing through a segment of 1000 km along the current sheath, in the direction perpendicular to the current vector, is 16 kA, and 1000 km is after all a small distance in this context. As we move tail-wise along the magnetosheath the current density is slowly diminishing. A semi-realistic illustration of the Earth's magnetosphere is given in Fig. 10.11.

- **Example:** We can use the simplified model from Figs. 10.9 and 10.10 to obtain an estimate for the distance R from the Earth to the stagnation point between the Earth and the Sun (Walker & Russell 1995, Børve et al. 2011). We take the dipolar Earth magnetic field component $B_\theta = \mu_0 \mathcal{M}/(4\pi r^3)$ and derive the magnetic field pressure $B^2/2\mu_0$ at this position. With the additional magnetic field contribution from the image dipole (see Fig. 10.9), we find $2\mu_0 \mathcal{M}^2/(4\pi R^3)^2$. This magnetic pressure has to balance the pressure from the solar wind. With this pressure being the momentum received per sec per area (see Section 2.6), we estimate $p = U^2 M n$ (often called "ram pressure" to distinguish it from the thermodynamic pressure), assuming that the collisions between solar wind particles and the magnetopause are inelastic (a perfect reflection may be difficult to justify). We included only the directed momentum density of the solar wind nMU, with M being an average ion mass, and ignored a thermal velocity spread. This can be justified since U is large, as we found before in Section 10.4.1. By the same argument, we can

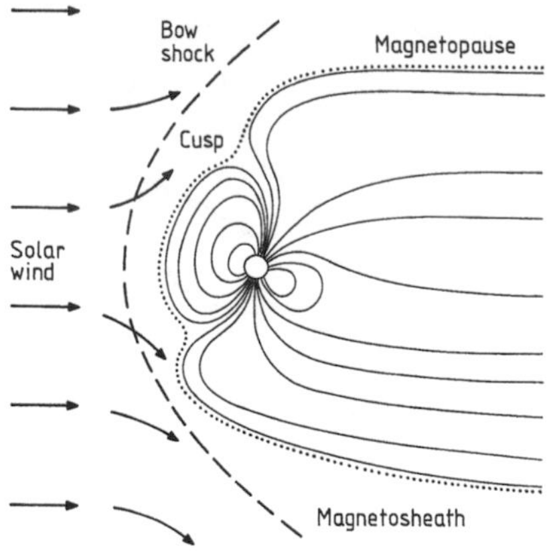

FIGURE 10.11: *Illustration of the Earth's magnetosphere, shown at winter conditions at the northern hemisphere.*

ignore also the electron contribution to the pressure at the magnetopause, at least as long as the ion and electron temperatures are comparable. (For isothermal conditions, the electron and ion thermal pressures are the same: ions are heavy and slow, while electrons are light but fast, so they transfer the same momentum to the surface per unit time.)

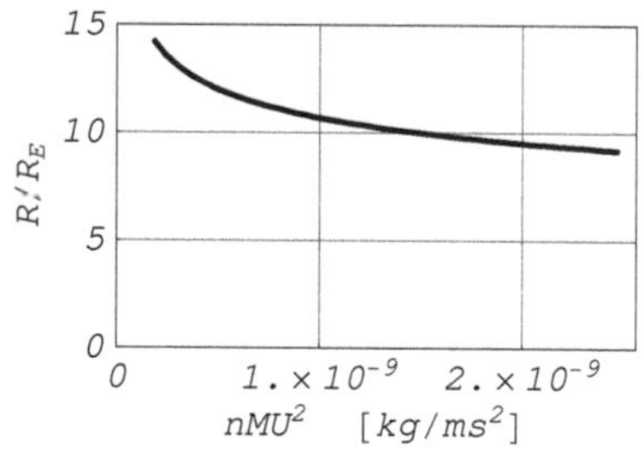

FIGURE 10.12: *The Earth-magnetopause distance (10.55), in units of the Earth radius R_E, for varying solar wind "ram pressures" $p = U^2Mn$.*

By equating the magnetic and solar wind pressures, we find the relation

$$R \approx \left(\frac{\mu_0 \mathcal{M}^2}{8\pi^2 U^2 nM} \right)^{1/6} \tag{10.55}$$

Inserting typical numbers as $U \approx 3 \times 10^5$ m/s, $n \approx 5 \times 10^6$ m^{-3} and hydrogen mass, $M = 1.66 \times 10^{-27}$ kg, we find $R \approx 72 \times 10^6$ m, or $R \approx 11.2R_E$, in terms of the Earth radius, $R_E = 6.4 \times 10^6$ m. The estimate for R is amazingly close to the generally accepted figure of $R \sim 10 - 15R_E$. In Fig. 10.12 we show the range of values obtained for R/R_E for varying solar wind pressures $p = U^2Mn$. Allowing for a small inclination of the Earth's dipole axis with the direction to the Sun, we have a slight reduction in the repulsive forces between the Earth's dipole and its mirror image, which acts to reduce the separation between the Earth and the magnetopause. Inclusion of the thermal pressure contribution would also have reduced the calculated distance between the Earth and magnetopause. The model (10.55) implies a scaling law for the distance to the magnetosheath boundary in terms of the solar wind velocity U and the solar wind mass density nM.

- **Exercise:** Calculate the magnetic field intensity just inside the surface separating the solar wind and the magnetosphere in Fig. 10.9, using the Earth magnetic dipole moment and the distance to the Earth from this model surface as obtained in the previous discussion.
- **Exercise:** Obtain analytical expressions for the magnetic field lines in the simple model illustrated in Fig. 10.9a.
- **Exercise:** The results summarized in Fig 10.12 give an equilibrium Earth-magnetopause distance, as illustrated in Fig. 10.9a. Demonstrate that this separation distance can be harmonically varying as a response to external impulse perturbations, like "gusts" in the solar wind (Børve et al. 2011). You will need to postulate a surface mass density on the interface.

Ⓢ **Exercise:** Assume that you have a laboratory experiment with the same number density as the solar Wind at say 1 AU, but at room temperature. What would the pressure be then? Ⓢ

The model outlined before represents a simplified argument for what is known as the *Chapman-Ferraro model* for the Earth's magnetosphere. Evidently, it is a *closed* model, where all magnetic field lines begin and end on the Earth. The Earth's magnetic field appears as an "obstacle", blocking the free flow of the solar wind (Chapman & Bartels 1940). The most important thing left out of the previous heuristic summary is the Sun's own magnetic field. Since the solar wind has a high but finite conductivity, it can support also a solar magnetic field. Under quiet conditions and depending on the actual phase of the solar cycle, it is of intensity in the range $1-5$ nT, as observed to be embedded in the solar wind at 1 AU. The intensity can be $3-4$ times this value under disturbed conditions. The polarity of this interplanetary magnetic field (IMF) with respect to the Earth's magnetic field is important for the dynamics of the coupling between the solar wind and the Earth's ionosphere and magnetosphere. The magnetosphere can become *open*, as a model proposed by Dungey (1963) suggests. Since the solar magnetic field is so weak, it will in most of Earth's near space introduce only minor corrections to the magnetic field illustrated on Fig. 10.11, apart from the vicinity of the cusp points where the Earth's magnetic field is weak. There the corrections can be important and depending on the polarity of the IMF it can give rise to a direct magnetic field line connection between the Earth's ionosphere and the solar wind. The discussion so far addressed static conditions, but dynamic conditions may be even more interesting (Dungey 1963, Stern 1973*b*, Børve et al. 2011) and can to some extent be discussed in terms of frozen-in field lines. The so-called magnetic reconnection problems will not be discussed here, and the reader is referred to the literature (Schmidt 1979, Kulsrud 2011).

We can introduce the magnetic field from the Sun into our model, by referring to *three* current systems: 1) the currents in the interior of the Earth, giving the approximately dipolar magnetic field, 2) the magnetosheath currents described before and 3) the distant solar and solar wind currents accounting for the locally homogeneous approximately 5 nT magnetic field in the vicinity of the Earth. In addition comes a weak "ring current" contribution. Following a magnetic field line from the Earth's interior we can encounter two basically different situations; either the field line is closing on the Earth, or connecting to the magnetic field of solar origin. The two cases are termed *closed* and *open* field lines, respectively. The open field lines will appear as a "bundle" emerging from the regions around the magnetic poles. At these positions, close to the cusp points, contributions to the magnetic field originating from the current systems 1) and 2) almost cancel, and the contribution from distant, weak solar system 3) becomes noticeable. The "footpoints" of field lines forming the boundary between open and closed magnetic field lines are associated with the *auroral oval*.

Ⓢ **Exercise:** Consider a simple magnetic dipole, with a homogeneous magnetic field $\mathbf{B}_0$ superimposed. Let the direction of this additional magnetic field be along the dipole axis. Show that the total magnetic field $\mathbf{B}_T$ has radial and angular components in the midplane perpendicular to the magnetic moment (Parks 2004)

$$B_r = \left(B_0 - \frac{\mu_0 \mathcal{M}}{2\pi r^3}\right)\sin\lambda \qquad \text{and} \qquad B_\lambda = \left(B_0 + \frac{\mu_0 \mathcal{M}}{4\pi r^3}\right)\cos\lambda .$$

where we introduced the angle λ, measured from the magnetic equator.

Write the expression for the total magnetic energy density, and demonstrate the possibility of having a magnetic neutral line, or discrete neutral points where $B_T = 0$, for the proper direction of $\mathbf{B}_0$. Show that if such a line exists, it is a circle. Give the analytical expression for the radius and the plane of this circle. In case we have neutral points, give an expression for their positions. This exercise also serves to illustrate the possibility for a boundary separating open and closed magnetic field lines in the Earth's near space. Ⓢ

Ⓢ **Exercise:** An unmagnetized object with radius R, a comet for instance, is exposed to the magnetized solar wind with velocity V. The object emits a flow of neutral gas at a steady rate of Q s^{-1} uniformly distributed over 4π with spherical symmetry. The average velocity of these neutral particles is U at $r = R$. Due to ion-neutral collisions this neutral flow acts with a pressure on the plasma component. Assume idealized conditions with a sharp locally plane interface between the neutral gas flow and the magnetized solar wind plasma. Estimate the equilibrium distance between the object and the stagnation point of the solar wind expressed in terms of Q, and solar wind plasma mass density ρ_s and a highly supersonic velocity V. Ignore ionization and recombination here (Cravens 1986). Does the local magnetic field intensity B enter the problem? Give a qualitative discussion of the effects of ionization and recombination on the estimated distance. Ⓢ

10.5 Alfvén waves

The foregoing section emphasized *static* plasma conditions. To study the *dynamic* properties of a plasma as described by the MHD equations we look for wave-like solutions. We might as well linearize the equations from the outset, since we will be used to doing so in later, and more general, discussions of waves in plasmas.

10.5.1 Alfvén waves in incompressible plasmas

We assume an ideal incompressible MHD model and with an unperturbed state $u = 0$, $\rho = \rho_0 =$ const., $p = p_0 =$ const. and an initially homogeneous magnetic field $\mathbf{B} = \mathbf{B}_0 =$ const. Perturbing this equilibrium with small quantities $\widetilde{\rho}$, $\widetilde{\mathbf{u}}$, etc., we have for the perturbations

$$\frac{\partial}{\partial t}\widetilde{\mathbf{B}} = \nabla \times (\widetilde{\mathbf{u}} \times \mathbf{B}_0), \tag{10.56}$$

$$\frac{\partial}{\partial t}\widetilde{\rho} = 0, \tag{10.57}$$

$$\nabla \cdot \widetilde{\mathbf{u}} = 0, \tag{10.58}$$

$$\rho_0 \frac{\partial}{\partial t}\widetilde{\mathbf{u}} = -\nabla \widetilde{p} + \frac{1}{\mu_0}(\nabla \times \widetilde{\mathbf{B}}) \times \mathbf{B}_0, \tag{10.59}$$

where we ignored small terms like $\widetilde{\mathbf{u}} \cdot \nabla \widetilde{\mathbf{u}}$.

We then make the assumption that the plasma supports a wave-like motion, characterized by a variation $\exp(-i(\omega t - \mathbf{k} \cdot \mathbf{r}))$ of all quantities, $\widetilde{p}$, $\widetilde{\rho}$, $\widetilde{\mathbf{B}}$ and $\widetilde{\mathbf{u}}$, where the frequency is ω and the wave vector is $\mathbf{k}$. We *choose* $\mathbf{k}$ to be real, and anticipate that ω just *might* turn out to be complex (although this will not be the case here). We then have

$$\left\{\widetilde{\mathbf{B}}(\mathbf{r},t), \widetilde{\mathbf{u}}(\mathbf{r},t), \widetilde{\rho}(\mathbf{r},t), \widetilde{p}(\mathbf{r},t)\right\} = \left\{\mathbf{B}_1, \mathbf{u}_1, \rho_1, p_1\right\} e^{-i(\omega t - \mathbf{k}\cdot\mathbf{r})}, \tag{10.60}$$

We can choose, for instance, $\mathbf{u}_1$ and then the equations will determine all the other perturbed quantities. Similarly, we can choose $\mathbf{k}$ and then the equations will determine ω for that given wave-number: this is the dispersion relation. By (10.57) we have $\widetilde{\rho} = \text{const.}$ and since $\widetilde{\rho}$ was the deviation from the unperturbed density ρ_0 we must have $\widetilde{\rho} = 0$. We then readily find

$$\begin{aligned}
-i\omega\mathbf{B}_1 &= i\mathbf{k}\times(\mathbf{u}_1\times\mathbf{B}_0) = i(\mathbf{k}\cdot\mathbf{B}_0)\mathbf{u}_1 - i(\overbrace{\mathbf{k}\cdot\mathbf{u}_1}^{=0})\mathbf{B}_0 && (10.61)\\
i\mathbf{k}\cdot\mathbf{u}_1 &= 0 && (10.62)\\
-i\omega\rho_0\mathbf{u}_1 &= -i\mathbf{k}p_1 + \frac{i}{\mu_0}(\mathbf{k}\times\mathbf{B}_1)\times\mathbf{B}_0\,. && (10.63)
\end{aligned}$$

It is interesting to note that the background pressure p_0 does not enter the expressions.

We have $\mathbf{k}$ being a real vector, by definition. The incompressibility condition $\nabla\cdot\mathbf{u} = 0$ then implies that $\mathbf{u}_1 \perp \mathbf{k}$, giving $\omega\mathbf{B}_1 = -\mathbf{u}_1(\mathbf{k}\cdot\mathbf{B}_0)$, which states that the perturbations in the magnetic and velocity vectors are parallel, and that $|B_1/u_1| = |\mathbf{k}\cdot\mathbf{B}_0/\omega|$. The magnetic field vector oscillates in proportion to the velocity vector.

Using $\mathbf{B}_0\times(\mathbf{k}\times\mathbf{B}_1) = (\mathbf{B}_0\cdot\mathbf{B}_1)\mathbf{k} - (\mathbf{B}_0\cdot\mathbf{k})\mathbf{B}_1$, the magnetic field vector $\mathbf{B}_1$ can now in part be eliminated from the equations, and we have

$$\left(\omega\rho_0 - \frac{1}{\mu_0\omega}(\mathbf{k}\cdot\mathbf{B}_0)^2\right)\mathbf{u}_1 = \left(p_1 + \frac{1}{\mu_0}\mathbf{B}_0\cdot\mathbf{B}_1\right)\mathbf{k}. \qquad (10.64)$$

We have already chosen $\mathbf{k}$ to be a real vector. We have the freedom to take $\mathbf{u}_1$ to be a real vector as well, but then both sides of the relation must equal zero in (10.64), since we found $\mathbf{u}_1 \perp \mathbf{k}$ from the incompressibility condition. The vanishing of the left side implies the dispersion relation

$$\omega^2 = \frac{1}{\mu_0\rho_0}(\mathbf{k}\cdot\mathbf{B}_0)^2. \qquad (10.65)$$

Provided (10.65) is fulfilled, it is possible to take $\mathbf{u}_1$ to be a real vector, and the corresponding wave is linearly polarized. Section 10.7.1 will demonstrate that this is not an obvious result.

One consequence of (10.65) is that from (10.61) we have the simple relation $\mathbf{u}_1 = \mp\mathbf{B}_1/\sqrt{\rho_0\mu_0}$, between the fluctuations in velocity and magnetic field, for the present ideal MHD model. This result can be seen as a consequence of "frozen-in" field lines. For a plasma with *finite* conductivity, this relation is modified.

> We can introduce a characteristic velocity, the *Alfvén velocity*, as
>
> $$V_A \equiv \frac{B_0}{\sqrt{\mu_0\rho_0}}.$$

This velocity characterizes the low frequency dynamics of many phenomena in astrophysics and magnetospheric physics. The discovery of the existence of this wave-mode was a major result, as also mentioned in Section 1.4.1.

The wave dispersion relation (10.65) can be written as $\omega^2 = V_A^2(\mathbf{k}\cdot\widehat{\mathbf{b}})^2$, where $\widehat{\mathbf{b}}$ is a unit vector in the direction of $\mathbf{B}_0$. It follows that the phase velocity of these *Alfvén waves* is $\omega/k = V_A\cos\Theta$, where Θ is the angle between the wave-vector and the initial or unperturbed magnetic field $\mathbf{B}_0$. The waves are *transverse* with $\mathbf{u}_1 \perp \mathbf{k}$. Here they are linearly polarized. It is possible to obtain a *circularly* polarized wave by addition of two orthogonally and linearly polarized waves.

The assumption of linearization requires $\widetilde{B} \ll B_0$. By an order of magnitude estimate we have $\widetilde{u}/V_A \sim \widetilde{B}/B_0$, i.e., $\widetilde{u}/V_A \ll 1$. There are thus no direct restrictions on the ratio $\widetilde{u}/u_{The}$, where u_{The} is a thermal plasma velocity. Since in general $V_A \gg u_{The}$, we *can* have $\widetilde{u} \sim u_{The}$ even for linear MHD waves.

We are now in a position to make a check on the consistency of our simple model; we are free to ignore Maxwell's displacement current when the characteristic velocity V_A is much less than the speed of light. Since also the right-hand side of (10.64) must vanish we have

$$p_1 = -\frac{1}{\mu_0}(\mathbf{B}_0 \cdot \mathbf{B}_1) = \frac{1}{\mu_0 \omega}(\mathbf{B}_0 \cdot \mathbf{k})(\mathbf{B}_0 \cdot \mathbf{u}_1), \tag{10.66}$$

which, in effect, corresponds to an equation of state in the present incompressible model, by relating fluctuations in pressure to fluctuations in another quantity, here either magnetic field or velocity.

For waves propagating along the unperturbed magnetic field lines, $\mathbf{k} \parallel \mathbf{B}_0$, the relation (10.66) is trivially fulfilled since $\mathbf{B}_0 \cdot \mathbf{u}_1 = 0$ here, and the pressure perturbation vanishes. For wave propagation in other directions, we find basically two different cases, depending on the polarization of the waves. If the fluctuating velocity and magnetic field vectors, $\mathbf{u}_1$ and $\mathbf{B}_1$, are perpendicular to the plane defined by $\mathbf{k}$ and $\mathbf{B}_0$, the fluctuations in pressure p_1 vanish. For this polarization, the waves are often called *shear* Alfvén waves. If, on the other hand, $\mathbf{u}_1$ and $\mathbf{B}_1$ are *in* this plane, we have $p_1 \neq 0$.

Since the model is linear, we can add any suitable linear combination of such waves to obtain an arbitrary polarization. With the assumed incompressibility of the plasma fluctuations, we have no explicit equation relating pressure, density and temperature; see, for instance, the discussion in Section 10.1.

From the dispersion relation (10.65), we can find the group velocity

$$\mathbf{u}_g \equiv \nabla_k \omega(\mathbf{k}) = \pm V_A \widehat{\mathbf{b}}, \tag{10.67}$$

where again $\widehat{\mathbf{b}}$ is a unit vector along $\mathbf{B}_0$. Even though the direction of wave propagation as given by $\mathbf{k}$ can have any angle with the unperturbed magnetic field lines (with the exception of the direction $\perp \mathbf{B}_0$), the direction of the group velocity is always *along* $\mathbf{B}_0$. If two observers are to communicate by sending messages in the form of shear Alfvén pulses, they have to be located on the same $\mathbf{B}_0$-magnetic field lines.

It is important to note that Alfvén waves are *transverse*, and for a homogeneously magnetized medium the waves are incompressible with $p_1 = 0$ for the polarization $\mathbf{B}_0 \cdot \mathbf{u}_1 = 0$ and $\mathbf{B}_0 \cdot \mathbf{B}_1 = 0$, irrespective of the relative magnitudes of the sound speed and the phase and group velocities entering the problem. We need not here go through the same discussions as in Section 2.5. The other polarization with $\mathbf{B}_0 \cdot \mathbf{B}_1 \neq 0$ has $p_1 \neq 0$. The dispersion relation of these two modes is here found to be the same, i.e., (10.65). The properties of the mode with $p_1 \neq 0$ will be more clear when we discuss compressional modes in Section 10.6.

Having derived the Alfvén velocity, we can present some interpretations of the plasma β as defined in (10.38). We have

$$\beta \equiv \frac{n\kappa T}{B^2/2\mu_0} = \frac{\rho\kappa T}{MB^2/2\mu_0} \approx 2\frac{C_s^2}{V_A^2} \approx 2\frac{r_L^2}{c^2}\Omega_{pi}^2, \tag{10.68}$$

where $C_s^2 \equiv \kappa T/M$ is the square of an isothermal sound velocity and $r_L = \sqrt{\kappa T/M}/\Omega_{ci}$ is the Larmor radius of the ions.

The assumption of incompressibility is tantamount to assuming all velocities of propagation well below the sound speed; see discussion in, for instance, Section 2.5. We will find later that the shear Alfvén wave, propagating with phase velocity V_A, remains for a particular wave polarization also without this incompressibility assumption. Results requiring $C_s \gg V_A$ generally imply $\beta \gg 1$, a condition rarely met in nature, not even in the solar wind.

A possible interpretation of transverse Alfvén waves is offered by noting that the dispersion relation resembles that for transverse waves on a string (Symon 1960, Elmore & Heald 1969), where the dispersion relation is $(\omega/k)^2 = \tau/\sigma$, where τ is the tension of the string and σ its density, i.e., mass per unit length. By comparison with the similar relation for transverse Alfvén waves, we can thus interpret B_0^2/μ_0 as a magnetic tension (see also Section 10.3.2) and ρ_0 is (more obviously) the plasma mass density, which acts as the mass loading on the "magnetic strings". The similarity is best appreciated for wave propagation along $\mathbf{B}_0$ and may in general be of limited value, after all.

- **Exercise:** Show that the Alfvén velocity can be expressed as $V_A = c\Omega_{ci}/\Omega_{pi}$, in terms of ion cyclotron and ion plasma frequencies. In order to ignore Maxwell's displacement current when obtaining (10.4), we assumed that characteristic propagation velocities were much smaller than the speed of light. The present result consequently implies $\Omega_{ci} \ll \Omega_{pi}$ for the validity of ideal MHD.

Ⓢ **Exercise:** In (8.48) we obtained an expression for the dielectric constant for a magnetized plasma. Express this ε_r in terms of the Alfvén velocity and other relevant quantities. Ⓢ

- **Exercise:** Demonstrate that a circularly polarized Alfvén wave, $B_0\widehat{\mathbf{z}} + \widetilde{B}(\widehat{\mathbf{x}} + i\widehat{\mathbf{y}})\exp(ikz - i\omega t)$, is an *exact* solution to the full set of the ideal nonlinear MHD equations in homogeneous plasmas.

Ⓢ **Exercise:** Derive the dispersion relation for the case where Alfvén waves are damped by finite conductivity as well as ion-neutral collisions. Consider $\mathbf{B}_0$-parallel motion. The basic linearized equations can be taken as

$$\nabla\cdot\mathbf{u} = 0, \qquad \rho_0\frac{\partial}{\partial t}\mathbf{u} = \mathbf{J}\times\mathbf{B}_0 - \rho_0\nu_{in}(\mathbf{u}-\mathbf{u}_n),$$
$$\nabla\times\mathbf{E} = -\frac{\partial}{\partial t}\mathbf{B}, \qquad \nabla\times\mathbf{B} = \mu_0\mathbf{J},$$

where $\rho_0 =$ const. is the plasma mass density, assuming homogeneous conditions, and ν_{in} is the ion-neutral collision frequency. The linearized equations governing the dynamics of the incompressible neutral component are taken to be

$$\nabla\cdot\mathbf{u}_n = 0, \qquad \text{and} \qquad \rho_n\frac{\partial}{\partial t}\mathbf{u}_n = -\rho_n\nu_{ni}(\mathbf{u}_n - \mathbf{u}),$$

with $\rho_n =$ const. being the homogeneous neutral density and $\mathbf{u}_n$ the velocity of the neutral component, and ν_{ni} being the neutral-ion collision frequency. Assume for simplicity that the atomic mass of the ions and neutrals is the same, and similarly for the ion and neutral temperatures. We have $\rho_0\nu_{in} = \rho_n\nu_{ni}$, so in general we have $\nu_{ni} \neq \nu_{in}$. A finite conductivity, σ, is retained by $\mathbf{J} = \sigma(\mathbf{E} + \mathbf{u}\times\mathbf{B}_0)$. The problem has been analyzed in some detail by, for instance, Pécseli and Engvold (2000) with further references given therein. Ⓢ

- **Exercise:** A shear Alfvén wavepacket with an initial magnetic field perturbation $\widetilde{\mathbf{B}} = \{A_1\cos(\mathbf{k}\cdot\mathbf{r}), 0, A_2\cos(\mathbf{k}\cdot\mathbf{r})\}\exp(-(x^2+y^2+z^2)/\Delta)$ propagating in a homogeneous magnetic field $\mathbf{B}_0 = \{0,0,B_0\}$. Determine the relation between A_1 and A_2 for given $\mathbf{k} = \{k_x, 0, k_z\}$. What is the pressure perturbation associated with this wavepacket? Give the governing differential equations for the magnetic field lines. Try to solve these equations. Consider the limit of small wave amplitudes.

  Change now the polarization to $\{\widetilde{B}_x, \widetilde{B}_y, \widetilde{B}_z\} = \{A_1\cos(\mathbf{k}\cdot\mathbf{r}), A_2\cos(\mathbf{k}\cdot\mathbf{r}), 0\}\exp(-(x^2+y^2+z^2)/\Delta)$. Careful here: what is now the relation between A_1 and A_2 for given $\mathbf{k}$? Express again the pressure perturbation associated with this wavepacket. Set-up the differential equations for the magnetic field lines valid a any time. Discuss the solutions also here.

10.5.2 Energy density of shear Alfvén waves

Taking the **B**-parallel wave propagation as an illustration, the ratio of magnetic field energy density and electric field energy density associated with a shear Alfvén wave can be estimated as

$$\frac{W_E}{W_B} = \frac{\frac{1}{2}\varepsilon_0|\widetilde{E}|^2}{\frac{1}{2}|\widetilde{B}|^2/\mu_0} = \frac{V_A^2}{c^2} \ll 1, \tag{10.69}$$

using $|\widetilde{E}/\widetilde{B}| = \omega/k = V_A$ (from Faraday's law since $\widetilde{\mathbf{E}} \perp \mathbf{k}$ here), where again V_A is the Alfvén velocity. When deriving the MHD equations, we ignored Maxwell's displacement current by arguing that relevant propagation velocities are much smaller than the speed of light. Consequently we must have $V_A \ll c$. We can therefore conclude that for plasmas where a simple MHD description is applicable, the electric wave energy density is negligible as compared to the magnetic energy density. In a sense, the electric field only acts as a sort of "catalyst" for mediating the transfer of magnetic field energy to particle kinetic energy. Shear Alfvén waves are not electromagnetic if we are to take the term "electro" literally; the term *magnetokinetic* may be appropriate?

By Faraday's law (5.58) and Ampere's law (10.4) we have the simple MHD equivalent of Poynting's theorem (see Section 5.4)

$$\frac{1}{2\mu_o}\frac{\partial}{\partial t}\widetilde{B}^2 + \frac{1}{\mu_o}\nabla\cdot(\widetilde{\mathbf{E}}\times\widetilde{\mathbf{B}}) = -\widetilde{\mathbf{E}}\cdot\widetilde{\mathbf{J}}. \tag{10.70}$$

Assuming that $\widetilde{B}$ varies on a fast time scale, $\sim 1/\omega$, as well as on a slow time scale, $\tau \gg 1/\omega$, we can average over $1/\omega$, and have

$$\frac{\partial}{\partial \tau}\overline{W_B} + \frac{1}{\mu_o}\nabla\cdot\overline{(\widetilde{\mathbf{E}}\times\widetilde{\mathbf{B}})} = -\overline{\widetilde{\mathbf{E}}\cdot\widetilde{\mathbf{J}}} \tag{10.71}$$

where $\overline{W_B} \equiv \frac{1}{2}\overline{\widetilde{B}^2}/\mu_o$, with the overline denoting average over the fast time scale. The magnitude of the Poynting flux is $V_A\overline{W_B}$. The discussion of (10.70) and (10.71) is completed as in Section 5.4.

- **Exercise:** Prove that $\frac{1}{2}\rho_0\overline{\widetilde{u}^2} = \frac{1}{2}\overline{\widetilde{B}^2}/\mu_0$, i.e., the average kinetic energy density, equals the magnetic energy density for an ideal shear Alfvén wave.

- **Exercise:** Prove the relation (10.70). What is the difference between this result and the "standard" version on Poynting's theorem in Section 5.4?

Ⓢ **Exercise:** Based on the analysis of a simple incompressible MHD description, the force density exerted by the fluctuations on the medium is given by $\mathbf{f} = \mathbf{J}\times\mathbf{B}$. We then have trivially that

$$\mathbf{f}(\mathbf{r},t) = \frac{1}{\mu_0}[\nabla\times\mathbf{B}(\mathbf{r},t)]\times\mathbf{B}(\mathbf{r},t). \tag{10.72}$$

This force will in general be fluctuating in time, and the only part that really matters is its time average, $\bar{\mathbf{f}} \equiv \frac{1}{T}\int^T \mathbf{f}(\mathbf{r},t)dt$. Obtain analytical results for this average force. In order to make the analysis general, include also finite conductivity as well as the effect of ion-neutral collisions in the ion momentum equation. Ⓢ

10.6 Compressional Alfvén waves

The simple Alfvén waves are modified somewhat in case we allow for the plasma being compressible, $\nabla\cdot\mathbf{u} \neq 0$, and introduce an explicit equation of state, $p = p(\rho)$. We assume isothermal motion, with $p = n\kappa T = \rho\kappa T/M$ as for ideal gases, noting that in a linearized analysis, adiabatic dynamics can be introduced "a posteriori" by a factor γ on the temperature. We have the linearized equations

$$\begin{aligned} \frac{\partial}{\partial t}\widetilde{\mathbf{B}} &= \nabla\times(\widetilde{\mathbf{u}}\times\mathbf{B}_0) &\quad (10.73)\\ \frac{\partial}{\partial t}\widetilde{\rho}+\rho_0\nabla\cdot\widetilde{\mathbf{u}} &= 0 &\quad (10.74)\\ \rho_0\frac{\partial}{\partial t}\widetilde{\mathbf{u}} &= -\nabla\widetilde{p}+\frac{1}{\mu_0}(\nabla\times\widetilde{\mathbf{B}})\times\mathbf{B}_0 &\quad (10.75)\\ \widetilde{p} &= \frac{\kappa T}{M}\widetilde{\rho}. &\quad (10.76) \end{aligned}$$

Assuming again a plane wave solution as in (10.60), we find

$$\begin{aligned} -i\omega\mathbf{B}_1 &= i\mathbf{k}\times(\mathbf{u}_1\times\mathbf{B}_0) = i(\mathbf{k}\cdot\mathbf{B}_0)\mathbf{u}_1 - i(\mathbf{k}\cdot\mathbf{u}_1)\mathbf{B}_0 &\quad (10.77)\\ -i\omega\rho_1+i\rho_0\mathbf{k}\cdot\mathbf{u}_1 &= 0 &\quad (10.78)\\ -i\omega\rho_0\mathbf{u}_1 &= -i\mathbf{k}p_1+\frac{i}{\mu_0}(\mathbf{k}\times\mathbf{B}_1)\times\mathbf{B}_0 &\\ &= -i\mathbf{k}p_1-\frac{i}{\mu_0}[(\mathbf{B}_0\cdot\mathbf{B}_1)\mathbf{k}-(\mathbf{k}\cdot\mathbf{B}_0)\mathbf{B}_1] &\quad (10.79)\\ p_1 &= \frac{\kappa T}{M}\rho_1. &\quad (10.80) \end{aligned}$$

From (10.77) we find that $\mathbf{B}_1 \perp \mathbf{k}$, even when compressibility is allowed for, while (10.78) demonstrates that the fluctuating velocity $\mathbf{u}_1$ has a component along $\mathbf{k}$ when the mass density fluctuates. The fluctuating electric field is given by $\mathbf{E}_1 = -\mathbf{u}_1\times\mathbf{B}_0$ or $\mathbf{E}_1(\mathbf{k}\cdot\mathbf{B}_0) = \omega\mathbf{B}_1\times\mathbf{B}_0$ obtained by (10.77). For ideal MHD waves, we have $\mathbf{E}_1 \perp \mathbf{B}_1$ also with compressional motions allowed for.

It is now a simple matter to eliminate $\mathbf{B}_1$, p_1 and ρ_1 to obtain the dispersion relation

$$\left(\frac{\omega^4}{k^4}-\frac{\omega^2}{k^2}\left(C_s^2+V_A^2\right)+C_s^2V_A^2\cos^2\Theta\right)\left(\frac{\omega^2}{k^2}-V_A^2\cos^2\Theta\right)=0, \quad (10.81)$$

where $C_s = \sqrt{\kappa T/M}$ is the fluid sound speed for the present isothermal model. Corrections for adiabatic sound dynamics are, as already mentioned, readily incorporated in the present linear analysis.

We recover the foregoing dispersion relation for the shear Alfvén waves by putting the second parenthesis in (10.81) equal to zero, but in addition we find two new modes of oscillation, the *slow* and *fast* magnetosonic waves with dispersion relation

$$\frac{\omega^2}{k^2}=\frac{1}{2}\left(C_s^2+V_A^2\right)\pm\frac{1}{2}\sqrt{\left(C_s^2+V_A^2\right)^2-4C_s^2V_A^2\cos^2\Theta}. \quad (10.82)$$

This dispersion relation remains nontrivial also in the limit where the plasma is assumed to be cold, $T=0$, implying $C_s=0$. The "stiffness" of the magnetic field lines is sufficient to ensure the existence of the magnetosonic mode.

For propagation along the magnetic field, $\Theta=0$, we have $(\omega/k)^2 = C_s^2$ and $(\omega/k)^2 = V_A^2$, i.e., sound waves and Alfvén waves, while for $\Theta=\pi/2$ only the fast mode exists, with $(\omega/k)^2 = C_s^2+V_A^2$. When $V_A^2 \gg C_s^2$, i.e., for low β plasmas, we have

$$\begin{aligned} \frac{\omega^2}{k^2} &= \frac{1}{2}(V_A^2+C_s^2)\pm\frac{1}{2}(V_A^2+C_s^2)\sqrt{1-4\frac{C_s^2V_A^2}{(V_A^2+C_s^2)^2}\cos^2\Theta}\\ &\approx \frac{1}{2}(V_A^2+C_s^2)\pm\frac{1}{2}(V_A^2+C_s^2)\left(1-2\frac{C_s^2}{V_A^2+C_s^2}\cos^2\Theta\right). \end{aligned}$$

Here, the mode with phase velocity $\omega/k \approx C_s \cos\Theta$ is directionally anisotropic, just like the shear Alfvén wave with phase velocity $\omega/k \approx V_A \cos\Theta$. The other mode, with phase velocity $\omega/k \approx V_A(1 + \frac{1}{2}(C_s^2/V_A^2)\sin^2\Theta)$, is in this limit only weakly affected by variations in direction of the wave-vector.

The shear Alfvén waves remain incompressible, with $\rho =$ const., even though we allowed for compressibility of the plasma in the analysis in this section. It is important to note *how* this incompressibility arises: by taking the scalar product with $\mathbf{k}$ of (10.79) we readily see that it requires $\mathbf{B}_1 \cdot \mathbf{B}_0 = 0$, implying that the fluctuating component of the magnetic field is perpendicular to the plane spanned by $\mathbf{k}$ and $\mathbf{B}_0$. This result is consistent with what we found in (10.64), giving $p = 0$ in agreement also with the proposed equation of state (10.76). More generally, we expect that the incompressibility condition is satisfied for *any* polarization of $\mathbf{B}_1$, provided we let $C_s \to \infty$; see the discussion in Section 2.5. We find that in the limit $C_s^2 \gg V_A^2$, the expression (10.82) reduces to

$$\begin{aligned}
\frac{\omega^2}{k^2} &= \frac{1}{2}(V_A^2+C_s^2) \pm \frac{1}{2}(V_A^2+C_s^2)\sqrt{1-4\frac{C_s^2V_A^2}{(V_A^2+C_s^2)^2}\cos^2\Theta} \\
&\approx \frac{1}{2}(V_A^2+C_s^2) \pm \frac{1}{2}(V_A^2+C_s^2)\left(1-2\frac{V_A^2}{V_A^2+C_s^2}\cos^2\Theta\right) \\
&= \begin{cases} V_A^2+C_s^2-V_A^2\cos^2\Theta \equiv C_s^2+V_A^2\sin^2\Theta \\ V_A^2\cos^2\Theta, \end{cases}
\end{aligned}$$

using $V_A^2\cos^2\Theta = V_A^2 - V_A^2\sin^2\Theta$. In the limit considered here, the fast mode (+ sign) becomes acoustic waves with phase velocity $\omega/k \approx C_s(1+\frac{1}{2}(V_A^2/C_s^2)\sin^2\Theta)$, while the slow mode (−sign) dispersion relation becomes identical to that of the incompressible shear Alfvén wave. This limit is applicable for high β plasmas, recalling here that $\beta \approx C_s^2/V_A^2$. Note that $C_s \to \infty$ is a sufficient but not *necessary* condition for having incompressible waves in MHD!

- **Exercise:** Derive the dispersion relation (10.81) for the general compressible case, $\widetilde{p} \neq 0$.

Ⓢ **Exercise:** Consider a compressible cold plasma, where we take $p = 0$. Derive the relation

$$(\mathbf{k}\cdot\mathbf{B}_0)^2\mathbf{B}_1 + k^2(\mathbf{B}_1\cdot\mathbf{B}_0)\mathbf{B}_0 - (\mathbf{k}\cdot\mathbf{B}_0)(\mathbf{B}_1\cdot\mathbf{B}_0)\mathbf{k} = \omega^2\rho_0\mu_0\mathbf{B}_1\,. \tag{10.83}$$

Demonstrate that *either* we have $\mathbf{B}_1 \cdot \mathbf{B}_0 = 0$, in which case the dispersion relation is $\omega^2 = (\mathbf{k}\cdot\mathbf{B}_0)^2/(\rho_0\mu_0)$, *or* we have $\mathbf{B}_1 \cdot \mathbf{B}_0 \neq 0$, giving $\omega^2 = k^2B_0^2/(\rho_0\mu_0)$. What is the mass density perturbation in the two cases? Discuss the two polarizations for the magnetic field fluctuations. Ⓢ

We note that once the direction of propagation Θ is fixed, the phase velocity is constant, independent of k, in all cases. This invites a polar diagrammatic representation of the phase velocity, as the one shown in Fig. 10.13. The horizontal axis is along $\mathbf{B}_0$, and a straight dashed line indicates a possible direction of the phase velocity. The length of the distance from the origin to the intersection with one of the curves gives the phase velocity of the appropriate wave type for that direction. The diagrams in Fig. 10.13 justify the terms "fast" and "slow" waves. Note that the direction of propagation with $\mathbf{k} \parallel \mathbf{B}_0$ is degenerate, since two dispersion curves are tangential there, irrespective of the ratio V_A/C_s. For one case it is the Alfvén wave and the fast mode that merge, in the other case the Alfvén wave and the slow mode. Note that the case with vanishing temperature, $C_s = 0$, is degenerate. As $C_s \to 0$, we see the polar diagram for the slow mode shrinking into a point. The topology of Fig. 10.13b) and 10.13c) characterize low and high β-plasmas, respectively; see (10.68). Ignoring the dispersion of the wavepacket, we find that a sound pulse propagates without change in shape, since phase-front velocities and group velocities are the same. It is interesting to note that even though group velocities and phase velocities differ for the shear Alfvén waves, a

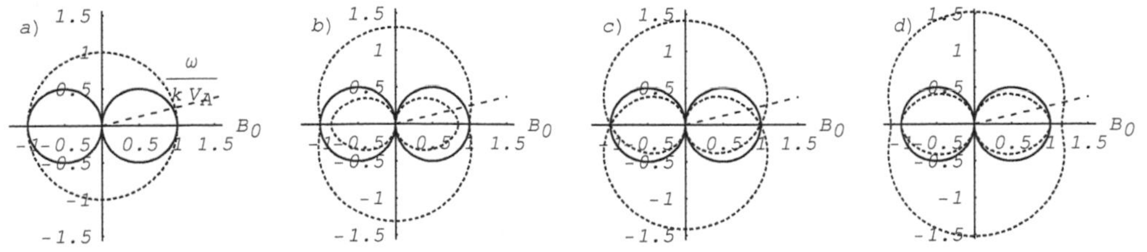

FIGURE 10.13: *Polar diagrams for the normalized phase velocity, $(\omega/k)/V_A$. The three cases shown are a) cold plasma with $C_s = 0$. In b) we have $C_s/V_A = 0.85$, and in c) $C_s/V_A = 1$, while $C_s/V_A = 1.15$ in d). The solid line corresponds to the shear Alfvén wave in all cases. In a) we formally include also $(\omega, \mathbf{k}) = (0,0)$ as a "dispersion relation", as explained in the text, so the "topology" of the three figures is the same. The horizontal axis is along the magnetic field $\mathbf{B}_0$ as indicated. These simple polar diagrams assume that a single characteristic phase velocity, independent of frequency, can be assigned any given direction.*

wavepacket propagates without change in shape also in this case, but now in the direction parallel to the magnetic field $\mathbf{B}_0$.

Shear as well as compressional, or magneto-acoustic, Alfvén waves are damped by finite resistivity, and for relevant conditions in, for instance, the solar plasma, also by ion-neutral collisions. Inclusion of such terms makes the MHD-equations time irreversible.

Some limitations of the standard MHD model were noted by Lighthill (1960). Basically, he argued that because of their small inertia the electrons will follow $\mathbf{E} = -\mathbf{u}_e \times \mathbf{B}$, while the corresponding expression for the ion dynamics has to contain inertial ion forces. The resulting extension of the standard MHD equations as summarized in Chapter 10 is often called the Hall-MHD model (Shivamoggi 2009); see also Section 12.4.3.

10.7 Ideal electron MHD

Basically, in ideal MHD we assumed that the plasma dynamics in the direction perpendicular to the local magnetic field was controlled by the bulk plasma $\mathbf{E} \times \mathbf{B}/B^2$-velocity. It is tempting to consider the possibility of describing the *electron* dynamics in the same way, but assuming that the ions form an immobile background (Kingsep et al. 1990). It turns out that such a description *is* indeed meaningful, and we shall summarize its basic characteristics here. Assuming the frequencies associated with the electron motion to be high, we can consider the ions to be an immobile background of positive charge. The only role of the ions is here to provide a charge neutralizing background for the electron motion. We ignore electron pressure forces and also resistivity. For frequencies much below the electron cyclotron frequency, $\omega \ll \omega_{ce}$ (but still high enough to allow the ion dynamics to be ignored), the electron velocity across magnetic field lines is taken to be the $\mathbf{E} \times \mathbf{B}/B^2$-velocity. With this inertia free description, the electron momentum equation becomes

$$-en\mathbf{E} + \mathbf{J} \times \mathbf{B} = 0. \tag{10.84}$$

With $\mathbf{J} \times \mathbf{B} \perp \mathbf{B}$, this relation implies that $\mathbf{E} \perp \mathbf{B}$ at all times, in the present limit. We can also obtain (10.84) directly from (10.6) by multiplying by $-en$. We ignore charge separations, and assume that the only source of electric fields is time-varying magnetic fields as given by Maxwell's equation

$$\nabla \times \mathbf{E} = -\frac{\partial}{\partial t}\mathbf{B}. \tag{10.85}$$

Finally, we again ignore Maxwell's displacement current just as when we discussed the MHD equations, and find Ampere's law also for this case

$$\nabla \times \mathbf{B} = \mu_0 \mathbf{J}. \tag{10.86}$$

We can now eliminate the current from (10.84). Note that the electron mass does not appear explicitly in these basic equations, consistent with the assumption of negligible electron inertia in (10.84). By the divergence of (10.86) we find $\nabla \cdot \mathbf{J} = 0$. It is thus implicitly assumed that the electron motion is incompressible. We of course still have an electron continuity equation, but it is here redundant, and does not enter the basic set of equations. In case the ion density, n_0, is uniform and the initial electron density is n_0 also, then the basic equations assume that the density remains the same constant for all later times. We would like the analysis to be valid for inhomogeneous plasmas as well, and consistency with the basic model then implies the restriction that density gradients must be weak. Otherwise, the fluctuating electron velocity would build up nontrivial charges at the "slopes" of the plasma density. The information in the electron continuity equation can then no longer be ignored.

It is an easy matter to combine these simple electron MHD equations into one. It is, however, an advantage to keep them as two separate equations in case they need modifications, in terms of resistivity, finite electron pressure, etc. If we choose to reduce the two equation into one, the result will be simple only for constant plasma density n.

Basic equations of simple electron MHD

$$\nabla \times \mathbf{E} = -\frac{\partial}{\partial t}\mathbf{B} \qquad \text{and} \qquad -en\mathbf{E} + \frac{1}{\mu_0}(\nabla \times \mathbf{B}) \times \mathbf{B} = 0$$

10.7.1 Whistlers Ⓚ

A basic dynamical waveform found in electron MHD is the whistler wave (Stenzel 1999). Historically, these waves have been important, since they were detected in the beginning of 20th century, in spite of the rather primitive radio equipment available at that time. In order to find the dispersion relation for these waves, we assume that the ion density, n_0, is uniform and that the initial, unperturbed magnetic field, $\mathbf{B}_0$, is homogeneous. We linearize the basic equations and obtain with little effort the relation for the fluctuating magnetic field

$$\nabla \times \left[(\nabla \times \widetilde{\mathbf{B}}) \times \mathbf{B}_0 \right] = -e\mu_0 n_0 \frac{\partial}{\partial t} \widetilde{\mathbf{B}}. \tag{10.87}$$

With simple manipulations involving the general relation $\nabla \times (\mathbf{X} \times \mathbf{Y}) = \mathbf{X}(\nabla \cdot \mathbf{Y}) - \mathbf{Y}(\nabla \cdot \mathbf{X}) + (\mathbf{Y} \cdot \nabla)\mathbf{X} - (\mathbf{X} \cdot \nabla)\mathbf{Y}$ (see, e.g., Appendix D.7), we find

$$\mathbf{B}_0 \cdot \nabla(\nabla \times \widetilde{\mathbf{B}}) = -e\mu_0 n_0 \frac{\partial}{\partial t} \widetilde{\mathbf{B}}. \tag{10.88}$$

After a Fourier transform with respect to the temporal as well as the spatial variables we have

$$B_0 k_{\|}(\mathbf{k} \times \mathbf{B}_1) = -ie\mu_0 n_0 \omega \mathbf{B}_1. \tag{10.89}$$

Note that $\mathbf{B}_1 \cdot \mathbf{k} = 0$ as it should be. Note also that $\mathbf{k}$ need not be in the direction of $\mathbf{B}_0$. We choose $\mathbf{k}$ to be a real vector.

Since $\mathbf{k} \times \mathbf{B}_1 \perp \mathbf{B}_1$, the relation (10.89) seems to state that the magnetic field vector $\mathbf{B}_1$ is perpendicular to itself! First we might expect that this can only be so for $B_1 = 0$, but on second thought we recall that these Fourier transformed fields are complex, see also Section 3.1.5. According to (10.89), it is not possible to choose $\mathbf{B}_1$ to be a real vector, otherwise we would have the left side real, while the right-hand side was imaginary. With this hindsight we insert the general expression for a complex vector $\mathbf{B}_1 = \mathbf{a} + i\mathbf{b}$ in (10.89), with two real vectors **a** and **b**, and obtain

$$\zeta \mathbf{k} \times \mathbf{a} = \mathbf{b} \qquad \text{and} \qquad \zeta \mathbf{k} \times \mathbf{b} = -\mathbf{a} \tag{10.90}$$

where we introduced the abbreviation $\zeta \equiv B_0 k_\parallel/(e\mu_0 n_0 \omega)$. From (10.90) we find trivially $\mathbf{a}\cdot\mathbf{b} = 0$, i.e., $\mathbf{a} \perp \mathbf{b}$, assuming $\mathbf{k} \neq 0$. By taking the scalar product of (10.90) with **k** we find $\mathbf{k}\cdot\mathbf{a} = 0$ and $\mathbf{k}\cdot\mathbf{b} = 0$, giving $\mathbf{k} \perp \mathbf{a}$ and $\mathbf{k} \perp \mathbf{b}$. Similarly we have

$$\zeta(\mathbf{k} \times \mathbf{a})\cdot\mathbf{b} = b^2 \qquad \text{and} \qquad \zeta(\mathbf{k} \times \mathbf{b})\cdot\mathbf{a} = -a^2 \tag{10.91}$$

Since $(\mathbf{k}\times\mathbf{a})\cdot\mathbf{b} = -(\mathbf{k}\times\mathbf{b})\cdot\mathbf{a}$, the result (10.91) implies $a^2 = b^2$, or $a = \pm b$. Note that $|\mathbf{a}| = |\mathbf{b}|$, but the two vectors have different and perpendicular directions! With $\zeta > 0$ and $k > 0$, the three vectors **k**, **a** and **b** form the basis of a right-hand orthogonal coordinate system, as evident from (10.90). Consequently we have to use the + sign in this case, giving $a = b$.

To understand the physical implications of the signs appearing in the vector $\widetilde{\mathbf{B}} = \mathbf{a} + i\mathbf{b}$, we recall that in configuration space the magnetic field variation associated with the wave is given as $\Re\{\widetilde{\mathbf{B}}\exp(-i(\omega t - \mathbf{k}\cdot\mathbf{r}))\}$. Assume for simplicity that **a** is in the $\widehat{\mathbf{x}}$-direction. Then, according to the foregoing results, **b** is in the $\widehat{\mathbf{y}}$-direction and **k** along $\widehat{\mathbf{z}}$, giving $\Re\{a(\widehat{\mathbf{x}} + i\widehat{\mathbf{y}})\exp(-i(\omega t - \mathbf{k}\cdot\mathbf{r}))\} = a\cos(\omega t - kz)\widehat{\mathbf{x}} + a\sin(\omega t - kz)\widehat{\mathbf{y}}$ for the fluctuating magnetic field. The expression represents a circularly polarized wave. Take $\mathbf{k} \parallel \mathbf{B}_0$. In case the real part of $\widetilde{\mathbf{B}}$ is along the $\widehat{\mathbf{x}}$-axis at a certain time t_1 at, say, position z_1, then the magnetic field vector will be along the $\widehat{\mathbf{y}}$-direction at a later time $t_1 + \frac{1}{2}\pi/\omega$, at the same z-position. At an even later time $t_1 + \pi/\omega$, the magnetic field vector points in the $-\widehat{\mathbf{x}}$-direction. We have a *right-hand rotation*, i.e., seeing toward the wave-vector from a fixed spatial position we will experience the fluctuating magnetic field vector rotating *against* the clock. If we take t constant, and let z vary, we find that for $\widetilde{\mathbf{B}} \parallel \widehat{\mathbf{x}}$ at position, say, z_1, then $-\widetilde{\mathbf{B}} \parallel \widehat{\mathbf{y}}$ at $z_1 + \frac{1}{2}\pi/k$. A wave with *left-hand rotation* would have the representation $\Re\{a(\widehat{\mathbf{x}} - i\widehat{\mathbf{y}})\exp(-i(\omega t - \mathbf{k}\cdot\mathbf{r}))\}$, but this is not a solution to the electron MHD equations. It is thus not possible here to obtain a linearly polarized wave by a superposition of left and right circularly polarized waves.

Using (10.89) we take the cross product with **k** on both sides of the equations, and find, for instance, $\zeta^2\mathbf{k}\times(\mathbf{k}\times\mathbf{a}) = -\mathbf{a}$. With $\mathbf{k}\times(\mathbf{k}\times\mathbf{a}) = -k^2\mathbf{a}$ we find $\zeta^2 k^2 \mathbf{a} = \mathbf{a}$, and obtain the whistler wave dispersion relation for the electron MHD limit as

$$\omega^2 = \left(\frac{\mathbf{B}_0\cdot\mathbf{k}}{e\mu_0 n_0}k\right)^2 \equiv \left(\frac{B_0}{e\mu_0 n_0}k_\parallel k\right)^2 \equiv \left(\frac{B_0}{e\mu_0 n_0}k^2\cos\Theta\right)^2, \tag{10.92}$$

with Θ being the angle between **k** and **B**. The dispersion relation (10.92) is illustrated in Fig. 10.14 using a convenient normalization of the variables obtained by introducing the electron mass both in the plasma frequency and cyclotron frequency.

For whistlers we have $\widetilde{\mathbf{J}} \equiv -en_0\mathbf{u}_e \perp \mathbf{k}$ from (10.86), i.e., whistlers are *transverse* or *shear* waves, just like Alfvén waves. Consistency of the result (10.92) requires that $\Omega_{ci} \ll \omega \ll \omega_{ce}$.

With the right-hand rotating circularly polarized waves in the present case, where we took $B_0 > 0$, the rotation of the fluctuating magnetic field vector is in the same direction as the electron gyro motion (this is a good way to remember the sign). If we change the direction of $k_\parallel$, nothing else is changed since (10.92) is quadratic in k. The electric field at a fixed position rotates in the right-hand sense, irrespective of the direction of whistler wave propagation with respect to the ambient magnetic field vector. If we change the direction of $\mathbf{B}_0$, the rotation of the fluctuating magnetic field vector is reversed, but so is the direction of the electron gyro motion. The electron MHD equations do not recognize any left-hand polarized waves, i.e., solutions with field vectors rotating in the direction opposite to the electron gyration. Left-hand polarized waves, for $\omega \ll \omega_{ce}$, can be found when the ion motion is also included (Yeh & Liu 1972).

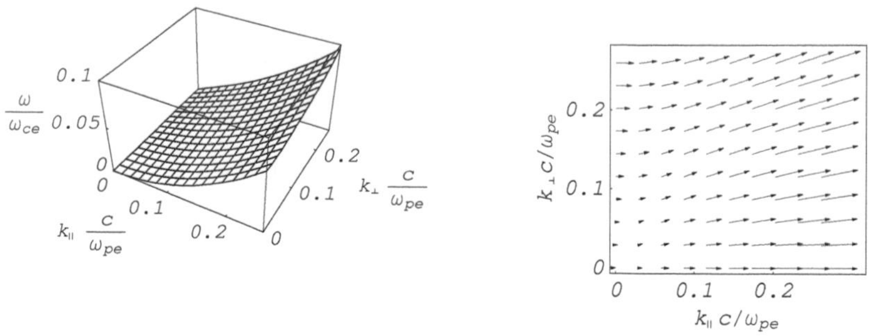

FIGURE 10.14: *Dispersion relation (left) for whistler waves as described by electron magnetohydrodynamics. The simple polar diagrams in Fig. 10.13 are of no use when the phase velocity depends on frequency as in the present case. The figure to the right shows the variation of the group velocity vectors with varying wave-numbers. The group velocity has a large* $\mathbf{B}_0$*-parallel component even for wavevectors directed at a large angle to* $\mathbf{B}_0$.

For waves propagating with the wave-vector parallel to the magnetic field, we find $\omega \sim k^2$. If you listen to these waves using a radio receiver (and that was the way they were discovered originally) you hear a characteristic whistling sound, with high frequencies coming first. When no single characteristic phase velocity can be assigned, we can no longer use simple polar phase velocity diagrams like those shown in Fig. 10.13. Whistlers are also observed in metals and semiconductors (in which case the ions can *definitely* be assumed to be at rest on the relevant time scales). Traditionally, the waves are termed *helicons* in this case, but it is basically the same thing!

Seemingly, whistlers in the lower parts of the Earth's ionosphere are often excited by lightning (recall that lightning frequently strikes *upwards* from clouds to the lower parts of the ionosphere). The excitation therefore appears almost as a δ-function perturbation, which contains all frequencies. We will therefore observe a wide frequency band signal. Since the high frequencies have the largest propagation velocity they arrive first, low frequencies later. The term "whistler" originates from this observation. Lightning is most frequent in the equatorial regions, and naturally occurring whistlers are relatively seldom in the polar ionospheric regions.

The electron whistler branch continues to $\omega \to 0$. A general theory takes into account also the ion motion for frequencies close to or below the ion cyclotron frequency, and in the magnetospheric plasma also the ion composition is important (Yeh & Liu 1972).

The interest in whistler waves had a "revival" when it was discovered that discharges based on these waves were effective in production of dense plasmas (Boswell 1984, Chen & Boswell 1997).

Ⓢ **Exercise:** Include finite resistivity in the electron MHD equations by including collision between the electrons and the immobile ion background, and demonstrate how the whistler waves become damped (Kingsep et al. 1990). Ⓢ

11

Plasma as a Mixture of Charged Gases

The summary of MHD theory presented in Chapter 10 was heuristic and based on some common sense arguments. For some strange reason, such an approach is at times being discredited, so we shall here attempt a more rigorous derivation (Lighthill 1960, van Kampen & Felderhof 1967, Boyd & Sanderson 1969). What we *should* take into account is the nature of the high electrical conductivity of a plasma, namely, the possibility for the electrons to move freely, or almost freely, through an ion background, which can have dynamics of its own. Complex plasma states with many different types of ions, negative ions, partial ionization, etc., can be seen as generalizations of this case, so to be specific we consider in the following a mixture of an electron and an ion gas, where the ion charge is of the same magnitude as the elementary electron charge but with opposite sign. The obvious example is a fully ionized hydrogen gas.

11.1 Multi-component plasmas

The multi-component plasma model has to follow continuity equations, one for each component. We ignore sources and sinks, i.e., ionization of a neutral background and recombination. We then have, using subscript e and i for electron and ion densities

$$\frac{\partial}{\partial t} n_e + \nabla \cdot (n_e \mathbf{u}_e) = 0 \tag{11.1}$$

and

$$\frac{\partial}{\partial t} n_i + \nabla \cdot (n_i \mathbf{u}_i) = 0 . \tag{11.2}$$

The system also follows Newton's second law, here in the form of momentum equations

$$m n_e \left(\frac{\partial}{\partial t} \mathbf{u}_e + \mathbf{u}_e \cdot \nabla \mathbf{u}_e \right) = -\nabla p_e - e n_e (\mathbf{E} + \mathbf{u}_e \times \mathbf{B}) + \mathbf{P} , \tag{11.3}$$

where we introduced a term $\mathbf{P}$ which accounts for a friction, or momentum transfer *from* the electron gas *to* the ion gas by collisions. Similarly we have an equation for the ions, where we now take into account that momentum *loss* from electrons is momentum *gain* for the ions, and vice versa.

$$M n_i \left(\frac{\partial}{\partial t} \mathbf{u}_i + \mathbf{u}_i \cdot \nabla \mathbf{u}_i \right) = -\nabla p_i + e n_i (\mathbf{E} + \mathbf{u}_i \times \mathbf{B}) - \mathbf{P} . \tag{11.4}$$

It is implicitly understood here that all varying quantities depend on the spatial coordinate as well as time, in general, i.e., $\mathbf{u}_{e,i} = \mathbf{u}_{e,i}(\mathbf{r}, t)$, etc.

We assume that the momentum exchange term can be written as

$$\mathbf{P} = \vartheta n_e n_i (\mathbf{u}_i - \mathbf{u}_e) . \tag{11.5}$$

The constant of proportionality ϑ is positive, but otherwise unspecified for the moment. We proposed the expression (11.5) without demonstrating any detailed insight into the collisional interactions mediating the momentum exchange. The justification of (11.5) is that it is the simplest expression obeying the requirement that **P** must be a vector in the direction of the relative velocity, and that it vanishes where the two velocities are equal, and vanishes also when either of the particle densities vanishes. The momentum exchange is mediated via electron-ion collisions with an average frequency $\nu_{e,i}$. The momentum transfer per collision is of the order of $m|\mathbf{u}_i - \mathbf{u}_e|$. The number of collisions per sec per m^{-3} is $\nu_{e,i}n_e$, with $\nu_{e,i}$ being the number of electron collisions per sec. Hence, $P \sim \nu_{e,i}n_e m|\mathbf{u}_i - \mathbf{u}_e|$, giving $\vartheta \sim \nu_{e,i}m/n_i$. Note that ϑ has a dimension $mass \times length^3 \times time^{-1}$. We recall here, from Section 8.8, the plasma conductivity $\sigma \approx e^2 n/(m\nu_{e,i})$, implying $\sigma \sim e^2/\vartheta$ or $\xi \sim \vartheta/e^2$.

We have not yet introduced any equations of state for p_e and p_i. For the most general case, this will involve a dynamic equation for the temperatures as given by, for instance, Braginskiı́ (1965). The relation (10.19) serves as a particularly simple example. Even though the two components, electrons and ions, are in thermal contact as mediated by the momentum exchange term **P** in (11.3), the energy inputs by dissipation can be different and we will have $T_e = T_i$ only in thermal equilibrium conditions. As a simple ad hoc model we can here complete the set of equations with an equation of state connecting pressure and density

$$p_e = f_e(n_e), \qquad p_i = f_i(n_i), \tag{11.6}$$

where details will not be discussed further here.

The equations (11.1)–(11.5) are quite general, and have at times been used directly, for instance, for numerical simulations. They show, however, little if any relation to the foregoing MHD equations in Chapter 10, and are therefore of little help in estimating the credibility of the standard MHD equations. We therefore introduce some quantities for describing bulk plasma properties. First we have the plasma mass density

$$\rho \equiv mn_e + Mn_i\,,$$

and average momentum density $n_e m\mathbf{u}_e + n_i M\mathbf{u}_i \equiv (n_e m + n_i M)\mathbf{U}$, giving the average velocity

$$\mathbf{U} \equiv \frac{n_e m\mathbf{u}_e + n_i M\mathbf{u}_i}{n_e m + n_i M}\,,$$

and obtain the bulk plasma continuity equation (10.1) reproduced here

$$\frac{\partial}{\partial t}\rho + \nabla\cdot(\mathbf{U}\rho) = 0\,.$$

By the self-consistent bulk plasma charge density $\zeta \equiv e(n_i - n_e)$ and current density $\mathbf{J} \equiv e(n_i\mathbf{u}_i - n_e\mathbf{u}_e)$ we find the continuity equation for charge

$$\frac{\partial}{\partial t}\zeta + \nabla\cdot\mathbf{J} = 0\,.$$

Both the equations of continuity for plasma matter density and charge density are exact, as far as creation and annihilation of charges are ignored. It turns out that the remaining dynamic equations can only be given in an approximate form. The combination of the momentum equations (11.3) and (11.4) is less straightforward because of the nonlinear terms that appear. Anticipating that the final equations will only be used for minor deviations from a static equilibrium, some small terms can be ignored. Thus we neglect quadratic terms in the velocities, and obtain, after addition of (11.3) and (11.4),

$$\rho\frac{\partial}{\partial t}\mathbf{U} = -\nabla(p_e + p_i) + \zeta\mathbf{E} + \mathbf{J}\times\mathbf{B}\,, \tag{11.7}$$

noting the cancellation of the terms containing **P** for the momentum exchange. It is tempting here to introduce a net pressure $p \equiv p_e + p_i$, but it turns out that we later on need the electron pressure explicitly, so it is best to retain both p_e and p_i separately. Nonetheless, we note that what we interpreted as a temperature T in the brute force approach to derivation of the MHD equations in Chapter 10 in reality was the *sum* of ion and electron temperatures, as long as we can assume quasi-neutrality and let $n_e \approx n_i = n$, so that $p_e + p_i = \kappa T_e n_e + \kappa T_i n_i \approx \kappa (T_e + T_i) n$.

By subtraction of (11.3) from (11.4), and again ignoring quadratic velocity terms, we also find

$$\frac{\partial}{\partial t}\mathbf{J} = \nabla\left(\frac{e}{m}p_e - \frac{e}{M}p_i\right) + \left(\frac{n_e e^2}{m} + \frac{n_i e^2}{M}\right)\mathbf{E} + \left(\frac{n_e e^2}{m}\mathbf{u}_e + \frac{n_i e^2}{M}\mathbf{u}_i\right)\times\mathbf{B} - \left(\frac{e}{m} + \frac{e}{M}\right)\mathbf{P}. \tag{11.8}$$

Next, we make use of the fact that the electron mass is small in comparison with the ion mass. In the right-hand side of (11.8) we thus approximate

$$n_i \approx \rho/M \qquad \text{and} \qquad n_e \approx \rho/M - \zeta/e\,.$$

By this we associate the plasma mass density with the ion mass density (which is very reasonable, after all) and let the electron density be the difference between the charge density divided by e and the ion density, which is a logical consequence of the first approximation. We also have $\mathbf{U} \approx \mathbf{u}_e n_e m/(n_i M) + \mathbf{u}_i$, giving, after some algebra, where we ignore terms of relative order $O(m/M)$ in the expression for the ion velocity

$$\mathbf{u}_e \approx \mathbf{U} n_i/n_e - \mathbf{J}/(n_e e) \qquad \text{and} \qquad \mathbf{u}_i \approx \mathbf{U} + \mathbf{J}m/(\rho e)\,.$$

The first one of these expressions means that the electron current contribution $e n_e \mathbf{u}_e$ is taken as the difference between the bulk plasma flow $e n_i \mathbf{U}$ and the net current $\mathbf{J}$. Using the expressions given before, the sum $(n_e m \mathbf{u}_e + n_i M \mathbf{u}_i)/(n_i M)$ will reproduce $\mathbf{U}$ to the given accuracy. We note that although the electron mass m is small, we anticipate that the electron velocity $\mathbf{u}_e$ can be large, so we cannot off-hand ignore $m n_e \mathbf{u}_e$ as compared to $M n_i \mathbf{u}_i$.

Although the electron and ion pressures need not be identical, they are assumed to be of the same order of magnitude. With these approximations, we then obtain

$$\frac{\partial}{\partial t}\mathbf{J} = \nabla\frac{e}{m}p_e + \frac{n_e e^2}{m}\mathbf{E} + \left(\frac{\rho e^2}{mM}\mathbf{U} - \frac{e}{m}\mathbf{J}\right)\times\mathbf{B} - \frac{e}{m}\mathbf{P}. \tag{11.9}$$

As a final approximation in (11.7) and (11.9) we assume quasi-neutrality by omitting terms containing the charge density ζ at places where it is multiplied by a quantity which vanishes in equilibrium. The term $\zeta\mathbf{E}$ will then vanish from (11.7). The coefficient n_e for $\mathbf{E}$ in (11.9) is replaced with $n_e \approx \rho/M = n_i$, because the electron and ion densities must be close for ζ to be small. By this assumption, we state, as readily seen, that current densities are mainly due to velocity differences between the electrons and ions; we *could* in principle have the two components moving with the same velocity, but with different densities, and thus creating a current, but this we assume to be a process of minor importance. Electron and ion densities will usually be approximately the same as long as we are only interested in length scales much larger than λ_D, the Debye length. In the same limit, the term with the momentum exchange $\mathbf{P}$ is reduced to

$$-\frac{e}{m}\mathbf{P} = -\vartheta n_e n_i \frac{e}{m}\frac{\mathbf{J}}{n_e e} = -\vartheta\frac{\rho}{mM}\mathbf{J},$$

and we obtain a modified Ohm's law in the form

$$\frac{\partial}{\partial t}\mathbf{J} = \frac{e}{m}\nabla p_e + \frac{\rho e^2}{mM}\left(\mathbf{E} - \frac{1}{\sigma}\mathbf{J} + \mathbf{U}\times\mathbf{B}\right) - \frac{e}{m}\mathbf{J}\times\mathbf{B}, \tag{11.10}$$

where we introduced the plasma conductivity as $\sigma \equiv e^2/\vartheta$. The last term on the right-hand side of (11.10) is the *Hall term*. The most important new term in Ohm's equation is $\partial\mathbf{J}/\partial t$, which makes the equation non-local in time. The pressure term makes the equation non-local in space.

We find that the MHD continuity equation is trivially recovered here, by the proper interpretation of the bulk plasma flow velocity. To the given accuracy, also the MHD momentum equation is recovered, while the simple Ohm's law postulated in Section 10.1.3 differs significantly from the present result (11.10), although elements can be recognized right away. For certain parameter ranges, the simple expression (10.5) can be recovered, as argued in the following. For slow and large scale phenomena, we might argue right away that the left side of (11.10) and the pressure gradient terms might be neglected, but the $\mathbf{J}\times\mathbf{B}$-term requires particular attention.

In order to make a relative order of magnitude estimate of the all terms in (11.10) we first note that

$$\frac{E}{B} \sim \omega\mathcal{L},$$

obtained from (5.58), as already noted in Section 10.1, with $\mathcal{L}$ being a characteristic length scale for the electromagnetic fluctuations, and ω a characteristic frequency. Since magneto-hydrodynamics implies a strong interaction between the flow and the electromagnetic fields, we are primarily interested in cases where

$$U \sim \omega\mathcal{L},$$

where U is a characteristic plasma flow velocity. We now revert to the basic MHD assumption stating

$$\varepsilon_0\mu_0\left|\frac{\partial\mathbf{E}}{\partial t}\right| \bigg/ |\nabla\times\mathbf{B}| \sim \frac{\omega}{c^2}\frac{E}{B}\mathcal{L} \sim \left(\frac{\omega\mathcal{L}}{c}\right)^2 \ll 1,$$

giving (10.4), i.e., $\nabla\times\mathbf{B} = \mu_0\mathbf{J}$. As an estimate we then find

$$J \sim \frac{B}{\mu_0\mathcal{L}}.$$

We now divide all terms in (11.10) by $E\rho e^2/(mM)$ and find for the respective orders of magnitude

$$\left(\frac{c}{\omega\mathcal{L}}\right)^2\left(\frac{\omega}{\omega_{pe}}\right)^2 = \left(\frac{C_s}{\omega\mathcal{L}}\right)^2\frac{\omega}{\Omega_{ci}} : 1 : \frac{\varepsilon_0\omega}{\sigma}\left(\frac{c}{\omega\mathcal{L}}\right)^2 : 1 : \left(\frac{c}{\omega\mathcal{L}}\right)^2\frac{\Omega_{ci}\,\omega}{\omega_{pe}^2}\frac{M}{m}. \tag{11.11}$$

Again with $U \sim \omega\mathcal{L}$ we find that the $\partial\mathbf{J}/\partial t$ term can be neglected in the case $(c/U)^2 \ll (\omega_{pe}/\omega)^2$. Similarly, the Hall term is ignored if $(c/U)^2 \ll (m/M)(\omega_{pe}^2/\omega\Omega_{ci})$, which can be a rather strict condition. The pressure term is negligible in the case $(C_s/U)^2 \ll \Omega_{ci}/\omega$. With these inequalities satisfied, we recover (10.5). If, in addition, we also have $(c/U)^2 \ll \sigma/(\omega\varepsilon_0)$, we can assume ideal MHD by ignoring the plasma resistivity as in (10.6). Note that in all cases we find that the inequalities can be fulfilled for sufficiently low frequencies, $\omega \to 0$, all other parameters constant, which implies $\mathcal{L} \to \infty$ in such a way that $c/(\omega\mathcal{L}) = \text{const}$. The inequalities stated before can be seen as determining the range of validity for the magneto-hydrodynamic equations; they apply for very low frequencies and very large scale lengths.

To find a more general scaling, we note that in reality (11.10) is a *vector* relation, so that each component might have a scaling of its own. We might have (11.11) applying for the $\mathbf{B}$-perpendicular components, while a different relation applies for the **B**-parallel part, here along the local z-axis. In that case we can retain

$$m\frac{\partial}{\partial t}J_{\|} = e\frac{\partial}{\partial z}p_e + \frac{\rho e^2}{M}\left(E_{\|} - \frac{1}{\sigma}J_{\|}\right) \tag{11.12}$$

explicitly. This relation allows a generalization of the ideal MHD equations, where we now also have electric field components parallel to **B**. In reality (11.12) is nothing but an equivalent of the **B**-parallel component of the electron momentum equation. The reader might demonstrate this.

Ⓢ **Exercise:** The purpose of this exercise is to give some insight into the properties of pinches in a two fluid model. Set up the fluid equations for a two component plasma. Consider cylindrical symmetry with no z-variation and find a steady state quasi-neutral and isothermal solution (Bennett 1934) in a frame of reference where the ions are at rest and the current is due to electron motion along the symmetry axis. The radial velocities are vanishing. The equations can be made elegant by introducing a vector potential for the magnetic field, but it has no importance for the steady state solution. Give the direction and magnitude of the magnetic field. Ⓢ

11.1.1 Plasma diamagnetism

We found that the MHD equations imply that under steady state conditions a gradient in plasma pressure gives rise to a reduction in magnetic field; see Section 10.3.3. The current that gives rise to this change in magnetic field is then obtained by $\nabla p = \mathbf{J}\times\mathbf{B}$, giving $\mathbf{J} = \mathbf{B}\times\nabla p/B^2$.With the multi-fluid model together with the results for single particle motions in Chapter 7, we are in a position to explain this plasma diamagnetism from first principles.

The motion of charged particles in a given magnetic field is such as to produce a magnetic perturbation in the opposite direction to the imposed field; this is at times called *Lenz's law*. A plasma is a diamagnetic medium due to this effect. Note that the magnetic moments for electrons and ions having the same energy, in a given magnetic field, are the same: the Larmor radius of the electrons is small, but their velocity is large, and vice versa.

- **Density gradient:** In the present multi-component model of a plasma we can give a simple interpretation of the plasma diamagnetism discussed in Section 10.3.3. In Fig. 11.1 we illustrate a plasma embedded in a homogeneous magnetic field perpendicular to the plane of the paper. In Fig. 11.1a the density is increasing downward, while the temperature is constant. Assume that an observer counts the number of charges passing through a small area element $A-A'$. With the geometry implied in Fig. 11.1a, more positive particles pass through $A-A'$ from left to right than from right to left. Similarly, more negative particles are passing from right to left than vice versa. Consequently, there is a net current along the equi-density contour, as indicated.

- **Temperature gradient:** In Fig. 11.1b the density is constant, while the temperature is increasing downward. We note that in Fig. 11.1b we find a current in the same direction as in Fig. 11.1a, in this case due to the temperature gradient. Each of the charged particles crosses the line $A-A'$ with a time interval of $2\pi/\omega_c$. In Fig. 11.1a the net difference in the number of charges crossing $A-A'$ is due to the difference in particle density between the top and bottom of Fig. 11.1a. In Fig. 11.1b, with constant plasma density, the particles on larger circles have larger velocities as compared to the smaller ones and hence contribute more to the current through $A-A'$. Particles pass the area element once per gyro-period, but there are more particles distributed along the large circles as compared to the smaller ones on Fig. 11.1b.

These currents are the *diamagnetic currents*. They are just as good as any other currents, but in some sense peculiar because they are associated with the motion of particles which are not being displaced *on average*, i.e., the gyro-centers are assumed to be fixed in both Fig. 11.1a) and 11.1b).

In particular in the early days of plasma physics it was seen as a paradox that we can have a net current without average motion of particles. There is another related paradox for the case of a homogeneous plasma (i.e., one with constant density and temperature) confined by an *inhomogeneous* magnetic field, as briefly mentioned in Section 7.1.6; see in particular Fig. 7.3. In this case all particles move on average, but there need not be any net current density! If we, in this case, take a small area element perpendicular to plane of the paper in Fig. 7.3 and parallel with $\nabla|B|$, there will be an equal number of particles passing through it in both directions: half of them will have their gyro-radius above and half below the area element, and the velocities of the particles are entirely given by the temperatures, which are assumed to be spatially constant and the magnetic field at the position of the area element. A net current for these conditions with $\nabla|\mathbf{B}| \neq 0$ requires inhomogeneities in plasma density and/or temperature.

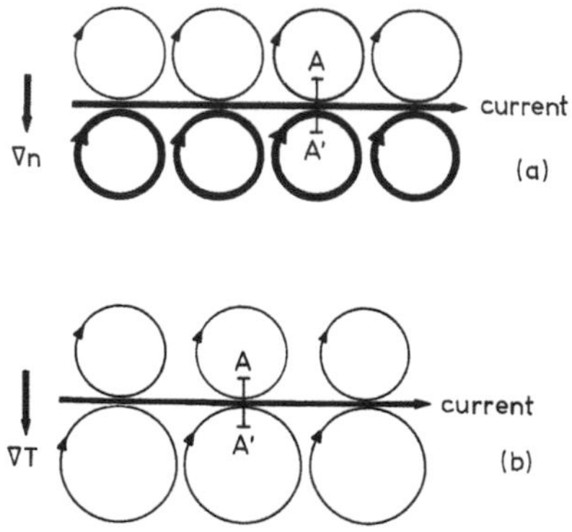

FIGURE 11.1: *Schematic illustration of the origin of the diamagnetic current, showing the gyration orbit of positively charged particles in a magnetic field perpendicular to the plane of the paper. In a) we have a gradient in density, but constant temperature, while b) illustrates the case with constant plasma density, but with a temperature gradient. Even though the individual particles are not displaced* on average, *an observer will find a net current through a cut* $A - A'$. *This is the diamagnetic current.*

We note that diamagnetic current is necessary in order to fulfill Newton's second law for stationary conditions with inhomogeneous magnetized plasmas. In that case we have a pressure gradient, which gives rise to a net force. This force has to be balanced to obtain a stationary condition. This is achieved by the $\mathbf{J} \times \mathbf{B}$-force, with $\mathbf{J}$ being the diamagnetic current. We have $\nabla(n\kappa(T_e + T_i)) = \mathbf{J} \times \mathbf{B}$. With $n_e = n_i \equiv n$ for local charge neutrality, we have $\mathbf{J} = en(\mathbf{u}_i - \mathbf{u}_e)$, and find for homogeneous temperatures $en(\mathbf{u}_i - \mathbf{u}_e)B^2 = -\kappa(T_e + T_i)\nabla n \times \mathbf{B}$. Evidently, we can have for each plasma species, here denoted by ℓ with $\ell = i$ or $\ell = e$, an expression $\mathbf{J}_\ell = -(\kappa T_\ell / B^2)\nabla n \times \mathbf{B}$. The electron and ion diamagnetic currents are adding up, $\mathbf{J} = \mathbf{J}_e + \mathbf{J}_i$. We often have $T_e \gg T_i$, and in such cases only the electron diamagnetic drift is important. Including the plasma temperature gradients, we have the diamagnetic drift velocity for each species

$$\mathbf{u}_{D\ell} = -\frac{\kappa T_\ell}{q_\ell B^2}\left(\frac{\nabla n}{n} + \frac{\nabla T_\ell}{T_\ell}\right) \times \mathbf{B}\,.$$

If we take the partial pressure for each species to be $p_\ell = n\kappa T_\ell$, we have $\mathbf{u}_{D\ell} = -\nabla p_\ell \times \mathbf{B}/(nq_\ell B^2)$.

Note that the diamagnetic velocity $\mathbf{u}_\perp \equiv -\kappa T(\nabla \ln n) \times \mathbf{B}/qB^2$ is divergence free *only* if $\mathbf{B} =$ const. For low-β plasmas, we can often ignore the variations in magnetic field intensity that might be induced by currents in the plasma, and if the field is homogeneous, we might assume it to be constant as well, also in the presence of the plasma.

- **Example:** There is a well known theorem in statistical mechanics, the Bohr-van Leeuwen theorem (Pathria 1998), stating that a classical system of charged particles moving in a magnetic

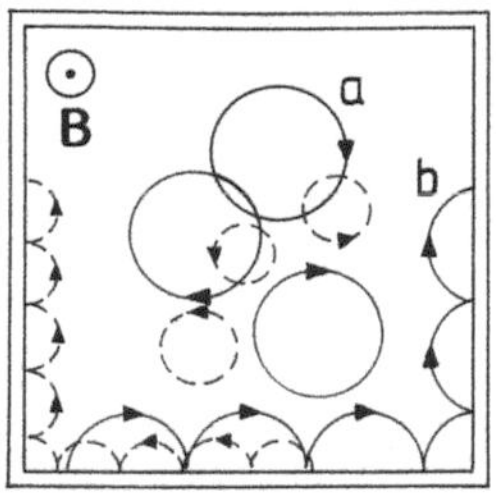

FIGURE 11.2: *Schematic illustration of particles moving in a magnetized box. Two types of particle orbits are distinguished, those which bounce off the walls, and those performing the closed cyclotron motion.*

field cannot be diamagnetic. This observation seems to be at variance with the claimed diamagnetism of a magnetized plasma. It is, however, important to note that such a plasma is implicitly assumed to be completely isolated from its surroundings. If we let the plasma completely fill a box with the walls at some fixed temperature, we can distinguish two types of particle orbits, as illustrated in Fig. 11.2. One, type a), are closed, while those of type b) are bouncing off the walls. It can be demonstrated that the current contribution from the latter type is exactly canceling the diamagnetic currents from type a) when the plasma density is uniform. If it is *not* uniform, we *can* have diamagnetism, but then the system is no longer in thermal equilibrium!

- **Exercise:** Satellite magnetometers sometimes pass through regions of plasma, such as the terrestrial ionosphere, where the plasma is sufficiently dense to reduce the original ambient magnetic field as compared to the intensity it would have *without* the plasma. This reduction of field inside the plasma region is due to the diamagnetic effect of the charged particles in their helical trajectory around the magnetic field lines. The container of the magnetometer will exclude the plasma, and a simple-minded approach, treating the ionosphere in the same way as for a conventional diamagnetic fluid, predicts that the field seen by the magnetometer will be somewhat larger than the (reduced) field in the plasma. Is this argument correct? If not, why not?

11.2 Quasi-neutrality

As an approximation it can at times be assumed (as we have seen) that the density fluctuations are quasi-neutral, i.e., make what is often called "the plasma approximation" (Chen 2016). This means that in the analysis we let $n_e = n_i$. Now, this cannot generally be true since it renders the right-hand side of Poisson's equation identically zero. We have seen that on scales comparable to or smaller than the Debye length we *can* have large deviations from charge neutrality. If, however, we restrict the analysis to scale lengths much larger than the Debye length, it is often safe to assume that no significant differences between n_e and n_i can occur; if by chance a large plasma volume lost even a small fraction of, say, the electrons, very large electric fields would be set up, which immediately would correct the imbalance. If the length scales are large, as assumed, this implies then that the gradients of the electric field components are small, even if the fields as such are nonzero. By the plasma approximation we can argue that Poisson's equation becomes redundant in the sense that, to the given approximation, it simply states $0 \approx 0$.

The arguments outlined here can be substantiated by considering Poisson's equation expressed in terms of ion and electron densities

$$\nabla\cdot\mathbf{E} = -\nabla^2\phi = \frac{e}{\varepsilon_0}(n_i - n_e)\,, \tag{11.13}$$

to be rewritten in normalized form. We introduce the normalized densities $\eta_e \equiv n_e/n_0$ and $\eta_i \equiv n_i/n_0$, assuming that we have a suitable normalizing reference density n_0, which is the same for electrons as well as ions. The electrostatic potential is normalized as $\psi \equiv e\phi/\kappa T_e$, whereby a local thermal equilibrium for the electrons is implicitly assumed. We hereby restrict the arguments to apply for low frequency phenomena; only then can we hope for the assumption of local thermal equilibrium for the electrons to be accurate. Finally, it is assumed that the perturbations have a well defined scale length, $\mathcal{L}$. Writing Poisson's equation (11.13) in terms of new variables, with $\xi \equiv r/\mathcal{L}$, we readily find

$$\left(\frac{\lambda_D}{\mathcal{L}}\right)^2 \frac{\partial^2}{\partial\xi^2}\psi = \eta_i - \eta_e\,. \tag{11.14}$$

If we now assume that all the normalized variables ψ, η_i and η_e are of the same order of magnitude, while $\mathcal{L} \gg \lambda_D$, we argue that to lowest order accuracy we must have $\eta_i \approx \eta_e$ to have both sides of (11.14) equally small. (Note that by normalizing x by $\mathcal{L}$, we ensure that $\partial^2\psi/\partial\xi^2$ is also of order unity, when ψ is near unity.) This is the analytical basis for the quasi-neutral assumption. It allows us to consider the space-time varying bulk plasma density $n(\mathbf{r},t)$, rather than the individual electron and ion densities. This can in some cases be a great simplification. We reduce the number of unknowns by one when setting $n_e \approx n_i \equiv n$ and we reduce the number of equations by one when we omit Poisson's equation.

In the quasi-neutral limit we have $\nabla_\xi^2\psi$ being equal to a product of a very small and a very large quantity, $\eta_i - \eta_e$ and $(\mathcal{L}/\lambda_D)^2$, respectively. In that case we can let $\nabla_\xi^2\psi$ be unspecified by Poisson's equation, and to be determined by another equation. It is thus important to emphasize that the assumption of quasi neutrality does not impose any conditions on $\nabla^2\phi$.

If we take $n_e \approx n_i \equiv n$ in the continuity equations for electrons and ions, $\partial n_{e,i}/\partial t + \nabla\cdot(n_{e,i}\mathbf{u}_{e,i}) \approx 0$, and take their difference, the result is $\nabla\cdot n(\mathbf{u}_e - \mathbf{u}_i) \approx 0$, or in terms of current, $\nabla\cdot\mathbf{J} \approx 0$. The currents in the plasma are divergence free in this limit. Note that the quasi-neutral assumption is not based on any direct assumption of linearization; it can be applicable also for nonlinear or large amplitude conditions.

Ignoring Poisson's equation, we note, just as in the discussion of the magneto-hydrodynamic equations in Chapter 10, that we can no longer obtain the continuity equation for current from (5.57). For consistency with the quasi-neutral approximation, or the "plasma approximation," we have to omit Maxwell's displacement current from (5.57), and use the original form of Ampere's law. This assumption implies again that we assume $\omega\mathbf{E}_\perp/c^2$ negligible as compared to $\mu_0\mathbf{J}$. Note that the assumption of quasi-neutrality has no association with the waves being electromagnetic or electrostatic! The assumption can be argued also without explicitly introducing the electrostatic potential as in (11.14).

When measured on the relevant slow time scale, the cancellation of charges will be essentially instantaneous; for high frequency phenomena this assumption will no longer be applicable, and the quasi-neutral approximation cannot be used, for instance, for electron plasma waves, even when their wavelength is large compared to the Debye length. For Langmuir waves, $\omega\mathbf{E}_\perp/c^2$ is *not* negligible: on the contrary, we have the plasma currents and Maxwell's displacement current balancing each other, $\omega\mathbf{E}_\perp/c^2 \approx \mu_0\mathbf{J}$.

As a heuristic reminder of the plasma approximation we may use that Poisson's equation becomes redundant in the limit where we, formally, let $\varepsilon_0 \to 0$. To avoid infinite electric fields in that case, we must have that the local charge density vanishes to the same approximation, i.e., $n_e \approx n_i$. The ratio of two numbers both going to zero can be anything, and we can assume that Poisson's equation in this limit simply adjusts to be consistent with the electric fields found from other equations. If, by chance, we come over a general solution to a plasma problem, derived without making the plasma approximation from the outset, we can find the simplified quasi-neutral result by letting $\varepsilon_0 \to 0$ everywhere. This means that we let λ_{De} and λ_{Di} both go to zero, while at the same time ω_{pe} and Ω_{pi} both go to infinity; the plasma approximation is valid for large scales and small frequencies. Another way of seeing this can be found by recalling that $\lambda_D\omega_p = u_{The}$. The thermal velocity of either species is unaffected by the assumption of quasi-neutrality, so *if* we assume the Debye length to be negligible we *must* assume the corresponding plasma frequency to be very large, i.e., much larger than any frequency we want to include in the analysis. This latter argument holds for both electrons and ions.

Ⓢ **Exercise:** Consider an infinitely wide and 10 cm thick slab of immobile singly charged ions with density $n_0 = 10^{14}$ cm^{-3}. Assume for the sake of argument that the electron density is $n_e \neq n_0$. The two surfaces of the slab are at the same potential, and the potential difference between the surface and the center of the slab is $\phi = 5$ kV. Calculate n_e. Ⓢ

11.3 Collisional diffusion in two component, magnetized plasmas

We assume collisional diffusion to be slow process, which is characterized by time scales longer than the collision time $1/\nu$, where ν is the collision frequency as before. In that case we can assume $D\mathbf{u}/Dt \ll \nu\mathbf{u}$, and write the fluid momentum equations for electrons and ions in a quasi-stationary form. This approximation amounts to ignoring inertia terms. Due to the smallness of the electron mass the approximation is more easily justified for electrons than for ions. Note that by the present approximation, we do not *necessarily* assume steady state conditions, since the time derivative in the *continuity equations* can still be different from zero!

11.3.1 Diffusion in fully ionized plasmas Ⓚ

First we consider the case where the collisions are mediated by electron-ion interactions, and ignore neutrals. Ions are assumed to be singly charged. The quasi steady state momentum equations are

$$\begin{aligned} -\nabla p_e - en_e(\mathbf{E} + \mathbf{u}_e \times \mathbf{B}) + \mathbf{P} &= 0, \\ -\nabla p_i + en_i(\mathbf{E} + \mathbf{u}_i \times \mathbf{B}) - \mathbf{P} &= 0, \end{aligned} \tag{11.15}$$

with $\mathbf{P} = \vartheta n_e n_i(\mathbf{u}_i - \mathbf{u}_e) = \xi e^2 n_e n_i(\mathbf{u}_i - \mathbf{u}_e)$ in terms of the plasma resistivity $\xi \equiv 1/\sigma$; see also (11.5). The coefficient $\vartheta = \nu_{e,i} m/n_i$ measures the electron-ion collision frequency, $\nu_{e,i}$, as before. Recall that $\xi = m\nu_{e,i}/(e^2 n) = \nu_{e,i}/(\varepsilon_0\omega_{pe}^2)$. The relations (11.15) state that momentum lost by one component is gained by the other component. We could also write the continuity equations, but we will not need these yet. We allow formally for a density dependence of the resistivity, $\xi = \xi(n)$, which gives an implicit dependence of temporal and spatial variables. This means that we, at least in principle, have to take care when differentiating the last terms in (11.15) with respect to the spatial variable. For resistivity due to electron-ion collisions we know, however, that the density dependence is weak, and can be ignored to logarithmic accuracy (i.e., ignoring the Spitzer logarithm).

Assuming quasi-neutrality (see Section 11.2), we simplify the equations (11.15) to

$$\begin{aligned} -\nabla p_e - en(\mathbf{E} + \mathbf{u}_e \times \mathbf{B}) + \xi e^2 n^2(\mathbf{u}_i - \mathbf{u}_e) &= 0, \\ -\nabla p_i + en(\mathbf{E} + \mathbf{u}_i \times \mathbf{B}) - \xi e^2 n^2(\mathbf{u}_i - \mathbf{u}_e) &= 0. \end{aligned} \tag{11.16}$$

We will assume that the plasma density varies with the x-coordinate, $n = n(x,t)$, i.e., we have a density gradient perpendicular to the magnetic field which is in the z-direction. By adding the two equations in (11.16), and then taking the cross product with **B** we obtain

$$en(\mathbf{u}_i - \mathbf{u}_e) = -\frac{\nabla(p_e + p_i)\times\mathbf{B}}{B^2}. \tag{11.17}$$

By taking the cross product with **B** on both sides of (11.17) and noting $\mathbf{J} = en(\mathbf{u}_i - \mathbf{u}_e)$, we find the relation $\mathbf{J}\times\mathbf{B} = \nabla(p_e + p_i) \equiv \nabla p$, which is simply expressing the pressure force being balanced by the $\mathbf{J}\times\mathbf{B}$-force by the diamagnetic current under the present quasi-equilibrium conditions.

Referring to the discussion in Section 8.8.1, we expect no charge separations to build up during diffusion processes, so the electric fields vanish, $\mathbf{E} = 0$. A necessary requirement is then that the electron and ion velocities along the density gradient are equal, which is consistent with an assumption of quasi-neutrality. With these conditions fulfilled, we can set $u_{ix} = u_{ex} \equiv u_x$. A detailed calculation will give the same results, but we might as well use the little a priori knowledge we have! With isothermal conditions for the slow diffusion processes, we use $p_{e,i} = \kappa T_{e,i} n$, and find, after a little algebra, left to the reader as an exercise, that

$$u_{ey} = -\frac{\kappa T_e}{eB}\frac{1}{n}\frac{\partial n}{\partial x},$$

and

$$u_{iy} = \frac{\kappa T_i}{eB}\frac{1}{n}\frac{\partial n}{\partial x}.$$

The velocities u_{ey} and u_{iy} obtained here are the diamagnetic drifts, which exist irrespective of the collisions.

The equations (11.16) are simple: we have $\mathbf{E} = 0$, and also the x-component of the last term (containing the resistivity ξ) vanishes. Taking the y-component of any of the two equations in (11.16) we find

$$u_x = -\xi\frac{\kappa(T_e + T_i)}{B^2}\frac{\partial n}{\partial x}.$$

It is implicitly assumed that we are dealing with a low-β plasma, so that the diffusion currents do not modify the magnetic field appreciably.

Inserting the velocity components obtained before into the electron or the ion continuity equations, we find that the y-components of the velocities do not contribute, and we end up with the diffusion equation

$$\frac{\partial}{\partial t}n - \frac{\partial}{\partial x}D_\perp\frac{\partial}{\partial x}n = 0, \tag{11.18}$$

where $D_\perp \equiv \xi n\kappa(T_e + T_i)/B^2$ is proportional to the plasma resistivity to the assumed logarithmic accuracy. We note a rather favorable magnetic field dependence (that is, favorable in case we want to confine a slightly collisional plasma by a magnetic field). The expression for the Spitzer resistivity (8.30) states that ξ is only weakly density dependent (via $\log\Lambda$), and scales with temperature as $T^{-3/2}$, so by taking electron and ion temperatures comparable, $T_i \approx T_e \equiv T$, we have

$$D_\perp \sim \frac{n}{B^2\sqrt{\kappa T}}, \tag{11.19}$$

indicating that by increasing an externally imposed magnetic field we can obtain a significant reduction in the plasma diffusion coefficient. Note that the diffusion equation (11.18) was obtained without linearizing the basic equations (11.16).

By dimensional reasoning and a little common sense, we might estimate the diffusion coefficient due to electron-ion collisions by using the analogy to a random walk (MacDonald 1962, Pécseli 2000). We recall that a typical step length for an electron-ion pair per collision is the electron Larmor radius r_{Le} (see Section 8.8.1), with a time between collisions being $1/\nu_{ei}$. We have then the estimate $D_\perp \approx r_{Le}^2\nu_{ei}$. Now, $r_{Le}^2 = (mu_{The})^2/(eB)^2$ while we in Section 8.8 found $\xi = 1/(N_D\omega_{pe}\varepsilon_0)$ with the plasma parameter being $N_p = \omega_{pe}/\nu_{e,i}$. This gives $\nu_{ei} \approx \xi n e^2/m$, resulting in $D_\perp \approx \xi \kappa T_e n/B^2$, which is essentially the previous result.

Considering (11.18) we recall the form of the continuity equation (2.1) or (11.1) used several times. We see that the diffusion equation will have the form of a continuity equation if we introduce a *diffusion velocity* $-(D_\perp/n)\partial n/\partial x$. Introducing a length scale $L \equiv n/(\partial n/\partial x)$, we can write the diffusion velocity as the ratio $-D_\perp/L$. The diffusion flux is $-D_\perp \partial n/\partial x$. For steady state conditions we have a constant particle flux given by $-D_\perp \partial n/\partial x$ which we recognize as being a result from Fick's law.

- **Exercise:** In (11.15) we ignored the terms $\partial \mathbf{u}_{e,i}/\partial t$ as well as $\mathbf{u}_{e,i} \cdot \nabla \mathbf{u}_{e,i}$. Make a test on the consistency of omitting these terms. Will the assumption hold for all times, also if we take the initial condition to be a step function in density?

Ⓢ **Exercise:** Consider a Q-machine plasma with density $n \approx 10^{16}$ m^{-3}, temperatures $T_e \approx 2500$ K and $T_i \approx 1500$ K, and a magnetic field of 0.1 T, operated in single ended conditions. Obtain the basic plasma parameters, such as Debye length, etc. Determine in particular the diffusion coefficient $D_\perp$. Assume a density gradient at the plasma edge so that $n/(dn/dr) \approx 5$ mm. Determine the diffusion velocity. Consider cesium ions (take the mass to be 133 AMU). What is an average lifetime for an ion in the device? Discuss the effects of a finite length of the device. Ⓢ

- **Exercise:** Estimate the thermal conductivity (see Section 2.7) in a fully ionized collisional magnetized plasma as well as in an unmagnetized plasma.

Generally, we expect that for a magnetically confined warm plasma, the plasma density will decrease as we approach the walls of the device, seemingly implying $D \to 0$, according to (11.19), which might lead us naively to expect good confinement. Unfortunately, at the same time we also have decreasing temperatures at the plasma edges, and this will tend to *increase* D, and the diffusion coefficient remains finite everywhere. Nevertheless, it is important to keep in mind that in fully ionized plasmas, the diffusion coefficient will depend on the spatial variables, in general.

It might be instructive to discuss another view of collisional diffusion in fully ionized magnetized plasma, here in terms of an $\mathbf{f} \times \mathbf{B}_0/(qB_0^2)$-drift where $\mathbf{f}$ is some force density (see (7.8)), assuming here that the magnetic field is homogeneous. As argued before, we have a term $\mathbf{P}_{e,i} \equiv \vartheta n_e n_i(\mathbf{u}_i - \mathbf{u}_e) = \xi e^2 n_e n_i(\mathbf{u}_i - \mathbf{u}_e) \approx e\xi n\mathbf{J}$ in terms of the plasma resistivity $\xi \equiv 1/\sigma$ (see also (11.5)), where we here assumed quasi-neutrality, $n_e \approx n_i = n$. This terms enters (11.15) as a frictional force $\mathbf{f}$ on the electrons due to their collisions with ions. Similarly we have a frictional term $\mathbf{P}_{i,e} \approx -e\xi n\mathbf{J}$, acting on the ions. If we insert these forces into (7.8), we obtain the drift velocity $\mathbf{u}_{d\perp} = -\xi \mathbf{J} \times \mathbf{B}_0/B^2$, which turns out to be same for both electrons and ions. Since $\mathbf{J} \times \mathbf{B}_0 = \nabla p = \nabla(n\kappa(T_e + T_i))$ for quasi-equilibrium processes, we find the drift velocity $\mathbf{u}_{d\perp} = -\xi \nabla p/B^2$. With $D_\perp \equiv \xi n\kappa(T_e + T_i)/B_0^2$ found before, and assuming isothermal processes with T_e and T_i constant, we readily demonstrate that $\mathbf{u}_{d\perp}$ obtained here is the diffusion velocity we found before for the direction perpendicular to $\mathbf{B}$.

11.3.2 Diffusion in partially ionized plasmas Ⓚ

In some respects, the diffusion process becomes somewhat more interesting but also more complicated when we consider collisions between neutrals and charged particles. In this case we can write the quasi-stationary fluid momentum equations as

$$
\begin{aligned}
-\nabla p_e - en(\mathbf{E} + \mathbf{u}_e \times \mathbf{B}) - nm\nu_e \mathbf{u}_e &= 0, \\
-\nabla p_i + en(\mathbf{E} + \mathbf{u}_i \times \mathbf{B}) - nM\nu_i \mathbf{u}_i &= 0,
\end{aligned}
$$

where we again assumed quasi-neutrality, and introduced the electron and ion neutral collision frequencies as ν_e and ν_i, respectively. A constant magnetic field $\mathbf{B}$ is along the z-axis. The neutral gas is taken to be at rest. We also assume that the electric fields are electrostatic, $\mathbf{E} = -\nabla\phi$, which can hardly be considered as a restriction. Assuming again plasma conditions to vary only with the x-coordinate, we have $\nabla n = \widehat{\mathbf{x}}\partial n/\partial x$ and $\nabla\phi = \widehat{\mathbf{x}}\partial\phi/\partial x$. From the y-component of the equation for the ions we readily obtain

$$u_{iy} = -\frac{eB}{M\nu_i}u_{ix},$$

and by using the x-component

$$u_{ix} = -\frac{1}{nM}\frac{\kappa T_i\partial n/\partial x + ne\partial\phi/\partial x}{\nu_i(1+(\Omega_{ci}/\nu_i)^2)}. \tag{11.20}$$

From this it is trivial to obtain an electron velocity as

$$u_{ex} = -\frac{1}{nm}\frac{\kappa T_e\partial n/\partial x - ne\partial\phi/\partial x}{\nu_e(1+(\omega_{ce}/\nu_e)^2)}. \tag{11.21}$$

In Section 11.3.1 we could argue from first principles that there were no electric fields. Since the diffusion coefficients for electrons and ions are different when the diffusion is due to collisions with neutrals as here, we expect electric fields to build up. By the quasi-neutrality assumption we have, however, excluded Poisson's equation, and have to determine the electric field by other means.

We can now encounter two basically different cases: one where the electric field is obtained self-consistently from the equations, and another where we can assume that the electrostatic potential is short-circuited by some "external" agency, i.e., we have ϕ = const. In this latter case we will in general have an electron flow along magnetic field lines *into* and *out* from conducting plates placed perpendicular to magnetic field lines, terminating the device assumed to be long. A very small electron flow velocity will suffice. This flow need not be included in the analysis.

1) We consider the latter case first, since it is the simplest. In this case we obtain a diffusion flux (i.e., *velocity* $\times$ *density*) for ions as

$$n\,u_{ix} = -\frac{\kappa T_i/M}{\nu_i(1+(\Omega_{ci}/\nu_i)^2)}\frac{\partial n}{\partial x},$$

and for electrons as

$$n\,u_{ex} = -\frac{\kappa T_e/m}{\nu_e(1+(\omega_{ce}/\nu_e)^2)}\frac{\partial n}{\partial x}.$$

From these results we trivially identify diffusion coefficients as

$$D_i \equiv \frac{\kappa T_i/M}{\nu_i(1+(\Omega_{ci}/\nu_i)^2)},$$

and

$$D_e \equiv \frac{\kappa T_e/m}{\nu_e(1+(\omega_{ce}/\nu_e)^2)}.$$

With $\nu_{e,i}$ being constants, we have $D_{e,i}$ being constants as well, in contrast to the case where we discussed electron-ion collisions. In particular for weakly collisional plasmas, where $\omega_{ce,i} \gg \nu_{e,i}$, we find

$$\frac{D_i}{D_e} \approx \frac{T_i m}{T_e M}\left(\frac{\omega_{ce}}{\Omega_{ci}}\right)^2\frac{\nu_i}{\nu_e} \approx \left(\frac{T_i}{T_e}\right)^{3/2}\left(\frac{M}{m}\right)^{1/2}\frac{\ell_{ce}}{\ell_{ci}},$$

in terms of collisional mean free paths $\ell_{ci} = u_{Thi}/\nu_i \approx \sqrt{\kappa T_i/M}\Big/\nu_i$ and $\ell_{ce} = u_{The}/\nu_e \approx \sqrt{\kappa T_e/m}\Big/\nu_e$. In case $T_e \approx T_i$ and $\ell_{ci} \sim \ell_{ce}$, we find as an estimate

$$\frac{D_i}{D_e} \approx \sqrt{\frac{M}{m}} \gg 1,$$

implying that ions diffuse much faster than electrons in this case. Physically, we argue that the ions collide more seldomly than electrons because of their smaller average velocity, but *when* they collide they "jump" a much larger distance in comparison, since the ion Larmor radius is much larger than that of the electrons. This type of diffusion is sometimes called "Simon diffusion" after the scientist noting the phenomena first (Simon 1955).

2) We now retain the electric field, $\phi \neq$ const. In order to retain a quasi-neutral plasma, we must require that $u_{ix} = u_{ex}$ in (11.20) and (11.21). This gives

$$\frac{1}{nM}\frac{\kappa T_i \partial n/\partial x + ne\partial\phi/\partial x}{\nu_i(1+(\Omega_{ci}/\nu_i)^2)} = \frac{1}{nm}\frac{\kappa T_e \partial n/\partial x - ne\partial\phi/\partial x}{\nu_e(1+(\omega_{ce}/\nu_e)^2)}.$$

In terms of the diffusion coefficients $D_{e,i}$ introduced before, the results can be written as

$$D_i\left(\frac{1}{n}\frac{\partial n}{\partial x} + \frac{e}{\kappa T_i}\frac{\partial\phi}{\partial x}\right) = D_e\left(\frac{1}{n}\frac{\partial n}{\partial x} - \frac{e}{\kappa T_e}\frac{\partial\phi}{\partial x}\right),$$

or

$$e\frac{\partial\phi}{\partial x} = (D_e - D_i)\frac{1}{n}\frac{\partial n}{\partial x} \Big/ \left(\frac{D_e}{\kappa T_e} + \frac{D_i}{\kappa T_i}\right).$$

In principle we have solved the problem, since the potential can now be eliminated, to give a diffusion type equation for the density evolution. The resulting diffusion coefficient can be made neater for the case where the plasma is isothermal, i.e., $T_e = T_i \equiv T_0$. Then

$$\frac{e}{\kappa T_0}\frac{\partial\phi}{\partial x} = \frac{D_e - D_i}{D_e + D_i}\frac{1}{n}\frac{\partial n}{\partial x} \approx \left(2\frac{D_e}{D_i} - 1\right)\frac{1}{n}\frac{\partial n}{\partial x}. \tag{11.22}$$

Since generally $D_e/D_i \ll 1$, we have $\partial\phi/\partial x$ and $\partial n/\partial x$ to be in opposite directions consistent with our expectation that ions are moving faster than electrons in the direction $\perp \mathbf{B}_0$. By insertion into the expressions for the velocities (11.20) and (11.21), we find here

$$u_{ix} = u_{ex} = -2\frac{D_e}{D_i}\frac{\kappa T_0}{nM}\frac{\partial n/\partial x}{\nu_i(1+(\Omega_{ci}/\nu_i)^2)} = -2D_e\frac{1}{n}\frac{\partial n}{\partial x}.$$

From this we find the diffusion equation for the ion density (which is the same as for the electron density by the assumption of quasi-neutrality)

$$\frac{\partial}{\partial t}n = 2D_e\frac{\partial^2}{\partial x^2}n.$$

The assumption of quasi-neutrality "locks" the ion and electron diffusion equations together, but the electric field induced by the large ion diffusion enhances the effective diffusion coefficient by a factor of 2 in comparison to what we would get if the electrons alone were to control the process. It is in this sense interesting that it is the *ion* pressure that gives rise to the ambipolar electric field. If the electron temperature was very large, $T_e > T_i\sqrt{M/m}$, we see from (11.22) that the electric field would change sign. Such large temperature ratios are, however, entirely unrealistic for plasmas occurring in nature, although not necessarily in laboratories where various types of discharge plasmas can have very high temperature ratios (Takahashi et al. 1998).

The two sections, 11.3.1 and 11.3.2, describe collisional diffusion as seen from a macroscopic continuum fluid model. The problem can be studied also by microscopic models where the particles undergo a random walk (MacDonald 1962, Pécseli 2000) as also discussed earlier in this section. It is again illustrative to see how a diffusion equation can be derived starting from such microscopic models. To substantiate the dependence of the diffusion coefficient D with the particle mass, we can estimate D in terms of a "step length" Δ and a time τ between steps, so that $D \sim \Delta^2/\tau$. We take τ to be the average time between collisions, $\tau_c = \ell_c/u_{The}$, i.e., the collisional mean free path ℓ_c divided by a thermal velocity $u_{The} = \sqrt{\kappa T/M}$. For a magnetized plasma it is reasonable to take the step length to be the Larmor radius, so that $\Delta = r_L = u_{The}/\Omega_c$, giving

$$D \sim \left(\sqrt{\frac{\kappa T}{M}} \frac{M}{|q|B}\right)^2 \frac{\sqrt{\kappa T/M}}{\ell_c}.$$

It then follows that for particles with similar temperatures and mean free paths, the diffusion coefficient will scale with $\sqrt{M}$ as we found for diffusion caused by collisions between the plasma species and neutrals. For ion-electrons collisions the argument no longer applies since the step length is the same for both species due to momentum conservation, see Section 8.8.1. Evidently, the electron-ion and ion-electron collision times are also identical, so in this case we have Δ and τ being the same for both species and their diffusion coefficients are also the same.

- **Exercise:** Demonstrate that

$$n(x,t) = \frac{e^{-x^2/2\sigma^2(t)}}{\sigma(t)\sqrt{2\pi}},$$

is a solution to the generalized one-dimensional diffusion equation

$$\frac{\partial}{\partial t} n(x,t) = \frac{1}{2} \frac{d\sigma^2(t)}{dt} \frac{\partial^2}{\partial x^2} n(x,t), \tag{11.23}$$

with the initial condition $n(x,0) = \delta(x)$. Generalize the result to two and three spatial dimensions. We note, by inspection, the relation between a diffusion coefficient and $d\sigma^2(t)/dt$.

Ⓢ **Exercise:** Consider an infinitely extended plasma with constant density n_0, and assume that at $t = 0$ the density in the half-space $x < 0$ is suddenly increased to $n_0 + \Delta n$. We can have two scenarios: either this represents an initial condition for the plasma so that the density evolves for all $-\infty < x < \infty$, or some external agency maintains the plasma density at the level $n_0 + \Delta n$ for $x < 0$, so that we have a boundary condition where the plasma density evolves for $x > 0$. Solve both cases by assuming that the evolution of $n(x,t)$ is well described by a diffusion equation with a known diffusion coefficient. The plasma need not be magnetized, but if it is, we assume the magnetic field to be perpendicular to $\widehat{\mathbf{x}}$. Give a simple physical relation between the two solutions *before* you solve the basic equations! Ⓢ

- **Exercise:** Solve the one-dimensional diffusion equation for the case where we have perfectly absorbing boundaries at $x = \pm L$, and a constant source, $a\delta(x)$, of particles at $x = 0$. Give first the steady state solution, and then the full time evolution. (Hint: the problem can be made easy by using a generalization of the method of images, see Section 5.7.1). Give the steady state solutions for the two- and three-dimensional generalization of the problem, assuming cylindrical and spherical symmetries, respectively.

12

Waves in Cold Plasmas

We take the simplest possible version of the two-fluid model for describing waves in homogeneous and isotropic plasmas. Here, the plasma is described as two fluids, electrons and ions, solely interacting by the electric and magnetic fields they set up. Collisional interactions are ignored. This simple model can be further simplified by ignoring pressure terms. On one hand it can be argued that in many cases the model can become oversimplified to such a degree that it is futile to expect any reasonable agreement with plasma processes occurring in nature. On the other hand it is experienced that the resulting model has a great potential by giving an overview of possible wave-types, and giving a convenient representation of the topology of an $(\omega, \mathbf{k})$-dispersion diagram (Allis et al. 1963, Bekefi 1966).

The simplest analysis assumes that the ion as well as the electron temperatures vanish. Evidently, this assumption can never be exactly fulfilled, but the results obtained by it have nonetheless proved most valuable, in particular by offering a convenient classification of electromagnetic waves in plasmas. A heuristic argument will give that thermal effects are negligible when the phase velocity of waves in a linear model is much larger than the thermal particle velocity. In this limit we might as well set $T_e = T_i = 0$.

12.1 Waves in unmagnetized plasmas

The basic equations are here the simple fluid equations for plasma continuity (11.1) and (11.2) and the momentum equations (11.3) and (11.4), together with Maxwell's equations. First we analyze the case with a homogeneous unperturbed plasma density. The basic equations for this simple limit are, in their linearized form,

$$m\frac{\partial}{\partial t}\widetilde{\mathbf{u}}_e = -e\widetilde{\mathbf{E}} \quad \text{and} \quad M\frac{\partial}{\partial t}\widetilde{\mathbf{u}}_i = e\widetilde{\mathbf{E}} \tag{12.1}$$

from the momentum equations. We note that there is no contribution from the Lorentz force $\mathbf{u}\times\mathbf{B}$ to these linear equations, since we have $u_0 = 0$ and $B_0 = 0$ here. The plasma density cancels in (12.1) by the assumption of vanishing pressure. This would be true even without linearizing the equations.

The set of equations (12.1) is completed by the full set of Maxwell equations. It turns out that we will not need the continuity equations: note, for instance, that perturbations of the plasma density do not enter (12.1). We look for electrostatic as well as electromagnetic waves here. The plasma current density is to the present accuracy given by $\widetilde{\mathbf{J}} = en_0(\widetilde{\mathbf{u}}_i - \widetilde{\mathbf{u}}_e)$, or

$$\frac{\partial}{\partial t}\widetilde{\mathbf{J}} = e^2 n_0 \frac{m+M}{mM}\widetilde{\mathbf{E}} \equiv \frac{e^2 n_0}{\mathcal{M}}\widetilde{\mathbf{E}}, \tag{12.2}$$

where for plasmas relevant here, we obviously have for the reduced mass $\mathcal{M} \equiv mM/(M+m) \approx m$. Bear in mind, however, that electron-positron plasmas, with identical masses for negative and positive charge carriers, i.e., $m = M$, are relevant, for instance, for astrophysics in some cases. In other cases we can have negative ions in abundance, with a mass greatly exceeding the electron mass, m. In such cases we need the full expressions for the reduced mass, $\mathcal{M}$. Generalizations to include negative ions, several ions species, etc. are straightforward.

From the Maxwell equations we have generally, i.e., without need of linearization,

$$\nabla \times (\nabla \times \widetilde{\mathbf{E}}) = \nabla(\nabla \cdot \widetilde{\mathbf{E}}) - \nabla^2 \widetilde{\mathbf{E}} = -\mu_0 \frac{\partial}{\partial t} \widetilde{\mathbf{J}} - \frac{1}{c^2} \frac{\partial^2}{\partial t^2} \widetilde{\mathbf{E}}, \tag{12.3}$$

giving in the present case

$$c^2 \nabla \times (\nabla \times \widetilde{\mathbf{E}}) + \frac{e^2 n_0}{\mathcal{M} \varepsilon_0} \widetilde{\mathbf{E}} + \frac{\partial^2}{\partial t^2} \widetilde{\mathbf{E}} = 0, \tag{12.4}$$

where the coefficient of the second term is recognized as the square of the plasma frequency (8.1) corresponding to the reduced mass $\mathcal{M}$. We use also $1/c^2 = \varepsilon_0 \mu_0$. Considering a plane wave in the form $\widetilde{\mathbf{E}} = \mathbf{E}_1 \exp\big(-i(\omega t - \mathbf{k} \cdot \mathbf{r})\big)$ we find

$$c^2 \mathbf{k} \times (\mathbf{k} \times \mathbf{E}_1) = c^2 \left(\mathbf{k}\mathbf{k} \cdot \mathbf{E}_1 - k^2 \mathbf{E}_1\right) = \left(\omega_{pe}^2 - \omega^2\right) \mathbf{E}_1. \tag{12.5}$$

We introduced the electron plasma frequency ω_{pe} since $\mathcal{M} \approx m$.

So far nothing was said concerning the electric field polarization, i.e., the direction of $\mathbf{E}_1$ with respect to $\mathbf{k}$. For transverse electromagnetic waves, $\mathbf{k} \perp \mathbf{E}_1$, we find the dispersion relation

$$\omega^2 = \omega_{pe}^2 + c^2 k^2, \tag{12.6}$$

as the criterion for the existence of nontrivial solutions for (12.5). For longitudinal electrostatic waves with $\mathbf{k} \parallel \mathbf{E}_1$ we have

$$\omega^2 = \omega_{pe}^2.$$

For the transverse waves we have $\mathbf{k} \cdot \mathbf{E}_1 = 0$, or $\nabla \cdot \widetilde{\mathbf{E}} = 0$, implying by Poisson's equation that there are no perturbations of the charge density, only fluctuating currents. These have to be divergence free, $\nabla \cdot \mathbf{J} = 0$ as found from the charge continuity equation. The transverse waves reduce to simple electromagnetic waves, i.e., light with $E_1/B_1 = c$, when the plasma density vanishes. For plasma conditions we have $B_1/E_1 = k \Big/ \sqrt{\omega_{pe}^2 + c^2 k^2}$, or $E_1/B_1 \approx \omega_{pe}/k$ for $k \ll \omega_{pe}/c$.

For the longitudinal waves, we find that the plasma current and the Maxwell displacement current exactly cancel, and there are no magnetic perturbations associated with these waves. They are consequently termed *electrostatic*, since the electric fields can be derived from Poisson's equation, just as in classical electrostatics.

Formally the equations contain one more solution, namely, one where $\mathbf{E} = 0$, but the current density is constant, $\mathbf{J} \neq 0$. Since we have no dissipation in the system, a dc current which is present initially will prevail for all later times in this model and give rise to a stationary magnetic field.

For waves in unmagnetized cold plasmas we have a significant difference for electromagnetic and electrostatic waves: in the first case we have a characteristic frequency ω_{pe} and a characteristic velocity c, and thereby also a characteristic length, c/ω_{pe}. For electrostatic waves we only have a characteristic frequency ω_{pe}.

- **Exercise:** In the derivations of this section we did not explicitly use the continuity equations for electrons or for the ions, and neither did we need Poisson's equation. Why not?

Ⓢ **Exercise:** Derive and discuss an equation similar to (12.3), this time for the magnetic field fluctuations. Ⓢ

Ⓢ **Exercise:** Assume a given relative density variation $\widetilde{n}/n_0$ for electrostatic electron waves in a cold plasma. What is the electric field and the electrostatic potential associated with these waves for varying wavenumber? Can you normalize the result? As a numerical example you can take $\widetilde{n}/n_0 = 10^{-4}$ and $n_0 = 10^{12}$ m^{-3}, together with $\lambda = 10$ m. Ⓢ

- **Exercise:** Demonstrate that for the transverse waves with dispersion relation $\omega^2 = \omega_{pe}^2 + c^2k^2$, we have $u_f u_g = c^2$, in terms of the phase velocity $u_f \equiv \omega/k$ and the group velocity $u_g \equiv d\omega/dk$.

Ⓢ **Exercise:** Analyze electromagnetic waves from a moving frame of reference, where we can take the unperturbed uniform electron velocity to be $U \ll c$. Let the ions be moving with velocity U as well, so there is no net dc current in this system, but the ion motion is not otherwise included in the analysis. Comment on the limit $|k| \to \infty$ for $U \neq 0$. Ⓢ

12.1.1 Penetration depth

We can try to force an oscillation into a plasma so that the frequency is sub-critical, $\omega_0 < \omega_{pe}$. Imagine, for instance, a semi-infinite plasma and shine an electromagnetic wave from the vacuum side normally onto the plasma surface. Since the tangential electric field component is the same on both sides of the interface, we can manage to impose an oscillation with arbitrary frequency, in principle. However, in order for this to be a *propagating* wave, it must fulfill the dispersion relation, while the space-time variation of the waveform has the form $\exp(-i(\omega_0 t - k_0 z))$. With (12.6), we find that with $\omega_0 < \omega_{pe}$, the wave must have the form

$$\exp\big(-i(\omega_0 t - k_0 z)\big) = \exp\left(-i\omega_0 t + i\frac{z}{c}\sqrt{\omega_0^2 - \omega_{pe}^2}\right) = \exp(-i\omega_0 t)\exp\left(-\frac{z}{c}\sqrt{\omega_{pe}^2 - \omega_0^2}\right).$$

The conclusion is that the electromagnetic wave is exponentially damped in the plasma, when the imposed frequency is sub-critical, $\omega_0^2 < \omega_{pe}^2$. The damping distance is $\delta = c\big/\sqrt{\omega_{pe}^2 - \omega_0^2} \to c/\omega_{pe}$ for $\omega_0 \to 0$. In this low frequency limit δ is often called the *London penetration depth*, or the *skin depth*, although this latter designation is not quite correct. We can consider δ as a *dynamic shielding*, in contrast to the static shielding, characterized by the Debye length. Recall that we found a skin depth also for media with finite conductivity (see Section 5.6) so to distinguish them we often use the terminology "resistive skin effect" and "collisionless skin effect."

A fluctuation of the form $\exp(-i\omega t)\exp(-z/\delta)$ is not a wavelike motion, and an effort to define a phase velocity by counting the number of wave-tops passing by per sec will be futile.

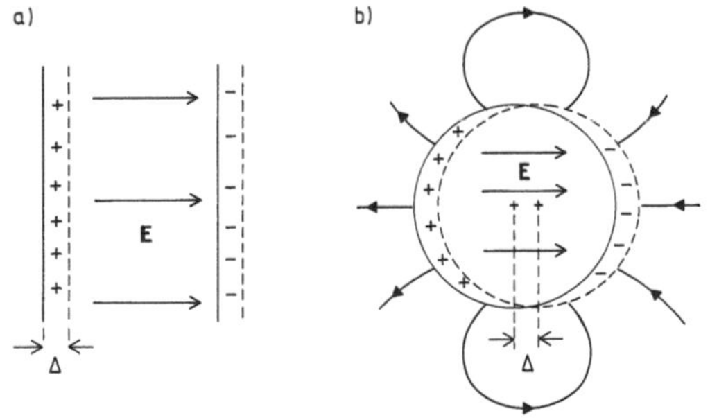

FIGURE 12.1: *Illustration of the effect of geometry on electron plasma oscillations with immobile ions.*

- **Example:** The case with *in*homogeneous plasma densities offers a surprisingly rich modification in the description of linear Langmuir oscillations in cold plasmas (Barston 1964). To illustrate the problem it suffices to point to the problem illustrated in Fig. 12.1, showing the cold electron component displaced a distance Δ, with respect to the immobile ions. Since the electrons' relative motion is ignored, we have Δ being the displacement of *all* electrons, and we have $md^2\Delta/dt^2 = -eE$; see also Section 5.9.1. With n_0 being the plasma density, the resulting surface charge is in a) given by $\sigma = en_0\Delta$ and we have a homogeneous electric field

$E = \sigma/\varepsilon_0 = en_0\Delta/\varepsilon_0$. In case b) the electric field is still homogeneous inside the sphere, but now with an intensity $E = \frac{1}{3}en_0\Delta/\varepsilon_0$; see the exercise on page 72. The natural frequency of oscillations is then $\omega = \omega_{pe}$ in case a), while it is $\omega = \omega_{pe}/\sqrt{3}$ in case b). More generally, we expect that the spatial variation of the plasma density will be important for the characteristic frequencies of the electron oscillations.

Ⓢ **Exercise:** Consider a geometry where the (cold) plasma density is constant inside a confining surface, and vanishes outside, as in Fig. 12.1, but in this case let the surface be a cylinder perpendicular to the plane of the paper. Figure 12.1b) can then show a cut perpendicular to the cylinder axis. Determine the characteristic frequency of plasma oscillations for this case. Ⓢ

12.2 Waves in magnetized plasmas

Inclusion of an externally imposed magnetic gives rise to a nontrivial complication of the basic equations for describing the dynamics of plasmas. The complete analysis tends to become rather lengthy, while individual cases (selected direction or certain polarizations) can be made simpler. Often only such cases are discussed but the expense of the simplification is that we lose the general picture. Here we choose the alternative disposition by deriving a general formalism, and make the simplifications later.

Take first the momentum equation

$$nm\left(\frac{\partial}{\partial t}\mathbf{u} + \mathbf{u}\cdot\nabla\mathbf{u}\right) = qn\mathbf{E} + qn\mathbf{u}\times\mathbf{B}, \tag{12.7}$$

for particles with charge q and mass m. There is no pressure term because the plasma is assumed to be cold, so in this case the density cancels in all terms. As an equilibrium solution we again take the case where $u_0 = 0$, $\mathbf{E}_0 = 0$, and $\mathbf{B}_0 = B_0\widehat{\mathbf{z}}$. Linearizing the equations, we obtain for the fluctuating quantities

$$m\frac{\partial}{\partial t}\widetilde{u}_\parallel = q\widetilde{E}_\parallel, \tag{12.8}$$

$$m\frac{\partial}{\partial t}\widetilde{\mathbf{u}}_\perp = q\widetilde{\mathbf{E}}_\perp + q\widetilde{\mathbf{u}}_\perp\times\mathbf{B}_0, \tag{12.9}$$

for the components parallel to the magnetic field, and perpendicular to it. After a Fourier transform with respect to the temporal and spatial variables we find

$$u_\parallel = i\frac{q}{m\omega}E_\parallel, \tag{12.10}$$

$$-i\omega\mathbf{u}_\perp = \frac{q}{m}\mathbf{E}_\perp + \frac{q}{m}\mathbf{u}_\perp\times\mathbf{B}_0. \tag{12.11}$$

We omitted the indexes $_1$ here, since they would be mixed up with the subscripts like $_\parallel$, etc. After some simple algebra we find

$$u_y = i\frac{q}{m}\omega\left(E_y - i\frac{\Omega_c}{\omega}E_x\right)\frac{1}{\omega^2-\Omega_c^2}, \tag{12.12}$$

and

$$u_x = i\frac{q}{m}\omega\left(E_x + i\frac{\Omega_c}{\omega}E_y\right)\frac{1}{\omega^2-\Omega_c^2}, \tag{12.13}$$

where we take the sign of Ω_c equal to the sign of the charge q. Given the fluctuating velocity of the plasma species, we can now to the present accuracy, as obtained by the linearization, calculate the contribution to the plasma current originating from the relevant species as $\mathbf{J}_l = q_l n_l \mathbf{u}_l$. The present analysis is similar to that in Section 8.9. The resulting expression for the total current density, $\mathbf{J} = \sum_l \mathbf{J}_l$, is subsequently inserted into the general form (12.3)

$$\mathbf{k} \times (\mathbf{k} \times \mathbf{E}_1) = \left(\mathbf{kk} \cdot \mathbf{E}_1 - k^2 \mathbf{E}_1\right) = -i\mu_0 \omega \mathbf{J}_1 - \frac{1}{c^2}\omega^2 \mathbf{E}_1.$$

- **Exercise:** Show that (12.11) can be written in the form

$$\mathbf{u}_\perp = \frac{i\omega \mathbf{E}_\perp (q/m) - \mathbf{E} \times \mathbf{B}(q/m)^2}{\omega^2 - \Omega_c^2}.$$

12.2.1 High frequency waves

To simplify the analysis, we begin with high frequency waves in the limit where the ions can be assumed to be an immobile background of positive charge, while the electron charge is now $q = -e$. We have in tensor notation again

$$\mathbf{J} = \underline{\sigma} \cdot \mathbf{E},$$

with the conductivity tensor σ given as

$$\underline{\sigma} \equiv \begin{pmatrix} \sigma_P & \sigma_H & 0 \\ -\sigma_H & \sigma_P & 0 \\ 0 & 0 & \sigma_\parallel \end{pmatrix} \tag{12.14}$$

with

$$\sigma_P = i\varepsilon_0 \frac{\omega\omega_{pe}^2}{\omega^2 - \omega_{ce}^2}, \qquad \sigma_H = \varepsilon_0 \frac{|\omega_{ce}|\omega_{pe}^2}{\omega^2 - \omega_{ce}^2}, \qquad \sigma_\parallel = i\varepsilon_0 \frac{\omega_{pe}^2}{\omega}.$$

See also Section 8.9.1.

In an almost universally adopted notation (Yeh & Liu 1972) one introduces $X \equiv (\omega_{pe}/\omega)^2$ and the vector $\mathbf{Y}$ is in the direction of $\mathbf{B}$ with $|\mathbf{Y}| \equiv \omega_{ce}/\omega$. Here we note that for a fixed frequency, X is proportional to the plasma density, and $|\mathbf{Y}|$ to the magnetic field, assumed locally homogeneous. We then have

$$\sigma_P = i\varepsilon_0\omega \frac{X}{1 - Y^2}, \qquad \sigma_H = \varepsilon_0\omega \frac{|Y|X}{1 - Y^2}, \qquad \sigma_\parallel = i\varepsilon_0\omega X,$$

where we took the absolute values $|Y|$, and $|\omega_{ce}|$ to emphasize that the signs are already taken care of by using the proper sign for the electron charge when defining $q = -e$, taking $e > 0$.

We have an expression for the conductivity tensor $\underline{\sigma}$, but it is an advantage to introduce the dielectric tensor instead. Consider a harmonically varying electric field with phase factor $\exp(-i(\omega t - \mathbf{k} \cdot \mathbf{r}))$. With $\mathbf{J} = \partial \mathbf{P}/\partial t$ we find from Section 5.7 the expression

$$\mathbf{D} = \varepsilon_0 \mathbf{E} + \frac{i}{\omega}\mathbf{J},$$

for the dielectric displacement, $\mathbf{D}$. If we now have $\mathbf{J} = \underline{\sigma} \cdot \mathbf{E}$, we can write

$$\mathbf{D} = \varepsilon_0 \left(\underline{1} + \frac{i}{\omega\varepsilon_0}\underline{\sigma}\right) \cdot \mathbf{E},$$

where also the unit tensor $\underline{1}$ was introduced. We can now define an effective relative dielectric tensor

$$\underline{\varepsilon} = \left(\underline{1} + \frac{i}{\omega\varepsilon_0}\underline{\sigma}\right), \tag{12.15}$$

noting that σ may, in the general case, be complex. The tensor $\underline{\varepsilon}$ contains no information of wave polarizations. Given a solution for a dispersion relation, the corresponding electric field polarization has to be determined by the basic equations.

To make the basic equations for the present case look neater, it is an advantage to introduce the relative dielectric tensor (12.15), giving

$$\underline{\varepsilon} \equiv \begin{pmatrix} 1-\dfrac{X}{1-Y^2} & i\dfrac{|Y|X}{1-Y^2} & 0 \\ -i\dfrac{|Y|X}{1-Y^2} & 1-\dfrac{X}{1-Y^2} & 0 \\ 0 & 0 & 1-X \end{pmatrix} \tag{12.16}$$

where the externally imposed magnetic field $\mathbf{B}_0$ is assumed to be along the z-axis. We can now write (12.5) as appropriate for the present case as

$$\left(k^2\underline{1}-\mathbf{kk}-\left(\frac{\omega}{c}\right)^2\underline{\varepsilon}\right)\cdot\mathbf{E}=0. \tag{12.17}$$

In principle we could simply have written (12.17) in terms of the conductivity tensor $\underline{\sigma}$, and omitted any reference to the dielectric displacement $\mathbf{D}$. The physical interpretation of the result becomes more transparent, however, in terms of the dielectric tensor.

For the following discussion it is important to note that $\underline{\varepsilon}$ depends on frequency ω only, in addition to the plasma parameter dependencies through X and Y. All of the wave-number dependence is contained in $k^2\left(\underline{1}-\widehat{\mathbf{kk}}\right)\cdot\mathbf{E}$.

It is an advantage here to distinguish two cases: longitudinal waves and transverse waves.

12.2.1.1 Longitudinal or electrostatic waves

For longitudinal waves we have, by definition, that $\mathbf{k} \parallel \mathbf{E}$, in which case $\mathbf{k}\cdot\mathbf{E}=kE$. These waves are often termed *electrostatic*. For this special case two first terms in (12.17) cancel, and that equation becomes simply $\underline{\varepsilon}\cdot\mathbf{E}=0$. The requirement for this equation to have nontrivial solutions is that the determinant $\det|\underline{\varepsilon}|=0$. However, the relation $\left(k^2\underline{1}-\mathbf{kk}\right)\cdot\mathbf{E}=k^2\left(\underline{1}-\widehat{\mathbf{kk}}\right)\cdot\mathbf{E}=0$ can be fulfilled also for $k=0$ and this limit should not properly be considered a propagating *wave*. If $k=0$ while $\omega\neq 0$, we have by Faraday's law (5.58) that the fluctuating part of the magnetic field vanishes, so this limit is also purely electric. For the cases listed, we have by (12.16) one of the following conditions fulfilled.

1. $1-X=0$, giving $\omega=\omega_{pe}$. Inserting $X=1$ into (12.16) we find solutions for the electric field components so that $E_z\neq 0$, while $E_x=E_y=0$. The electric field is polarized along the externally imposed magnetic field. We can express this observation as $E_x:E_y:E_z=0:0:1$. We have $\mathbf{k}\parallel\mathbf{E}$ irrespective of the magnitude of the wave-vector.

2. $1+|Y|=X$, giving $\omega=\omega_L\equiv-\omega_{ce}/2+\sqrt{\omega_{pe}^2+\omega_{ce}^2/4}$. Inserting $X=1+|Y|$ in (12.16) we find in this case the relative electric field magnitudes $E_x:E_y:E_z=i:1:0$, i.e., the electric field component along the magnetic field vanishes, and the electric field vector performs a circular motion in the plane $\perp\mathbf{B}_0$ with E_y leading. The E_x and E_y components of the electric field are of equal magnitude but 90° out of phase. We have $|\mathbf{k}|=0$. See Fig. 12.2.

3. $1-|Y|=X$, giving $\omega=\omega_R\equiv\omega_{ce}/2+\sqrt{\omega_{pe}^2+\omega_{ce}^2/4}$. In this case we have for the relative magnitudes of the electric field $E_x:E_y:E_z=-i:1:0$, i.e., the electric field component along the dc-magnetic field vanishes, and the electric field vector performs a circular motion in the plane $\perp\mathbf{B}_0$ with E_x leading. We have $|\mathbf{k}|=0$ also here.

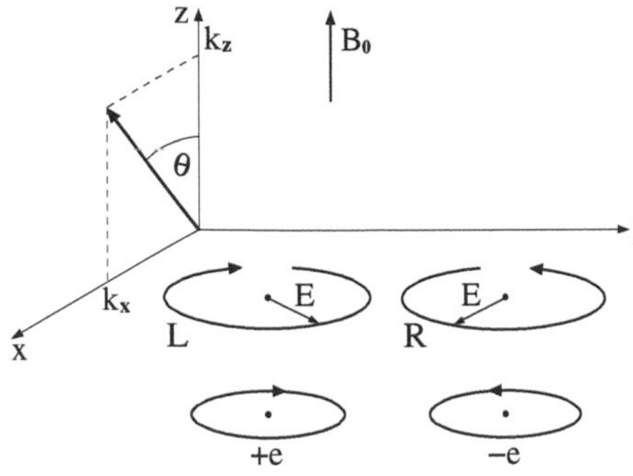

FIGURE 12.2: *Illustration of left (L) and right (R) wave polarizations in a right-hand coordinate system with* $\mathbf{B}_0 \parallel \widehat{\mathbf{z}}$*. The circular motions of magnetized particles with positive and negative charges are also shown, taking* $e > 0$*. We also show the definition of the angle* θ *between* $\mathbf{B}_0$ *and* $\mathbf{k}$ *used for discussing the Appleton-Hartree relation.*

Note that $\omega_R - \omega_L = \omega_{ce}$. The subscripts R and L on the frequencies denote *right* and *left* circular polarizations, and are introduced for later reference. A left circularly polarized **E**-vector rotates clockwise, a right polarized field counter-clockwise. Both of these definitions of rotations assume that the observer is facing the (x,y)-plane with the z-axis also the **B**-field direction pointing towards the observer; see Fig. 12.2.

These are the only *exact* results for the case $\left(k^2\underline{1} - \mathbf{kk}\right)\cdot\mathbf{E} = 0$. As demonstrated later on, there are cases, typically at large wave-numbers, i.e., short wavelengths, where waves can be *almost* electrostatic, in the sense that the electric field component along $\mathbf{k}$ is by far the largest one. We can argue that *if* a mode of wave propagation exists such that ω is finite as $k \to \infty$, then also $\underline{\varepsilon}$ is constant and (12.17) consequently implies that $\left(\underline{1} - \widehat{\mathbf{k}}\widehat{\mathbf{k}}\right)\cdot\mathbf{E} \to 0$ in this limit, with $\widehat{\mathbf{k}} \equiv \mathbf{k}/k$. This requirement coincides with our interpretation of longitudinal waves. This limit should be considered with some care since a cold plasma model is bound to become invalid in a limit where $\omega/k \to 0$, or more precisely, when the phase velocity of the wave becomes comparable to the particle thermal velocities.

- **Exercise:** Show that after linearization the cold electron momentum expression

$$n(\mathbf{r},t)m\left(\frac{\partial \mathbf{u}(\mathbf{r},t)}{\partial t} + \mathbf{u}(\mathbf{r},t)\cdot\nabla\mathbf{u}(\mathbf{r},t)\right) = -en(\mathbf{r},t)\left(\mathbf{E}(\mathbf{r},t) + \mathbf{u}(\mathbf{r},t)\times\mathbf{B}_0\right)$$

can be expressed in the form

$$\widetilde{\mathbf{u}} = i\frac{e}{m}\frac{\omega}{\omega^2 - \omega_{ce}^2}\left(\nabla\widetilde{\phi} + i\frac{\omega_{ce}}{\omega}\widehat{\mathbf{b}}\times\nabla\widetilde{\phi} - \frac{\omega_{ce}^2}{\omega^2}\widehat{\mathbf{b}}\frac{\partial}{\partial z}\widetilde{\phi}\right),$$

for harmonic time variations as, e.g., $\widetilde{\phi}(\mathbf{r})\exp(-i\omega t)$.

12.2.1.2 Transverse waves

We now consider the general case where the electric field has a component in the direction perpendicular to the wave-number $\mathbf{k}$. Note that we are *not* requiring that we *only* have components $\mathbf{E} \perp \mathbf{k}$. Because of the cylindrical symmetry with respect to the magnetic field, the problem is specified once the direction of the wave-number $\mathbf{k}$ is given with respect to the magnetic field, as well as its magnitude $|\mathbf{k}|$. Taking again $\mathbf{B}_0$ to be along the $\widehat{\mathbf{z}}$-axis, we use $k_x = k\sin\theta, k_y = 0$ and $k_z = k\cos\theta$, where θ is the angle between $\mathbf{k}$ and $\mathbf{B}_0$; see Fig. 12.2. We then rewrite $k^2\underline{1}-\mathbf{kk}$ in terms of the wave-vector components. It is customary to introduce the refractive index, n, as the ratio between the speed of light in a vacuum and the phase velocity of the wave, $n\equiv c/(\omega/k)$, i.e., $k=n\omega/c$ which can also be interpreted as the ratio of the wave-number $k_0\equiv\omega/c$ an electromagnetic wave would have at the same frequency in a vacuum as compared to the actual wave-number, i.e., $n=k/k_0$. We find

$$\begin{pmatrix} n^2\cos^2\theta-\left(1-\frac{X}{1-Y^2}\right) & -i\frac{|Y|X}{1-Y^2} & -n^2\cos\theta\sin\theta \\ i\frac{|Y|X}{1-Y^2} & n^2-\left(1-\frac{X}{1-Y^2}\right) & 0 \\ -n^2\cos\theta\sin\theta & 0 & n^2\sin^2\theta-(1-X) \end{pmatrix}\cdot\mathbf{E}=0. \quad (12.18)$$

The criterion for nontrivial solutions of (12.18) is that the determinant of the coefficient tensor vanishes. Simple calculations give, in terms of the refractive index, the dispersion relation

$$a_4 n^4 + a_2 n^2 + a_0 = 0 \quad (12.19)$$

with

$$\begin{aligned} a_4 &\equiv \left(1-\frac{X}{1-Y^2}\right)\sin^2\theta+(1-X)\cos^2\theta \\ a_2 &\equiv -\left(1-\frac{X}{1-Y^2}\right)(1-X)(1+\cos^2\theta) \\ &\quad -\left(\left(1-\frac{X}{1-Y^2}\right)^2-\left(\frac{|Y|X}{1-Y^2}\right)^2\right)\sin^2\theta \\ a_0 &\equiv (1-X)\left(\left(1-\frac{X}{1-Y^2}\right)^2-\left(\frac{|Y|X}{1-Y^2}\right)^2\right). \end{aligned}$$

It turns out to that it is convenient to retain the two terms in a_4 as they stand, without using the identity $\cos^2\theta+\sin^2\theta=1$. We note that the coefficients a_4, a_2 and a_0 are functions only of the polar angle θ in addition to the plasma parameter dependencies through X and Y. In particular, a_0 is the determinant of the cold plasma relative dielectric tensor. The coefficients in the dispersion relation (12.19) are functions of the *direction* of the wave propagation as given by the wave-vector $\mathbf{k}$, but not its *magnitude*, k. Further the coefficients are unchanged if θ is replaced by $\pi-\theta$. The refractive index has a symmetry with respect to the plane perpendicular to the magnetic field.

For very large n^2, we have by (12.19) the approximation $a_4 n^2 + a_2 \approx 0$. For the limit $n^2\to\infty$ to exist we require $a_4=0$ for some directions of propagation, giving the form

$$\tan^2\theta = -\frac{(1-Y^2)(1-X)}{1-Y^2-X}. \quad (12.20)$$

The right-hand side has to be positive for physically acceptable solutions, implying constraints on Y^2 and X. If a solution for (12.20) is found for a value of θ, this angle will define a cone of directions with respect to the magnetic field. In general $\theta=\theta(\omega)$ for these directions. Since $n^2\to\infty$ corresponds to phase velocities $\omega/k\to 0$, i.e. resonances, this cone is called a "resonance cone" (Fisher & Gould 1969, Piel 2010). Since the structure of the resonance cone depends on ω_{pe} and ω_{ce}, a careful measurement of cone characteristics will contain information of local plasma parameters such as density and magnetic field (assumed locally homogeneous) (Rohde et al. 1993). Keep in mind, though, that the assumption of a cold plasma is not accurate for phase velocities of the order of the thermal velocities, so temperature effects are likely to play a role for resonance cones (Piel & Oelerich 1985). These non-zero temperature effects are not accounted for by (12.20).

The solutions of (12.19) can be written as

$$\begin{aligned} n^2 &= \frac{-a_2 \pm \sqrt{a_2^2 - 4a_4a_0}}{2a_4} \\ &= 1 - \frac{2(a_4 + a_2 + a_0)}{2a_4 + a_2 \pm \sqrt{a_2^2 - 4a_4a_0}} . \end{aligned} \tag{12.21}$$

Through the ω dependence of X and Y, the expression (12.21) relates the phase velocity to frequency (albeit in a rather intricate way), and (12.21) is therefore equivalent to a dispersion relation $\omega = \omega(\mathbf{k})$. It is often called the *Appleton-Hartree* equation. The dispersion relation is simple to use in representing wave-number as a function of frequency, once a direction of propagation is selected. It is much more difficult to "invert" (12.21) with $n \equiv c/(\omega/k)$ to give the more standard formulation with frequency as a function of wave-vector for arbitrary directions. The two scientists Appleton and Hartree have received much credit for studies of wave propagation in the ionosphere, but it may be proper to mention significant contributions also from the Danish scientist P. O. Pedersen, who presented very detailed and closely related results (Pedersen 1927).

Illustrations of the dispersion relations given by the Appleton-Hartree equation are shown in Fig. 12.3 for two sets of parameters. For small angles θ we note that the some of the dispersion relations become close and merging for $\theta = 0°$. To show some of the details we have in Fig. 12.4 dispersion relations for a selected direction $\theta = 10°$. The transition $\theta \to 0°$ is numerically challenging.

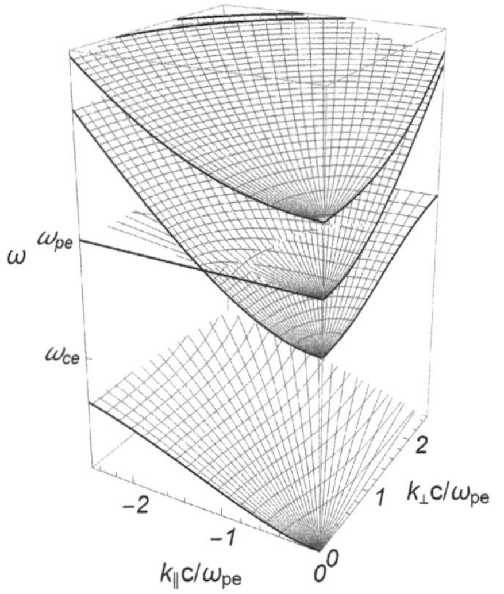

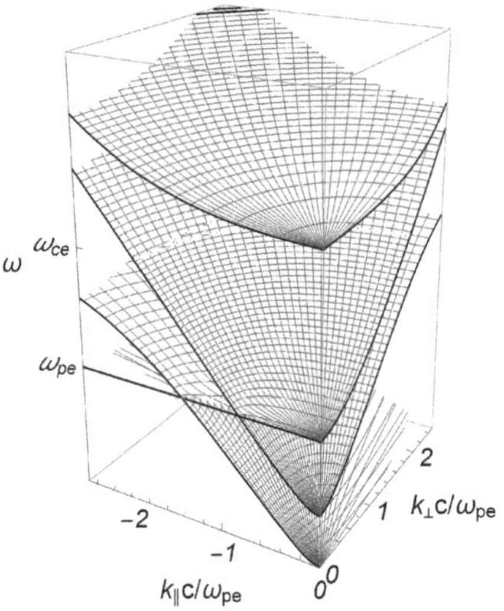

FIGURE 12.3: *Illustration of the dispersion relations given by the Appleton-Hartree equation, here shown for two cases,* $\omega_{pe} = 2\omega_{ce}$ *and* $\omega_{pe} = \omega_{ce}/2$. *Near the ionospheric F-region maximum you will usually find* $\omega_{pe} > \omega_{ce}$. *Neither one of these two figures, nor the following Fig. 12.4, distinguish the wave polarizations.*

The version (12.21) of the Appleton-Hartree equation is most convenient for generalizations to include one or more ion species (Yeh & Liu 1972). For the case where only the effects of the electron dynamics on the wave propagation are considered, we can, after some algebra, simplify (12.21) to become

$$n^2 = 1 - \frac{X}{1 - \dfrac{Y^2 \sin^2\theta}{2(1-X)} \pm \sqrt{\dfrac{Y^4 \sin^4\theta}{4(1-X)^2} + Y^2\cos^2\theta}}. \tag{12.22}$$

This formulation is particularly convenient for analyzing cases where either $\theta \approx 0^\circ$ or $\theta \approx 90^\circ$.

In order to have wave propagation, we require $n^2 \geq 0$, i.e., the right-hand side of (12.21) or (12.22) is positive. If this is the case for some ω, it is always possible to find a value of k which, inserted into $n^2 \equiv c^2k^2/\omega^2$, fulfills (12.21). Frequency intervals where $n^2 \geq 0$ are called "propagation bands," while regions with $n^2 < 0$ are "stop bands" since wave propagation is not possible here.

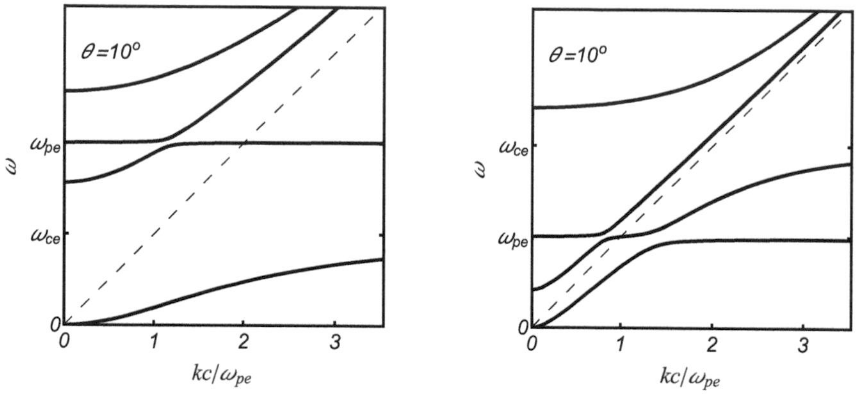

FIGURE 12.4: *Illustration of the dispersion relations given by the Appleton-Hartree equation, here shown for a selected direction* $\theta = 10^\circ$ *with frequency ratios* ω_{pe}/ω_{ce} *as in Fig. 12.3.*

Since free charges ρ are absent, we note that according to (5.7) and (5.8) both **D** and **B** fields are *transverse*, i.e., perpendicular to **k**, while for **E** it is not necessarily so!

- **Exercise:** Consider the limit of (12.14) where $\omega^2 \ll \omega_c^2$, and $\omega^2 \gg \omega_c^2$, respectively, and discuss wave propagation $\parallel$ **B** and $\perp$ **B**. Interpret the results in terms of electron motion $\parallel$ **B** and $\perp$ **B**.
- **Exercise:** Write out the tensor $k^2\underline{1} - \mathbf{kk}$ in all its components, assuming the geometry shown in Fig. 12.2, and derive (12.18).
- **Exercise:** Demonstrate that the two expressions in (12.21) are indeed identical.
- **Exercise:** Derive the relation (12.22).
- **Exercise:** Compare the results for whistler waves from electron MHD, Section 10.7.1, with the cold plasma Appleton-Hartree results shown in Fig. 12.3. Comment on the range of validity for electron MHD.

12.2.2 Wave propagation perpendicular to $\mathbf{B}_0$

The full dispersion relation, as given by (12.21), is rather unhandy, and it is an advantage to consider some special directions of wave propagation separately. In Fig. 12.4 we have such an illustration, but we would like to have cases that are tractable also analytically. First we take the case where the wave propagates in the direction perpendicular to the externally imposed homogeneous magnetic field, $\mathbf{k} \perp \mathbf{B}_0$. After some simple manipulations of (12.19) we find two cases. One is the simple result

$$\omega^2 = \omega_{pe}^2 + c^2k^2, \tag{12.23}$$

which is the same wave mode we found in an *unmagnetized* plasma. This wave has its electric field polarized *along* the magnetic field $\mathbf{B}_0$, i.e., $E_z \neq 0$ while $E_x = E_y = 0$, and in this case the electron dynamics is unaffected by $\mathbf{B}_0$. Consequently, we find the same result as without a magnetic field. This is the plane polarized *ordinary wave*.

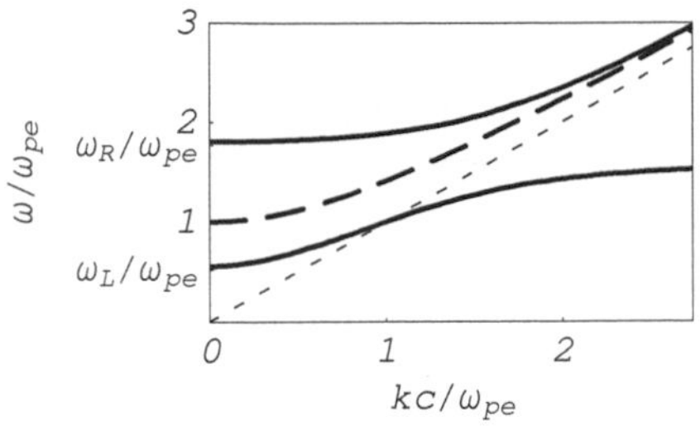

FIGURE 12.5: *Illustration of the dispersion relations for the case of* $\mathbf{B}_0$ *perpendicular propagation,i.e.,* $\theta = 0°$*, for* $\omega_{ce} = 1.25\,\omega_{pe}$*, implying that the two cut-off frequencies are* $\omega_R \approx 1.8\,\omega_{pe}$ *and* $\omega_L \approx 0.55\,\omega_{pe}$ *here, while* $\omega_{uh} \approx 1.6\omega_{pe}$*. The asymptotic limit,* $k \to \infty$ *for the lowest branch is* ω_{uh}*. The dashed line gives the* ordinary *wave. The lower extraordinary branch crosses the thin dashed line that is indicating the speed of light at* $\omega = \omega_{pe}$ *with* $k = \omega_{pe}/c$*.*

For the electric field polarized $\perp \mathbf{B}_0$ we find

$$\frac{c^2k^2}{\omega^2} = 1 - \left(\frac{\omega_{pe}}{\omega}\right)^2 \frac{\omega^2 - \omega_{pe}^2}{\omega^2 - \omega_{pe}^2 - \omega_{ce}^2}. \tag{12.24}$$

This relation has two solutions $\omega = \omega(k)$; they are the *extraordinary waves*. Evidently it is easy to find an expression $k = k(\omega)$, but difficult to find a closed analytical expression $\omega = \omega(k)$. The dispersion relations are illustrated by Fig. 12.5. For these waves we have $E_x \neq 0$ $E_y \neq 0$, while $E_z = 0$.

For $k = 0$, we have $\omega^2 - \omega_{pe}^2 - \omega_{ce}^2 = (\omega_{pe}^2/\omega)^2(\omega^2 - \omega_{pe}^2)$ being a second order equation in ω^2 which is readily solved to give two solutions for ω^2. With a few tricky steps (Chen 2016) we reduce the expression to $\omega^2 \mp \omega\,\omega_{ce} - \omega_{pe}^2 = 0$, giving the two roots

$$\omega_R = \frac{1}{2}\left(\omega_{ce} + \sqrt{\omega_{ce}^2 + 4\omega_{pe}^2}\right) > \omega_{pe} \quad \text{and} \quad \omega_L = \frac{1}{2}\left(-\omega_{ce} + \sqrt{\omega_{ce}^2 + 4\omega_{pe}^2}\right) < \omega_{pe},$$

being the right and left polarized waves we found before with $\omega_R - \omega_L = \omega_{ce}$.

For $k \to \infty$ the *left* side of (12.24) becomes infinite for finite frequencies. The only way the *right*-hand side can become infinite is if $\omega^2 = \omega_{pe}^2 + \omega_{ce}^2$, which *is* a finite constant frequency. The frequency of the lowest ordinary branch approaches a constant value, and in this limit the phase as well as the group velocity vanishes in the present cold plasma formulation. This frequency, the *upper-hybrid frequency*, $\omega_{uh} \equiv \sqrt{\omega_{ce}^2 + \omega_{pe}^2}$, is an example of a *resonance frequency*. With reference to Faraday's law (see Section 5.2), we have in general *electrostatic waves*, $\mathbf{E}\cdot\mathbf{k} \neq 0$ with $\mathbf{E}\times\mathbf{k} = 0$, provided the limit exists where $\omega/k \to 0$ for $k \to \infty$, while at the same time $\omega \neq 0$.

The polarizations of the electric fields for the case with $E_z = 0$ and $\mathbf{k} \parallel \widehat{\mathbf{x}}$ are obtained, for instance, from (12.18) to give

$$\begin{aligned} -\left(1 - \frac{X}{1-Y^2}\right)E_x - i\frac{|Y|X}{1-Y^2}E_y &= 0 \\ i\frac{|Y|X}{1-Y^2}E_x + \left(\frac{c^2}{(\omega/k)^2} - \left(1 - \frac{X}{1-Y^2}\right)\right)E_y &= 0, \end{aligned} \tag{12.25}$$

again with the notation $X \equiv \omega_{pe}^2/\omega^2$ and $|Y| \equiv \omega_{ce}/\omega$. Using the dispersion relation (12.24) for $c^2/(\omega/k)^2$ we find

$$E_x : E_y : E_z = iX|Y| : -(1 - X - Y^2) : 0,$$

implying, for instance, that the electric field becomes longitudinal, $E_y \to 0$, when $\omega \to \omega_{uh}$, as expected for electrostatic waves at a resonance.

A word of caution might be appropriate here: we often see the statement that for electrostatic waves, Maxwell's displacement current is exactly compensating the plasma current. For upper-hybrid waves this statement deserves some scrutiny. Indeed we have $\varepsilon_0 \partial E_x/\partial t = -en_0 u_x$ in the present linear description, but from the electron momentum equation we also have $\partial u_y/\partial t = -eu_x B_0$, implying that we have an electron current $-en_0 u_y$ in the $\widehat{\mathbf{y}}$-direction, which is *not* compensated by any displacement current since we have $E_y = 0$ in this electrostatic limit. The plasma current in the $\widehat{\mathbf{y}}$-direction gives a contribution to $\nabla \times \mathbf{B}$, but this observation remains consistent with the results we obtained from Faraday's law, namely, that $E_\perp/B \to 0$ for electrostatic waves, the waves at a resonance in particular. We have the electric field component $\perp$ **k** vanishing, i.e., $E_\perp \to 0$. With $E_y \to 0$ as $\omega \to \omega_{uh}$ and Maxwell's displacement current compensated by plasma currents in the $\widehat{\mathbf{x}}$-direction, we have from the y-component of Ampere's law (even though the problem here is not static) $\partial B_z/\partial x = \mu_0 e n_0 u_y$, giving $ikB_z = \mu_0 e n_0 u_y$, since all quantities are assumed to vary with the x-coordinate only. On the other hand we have by Faraday's law for the same conditions $\nabla \times \mathbf{E} = 0$ implying $\partial \mathbf{B}/\partial t = 0$, which together with the previous result from Ampere's law would imply that $u_y = 0$. The seeming inconsistency is resolved by emphasizing that the strict electrostatic limit is obtained only when $k \to \infty$, so we have simultaneously $B_z \to 0$ in such a way that the product ikB_z remains finite and equal to $\mu_0 e n_0 u_y$.

- **Exercise:** Express (12.24) in terms of X and Y.

- **Exercise:** Derive (12.23), (12.24) and (12.25) from the Appleton-Hartree equation.

Ⓢ **Exercise:** Consider a model ionosphere with a constant magnetic field $B = 50\ \mu$T and plasma densities relevant for the sunward side of the Earth's ionosphere. Illustrate the density variation of the two cut-off frequencies ω_L and ω_R. Ⓢ

12.2.3 Wave propagation parallel to $\mathbf{B}_0$

We now consider the case where the wave propagates in the direction parallel with the externally imposed homogeneous magnetic field, $\mathbf{k} \parallel \mathbf{B}_0$. We find two relations for circularly polarized waves

$$\frac{c^2 k^2}{\omega^2} = 1 - \left(\frac{\omega_{pe}}{\omega}\right)^2 \frac{\omega}{\omega - \omega_{ce}}, \tag{12.26}$$

called the R-wave rotating in the direction of the electron gyro-rotation, and

$$\frac{c^2 k^2}{\omega^2} = 1 - \left(\frac{\omega_{pe}}{\omega}\right)^2 \frac{\omega}{\omega + \omega_{ce}}, \tag{12.27}$$

called the L-wave. The Appleton-Hartree relation depends on k^2 only, so it does not make any difference to change the direction of the wave-vector.

We recognize what is defined as the two *cut-off frequencies* ω_R and ω_L in the limit of $k \to 0$ for $\mathbf{k} \parallel \mathbf{B}_0$ as well as $\mathbf{k} \perp \mathbf{B}_0$ in (12.24), (12.26) and (12.27). In a way this is natural: imagine that we start out with a circularly polarized electric field in the plane perpendicular to $\mathbf{B}_0$ with constant phase for all space (corresponding to $k = 0$). We would be surprised to see it making any difference for these characteristic frequencies, ω_R or ω_L, whether we then choose to increment **k** *along* $\mathbf{B}_0$ or *perpendicular* to $\mathbf{B}_0$. The cut-off frequencies can be noted also in Fig. 12.3.

At any instant we can construct a linearly polarized wave from a left and a right circularly polarized component. The resulting wave will remain linearly polarized at all later times. Since, however, the two circular constituents propagate with different velocities along the magnetic field, we will experience the direction of the linear polarization to change with time: this is the Faraday rotation (Yeh & Liu 1972, Chen 2016, Piel 2010).

We can have two basically different cases: $\omega_L < \omega_{ce}$ or $\omega_L > \omega_{ce}$. The nontrivial difference is the presence of a *stop band* $\{\omega_L; \omega_{ce}\}$ in the latter case. We find by elementary calculations the criterion $\omega_{pe}^2 > 2\omega_{ce}^2$ for having $\omega_L > \omega_{ce}$. In general, by the arguments given before, we expect cut-offs to be *global*, independent of the polarization of the electric field for $k = 0$. *Resonances*, on the other hand, may be local, i.e., referring to the direction of $\mathbf{k}$, as $k \to \infty$.

The R-wave gives two branches, where the one confined to the interval $\omega \in \{0, \omega_{ce}\}$ is called the *whistler mode*; see also Section 10.7.1. Whistlers are important also historically by being some of the first naturally occurring waves in plasma media to be observed. The dispersion relations are illustrated by Fig. 12.6. As in Section 10.7.1, we have that irrespective of the direction of whistler wave propagation with respect to the ambient magnetic field vector, the electric field rotates in the same direction as the electrons do in their cyclotron motion. As $k \to \infty$, the whistler frequency approaches a limiting value $\omega = \omega_{ce}$, which is another example of a resonance frequency. This limit should be considered with some care, since a cold plasma model is bound to become invalid in a limit where $\omega/k \to 0$. For *any* finite temperature, thermal effects will, in general, become important near resonances.

The polarization of the waves for $\mathbf{B}_0$-parallel propagation is determined by the equations

$$
\begin{aligned}
(\omega^2 - c^2k^2 - \alpha)E_x + i\alpha\frac{\omega_{ce}}{\omega}E_y &= 0 \\
-i\alpha\frac{\omega_{ce}}{\omega}E_x + (\omega^2 - c^2k^2 - \alpha)E_y &= 0
\end{aligned}
$$

with $\alpha \equiv \omega_{pe}^2/(1 - \omega_{ce}^2/\omega^2)$. We readily find that for the L-wave, we have $E_x : E_y : E_z = i : 1 : 0$ for all frequencies, corresponding to a left-hand polarized wave, i.e., the wave starts out being left-handed at $k = 0$ and remains so for all frequencies. For the R-wave we find $E_x : E_y : E_z = -i : 1 : 0$, corresponding to a right-hand polarized wave, except for the whistler wave resonance at $\omega = \omega_{ce}$, where the polarization is not determined, since α diverges there.

We have here assumed ions to form an immobile background neutralizing the electron charge. If also ion motion is allowed, we will find *ion* resonances, this time for the left-hand polarized waves, just as we have electron resonances for the right-hand polarized waves. The left-hand polarization rotates in the direction of a positive ion gyro motions.

The electron MHD limit of the whistler branch (10.92) is recovered if we in (12.26) assume $\omega \ll \omega_{ce}$ and in the resulting expression $c^2k^2/\omega^2 = 1 + \omega_{pe}^2/(\omega\omega_{ce})$ take also $\omega_{pe}^2/(\omega\omega_{ce}) \gg 1$ for small frequencies.

- **Exercise:** Demonstrate that in the high frequency limit, $\omega \gg \omega_{pe}$ and $\omega \gg \omega_{ce}$, the change in angle of the plane of polarization due to the Faraday rotation is $\Delta\theta_F \approx \frac{1}{2}(\omega_{pe}^2\omega_{ce}/\omega^2)(L/c)$, where L is the distance of wave propagation. Since $\Delta\theta_F$ is proportional to the product of density and magnetic field, it can be used as a diagnostic for one of the two if the other one is known. Usually we know the magnetic field relatively accurately, and will then be able to estimate an average value of the plasma density along L by measuring $\Delta\theta_F$.

- **Exercise:** Demonstrate that the whistler branch has its maximum phase velocity at $\frac{1}{2}\omega_{ce}$. What is the magnitude of the phase velocity as well as the group velocity at this frequency? Compare with the speed of light.

- **Exercise:** Write out the linearized equation of electron motion for the special cases where $\mathbf{k} \perp \mathbf{B}_0$ and $\mathbf{k} \parallel \mathbf{B}_0$, respectively. Derive the dispersion relation by this approach, and in particular determine the electron velocities and electron orbits for these cases.

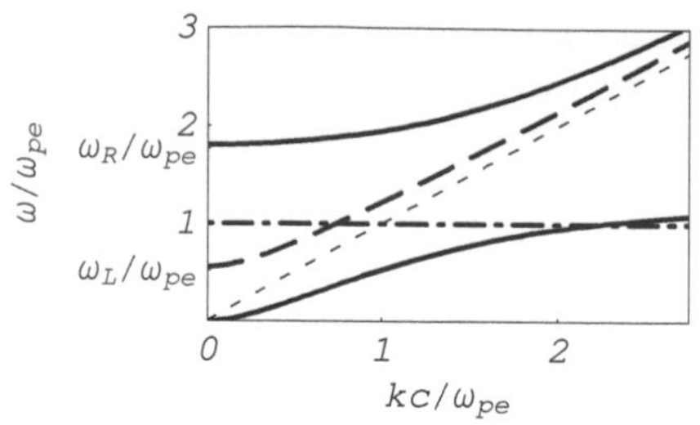

FIGURE 12.6: *Illustration of the dispersion relations for the case of* $\mathbf{B}_0$*-parallel propagation, i.e.,* $\theta = 90^\circ$*, for* $\omega_{ce} = 1.25\,\omega_{pe}$*; see also Fig. 12.5. The dashed line gives the left-hand polarized wave. The lowest branch (the one that starts out as a parabola from* $(\omega, k) = (0, 0)$*) is the right-hand polarized whistler branch. In the asymptotic,* $k \to \infty$*, electrostatic limit, the whistler wave frequency approaches* ω_{ce}*. The electrostatic Langmuir wave, or electron plasma wave, is here a horizontal dot-dashed line at* $\omega/\omega_{pe} = 1$*. A thin dashed line with a slope giving the speed of light is inserted as a reference.*

12.2.4 Wave propagation at an arbitrary angle to $\mathbf{B}_0$

Sections 12.2.2 and 12.2.3 refer to particular directions of propagation with respect to the magnetic field. The problem of *arbitrary* directions becomes much more complicated, unfortunately. Within the present cold plasma approximation, the dispersion relation is of course still described by the Appleton-Hartree relation (12.21). It is not too difficult to make a plot which is effectively the same as a dispersion relation: by use of (12.21), we can select a direction as defined by θ and then find $k = k(\omega)$ for a given set of parameters. This relation can readily be plotted given some program package, and the result is just as good as the more conventional presentation in the form $\omega = \omega(k)$.

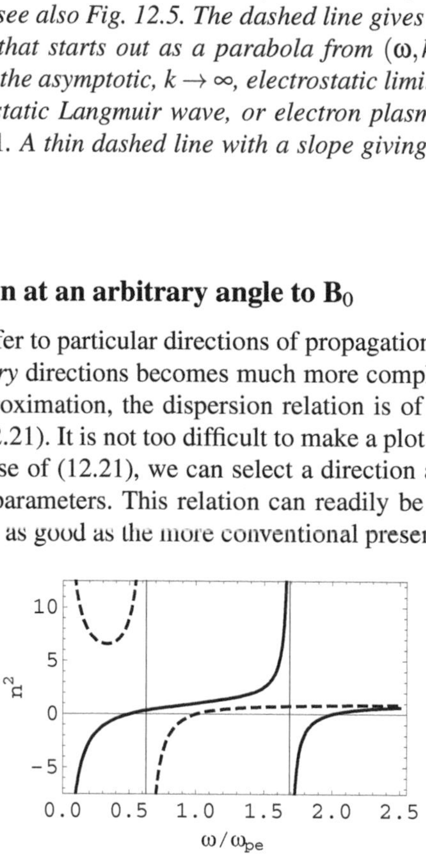

FIGURE 12.7: *Illustration of stop and pass band for the case with* $\theta = 45^\circ$ *and* $\omega_{ce}/\omega_{pe} = 1.25$ *as in Figs. 12.5 and 12.6. The two polarizations are shown by solid and dashed lines, respectively. Vertical lines show frequencies where* $n^2 \to \infty$.

A convenient, although not so detailed overview of the conditions for wave propagation can be obtained by determining the "stop" and "pass" bands for a given choice of direction θ. We use that n^2 has to be positive for wave propagation. *If* we have $n^2 > 0$, it will always be possible to find a combination of ω and k which satisfies the dispersion relation. On the other hand, a negative value for n^2 indicates that no wave propagation is possible for the given choice of plasma parameters and wave polarization. The sign of n^2 is readily determined by (12.21). We show an illustrative case in Fig. 12.7 for the two signs of the square root in n^2, i.e., the two wave polarizations, indicated by solid and dashed lines, respectively. Note that $n^2 \to 1$ for $\omega \to \infty$ for either polarization, as expected. Vertical lines show the asymptotes where $n^2 \to \infty$ at resonances where the phase velocity

$\omega/k \to 0$. A zero crossing of n^2 corresponds to a cut-off. The sign of n^2 is easily recognized in this presentation. Note the large value of n^2 for small frequencies: these are the slow whistler modes.

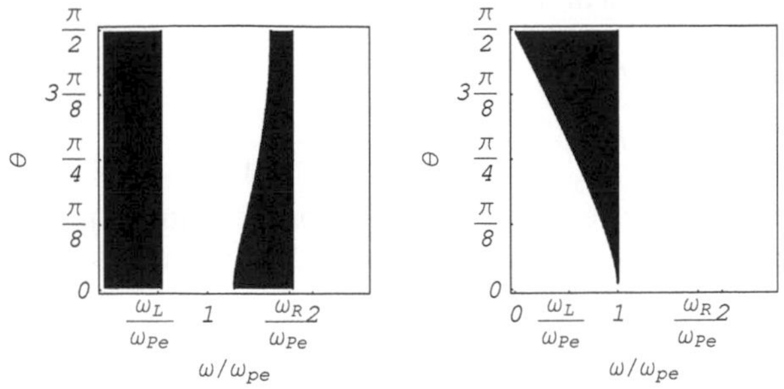

FIGURE 12.8: *The sign of n^2 is positive in the white areas, negative in the dark areas. The case illustrated has $\omega_{ce}/\omega_{pe} = 1.25$ as in Figs. 12.5 and 12.6. The two figures correspond to the two wave polarizations.*

We can generate new figures like Fig. 12.7 for every new choice of θ. To present the information in a more handy form, we show the sign of n^2 in Figs. 12.8 and 12.9 by the white color for positive (pass-band) and dark color for negative (stop-band). We can distinguish two basically different cases, one where $\omega_{ce}/\omega_{pe} > 1$ and another where $\omega_{ce}/\omega_{pe} < 1$. Examples for both cases are shown. By comparison with Figs. 12.5 and 12.6 we can identify the two polarizations. We find also the explanation for the disappearance of one wave mode when we compare Figs. 12.5 and 12.6: we have four wave modes in Fig. 12.6 and three modes in Fig. 12.5! From Figs. 12.8 and 12.9 we find that one pass-band disappears as $\theta \to \pi/2$, so that the line $\omega = 0$ should in reality be included in Fig. 12.5. Then we have four modes in both Figs. 12.5 and 12.6.

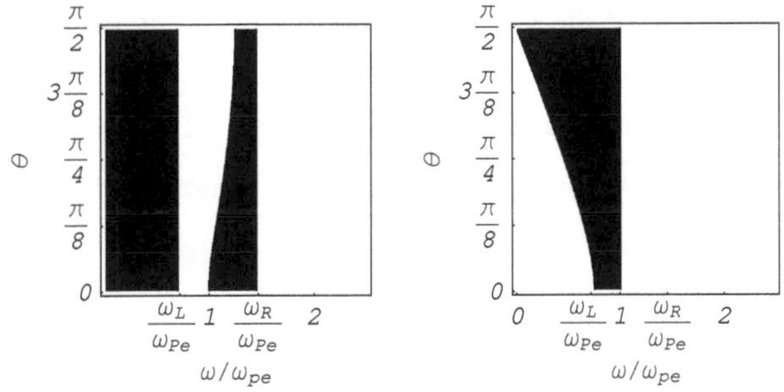

FIGURE 12.9: *The sign of n^2 is positive in the white areas, negative in the dark areas. The case illustrated has $\omega_{ce}/\omega_{pe} = 0.75$. For this case we have $\omega_R \approx 1.44\,\omega_{pe}$, $\omega_L \approx 0.69\,\omega_{pe}$ and $\omega_{uh} = 1.25\,\omega_{pe}$. The two figures correspond to the two wave polarizations. Compare with Fig. 12.8 and also Fig. 12.3.*

Some approximate results can be obtained in certain limits, when the angle is either $\theta \approx 0^\circ$ or $\theta \approx 90^\circ$, i.e., for quasi-normal and quasi-parallel wave propagation with respect to the given homogeneous magnetic field.

12.2.4.1 Quasi-normal wave propagation

The expression (12.22) can be simplified when $\cos^2\theta \ll \frac{1}{4}\sin^4\theta(Y^2/(1-X)^2)$. In this case we have the two solutions (one for each sign in front of the $\sqrt{}$-sign)

$$n_0^2 = 1-X \tag{12.28}$$

$$n_x^2 = 1-\frac{X(1-X)}{1-X-Y^2\sin^2\theta}. \tag{12.29}$$

Within the present limit, we have no θ dependence for the ordinary polarization. For one of the branches of the extraordinary wave (see Figs. 12.3 and 12.5), we find that the resonance frequency ω_{uh} at **B**-normal propagation is modified to $\sqrt{\omega_{pe}^2+\omega_{ce}^2\sin^2\theta}$, for θ close to 90°.

12.2.4.2 Quasi-parallel wave propagation

The expression (12.22) can also be simplified when $\cos^2\theta \gg \frac{1}{4}\sin^4\theta(Y^2/(1-X)^2)$. In this case we have the two solutions (again one for each sign in front of the $\sqrt{}$-sign)

$$n^2 = 1-\frac{X}{1\pm Y\cos\theta}. \tag{12.30}$$

In particular we can analyze whistler wave propagation in the quasi-parallel approximation by using $n_R^2 \approx 1-X/(1-Y\cos\theta)$. The electron cyclotron resonance found at **B**-parallel whistler propagation (see Fig. 12.6) is modified as $\omega_{ce}\cos\theta$ for propagation at an angle θ with respect to **B**: at the resonance frequency where $n^2 \to \infty$, we find a *resonance cone* (Yeh & Liu 1972, Piel 2010), see also (12.20). Strictly speaking, the result found for the resonance cone is not quite correct: the phase velocities become comparable to thermal particle velocities, and corrections due to this need to be included (Piel & Oelerich 1985). The cold-plasma results are, however, useful as guidelines.

- **Exercise:** Express the conditions for quasi-normal and quasi-parallel wave propagation in terms of θ, ω, ω_{pe} and ω_{ce}.
- **Exercise:** Find expressions for the wave polarizations in the quasi-normal and quasi-parallel limits.
- **Exercise:** Identify the frequencies ω_R, ω_L and ω_{uh} in Figs. 12.8 and 12.9.
- **Exercise:** Demonstrate that with immobile ions, the Appleton-Hartree relation (12.21) can be written in the form

$$Ak^4 - B\frac{\omega^2}{c^2}k^2 + C\frac{\omega^4}{c^4} = 0,$$

where (Schmidt 1979) $A = \kappa_T\sin^2\theta + \kappa_\parallel\cos^2\theta$, $B = \kappa_R\kappa_L\sin^2\theta + \kappa_\parallel\kappa_T(1+\cos^2\theta)$ and $C = \kappa_\parallel\kappa_R\kappa_L$, with

$$\kappa_L = 1-\frac{\omega_{pe}^2/\omega^2}{1+\omega_{ce}/\omega}, \qquad \kappa_R = 1-\frac{\omega_{pe}^2/\omega^2}{1-\omega_{ce}/\omega}, \qquad \text{and} \qquad \kappa_\parallel = 1-\frac{\omega_{pe}^2}{\omega^2},$$

introducing also $\kappa_T \equiv \frac{1}{2}(\kappa_R+\kappa_L)$ and $\kappa_H \equiv \frac{1}{2}(\kappa_R-\kappa_L)$.

Demonstrate that we have

$$\tan^2\theta = -\kappa_\parallel\frac{(n^2-\kappa_R)(n^2-\kappa_L)}{(n^2-\kappa_\parallel)(\kappa_T n^2-\kappa_R\kappa_L)}, \tag{12.31}$$

in terms of the index of refraction $n \equiv kc/\omega$. What is the physical distinction between κ_L and κ_R?

Write up the solution for (12.31) for the special case $\theta = 0$.

Write the expressions for the resonance cones in terms of κ_T, κ_L and κ_R. In case one of the components κ_T, κ_L and κ_R are absent for resonance cones, explain why.

12.2.5 Wave propagation in stratified plasmas

We can give a simple illustration of the propagation of wave in an inhomogeneous plasma. The simplest case we can imagine is stratified in a sense that the plasma (or any other medium for that matter) is layered so it consists of a stack of "plates" of uniform thickness, which might vary from one layer to the next; see Fig. 12.10. We assume that the index of refraction for each plate is given as n_j, with $j = 1, 2, \ldots$. Each separating surface between layers j and $j+1$ can be associated with an incidence angle and a transmission angle for the ray, as indicated at the first interface on Fig. 12.10.

We consider a wave ray in the sense discussed in Section 3.2.2 entering layer number 1. We use Snell's law (see Section 3.2.1) to determine the direction of the ray in layer 2, etc. The angle for the ray leaving the interface j is the same as the angle of incidence on the interface $j+1$. We find

$$n_1 \sin\theta_1 = n_2 \sin\theta_2 = \ldots = n_j \sin\theta_j \,, \tag{12.32}$$

or $\sin\theta_1 = (n_j/n_1)\sin\theta_j$. In case the wave is completely reflected at a certain layer j, we must necessarily have $\theta_j = \pi/2$. This can be interpreted as a criterion for total reflection by writing

$$\sin\theta_1 = n_j/n_1 \,, \tag{12.33}$$

as the criterion for an entrance angle θ_1 for having reflection. In other words, if, for the given angle of incidence θ_1, we somewhere at a layer j have an index of refraction satisfying (12.33), the ray is reflected.

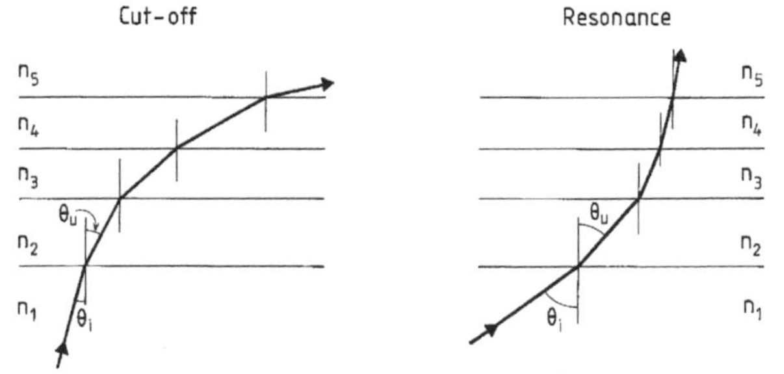

FIGURE 12.10: *Illustration of ray propagation in stratified media. One figure shows propagation toward a cut-off, the other toward a resonance.*

An obvious example for electromagnetic waves propagating toward a cut-off is found in the Earth's ionosphere, where electromagnetic waves launched from the ground propagates from air, with index of refraction $n \approx 1$, toward a plasma where $n \ll 1$. At normal propagation these waves are reflected in case their frequency matches the local plasma frequency at some altitude. If the wave frequency is larger than the largest plasma frequency (typically found in the F-region; see Fig. 6.3), the waves propagate into space. We recall that Snell's law, in effect, states that the wave-vector component *parallel* to the interface is the same in both layers, as easily shown. The "horizontal" wave-vector component $k_\parallel$ is thus constant, and at the reflection point, the wave-vector is equal to the incoming value of $k_\parallel$. With the index of refraction being $n \equiv \sqrt{1-(\omega_{pe}/\omega)^2}$ for a given constant frequency ω in an unmagnetized plasma, we find that an electromagnetic wave is reflected at the position where the local electron plasma frequency and the wave frequency are related by $\sqrt{\omega^2 - \omega_{pe}^2} = ck_\parallel$.

The discussion is entirely similar when it comes to a resonance. In this case we have a region in space where the index of refraction goes to infinity, $n_j \to \infty$, as the phase velocity goes to zero. In this case we must have $\theta_j \to 0$ in order to keep the product $n_j \sin\theta_j$ finite, and the wave-vector is turning to become perpendicular to the stratifications. Electromagnetic waves propagating toward a resonance can be found, for instance, for extraordinary polarization of waves in a *magnetic beach*, where $\mathbf{B} \equiv \{B_x, B_y, B_z\} = \{0, 0, B_0(y)\}$, and the wave-vector is in the (x,y)-plane for Fig. 12.10 to apply.

Note that Snell's law is based on a geometrical observation concerning phase velocities on the two sides of a locally plane interface. We need not know the polarization state of the waves. Discussions of waves at interfaces are often supplemented with expressions for transmission and reflection coefficients, but for these results it is important to know the direction of, for instance, the electric field with respect to the wave-vector (Jackson 1975).

Ⓢ **Exercise:** Consider a horizontally stratified plasma, where we have vacuum in the interval $0 < z < h_0$, and then a plasma with increasing density so that $n = az^2$. An electromagnetic wave is propagating vertically along the z-direction. The frequency ω_0 of the wave is chosen so that it matches the plasma frequency at $z = 2h_0$. The wavelength of the wave in a vacuum is much smaller than h_0. Illustrate the local phase and group velocities as a function of a normalized z-variable. How will you normalize the velocities? Ⓢ

12.2.6 Electrostatic waves in a strongly magnetized waveguide

If we want to verify analytical results concerning wave propagation in plasmas we encounter the problem that most of the simple results refer to infinitely extended homogeneous plasmas. It can be difficult to justify such assumptions in a laboratory, and we have to compromise by defining conditions which are experimentally accessible, at least to some approximation, and then modify our analysis accordingly.

From a laboratory experimental point of view, one particular problem has been of interest. We are here concerned with an experimental situation, where a long (assumed *infinitely* long) waveguide, with circular cross section of radius R, is magnetized with a strong homogeneous magnetic field, $\mathbf{B}_0$, along its axis, taken to be the z-direction. The waveguide is filled with a plasma, where we for the moment assume the temperature to be vanishing. The density may be varying in the radial direction; this will be the case in most practical realizations of the problem. Azimuthal density variations are ignored from the outset. We will later on simplify the problem by assuming the density to be radially uniform as well, but will illustrate where the complications of inhomogeneous densities arise. We consider only high frequency waves at, or below, the electron plasma frequency, and assume also here that the ions form an immobile neutralizing background of positive charge. The metal forming the waveguide is assumed to be an ideal conductor, with infinite conductivity, implying that the electric field inside it is vanishing. The proper boundary condition is then that the tangential part of the electric field in the plasma is vanishing at $r = R$.

The fluctuating electric field can drive currents along as well as across the magnetic field lines, where the latter contribution comes solely from an azimuthal $\mathbf{E}_\perp \times \mathbf{B}_0/B_0^2$-drift of the electrons, with the implied assumption of $\omega_{pe} \ll \omega_{ce}$. By a strong magnetic field, we understand the limit where the azimuthal currents in the plasma are negligible. For time-harmonic oscillations, we find the plasma current in a linearized model to be $\widetilde{\mathbf{J}} = \widetilde{J}\widehat{\mathbf{z}} = -ien_0(r)\widetilde{E}_z/(m\omega)\widehat{\mathbf{z}}$, with $\widetilde{E}_z(r = R) = 0$. Within the present model, it is thus only the z-component of (12.3) that is affected by the presence of the plasma. Since the problem is homogeneous as far as the z-variable is concerned, we can consider a variation like $\exp(-i(\omega t - k_\parallel z))$ from the outset, retaining only the explicit radial variation. Our only concern is then the variables perpendicular to $\mathbf{B}$, since we make a "usual" Fourier transform with respect to the z-variable. It is then an advantage to separate the ∇-operator as $\nabla \to \nabla_\perp + ik_\parallel \widehat{\mathbf{z}}$, where $\nabla_\perp$ refers to the coordinates perpendicular to $\mathbf{B}$. By considering very strong magnetic fields, we imply $\omega_{ce} \to \infty$, or $Y \to \infty$ in (12.16), giving

$$\underline{\varepsilon} \approx \begin{pmatrix} 1 & 0 & 0 \\ 0 & 1 & 0 \\ 0 & 0 & 1 - X \end{pmatrix} \tag{12.34}$$

For this case we can write two equations, one for the z-component of the electric field and one for the **B**-perpendicular component

$$\begin{aligned} -ik_{\parallel}\nabla_{\perp}\cdot\mathbf{E}_{\perp} - \nabla_{\perp}^2 E_z &= \frac{\omega^2}{c^2}(1-X)E_z\,, \\ \nabla_{\perp}\nabla_{\perp}\cdot\mathbf{E}_{\perp} - ik_{\parallel}\nabla_{\perp}E_z - \nabla_{\perp}^2\mathbf{E}_{\perp} + k_{\parallel}^2\mathbf{E}_{\perp} &= \frac{\omega^2}{c^2}\mathbf{E}_{\perp}\,. \end{aligned}$$

Taking the scalar product $\nabla_{\perp}\cdot$ of the second equation, and inserting the result into the first one, we find (Trivelpiece & Gould 1959, Schmidt 1979)

$$\nabla_{\perp}^2 E_z + \left(\frac{\omega^2}{c^2} - k_{\parallel}^2\right)\left(1 - \frac{\omega_{pe}^2(r)}{\omega^2}\right)E_z\,, = 0\,, \tag{12.35}$$

to be solved with the boundary condition $E_z(R) = 0$. The plasma frequency has a radial dependence due to the radial density variation.

With reference to the proposed geometry of the problem we write $\nabla_{\perp}$ in cylindrical geometry. Since we have assumed that the density variation in the plasma column has no poloidal variation, we can assume that the electric field varies as $E_1(r)\exp(-i(\omega t + m\theta - k_{\parallel}z))$. Because of the periodicity of the θ-variation, we have the mode number m as an integer. We write (12.35) in the form

$$\frac{1}{r}\frac{\partial}{\partial r}\left(r\frac{\partial}{\partial r}E_z\right) - \frac{m^2}{r^2}E_z + \left(\frac{\omega^2}{c^2} - k_{\parallel}^2\right)\left(1 - \frac{\omega_{pe}^2(r)}{\omega^2}\right)E_z = 0. \tag{12.36}$$

The relation (12.36) can be seen as an eigenvalue equation for ω with corresponding eigenfunctions $E_z = E_z(r)$, and is in general rather difficult to solve. To simplify the problem, we now assume that the plasma density is constant inside the waveguide, $\omega_{pe} =$ constant and introduce the quantity

$$T^2 \equiv \left(\frac{\omega^2}{c^2} - k_{\parallel}^2\right)\left(1 - \frac{\omega_{pe}^2}{\omega^2}\right),$$

and write, using (12.36), a general solution for the electric field as

$$E(r,z,t) = AJ_m(Tr)\exp(-i(\omega t + m\theta - k_{\parallel}z))\,,$$

where J_m is the m-th order Bessel function of the first kind.

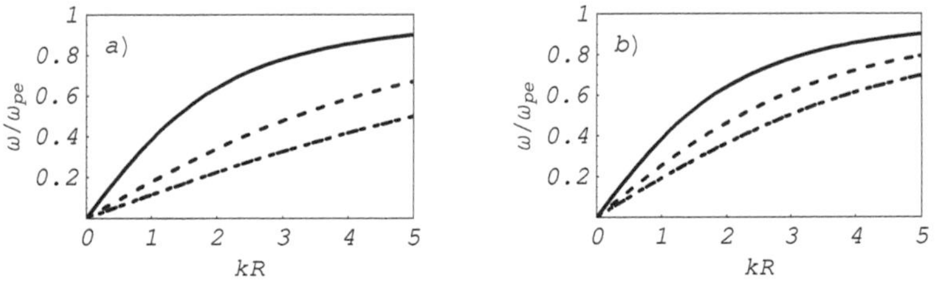

FIGURE 12.11: *Dispersion relation for the Trivelpiece-Gould waves, in a) for* $p_{0,1}$ *(solid line),* $p_{0,2}$ *(dashed line) and* $p_{0,3}$ *(dot-dashed line), and in b) for* $p_{0,1}$ *(solid line),* $p_{1,1}$ *(dashed line) and* $p_{2,1}$ *(dot-dashed line).*

To fulfill the boundary condition at $r = R$ we require $J_m(TR) = 0$, giving the dispersion relation for the guided waves as $TR = p_{m,\nu}$ in terms of the Bessel function zeroes $p_{m,\nu}$, or

$$(kR)^2 = \left(\frac{\omega R}{c}\right)^2 - \frac{p_{m,\nu}^2}{1 - \omega_{pe}^2/\omega^2}\,. \tag{12.37}$$

This relation has two solutions, one corresponding to the high frequency electromagnetic mode, and a low frequency electrostatic mode. The high frequency branch can be approximated by letting $\omega_{pe}^2 \ll \omega^2$ in (12.37), to give

$$\omega^2 = \frac{p_{m,\nu}^2}{R^2} + c^2k^2 .$$

In the present limit, we have the "usual" waveguide mode (Jackson 1975), unaffected by the presence of the plasma, when $\omega_{pe}^2 \ll p_{m,\nu}^2/R^2$.

To "isolate" the lower frequency electrostatic mode, we formally let $c \to \infty$, and find the simplified dispersion relation

$$\omega^2 = \omega_{pe}^2 \frac{(kR)^2}{p_{m,\nu}^2 + (kR)^2}, \tag{12.38}$$

where m is the order of Bessel function we consider, and ν is the actual number of the zero for that function (there are infinitely many for each Bessel function!). The index ν refers to the radial structure of the mode, and m to its azimuthal variation. Examples are shown in Fig. 12.11a) and 12.11b). For fixed wave-number k, the mode corresponding to the J_0-solution and its first zero has the largest phase velocity. Note that for $kR \leq 3$, approximately, the mode corresponding to $p_{0,1}$ has a phase velocity which is larger than for any other mode.

The wave mode (12.38) is often called the *Trivelpiece-Gould* mode. More correctly, it is one limiting case of a general waveguide mode analyzed by Trivelpiece and Gould (1959). It turns out that this particular dispersion relation (12.38) is representative for a large number of laboratory experiments (in Q-machines, in particular) on high frequency electrostatic waves. The physical interpretation of the waves is given by, for instance, Trivelpiece and Gould (1959) and Schmidt (1979).

In the limit of long wavelengths, $k \to 0$, the dispersion relation (12.38) has the form

$$\omega^2 = k^2 \left(\frac{\omega_{pe} R}{p_{m,\nu}} \right)^2, \tag{12.39}$$

which resembles the dispersion relation for an acoustic wave, where now $\omega_{pe}R/p_{m,\nu}$ takes the role of a sound speed.

12.3 Collisional losses

The foregoing discussions explicitly assumed the dielectric tensor to be real, corresponding to conditions without losses. It is a relatively simple matter to include the effect of collisions in the model for the electron dynamics. As long as the ions are considered as an immobile background of positive charge, it does not really matter whether we study electron-ion collisions or electron-neutral collisions; in both cases we simply supplement the present cold plasma equation of motion (12.7) with a term $-\nu mn\mathbf{u}$, accounting for an effective "friction," with ν being the effective collision frequency.

We can combine the terms $\partial\mathbf{u}/\partial t \to -i\omega\mathbf{u}$ and $-\nu\mathbf{u}$ from the two sides of the momentum equation and find that the collisional effects are most easily taken into account by introducing a scalar quantity $U \equiv 1 + i\nu/\omega$, which enters as a multiplier on ω all places where this variable originates from the momentum equations; see also the discussion in Section 8.9.1. This is easy, as long as we use a cold plasma model, since in this case the continuity equations do not enter the discussions of the time derivations of the fluctuating currents in, e.g., (12.3). Since the collision frequency does not enter the continuity equation explicitly, we would make an error by making the replacement $\omega \to \omega + i\nu$ there. To take into account collisional losses (Yeh & Liu 1972) for cold plasmas, we can simply replace ωX by $\omega X/U$, $\mathbf{Y}$ by $\mathbf{Y}/U$ and $1-Y^2$ by $(U^2-Y^2)/U^2$ in the expression for $\underline{\sigma}$. The conductivity tensor for this case becomes

$$\underline{\sigma} \equiv \frac{i\varepsilon_0 \omega X}{U^2 - Y^2} \begin{pmatrix} U & -iY & 0 \\ iY & U & 0 \\ 0 & 0 & (U^2-Y^2)/U \end{pmatrix}. \tag{12.40}$$

From this result, we can obtain the generalized Appleton-Hartree equation by the same procedure as used before, at least in principle.

As an illustration, we consider the extraordinary mode (or, what amounts to give here the same analysis, an unmagnetized cold plasma). In this case we find the index of refraction to be given by

$$n^2 = 1 - \frac{X}{1+iZ}, \tag{12.41}$$

with $Z \equiv \nu/\omega$. The refractive index is now complex, $n \equiv n_1 + in_2$. Taking, as before, the frequency to be real, and $\mathbf{k} \parallel \widehat{\mathbf{z}}$, we easily verify that $\exp(-i(\omega t - \mathbf{k}\cdot\mathbf{r})) \equiv \exp(-i(\omega t - kz)) = \exp(-i\omega(t - n_1 z/c))\exp(-n_2 z/c)$, so the imaginary part n_2 gives rise to an imaginary part of the wave-number k, which for $n_2 > 0$ gives a damping of the wave. This damping is of a different nature than the one discussed in Section 12.1.1, where the penetration depth was associated with a spatially aperiodic variation, i.e., no wave propagation. In the present case the damping is due to loss of wave energy, and the waves propagate while they are damped.

We find the expressions for the real and imaginary parts of the refractive index as

$$(n_1 + in_2)^2 = 1 - \frac{X}{1+Z^2} + i\frac{XZ}{1+Z^2}, \tag{12.42}$$

giving $n_1^2 - n_2^2 = 1 - X/(1+Z^2)$ and $2n_1 n_2 = XZ/(1+Z^2)$. For weak dampings, $Z \ll 1$, we find the approximate relations

$$n_1 \approx 1 - \frac{X}{1+Z^2} \qquad \text{and} \qquad n_2 \approx \frac{1}{2n_1}\frac{XZ}{1+Z^2}, \tag{12.43}$$

where the range of application is limited by the assumption $n_2 \ll n_1$. We let $X = 1$, which would correspond to a cut-off for the case without losses. For the present collisional case we then find $n_1 \approx Z^2$, which implies that the wave now has a finite wave-number, representing a propagating wave. As a consequence, we can have $X < 1$ and *still* have wave propagation, in contrast to the collisionless result, where $X < 1$ corresponds to a "stop-band." Before getting too enthusiastic (for once we might think to get something for nothing), we note that in this case we have $n_2 \approx 1/(2Z) \gg 1$, so, although we *do* have sub-critical wave propagation due to the finite collision frequency, the waves are heavily damped, and we did not gain anything, really.

As an illustration for *magnetized* plasmas, we consider waves propagating *along* the magnetic field lines. After a little algebra, we find the (complex) refractive index to be given by

$$n^2 = 1 - \frac{X}{1 \pm Y_L + iZ}, \tag{12.44}$$

with $Y_L \equiv Y\cos\theta$. For the unmagnetized case, $Y_L = 0$, we have (12.44) reproducing (12.41). The result (12.44) turns out to be a good approximation also for $\theta \neq 0$ as long as we have $\theta \ll \pi/4$. We have for this case

$$n_1 \approx 1 - \frac{X(1\pm Y_L)}{(1\pm Y_L)^2 + Z^2} \qquad \text{and} \qquad n_2 \approx \frac{1}{2n_1}\frac{XZ}{(1\pm Y_L)^2+Z^2}. \tag{12.45}$$

Collisional absorption is most effective for the sign combination $1 - Y_L$ in the denominator. We note quite generally that absorption is proportional to the product of collision frequency and density; this we could have guessed right away.

The present brief summary gives an indication of the consequences of including collisional losses in the cold plasma model. Such effects can be important for naturally occurring plasma, in the Earth's ionosphere, for instance. A general or universal description becomes very lengthy, and it may be advisable to consider selected problems individually.

- **Exercise:** Demonstrate that the dielectric function for a cold collisional electron plasma with immobile ions becomes

$$\varepsilon(\omega, k) = 1 - \frac{\omega_{pe}^2}{\omega(\omega + i\nu)}, \qquad (12.46)$$

where ν is a collision frequency, here taken to be a constant; see also Section 8.10.1. Express the refractive index for this case.

- **Exercise:** Rewrite the Appleton-Hartree equation in the form given in (12.22), but now with collisional effects included.

- **Exercise:** Explain by physical arguments why collisional damping is most effective for the combination $1 - Y_L$ in (12.45).

12.4 Waves including the ion dynamics

As the frequency decreases, the ion dynamics can no longer be ignored. It is a particularly interesting question exactly *how* low the frequency has to be in order for the ion dynamics to be significant. In principle, this problem can be solved by a straightforward generalization of the results in Section 12.2.1. We can include a subscript $_e$ for electrons on ε in (12.15), and obtain the corresponding ion contribution ε_i by the substitution $X \rightarrow (\Omega_{pi}/\omega)^2$ and $\mathbf{Y} \rightarrow \Omega_{ci}/\omega$. Note that the *sign* of Ω_{ci} is important, as emphasized by giving it a bold face, vector representation. The result for the dielectric tensor is then $\underline{\varepsilon} = \underline{\varepsilon}_e + \underline{\varepsilon}_i - \underline{1}$; see, for instance, (5.90). The motion of the ions allows new low frequency waves to propagate. The analysis for cold unmagnetized homogeneous plasmas is relatively trivial, and is left to the reader as an exercise. The ion dynamics is particularly important for magnetized plasmas, and this case will be discussed in some detail here. We only consider some particular directions of propagation, so we might as well solve the equations of motion directly, instead of re-deriving the entire Appleton-Hartree equations, including the ion dynamics.

12.4.1 Lower-hybrid waves

A particularly important wave type is the *lower-hybrid* wave. It is electrostatic, $\mathbf{E} \parallel \mathbf{k}$, and propagates in the direction perpendicular to the magnetic field. First we take a very simple approach, and postulate without much explanation that an electrostatic mode of oscillation exists where both the ion and electron dynamics are important, and associated with a wave propagating in the direction $\perp \mathbf{B}_0$. Take it as an inspired guess, if you like! Since the frequency has to be relatively low for the ion dynamics to be important, it is safe to assume characteristic frequencies to be well below the electron cyclotron frequency. In that case we can take the electron velocity to be $\widetilde{\mathbf{u}}_{e\perp} \approx \widetilde{\mathbf{E}} \times \mathbf{B}_0/B_0^2 - (1/\omega_{ce}B_0)\partial\widetilde{\mathbf{E}}/\partial t$, including the polarization drift as a small correction to the $\widetilde{\mathbf{E}} \times \mathbf{B}_0/B_0^2$-velocity. On the other hand, for electron inertia to be important (as included in the electron polarization drift), we expect that the frequency is not very low either, presumably well above the ion cyclotron frequency. We can then take the velocity of singly charged ions to be $\mathbf{u}_{1i\perp} \approx ie\mathbf{E}_1/(M\omega)$ as for unmagnetized plasmas. The index $_1$ denotes Fourier transformed fields. Since we assumed the waves to be electrostatic, and propagating $\perp \mathbf{B}_0$, we have $\mathbf{E}_1 \perp \mathbf{B}_0$. We assume quasi-neutrality, and have for homogeneous plasma densities that

$$n_0 \nabla \cdot (\mathbf{u}_{1e\perp} - \mathbf{u}_{1i\perp}) = 0, \tag{12.47}$$

by linearizing the continuity equation. Since the $\mathbf{E}_1 \times \mathbf{B}_0/B_0^2$-drift is divergence free, the only contributions to (12.47) come from the electron polarization drift and the ion velocity, implying $\mathbf{E}_1\omega/(\omega_{ce}B_0) = e\mathbf{E}_1/(M\omega)$, or

$$\omega^2 = \Omega_{lh}^2 \equiv \omega_{ce}\Omega_{ci}, \tag{12.48}$$

defining the lower-hybrid frequency, Ω_{lh}. Due to the combination of symbols we made the ion cyclotron frequency appear explicitly, although we assumed from the outset that the ions could be considered in effect unmagnetized! The expense for the assumed quasi-neutrality (or the plasma approximation) is, as we know, that for consistency the result is applicable only as long as the frequencies are well below both the electron and the ion plasma frequencies.

We can estimate the kinetic energy density of electrons and ions, $W_{e,i}$, when they participate in the lower-hybrid oscillation by taking $W_e \equiv \frac{1}{2}mn_0u_{e\perp}^2 \approx \frac{1}{2}mn_0(E/B_0)^2$ (apart from a small correction due to the polarization drift) and $W_i \equiv \frac{1}{2}Mn_0u_{i\perp}^2 \approx \frac{1}{2}Mn_0(eE/M\omega)^2$. We readily find that $W_e = W_i$ when $\omega \equiv \Omega_{lh}$ as given by (12.48). Lower-hybrid waves are particularly effective when we want to impart energy to electrons as well as ions! The electron/ion energy density ratio, W_e/W_i, found here has to be modified when $\Omega_{lh} \geq \Omega_{pi}$. Lower-hybrid waves have been considered for heating in plasma fusion experiments, Tokamaks for instance.

To relax the condition of quasi-neutrality, we retain Poisson's equation, and have by explicit use of the ion and electron continuity equations that

$$k^2\phi_1 = \frac{en_0}{\varepsilon_0\omega}(\mathbf{k}\cdot\mathbf{u}_{1i\perp} - \mathbf{k}\cdot\mathbf{u}_{1e\perp}),$$

using $n_1 = n_0\mathbf{k}\cdot\mathbf{v}_{1\perp}/\omega$ from the continuity equations, and $\mathbf{E}_1 = -i\mathbf{k}\phi_1$. With a little algebra, we obtain the handy formula

$$\frac{1}{\omega^2} = \frac{1}{\Omega_{lh}^2} \equiv \frac{1}{\omega_{ce}\Omega_{ci}} + \frac{1}{\Omega_{pi}^2}, \tag{12.49}$$

which reproduces (12.48) in the quasi-neutral limit, $\Omega_{pi} \to \infty$. The result (12.49) is found by using the electron polarization current as the correction to the $\mathbf{E} \times \mathbf{B}_0/B_0^2$-drift as before. Note that the lower-hybrid frequency is independent of density in the quasi-neutral limit, while Poisson's equation introduces an explicit density dependence through Ω_{pi}^2. This correction is important in, for instance, most parts of the Earth's upper ionosphere (Pécseli et al. 1996, Schuck et al. 2003), in particular also for ionospheric heating experiments (Kuo 2015). The waves are also important for heating of plasmas in fusion experiments (Porkolab 1977): in this case the near equal distribution of energy between electrons and ions can be of importance.

The results (12.48) and (12.49) assume waves propagating in the direction *exactly* perpendicular to $\mathbf{B}_0$. If we allow even a slight wave-number component along $\mathbf{B}_0$ (implying an electric field component $\parallel \mathbf{B}_0$), the electron velocity can be large in this direction, much larger than the $\perp \mathbf{B}_0$ component, while the ion velocity is along $\mathbf{E}$ in either case, since we assumed the ions to be in effect unmagnetized. The critical angle, where the magnitude of the $\mathbf{B}_0$-parallel electron velocity component equals the $\mathbf{B}_0$-perpendicular component is approximately $\theta \approx \sqrt{m/M}$, as easily demonstrated. To obtain a dispersion relation of the electrostatic lower-hybrid waves with $k_\parallel \neq 0$, we take the full expression for the (cold) magnetized electron dynamics, retaining an unmagnetized ion model, and find

$$\omega^2 = \frac{\Omega_{lh}^2}{k^2}\left(k^2 + \frac{M}{m}k_\parallel^2\right) \equiv \Omega_{lh}^2\left(1 + \frac{M}{m}\cos^2\theta\right), \tag{12.50}$$

where θ is again the angle between $\mathbf{k}$ and the magnetic field $\mathbf{B}$. For fixed direction of propagation, as given by $\hat{\mathbf{k}}$, the wave frequency is constant for this cold plasma limit, but the waves are strongly *directionally* dispersive (Verdon et al. 2009).

As the wavelengths are increased, the lower-hybrid waves acquire an increasing electromagnetic component, and for $k \ll \omega_{pe}/c$ they continue into the dispersive part of the magnetosonic branch (Stringer 1963).

Ⓢ **Exercise:** Illustrate the variation of the lower-hybrid wave frequency for varying ion plasma frequency. The exercise seems nearly trivial, but the point is to decide which normalization is the most appropriate. Ⓢ

- **Exercise:** Demonstrate that the lower-hybrid frequency for a multi ion species plasma is obtained as the larger solution of

$$1+\left(\frac{\omega_{pe}}{\omega_{ce}}\right)^2 = \sum_j \frac{\Omega_{pj}^2}{\Omega_{lh}^2 - \Omega_{cj}^2}$$

assuming $\Omega_{lh}^2 \ll \omega_{ce}^2$, where we labeled the various ion species by j. The result has direct applications to the multi-component plasma in the Earth's upper ionosphere, where lower-hybrid waves are often observed by instrumented satellites.

- **Exercise:** Derive the dispersion relation (12.50).

Ⓢ **Exercise:** The expression (12.49) accounts for the frequency of electrostatic waves propagating $\perp$ $\mathbf{B}$ in a cold plasma. We have another wavetype, the upper-hybrid wave at the resonance frequency $\omega_{uh} \equiv \sqrt{\omega_{ce}^2 + \omega_{pe}^2}$. Why did it disappear in the derivation of (12.49)? Our analysis includes, after all, also the electron dynamics. Derive a general expression containing both ω_{uh} and Ω_{lh}. Ⓢ

12.4.2 Alfvén waves

To investigate low frequency electromagnetic waves in a magnetized plasma we assume (again) that the relevant frequencies are much smaller than the ion cyclotron frequency. We let the wave-vector be along $\mathbf{B}_0$ and the electric field be polarized in the direction perpendicular to $\mathbf{B}_0$, and have the electron velocity as $\mathbf{u}_{e\perp} \approx \mathbf{E}_1 \times \mathbf{B}_0/B_0^2$. The corresponding ion velocity is then $\mathbf{u}_{i\perp} \approx \mathbf{E}_1 \times \mathbf{B}_0/B_0^2 + (1/\Omega_{ci})\partial(\mathbf{E}_1/B_0)/\partial t$, including the polarization drift as a small correction to the $\mathbf{E}_1 \times \mathbf{B}_0/B_0^2$-velocity. The net plasma current density is then $\mathbf{J}_1 = (en_0/\Omega_{ci}B_0)\partial\mathbf{E}_1/\partial t$. The shear wave is incompressible, so there are no density perturbations in homogeneous plasmas.

By use of (12.3) we find after Fourier transform with respect to spatial as well as temporal variables

$$\varepsilon_0(\omega^2 - c^2k^2)E_1 = -i\omega n_0 e\left(\frac{-i\omega}{\Omega_{ci}B_0}E_1\right),$$

which gives the dispersion relation

$$\omega^2 = k^2\frac{c^2}{1+(n_0M/\varepsilon_0 B_0^2)} \equiv k^2\frac{c^2}{1+c^2/V_A^2} \xrightarrow{c^2 \gg V_A^2} k^2V_A^2, \tag{12.51}$$

where we introduced the Alfvén velocity $V_A \equiv B_0/\sqrt{\mu_0 n_0 M}$. In the limit $V_A^2 \ll c^2$, the result (12.51) reduces to the familiar Alfvén wave dispersion relation $(\omega/k)^2 = V_A^2$ for waves propagating along $\mathbf{B}_0$. This result was known already from Section 10.5.1, but we have now learned that the fluctuating current associated with Alfvén waves is due to the *ion* polarization drift. A detailed and more general analysis of low frequency wave modes in magnetized plasmas is presented, for instance, by Stringer (1963).

12.4.3 The Hall-MHD model

We can use elements of Section 11.1 to introduce a generalization of the simple compressible MHD equations, using here a cold plasma model for simplicity. Retaining an idealized plasma model with infinite conductivity, $\sigma \to \infty$, we can generalize the ideal MHD model (Cramer 2001) by including the Hall term, giving

$$\mathbf{E} + \mathbf{u} \times \mathbf{B} = \frac{M}{\rho e} \mathbf{J} \times \mathbf{B}, \tag{12.52}$$

see (11.10), where we assume slow temporal variations and ignore $\partial \mathbf{J}/\partial t$, and large spatial scales to ignore ∇p_e. Evidently, the $\mathbf{B}$-parallel part of (12.52) is trivial, while the Hall term gives a significant modification of the $\mathbf{B}$-perpendicular part of the ideal MHD model as the frequency increases to become comparable to the ion cyclotron frequency. This is readily verified by scaling arguments as in Section 11.1.

Using (12.52) to eliminate $\mathbf{E}$, the basic compressible Hall-MHD equations become

$$\frac{\partial}{\partial t}\mathbf{B} = \nabla \times (\mathbf{u} \times \mathbf{B}) - \nabla \times \left(\frac{M}{\rho e \mu_0} (\nabla \times \mathbf{B}) \times \mathbf{B} \right) \tag{12.53}$$

$$\frac{\partial}{\partial t}\rho + \mathbf{u} \cdot \nabla \rho = -\rho \nabla \cdot \mathbf{u} \tag{12.54}$$

$$\rho \left(\frac{\partial}{\partial t}\mathbf{u} + \mathbf{u} \cdot \nabla \mathbf{u} \right) = \frac{1}{\mu_0} (\nabla \times \mathbf{B}) \times \mathbf{B}, \tag{12.55}$$

where a pressure term in (12.55) is omitted in the present cold plasma model. Note that we have retained Ampere's law in its form (10.4), since this equation only assumes phase velocities for transverse waves being much smaller than the speed of light in a vacuum, irrespective of frequency, while (12.52) refers to an explicit frequency scaling.

Linearizing the equations, we find after Fourier transform and some simple algebra the relation

$$\omega^2 \rho_0 \mu_0 \mathbf{B} = \mathbf{k} \times [((\mathbf{k} \times \mathbf{B}) \times \mathbf{B}_0) \times \mathbf{B}_0] + i\omega \frac{M}{e} \mathbf{k} \times ((\mathbf{k} \times \mathbf{B}) \times \mathbf{B}_0), \tag{12.56}$$

where M is the ion mass. For small ω, the first term on the right-hand side of (12.56) is dominating, and we recover the results of ideal cold compressible MHD in that limit, the dispersion relation for linearly polarized waves in particular. For larger frequencies, we see that the last term in (12.56) prohibits a simple solution in terms of linearly polarized waves. As an illustration, we consider the particular case where $\mathbf{k} \parallel \mathbf{B}_0 \parallel \widehat{\mathbf{z}}$, and introduce a circularly polarized magnetic field $\mathbf{B} = a\widehat{\mathbf{x}} \pm ia\widehat{\mathbf{y}}$ to find the dispersion relation

$$\omega^2 = V_A^2 k^2 \pm \frac{\omega}{\Omega_{ci}} V_A^2 k^2, \tag{12.57}$$

where we again introduced the Alfvén velocity V_A. The two signs give rise to a *fast* (+ sign) and a *slow* ($-$sign) wave. The fast mode is right-hand polarized, the slow left-hand polarized. In the latter case, the electric field vector is rotating in the same direction as the ions with respect to the magnetic field $\mathbf{B}_0$.

An approximate solution of (12.57) is found as

$$\omega^2 \approx V_A^2 k^2 \left(1 \pm \frac{1}{\Omega_{ci}} V_A k \right). \tag{12.58}$$

If for $\mathbf{k} \parallel \mathbf{B}_0$ we continue along the fast and slow wave dispersion relations toward higher frequencies, the analysis breaks down. A more detailed analysis reveals that the slow mode approaches ion cyclotron frequency, in the present cold plasma model. The slow mode has a resonance at Ω_{ci}, where it becomes electrostatic. The fast mode crosses the ion cyclotron frequency unaffected by it, and ends up as the whistler branch, where frequencies become so large that the electron contribution becomes dominant, see e.g., Section 10.7.1.

For arbitrary wave-vector directions, Hall-MHD gives a dispersion relation in the form (Cramer 2001)

$$\left(\omega^2 - V_A^2 k_\parallel^2\right)\left(\omega^2 - V_A^2 k^2\right) = \left(\frac{\omega}{\Omega_{ci}}\right)^2 V_A^4 k_\parallel^2 k^2 . \tag{12.59}$$

The two parentheses on the left side set equal to zero give the shear and the compressional Alfvén waves, respectively. The right-hand side of (12.59) is negligible in ideal MHD, but for finite frequencies it couples the two branches. The compressional becomes the fastest mode, the shear Alfvén wave the slowest of the two modes. For propagation $\perp$ **B** the compressional magnetosonic mode becomes the lower-hybrid wave as the frequency ω becomes large, so also this mode ends up being electrostatic for large frequencies.

12.4.4 Multi ion species

The presence of several ion species can open some particular modes of wave propagation which do not exist in plasmas with only one ion component. Qualitatively, we can argue as follows: the high frequency modes are in general supported by light negatively charged particles (electrons) moving in the background of positively charged heavy species. If we now have *two* ion masses present with a large mass separation, we can anticipate that the light ion component can oscillate with the other one, constituting a charge neutralizing background for the electrons and the light ions. In particular, we emphasize that we *might* have negative ions present. For instance, chlorine has a large electron affinity.

For a generalization of (12.26) and (12.27) for inclusion of ions, or more generally, several ion species, we have for waves with $\mathbf{k} \parallel \mathbf{B}_0$ (Yeh & Liu 1972)

$$n_R^2 \equiv \frac{c^2k^2}{\omega^2} = 1 - \frac{1}{\omega}\sum_\ell \frac{\omega_{p\ell}^2}{\omega + \omega_{c\ell}} \tag{12.60}$$

$$n_L^2 \equiv \frac{c^2k^2}{\omega^2} = 1 - \frac{1}{\omega}\sum_\ell \frac{\omega_{p\ell}^2}{\omega - \omega_{c\ell}}, \tag{12.61}$$

in terms of the appropriate index of refraction n, and the plasma frequencies, and cyclotron frequencies, for the various species, i.e., electrons and ion components, labeled by ℓ. The fraction of the corresponding species with respect to the total plasma density enters through the definition of the appropriate plasma frequency $\omega_{p\ell} \equiv \sqrt{q_\ell^2 n_\ell/\varepsilon_0 m_\ell}$. The signs of the appropriate charges enter through $\omega_{c\ell} \equiv q_\ell B_0/m_\ell$. For high frequencies, we can assume the ions to be infinitely massive, and recover the previous results, e.g., (12.26) and (12.27), for right and left-hand polarized waves, and we indicated this by the subscripts R and L on n. Note that the right-hand polarized whistler branch has a resonance at the electron cyclotron frequency, but it passes through any resonance frequency associated with positive ions, as expected for this polarization.

Assume now for illustration only that we have two positive ion species denoted by A and B, with a large mass ratio, $m_A/m_B \ll 1$. (This may not be so readily found in nature; the mass ratio of hydrogen and oxygen, for instance, is ~ 16, which is hardly "much larger than unity.") Assuming that we consider frequencies $\omega \gg \omega_{cA}$ but $\omega \ll \omega_{cB} \ll \omega_{ce}$, we can make an approximation of

(12.61) in the form

$$\begin{aligned}\frac{c^2k^2}{\omega^2} &= 1-\frac{1}{\omega}\sum_{\ell}^{3}\frac{\omega_{p\ell}^2}{\omega-\omega_{c\ell}} \\ &\approx 1-\frac{\omega_{pe}^2}{\omega\omega_{ce}}+\frac{\Omega_{pB}^2}{\omega\Omega_{cB}}-\frac{\Omega_{pA}^2}{\omega^2} \\ &= 1-\frac{\Omega_{pA}^2}{\omega\Omega_{cA}}-\frac{\Omega_{pA}^2}{\omega^2}=1-\frac{\Omega_{pA}^2}{\omega^2\Omega_{cA}}(\omega+\Omega_{cA})\approx 1-\frac{\Omega_{pA}^2}{\omega\Omega_{cA}},\end{aligned} \tag{12.62}$$

where we reintroduced capital Greek letters for the characteristic ion frequencies for clarity. We also used that for a charge neutral plasma we have $\sum_\ell \omega_{p\ell}^2/\omega_{c\ell}=0$, giving the relation $-\omega_{pe}^2/\omega_{ce}+\Omega_{pB}^2/\Omega_{cB}=-\Omega_{pA}^2/\Omega_{cA}$. The result (12.62) describes whistler-like waves, but here the light and heavy particles are two ion species rather than electrons and ions. The validity of the result is by its derivation restricted to $\Omega_{cA}\ll\omega\ll\Omega_{cB}$.

Ⓢ **Exercise:** Demonstrate that the relations (12.60) and (12.61) for plasmas with multi ion species can be written as

$$\frac{c^2k^2}{\omega^2} = 1+\sum_\ell\frac{\omega_{p\ell}^2}{\omega_{c\ell}(\omega+\omega_{c\ell})} \tag{12.63}$$

$$\frac{c^2k^2}{\omega^2} = 1+\sum_\ell\frac{\omega_{p\ell}^2}{\omega_{c\ell}(\omega-\omega_{c\ell})}, \tag{12.64}$$

for a charge neutral plasma. Ⓢ

Ⓢ **Exercise:** Derive by use of (12.60)–(12.61), or by (12.63)–(12.64), the results for the two circular polarizations of Alfvén waves propagating along $\mathbf{B}_0$, as obtained from Hall-MHD; see Section 12.4.3. Ⓢ

• **Exercise:** Determine how the energy density of a whistler wave is distributed on electric, magnetic and electron kinetic energy densities. Consider in particular also the limit of electron MHD.

12.5 Quasi-electrostatic approximation

We have seen that a magnetized plasma allows the electric field polarization to change from transverse (electromagnetic) to longitudinal (electrostatic) when we follow a dispersion relation from some finite wave-number to large wave-numbers. Strictly speaking, we have electrostatic waves only in the limit $k\to\infty$. We ask for what *finite* wave-number can we consider the waves to be electrostatic without making any significant error (Chen 1987)?

We use the "wave equation" in the form (12.17)

$$\left(k^2\underline{1}-\mathbf{kk}-\left(\frac{\omega}{c}\right)^2\underline{\varepsilon}\right)\cdot\mathbf{E}=0,$$

where the electric field can be separated into a longitudinal $\mathbf{E}_\ell$ and a transverse $\mathbf{E}_t$ part as

$$\mathbf{E}=E_\ell\widehat{\mathbf{k}}+E_t\widehat{\mathbf{k}}_\perp$$

where $\widehat{\mathbf{k}}\cdot\widehat{\mathbf{k}}_\perp = 0$. Taking the scalar product of the unit vector $\widehat{\mathbf{k}}$ and (12.17) as reproduced before, we have

$$E_\ell \widehat{\mathbf{k}}\cdot\underline{\varepsilon}\cdot\widehat{\mathbf{k}} + E_t \widehat{\mathbf{k}}\cdot\underline{\varepsilon}\cdot\widehat{\mathbf{k}}_\perp = 0. \tag{12.65}$$

A similar manipulation of (12.17), this time with the transverse unit vector $\widehat{\mathbf{k}}_\perp$, gives

$$E_t\left(\left(\frac{ck}{\omega}\right)^2 - \widehat{\mathbf{k}}_\perp\cdot\underline{\varepsilon}\cdot\widehat{\mathbf{k}}_\perp\right) - E_\ell \widehat{\mathbf{k}}_\perp\cdot\underline{\varepsilon}\cdot\widehat{\mathbf{k}} = 0. \tag{12.66}$$

Recall that we have the index of refraction as $n \equiv ck/\omega$.

We now anticipate that we can define a frequency ω_ℓ which gives a purely electrostatic wave where

$$\widehat{\mathbf{k}}\cdot\underline{\varepsilon}(\omega_\ell)\cdot\widehat{\mathbf{k}} = 0.$$

The actual wave frequency is then written as $\omega = \omega_\ell + \delta\omega_\ell$, anticipating that for *almost* longitudinal waves we have $\delta\omega_\ell/\omega_\ell \ll 1$. For this case we find the approximation

$$E_\ell\, \delta\omega_\ell \frac{\partial}{\partial\omega} \widehat{\mathbf{k}}\cdot\underline{\varepsilon}(\omega)\cdot\widehat{\mathbf{k}}\Big|_{\omega=\omega_\ell} + E_t \widehat{\mathbf{k}}\cdot\underline{\varepsilon}(\omega_\ell)\cdot\widehat{\mathbf{k}}_\perp = 0, \tag{12.67}$$

and

$$E_t = \frac{E_\ell \widehat{\mathbf{k}}_\perp\cdot\underline{\varepsilon}(\omega_\ell)\cdot\widehat{\mathbf{k}}}{(ck/\omega_\ell)^2 - \widehat{\mathbf{k}}_\perp\cdot\underline{\varepsilon}(\omega_\ell)\cdot\widehat{\mathbf{k}}_\perp}. \tag{12.68}$$

Inserting (12.68) into (12.67) we can cancel E_t and find the criterion for quasi-electrostatic polarizations to be (Chen 1987)

$$\left|\frac{\delta\omega_\ell}{\omega_\ell}\right| = -\left|\frac{\widehat{\mathbf{k}}\cdot\underline{\varepsilon}(\omega_\ell)\cdot\widehat{\mathbf{k}}_\perp\, \widehat{\mathbf{k}}_\perp\cdot\underline{\varepsilon}(\omega_\ell)\cdot\widehat{\mathbf{k}}}{\left((ck/\omega_\ell)^2 - \widehat{\mathbf{k}}_\perp\cdot\underline{\varepsilon}(\omega_\ell)\cdot\widehat{\mathbf{k}}_\perp\right)\omega_\ell \partial\, \widehat{\mathbf{k}}\cdot\underline{\varepsilon}(\omega)\cdot\widehat{\mathbf{k}}\Big/\partial\omega\Big|_{\omega=\omega_\ell}}\right| \ll 1. \tag{12.69}$$

We can make the denominator look nicer by noting that here

$$\omega_\ell \frac{\partial}{\partial\omega}\widehat{\mathbf{k}}\cdot\underline{\varepsilon}(\omega)\cdot\widehat{\mathbf{k}}\Big|_{\omega=\omega_\ell} = \frac{\partial}{\partial\omega}\omega\widehat{\mathbf{k}}\cdot\underline{\varepsilon}(\omega)\cdot\widehat{\mathbf{k}}\Big|_{\omega=\omega_\ell}.$$

12.5.1 Upper-hybrid waves

Considering, in particular, electrostatic waves in magnetized plasmas, we encounter the case for propagation along the magnetic field, where the wave at frequency ω_{pe} is electrostatic for all wavenumbers, while for propagation in the **B**-transverse direction the waves are electrostatic only for $k_\perp \to \infty$, at the upper-hybrid resonance $\omega = \omega_{uh}$. The question naturally arises: what are the constraints on the wave-vector allowing the wave to be treated as electrostatic? A detailed analysis (Dysthe et al. 1978) gives that within the present cold fluid model taking $\omega_{pe} > \omega_{ce}$, the electron waves can be taken as electrostatic provided

$$\frac{\omega}{\omega_{ce}}\left(\frac{\omega_{pe}}{kc}\right)^2 \sin\theta \ll 1, \tag{12.70}$$

where θ is again the angle between the magnetic field and the wave-vector $\mathbf{k}$, i.e., $\sin\theta = k_\perp/k$, and ω is given by the dispersion relation via k and θ. In the limit (12.70) we have the electric field component along $\mathbf{k}$ to be larger than in the direction $\perp \mathbf{k}$. The waves become electromagnetic for $k \to 0$, except when $\theta = 0$

12.5.2 Lower-hybrid waves

Lower-hybrid waves (see Section 12.4.1) are electrostatic when $k_\perp \gg k_\parallel$ and $\omega \sim \Omega_{pi} \ll \omega_{ce}, \omega_{pe}$, where the subscripts $\perp$ and $\parallel$ here refer to the wave vector components taken with respect to the ambient magnetic field. The dispersion relation is here given by

$$\widehat{\mathbf{k}} \cdot \underline{\varepsilon}(\omega) \cdot \widehat{\mathbf{k}} \approx 1 - \left(\frac{\Omega_{pi}}{\omega}\right)^2 + \left(\frac{\omega_{pe}}{\omega_{ce}}\right)^2 - \left(\frac{\omega_{pe} k_\parallel}{\omega k}\right)^2 = 0 .$$

Since $\widehat{\mathbf{e}}_k \cdot \widehat{\mathbf{e}}_\parallel = k_\parallel / k$, we find

$$\widehat{\mathbf{k}}_\perp \cdot \underline{\varepsilon}(\omega) \cdot \widehat{\mathbf{k}} \approx \widehat{\mathbf{k}} \cdot \underline{\varepsilon}(\omega) \cdot \widehat{\mathbf{k}}_\perp \approx -\left(\frac{\omega_{pe}}{\omega}\right)^2 \frac{k_\parallel}{k} .$$

After some calculations assuming $(ck)^2 \gg \omega_{pe}^2$ we find (Chen 1987) the criterion for quasi-electrostatic lower-hybrid waves to be

$$\left|\frac{\delta\omega_\ell}{\omega_\ell}\right| \approx \frac{1}{2} \frac{\left(\omega_{pe} k_\parallel / \omega k\right)^4}{1 + (\omega_{pe}/\omega_{ce})^2} \left(\frac{\omega}{c k_\parallel}\right)^2 \ll 1 ,$$

where we have to insert $\omega = \omega(\mathbf{k})$ as determined from the dispersion relation.

12.6 Quasi-transverse approximation

We extend the exposition of Section 12.5 to ask when wave polarizations can be taken to be *transverse* with good approximation. We introduce a frequency $\omega = \omega_t + \delta\omega_t$, where the subscript now indicates transverse waves. Let ω_t be given by

$$\left(\frac{ck}{\omega_t}\right)^2 - \widehat{\mathbf{k}}_\perp \cdot \underline{\varepsilon}(\omega_t) \cdot \widehat{\mathbf{k}}_\perp = 0 .$$

Following the procedure from Section 12.5, we find the criterion for quasi-transverse polarizations to be (Chen 1987)

$$\left|\frac{\delta\omega_t}{\omega_t}\right| = -\left|\frac{\widehat{\mathbf{k}} \cdot \underline{\varepsilon}(\omega_t) \cdot \widehat{\mathbf{k}}_\perp \; \widehat{\mathbf{k}}_\perp \cdot \underline{\varepsilon}(\omega_t) \cdot \widehat{\mathbf{k}}}{\widehat{\mathbf{k}} \cdot \underline{\varepsilon}(\omega_t) \cdot \widehat{\mathbf{k}} \; \omega_t \; \partial\left((ck/\omega)^2 - \widehat{\mathbf{k}}_\perp \cdot \underline{\varepsilon}(\omega) \cdot \widehat{\mathbf{k}}_\perp\right) \Big/ \partial\omega \Big|_{\omega=\omega_t}}\right| \ll 1 . \tag{12.71}$$

We can rewrite the denominator also in this case, but this is, after all, merely a cosmetic change.

- **Example:** Assume that a spacecraft, a rocket or a satellite has detected some low frequency wave phenomenon, where variations of electric and magnetic fields are measured. Consider this as a point measurement, so you know little if anything about the relevant wavelengths, etc. Often our knowledge about the plasma conditions leaves much to be desired, so it may not offhand be evident what wave type we are dealing with. It is not easy to guess the nature of the waves simply from the satellite frame frequency. One feasible idea is to take the ratio between the root-mean-square (RMS) absolute values of the electric and magnetic fields. Comparing this ratio with an estimate of the Alfvén velocity may give a hint (Eriksson et al. 1994). If we happen to find $|E|/|B| \gg V_A$, or even $|E|/|B| \gg c$, it may be argued that the waves are unlikely to be of electromagnetic origin, since at low frequencies it is unlikely to find electromagnetic waves faster that an Alfvén wave or the speed of light. According to Faraday's law, the phase velocity is given by the ratio of the fluctuating electric field component perpendicular to **k** and the fluctuating magnetic field that is always perpendicular to **k**. Any "excess" electric field must be parallel to the wave-vector, and consequently we can have $|E|/|B| \gg V_A$, or even $|E|/|B| \gg c$ indicating electrostatic waves.

12.7 Instabilities

Significant modifications of the results of this section result in case the various species are moving with respect to each other. The most significant change will be that waves can become *unstable*. These types of problems will be discussed in more detail in a section concerned with kinetic plasma models, but a simple case can be outlined here. We can, for instance, let the electron component be moving with respect to the ions. We retain a cold plasma model, and implied in the basic assumptions is then that the electron drift velocity u_0 must be much larger than the thermal velocity, a non-trivial implication for realistic plasmas. The plasma carries a dc-current in the present case, and there is *free energy* in the system. If the ions are to be considered as an immobile background of positive charge, the result is of minor consequence, basically resulting in a change in frame of reference. If the ion dynamics are included in the analysis, the free energy in the electron flow can be converted into organized motion in the form of waves increasing in amplitude.

Inclusion of an electron drift is at least formally quite simple: in the electron contribution to σ we simply replace ω with $\omega - \mathbf{k} \cdot \mathbf{u}_0$ everywhere. It is assumed implicitly that the current associated with the electron flow is so small that we can keep the magnetic field unchanged, as maintained by some "external" currents, flowing in coils or elsewhere. As far as the ion contribution to ε is concerned, we keep ω there without a Doppler shift, corresponding to stationary ions in the unperturbed state.

12.7.1 Buneman instability Ⓚ

We consider here the special case with electrons flowing along the magnetic field lines (we *might* have electron flow also across magnetic field lines in some cases), and restrict the analysis to a one-dimensional case with waves propagating along the magnetic field. We assumed singly charged ions, so that both species have density n_0. The relative drift velocity u_0 is assumed to be much larger than any thermal velocity, so we can ignore temperature effects also here. This is not a trivial restriction.

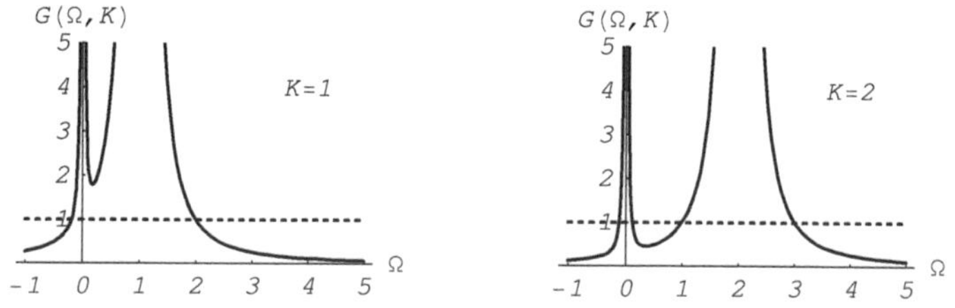

FIGURE 12.12: *Diagram for analyzing stability conditions for the Buneman instability. The figure assumes an artificial mass ratio of $m/M = 10^{-2}$; otherwise the curve would be extremely narrow in the vicinity of $\Omega = 0$. The figure to the left, for $K = 1$, corresponds to unstable conditions, the one with $K = 2$ to stable conditions.*

It is a simple matter (left to the reader as an exercise) to derive the dispersion relation for electrostatic waves in a one-dimensional case, $\mathbf{k} \parallel \mathbf{u}_0$, as

$$1 = \omega_{pe}^2 \left(\frac{m/M}{\omega^2} + \frac{1}{(\omega - k u_0)^2} \right). \tag{12.72}$$

For discussing this relation, it is an advantage to introduce the normalized variables $\Omega \equiv \omega/\omega_{pe}$ and $K \equiv k u_0/\omega_{pe}$. We then have

$$1 = \frac{m/M}{\Omega^2} + \frac{1}{(\Omega - K)^2}. \tag{12.73}$$

This relation can be rewritten as a fourth order polynomium, and will have four roots, where some might be double roots. Since all coefficients are real, complex roots will appear in complex conjugate pairs, i.e., if we have one complex solution, also its complex conjugate will be a solution.

It is an advantage to introduce a new function

$$G(\Omega, K) \equiv \left(\frac{m/M}{\Omega^2} + \frac{1}{(\Omega - K)^2} \right).$$

Since m/M is generally a small number, we have $G(\Omega, K) \approx 1$ for $\Omega - K = \pm 1$, unless Ω is close to zero, in which case also the term with coefficient m/M becomes important. To analyze the stability conditions of (12.73), it is best to use a graphic representation as the one in Fig. 12.12. Evidently, $G(\Omega, K) \to 0$ for $\Omega \to \pm\infty$ and $G(\Omega, K) \to \infty$ for $\Omega \to 0$ and $\Omega \to K$, so we must have at least two real solutions for $G(\Omega, K) = 1$. If the line $G(\Omega, K) = 1$ is crossing the curve *four* places we have four real solutions, corresponding to stability. If we have only *two* crossings, these correspond to the two real oscillatory solutions. We *must* have two more solutions, these being complex conjugate. One of these solutions then corresponds to *instability*. Small wave-numbers, K, are those giving unstable conditions. This can happen when either k or u_0 is small. The condition of marginal stability, with one double root at the coordinate where $\partial G(\Omega, K)/\partial\Omega = 0$, is readily found to be $\Omega = K(m/M)^{1/3}/(1 + (m/M)^{1/3})$, which, inserted into the marginal stability condition $G(\Omega, K) = 1$, gives $K_c = \sqrt{1 + 3(m/M)^{1/3} + 3(m/M)^{2/3} + m/M}$. We have $K_c \approx 1$, but note that even though $m/M \ll 1$, then $(m/M)^{1/3}$ might be a nontrivial correction in K_c. We find that wave-numbers $K < K_c$ are unstable. It can appear counter-intuitive that any small value of u_0 leads to instability for some k, but we should keep in mind that the analysis breaks down as soon as u_0 approaches the thermal velocity of the electrons.

The instability is often called the *Buneman instability*. Also in this case we are dealing with a "semi-fluid type" instability, just as the case with two cold counter-streaming electron beams; see Section 20.3.1. The problem can be analyzed equally well by a simple cold fluid model as with the kinetic plasma dispersion relation to be discussed later, e.g., (21.1), which readily gives (12.72).

The Buneman instability can be understood in terms of positive and negative energy waves (Chen 2016) (see also Section 5.10.4). In the present case the electron plasma waves supported by the streaming electron component can have a negative energy density. *With* the wave the system is in a state with less energy than *without* it. The dispersion relation for the electron plasma waves is in this case $(\omega - u_0 k)^2 = \omega_{pe}^2$ or in the present normalized units $(\Omega - K)^2 = 1$.

Ⓢ **Exercise:** Give an estimate (i.e., not an exact expression) for the growth rate of the Buneman instability for small (Ω, K) and its scaling with the mass ratio m/M. Ⓢ

- **Exercise:** Give an estimate (i.e., not an exact expression) for the most unstable wave-number for the Buneman instability, and for the corresponding growth rate.

12.8 Conclusions

Intuitively we expect that the cold wave results will be applicable for any wave type having phase velocities much larger than any relevant particle thermal velocity (being aware that the results discussed before explicitly refer to non-relativistic particles). Finite temperature effects are supposedly relevant for small phase velocities, but in this limit we expect waves to be electrostatic, as already mentioned. For this reason, most studies of slow waves in plasmas are emphasizing the electrostatic waves. A variety of plasma waves is observed, for instance, in the Earth's magnetosphere (Shawhan 1979), and a detailed knowledge of the dispersion relations is often necessary in order to sort out the observations.

We may illustrate some basic properties of various distinct modes of oscillations in plasmas as in Fig. 1.1. The energy density of wave motion is associated with two or more "reservoirs" of energy, just as the energy of a simple pendulum is oscillating between potential energy and kinetic energy. Wave-like motion is associated with the "sloshing" of energy between two or more such energy reservoirs. We might argue that for mechanical systems, oscillatory motion is a consequence of a restoring force and inertia: as a response to an initial perturbation, the restoring force sets a perturbed system into motion, but due to a finite inertia, the motion continues also when the force vanishes at an equilibrium position. Such a heuristic model applies, for instance, for the Langmuir waves and for Alfvén waves. For Langmuir waves the restoring force is the electrostatic electric field; for Alfvén waves it is the $\mathbf{J} \times \mathbf{B}$-force. The nature of electromagnetic waves is different, however, as far as inertia effects are concerned.

- **Exercise:** Demonstrate the basic wave mode characteristics illustrated in Fig. 1.1 by analyzing a standing Alfvén wave, a standing Langmuir wave and a standing electromagnetic wave in a plasma. How would you extend the diagram in order to accommodate sound waves in *neutral* gases, for instance?

13

Electrostatic Waves in Warm Homogeneous and Isotropic Plasmas

Finite plasma temperatures are most important for electrostatic waves, which are generally those with the smallest phase velocities. Heuristically, we might argue that thermal effects are of minor importance when a wave propagates at a velocity much larger than the thermal speed. Care should be taken, though, to distinguish phase velocities and group velocities.

13.1 Electron plasma waves

Consider first high frequency electron oscillations in an unmagnetized plasma. The waves are often called *electron plasma waves* or *Langmuir waves*. The description in this section will be based on the so-called fluid equations, which will here simply be postulated by the, in principle, reasonable argument that a gas of charged particles is in many ways like any other gas, and is supposed to follow the same basic physical laws as other gases. First, only the dynamics of electrons will be discussed here by arguing that for very high frequencies, ion inertia makes the displacement of ions negligible, and we can safely assume that they are in effect immobile. The only role of the ions will be to constitute a background distribution of positive charge with density n_0 which neutralizes the electrons globally, i.e., locally we can have deviations from charge neutrality, but the net amount of charge vanishes. In the following it will be assumed that the ion density is uniform, i.e., $n_0 = \text{const.}$ Since the ion dynamics is not entering the analysis, we might as well omit the index e for electrons on the space and time varying density and velocity.

The basic equation for the analysis is the equation of continuity for the electron density, n,

$$\frac{\partial n(\mathbf{r},t)}{\partial t} + \nabla \cdot \Big(n(\mathbf{r},t)\mathbf{u}(\mathbf{r},t) \Big) = 0, \tag{13.1}$$

which simply states that electrons are neither being lost nor created on time scales of interest. In nontrivial cases the equation (13.1) has to be completed by recombination and ionization terms. The second basic equation is the momentum equation, which simply states Newton's second law

$$n(\mathbf{r},t)m\left(\frac{\partial \mathbf{u}(\mathbf{r},t)}{\partial t} + \mathbf{u}(\mathbf{r},t)\cdot\nabla\mathbf{u}(\mathbf{r},t)\right) = -\nabla p(\mathbf{r},t) - en(\mathbf{r},t)\mathbf{E}(\mathbf{r},t), \tag{13.2}$$

where m is the electron mass and p is the electron pressure. Since we assume from the outset that we are dealing with electrostatic fluctuations or waves, there is no magnetic contribution to the force on a volume element. The electric field is determined here solely by Poisson's equation

$$\nabla \cdot \mathbf{E}(\mathbf{r},t) = \frac{e}{\varepsilon_0}\Big(n_0 - n(\mathbf{r},t) \Big). \tag{13.3}$$

We assume that there are no magnetic fluctuations associated with these waves, i.e., $\mathbf{B} = 0$. From the first Maxwell equation, **MI** (see, for instance, the form (5.57)), we find that this constraint implies that currents are exactly canceled by Maxwell's displacement current. In many relevant cases this cancellation is only partial, but as an approximation we can disregard the magnetic fields anyway.

For the electron pressure we assume that the ideal gas law $p = n\kappa T_e$ applies, with T_e being the electron temperature. We assume the pressure fluctuations to be adiabatic. These are evidently postulates, but with reference to fluctuation phenomena in neutral gases the two assumptions seem physically quite plausible. Consequently

$$p = Cn^{\gamma}, \tag{13.4}$$

or

$$\nabla p = \gamma \frac{p}{n} \nabla n = \gamma \kappa T_e \nabla n, \tag{13.5}$$

where $\gamma = C_p/C_v$ in terms of the ratio of the specific heats obtained at fixed pressure and fixed volume, respectively. For an ideal gas we have generally $\gamma = (d+2)/d$ where d is the dimensionally of the problem. Here we have $d = 3$ and use $\gamma = 5/3$. The equations (13.1)–(13.5) form a closed set. A magnetic field is not included, which at first sight may seem incorrect as a time-varying electric field should give rise to fluctuating magnetic fields also. The accuracy of this omission will be justified later on.

As they stand, the equations (13.1)–(13.5) are of little use. They can be solved in special cases only. If the analysis is restricted to small amplitude fluctuations, the situation becomes simpler. In this limit a linearization of the equations can be argued. Take, for instance, the simplest possible solution of (13.1)–(13.5), which is $\mathbf{E}_0 = 0$, $\mathbf{u}_0 = 0$, $n = n_0$ and $p_0 =$ const. Assume now that this equilibrium is disturbed, but only a little. The perturbations $\widetilde{n}$, $\widetilde{\mathbf{u}}$, $\widetilde{\mathbf{E}}$ and $\widetilde{p}$ are assumed to be small compared to their unperturbed values, and nonlinear terms as $n\mathbf{u}$ in (13.1) can be approximated by $n_0\widetilde{\mathbf{u}}$, etc., since terms like $\widetilde{n}\widetilde{\mathbf{u}}$ will be of second order. The term $\mathbf{u} \cdot \nabla \mathbf{u}$ disappears altogether in this approximation, since we assumed $\mathbf{u}_0 = 0$. In their linearized version the basic equations become

$$\frac{\partial \widetilde{n}(\mathbf{r},t)}{\partial t} + n_0 \nabla \cdot \widetilde{\mathbf{u}}(\mathbf{r},t) = 0, \tag{13.6}$$

$$\frac{\partial \widetilde{\mathbf{u}}(\mathbf{r},t)}{\partial t} = -\frac{\gamma \kappa T_e}{m n_0} \nabla \widetilde{n}(\mathbf{r},t) - \frac{e}{m} \widetilde{\mathbf{E}}(\mathbf{r},t), \tag{13.7}$$

and

$$\nabla \cdot \widetilde{\mathbf{E}}(\mathbf{r},t) = -\frac{e}{\varepsilon_0} \widetilde{n}(\mathbf{r},t). \tag{13.8}$$

With a little algebra the new set of equations (13.6)–(13.8) can, in terms of, for instance, the normalized electron density $\eta \equiv \widetilde{n}/n_0$, be reduced to the single equation

$$\frac{\partial^2 \eta}{\partial t^2} + \omega_{pe}^2 \eta - \frac{5}{3} u_{The}^2 \nabla^2 \eta = 0, \tag{13.9}$$

where the electron plasma frequency was introduced and $u_{The} = \sqrt{\kappa T_e/m}$ is an electron thermal velocity. It is important that in principle the same differential operator, $D(\partial_t, \nabla) \equiv \partial_t^2 + \omega_{pe}^2 - \frac{5}{3} u_{The}^2 \nabla^2$, applies also for the electric field and the velocity fluctuations since $\eta = -(\varepsilon_0/en_0)\nabla \cdot \widetilde{\mathbf{E}}$, or $\partial_t \eta = -\nabla \cdot \widetilde{\mathbf{u}}$ can be inserted into (13.9) as well. Consequently the entire dynamics of the electron fluctuations is determined by the operator $D(\partial_t, \nabla)$ in the limit of small amplitudes. A more convenient representation of this result is in terms of a dispersion relation $\omega = \omega(\mathbf{k})$ relating a spatial Fourier component denoted by $\mathbf{k}$ to a frequency ω

$$\omega^2 = \omega_{pe}^2 + \frac{5}{3} u_{The}^2 k^2 = \omega_{pe}^2 \left(1 + \frac{5}{3}(k\lambda_D)^2\right). \tag{13.10}$$

This is often called the Bohm–Gross dispersion relation. Note that the information contained in the dispersion relation and the operator $D(\partial_t, \nabla) = \partial_t^2 + \omega_{pe}^2 - \frac{5}{3}u_{The}^2\nabla^2$ are identical; one is related to the other by the substitution $\partial_t \to -i\omega$ and $\nabla \to i\mathbf{k}$. Given the dispersion relation we can obtain the governing differential equation for small amplitude fluctuations right away.

We are usually interested in relatively long wavelength solutions for (13.10), and have a standard form for this limit

$$\omega \approx \omega_{pe}\left(1 + \frac{5}{6}(\lambda_D k)^2\right), \tag{13.11}$$

where we introduced the electron Debye length $\lambda_D = u_{The}/\omega_{pe}$. We note that the numerical coefficient 5/6 in (13.11) originates in part from the ratio of specific heats for three-dimensional motions. This turns out to be a nontrivial observation when we deal with the same waves in a kinetic model.

We note that by perturbing the electron density we essentially move some of the electrons from one place to another, thus creating a charge separation, which gives the electric fields acting as "restoring forces." These move electrons from enhanced electron density regions toward electron density depletions. The electron pressure acts to enhance this restoring force. It is therefore reasonable that the oscillation frequency is increased from ω_{pe} by a finite electron temperature. On the other hand, it is also clear that the thermal correction becomes smaller for long wavelengths, since the pressure gradient decreases. The two forces, pressure gradient and electrostatic electric fields, enter as $-\nabla\widetilde{p} + en_0\nabla\widetilde{\phi}$ in a linearized analysis. To facilitate the comparison of the two terms, we can take the divergence of the force to find $-\nabla^2\widetilde{p} + en_0\nabla^2\widetilde{\phi} = -\nabla^2\widetilde{p} + e^2n_0\widetilde{n}$, where we used Poisson's equation. For a plane wave with wave-number k the previous expression gives $k^2\widetilde{p} + e^2n_0\widetilde{n}$. For a given density perturbation with $\widetilde{p} = \kappa T_e\widetilde{n}$, we find that the relative magnitude of the pressure and electric forces decreases with the square of the inverse wavelength $k \sim 1/\lambda$, as expected.

The result (13.10) contains the basic physics of high frequency electron plasma waves, but does not exhibit any damping of these waves, in contradiction to experimental observations. Also it is found that a numerical factor 3 gives a better agreement with observations than the 5/3 in (13.10). The remedy for these two objections is given by kinetic theory. Sometimes it was argued that the motion was of such a high frequency that the electrons "never discovered that they were allowed to move in full three dimensions," so one has to use the one-dimensional value for the ratio of the specific heats, a rather far fetched way of making observations fit a theory! Sometimes you may see a definition of u_{The} that incorporates the numerical factor, in which case the expression for the dispersion relation becomes somewhat neater.

We note that an acoustic-like dispersion relation $\omega^2 = (5/3)k^2u_{The}^2$ remains even in the limit where the plasma density vanishes in (13.10), which is, in a sense, an unphysical result. Here this dispersion relation is a "false solution," but acoustic electron oscillations are found for cases where the electron population consists of two components with significantly different temperatures.

Note that the derivation of the dispersion relation for electrostatic electron plasma waves explicitly uses the assumption of *linear* waves; we have no a priori reason to expect *nonlinear* electron plasma waves to be purely electrostatic!

Langmuir waves can be generated in the Earth's near space plasma environment. In local thermal equilibrium, low level plasma waves can be thermally excited as described by the fluctuation-dissipation theorem (Bekefi 1966, Pécseli 2000). Advanced modern radar scattering techniques allow detection of this electron plasma line (Vierinen et al. 2017). The electron plasma frequency can then be estimated to good accuracy. Thereby also the local plasma density in the ionosphere can be obtained by remote techniques, i.e., as an alternative to in situ rocket or satellite observations.

Ⓢ **Exercise:** Assume a given relative density variation $\widetilde{n}/n_0$ for electrostatic electron waves in a warm plasma at temperature T_e. What is the electric field and the electrostatic potential associated with these waves for varying wavenumber? How will you normalize the result? Compare your expressions with the similar problem for cold plasma waves. Ⓢ

Ⓢ **Exercise:** Consider a plasma that can be described as a mixture of two electron components with different temperatures, T_{e1} and T_{e2}, with densities n_{01} and n_{02}, respectively. Let the ions be immobile. Demonstrate the existence of a branch of the dispersion relation that has properties similar to acoustic waves. Discuss the general properties of the dispersion relation. Such waves have been observed in a laboratory plasma (Chowdhury et al. 2017).Ⓢ

Ⓢ **Exercise:** In a two-fluid model for a two-electron temperature plasma we have two dispersion relations, one Langmuir-like and one acoustic-like. How will you design an initial condition to excite a plane wave on only *one* of the branches of these dispersion relations? Ⓢ

13.1.1 Radiation of Langmuir waves from a moving charge

The concept of *Čerenkov radiation* is assumed to be well known; when a particle happens to move with a velocity exceeding the speed of light in the actual medium (of course not exceeding the speed of light in a vacuum!) we observe small "blips" of light being emitted. These small bursts of light have a short duration because the particle is decelerated by the energy loss. In case we could keep its velocity constant, we would expect it to act as a constant light source, the necessary energy being supplied by the agency maintaining the particle's velocity. Intuitively, one would expect that *any* particle moving at a speed exceeding a local characteristic wave velocity would emit waves of the appropriate sort. If several wave types are candidates, then they will all be generated, possibly with different intensities. One such example is *ship waves*; the rather intricate form of the waves trailing a ship moving at constant speed is due to the fact that two different wave forms are being "radiated."

In the present discussion we will provide qualitative arguments for the appearance of the wave form of electron plasma waves being radiated by a charge moving with constant velocity. The analysis will be kept in general terms, allowing the arguments to be generalized also to other related problems. We make the following simplifying assumptions: 1) only one relevant wave-type is present, as described by the dispersion relation (13.10), 2) we consider only the "far field" of the wave train, i.e., restrict the analysis to the region several wavelengths from the source and 3) we will be concerned only with the qualitative waveform, and not in the precise amplitude variation. Figure 13.1 is used for illustrating the arguments. The charged particle is assumed to be at the point A, and the analysis will refer to the co-moving coordinate system, where the wave field is stationary. The plasma will then be streaming backwards in this frame, as illustrated by the arrow $-\mathbf{U}$ in Fig. 13.1a. We then take an arbitrary position P, where the local wave-vector $\mathbf{k}$ is drawn, and define two angles θ and ψ.

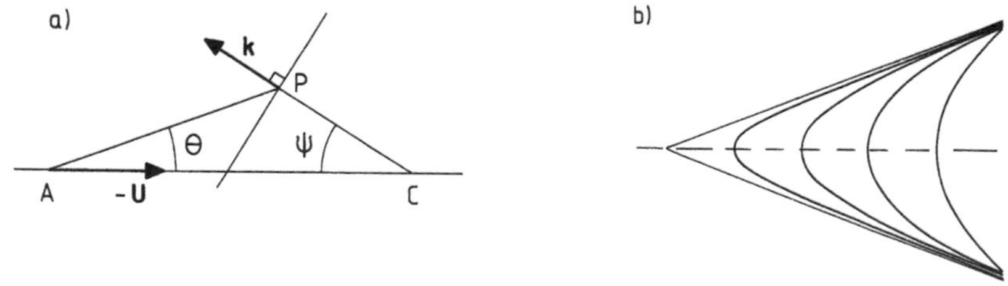

FIGURE 13.1: *Illustration of the angles and reference points entering the discussion of radiation of electron plasma waves by a charge moving with constant velocity in a). The local phase front is shown perpendicular to* **k***. Figure b) shows equipotential contours schematically.*

Since we assume the wave field to be constant in the co-moving frame of reference, we must have $\Omega(\mathbf{k}) = \omega(k) - \mathbf{U}\cdot\mathbf{k} = 0$, with $\omega(k)$ given by (13.10), and $\Omega(\mathbf{k})$ is the appropriate dispersion relation in the co-moving frame of reference. In particular, the group velocity $d\Omega/dk$ is different from $u_g \equiv d\omega/dk$. We have $\nabla_{\mathbf{k}}\Omega(\mathbf{k}) = \mathbf{u}_g - \mathbf{U} = u_g\mathbf{k}/k - \mathbf{U}$, making explicit use of the isotropy of the dispersion relation (13.10). Note our distinction in the notation; $\omega(k)$ is a function of the *magnitude* of k only, while $\Omega(\mathbf{k})$ depends also on the *direction* of $\mathbf{k}$. Referring to Fig. 13.1a) we have the horizontal component of $\nabla_{\mathbf{k}}\Omega$, along the line AC, to be $u_g \cos\psi - U$, while the vertical (or perpendicular) component is $u_g \sin\psi$.

Since the wave field in the position P is induced by the source in position A, the group velocity vector $\nabla_{\mathbf{k}}\Omega(\mathbf{k})$ must be parallel to the direction of the line AP. This implies

$$\tan\theta = \frac{u_g \sin\psi}{U - u_g \cos\psi}.$$

We have not yet used the actual form of the dispersion relation, only its isotropy.

We can determine the range of variability of ψ and $\mathbf{k}$ from the relation $\Omega(\mathbf{k}) = \omega(k) - Uk\cos\psi = 0$ and find

$$\frac{\sqrt{\omega_{pe}^2 + \frac{5}{3}u_{The}^2 k^2}}{k} = U\cos\psi.$$

Since the maximum value for the right-hand side is U, obtained when $\psi = 0$, we have a lower limit for the wave-number given by

$$k_0 = \frac{\omega_{pe}}{\sqrt{U^2 - \frac{5}{3}u_{The}^2}}.$$

It is obvious that no wave is found if $U^2 < \frac{5}{3}u_{The}^2$, which is rather self-evident in terms of the Čerenkov radiation arguments presented before. For given $U > u_{The}\sqrt{5/3}$, we have the maximum value for ψ_m given by $\cos\psi_m \equiv \sqrt{\frac{5}{3}u_{The}^2/U^2}$, obtained in the limit $k \to \infty$. To find the maximum value for θ, we use that in the limit of $k \to \infty$ we must have $u_g = u_{The}\sqrt{5/3}$. We then have $\cot\theta_m = \tan\psi_m$. This implies that APC becomes a right angle in this limit, meaning that at θ_m, the local phase fronts are parallel to the limiting line obtain for $k \to \infty$, i.e., at $\lambda \equiv 2\pi/k \to 0$. We also have

$$\tan\theta = \frac{1}{\frac{3}{5}U^2/u_{The}^2 - 1}\tan\psi,$$

obtained by use of $\omega/k = U\cos\psi$ as found before and the relation $u_g\omega/k = \frac{5}{3}u_{The}^2$, which is valid for Langmuir waves.

Selecting an arbitrary position, say P in Fig. 13.1a, we have one and only one wave-number that gives $\nabla_k\Omega$ parallel to AP. Recalling also the limiting angle θ_m, we can readily make a "free hand drawing" of the wave field trailing a charged particle moving at super-thermal energy; see Fig. 13.1b. In particular, the wavelength for the oscillations along the horizontal axis on Fig. 13.1b is obtained for $\theta = \psi = 0$ as $2\pi/k_0$, where k_0 is given before. Note that this particular wavelength is independent of position behind the moving particle.

The arguments work equally well for *any* dispersion relation of the form $\omega^2 = \omega_0^2 + C^2k^2$, with suitably defined characteristic frequency ω_0 and characteristic speed C. We meet several such candidates in plasmas. The arguments also are independent of the dimensionality of the problem; this does not sound right! Seemingly, the arguments are the same for a point charge moving in three-dimensional space or on a plane. However, is must be borne in mind that we have only obtained the variation of the equi-amplitude lines; the actual value of the amplitudes along these lines will indeed depend on the problem being formulated in two or three spatial dimensions. The present outline is correct for a construction of the wave-fronts, independent of the spatial dimensionality of the problem. The arguments are in fact also independent of the nature of the moving source: usually we imagine it to be a fast charged particle, but the arguments will hold also for a moving dipole.

It can be argued that for a plasma it is an artifact to have a charged particle moving with a constant velocity, and indeed it is an oversimplification to assume a complete similarity with the motion of a ship. The electric fields associated with the electrostatic electron plasma waves will act on the charge itself, and this will eventually be decelerated by the corresponding loss of energy. If the initial particle velocity is large, however, we can assume that at least for a nontrivial period of time, the particle speed can be taken to be constant. For the analogous problem of waves in the wake of a moving ship, the problem is simpler since in this case the engines are supposed to keep the velocity of the vessel constant by compensating for the energy lost in exciting the waves.

The foregoing discussions are based on a fluid model where the electron plasma waves are not damped. In a more general kinetic analysis, the present model will have to be modified by inclusion of Landau damping.

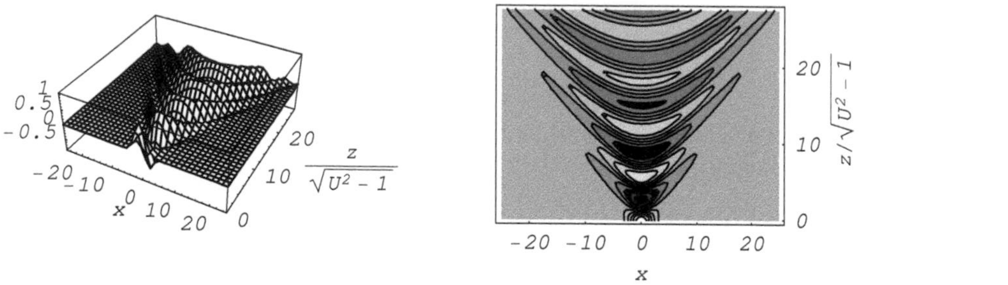

FIGURE 13.2: *Results of a numerical solution of (13.12) for the two-dimensional case, to be compared with the illustrative result in Fig. 13.1b. The δ-function boundary condition has here been replaced by a narrow Gaussian.*

- **Example:** We can analyze radiation of Langmuir waves by moving charges by (13.9), here written in normalized units, where time is normalized by $1/\omega_{pe}$, velocities by u_{The} and lengths by u_{The}/ω_{pe}, which is the electron Debye length, apart from a numerical constant which is here assumed to be included in the definition of u_{The}. We have $\partial^2\eta/\partial t^2 + \eta - \nabla^2\eta = 0$. If we change the frame of reference to one moving with the particle, we have a stationary pattern. We make the replacement $\partial/\partial t \to -\mathbf{U}\cdot\nabla$, with $\mathbf{U} = U\widehat{\mathbf{z}}$ being the charge velocity, and have $(U^2\partial^2/\partial z^2 + 1 - \nabla^2)\eta = 0$, or

$$\left((U^2-1)\frac{\partial^2}{\partial z^2} + 1 - \frac{\partial^2}{\partial x^2} - \frac{\partial^2}{\partial y^2}\right)\eta = 0. \tag{13.12}$$

  This result is written for three spatial dimensions. For a two-dimensional case (corresponding to a moving thin charged wire), we omit, for instance, $\partial^2/\partial y^2$. In either case, the resulting equation has to be solved for $\eta = \delta(r)$, corresponding to a disturbance at the origin. Note that we can normalize z by $\sqrt{U^2-1}$. Then only the sign of $U^2 - 1$ will be relevant, and not the magnitude of the velocity U. Results from a numerical solution of (13.12) are shown in Fig. 13.2.

- **Example:** The charged particle loses kinetic energy by exciting the wave pattern (Yeh & Liu 1972). The moving charge polarizes the surrounding plasma and this polarization field acts back on the particle. We find that this problem gives an interesting illustration of some of the shortcomings of the model based on a simple, and seemingly physically acceptable, dielectric function such as $\varepsilon(\omega) = 1 - (\omega_{pe}/\omega)^2$.

  To calculate the electric field at the position of the particle we have the Fourier transform of the electrostatic field as

$$\mathbf{E} = i\mathbf{k}\frac{q}{k^2\varepsilon_0\,\varepsilon(-\mathbf{k}\cdot\mathbf{U},k)} \tag{13.13}$$

with q being the charge on the particle. The potential associated with the "self-field" is $\phi_s = q/(\varepsilon_0 k)^2$ and the corresponding electric field is $iq\mathbf{k}/(\varepsilon_0 k)^2$. With $\mathbf{k}\cdot\mathbf{U} = k_\parallel U$ we determine the electric field at the particle position where we subtract the self-field

$$E_\parallel(\mathbf{r}=0) = i\frac{q}{\varepsilon_0}\int_{-\infty}^{\infty}\int_0^{\infty}\int_0^{2\pi}\left(\frac{k_\parallel}{k^2\varepsilon(-k_\parallel U,k)} - \frac{k_\parallel}{k^2}\right)k_\perp dk_\parallel dk_\perp d\theta\,.$$

Clearly, we have no E_θ or E_r components of the electric field here by symmetry arguments. It is evident that an imaginary part of ε is essential for expressing $E_\parallel$ correctly. The Kronig-Kramers relations (Champeney 1973, Pécseli 2000) ensure that any dispersive medium will be dissipative: taking (12.46) in the limit where $\nu \to 0$, we thus find

$$\varepsilon(\omega,k) \to 1 - \left(\frac{\omega_{pe}^2}{\omega}\right)^2 - i\frac{\pi\omega_{pe}^2}{\omega}\delta(\omega)\,,$$

by using $\lim_{\nu\to 0}\nu/(\omega^2+\nu^2) = \pi\delta(\omega)$. Although it will not be needed here, it might be worthwhile to recall that $\omega\delta'(\omega) = -\delta(\omega)$.

Needing only the real part of $E_\parallel(\mathbf{r}=0)$, we have

$$E_\parallel(\mathbf{r}=0) = -\frac{q2\pi^2}{\varepsilon_0}\int_{-\infty}^{\infty}\int_0^{\infty}\frac{k_\perp k_\parallel}{k^2}\frac{\varepsilon_2(-k_\parallel U,k)}{\varepsilon_1^2(-k_\parallel U,k)+\varepsilon_2^2(-k_\parallel U,k)}dk_\parallel dk_\perp\,,$$

in terms of real and imaginary parts of $\varepsilon \equiv \varepsilon_1 + i\varepsilon_2$. In the limit where $\varepsilon_2 \to 0$ we have

$$E_\parallel(\mathbf{r}=0) = -\frac{q2\pi}{\varepsilon_0}\int_{-\infty}^{\infty}\int_0^{\infty}\frac{k_\perp k_\parallel}{k^2}\,\delta(\varepsilon_1(-k_\parallel U,k))\,dk_\parallel dk_\perp\,. \tag{13.14}$$

We here recall the relation for Dirac's δ-function (Davydov 1965, Pécseli 2000)

$$\delta(f(x)) = \sum_j \frac{\delta(x-x_j)}{|df/dx|_{x=x_j}}\,,$$

where x_j is any one of the zeroes for the relevant function, i.e., $f(x_j)=0$. We obtain the stopping power, i.e., energy loss per time unit, as

$$\frac{dW}{dt} \equiv UqE_\parallel = -\frac{2q^2\pi}{\varepsilon_0}U\int_0^{\infty}\frac{k_\perp}{k_\perp^2+\omega_{pe}^2/U^2}\,dk_\perp\,, \tag{13.15}$$

using $\partial\varepsilon_1(-k_\parallel U)/\partial k_\parallel = 2\omega_{pe}^2/(U^2k_\parallel^3)$. We here took U to be a constant, which is acceptable for small losses, or for the case where the particle is moved by some external agency, in which case we obtain the energy deposited into the plasma per time unit. The integral in (13.15) is ill-posed due to a logarithmic divergence. In reality, however, we have to cut off the integration at wavelengths $k_\perp \sim 1/\lambda_D$, since the assumed model ceases to be valid there for any temperature, no matter how small (Yeh & Liu 1972). Consequently we can obtain a finite stopping power as

$$\frac{dW}{dt} \approx -\frac{q^2\pi\omega_{pe}^2}{U\varepsilon_0}\ln\left(1+\frac{U^2}{\lambda_D^2\omega_{pe}^2}\right) \approx -2\frac{q^2\pi\omega_{pe}^2}{U\varepsilon_0}\ln\left(\frac{U}{u_{The}}\right),$$

with $U \gg u_{The}$. This result turns out to be quite close to the generally accepted expression. The analysis can be generalized to have U varying with time, i.e., for a decelerating charge.

- **Exercise:** Generalize the discussion and results from the foregoing analysis of Čerenkov radiation of Langmuir waves to a more general anisotropic wave-type, where the directions of the group velocity $\mathbf{u}_g$ and the wave-vector $\mathbf{k}$ are different, but where the dispersion relation is otherwise unspecified. Make a model dispersion relation of your own choice for a detailed discussion.

13.2 Ion acoustic waves

When describing the low frequency plasma dynamics associated with ion acoustic waves, we extend the analysis by introducing two more equations: one for ion continuity

$$\frac{\partial n_i(\mathbf{r},t)}{\partial t} + \nabla \cdot (n_i(\mathbf{r},t)\mathbf{u}_i(\mathbf{r},t)) = 0, \tag{13.16}$$

and one for ion momentum

$$n_i(\mathbf{r},t)M\left(\frac{\partial \mathbf{u}_i(\mathbf{r},t)}{\partial t} + \mathbf{u}_i(\mathbf{r},t)\cdot\nabla\mathbf{u}_i(\mathbf{r},t)\right) = -\nabla p_i(\mathbf{r},t) + en_i(\mathbf{r},t)\mathbf{E}(\mathbf{r},t), \tag{13.17}$$

in terms of ion density n_i and ion velocity $\mathbf{u}_i$. The dynamics of the electrons and ions is for electrostatic waves coupled through Poisson's equation. We assume that the ions can adequately be described by the equation of state for ideal gases and have for adiabatic ion dynamics

$$p_i(\mathbf{r},t) = Cn_i^{\gamma}(\mathbf{r},t), \tag{13.18}$$

with $\gamma \equiv C_p/C_v$ as in (13.4).

A considerable simplification of the analysis is obtained by some physically plausible simplifying assumptions concerning the electron dynamics. Assuming that the fluctuations are electrostatic, we introduce the potential ϕ through $\mathbf{E} = -\nabla\phi$. The electron momentum equation gives

$$n_e m\left(\frac{\partial}{\partial t}\mathbf{u}_e + \mathbf{u}_e\cdot\nabla\mathbf{u}_e\right) = -\kappa T_e \nabla n_e + en_e\nabla\phi, \tag{13.19}$$

where the electron motion can be taken to be isothermal if we for ion acoustic waves assume from the outset that $\omega/k \ll u_{The}$. The electron inertia effects, i.e., the terms on the left-hand side of (13.19), can be neglected, since they are $(\omega/k)^2/u_{The}^2$ times smaller than those on the right-hand side, giving

$$\nabla \ln\left(\frac{n_e}{n_0}\right) = \frac{e}{\kappa T_e}\nabla\phi. \tag{13.20}$$

In the logarithm of n_e, we introduced the unperturbed density n_0 a normalizing dimensional quantity.

By integration of (13.20) we find a result corresponding to the assumption that the electrons maintain an isothermal Boltzmann equilibrium at all times in the electrostatic potential as in (8.5)

$$n_e(\mathbf{r},t) = n_0 \exp\left(e\phi(\mathbf{r},t)/\kappa T_e\right). \tag{13.21}$$

This assumption is, in effect, based on the observation that for relevant conditions, the electron thermal velocity is much larger than any velocity we might imagine being relevant for the ion dynamics. By this assumption, no equations for the electron dynamics are necessary, and the electron continuity as well as electron momentum equations become redundant, so (13.21) is inserted directly into Poisson's equation, giving

$$\nabla^2\phi(\mathbf{r},t) = -\frac{e}{\varepsilon_0}\left[n_i(\mathbf{r},t) - n_0\exp(e\phi(\mathbf{r},t)/\kappa T_e)\right]. \tag{13.22}$$

We have a steady state solution $\phi_0 = 0$, $n_i = n_0 =$ const. and $\mathbf{u}_0 = 0$. Linearizing the equations we obtain

$$\frac{\partial \widetilde{n}_i(\mathbf{r},t)}{\partial t} + n_0\nabla\cdot\widetilde{\mathbf{u}}_i(\mathbf{r},t) = 0, \tag{13.23}$$

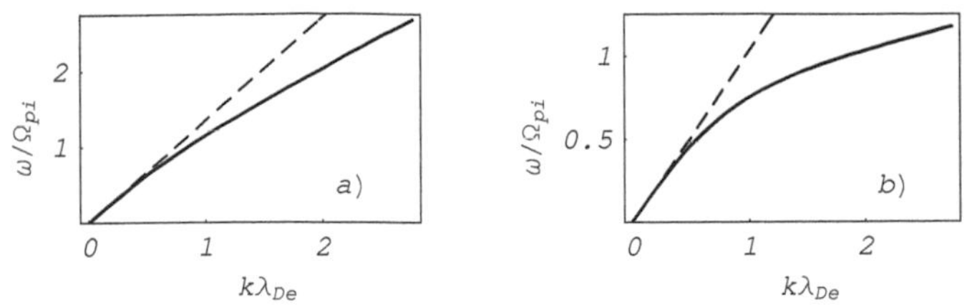

FIGURE 13.3: *Dispersion relation of ion acoustic waves within a fluid model for $T_e/T_i = 2$ in a), and $T_e/T_i = 25$ in b). A dashed line shows the quasi-neutral limit. Note the different scales on the vertical axes.*

$$\frac{\partial \widetilde{\mathbf{u}}_i(\mathbf{r},t)}{\partial t} = -\frac{\gamma\kappa T_i}{M n_0}\nabla \widetilde{n}_i(\mathbf{r},t) - \frac{e}{M}\nabla\widetilde{\phi}(\mathbf{r},t), \tag{13.24}$$

and

$$\nabla^2\widetilde{\phi}(\mathbf{r},t) = -\frac{e}{\varepsilon_0}\left(\widetilde{n}_i(\mathbf{r},t) - n_0\frac{e\widetilde{\phi}(\mathbf{r},t)}{\kappa T_e}\right). \tag{13.25}$$

With a little algebra, we obtain the partial differential equation

$$\left[\frac{\partial^2}{\partial t^2}\left(1 - \frac{\varepsilon_0\kappa T_e}{e^2 n_0}\nabla^2\right) - \frac{\gamma\kappa T_i}{M}\nabla^2\left(1 - \frac{\varepsilon_0\kappa T_e}{e^2 n_0}\nabla^2\right) - \frac{\kappa T_e}{M}\nabla^2\right]\frac{e\widetilde{\phi}}{\kappa T_e} = 0, \tag{13.26}$$

which, by Fourier transform, ultimately gives the dispersion relation

$$\omega^2 = \frac{\kappa T_e}{M}\frac{k^2}{1 + (k\lambda_{De})^2} + \frac{\gamma\kappa T_i}{M}k^2, \tag{13.27}$$

where we introduced $\lambda_{De} \equiv \sqrt{\kappa T_e \varepsilon_0/e^2 n_0}$, the electron Debye length. The dispersion relation of ion acoustic waves is illustrated in Fig. 13.3. Note that $\omega^2 \approx (\gamma\kappa T_i/M)k^2$ for $|k| \to \infty$, which is a finite ion temperature effect, independent of the electron temperature. On the other hand, for small k in the quasi-neutral limit we have $\omega^2 \approx k^2\kappa(T_e + \gamma T_i)/M$, which also has a contribution from finite ion temperatures. Later, we shall learn that ion acoustic waves become heavily damped by Landau damping for $|k| \to \infty$, so this limit turns out to be of minor physical relevance. Intuitively, we find that organized motion of plasmas on the Debye length are unlikely, because of the thermal motions, as outlined in, for instance, Sections 8.2, 8.4 and 8.5.

Including the first order correction due to wave dispersion, we find the ion acoustic dispersion relation as

$$\omega = k\sqrt{\frac{\kappa(T_e + \gamma T_i)}{M}} - \frac{1}{2}k^3\frac{\kappa T_e \lambda_{De}^2}{M\sqrt{\kappa(T_e + \gamma T_i)/M}}. \tag{13.28}$$

For cold plasmas, $T_e = T_i = 0$, the sound speed is vanishing, and we have no ion acoustic waves. A finite *electron* temperature, still with $T_i = 0$, is, however, sufficient for the existence of these sound waves. In this latter case the finite electron pressure attempts to "push" electrons away from the wave crests, thereby creating a slight deficit of electrons there as compared to the local ion density, while the wave troughs, on the other hand, will be slightly "overpopulated" by electrons. This process continues until electric fields have been built up, which arrests further electron motion. These electric fields are usually called *ambipolar* electric fields. We can estimate their intensity by arguing that the electric field force $-en_eE = en_e\nabla\phi$ must balance the electron pressure $-\kappa T_e\nabla n_e$, giving $e\phi/\kappa T_e \approx \ln n_e$, i.e., precisely the relation for Boltzmann distributed electrons. The ambipolar electric field sets the ions into motion, and thus ion sound waves can be propagated solely by the electron pressure, also for $T_i = 0$. With $T_i \ll T_e$ and $(k\lambda_D)^2 > 1$, we find a wave-number range where $\omega^2 \approx \Omega_{pi}^2 \equiv e^2 n/\varepsilon_0 M$, with Ω_{pi} being the *ion* plasma frequency.

For $T_e \approx T_i$ we no longer have the sound speed being much larger than the ion thermal velocity. In this case it is no longer obvious that we should take the ion equation of state to be adiabatic. The needed modifications will, however, not imply dramatic changes within a linearized fluid model.

- **Exercise:** Demonstrate that the inertial term (left side of electron momentum equation (17.4)) is indeed of the order of $(\omega/k)/u_{The}$ as compared to the terms on the right-hand side.

13.2.1 The quasi-neutral limit

For $(k\lambda_D)^2 \ll 1$, we find $\omega^2/k^2 \approx \kappa(T_e + \gamma T_i)/M$. This is the *quasi-neutral* limit (obtained by the *plasma approximation* (Chen 2016)), as discussed in Section 11.2. This quasi-neutral limit is shown with a dashed line in Fig. 13.3. We note that, as anticipated, we have $\omega^2 \ll \Omega_{pi}^2$ here. It is important to note that with the assumption of quasi-neutrality, the Debye length no longer appears in the dispersion relation. The problem contains no other physical length scale, and as far as relevant lengths are concerned, quasi-neutral ion-acoustic waves in a linearized model are characterized solely by the length scales of the initial conditions in the present linearized model.

Ⓢ **Exercise:** Assume that the plasma is composed by a mixture of electrons, singly charged ions and multi-charged ions with charge Ze, with different ion species having different temperatures. Find the dispersion relation for electrostatic ion-acoustic waves propagating in this plasma, assuming again Boltzmann distributed electrons. Consider, in particular, the quasi-neutral limit, and also the case where the ion temperatures are vanishing. Ⓢ

Ⓢ **Exercise:** In (13.27) we find that we have propagating ion-acoustic waves even in the case where we let $T_i \to 0$. Now isn't this strange? We learned in Section 8.3 that charges are shielded at an effective shielding distance determined through $1/\lambda_{ef}^2 = 1/\lambda_{De}^2 + 1/\lambda_{Di}^2$ for a two component plasma. If we let $T_i \to 0$ we find $\lambda_{Di} \to 0$ and consequently $\lambda_{ef} \to 0$, so all charges are completely shielded, but ion sound waves exist nonetheless. How come? Ⓢ

Ⓢ **Exercise:** Discuss the equivalent of Čerenkov radiation for the case of a charged particle propagating at velocities close to the ion sound speed (Guio et al. 2008). Assume $T_i \approx 0$. Take first the case where you assume quasi-neutrality from the outset and then relax this assumption. Ⓢ

- **Exercise:** Derive the dispersion relation for the case where you have a component of *negative* ions, i.e., a density n_1 of positive ions with charge e, mass M_1, and a component of negative ions with charge $-e$ of density $n_2 < n_1$ and mass M_2, so that the electron density gives $n_1 = n_e + n_2$ for charge neutrality. Consider both the high frequency electron branch and the low frequency ion branch, where you can assume the electrons to be Boltzmann distributed.

14

Fluid Models for Nonlinear Electrostatic Waves: Isotropic Case

This chapter describes nonlinear wave phenomena in a fluid model for the plasma dynamics in homogeneous, isotropic plasmas (in reality this means unmagnetized plasmas). We consider here weakly nonlinear Langmuir waves and also ion acoustic waves.

14.1 Weakly nonlinear Langmuir waves

Finite amplitude Langmuir waves have two sorts of contributions to their nonlinear evolution. To discuss these we can make a brief analogy to a simple oscillating system, such as a pendulum, as in Section 4.3.1. We discussed there the nonlinearity arising from a finite displacement of the pendulum mass. An additional effect can arise in the case that the displacement of the pendulum somehow affects the *length* of the supporting string (an effect which was not discussed in Section 4.3.1). Translated to the case of a Langmuir wave, this would correspond to a situation where the wave is affecting its own natural frequency of oscillation. That frequency is the plasma frequency, and it contains only one parameter which can be varied, the plasma density. We allow for the case where the wave by some, for the time unspecified, manner is able to change the plasma density locally, and then attempt to investigate the relative magnitude of the two linearities.

14.1.1 Cold electrons with immobile ions

First we consider the simplest case, where the ions are considered to be an immobile background of positive charge. After all, this is the most obvious generalization of the linear model, although, as we will see in a moment, it may not be the problem we find to be the most attractive or relevant. It turns out here that the assumption of vanishing electron temperature gives a great simplification. We will be interested mainly in the nonlinear dynamics of plane electron plasma waves, and consider the problem in one spatial dimension. In this limit we have the basic equations to be continuity of the electrons

$$\frac{\partial}{\partial t}n + u\frac{\partial}{\partial x}n = -n\frac{\partial}{\partial x}u. \tag{14.1}$$

Since the ions are immobile, there is no reason to add subscripts on the electron density $n = n(x,t)$, nor the electron velocity $u = u(x,t)$. For electrostatic electron plasma waves we have the Maxwell displacement current canceling the electron current, $\varepsilon_0 \partial E/\partial t = enu$, which combined with Poisson's equation $\varepsilon_0 \partial E/\partial x = e(n_0 - n)$, gives

$$\frac{\partial}{\partial t}E + u\frac{\partial}{\partial x}E = \frac{en_0}{\varepsilon_0}u. \tag{14.2}$$

Implicitly we have assumed that electron plasma waves that are electrostatic in the linearized analysis will remain electrostatic also in a nonlinear model. We have no a priori guarantee that this will be so, and in principle we have to be prepared that no solution can be found. Fortunately, it turns out that the implied electrostatic assumption is correct.

Finally we have the electron momentum equation in the form

$$\frac{\partial}{\partial t}u + u\frac{\partial}{\partial x}u = -\frac{e}{m}E, \tag{14.3}$$

with $-e$ being the electron charge. Electron pressure forces are absent by the assumption of vanishing temperatures.

It is now a simple matter to obtain one closed equation for the electron velocity

$$\left(\frac{\partial}{\partial t} + u\frac{\partial}{\partial x}\right)^2 u = -\omega_{pe}^2 u, \tag{14.4}$$

where the interpretation of the operator on the left side is evident. For the problem of linear oscillations, the result is a harmonic oscillation at the electron plasma frequency ω_{pe}. To solve the expression (14.4) without making the linear approximation is no simple task! We recognize the operator $\partial/\partial t + u\partial/\partial x$ as the time derivative we denoted D/Dt previously, and the one that gives the time derivative in a frame of reference following the flow with the local velocity. It turns out to be an advantage to consider the problem in such a frame. In imagination, we follow a small volume element of the electron gas which is at a position x_0 at time t_0, and find it at a position x at a later time t. The electron fluid quantities as density or velocity are functions of x and t, e.g., $u(x,t)$. We might, however, just as well select a set of coordinates (x,t), and ask from which set (x_0,t_0) did the volume element arrive? Traditionally, the variables x,t are denoted Eulerian and x_0,t_0 the Lagrangian variables (Davidson & Schram 1968, Davidson 1972). The transformation between the two is given by

$$x = x_0 + \int_0^\tau u(x_0,t')dt' \qquad \text{with} \qquad \tau = t - t_0. \tag{14.5}$$

The Eulerian and Lagrangian times are the same, as measured after the time t_0 where we establish the initial conditions.

We are dealing with partial differential equations and have to determine how to deal with differential operators, that is, when having, for instance, the velocity $u(x,t)$, and decide to consider x,t as functions of x_0,τ, how do we differentiate with respect to x_0 and τ? We readily find the relation

$$\frac{\partial}{\partial x} = \frac{1}{1+\int_0^\tau \frac{\partial}{\partial x_0}u(x_0,t')dt'}\frac{\partial}{\partial x_0}, \tag{14.6}$$

and with a little more algebra

$$\frac{\partial}{\partial t} = \frac{\partial}{\partial \tau} - \frac{u(x_0,\tau)}{1+\int_0^\tau \frac{\partial}{\partial x_0}u(x_0,t')dt'}\frac{\partial}{\partial x_0}. \tag{14.7}$$

We then trivially find

$$\frac{\partial}{\partial t} + u\frac{\partial}{\partial x} = \frac{\partial}{\partial \tau},$$

which, after all, is no surprise. We can now rewrite (14.4) as

$$\frac{\partial^2}{\partial \tau^2}u(x_0,\tau) = -\omega_{pe}^2 u(x_0,\tau), \tag{14.8}$$

with solutions

$$u(x_0,\tau) = U(x_0)\cos\omega_{pe}\tau + \omega_{pe}X(x_0)\sin\omega_{pe}\tau, \tag{14.9}$$

with $U(x_0)$ and $X(x_0)$ determined by the initial conditions. In this Lagrangian frame all oscillations, linear or nonlinear, have the same frequency, ω_{pe}. In the new variables, the momentum equation also becomes simple and gives

$$E(x_0,\tau) = \frac{m}{e}\omega_{pe}U(x_0)\sin\omega_{pe}\tau - \frac{m}{e}\omega_{pe}^2 X(x_0)\cos\omega_{pe}\tau.$$

We have an explicit expression for the electron fluid velocity for the present case, and can therefore by use of (14.5) write the continuity equation for the electron density in Lagrangian variables

$$\frac{\partial}{\partial\tau}\left(n(x_0,\tau)\left(1+\int_0^\tau \frac{\partial}{\partial x_0}u(x_0,t')dt'\right)\right) = 0,$$

giving the expression for the density in the form

$$n(x_0,\tau) = \frac{n(x_0,0)}{1+\dfrac{1}{\omega_{pe}}\dfrac{dU(x_0)}{dx_0}\sin\omega_{pe}\tau + \dfrac{dX(x_0)}{dx_0}(1-\cos\omega_{pe}\tau)}. \tag{14.10}$$

- **Exercise:** Demonstrate that the initial displacement $X(x_0)$ is related to the initial density through $dX(x_0)/dx_0 = n(x_0,0)/n_0 - 1$.

- **Exercise:** Demonstrate that the requirements of $n(x_0,\tau) \geq 0$ for all times and the requirement for the transformation from x_0 to x being unique (i.e., meaning that a position (x,t) can be reached from only one $(x_0,0)$ position) impose the conditions $n(x_0,0) \geq n_0/2$ and also

$$\frac{1}{\omega_{pe}}\left|\frac{dU(x_0)}{dx_0}\right| < \sqrt{2\frac{n(x_0,0)}{n_0} - 1}.$$

  Give a physical interpretation of these conditions.

- **Exercise:** The transformation (14.5) can readily be expressed in terms of $X(x_0)$ and $U(x_0)$, together with $\cos\omega_{pe}\tau$ and $\sin\omega_{pe}\tau$. Demonstrate this.

We will now apply the foregoing results to a specific example (Davidson & Schram 1968, Davidson 1972), where we have initially a density perturbation in the form of a plane wave

$$n(x_0,0) = n_0(1+\Delta\cos kx_0),$$

while $u(x_0,0) = 0$. The magnitude of the initial perturbation is Δ. For the linearized counterpart of the problem, we know that this condition gives rise to a standing wave, or if you like two counter-propagating plane waves. We note that $n(x_0,0) > 0$ trivially imposes the condition $\Delta < 1$, but we will find a more stringent condition in a moment.

With little effort we now find

$$u(x_0,\tau) = \frac{\omega_{pe}}{k}\Delta\sin kx_0\sin\omega_{pe}\tau, \tag{14.11}$$

$$E(x_0,\tau) = -\frac{m\omega_{pe}^2}{ek}\Delta\sin kx_0\cos\omega_{pe}\tau, \tag{14.12}$$

$$n(x_0,\tau) = n_0\frac{1+\Delta\cos kx_0}{1+\Delta\cos kx_0(1-\cos\omega_{pe}\tau)}. \tag{14.13}$$

The Eulerian-Lagrangian coordinate transformation here becomes $\tau = t$ and $kx = kx_0 + \alpha(\tau)\sin kx_0$, with $\alpha(\tau) = 2\Delta\sin^2(\omega_{pe}\tau/2)$. We thus find that the requirement $n(x_0,\tau) > 0$ for all times implies the restriction $\Delta < 1/2$. Note that the transformation from (x_0,τ) to (x,t) is by no

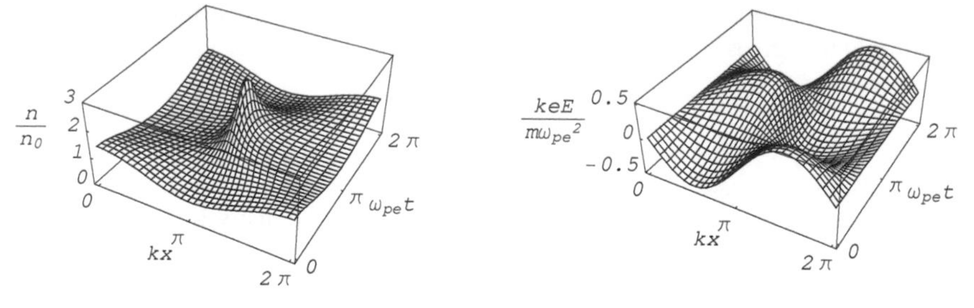

FIGURE 14.1: *Space-time evolution of a nonlinear standing electron wave. We have $\Delta = 0.4$. We show results for the relative electron density variation as well as the normalized electric field.*

means trivial (Davidson 1972, Shivamoggi 1988). For small amplitudes Δ in the initial perturbation, we note that the difference between the Eulerian and Lagrangian coordinates x and x_0 is immaterial, and the results from a linearized analysis will apply.

Plotting the results as in Fig. 14.1, it is best to avoid the inversion of x_0 to x, and it is easiest to make a "parametric plot," with x_0 a parameter, and calculate first $n(x_0, \tau)$ and then the corresponding (x,t). If we let $\Delta \to 0.5$ we find that the "spike" in the density evolution becomes a singularity. The electrons become trapped in the potential well associated with the wave, and the electron velocity becomes multi-valued: this is the physical reason for the limitation on the amplitude, $\Delta < 0.5$. When a spike develops in the electron density we see a corresponding steepening of the electric field as well as the electron velocity. We note that the transformation $x_0 \to x$ brings x_0 to the position x at time $t = 0$ (since this is the initial condition) and again at any later time $t = m2\pi/\omega_{pe}$ with $m = 1,2,\ldots$ where $\alpha(t) = 0$. The oscillation period is thus $2\pi/\omega_{pe}$, irrespective of the wave amplitude; there are no nonlinear frequency shifts. All that nonlinear effects are doing here is to distort the time evolutions of u, E and n as compared to the linear results. According to the Lighthill criterion, the waves are modulationally stable in this model; see Section 4.3.3.

The analysis summarized in the present section explicitly assumes the electron temperature to be vanishing. When the solutions exist, the electron velocity distribution is given at all times as $f(x,u,t) = n(x,t)\delta(u - u(x,t))$. If the initial wavelength is very large, corresponding to high phase velocities in the linear dispersion relation, we expect that a slight "broadening" of the electron velocity distribution function due to finite temperatures will give only minor corrections.

- **Exercise:** The results for $n(x,t)$, $E(x,t)$ and $u(x,t)$ can be expressed in terms of Bessel functions J_n of the first kind and order n. Demonstrate this by showing first

$$\sin kx_0(x,t) = \sum_{n=1}^{\infty} a_n(t) \sin nkx$$

where

$$a_n(t) = \frac{k}{\pi} \int_0^{2\pi/k} \sin nkx \sin kx_0(x,t) dx.$$

Show that this result gives

$$a_n(t) = (-1)^{n+1} \frac{2}{n\alpha(t)} J_n(n\alpha(t)).$$

With this, you can obtain a series expansion of $n(x,t)$, $E(x,t)$ and $u(x,t)$, where the individual terms with $a_n(t)$ give the time evolution of the Bessel function components (Davidson 1972, Shivamoggi 1988).

14.1.2 Mobile ions

In a more general, and as it turns out maybe also more interesting, analysis, we assume that the ions are free to move also, although we expect this motion to be characterized by much longer time scales than the electron motion.

The basic equations for the problem are still (13.1), (13.2), (13.3) and (13.5). We introduce rapidly and slowly varying quantities as $\widetilde{\mathbf{u}}$, $\widetilde{n}$, $\widetilde{\phi}$, and $\overline{\mathbf{u}}$, $\overline{n}$, $\overline{\phi}$, respectively. First we normalize these variables as

$$\begin{aligned} \widetilde{N} &\equiv \widetilde{n}/n_0 & \overline{N} &\equiv \overline{n}/n_0, \\ \widetilde{\mathbf{U}} &\equiv \widetilde{\mathbf{u}}/u_{The} & \overline{\mathbf{U}} &\equiv \overline{\mathbf{u}}/u_{The}, \\ \widetilde{\Phi} &\equiv e\widetilde{\phi}/(mu_{The}^2) & \overline{\Phi} &\equiv \overline{\phi}/(mu_{The}^2), \end{aligned}$$

where n_0 is the unperturbed plasma density, assumed homogeneous, and $u_{The}^2 = 3T_e/m$, where it turns out to simplify the notation to introduce the factor 3 here. Since we have no zero order drifts or electric fields, we can assume the quantities $\overline{N}$, $\overline{\mathbf{U}}$ and $\overline{\Phi}$ to be of second order in the wave amplitude. Note that we have *not* normalized spatial and temporal variables.

Following Dysthe and Pécseli (1977), the basic equations (13.1)–(13.5) can now be rewritten, after elimination of the pressure p, as

$$\begin{aligned} \frac{1}{u_{The}}\frac{\partial}{\partial t}\widetilde{\mathbf{U}} + \nabla\widetilde{N} - \nabla\widetilde{\Phi} &= \frac{1}{u_{The}}\frac{\partial}{\partial t}\overline{\mathbf{U}} - \nabla(\widetilde{\mathbf{U}}\cdot\overline{\mathbf{U}}) + \frac{1}{3}\widetilde{N}\nabla\overline{N} + \overline{N}\nabla\widetilde{N} \\ &\quad -\nabla\left(\frac{1}{3}\overline{N} - \overline{\Phi} + \frac{1}{2}\widetilde{U}^2 + \frac{1}{2}\widetilde{N}^2\right), \qquad (14.14) \\ \frac{1}{u_{The}}\frac{\partial}{\partial t}\widetilde{N} + \nabla\cdot\widetilde{\mathbf{U}} &= -\frac{1}{u_{The}}\frac{\partial}{\partial t}\overline{N} \\ &\quad -\nabla\cdot(\widetilde{N}\widetilde{\mathbf{U}} + \overline{\mathbf{U}}) - \nabla(\overline{N}\widetilde{\mathbf{U}} + \overline{\mathbf{U}}\widetilde{N}), \qquad (14.15) \end{aligned}$$

where fourth order terms have been neglected, and we introduced the Debye wave-number $k_D \equiv \sqrt{3\omega_{pe}^2/u_{The}^2}$.

Assuming quasi-neutrality in the slow motion that involves the ions, we have Poisson's equation in the form

$$\nabla^2\widetilde{\Phi} - \frac{1}{3}k_D^2\widetilde{N} = 0. \qquad (14.16)$$

Averaging (14.14) over the fast time scale and ignoring also here electron inertia for the slow variations, we find

$$\nabla\left(\frac{1}{3}\overline{N} - \overline{\Phi} + \frac{1}{2}\overline{\widetilde{U}^2} + \frac{1}{2}\overline{\widetilde{N}^2}\right) = 0. \qquad (14.17)$$

Combining (14.14)–(14.17) we find

$$\frac{1}{u_{The}^2}\frac{\partial^2}{\partial t^2}\widetilde{N} - \nabla^2\widetilde{N} + \frac{1}{3}k_D^2\widetilde{N} = Q_2 + Q_3 - \overline{Q}_2 - \overline{Q}_3, \qquad (14.18)$$

where

$$Q_2 \equiv \nabla^2\widetilde{N}^2 + \nabla\nabla : \widetilde{\mathbf{U}}\widetilde{\mathbf{U}} - \nabla\cdot(\widetilde{N}\nabla\widetilde{\Phi}),$$

and

$$Q_3 \equiv \frac{1}{3}\nabla^2\widetilde{N}^3 + \nabla\nabla : \left(\widetilde{N}\widetilde{\mathbf{U}}\widetilde{\mathbf{U}} + \widetilde{\mathbf{U}}\overline{\mathbf{U}} + \overline{\mathbf{U}}\widetilde{\mathbf{U}}\right) - \nabla\cdot\left(\overline{N}\nabla\widetilde{\Phi} + \widetilde{N}\nabla\overline{\Phi}\right).$$

Note that we must have the terms $\overline{Q}_2$ and $\overline{Q}_3$ in (14.18) in order to have both the left- and the right-hand side vanishing when averaged over the fast time scale!

Using complex notation, the rapidly oscillating quantities $\widetilde{\mathbf{U}}$, $\widetilde{N}$, $\widetilde{\Phi}$ are now expanded as

$$\begin{pmatrix} \widetilde{N} \\ \widetilde{\Phi} \\ \widetilde{\mathbf{U}} \end{pmatrix} = \begin{pmatrix} \widetilde{N}_1 \\ \widetilde{\Phi}_1 \\ \widetilde{\mathbf{U}}_1 \end{pmatrix} e^{-i\omega t} + \begin{pmatrix} \widetilde{N}_2 \\ \widetilde{\Phi}_2 \\ \widetilde{\mathbf{U}}_2 \end{pmatrix} e^{-i2\omega t} + \ldots + \text{c.c.} \tag{14.19}$$

where $\widetilde{N}_2$, $\widetilde{\Phi}_2$ and $\widetilde{\mathbf{U}}_2$ are of second order in the wave amplitudes $\widetilde{N}_1$, $\widetilde{\Phi}_1$ and $\widetilde{\mathbf{U}}_1$. We now wish to estimate the order of magnitudes of the various terms in Q_2 and Q_3. Since the final goal is an equation for the slow variation of the field quantities $\widetilde{N}_1$, $\widetilde{\Phi}_1$ and $\widetilde{\mathbf{U}}_1$, we are particularly interested in the terms in Q_2 and Q_3 which vary like $\exp(-i\omega t)$. There are quite a few of these, as listed here

$$\begin{aligned} 2\nabla^2 \widetilde{N}_1^* \widetilde{N}_2 + \nabla\nabla : \left(\widetilde{\mathbf{U}}_1^* \widetilde{\mathbf{U}}_2 + \widetilde{\mathbf{U}}_2 \widetilde{\mathbf{U}}_1^* \right) - \nabla \cdot \left(\widetilde{N}_2 \nabla \widetilde{\Phi}_1^* + \widetilde{N}_1^* \nabla \widetilde{\Phi}_2 \right) \\ + \nabla^2 \widetilde{N}_1 |\widetilde{N}_1|^2 + \nabla\nabla : \left(2\widetilde{N}_1 |\widetilde{\mathbf{U}}_1|^2 + \widetilde{N}_1^* \widetilde{\mathbf{U}}_1 \widetilde{\mathbf{U}}_1 + 2\widetilde{\mathbf{U}}_1 \overline{\mathbf{U}} \right) \\ - \nabla \cdot \left(\widetilde{N}_1 \nabla \overline{\Phi} \right) - \nabla \cdot \left(\overline{N} \nabla \widetilde{\Phi}_1 \right), \end{aligned} \tag{14.20}$$

where the asterisk denotes complex conjugate, as usual. For an order of magnitude estimate, we now use a plane wave as a "test-function," $\widetilde{N}_1 \sim \exp(i\mathbf{k} \cdot \mathbf{r})$, recalling that we have taken the exponential time variation separately already in (14.19). To lowest order we then have from (14.14)–(14.17) by neglecting the right-hand sides

$$\mathbf{k} \cdot \widetilde{\mathbf{U}}_1 = \frac{\omega}{u_{The}} \widetilde{N}_1, \quad \text{and} \quad \widetilde{\Phi}_1 = -\frac{1}{3} \left(\frac{k_D}{k} \right)^2 \widetilde{N}_1.$$

The second order quantities are now obtained by iteration. We insert the first order quantities on the right-hand sides of (14.14)–(14.17) and solve to obtain

$$\widetilde{N}_2 = 2\left(1 + 4(k_D/k)^2\right) \widetilde{N}_1^2$$

$$\mathbf{k} \cdot \widetilde{\mathbf{U}}_2 = \frac{\omega}{u_{The}} \widetilde{N}_1^2, \quad \text{and} \quad \widetilde{\Phi}_2 = -\frac{1}{6} \left(\frac{k_D}{k} \right)^2 \widetilde{N}_1^2.$$

Inserting these expressions into (14.20) we find that all the terms here can be expressed in terms of $\widetilde{N}_1 |\widetilde{N}_1|^2$. It is then found (the proof is left to the reader as an exercise) that the dominating term is the last one, which is greater than all the others by a factor $(k_D/k)^2$ or more. Neglecting smaller terms we find from (14.18) an equation for $A \equiv \widetilde{\Phi}_1 \exp(-i\omega t)$ correct to the third order in A

$$\left(\frac{\partial^2}{\partial t^2} - u_{The}^2 \nabla^2 + \omega_{pe}^2 \right) \nabla^2 A = -\omega_{pe}^2 \nabla \cdot \left(\overline{N} \nabla A \right),$$

with ω_{pe} here being a constant "reference" electron plasma frequency obtained from n_0. We can rewrite the equation as

$$\nabla \cdot \left(\frac{\partial^2}{\partial t^2} \nabla A - u_{The}^2 \nabla \nabla^2 A + \omega_{pe}^2 \nabla A \right) = -\omega_{pe}^2 \nabla \cdot \left(\overline{N} \nabla A \right). \tag{14.21}$$

Equation (14.21) serves as a basis for describing weakly nonlinear Langmuir waves. Note that we have not yet specified $\overline{N}$; we anticipate equations to be derived which express $\overline{N}$ in terms of A. The relation (14.21) can, however, also be used for describing Langmuir waves in *prescribed* variations of the plasma density, as long as these variations are consistent with the ordering of the variables. In this latter case we usually denote the process *scattering* of Langmuir waves on plasma density perturbations. In case the Langmuir waves themselves are generating the plasma density perturbations, the process is called *stimulated* scattering.

Physically, the term on the right-hand side of (14.21) can be traced back to the nonlinear term $\nabla \cdot (n\mathbf{u})$ in the continuity equation (13.1). Our analysis has thus demonstrated that the nonlinearities in the adiabatic pressure variation (13.4) and the term $\mathbf{u} \cdot \nabla \mathbf{u}$ in (13.2) are negligible in comparison. Physically, the result (14.21) can be argued (Pécseli 1985) by first considering a plasma slab where the electrons are displaced by a distance Δ with respect to the ions as in Fig. 12.1a. If the thickness of the slab $\mathcal{L}$ is much larger than Δ, it seems intuitively clear that the nonlinear correction to the oscillation frequency of the electron component is $\sim (\Delta/\mathcal{L})^a$. The power a must be an even number, since it cannot possibly matter whether the electrons are displaced to the left or to the right, i.e., $a \geq 2$. For large $\mathcal{L}$ the oscillation frequency is $\omega_{pe} = \sqrt{e^2 n/(\varepsilon_0 m)}$ and depends only weakly on the amplitude as $(\Delta/\mathcal{L})^a \ll 1$ so nonlinear effects are small in this limit. Significant changes in ω_{pe} can be induced only by a mechanism that varies the plasma density n.

As a conclusion we argue that to lowest order in wave amplitude we have oscillations at the plasma frequency, ω_{pe}; to next order we have small corrections due to thermal effects and nonlinearity! We emphasize here again that if we insist on keeping $\overline{N} = 0$, by assuming immobile ions, for instance, then we can still have nonlinear Langmuir oscillations as discussed before, e.g., in Section 14.1.1, but this nonlinearity will be comparatively weaker.

14.1.3 The ponderomotive force

We have derived an equation (14.21) for the Langmuir waves in terms of a yet unspecified perturbation of the bulk plasma density $\overline{N}$, and it remains to find a mechanism by which the Langmuir waves can induce such a perturbation, i.e., modify the index of refraction of the plasma in which they propagate. Physically, this mechanism can be identified as the *ponderomotive force*, which is induced by spatial variation in the amplitude of the high frequency electric field. Ponderomotive forces are sometimes called *Miller forces* (Gaponov & Miller 1958). See also Section 5.11.

To find the slowly varying part we average every term over an electron oscillation time, and indicate this averaging by an overline as usual. Since we will need this equation for describing *low frequency* phenomena, we take here the electron equation of state to be isothermal. We write the electron velocity as a sum of a slowly and a rapidly varying part, $\mathbf{u}_e = \overline{\mathbf{u}}_e + \widetilde{\mathbf{u}}_e$ with $\overline{\widetilde{\mathbf{u}}}_e = 0$, and similarly for other time varying quantities.

To illustrate the ponderomotive forces we solve a one dimensional model equation for the motion of a single particle (which is, you will see, is not the same as a fluid model) moving in a standing wave. We thus take $d^2x(t)/dt^2 = (e/m)E\sin(\omega t)\sin(Kx(t))$, normalize it and show the results in Fig. 14.2. It is most convenient to show results in a phase space. We find the oscillations of the particle position as well as its velocity. These would be seen for the motion in a plane wave as well. In addition we find for the present case a trapped solution and a drifting solution, for two different initial conditions. These drifts are caused by the ponderomotive forces associated with the spatial variation of the wave form. The variation on the slow time scale is here found by a time averaging as $(\omega/2\pi)\int_t^{t+2\pi/\omega} d\tau$ over the fast time scale. We note the acceleration and deceleration in the averaged phase space trajectory due to the ponderomotive forces. The results in Fig. 14.2 show in particular that ponderomotive forces can also trap charged particles.

To give an analytical expression for the ponderomotive force in a fluid model, we take the momentum equation (11.3) for the electrons, average it over the rapid time scale and have

$$m\left(\frac{\partial}{\partial t}\overline{\mathbf{u}}_e + \overline{\mathbf{u}}_e \cdot \nabla \overline{\mathbf{u}}_e + \overline{\widetilde{\mathbf{u}}_e \cdot \nabla \widetilde{\mathbf{u}}_e}\right) = -\frac{\overline{\nabla p_e}}{n_e} + e\nabla\overline{\phi}. \tag{14.22}$$

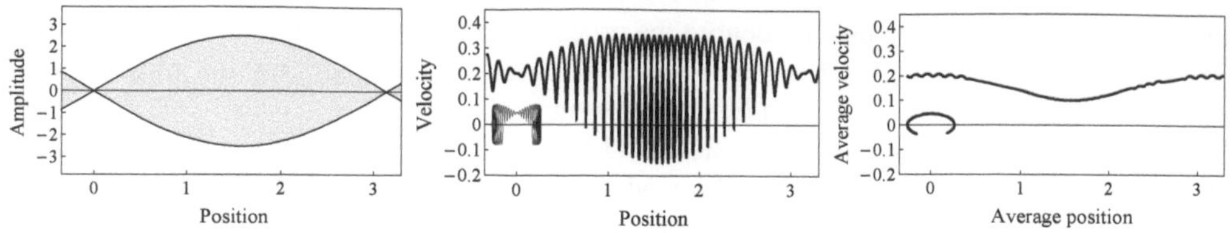

FIGURE 14.2: *Illustrations for the numerical solution of* $d^2\xi(t)/dt^2 = A\sin(\omega t)\sin(\xi(t))$ *with* $\xi \equiv Kx$*, for the parameters* $A = 2.5$ *and* $\omega = 10$*. The first frame show the variation range of the standing wave, the second frame a phase space* $\{\xi(t), d\xi(t)/dt\}$ *and the third frame an averaged phase space, where the averaging is over the period* $2\pi/\omega$*. Physically unrealistic parameters have been chosen for the sake of illustration.*

Ignoring again terms of the relative order of $(\lambda_D/L)^2$ or smaller, we have the high frequency fluctuating (complex) electron velocity $\widetilde{\mathbf{u}}_e = -i\widetilde{\mathbf{E}}e/(m\omega) \approx ie\nabla\widetilde{\phi}/(m\omega_{pe})$. We use this approximation for $\widetilde{\mathbf{u}}_e$ in the expression for $\overline{\widetilde{\mathbf{u}}_e \cdot \nabla\widetilde{\mathbf{u}}_e}$. Since the quantity $\overline{N}$ was assumed to be of second order when deriving (14.21), we linearize the equations with respect to the density variations $\overline{n}/n_0 \equiv 1 + \overline{N}$.

Introducing the real valued electron velocity and using the relation $\nabla\phi \cdot \nabla\nabla\phi = \frac{1}{2}\nabla(\nabla\phi)^2$ we find

$$mn_0\left(\frac{\partial}{\partial t}\overline{\mathbf{u}}_e + \overline{\mathbf{u}}_e \cdot \nabla\overline{\mathbf{u}}_e\right) + \frac{1}{2}\varepsilon_0\nabla\overline{(\nabla\widetilde{\phi})^2} = -\kappa T_e\nabla\overline{n} + en_0\nabla\overline{\phi}. \tag{14.23}$$

As apparent from the notation, we are here considering the forces on fluid elements, not on individual particles. This distinction is not trivial, as evident also from discussions in Section 19.2. Detailed and more general discussions of ponderomotive forces can be found in the literature (Karpman & Shagalov 1982).

On the slow time scale we assumed quasi-neutrality, $\overline{n}_e \approx \overline{n}_i = \overline{n}$. Ignoring once more electron inertia on the slow time scale, we omit the first term in (14.23), and find

$$\nabla\left(\frac{1}{2}\frac{\varepsilon_0}{n_0}\overline{(\nabla\widetilde{\phi})^2} + \kappa T_e\overline{N} - e\overline{\phi}\right) = 0. \tag{14.24}$$

Physically, this relation means that the electrons are isothermally Boltzmann distributed in the combination of the slowly varying electrostatic potential $\overline{\phi}$ and a potential $\frac{1}{2}(\varepsilon_0/n_0)\overline{(\nabla\widetilde{\phi})^2}$ associated with the ponderomotive force. A positive potential $\overline{\phi}$ attracts electrons, while a locally large amplitude, high frequency wave field with an associated ponderomotive potential expels electrons. The ponderomotive force associated with a spatially modulated high frequency wave train can be interpreted as a "radiation pressure" acting on the light electrons, but not on the heavy ions. The ions are set into motion by the electrostatic fields arising from the charge separations set up by, in part, the ponderomotive force.

The ponderomotive force can be argued physically by following an individual electron as it oscillates in the high frequency electric field. In the half period it spends in the larger amplitude part it is accelerated slightly more than in the lower amplitude part. This gives rise to a slight imbalance, which makes the electron drift slowly, while it oscillates in the wave field.

The final equation for this problem accounts for the low frequency ion dynamics. In their linearized form, the ion continuity and the ion momentum equations are easily combined into

$$\frac{\partial^2}{\partial t^2}\overline{N} - \frac{T_e}{M}\nabla^2\overline{\phi} = 0, \tag{14.25}$$

where we assumed cold ions, $T_i = 0$, for simplicity. Using (14.24) and (14.25) we finally have

$$\frac{\partial^2}{\partial t^2}\overline{N} - C_s^2\nabla^2\overline{N} = \frac{\varepsilon_0}{2Mn_0}\nabla^2\overline{(\nabla\widetilde{\phi})^2}. \tag{14.26}$$

We should emphasize that the ponderomotive force is *not* the only way the Langmuir waves can modify the local plasma density. We might readily imagine a case where externally maintained waves are dissipated slightly, for instance, by collisions, so as to heat the plasma locally (Gurevich 1978, Dysthe et al. 1985*b*). The enhanced pressure gives rise to a slight expansion of the plasma, and consequently to a slight local depletion of the plasma density and a corresponding change in the index of refraction. The density depletion acts back on the high frequency wave in the same way as the density depletion caused by a ponderomotive force. In the case of electromagnetic waves propagating in the lower parts of the Earth's ionosphere, this mechanism may actually be the most important one (Gurevich 1978).

14.1.3.1 Experimental observations

Ponderomotive forces have been studied in many experiments where large amplitude plasma waves were excited, e.g., by Ikezi et al. (1974) and Kim et al. (1974). Here we shall show a particularly simple illustrative example from an experiment carried out in a double ended Q-machine (Michelsen et al. 1977). In Fig. 14.3 we show an experimental set up for exciting a standing electron plasma wave, by two circular antennas, separated by a distance of $L = 26$ cm.

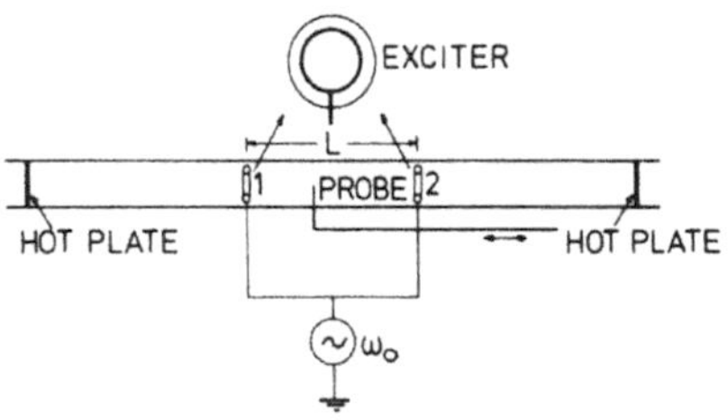

FIGURE 14.3: *Experimental set up used by Michelsen et al. (1977) for exciting a standing electron wave in a double ended Q-machine, reprinted with permission from the American Institute of Physics, copyright 1977.*

The plasma density was detected by a movable thin-wire Langmuir probe, and the dc-plasma density was measured. Typical results are shown in Fig. 14.4. A slight density enhancement, of the order of a few %, can be observed at the nodes of the standing electron plasma wave.

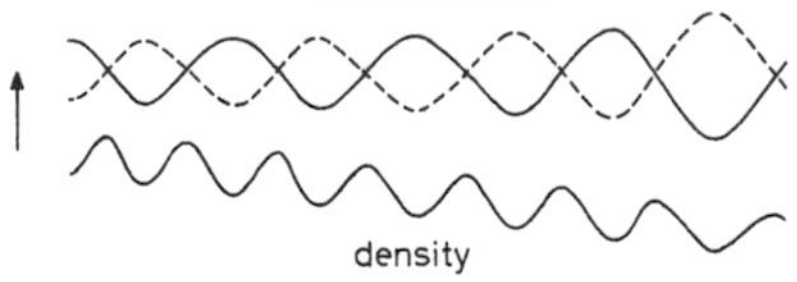

FIGURE 14.4: *Experimental results from the experiment of Michelsen et al. (1977), reprinted with permission from the American Institute of Physics, copyright 1977. The standing electron plasma wave is shown at the top for two times, 1/2 period of oscillation apart. The lower trace shows a stationary plasma density variation as detected by a movable Langmuir probe. Two arrows indicate the approximate positions of the exciter. The slight overall drift of the plasma density in the lower trace is likely to be caused by a slow coating of the probe during its axial motion.*

We have from (14.24) the expression $(\varepsilon_0/2n_0)\overline{(\nabla\widetilde{\phi})^2} + \kappa T_e\overline{N} = e\overline{\phi}$. For stationary conditions we assume the ions to be in a local isothermal Boltzmann equilibrium, with a linearized expression $n/n_0 = -e\phi/\kappa T_i$. For quasi-neutral density variations we find the analytical expression for the plasma density in the form $\kappa(T_e+T_i)\overline{N} = -(\varepsilon_0/2n_0)\overline{(\nabla\widetilde{\phi})^2}$. This relation is qualitatively correct in the limit where the plasma density in the Q-machine is very large and the bulk ion flow velocity is negligible; see Section 9.1. For smaller densities, we have to take into account an ion flow, and will distinguish supersonic and subsonic flows. Details of these discussions are given by Michelsen et al. (1977). For a non-vanishing ion flow or by a slight frequency mismatch between the forward and backward propagating electron waves, we can have ion waves being excited and to be detected also outside the region of occupied by the standing electron plasma wave, as anticipated also by Märk and Sato (1977).

In some ways, the experiment of Michelsen et al. (1977) can be seen as an analog of Kundt's tube, which was, at least for a time, a very popular high school experiment for illustrating standing sound waves; see Fig. 14.5. Some dust or lycopodium powder, distributed in the bottom of the tube, was found to settle near the nodes of the standing sound wave, where the motion of the air is minimum. The tube can be constructed with an open or a closed end: in any case a standing sound wave can be excited. We shall not overstate the analogy of the two experiments, but it may serve as an illustration of related phenomena.

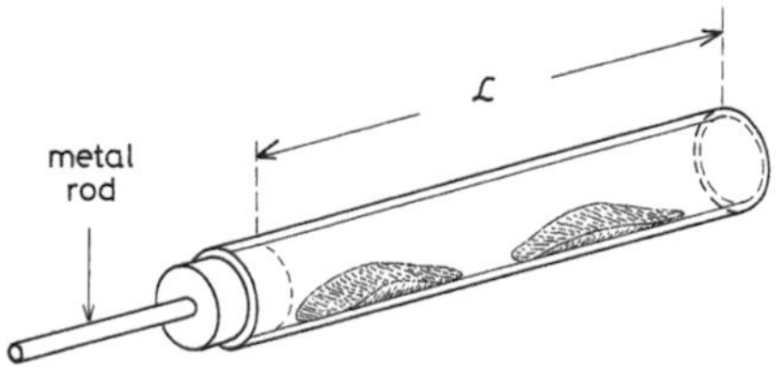

FIGURE 14.5: *Illustration of what is known as "Kundt's tube." By adjusting the length $\mathcal{L}$ of the tube and the length of the metal rod, a standing sound wave can be excited inside the tube when "tapping" the end of the rod. Some light powder (lycopodium powder, for instance) will collect at the nodes of the sound wave, as indicated on the figure.*

14.1.4 Nonlinear wave equations

Formally the set of equations (14.21) and (14.26) forms a complete set with $A = \widetilde{\phi}$. To put these equation into a more suitable form we make some intuitively simple assumptions, writing the Langmuir wave potential as the product of a slowly varying amplitude and a rapidly varying phase factor, i.e.,

$$\widetilde{\phi} \equiv \frac{1}{2}\phi(t,\mathbf{r})\exp(-i\omega t) + c.c., \tag{14.27}$$

where the wave frequency ω is assumed to be close to the electron plasma frequency. The exponential factor accounts for the variation on the time scale of the plasma period. The time variation of the coefficient $\phi(t,\mathbf{r})$ accounts for variations much slower than this.

Although seemingly reasonable, this model is not without controversies (Pécseli 1985); the notion of a sinusoidal oscillation with a slowly varying amplitude is about as meaningful as a "slightly bent straight line" (Vakman & Vaĭnshteĭn 1977). Different definitions of frequency and amplitude of a given signal can give rather different results! In fact, one cannot unambiguously discuss frequency and amplitude without knowing the *entire* time history of the process (Vakman & Vaĭnshteĭn 1977). Here we will (as so many others do) simply sweep all these arguments "under the rug" and assume that we have somehow made this separation in phase factor and amplitude. We can then write $\overline{(\nabla\widetilde{\phi})^2}$

as $\frac{1}{2}|\nabla\phi|^2$ in (14.26). Again using (14.27), with $\overline{N} \equiv \overline{n}/n_0$, we rewrite (14.26) as

$$\frac{\partial^2}{\partial t^2}\overline{N} - C_s^2\nabla^2\overline{N} = \frac{\varepsilon_0}{4Mn_0}\nabla^2|\nabla\phi|^2, \tag{14.28}$$

and have a closed set of equations (14.30) and (14.28).

Inserting (14.27) into (14.21) we find

$$\left(i2\omega\frac{\partial}{\partial t} - u_{The}^2\nabla^2 + (\omega_{pe}^2 - \omega^2)\right)\nabla^2\phi = -\omega_{pe}^2\nabla\cdot\left(\overline{N}\nabla\phi\right) \tag{14.29}$$

where we ignored a term $\partial^2\phi/\partial t^2$ as compared to $\omega\partial\phi/\partial t$, consistent with our assumption of a separation between the high frequency wave field and a slowly varying amplitude. Finally, we can make one additional assumption by setting $\omega = \omega_{pe}$. By this, we include the phase factor $\exp(-i\Delta\omega t)$ of the difference frequency $\Delta\omega \equiv \omega - \omega_{pe} \ll \omega$ into ϕ, which is complex anyhow. By this we impose a rather modest additional restriction on the relevant frequency range. We then have

$$\left(i\frac{\partial}{\partial t} + \frac{1}{2}\frac{u_{The}^2}{\omega_{pe}}\nabla^2\right)\nabla^2\phi = \frac{1}{2}\omega_{pe}\nabla\cdot\left(\overline{N}\nabla\phi\right)$$

or

$$\nabla\cdot\left(i\frac{\partial}{\partial t}\nabla\phi + \frac{1}{2}\frac{u_{The}^2}{\omega_{pe}}\nabla\nabla^2\phi\right) = \frac{1}{2}\omega_{pe}\nabla\cdot\left(\overline{N}\nabla\phi\right). \tag{14.30}$$

The relation (14.30) describes the slow dynamics of the wave potential ϕ coupled to the slow bulk plasma density variation, $\overline{N}$. We have thus introduced two time scales, a fast one on the plasma frequency scale and a slow variation accounted for by (14.30). As far as the spatial variation is concerned, we have *all* scales included in (14.26) and (14.30), subject to the restriction that the assumption of quasi-neutrality is justified, implying that all spatial scales are longer than the electron Debye length. The situation is thus somewhat different from the case discussed in Section 4.3.2, where we distinguished two time scales and also long and short spatial scales. The reason for the present problem being slightly simple is that Section 4.3.2 describes a local parabolic approximation to a dispersion relation (this can be done quite generally), where for the present case the Langmuir wave dispersion relation is parabolic by nature for $k\lambda_D \ll 1$.

The set of equations (14.26) and (14.30) represents a model for studying a number of important nonlinear plasma wave phenomena. It is often found to be an advantage to use a normalized version, obtained by the replacements

$$\frac{2}{3}\sqrt{\frac{m}{M}}\frac{\mathbf{r}}{\lambda_D} \rightarrow \mathbf{r}, \qquad \frac{2}{3}\frac{m}{M}\omega_{pe}t \rightarrow t,$$

$$\frac{\mathbf{E}}{4}\sqrt{\frac{3\varepsilon_0 M}{n_0 m\kappa T_e}} \rightarrow \mathbf{E}, \qquad \frac{3}{4}\frac{M}{m}\frac{\overline{n}}{n_0} \rightarrow n.$$

Assuming a hydrogen plasma, we find that the normalized unit of time is ~ 400 electron plasma periods, and the length unit is ~ 60 electron Debye lengths. The ion temperature was implicitly set to zero, and in the normalized variables the ion sound speed is unity. The square of the normalized electric field can be written as $(\frac{1}{2}\varepsilon_0 E^2/n_0 T_e)\frac{3}{8}M/m$, implying that a normalized electric field of order unity corresponds to the case where the ratio of the electric field energy density to the electron thermal energy density in the plasma is $\frac{1}{2}\varepsilon_0 E^2/(n_0 T_e) = \frac{8}{3}m/M$, i.e., the analysis refers to modest electric field energy densities.

With $\mathbf{E} = -\nabla\phi$ we have the normalized equation

$$i\nabla\cdot\frac{\partial}{\partial t}\mathbf{E} + \nabla^2\nabla\cdot\mathbf{E} = \nabla\cdot(n\mathbf{E}). \tag{14.31}$$

Ignoring the $\nabla\cdot$ operator, we then find the standard form

$$i\frac{\partial}{\partial t}\mathbf{E} + \nabla\nabla\cdot\mathbf{E} = n\mathbf{E}, \tag{14.32}$$

where a slight change in the numerical coefficients can place a factor $\frac{1}{2}$ in front of the second term on the left-hand side. We can argue that any solution of (14.32) is a solution of (14.30), but a solution of (14.31) need not be a solution for (14.32), since we can add the rotation of an arbitrary vector field to the electric field in (14.32), with a contribution which will disappear by insertion into (14.30). The question is of course relevant only in two or three spatial dimensions: in one dimension, the transition from (14.30) to a simpler version of (14.32) is trivial, given some modest requirements on E at $x \to \pm\infty$.

It should be emphasized that the seemingly innocent cancellation of the $\nabla\cdot$ operator on both sides of (14.30) or (14.31) has the rather drastic implication that instead of a scalar equation for ϕ we have a vector equation for $\mathbf{E}$, which in reality amounts to *three* identical differential equations for the three electric field components! We have the general relation $\nabla^2\mathbf{A} = \nabla(\nabla\cdot\mathbf{A}) - \nabla\times\nabla\times\mathbf{A}$, which applies for $\mathbf{E}$ in (14.32) as well. Since we, however, explicitly assumed electrostatic fields, $\mathbf{E} = -\nabla\phi$, in the derivation of (14.32) and (14.34), we have implied $\nabla\times\nabla\times\mathbf{E} \approx 0$, and therefore we cannot distinguish $\nabla^2\mathbf{E}$ and $\nabla(\nabla\cdot\mathbf{E})$, replacing (14.32) by

$$i\frac{\partial}{\partial t}\mathbf{E} + \nabla^2\mathbf{E} = n\mathbf{E}. \tag{14.33}$$

Consistency requires that $\nabla\times(n\mathbf{E}) = n(\nabla\times\mathbf{E}) + (\nabla n)\times\mathbf{E} \approx 0$: if not, the term on the right-hand side of (14.32) or (14.33) will act as a source of an electromagnetic component. If we have $\nabla\times\mathbf{E}$ small, we require either ∇n parallel to $\mathbf{E}$, approximately, or that $\nabla n/n$ is small, which means that the wave envelope varies slowly with the spatial coordinate, as compared to the wavelength of the Langmuir waves. This latter assumption implies that we deal with narrow-band processes, which, as you may recall, was precisely what we assumed by using a plane wave as a "test function" when deriving (14.21).

With the variable changes listed before, we find the self-consistent equation for the slowly varying normalized plasma density in the form

$$\frac{\partial^2}{\partial t^2}n - \nabla^2 n = \nabla^2|\mathbf{E}|^2. \tag{14.34}$$

We can make a few general statements on (14.32) and (14.34). Taking first (14.31) and then its complex conjugate version, we multiply the first one of these by $\mathbf{E}^*$ and the second one by $\mathbf{E}$, subtract the two and find after an integration that

$$\frac{d}{dt}N = 0 \qquad \text{with} \qquad N \equiv \int_{\mathcal{V}} |E|^2 d\mathbf{r}, \tag{14.35}$$

where $\mathcal{V}$ is the integration volume. The relation (14.35) is often interpreted as conservation of the "number of Langmuir wave quanta" (admittedly, this is a bit sloppy, since we have nowhere made any reference to quantum mechanics). The relation (14.34) was not used here.

Using both (14.31) and (14.34), a second conservation law is

$$\frac{d}{dt}H = 0 \qquad \text{with} \qquad H \equiv \int_{\mathcal{V}} \left(|\nabla\cdot\mathbf{E}|^2 + n|E|^2 + \frac{1}{2}(n^2 + u^2) \right) d\mathbf{r} \tag{14.36}$$

where the velocity fulfills the linearized and normalized plasma continuity equation $\partial n/\partial t + \nabla\cdot\mathbf{u} = 0$. Interpreting H as an energy, we have (14.36) being an energy conservation law.

An equivalent to a momentum conservation conservation law can be obtained also by use of (14.32) and (14.34) as

$$\frac{d}{dt}\mathbf{P} = 0 \qquad \text{with} \qquad \mathbf{P} \equiv \frac{1}{2}\int_{\mathcal{V}} \Big(i((\nabla \cdot \mathbf{E})\mathbf{E}^* - (\nabla \cdot \mathbf{E}^*)\mathbf{E}) + 2n\mathbf{u} \Big) d\mathbf{r}. \tag{14.37}$$

The conservation laws listed here are a consequence of the Hamiltonian nature of the problem, which can be derived from a principle of minimum action (Rudakov & Tsytovich 1978).

- **Exercise:** Consider the basic equations (14.31) and (14.34) and the conservation laws (14.35), (14.36) and (14.37). Demonstrate that we have an equation for the velocity u in the form

$$\frac{\partial}{\partial t}u = -\frac{\partial}{\partial x}n - \frac{\partial}{\partial x}|E|^2.$$

What is the physical interpretation of this relation? Can you write it in two or three spatial dimensions?

Ⓢ **Exercise:** Consider the one-dimensional normalized equation

$$i\frac{\partial}{\partial t}E + \frac{\partial^2}{\partial x^2}E = nE.$$

The model is here assumed to be linear in E, with the spatial and temporal variations of n are given, with n measuring the density difference from a reference background level.

Use the model to describe a plane electron plasma wave interacting with a time-stationary thin "step-like" plasma slab of width Δ and maximum density n_0. The level $n_0 = 0$ thus refers to the reference plasma frequency where the time-factor $\exp(-i\omega_{pe}t)$ has been removed. Note that n_0 can have both signs in general: we can have density enhancements and density depletions. In principle n could be time varying, but it is here assumed to vary with x only. The present problem is analogous to the "tunneling" problem encountered in wave mechanics.

Based on this result, describe the interaction of the electron plasma wave with a narrow slab-like density variation. Ⓢ

Ⓢ **Exercise:** Consider the one-dimensional normalized equation

$$i\frac{\partial}{\partial t}E + \frac{\partial^2}{\partial x^2}E = nE$$

and assume that the normalized plasma density n can be locally approximated by a parabolic well, $n \approx ax^2$. The potential well is deterministic and given; we have no nonlinearities here! Demonstrate that the resulting equation can be solved analytically, and give the solution in terms of Hermite polynomia, $H_n(x)$. These solutions correspond to standing Langmuir waves in a plasma density depletion. The problem is closely related to a textbook problem in quantum mechanics (Davydov 1965). There is a relation for Hermite polynomia that may come handy

$$H_n(\xi) = (-1)^n e^{\xi^2}\left(\frac{d}{d\xi}\right)^n e^{-\xi^2},$$

called the Rodrigues formula. Ⓢ

Ⓢ **Exercise:** Consider the one-dimensional normalized equation

$$i\frac{\partial}{\partial t}E + \frac{\partial^2}{\partial x^2}E = nE,$$

with n locally approximated by a parabolic well, as, for instance, $n \approx \frac{1}{4}\omega^2 x^2$. This is a *linear* problem, with n given. Consider the initial condition $E(x,0) = E_0 \exp(-\omega(x-b)^2)$. Demonstrate that this solution is related to the lowest order natural mode of the parabolic density depletion. Show that the solution for $E(x,t)$ for later times is

$$E(x,t) = E_0 \exp\left(-\frac{1}{4}\omega\left(x^2 + 2it + \frac{b^2}{2}\left(1 + e^{-2i\omega t}\right) - 2bxe^{-i\omega t}\right)\right).$$

(For your information, this problem is related to what is known as *Glauber states* in quantum mechanics.) Demonstrate that

$$|E|^2 = E_0^2 \exp\left(-\frac{1}{2}\omega\left(x - b\cos(\omega t)\right)^2\right),$$

and illustrate the space-time variation of $|E|^2$. Why is the result surprising? ⓢ

ⓢ **Exercise:** You may feel uneasy about the use of a complex equation like (14.33) for describing real physical phenomena. (This unhappiness can be transformed to quantum mechanics as well.) Demonstrate how to construct a real equation out of (14.33), and explain the price for this seeming simplification. ⓢ

14.1.5 Langmuir wave decay

We have now a set of equations (14.32) and (14.34) describing coupled high and low frequency electrostatic waves, varying on the scale of the electron plasma frequency and on another frequency, which can be estimated as $\Omega_{pi}\sqrt{m/M}$. The reason for the frequency being much smaller than the ion plasma frequency Ω_{pi} is our assumption of quasi-neutrality (or "the plasma approximation"), which restricts the analysis to very long scales and correspondingly low frequencies. The equations (14.21) and (14.26) are some of the most basic ones in nonlinear plasma wave theory, and have been applied to describe a variety of phenomena. Here we will use them to analyze a phenomenon known as "Langmuir wave decay." The point is that a plane electron plasma wave is a solution to (14.21) and (14.26), but not necessarily a *stable* solution. If the equilibrium is perturbed slightly, we may find it to be unstable (this will depend on the detailed parameters), and the wave can decay into another electron plasma wave and an ion sound wave (Davidson 1972, Thornhill & ter Haar 1978, Kono & Škorić 2010). The analysis here considers only Langmuir wave decay, but the methods used can be generalized to many other nonlinear wave processes. Only the nonlinear dynamics of modulated plane waves is studied here. For laboratory experiments this condition is easily realized, but in nature we will usually encounter broad spectra. Studies of also the nonlinear evolution of wave spectra are reported in the literature (Bhakta & Majumder 1983, Pécseli 2014, Kono & Pécseli 2016).

In the normalized version of our basic equations (14.32) and (14.34), the dispersion relations for the Langmuir waves and ion sound waves are

$$\omega = k^2 \qquad \text{and} \qquad \Omega = |K|, \tag{14.38}$$

respectively. In the present model the Langmuir wave dispersion relation now includes the origin, since we in effect subtracted the electron plasma frequency ω_{pe} from the linear dispersion relation by the manipulations leading to (14.30); see also Fig. 14.6.

We now consider an initially given Langmuir wave

$$\mathbf{E}_0 \exp\left(-i(\omega_0 t - \mathbf{k}_0 \cdot \mathbf{r})\right) \qquad \text{with} \qquad n_0 = 0$$

where, according to (14.38), we have $\omega_0 = k_0^2$. We then introduce the (small) perturbations

$$\mathbf{E}_1(t) \exp\left(-i(\omega_1 t - \mathbf{k}_1 \cdot \mathbf{r})\right) \qquad \text{and} \qquad n(t)\cos(\omega t - \mathbf{K} \cdot \mathbf{r})$$

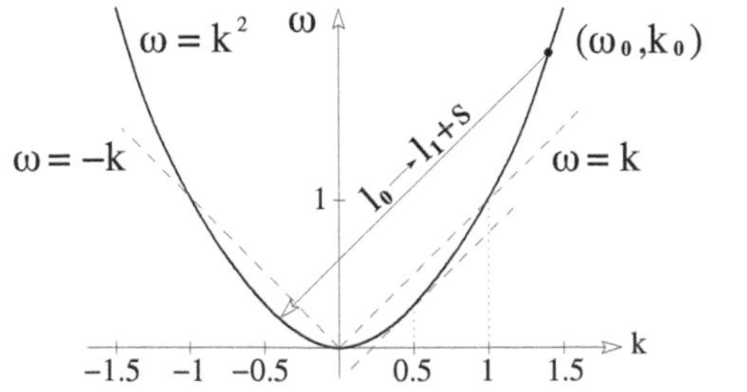

FIGURE 14.6: *Illustration of the dispersion relations entering the Langmuir wave decay process, $\ell_0 \to \ell_1 + s$. The figure is shown in one dimension for simplicity, and assumes normalized wavenumbers and frequencies.*

assuming $|\mathbf{E}_1| \ll |\mathbf{E}_0|$ for all times t, where again $\omega_1 = k_1^2$ and $\Omega = |K|$. The electric fields are assumed to be electrostatic and hence $\mathbf{E}_0 \parallel \mathbf{k}_0$ and $\mathbf{E}_1 \parallel \mathbf{k}_1$. We allow for a slow temporal variation of $\mathbf{E}_1$ and n, as indicated. No index is needed on n since it is a perturbation on $n_0 = 0$. We need not introduce a phase factor in the density perturbation, since we have E_0 and E_1 being complex quantities, in general, and these can accommodate any relative phase variations if needed. Introducing the angle θ between $\mathbf{k}_0$ and $\mathbf{k}_1$, we find by insertion into (14.26) and (14.30), followed by linearization

$$\begin{aligned} ie_1 \frac{\partial}{\partial t} E_1(t) &= \frac{1}{2} n(t) E_0 (e_+ + e_-) e_0 \cos\theta \\ \frac{1}{2}\left(\frac{\partial^2}{\partial t^2} n(t) - i2\Omega \frac{\partial}{\partial t} n(t)\right) e_+ &= K^2 E_1(t) E_0^* e_+ \cos\theta \end{aligned} \tag{14.39}$$

where we introduced the abbreviations $e_\pm \equiv \exp(\pm i(\omega t - \mathbf{K}\cdot\mathbf{r}))$ and $e_{0,1} \equiv \exp(-i(\omega_{0,1} t - \mathbf{k}_{0,1}\cdot\mathbf{r}))$, for simplicity (Thornhill & ter Haar 1978). Some terms cancel by use of the dispersion relations (14.38).

Assuming harmonic time variations of $E_t(t)$, etc., we introduce the complex variable ω by the assumption $\partial E_1(t)/\partial t \to -i\omega E_1$, which in turn implies $\partial n(t)/\partial t \to -i\omega n$. (By now there are many frequencies in this problem, so care should be taken not to confuse them!) Taking the resonance parts of (14.39), we impose the conditions

$$\omega_0 = \omega_1 + \Omega \qquad \text{and} \qquad \mathbf{k}_0 = \mathbf{k}_1 + \mathbf{K}, \tag{14.40}$$

which are usually interpreted as energy and momentum conservations, respectively. This interpretation would indeed be the most obvious if considering a quantum mechanical equivalent of the problem, where we had to multiply each term by $\hbar$. Eliminating n, we find the relation

$$\omega^3 + 2\Omega\omega^2 + K^2 |E_0|^2 \cos\theta = 0. \tag{14.41}$$

This expression has three roots for ω. At least one is real, while the two others can be either both real or both complex. In the latter case they form a complex-conjugate pair, where one of the roots corresponds to instability. Some limiting cases serve as illustrations of the properties of the solutions. If we take the weakly unstable case where $|\omega| \ll \Omega$ where the ion sound is only slightly modified, we find, by using $\Omega = |K|$ from (14.38), the approximate result for the growth rate as

$$\gamma \approx \sqrt{\frac{1}{2}|K|\,|E_0|^2 \cos\theta}, \tag{14.42}$$

while the real part is negligible to the same approximation, i.e., the instability is aperiodically increasing.

In the alternative limit, with $|\omega| \gg \Omega$, we find

$$\gamma \approx \frac{1}{2} K^{2/3} |E_0|^{2/3} (\cos\theta)^{1/3}. \tag{14.43}$$

In all cases we note that the maximum growth rate is obtained when $\theta = 0$ when the triangle formed by the vectors $\mathbf{k}_0$, $\mathbf{k}_1$ and $\mathbf{K}$ is reduced to a straight line ("slim triangles"), indicating that the Langmuir wave decay tends to give a strongly anisotropic wave spectrum. This result is often used as a justification for applying a one-dimensional model in, for instance, computer simulations, which would become excessively time consuming in a two- or three-dimensional treatment.

In the foregoing we have implicitly assumed that the wave-number and frequency matching conditions were fulfilled. This is, however, only possible for a limited range of wave-numbers $\mathbf{k}_0$. We consider (again) the triangle spanned by the three vectors $\mathbf{k}_0$, $\mathbf{k}_1$ and $\mathbf{K}$, and apply the "cosine" relation $k_1^2 = k_0^2 + K^2 - 2k_0 K \cos\theta_2$ to obtain

$$|K| = 2k_0 \cos\theta_2 - 1, \tag{14.44}$$

where θ_2 is the angle between $\mathbf{K}$ and $\mathbf{k}_0$. We eliminated k_1 by use of the frequency matching condition $k_0^2 = k_1^2 + |K|$, using $\omega_0 = k_0^2$, $\omega_1 = k_1^2$ and $\Omega = |K|$ in our present normalized units. Note that the constraint from the dispersion relations and the frequency and wave-number matching conditions are significant: we can choose $\mathbf{k}_0$ and the angle θ_2, but then both the direction and magnitude of $\mathbf{K}$ are determined by (14.44) and thereby also $\mathbf{k}_1$. The direction of $\mathbf{E}_1 \parallel \mathbf{k}_1$ is then also given since the electric field direction is the same as the wave-vector direction for electrostatic waves.

The relation (14.44) can be fulfilled only if $k_0 > 1/2$, or in dimensional units when

$$k_0 > \frac{1}{3} \frac{1}{\lambda_D} \sqrt{\frac{m}{M}} \equiv k_c. \tag{14.45}$$

This constraint can readily be understood by a geometrical consideration; see, for instance, Fig. 14.6. For $k_0 > 1$ in normalized units the decay Langmuir wave can propagate in the opposite direction of the initial wave and we can have back-scatter of Langmuir waves; see also Fig. 14.6. For $1/2 < k_0 < 1$ we can only have forward scattering of Langmuir waves.

In terms of non-normalized, or physical, variables we can express the maximum growth rate from (14.42) as

$$\gamma \approx \frac{1}{2\sqrt{2}} (k_0 \lambda_D)^{1/2} \left(\frac{m}{M}\right)^{1/4} \omega_{pe} \left(\frac{\varepsilon_0 |E_0|^2}{n_0 \kappa T_e}\right)^{1/2},$$

where the implied condition $\gamma \ll \Omega$ gives here $k_0 \lambda_D \gg (\sqrt{M/m}) \varepsilon_0 |E_0|^2 / (n_0 \kappa T_e)$. In the alternative limit (14.43), we find the maximum growth rate

$$\gamma \approx \frac{1}{2^{2/3}} (k_0 \lambda_D)^{2/3} \left(\frac{m}{M}\right)^{1/3} \omega_{pe} \left(\frac{\varepsilon_0 |E_0|^2}{n_0 \kappa T_e}\right)^{1/3}.$$

The condition $\gamma \gg \Omega$ gives here $k_0 \lambda_D \ll (\sqrt{M/m}) \varepsilon_0 |E_0|^2 / (n_0 \kappa T_e)$, which has to be satisfied simultaneously with $k_0 > k_c$ by (14.45). The foregoing analysis ignores damping of the waves. Indeed, for relevant plasma parameters in the ionosphere or in laboratory plasmas, the damping is small for a large range of wave-numbers.

We have here discussed the decay of an initially given Langmuir wave into another similar wave and an ion acoustic sound wave. However, the decay wave, which starts out with a small amplitude, will eventually reach a level where it decays into a new Langmuir wave and corresponding sound wave: we have a *cascade* of Langmuir decay.

From the derivation, it is evident that here $\omega_0 > \Omega$ as well as $\omega_0 > \omega_2$. If we make a quantum mechanical interpretation in terms of units of wave energy $\hbar\omega$, then this is a consequence of energy conservation in the wave interactions, $\hbar\omega_0 = \hbar\omega_2 + \hbar\Omega$; decay is possible only to other wave with frequencies below that of the initial wave. It is not admissible to make a quantum mechanical interpretation of a purely classical result, but it turns out that the frequency relation, and the ordering of the magnitudes of the frequencies has a classical analog, the *Manley-Rowe relations* (Sagdeev & Galeev 1969, Kono & Škorić 2010).

- **Exercise:** Express the constraint (14.45) in terms of the group velocity of the Langmuir waves and the ion sound speed. Explain the result by a diagram of the dispersion relation for Langmuir waves and ion sound waves.

- **Exercise:** Obtain the full solution for $\omega = \omega_r + i\gamma$ in (14.41), and recover the results (14.42) and (14.43) in the appropriate limits.

- **Exercise:** Consider a simple one-dimensional model system, with dispersion relations $\omega^2 = \omega_p^2 + C^2k^2$ and $\omega^2 = \omega_p^2 + U^2k^2$ with $C^2 < U^2$. Determine the frequencies and wave-numbers of the possible triplets for decay of waves ω_0, k_0 from the first of these branches, using $\omega_0 = \omega_1 + \omega_2$ and $k_0 = k_1 + k_2$.

Ⓢ **Exercise:** Consider a model system of three coupled oscillators (Sagdeev & Galeev 1969) with a Hamiltonian

$$H = \sum_{j=1}^{3} \frac{1}{2}\left(p_j^2 + \omega_j^2 x_j^2\right) + V x_1 x_2 x_3,$$

where p is the momentum and x the displacement of the oscillators, while V is a coupling coefficient. The oscillators are linear in the absence of the coupling term. This model can be seen as describing a sort of wave interaction without spatial variables. Analyze the stability of the motion of the three oscillators, for instance by numerical solutions. Ⓢ

Decay of Langmuir waves has been observed experimentally in ionospheric heating experiments, for instance, where intense radio waves are transmitted to the ionosphere, choosing the frequency to be resonant with the plasma frequency at an altitude somewhere in the ionospheric E- or F-region. The interaction of the electromagnetic wave and the plasma near the resonance layer produces an intense Langmuir wave, on a level much larger than those due to thermal fluctuations. One such experiment is described by Rietveld et al. (2000), where 200–300 MW effective radiated power was used at a frequency of ~ 5 MHz. The perturbations of the ionosphere were monitored by radar scattering. Laboratory experiments demonstrating decay of Langmuir waves have been reported by, for instance, Quon et al. (1974). In order to obtain agreement with observation in nature or in a laboratory it is often necessary to modify the basic equation (14.32) or its equivalent (14.33), together with (14.34) (Guio & Forme 2006).

The wave decay discussed in the present section can serve as an illustration for a general physical phenomenon when nonlinear media, plasma media in particular, carry a large amplitude wave, and a unified treatment can be formulated (Weiland & Wilhelmsson 1977, Stenflo 1994).

14.1.6 The nonlinear Schrödinger equation

The simplest possible model for inclusion of the low frequency dynamics of the plasma density, i.e., $\overline{N} = \overline{n}/n_0$, is obtained by assuming that the variation of the Langmuir wave amplitude is so slow that we can ignore $\partial^2\overline{N}/\partial t^2$ as compared to $C_s^2\nabla^2\overline{N}$ in (14.28). By this assumption, we ignore Langmuir wave decay involving a sound wave, where $\partial^2\overline{N}/\partial t^2$ and $C_s^2\nabla^2\overline{N}$ are of the same order of magnitude. If $k_0 < k_c$, we found that Langmuir wave decay is prohibited by geometric constraints, and the approximation in the following is easily justified.

Assuming the wave field to vanish at infinity, we let an integration constant be zero, and can readily integrate the simplified expression obtained from (14.28) to give $C_s^2\overline{N} = -(\varepsilon_0/4Mn_0)|\nabla\phi|^2$, which, inserted into (14.30), gives the simple nonlinear Schrödinger equation

$$\left(i\frac{\partial}{\partial t} + \frac{1}{2}\frac{u_{The}^2}{\omega_{pe}}\nabla\nabla\cdot\right)\mathbf{E} = -\frac{1}{8}\frac{\omega_{pe}\varepsilon_0}{\kappa T_e}|\mathbf{E}|^2\mathbf{E}, \tag{14.46}$$

where we re-introduced the electric field $\mathbf{E} = -\nabla\phi$ to make the equation look nicer, and also ignored the ion temperature in the expression for the sound speed. Since all three components of the electric field follow the same differential equations, there is no need to retain a vector notation for $\mathbf{E}$ in (14.46). It might seem that the notation could have been made simpler by keeping the electric field from the outset, rather than using the gradient of a potential, but by doing this we risk forgetting that the electric field is implicitly assumed to be *electrostatic* everywhere. We might indeed assume electrostatic waves in a linear analysis, but it is by no means self-evident that electrostatic waves remain so when nonlinear effects are involved. This problem has been discussed by, e.g., Kuznetsov (1974); see also Thornhill and ter Haar (1978).

When interpreting (14.46), we again recall that we have assumed an electric field variation of the form $\mathbf{E}(\mathbf{r},t)\exp(-i\omega_{pe}t)$. It is readily demonstrated that a plane wave as $\mathbf{E}(\mathbf{r},t) = a_0\exp(-i\omega t + ikx)$ is a solution to (14.46), since then $|E|^2 = a_0^2 = \text{const}$. This looks deceiving, but we should keep in mind that by obtaining (14.46) we assumed the wave field to vanish at infinity, in variance with the plane wave assumption. If we are bit cavalier concerning this point, and consider a plane wave as a limiting form of a very long wave-packet, we are nonetheless able to derive a dispersion relation for the assumed wave

$$\Omega = \frac{1}{2}\frac{u_{The}^2}{\omega_{pe}}k^2 - \frac{1}{8}\frac{\omega_{pe}\varepsilon_0}{n_0\kappa T_e}a_0^2, \tag{14.47}$$

emphasizing again that the Ω thus obtained is a correction to ω_{pe}, so that the wave frequency is $\omega = \omega_{pe} + \Omega$. Note that in the last term of (14.47) we have $\frac{1}{2}\varepsilon_0 a_0^2/(n_0\kappa T_e)$ being the ratio of the wave and the plasma thermal energy densities, respectively. If we let $a_0 \to 0$, we recover essentially the linear dispersion relation (13.11), apart from a difference in the numerical constant due to the definition $u_{The}^2 = 3\kappa T_e/m$ used here. The last term in (14.47) gives the nonlinear frequency shift, and we note that it is independent of k and that it is *negative*. The derivative of the group velocity is positive, so that Lighthill's condition, Section 4.3, is fulfilled and the Langmuir waves are modulationally unstable. This is an important difference from the cold electron model with immobile ions considered in Section 14.1.1. In the present case the ponderomotive forces deplete the local plasma density where the wave amplitude is large, and the lower density gives a lower local plasma frequency.

14.1.7 Nonlinear plasma waves in one, two and three spatial dimensions

The implications of the equation (14.46) have been discussed in Section 4.3, although that discussion was restricted to one spatial dimension. We will here give some heuristic arguments for the differences in the physics of weakly nonlinear electron plasma waves in one, two and three spatial dimensions.

First of all we note that irrespective of spatial dimensionality, the relation (14.46) is energy conserving, i.e., $\mathcal{W} \equiv \int |E|^2 dV = \text{const}$. The proof is trivial and independent of the spatial dimensionality of the problem. By order of magnitude estimates we compare the relative importance of the dispersive and the nonlinear terms in (14.46) and write their ratio as

$$\frac{\text{dispersion}}{\text{nonlinearity}} \sim \frac{E}{\mathcal{L}^2|E|^2E} \sim \frac{\mathcal{L}^{d-2}}{\mathcal{W}} \tag{14.48}$$

where we used $\nabla^2 E \sim E/\mathcal{L}^2$ and also $\mathcal{W} \sim |E|^2\mathcal{L}^d$, where d is the spatial dimensionality of the problem.

Since we assume $\mathcal{W}$ constant for the system, the relation (14.48) states that the relative importance of the dispersive and the nonlinear terms in (14.46) is proportional to $\mathcal{L}^{d-2}$. Consider a one-dimensional system, $d = 1$, and assume that we have a localized equilibrium solution where dispersion and nonlinearity are in balance. Perturb now this equilibrium by contracting the pulse slightly, i.e., decrease $\mathcal{L}$. By (14.48) with $d = 1$ this implies that the dispersion term becomes *more* important, and the pulse disperses back to its original wider shape. If the pulse is widened, the situation is reversed, and the dispersion term becomes *less* important, and nonlinearity tends to restore the pulse (Nishikawa 1984). The assumed equilibrium situation is stable; this is the soliton solution for the NLS equation. For the three-dimensional case, $d = 3$, the same argument gives the opposite result. If we state that an equilibrium solution exists with balance between nonlinearity and dispersion, a slight contraction of this state will let nonlinearity take over, while a slight expansion gives preference to the dispersion. The assumed equilibrium situation is unlikely to exist. For the two-dimensional case these arguments will not give any definite answer.

The foregoing discussion indicates that we may anticipate peculiar results in three spatial dimensions, but the arguments do not tell much about the *nature* of these results. A detailed investigation (Zakharov 1972, Rasmussen & Rypdal 1986) demonstrates that for the three-dimensional case the NLS equation has collapsing solutions, where a singularity in the electric field develops within a finite time. It turns out that such solutions can be obtained also in the spatially two-dimensional case. To account for the possibility of Langmuir wave collapse, we outline the derivation of a virial theorem, here based on the analysis of Goldman and Nicholson (1978), although the theorem had several predecessors (Rasmussen & Rypdal 1986). We take for simplicity a normalized version of the NLS equation, $i\partial\mathbf{E}/\partial t + \frac{1}{2}\nabla\nabla\cdot\mathbf{E} + |E|^2\mathbf{E} = 0$. The existence of at least two continuity equations can be demonstrated

$$\frac{\partial}{\partial t}|E|^2 + \nabla\cdot\mathbf{s} = 0 \tag{14.49}$$

$$\frac{\partial}{\partial t}\mathbf{p} + \nabla\cdot\underline{\mathsf{T}} = 0, \tag{14.50}$$

where $\mathbf{s}$ can be interpreted as a current density and $\mathbf{p}$ as a momentum density of the field. For the given NLS equation they are equal

$$\mathbf{s} = \mathbf{p} = \frac{1}{2i}\left(\mathbf{E}^*\nabla\cdot\mathbf{E} - \mathbf{E}\nabla\cdot\mathbf{E}^*\right).$$

The components of the stress tensor $\underline{\mathsf{T}}$ are given by

$$\mathsf{T}_{ij} = \Re\{(\nabla\cdot\mathbf{E})\partial_i E_j\} - \frac{1}{2}\delta_{ij}\left(|E|^4 + \nabla\cdot(\Re\mathbf{E}^*\nabla\cdot\mathbf{E})\right).$$

For fields vanishing sufficiently fast at $|\mathbf{r}| \to \infty$, we can integrate, for instance, (14.49) to give conservation of $N \equiv \int |E|^2 d\mathbf{r}$, which is often denoted "plasmon number." Similarly we can demonstrate conservation of $\mathbf{S} \equiv \int \mathbf{s} d\mathbf{r}$. In addition, a quantity interpreted as a total field energy $\mathcal{H} \equiv \frac{1}{2}\int\left(|\nabla\cdot\mathbf{E}|^2 - |E|^4\right)d\mathbf{r}$ is conserved as well. The integrals run over all space.

To analyze the space-time evolution of the localized wave field, we do not need to understand all possible details. We introduce some characteristic average quantities such as the average position $\langle\mathbf{r}\rangle \equiv \int \mathbf{r}|E|^2 d\mathbf{r}/N$ and a spread around this average as $\langle\Delta\mathbf{r}^2\rangle \equiv \int (r - \langle\mathbf{r}\rangle)^2|E|^2 d\mathbf{r}/N$, where the averaging $\langle\rangle$ here is over the spatial electric field distribution and should not be confused with an ensemble averaging.

- **Exercise:** Derive the equivalent of Ehrenfest's theorem (Davydov 1965) $d\langle\mathbf{r}\rangle/dt = \mathbf{S}/N = \text{const}$ for the present NLS equation.

It is now a trivial matter to obtain a virial theorem for the present problem as

$$\frac{d^2}{dt^2}\langle \Delta \mathbf{r}^2 \rangle = 2\left(\frac{1}{N}\int \mathrm{Tr}\{\underline{\mathrm{T}}\}d\mathbf{r} - \left(\frac{\mathbf{S}}{N}\right)^2\right),$$

where $\mathrm{Tr}\{\underline{\mathrm{T}}\}$ denotes the trace of the tensor $\underline{\mathrm{T}}$. After a little algebra, where we use the explicit forms of $\underline{\mathrm{T}}$ and $\mathcal{H}$ given before, we integrate the virial theorem to give

$$\langle \Delta \mathbf{r}^2 \rangle = At^2 + Bt + C + (2-d)\int_0^t \int_0^{t'} \int |E|^4 d\mathbf{r}dt''dt', \tag{14.51}$$

where B and C are integration constants, d is the spatial dimensionality of the problem, while $A \equiv 2\mathcal{H}/N - (\mathbf{S}/N)^2$.

We have both $\langle \Delta \mathbf{r}^2 \rangle \geq 0$ and $\int |E|^4 d\mathbf{r} \geq 0$, so from (14.51) we conclude that if $d \geq 2$ and $A < 0$, then the localized wave field must necessarily collapse to a point within a *finite* time: a most unexpected result when it was first demonstrated! The condition $A < 0$ implies, in terms of conserved quantities, $s\mathcal{H} < |S|^2/N$, which can be achieved by properly chosen initial conditions.

Since $N \equiv \int |E|^2 d\mathbf{r}$ is conserved, $\langle \Delta \mathbf{r}^2 \rangle \to 0$ must necessarily imply that $\mathbf{E} \to \infty$ at the collapse time. In reality, the process will of course never develop that far, since the basic NLS equation will be invalid long before that. A proper damping mechanism is lacking, for instance, and when very large density perturbations develop, we might expect that Langmuir wave breaking, as anticipated in Section 14.1.1, will become effective as well. By introducing a linear damping mechanism, we may arrest collapse if the initial energy density is close to threshold. If, however, we have conditions where the initial wave energy density is large, we intuitively expect a linear wave-damping mechanism to have little effect: the wave collapse proceeds faster than exponentially, while any physically realistic linear damping mechanism we might imagine gives rise to a damping being exponential with time. If we use (14.51) for a spatially one-dimensional case, $d = 1$, the positive last term will become dominating when $t \to \infty$ and any further approach to collapse will be arrested: wave collapse is a mechanism found in two or three spatial dimensions.

Even though many details have to be added to give a physically correct description, Langmuir wave collapse in its basic form as described here nevertheless represents an intriguing new phenomenon which is still being studied. A number of computer simulations and also some laboratory experiments have been interpreted by such processes (Wong & Quon 1975).

The discussion outlined in this and foregoing sections implicitly refers to cases where the initial plasma density is uniform and a wave field is introduced at some initial time. The variations in plasma density are then developing as the result of the nonlinear ponderomotive forces. Large amplitude wave fields can be obtained more easily when the initial plasma density has a gradient, e.g., as $n(x) \approx n_0 + x/\mathcal{L}$ in the vicinity of $x = 0$. Choosing the frequency ω_0 of an externally excited wave to match the plasma frequency ω_{pe} at $x = 0$, we obtain total reflection; see the discussion in Section 12.2.5. In the vicinity of the reflection point, the group velocity is small and vanishes at $x = 0$. At least with the WKB approximation we can assume the group velocity to be the velocity of energy transport (Brillouin 1960), and by the conservation of energy flux the wave energy density will increase where the group velocity decreases. The ponderomotive forces become particularly large here and a plasma cavity is formed near $x = 0$, as observed in, for instance, laboratory experiments (Kim et al. 1974). Analytical models for the process can be formulated in terms of modified NLS equations (Morales & Lee 1977). The most interesting results (and physically most relevant as well) may be obtained when a continuous external driving field is included. Such models have been used for studying ionospheric heating experiments, for instance (Hanssen et al. 1992).

14.1.8 Wave decay

The decay of Langmuir waves discussed before can be seen as an example of a general plasma wave phenomenon. A number of decay modes are illustrated in Fig. 14.7. The initial large am-

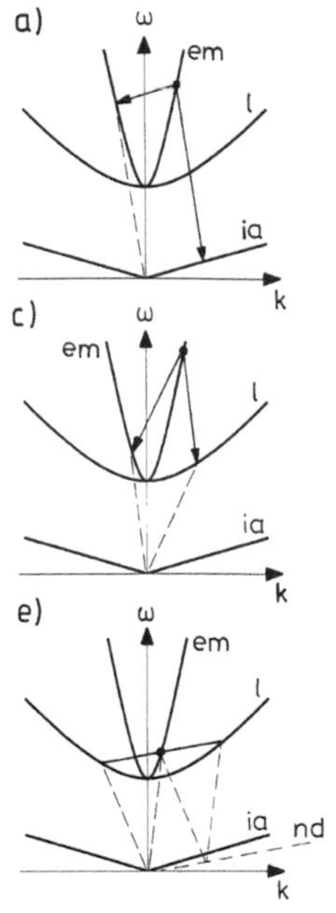

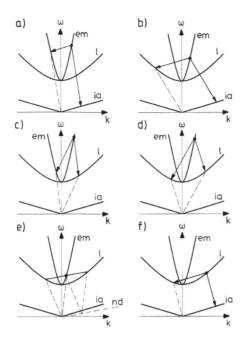

FIGURE 14.7: *Schematic representation of some nonlinear wave interactions. The abbreviations on the dispersion diagrams have the following meanings:* em → *electromagnetic,* l → *Langmuir,* ia → *ion-acoustic and* nd → *not a dispersion relation. Note that the figure is restrictive, for illustrative purposes, by showing wave-vectors in one plane only.*

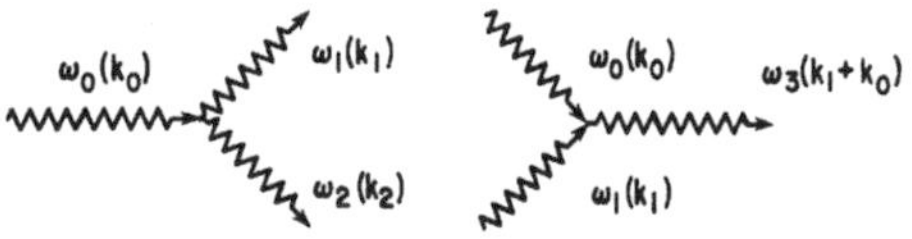

FIGURE 14.8: *Diagrammatic representation of parametric decay (left) and scattering (right). The notation* $\omega_j(\mathbf{k}_j)$ *indicates a wave that fulfills an appropriate dispersion relation.*

plitude wave, or "pump wave" can be electrostatic or electromagnetic. The process illustrated in Fig. 14.7a is often called "stimulated Brillouin scattering," the word "stimulated" emphasizing that we deal with an instability, where the sound wave is driven by the electromagnetic wave (Sjölund & Stenflo 1967*a*), in contrast to the standard "Brillouin scattering", where electromagnetic waves are scattered off "pre-existing" sound waves. Similarly, we have "stimulated Raman scattering" illustrated in Fig. 14.7c, where electromagnetic waves generate electron plasma waves (Sjölund & Stenflo 1967*b*). The names Brillouin and Raman refer to the scientists first studying these processes. Figure 14.7e), in particular, illustrates the "oscillating two stream instability" (Chen 2016). Often we see a diagrammatic representation of the processes as in Fig. 14.8, illustrating decay $\omega_0 \to \omega_1 + \omega_2$, $k_0 \to k_1 + k_2$ (left), and scattering (right). A magnetized plasma will be particularly rich in nonlinear wave interactions.

14.2 Weakly nonlinear ion acoustic waves

In order to discuss the lowest order corrections to the linear description of electrostatic ion acoustic waves, we choose a simplified model by assuming the ion component to be cold, $T_i = 0$. The sound waves are in this case solely propagated by the electron pressure, $T_e \neq 0$. Assuming the electrons to be isothermally Boltzmann distributed at all times, we find the basic equations for the problem in one spatial dimension to be

$$\frac{\partial n}{\partial t} + \frac{\partial}{\partial x}(nu) = 0, \tag{14.52}$$

$$\frac{\partial u}{\partial t} + u\frac{\partial}{\partial x}u = -\frac{e}{M}\frac{\partial}{\partial x}\phi, \tag{14.53}$$

$$n = n_0 \exp(e\phi/\kappa T_e) \quad \text{or} \quad \frac{e\phi}{\kappa T_e} = \ln\left(\frac{n}{n_0}\right), \tag{14.54}$$

where we can assume quasi-neutrality from the outset, i.e., $n_e \approx n_i \equiv n$. Alternatively, we can retain Poisson's equation and allow for a small charge separation to appear explicitly in the analysis. We assume that the plasma is initially homogeneous, $n_0 = \text{const}$.

14.2.1 Simple ion acoustic waves

By virtue of the Fourier transform of the continuity equation, $(n/n_0)(\omega/k) = u$, for *linear* ion acoustic waves in the quasi-neutral limit we have $n = n_0 u/C_s$ with $C_s = \sqrt{\kappa T_e/M}$, where again the ion mass is M. Evidently, the perturbation in plasma density n is linearly proportional to the perturbation in ion velocity, u in this case. For weakly nonlinear waves we now *assume* that there is still a one-to-one relation between density and velocity, and write $n(x,t) = n(u(x,t))$, without, however, specifying this functional relationship (at least not yet!). We simply imply that the plasma density somehow depends on x,t through the velocity dependence. From (14.53) we have, by first eliminating the potential ϕ and then using $\partial n/\partial x = (dn/du)\partial u/\partial x$,

$$\frac{\partial u}{\partial t} + u\frac{\partial u}{\partial x} + \frac{\kappa T_e}{M}\frac{1}{n}\frac{\partial n}{\partial x} = 0,$$

$$\frac{\partial u}{\partial t} + u\frac{\partial u}{\partial x} + C_s^2\frac{1}{n}\frac{dn}{du}\frac{\partial u}{\partial x} = 0. \tag{14.55}$$

By the continuity equation (14.52) we find

$$\frac{dn}{du}\left(\frac{\partial u}{\partial t} + u\frac{\partial u}{\partial x}\right) + n\frac{\partial u}{\partial x} = 0, \tag{14.56}$$

where we used, for instance, $\partial n/\partial t = (dn/du)\partial u/\partial t$. Using (14.55) and (14.56), we readily obtain

$$\left(\frac{dn}{du}\right)^2 = \frac{n^2}{C_s^2} \quad \text{or} \quad \frac{dn}{du} = \pm\frac{n}{C_s}. \tag{14.57}$$

This relation determines $n = n(u)$. Using (14.56) and (14.57) we find the amazingly simple result

$$\frac{\partial}{\partial t}u + (u \pm C_s)\frac{\partial}{\partial x}u = 0, \tag{14.58}$$

consistent with Section 4.1. By a trivial change in reference system moving with constant velocity C_s we have

$$\frac{\partial}{\partial t}u + u\frac{\partial}{\partial x}u = 0. \tag{14.59}$$

For a given initial condition $u(x,0) = F(x)$, we have the solution $u(x,t) = F(x - u(x,t)t)$. Ultimately, also these weakly nonlinear ion sound waves will exhibit a wave breaking, just like simple waves in a neutral gas; see Section 4.1. Before this time, however, the quasi-neutrality assumption in the basic equation is violated, i.e., the variations in plasma parameters occur on length scales smaller than or comparable to the Debye length, and the analysis has to be reformulated. For long initial wavelengths, the simple equation (14.59) will, on the other hand, remain adequate for a long time.

Notice how easy the analysis of this section is compared to the simple wave analysis of Section 4.1. This simplification was in part achieved by the assumption of cold ions and isothermal electrons. If we assume $T_i \neq 0$, and adiabatic ion dynamics, we will encounter the same complications as before.

- **Exercise:** Why did we assume $n(x,t) = n(u(x,t))$ in (14.55) and (14.56), and not $u = u(n(x,t))$?

- **Exercise:** Derive equation (14.59) for the case where $T_i \neq 0$. Discuss the case of isothermal as well as adiabatic ion dynamics.

14.2.2 The Korteweg-deVries model for ion acoustic waves

As we have seen in Section 14.2.1, the simple wave model for ion acoustic waves breaks down at a time where the basic assumption of quasi-neutrality is violated. It will be demonstrated that a remedy of this shortcoming in one spatial dimension leads to the derivation of a Korteweg-deVries equation for the ion acoustic wave dynamics. We have already found these waves to be weakly dispersive in the large wave-number limit; see (13.28).

A systematic method for the following analysis is the *reductive perturbation method* (Taniuti & Wei 1968, Washimi & Taniuti 1968). Physically, the idea is that to lowest order we have nondispersive ion acoustic waves, while nonlinearity and dispersion come as corrections on the same level to this basic model. The analytical derivation will make the implications of these arguments clear. Interested readers can find more details in the monograph by Nayfeh (1973), and in particular also in the special issue on "Reductive Perturbation Method for Nonlinear Wave Propagation", Supplement of the Progress in Theoretical Physics, (1974) Vol. 55, published by the Research Institute for Fundamental Physics and the Physical Society of Japan.

Consider again the set of equations

$$\frac{\partial}{\partial t}n_i + \frac{\partial}{\partial x}(n_i u_i) = 0, \tag{14.60}$$

$$\frac{\partial}{\partial t}u_i + u_i\frac{\partial}{\partial x}u_i = -\frac{e}{M}\frac{\partial}{\partial x}\phi, \tag{14.61}$$

$$n_e = n_0 \exp(e\phi/\kappa T_e) \qquad \text{or} \qquad \frac{e\phi}{\kappa T_e} = \ln(n_e/n_0), \tag{14.62}$$

to be completed with Poisson's equation

$$\frac{\partial^2}{\partial x^2}\phi = \frac{e}{\varepsilon_0}(n_e - n_i). \tag{14.63}$$

Note that we have added a subscript i for *ions* in (14.60)–(14.63) in order to distinguish electron and ion densities, since we relax the quasi-neutrality assumption, at least for the time being. In the linearized limit, these basic equations are readily solved to give the well-known linear wave dispersion relation $\omega^2 = C_s^2 k^2/(1 + (k\lambda_D)^2)$, which at least formally allows any phase velocity $0 < \omega/k \leq C_s$.This property is of some importance for understanding the results in the following sections. A finite ion temperature imposes a lower limit for this phase velocity.

Assume that the perturbations we are interested in can be characterized by a length scale $\mathcal{L}$ such as the width of a sound pulse. We normalize the variables as

$$\frac{n_i}{n_0} \equiv \eta, \quad \frac{n_e}{n_0} \equiv \eta_e, \quad u \equiv \frac{u_i}{C_s}, \quad \frac{e\phi}{\kappa T_e} \equiv \Phi, \quad \frac{tC_s}{\mathcal{L}} \equiv t', \quad \text{while} \quad \frac{x}{\mathcal{L}} \equiv x',$$

in terms of the sound speed $C_s \equiv \sqrt{T_e/M}$, and have, for instance, $\Phi = \Phi(x', t')$. Making a series expansion in the exponential factor in the assumption of Boltzmann distributed electrons, the basic equations then become

$$\frac{\partial}{\partial t'}\eta + \frac{\partial}{\partial x'}(\eta u) = 0, \tag{14.64}$$

$$\frac{\partial}{\partial t'}u + u\frac{\partial}{\partial x'}u = -\frac{\partial}{\partial x'}\Phi, \tag{14.65}$$

$$\eta_e = e^{\Phi} \approx 1 + \Phi + \frac{1}{2}\Phi^2 + \ldots, \tag{14.66}$$

$$\left(\frac{\lambda_D}{\mathcal{L}}\right)^2 \frac{\partial^2}{\partial x'^2}\Phi = (\eta_e - \eta), \tag{14.67}$$

see also (11.14).

We then expand the dependent variables as

$$\begin{aligned} \eta &= 1 + \varepsilon\eta_1 + \varepsilon^2\eta_2 + \ldots \\ u &= 0 + \varepsilon u_1 + \varepsilon^2 u_2 + \ldots \\ \Phi &= 0 + \varepsilon\Phi_1 + \varepsilon^2\Phi_2 + \ldots \end{aligned}$$

where we took into account that u_i and ϕ vanish in equilibrium. The quantity ε is a small expansion parameter, serving to identify the order of magnitude of the corresponding perturbations. We can use ε as a measure for the amplitude of the perturbation. In order to have dispersion entering as a small quantity we assign $(\lambda_D/\mathcal{L})^2 \equiv \varepsilon$. We now proceed by considering terms containing the same powers in ε.

- Zeroth order, ε^0: The equations are trivially satisfied.
- First order, ε^1: To this order we find

$$\begin{aligned} \frac{\partial}{\partial t'}\eta_1 + \frac{\partial}{\partial x'}u_1 &= 0, \\ \frac{\partial}{\partial t'}u_1 &= -\frac{\partial}{\partial x'}\Phi_1, \\ \eta_{e1} &= \Phi_1, \\ \eta_{e1} &= \eta_1, \quad \text{i.e., quasi-neutrality is recovered to this order!} \end{aligned}$$

By simple elimination we find the relation $\partial^2\eta_1/\partial t'^2 + \partial^2\eta_1/\partial x'^2 = 0$, giving the dispersion relation $\omega/k = \pm 1$, where we recall that velocities are normalized by the sound speed C_s. We also find $\eta_1 = \eta_{e1} = \Phi_1 = u_1$ to this order. To this accuracy, all quantities propagate with a constant speed without distortion. In particular, if we change the frame of reference to one moving with velocity $+1$ or -1 (i.e., $\pm C_s$), everything is stationary, again to the present accuracy. It is therefore convenient to introduce new variables

$$\xi \equiv x' - t' \quad \text{and} \quad \tau \equiv \varepsilon t', \tag{14.68}$$

where by the scaling of τ we anticipate that we can have time variations to higher order in this moving frame of reference, and have $\eta_1 = \eta_1(\xi, \tau)$, $\Phi_1 = \Phi_1(\xi, \tau)$, etc. Usually (14.68) is assumed from the outset, without the physical justification given here.

- Second order, ε^2: Using $\partial\eta_1/\partial t' = \varepsilon\partial\eta_1/\partial\tau - \partial\eta_1/\partial\xi$, etc., we obtain the following ε^2 order relations in the new frame of reference

$$\begin{aligned}
\frac{\partial}{\partial\tau}\eta_1 - \frac{\partial}{\partial\xi}\eta_2 + \frac{\partial}{\partial\xi}u_2 + \frac{\partial}{\partial\xi}(\eta_1 u_1) &= 0, \\
\frac{\partial}{\partial\tau}u_1 - \frac{\partial}{\partial\xi}u_2 + \frac{1}{2}\frac{\partial}{\partial\xi}u_1^2 &= -\frac{\partial}{\partial\xi}\Phi_2, \\
\eta_{e2} &= \Phi_2 + \frac{1}{2}\Phi_1^2, \\
\frac{\partial^2}{\partial\xi^2}\Phi_1 &= \eta_{e2} - \eta_2.
\end{aligned}$$

Eliminating η_1 and u_1 in favor of Φ_1 we find

$$2\frac{\partial}{\partial\tau}\Phi_1 + \frac{3}{2}\frac{\partial}{\partial\xi}\Phi_1^2 = \frac{\partial}{\partial\xi}\eta_2 - \frac{\partial}{\partial\xi}\Phi_2. \tag{14.69}$$

Similarly, we find

$$\frac{\partial^3}{\partial\xi^3}\Phi_1 = \frac{\partial}{\partial\xi}\Phi_2 + \frac{1}{2}\frac{\partial}{\partial\xi}\Phi_1^2 - \frac{\partial}{\partial\xi}\eta_2. \tag{14.70}$$

This equation can be used to eliminate $\partial\eta_2/\partial\xi$ from (14.69). We find that by this $\partial\Phi_2/\partial\xi$ vanishes also! Evidently, it is most fortunate that this "double" elimination is possible! Finally we have

$$\frac{\partial}{\partial\tau}\Phi_1 + \Phi_1\frac{\partial}{\partial\xi}\Phi_1 + \frac{1}{2}\frac{\partial^3}{\partial\xi^3}\Phi_1 = 0, \tag{14.71}$$

which is the Korteweg-deVries equation, written in the present normalized variables. We note that Φ_1 is a first order quantity, which varies slowly with time so as to make its time derivative of second order. To complete the discussion of the KdV model for weakly nonlinear ion acoustic waves, we refer to Section 4.2. From (4.30), we thus readily find a propagating soliton solution of (14.71) in the normalized form

$$\Phi_1 = a\,\mathrm{sech}^2\left(\frac{1}{\Delta}(\xi - U\tau)\right)$$

with $U = a/3$ and $\Delta = \sqrt{6/a}$, since the dispersion coefficient is trivially found as $1/2$ in the present normalized units.

- **Example:** You might like to have the KdV equation in physical units, in which case we find to the present accuracy, with cold ions

$$\frac{\partial}{\partial t}\phi_1 + \left(C_s + C_s\frac{e}{\kappa T_e}\phi_1\right)\frac{\partial}{\partial x}\phi_1 + \frac{1}{2}C_s\lambda_{De}^2\frac{\partial^3}{\partial x^3}\phi_1 = 0, \tag{14.72}$$

here with soliton solutions $\phi_1(x,t) = a\,\mathrm{sech}^2\left(\left(x - C_s t - \frac{1}{3}t\,aC_s e/\kappa T_e\right)\sqrt{ae/(6\kappa T_e\lambda_{De}^2)}\right)$.

- **Exercise:** Obtain the expression in physical units for the ion density and ion velocity for the ion acoustic soliton solution to the KdV equation, accurate to second order, ε^2.

14.2.3 Stationary nonlinear solutions

One spectacular result of the KdV-type equations seems to be the prediction of soliton solutions. It turns out, however, that such solutions can be found by a simplified procedure, without any reductive perturbation procedure, i.e., expansions in amplitude and stretching of time (Biskamp & Parkinson 1970). The *really* spectacular result of the KdV equation is the prediction of the inverse scattering method (which has not been derived here, only quoted), stating that these soliton solutions emerge unchanged (apart from a phase shift) from interactions; see Fig. 4.9. Some interesting and elegant discussions of steady state solutions for nonlinear equations (the KdV equation being one of them) were summarized by Nycander (1994).

To demonstrate that stationary solutions can be found for a wide class of problems we again consider ion acoustic waves, and take also here the simple limit with cold ions and isothermal Boltzmann distributed electrons. We postulate that a moving frame of reference exists, where such stationary solutions can be found. Let the constant velocity of this frame be U. This assumption implies that $\partial n/\partial t \to -U dn/dx$, and similarly for the ion velocity u.

From the one-dimensional continuity equation $\partial n/\partial t + \partial(nu)/\partial x = 0$ we find, after integration with respect to x, the relation

$$n(u-U) = -Un_0 \tag{14.73}$$

where we choose a reference for the electrostatic potential such that $n = 0$ when $\phi = 0$. We emphasize here that u is the ion velocity in the rest frame where we see the solitary structure propagate with velocity U.

Similarly we find from the momentum and Poisson equations the following relations

$$\frac{1}{2}(u-U)^2 = -\frac{e}{M}\phi + \frac{1}{2}U^2 \tag{14.74}$$

$$\frac{d^2}{dx^2}\phi = \frac{e}{\varepsilon_0}\left(n_0 \exp\left(e\phi/\kappa T_e\right) - n\right) \tag{14.75}$$

where we choose an integration constant by requiring $u = 0$ for $\phi = 0$.

To ease the notation it is also here an advantage to use normalized variables, and we take $\psi \equiv e\phi/\kappa T_e$, $\eta \equiv n/n_0$, and $\xi \equiv x/\lambda_D$, while velocities are normalized by the sound speed $C_s \equiv \sqrt{\kappa T_e/M}$ so that we have $C \equiv U/C_s$. Eliminating the ion velocity and ion density, we then find an expression for the normalized potential as

$$\frac{d^2}{d\xi^2}\psi = \left(e^{\psi} - \frac{C}{\sqrt{C^2 - 2\psi}}\right). \tag{14.76}$$

When solving for the density, we have $\pm$ in front of a square root, but taking $C > 0$ we choose $-$ by requiring $n = n_0$ when $u = 0$. After multiplication with $d\psi/d\xi$ and using

$$\frac{d\psi}{d\xi}\frac{d^2\psi}{d\xi^2} = \frac{1}{2}\frac{d}{d\xi}\left(\frac{d\psi}{d\xi}\right)^2,$$

we find after integration with respect to ξ again the form

$$\frac{1}{2}\left(\frac{d\psi}{d\xi}\right)^2 + V(\psi) = W, \tag{14.77}$$

here with the quasi-potential

$$V(\psi) \equiv -e^{\psi} - C\sqrt{C^2 - 2\psi},$$

while W is here an integration constant; see also (4.29). Examples of pseudo-potentials are shown in Fig. 14.9. The existence of stationary solutions requires that a local minimum exists for $V(\psi)$.

We readily find that $dV(\psi)/d\psi = 0$ at least for $\psi = 0$. This is a local minimum for $C < 1$. When $C > 1$ we have another local minimum at ψ_0 determined implicitly by $\exp\psi_0 = C/\sqrt{C^2 - 2\psi_0}$ for $\psi_0 > 0$. For the sonic case $C = 1$, we have that both the first and second derivatives of $V(\psi)$ are vanishing at $\psi = 0$.

In order to have a localized, "pulse-like" solution we require $\psi \to 0$ for $\xi \to \pm\infty$, which can be obtained when the position $\psi = 0$ corresponds to a local maximum, i.e., for the supersonic case when $C > 1$. However, for such solutions we have the additional constraint, namely, that the value $V(0) \le V(C^2/2)$, which implies $1 < C \le C_0 \approx 1.5852$. When this condition is satisfied, we have the maximum value ψ_m of the solitary pulse-like solution to be implicitly given by $V(0) = V(\psi_m)$, or $1 + C^2 = \exp\psi_m + C\sqrt{C^2 - 2\psi_m}$ for given C. This relation is probably best solved graphically.

For pulse velocities close to the sonic case we expect to have small ψ_m, and therefore $2\psi \ll C^2$, which allows a series expansion of the pseudo-potential as

$$V(\psi) \approx -\left(1 + C^2 + \frac{1}{2}\left(1 - \frac{1}{C^2}\right)\psi^2 + \frac{1}{6}\left(1 - \frac{3}{C^4}\right)\psi^3\right),$$

giving the implicit expression for ψ_m to be $3(1 - 1/C^2) + (1 - 3/C^4)\psi_m \approx 0$, as obtained by requiring $V(0) = V(\psi_m)$. Since, however, for this case we expect C to be only slightly larger than unity, we can approximate $3(C^4 - C^2)/(C^4 - 3) \approx 3(C - 1)$, which turns out to give precisely the KdV result, albeit in a different notation.

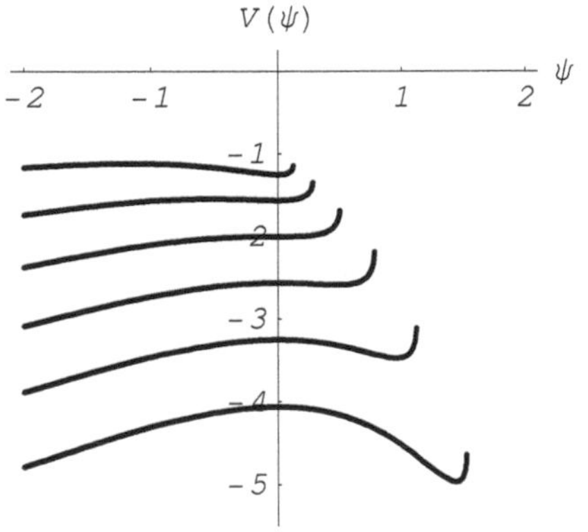

FIGURE 14.9: *Pseudo-potentials $V(\psi)$ for $C = 0.5, 0.75, 1.0, 1.25, 1.5$ and 1.75, the largest value of C being the lower-most curve. For the last case, there are no solitary solutions of (14.77), since $V(0) > V(C^2/2)$ here. Pulse-like localized solutions are found for the supersonic case $1 < C \le C_0 \approx$* 1.5852.

- **Exercise:** Demonstrate by inspection of the pseudo-potentials in Fig. 14.9 that the solitary solution becomes more "pointed" as compared to the soliton when C is increased. For the critical velocity, $C = C_0$, demonstrate that the solution has a "cusp" at the maximum.

We have thus succeeded in presenting a general formalism, which in the small amplitude limit contains the KdV soliton as a limiting case. We have not been able to present an exact analytical expression for this solitary solution, but in principle the differential equation (14.77) is just as good. It is, however, important to emphasize that we do *not* have any closed solvable *dynamic* equation which includes the KdV equation as a special limit. Even if we had, it would be unlikely that the more general equation could be demonstrated to have pulse-like solutions possessing the soliton property, i.e., emerging naturally from a large class of initial conditions and remaining unaffected by nonlinear interactions with other pulses. At times the scientific journals in plasma sciences were crowded with stationary solutions of complicated nonlinear equations, but, as stated, such solutions have little physical interest unless they are associated with a soliton property, at least in some limits. It is important to distinguish *solitons* from *solitary solutions*! For the present case we have the general set of equations (14.60)–(14.63) instead of the simpler KdV equation. These equations were solved numerically by Biskamp and Parkinson (1970) and the development of shock-like solutions was found for large initial amplitudes, while smaller amplitudes gave solutions with properties close to those predicted by the KdV equation.

We might emphasize here that the analysis outlined in the present section is correct only for the localized solutions obtained. For the periodic nonlinear wave-trains, for instance, we have to determine the constant n_0 in such a way that the total number of electrons and ions is the same at all times (Pécseli et al. 1984). For the localized solution this requirement is trivially satisfied with the integration constants chosen here. The basic arguments can be outlined as follows: overall charge neutrality can be expressed as $\int\int\int_{-\infty}^{\infty}(n_e(\mathbf{r},t) - n_i(\mathbf{r},t))dxdydz = 0$. With Boltzmann distributed electrons in one spatial dimension in a finite domain $x \in \{-L/2, L/2\}$ this requirement becomes

$$\int_{-L/2}^{L/2} (n_0 \exp(-e\phi(\mathbf{r},t)/\kappa T_e) - n_i(\mathbf{r},t))\, dx = 0, \tag{14.78}$$

where n_0 is independent of x but may depend on time. The relation (14.78) can be seen as a consistency expression determining n_0. For the solitary localized solutions (14.78) becomes immaterial as we let $L \to \infty$. For the periodic solutions (with period L) that also exist (the "cnoidal waves," for instance), the condition is important (Pécseli et al. 1984). Similar conditions can arise also for stationary nonlinear electron structures with immobile ions (Kako et al. 1971).

The fact that for nonlinear problems we generally can have $n_0 = n_0(t)$ is often ignored in the literature. In two- or three-dimensional models we might have n_0 depending on position; this will be particularly relevant in a magnetized plasma where different magnetic flux tubes can have different reference or unperturbed plasma densities. When describing the low frequency dynamics of a Q-machine (see Section 9.1), for instance, it is straightforward to assume "each magnetic field line to have its own n_0," by this accounting for the radial variation of the unperturbed density in the plasma column, $n_0 = n_0(r)$. Since the hot-plate of a Q-machine usually has a small radial temperature variation, it is consistent to allow for a corresponding variation of the electron temperature, $T_e = T_e(r)$.

14.2.4 Experimental results for soliton propagation

Weakly nonlinear ion acoustic waves have been investigated experimentally by, for instance, Ikezi et al. (1970), Cohn and MacKenzie (1973), and Ikezi (1973); see also the summary by Lonngren (1983). The formation and interaction of solitary structures were found to be reasonably well described by the KdV equation, although a correction in terms of a damping term was needed to improve the agreement with theory. This damping was mostly due to kinetic effects, not accounted for by a fluid model. It should be emphasized that the KdV equation is valid only to second order in the expansion parameter ε. It is futile to attempt a verification of the results beyond the range of validity of the equation. There are other cases, however (Lynov et al. 1979b), where the results from a KdV-type analysis are "incidentally" valid also for a wider amplitude range.

For illustration we here show selected experimental result by Ikezi et al. (1970), see Figs. 14.10 and 14.11. The experiment was carried out in a Double Plasma device, see Section 9.2, in an an Argon plasma where the electron-ion temperature ratio was $T_e/T_i \approx 15$. The results of Fig. 14.10a demonstrate a scaling of the normalized squared soliton width D/λ_D and normalized soliton amplitude as $D/\lambda_D \sim \sqrt{n_0/\delta n}$: large amplitude solitons are narrow. The results of Fig. 14.10b demonstrate that the soliton's velocity is linearly proportional to the soliton amplitude to a good

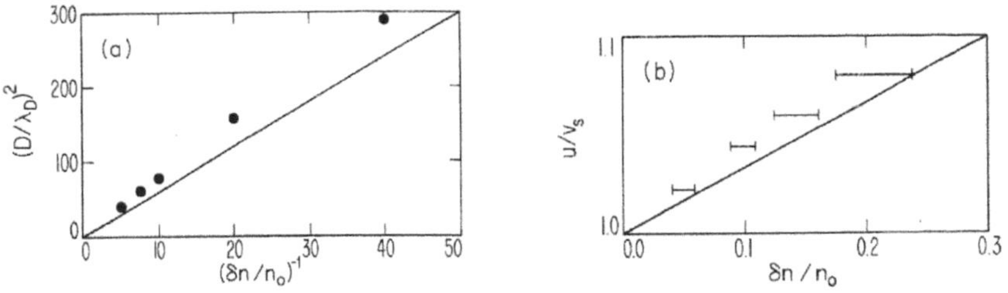

FIGURE 14.10: *Experimental results for the variation of the squared soliton width in a) and the soliton velocity in b), shown for different amplitudes, measured in normalized units* $(\delta n/n_0)^{-1}$ *and* $\delta n/n_0$*, respectively. The figures are reproduced with permission from Ikezi et al. (1970), copyright 1970 by the American Physical Society.*

approximation. Both results agree well with predictions based on the KdV-equation. Results with different plasma parameters were also shown by Ikezi (1973).

The results summarized in Fig. 14.11 show collisions between solitons. In Fig. 14.11a we see an example of a collision where a large amplitude soliton is overtaking one with a smaller amplitude. Although the waves are damped due to features not included in the Korteweg-deVries equation (notably ion Landau damping, to be discussed in Section 21) the main features are sufficiently convincing: a large amplitude fast soliton overtakes its small amplitude slow counterpart and, apart from the damping, the shape of the pulses are retained before and after the collision. See also Fig. 4.9 giving a presentation of analytical results for two colliding solitons (Drazin & Johnson 1989).

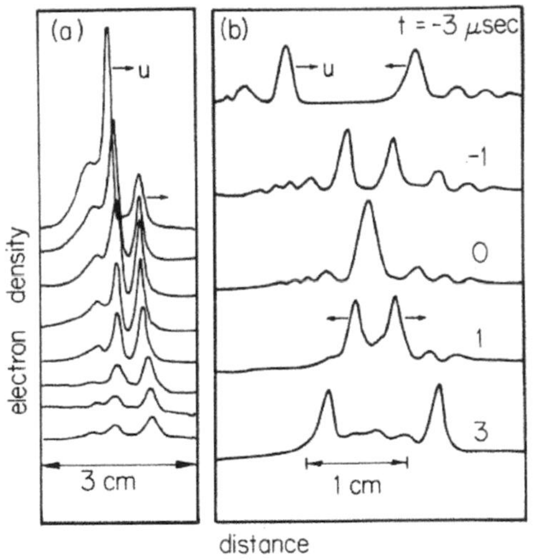

FIGURE 14.11: *Experimental results for colliding solitons. In a) we have an overtaking collision, while b) shows head-on collisions. This latter problem is not properly described the Korteweg-deVries equation. The figures are reproduced with permission from Ikezi et al. (1970), copyright 1970 by the American Physical Society.*

Also the break-up of an initial sound pulse into a sequence of several solitons with decreasing amplitudes have been observed experimentally (Ikezi 1973). The results were in qualitative agreement with the illustrative numerical solutions as shown in, e.g., Fig. 4.8.

Although some differences in detail can be noted, it can be argued that the agreement between the weakly nonlinear evolution of ion sound waves and predictions based on the KdV equation is robust.

By its derivation, the KdV equation can account only for waves or solitons propagating in the same direction, i.e., undergo overtaking collisions. Experimentally, it is feasible to arrange for head-on collisions as well, and such a case is shown in Fig. 14.11b). This nonlinear process can in principle be described by a nonlinear Boussinesq equation. See the relevant discussion in Section 4.2.3. The soliton damping seen in Fig. 14.11a) is not accounted for by the KdV-equation. This particular feature is discussed later in Section 23.2.

The results summarized here refer to one spatial direction. Also cylindrical and spherical solitons can be excited. For a one soliton case, these examples will, however, depend on time and one radial coordinate so the higher dimensionality of the problem is still accounted for by a temporal and one spatial variable. The full dimensionality of the problem is manifested when two such solitons are colliding. There is vast amount of literature on also this subject, with review papers available (Lonngren 1983).

The present section emphasizes weakly dispersive ion acoustic waves. It is found that a variety of wave phenomena can be accounted for by the Korteweg-deVries equation, not only plasma waves. The nature of the waves enters through the parameters accounting for the velocity, the dispersion and the coefficient for the nonlinear term. Detailed features have to be accounted for by additional perturbation terms in the KdV equation.

15

Small Amplitude Waves in Anisotropic Warm Plasmas

The most relevant form of anisotropy and inhomogeneity is encountered for plasmas in magnetic or gravitational fields, or combinations of the two. Even in a fluid model we *can* have anisotropic temperatures, for instance, but this case will not be discussed here. The plasma can be anisotropic but homogeneous, i.e., its properties depend on the selected *direction* in space, but not on absolute position. More generally, the plasma can be anisotropic as well as inhomogeneous. The example considered in Section 15.5 illustrates such a case.

First we consider the case where the plasma density is uniform, and the anisotropy is solely imposed by a homogeneous magnetic field $\mathbf{B}_0$. We found in Chapter 12 that electromagnetic waves had a large phase velocity, except in the cases where resonance frequencies were present. In the limit of large wave-numbers, the waves became electrostatic for such cases. We do not expect finite temperature effects to have any particular significance for the fast electromagnetic modes (maybe with an exception when relativistic particles are present). Our analysis of finite temperature effects will therefore be concentrated on the electrostatic modes, which are relatively slow. The notable exception is the Langmuir wave, which can in principle have arbitrarily large phase velocities.

15.1 High frequency electrostatic electron waves

For high frequency phenomena, the basic fluid equations are obtained with the assumption of immobile ions with uniform density n_0. We have "as usual" the electron continuity equation

$$\frac{\partial n(\mathbf{r},t)}{\partial t} + \nabla \cdot \Big(n(\mathbf{r},t)\mathbf{u}(\mathbf{r},t) \Big) = 0 \tag{15.1}$$

and the momentum equation

$$n(\mathbf{r},t)m\left(\frac{\partial \mathbf{u}(\mathbf{r},t)}{\partial t} + \mathbf{u}(\mathbf{r},t)\cdot\nabla\mathbf{u}(\mathbf{r},t) \right) = -\nabla p_e(\mathbf{r},t) - en(\mathbf{r},t)\left(\mathbf{E}(\mathbf{r},t) + \mathbf{u}(\mathbf{r},t)\times\mathbf{B}_0\right), \tag{15.2}$$

where m is the electron mass and p_e is the electron pressure. By the assumption of electrostatic waves, we have $\mathbf{E} = -\nabla\phi$, where the potential is determined by Poisson's equation, here with immobile ions

$$\nabla\cdot\mathbf{E}(\mathbf{r},t) = -\nabla^2\phi(\mathbf{r},t) = \frac{e}{\varepsilon_0}(n_0 - n(\mathbf{r},t)) \tag{15.3}$$

By the basic assumption here, we have the magnetic field being constant in time, and we take it to be independent of the spatial coordinate as well. Initially, the plasma is assumed to be charge neutral with vanishing electric fields.

Linearizing the equations for small amplitude perturbations, assuming an adiabatic electron equation of state $p = p_0 + \widetilde{p} = p_0(1 + \widetilde{n}/n_0)^\gamma$ (see Section 2.4), with $p_0 = \kappa T_e n_0$ and $\gamma = C_p/C_V$, we find the electron continuity equation

$$\frac{\partial}{\partial t}\widetilde{n} + n_0\nabla\cdot\widetilde{\mathbf{u}} = 0, \tag{15.4}$$

and the momentum equation

$$n_0 m \frac{\partial}{\partial t}\widetilde{\mathbf{u}} = -\gamma\kappa T_e \nabla\widetilde{n} + e n_0 \left(\nabla\widetilde{\phi} - \widetilde{\mathbf{u}} \times \mathbf{B}_0\right) \tag{15.5}$$

and Poisson's equation in terms of linearized quantities

$$\nabla^2\widetilde{\phi} = \frac{e}{\varepsilon_0}\widetilde{n}. \tag{15.6}$$

From (15.5) we find by taking the cross product with $\mathbf{B}_0$, followed by some simple manipulations

$$\frac{\partial}{\partial t}\nabla_\perp\left(e n_0\widetilde{\phi} - \frac{\gamma\kappa T_e\varepsilon_0}{e}\nabla^2\widetilde{\phi}\right) - n_0 m\frac{\partial^2}{\partial t^2}\widetilde{\mathbf{u}}_\perp =$$
$$\nabla_\perp\left(\frac{e^2 n_0}{m}\widetilde{\phi} - \frac{\gamma\kappa T_e\varepsilon_0}{m}\nabla^2\widetilde{\phi}\right) \times \mathbf{B}_0 + \frac{e^2 n_0 B_0^2}{m}\widetilde{\mathbf{u}}_\perp,$$

while the $\mathbf{B}_0$-parallel part of the equation gives

$$n_0 m\frac{\partial}{\partial t}\widetilde{u}_\parallel = \frac{\partial}{\partial z}\left(e n_0\widetilde{\phi} - \frac{\gamma\kappa T_e\varepsilon_0}{e}\nabla^2\widetilde{\phi}\right),$$

where we used Poisson's equation. Similarly, we find from the continuity equation

$$\frac{\varepsilon_0}{e n_0}\frac{\partial^2}{\partial t^2}\nabla^2\widetilde{\phi} + \frac{\partial^2}{\partial z\partial t}\widetilde{u}_\parallel + \frac{\partial}{\partial t}\nabla_\perp\cdot\widetilde{\mathbf{u}}_\perp = 0.$$

Eliminating the velocity components, we finally obtain a partial differential equation for the potential in the form

$$\frac{\partial^2}{\partial t^2}\nabla_\perp^2\left(e\widetilde{\phi} - \frac{\gamma\kappa T_e\varepsilon_0}{n_0 e}\nabla^2\widetilde{\phi}\right) =$$
$$-\left(\frac{e^2 B_0^2}{m} + m\frac{\partial^2}{\partial t^2}\right)\left[\frac{\varepsilon_0}{e n_0}\frac{\partial^2}{\partial t^2}\nabla^2\widetilde{\phi} + \frac{\partial^2}{\partial z^2}\left(\frac{e}{m}\widetilde{\phi} - \frac{\gamma\kappa T_e\varepsilon_0}{e n_0 m}\nabla^2\widetilde{\phi}\right)\right],$$

since $\nabla\cdot(\nabla\widetilde{\phi}\times\mathbf{B}_0) = 0$. The equation can be made neater by introducing characteristic frequencies ω_{ce}, ω_{pe} together with the Debye length, λ_D, to give

$$\frac{\partial^2}{\partial t^2}\nabla_\perp^2\left(\widetilde{\phi} - \lambda_D^2\nabla^2\widetilde{\phi}\right) =$$
$$-\left(\omega_{ce}^2 + \frac{\partial^2}{\partial t^2}\right)\left[\frac{1}{\omega_{pe}^2}\frac{\partial^2}{\partial t^2}\nabla^2\widetilde{\phi} + \frac{\partial^2}{\partial z^2}\left(\widetilde{\phi} - \lambda_D^2\nabla^2\widetilde{\phi}\right)\right].$$

By Fourier transforming the differential equation with respect to spatial and temporal variables, we find a dispersion relation in the form

$$\omega^2(\omega_{ce}^2 - \omega^2)k^2 - \omega_{pe}^2\omega_{ce}^2 k_\parallel^2(1 + (k\lambda_D)^2) + \omega^2\omega_{pe}^2 k^2(1 + (k\lambda_D)^2) = 0. \tag{15.7}$$

Two special limits are readily simplified. For $k_\parallel = 0$ we find $\omega^2 = \omega_{ce}^2 + \omega_{pe}^2(1 + (k\lambda_D)^2)$, where we recognize the upper-hybrid frequency, $\omega_{uh} \equiv \sqrt{\omega_{ce}^2 + \omega_{pe}^2}$. For $k_\perp = 0$, we find two solutions, $\omega^2 = \omega_{pe}^2(1 + (k\lambda_D)^2)$ and $\omega^2 = \omega_{ce}^2$. The former mode is the well-known Langmuir wave. The latter solution is the electrostatic resonance frequency associated with the whistler mode, which for finite

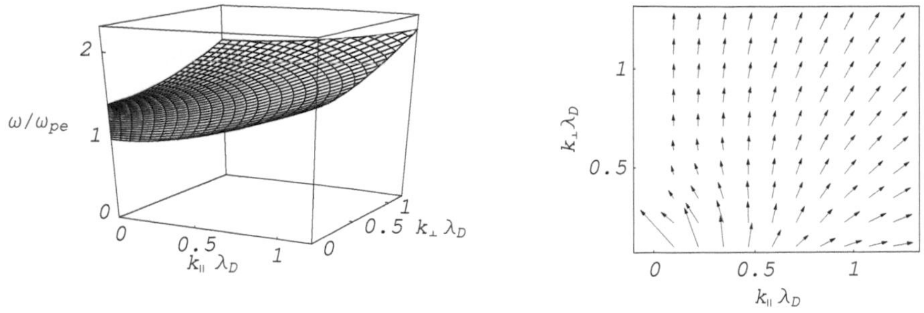

FIGURE 15.1: *High frequency branch of the electrostatic dispersion relation for electron waves in magnetized plasmas for $\omega_{pe} = \omega_{ce}$. Frequencies are normalized by ω_{pe} and wave-numbers with λ_D. The figure to the right shows the directions and relative magnitudes of the group velocities by small arrows.*

wave-numbers is electromagnetic. With the present basic assumption of electrostatic waves, we will *not* recover the full whistler mode, only its electrostatic resonance frequency at $k_\parallel \to \infty$.

We distinguish two topologically different cases, one where $\omega_{pe}^2 > \omega_{ce}^2$, and another where $\omega_{pe}^2 < \omega_{ce}^2$. We had the same situation in Sections 12.2.2 and 12.2.3. Note also that if we take $T_e = 0$ in the present electrostatic description, i.e., $\lambda_D \to 0$, the dispersion relation depends on the *direction* of propagation through $k_\parallel^2/k^2$, but not on the actual magnitude of k. We have a resonance cone (Piel 2010).

Illustrative examples are shown in Figs. 15.1 and 15.2 for the case $\omega_{pe} = \omega_{ce}$, i.e., the "cross-over" case. We note the thermal dispersion of the high frequency branch in all directions with respect to the magnetic field, except at the electron cyclotron resonance. In addition, we have a sharp "fringe" in the dispersion relation at small wave-numbers, which would be even more noticeable in the cold plasma limit.

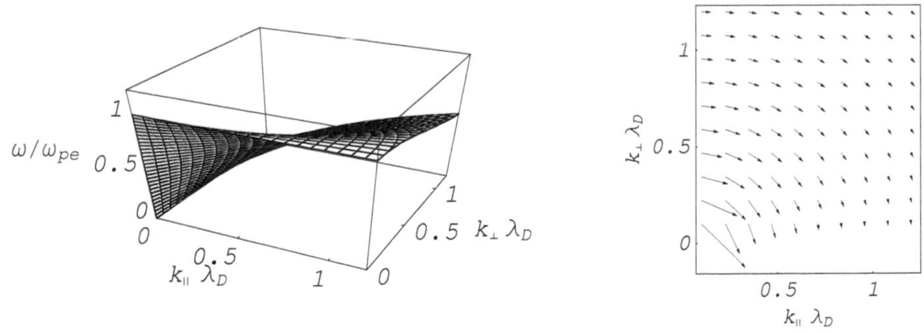

FIGURE 15.2: *Low frequency branch of the electrostatic dispersion relation for electron waves in magnetized plasmas for $\omega_{pe} = \omega_{ce}$. Frequencies are normalized by ω_{pe} and wave-numbers with λ_D. The figure to the right shows the directions and relative magnitudes of the group velocities by small arrows.*

To illustrate the complexity of the dispersion relations we show also the group velocities $\mathbf{u}_g(k_\perp, k_\parallel)$ in the appropriate figures. Note that for both Figs. 15.1 and 15.2 we avoid $k_\perp = 0$ as well as $k_\parallel = 0$ when presenting the group velocities. Note the large values of u_g close to the origin $(k_\perp, k_\parallel) \approx (0,0)$. It is interesting to note the "competition" between the cold plasma resonance cone and thermal effects in the group velocity for the high frequency branch: for small $|k\lambda_D|$ the dispersion is strongly anisotropic, while it becomes close to isotropic with $\mathbf{u}_g \parallel \mathbf{k}$ when $|k\lambda_D| > 1$. It should be emphasized that the electrostatic approximation breaks down for small k, as discussed in Section 12.2.4.

- **Exercise:** Discuss the dispersion relation (15.7) in the limit $k_\perp = 0$ for $\omega_{pe} < \omega_{ce}$ as well as $\omega_{pe} > \omega_{ce}$.

15.1.1 Electrostatic waves in a strongly magnetized wave guide

In Section 12.2.6 we discussed electron plasma waves propagating in a cold plasma, which was "embedded" in a strongly magnetized cylindrical wave guide. Many studies of high frequency electrostatic plasma waves are carried out in this configuration, but under conditions where corrections of the dispersion relation due to finite electron temperatures are important. In the limit of infinite magnetic fields, these corrections are easily introduced into (12.38), giving the result

$$\omega^2 = \omega_{pe}^2 \frac{(kR)^2}{p_{m\nu}^2 + (kR)^2} + k^2 u_{te}^2 . \tag{15.8}$$

A classical study of this wave-mode with emphasis on finite electron temperature effects was carried out by Malmberg and Wharton (1964).

15.2 Low frequency electrostatic ion waves

Analyzing low frequency waves, we include the dynamics of the ion plasma component. The basic fluid equations for this case are the ion continuity equation

$$\frac{\partial n(\mathbf{r},t)}{\partial t} + \nabla \cdot (n(\mathbf{r},t)\mathbf{u}(\mathbf{r},t)) = 0, \tag{15.9}$$

and the momentum equation

$$n(\mathbf{r},t)M\left(\frac{\partial \mathbf{u}(\mathbf{r},t)}{\partial t} + \mathbf{u}(\mathbf{r},t)\cdot\nabla\mathbf{u}(\mathbf{r},t)\right) = -\nabla p_i(\mathbf{r},t) + qn(\mathbf{r},t)\left(\mathbf{E}(\mathbf{r},t) + \mathbf{u}(\mathbf{r},t)\times\mathbf{B}_0\right), \tag{15.10}$$

where M is the ion mass and p_i is the ion pressure, taking the ion charge to be q, allowing for multi-charged ions $q = Ne$. As far as the electrons are concerned, we *could* use (15.1) and (15.2), but often the fluctuation frequencies are so low that the assumption of a local Boltzmann equilibrium is justified at all times. Physically, we assume that the electrons can flow along magnetic field lines, from wave crest to wave trough, to maintain this equilibrium. Of course, in case the wave-vector is *exactly* perpendicular to $\mathbf{B}_0$, this is not possible, so we have to exclude a small region of the dispersion diagram around $k_\parallel \approx 0$ from the analysis, but this is a minor restriction which will not be elaborated here. In the following we will explicitly assume that we are dealing with electrostatic waves, and let the magnetic field be constant.

We linearize the basic equations, with the assumption of a simple equilibrium state, with constant plasma density n_0 and homogeneous magnetic fields $\mathbf{B}_0$. The equations then reduce to

$$\frac{\partial}{\partial t}\widetilde{n} + n_0 \nabla \cdot \widetilde{\mathbf{u}} = 0, \tag{15.11}$$

and the momentum equation

$$n_0 M \frac{\partial}{\partial t}\widetilde{\mathbf{u}} = -\gamma T_i \nabla \widetilde{n} - e n_0 \left(\nabla \widetilde{\phi} - \widetilde{\mathbf{u}} \times \mathbf{B}_0\right), \tag{15.12}$$

for singly charged ions. We used again the ideal gas law for the ion equation of state, $p_i = \kappa T_i n$, and used a wiggle $\widetilde{}$ to indicate perturbations of the equilibrium. We also make the assumption of quasi-neutrality, $\widetilde{n}_e \approx \widetilde{n}_i = \widetilde{n}$. Ignoring the inertia term in the electron momentum equation, we obtain $\widetilde{n}/n_0 = e\phi/\kappa T_e$; see also (13.21). Introduce the sound speed as $C_s^2 \equiv \kappa(T_e + \gamma T_i)/M$ and the normalized density as $\eta \equiv \widetilde{n}/n_0$. After a Fourier transform with respect to the spatial as well as the temporal variable we find the ion continuity and the ion momentum equations in the form

$$\omega \eta_1 = \mathbf{k} \cdot \mathbf{u}_1 \qquad \text{and} \qquad \omega \mathbf{u}_1 = \mathbf{k} C_s^2 \eta_1 + i\frac{e}{M}\mathbf{u}_1 \times \mathbf{B}_0,$$

where we eliminated the potential. The subscript $_1$ indicates also here the coefficient to $\exp(-i(\omega t - \mathbf{k}\cdot\mathbf{r}))$. We then have $\omega u_{1\parallel} = k_\parallel C_s^2 \eta_1$ and

$$\mathbf{u}_{1\perp} = C_s^2 \eta_1 \frac{\omega \mathbf{k}_\perp + i e \mathbf{k}_\perp \times \mathbf{B}_0/M}{\omega^2 - \Omega_{ci}^2},$$

which by insertion into the ion continuity equation (where $\mathbf{k}_\perp \cdot \mathbf{k}_\perp \times \mathbf{B}_0 = 0$ makes one term disappear) finally gives the dispersion relation in the form

$$\omega^2(\omega^2 - \Omega_{ci}^2) - C_s^2\left(k_\perp^2 \omega^2 + k_\parallel^2(\omega^2 - \Omega_{ci}^2)\right) = 0. \tag{15.13}$$

In the limit $k_\parallel = 0$ we find the result

$$\omega^2 = \Omega_{ci}^2 + C_s^2 k_\perp^2, \tag{15.14}$$

which is usually called the dispersion relation for electrostatic ion cyclotron waves. The limit $k_\parallel \to 0$ has to be interpreted in a restrictive sense, as already mentioned, since the phase velocity component $\omega/k_\parallel$ parallel to $\mathbf{B}_0$ has to be smaller than the electron thermal velocity so the electrons can maintain Boltzmann equilibrium. The limit $k_\parallel \to 0$ requires a separate analysis, see also Section 15.3.1.

More generally, the dispersion relation has two branches, one for $\omega^2 < \Omega_{ci}^2$ and one for $\omega^2 > \Omega_{ci}^2$. The full dispersion relations are illustrated in Figs. 15.3 and 15.4. Ion cyclotron waves are found in the limit of small $k_\parallel$. An interesting feature of these dispersion relations is that the group velocity in general has a direction different from the wave-vector direction. The point $(\omega, k_\parallel) = (\Omega_{ci}, \Omega_{ci}/C_s)$ is singular, by the merging of two dispersion branches. The figures also show the direction and relative magnitude of the group velocity in the $(k_\perp, k_\parallel)$-plane. We note that by the quasi-neutrality assumption, the ion plasma frequency does not enter the expressions. It therefore does not have any meaning to ask for the ratio Ω_{ci}/Ω_{pi} in Figs. 15.3 and 15.4. It is important to emphasize that the assumption of quasi-neutrality in general restricts the applicability of results like (15.13) to relatively low frequencies as compared to the ion plasma frequency, i.e., $\omega \ll \Omega_{pi}$.

To obtain a more generally applicable, but also less transparent result, we have to include Poisson's equation explicitly, which gives the relation $n_1/n_0 = (e\phi/\kappa T_e)(1 + (k\lambda_{De})^2)$ for the perturbations in ion density, expressed in terms of the electron Debye length λ_{De}. By inspection we find that this is the only relation where the electron temperature remains explicitly when we have eliminated the expression for Boltzmann distributed electrons. All we have to do is take $T_e \to T_e/(1 + (k\lambda_{De})^2)$ in order to include the effects of Poisson's equation. Thus, by the replacement

$$C_s^2 \to \kappa\left(\gamma T_i + \frac{T_e}{1 + (k\lambda_{De})^2}\right)\frac{1}{M}$$

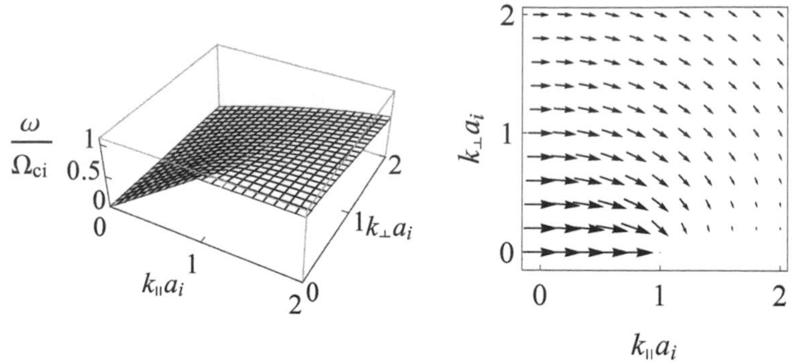

FIGURE 15.3: *Low frequency branch,* $\omega^2 < \Omega_{ci}^2$*, of the electrostatic dispersion relation in magnetized plasmas. Frequencies are normalized by* Ω_{ci} *and wave-numbers with* $a_i \equiv \Omega_{ci}/C_s$*. The right-hand part of the figure shows the direction and relative magnitude of the group velocity,* $\{\partial\omega/\partial k_\perp; \partial\omega/\partial k_\parallel\}$*, in the* $(k_\perp, k_\parallel)$*-plane.*

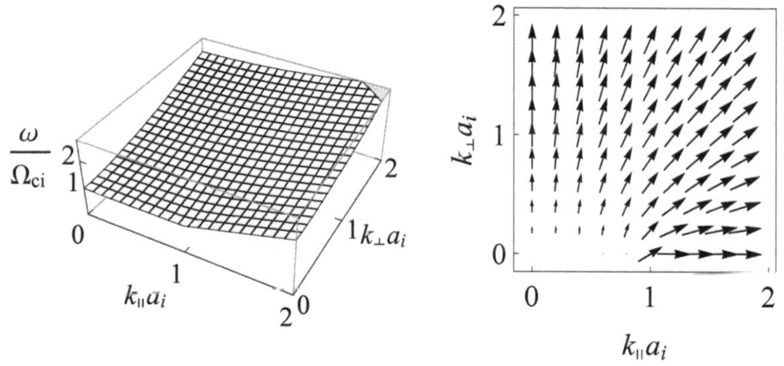

FIGURE 15.4: *High frequency branch,* $\omega^2 > \Omega_{ci}^2$*, of the electrostatic dispersion relation in magnetized plasmas. Frequencies are normalized by* Ω_{ci} *and wave-numbers with* $a_i \equiv \Omega_{ci}/C_s$*. The right-hand part of the figure shows the direction and relative magnitude of the group velocity,* $\{\partial\omega/\partial k_\perp; \partial\omega/\partial k_\parallel\}$*, in the* $(k_\perp, k_\parallel)$*-plane.*

in (15.13), we can relax the assumption of quasi-neutrality and find

$$\omega^2(\omega^2 - \Omega_{ci}^2) - \left(\gamma\frac{\kappa T_i}{M} + \frac{\kappa T_e}{M}\frac{1}{1+(k\lambda_{De})^2}\right)\left(k_\perp^2\omega^2 + k_\parallel^2(\omega^2 - \Omega_{ci}^2)\right) = 0. \tag{15.15}$$

This dispersion relation is understood best in the limit where $T_i \to 0$. In that case we have a stop-band for frequencies between Ω_{pi} and Ω_{ci}. This case is illustrated in Fig. 15.5, here shown for the example where $\Omega_{pi}/\Omega_{ci} = 0.5$.

Ⓢ **Exercise:** Write the full set of fluid equations describing the dynamics of a plasma on a plane $\perp$ **B**, when the Finite Larmor Radius (FLR)-corrections are taken into account for the ion dynamics (Knorr et al. 1988, Hansen et al. 1989). See also Section 7.1.4. Ⓢ

Ⓢ **Exercise:** Let a plasma consist of electrons and two types of singly charged ions with different ion masses, M_1 and M_2, and charges q_1 and q_2, with given relative densities n_1 and n_2, confined by a homogeneous magnetic field. Obtain a general expression for the dispersion relation, and consider the two limiting cases $k_\parallel \to 0$ and $k_\perp \to 0$ in detail. Assume that the fluctuations are

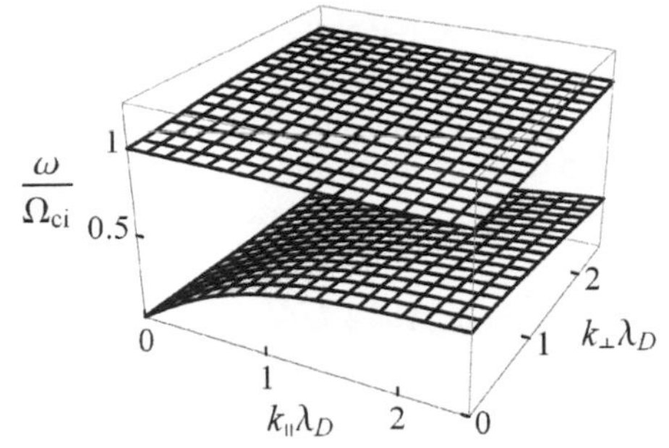

FIGURE 15.5: *The two branches for the dispersion relation (15.15) obtained for $T_i = 0$, without the quasi-neutral approximation, with a strong magnetic field, here with $\Omega_{ci}/\Omega_{pi} = 2$. Note the normalizations on the axes used here.*

quasi-neutral and the electrons are isothermally Boltzmann distributes at all times. You may let the ion temperatures vanish without too much loss of generality. Give attention to the case where $q_1 = q_2$, where we have isotopes of the same ions: do you see any possibility for isotope separation? Ⓢ

- **Exercise:** Derive the dispersion relation for a magnetized plasma for the case where you have a component of *negative* ions, i.e., a density n_1 of positive ions with charge e, mass M_1, and a component of negative ions with charge $-e$ of density $n_2 < n_1$ and mass M_2, so that the electron density is $n_e = n_1 - n_2$ for charge neutrality. Consider both the case of a weak and a strong magnetic field.

15.3 Lower-hybrid waves

For magnetized plasmas there is an intermediate frequency region, where the frequencies are so large that we can no longer assume electrons to be Boltzmann distributed, but the frequencies are still so low that the ion dynamics are of importance. More specifically, the phase velocity component along the magnetic field can be comparable to or even larger than the electron thermal velocity, $\omega/k_\parallel \geq u_{The}$. For these waves, we expect ion as well as electron inertia to be important, so we have to retain the full momentum equations for both components. Again it is advantageous to consider first the analysis using the quasi-neutral approximation, and then to see the limitations imposed by this simplification. In the cold plasma limit, these waves were analyzed in Section 12.4.1.

15.3.1 The quasi-neutral limit

The linear dispersion relation for the low frequency electrostatic waves is found here in the form

$$\left[\Omega^2 - \Omega_{ci}^2 \frac{k_\parallel^2}{k^2}\right]\left[\left(\Omega^2 - \omega_{ce}^2\right)\left(\Omega^2 - k_\parallel^2 u_{The}^2\right) - k_\perp^2 u_{The}^2 \Omega^2\right]$$
$$+\frac{u_{The}^2}{C_s^2}\left[\Omega^2 - \omega_{ce}^2 \frac{k_\parallel^2}{k^2}\right]\left[\left(\Omega^2 - \Omega_{ci}^2\right)\left(\Omega^2 - k_\parallel^2 \gamma u_{Thi}^2\right) - k_\perp^2 \gamma u_{Thi}^2 \Omega^2\right] = 0, \qquad (15.16)$$

where $C_s^2 \equiv T_e/M$. We have the adiabatic exponent $\gamma \equiv C_P/C_V \approx 5/3$, taking as usual the value for ideal gases. The relation (15.16) is general and includes also, for instance, ion cyclotron waves with $\omega/k_\| \geq u_{The}$, i.e., the limit where the Boltzmann distribution for electrons is no longer applicable, see the previous section. See Fig. 15.6 for an illustration.

In the limit $k_\| = 0$ in (15.16) we find the lower hybrid waves dispersion as

$$\omega^2 = \omega_{ce}\Omega_{ci} + k^2\frac{T_e + \gamma T_i}{M}, \tag{15.17}$$

where some small terms $O(m/M)$ have been ignored. Since the characteristic frequency $\sqrt{\omega_{ce}\Omega_{ci}} \gg \Omega_{ci}$, we might as well ignore the magnetization of the ions from the outset.

For $k_\perp = 0$ we find the cyclotron resonances $\omega^2 = \omega_{ce}^2$ and $\omega^2 = \Omega_{ci}^2$, together with the sound wave dispersion relation in the form $\omega^2 = k^2(mu_{The}^2 + M\gamma u_{Thi}^2)/M$.

For small angles θ, i.e., nearly normal to the magnetic field, see also (12.50), we have the approximate relation

$$\omega^2 = \Omega_{lh}^2(1 + \frac{M}{m}\cos^2\theta) + k^2\frac{T_e + \gamma T_i}{M}, \tag{15.18}$$

where $k^2 = k_\perp^2 + k_\|^2$ in the last term. Details of this expression have been debated (Verdon et al. 2009).

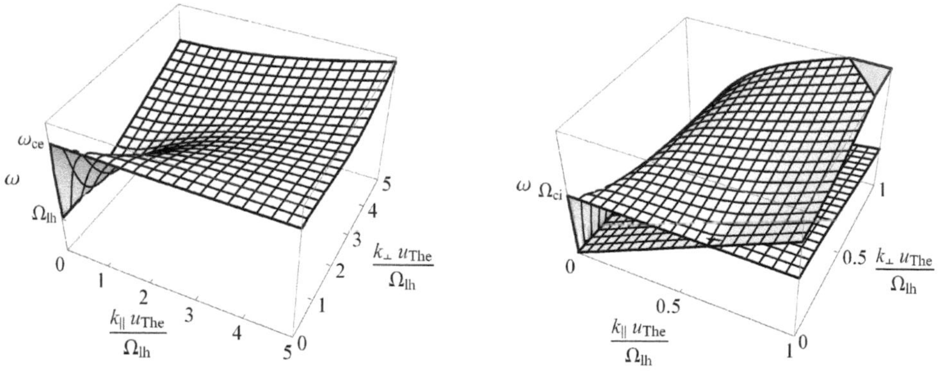

FIGURE 15.6: *High and low frequency parts of the dispersion relation obtained by (15.16), here shown with an artificially small mass ration $M/m = 10$ in order to obtain manageable figures. We have $T_e = T_i$.*

The result for the cold plasma limit can be derived from (15.16) for $k_\| u_{The} \ll \omega \ll \omega_{ce}$. It is important that for waves propagating strictly normal to **B** (and not only *almost* normal), the lower hybrid wave is the only solution found for ion waves: the electrostatic ion cyclotron waves found in (15.14) require a small but nonvanishing $k_\|$ for propagation. The only other electrostatic wave that can propagate strictly normal to **B** is the upper-hybrid wave, see (15.7), which does not involve ion motion.

Lower hybrid waves are found important for instance at plasma heating in nuclear fusion related experiments (Porkolab 1977) and ionospheric heating experiments (Kuo 2015). This wave-type is often found in nature as well (Pécseli et al. 1996, Schuck et al. 2003) and studies of excitation methods are interesting and relevant, but this problem is not considered in further detail here. Also lower hybrid waves have an electromagnetic component at long wavelengths (Hsu & Kuehl 1982, Verdon et al. 2009). This component is not included in the present summary either.

15.3.2 Deviations from quasi neutrality

For the limit where the quasi neutral assumption breaks down, we have to generalize the previous result (Verdon et al. 2009). As an order of magnitude estimate this will happen when $\Omega_{lh} \geq \Omega_{pi}$. After some rather lengthy algebra, one finds a partial differential equation for the relative density perturbation in lower-hybrid waves for warm plasmas as

$$\left[\frac{\partial^4}{\partial t^4}\nabla^2 + \Omega_{lh}^2\left(\nabla^2 + \frac{M}{m}\frac{\partial^2}{\partial z^2}\right)\frac{\partial^2}{\partial t^2}\right]\eta$$
$$-\frac{\omega_{pe}^2}{\omega_{pe}^2+\omega_{ce}^2}C_s^2\left(\omega_{ce}^2\frac{\partial^2}{\partial z^2}+\frac{\partial^2}{\partial t^2}\nabla^2\right)\nabla^2\eta$$
$$-\frac{1}{\omega_{pe}^2+\omega_{ce}^2}\left[u_{Thi}^2\omega_{ce}^2\nabla^2 + u_{The}^2\left(\omega_{ce}^2\frac{\partial^2}{\partial z^2}+\frac{\partial^2}{\partial t^2}\nabla^2\right)\right]\frac{\partial^2}{\partial t^2}\nabla^2\eta = 0, \quad (15.19)$$

where we introduced the abbreviations $C_s^2 \equiv u_{Thi}^2 + u_{The}^2 m/M$, and $\eta \equiv \widetilde{n}/n_0$, having $\Omega_{lh}^2 \equiv \Omega_{pi}^2/(1+\omega_{pe}^2/\omega_{ce}^2)$ consistent with (12.49), and ignored terms containing products of the ion and electron thermal velocities like $u_{Thi}^2 u_{The}^2$. Note how greatly simplified the differential equation becomes when thermal effects are ignored! In that case only the first term in (15.19) remains. There is no reason to show the full lower-hybrid wave dispersion relation here: this is left to the reader as an exercise.

- **Exercise:** Find the quasi neutral limit of (15.19) and compare with (15.18).
- **Exercise:** Use (15.19) to derive a full dispersion relation for the lower hybrid waves and illustrate the result by a figure.

15.4 Alfvén waves in warm plasmas

By a simple MHD model, we found the possibility of propagating low frequency electromagnetic modes in a plasma. The model was restrictive and applied to low frequency and long wavelengths only. We can extend the analysis somewhat, and include wavelengths comparable to an effective Larmor radius, a_i, which is derived from the electron temperature and the ion mass. The basic difference from the simple ideal MHD model is that we allow for the electric field to have a component *along* the magnetic field. Assuming quasi-neutrality, the linearized model equations are

$$Mn_0\frac{\partial}{\partial t}\widetilde{\mathbf{u}}_\perp = \left(\widetilde{\mathbf{J}}\times\mathbf{B}_0\right)_\perp - \nabla_\perp\widetilde{p}, \quad (15.20)$$

$$mn_0\frac{\partial}{\partial t}\left(\frac{\widetilde{J}_z}{en_0}\right) = en_0\widetilde{E}_z + \kappa T_e\frac{\partial}{\partial z}\widetilde{n}, \quad (15.21)$$

$$\frac{\partial}{\partial t}\widetilde{n} + n_0\nabla_\perp\cdot\widetilde{\mathbf{u}}_\perp = \frac{\partial}{\partial z}\left(\frac{\widetilde{J}_z}{e}\right), \quad (15.22)$$

$$\widetilde{\mathbf{E}}_\perp + \widetilde{\mathbf{u}}_\perp\times\mathbf{B}_0 = 0, \quad (15.23)$$

where we take the unperturbed magnetic field in the z-direction. We used that in the linear approximation the $\mathbf{B}_0$-parallel part of the current is basically due to the electron motion, with $\widetilde{J}_z \approx -en_0\widetilde{u}_z$. By the relation (15.23) we retain the ideal $\mathbf{B}_0$-perpendicular part of the MHD equation. In deriving the relation (15.21) we used (11.12), which in the present ideal linearized model without resistivity in effect corresponds to the electron momentum equation.

By the operation $\nabla_\perp \times$ on both sides of the equation, we find from (15.20)

$$Mn_0 \widehat{\mathbf{z}} \cdot \frac{\partial}{\partial t} \nabla \times \widetilde{\mathbf{u}}_\perp = B_0 \frac{\partial}{\partial z} \widetilde{J}_z ,$$

where we used that $\nabla_\perp \times \left(\widetilde{\mathbf{J}} \times \mathbf{B}_0\right)_\perp = \nabla_\perp \times (\widetilde{J}_y \widehat{\mathbf{x}} - \widetilde{J}_x \widehat{\mathbf{y}}) B_0 = -B_0 (\partial \widetilde{J}_x / \partial x + \partial \widetilde{J}_y / \partial y) = B_0 \partial \widetilde{J}_z / \partial z$, recalling that the quasi-neutrality assumption implies $\nabla \cdot \widetilde{\mathbf{J}} \approx 0$. From (15.22) we find

$$e\mu_0 n_0 \frac{\partial}{\partial t} \left(\frac{\widetilde{n}}{n_0} \right) - \frac{\partial}{\partial z} \left(\mu_0 \widetilde{J}_z \right) = 0 .$$

By the relations (15.21)–(15.23), we also find

$$B_0 \frac{\partial}{\partial z} (\nabla \times \widetilde{\mathbf{u}}_\perp) \cdot \widehat{\mathbf{z}} + \left(\frac{c^2}{\omega_{pe}^2} \nabla_\perp^2 - 1 \right) \frac{\partial}{\partial t} \left(\mu_0 \widetilde{J}_z \right) - \frac{\kappa T_e}{e} \nabla_\perp^2 \frac{\partial}{\partial z} \left(\frac{\widetilde{n}}{n_0} \right) = 0 .$$

By Fourier transforming the equation with respect to the spatial as well as the temporal variables, it is now a simple matter to obtain the dispersion relation in the form

$$\omega^2 = \frac{k_z^2 V_A^2 (1 + (k_\perp a_i)^2)}{1 + (ck_\perp / \omega_{pe})^2} , \tag{15.24}$$

where $a_i \equiv \sqrt{\kappa T_e / (M \Omega_{ci})}$ was introduced as an effective Larmor radius obtained by the ion mass but the *electron* temperature. In (15.24) we recognize also the skin depth, c/ω_{pe}, which represents the correction due to electron inertia. Often we find the term "inertial length" used for c/ω_{pe}, although this term is sometimes used for the quantity where the speed of light is replaced by the electron thermal velocity. We note that the MHD limit discussed in Chapter 10 assumes, in effect, $c \to \infty$, but the assumption of quasi-neutrality assumes $\omega_{pe} \to \infty$, so that ε_0 cancels in the denominator of (15.24). For the case where the Alfvén velocity exceeds the electron thermal velocity, $V_A \gg u_{The}$, the electrons could be taken to be cold from the outset. For this *inertial* limit, the electron skin depth is the relevant length scale. These waves have been observed experimentally (Kletzing et al. 2010).

Some limiting cases can be recognized in (15.24). First we take $\beta \ll m/M$ and find

$$\omega^2 = \frac{k_z^2 V_A^2}{1 + (ck_\perp / \omega_{pe})^2} \approx k_z^2 V_A^2 (1 - (ck_\perp / \omega_{pe})^2) , \tag{15.25}$$

where the approximation applies for $(ck_\perp / \omega_{pe})^2 \ll 1$. On the other hand, the opposite limit $\beta \gg m/M$ gives

$$\omega^2 = k_z^2 V_A^2 (1 + (k_\perp a_i)^2) . \tag{15.26}$$

The important difference between (15.25) and (15.26) is the change in sign of the curvature with respect to the $k_\perp$-variation. The summary in the present section is simplified, and does not account for the ion skin depth. A related more general, but also much more complicated, dispersion relation is given by Hirose et al. (2004).

An unfortunate notation has developed in denoting (15.24), the dispersion relation for *kinetic Alfvén waves*. As we have seen, the analysis is based on a fluid model (and quite simplified as such), and has nothing to do with the proper plasma kinetic theory.

- **Exercise:** Demonstrate how the plasma-β can be introduced in the general expression (15.24), to give the limiting cases (15.25) and (15.26) as indicated.

- **Exercise:** Determine the sign of rotation (right- or left-handed) of the waves corresponding to the two dispersion relations (15.25) and (15.26).

15.5 Ion acoustic waves in gravitational atmospheres ⓚ

In this section we consider a special case, where low frequency waves are propagating in an inhomogeneous gravitationally confined plasma. This example is important by giving an example for a "fake instability," i.e., a case where we have waves growing in amplitude, but without the presence of free energy to drive an instability. The apparent instability is due to the conservation of wave energy flux as the waves propagate in the vertical direction. As the plasma density decreases with altitude, the wave amplitude has to increase in order to keep the energy flux constant.

As the basic equations for the problem, we take the continuity equation for the ions, the momentum equation assuming cold ions, $T_i = 0$, and Boltzmann distributed electrons, i.e.,

$$\begin{aligned} \frac{\partial}{\partial t}n_i + \frac{\partial}{\partial z}(n_i u_i) &= 0, \\ n_i M\left(\frac{\partial}{\partial t}u_i + u_i\frac{\partial}{\partial z}u_i\right) &= -n_i e\frac{\partial}{\partial z}\phi - n_i Mg, \\ n_e &= n_0 \exp\left(\frac{e\phi}{\kappa T_e}\right), \end{aligned}$$

where g is the (constant) gravitational acceleration, and z is the vertical coordinate. The purpose of this section is to illustrate the problem of this "fake" instability, and not to solve the general problem of waves in gravitational atmospheres, so we restrict the analysis to the one-dimensional system. To simplify the problem further, we assume quasi-neutrality, $n_e \approx n_i \equiv n$.

First we obtain a stationary solution for the problem. This is readily found as

$$\overline{n} = n_0 \exp(-zMg/\kappa T_e) \qquad \text{and} \qquad \overline{\phi} = -zMg/e,$$

implying that we have a constant vertical electric field $\overline{E} = Mg/e$. Physically, this solution means that the electron pressure expands the electron gas in the vertical direction, until a steady state electric field has been built up, which exactly counterbalances the electron pressure. For an isothermal *neutral* gas atmosphere in a gravitational field, it is the force due to a pressure gradient and the gravitational force that are in balance.

If we now introduce one more ion of the same species with, say, velocity U_0 moving upwards at $z = 0$, then this ion will continue without acceleration or deceleration, since the gravitational force is exactly canceled by $e\overline{E}$. If we instead take another much lighter ion, this one will be *accelerated* upwards by the electric field, since the electric force is the same as before, but the gravitational force is reduced. This is the basic mechanism of the *polar wind*, where ions of light species, hydrogen and helium, are lost from the Earth's ionosphere to the magnetopause by propagating along the magnetic field lines in the magnetic polar regions. The polar wind has been studied in great detail, and a number of relevant publications can be found in a collection of papers in Journal of Atmospheric and Terrestrial Physics, Volume **69**, Issue: 16 (2007). A review paper can be found also (Ganguli 1996).

To investigate the propagation of small amplitude ion acoustic waves in the vertical direction for the given environment, we linearize the basic equations with $\partial\overline{n}/\partial z = -(Mg/\kappa T_e)\overline{n}$, and find easily

$$\begin{aligned} \frac{\partial}{\partial t}\widetilde{\eta} - \frac{Mg}{\kappa T_e}\widetilde{u} + \frac{\partial}{\partial z}\widetilde{u} &= 0, \\ \frac{\partial}{\partial t}\widetilde{u} &= -\frac{e}{M}\frac{\partial}{\partial z}\widetilde{\phi}, \\ \widetilde{\eta} &= \frac{e}{\kappa T_e}\widetilde{\phi}, \end{aligned}$$

where we introduced $\widetilde{n}/\overline{n} \equiv \widetilde{\eta}$ and omit the index i on the ion fluid velocity. We then obtain the dispersion relation

$$\omega^2 = C_s^2 k_z^2 \left(1 + i\frac{Mg}{\kappa T_e k_z}\right). \tag{15.27}$$

The problem is inhomogeneous, but we succeeded in Fourier transforming with respect to the z-variable nonetheless because the z-dependence of $\overline{n}$ does not appear explicitly in the linearized equations.

For upward propagating waves, the relation (15.27) has solutions which grow exponentially with time, where long wavelengths (small k_z) have the largest relative growth rates, $\Im\{\omega\}/\Re\{\omega\}$. Waves propagating downwards ($k_z < 0$) are damped. The plasma has here no source of free energy, so the observed wave growth cannot properly be called an instability (Dysthe et al. 1975). When the wavelength becomes much smaller than the vertical length scale $\kappa T_e/Mg$, the relative growth rate $\Im\{\omega\}/\Re\{\omega\}$ becomes small, and the waves can be assumed to propagate with the local dispersion relation, corresponding to the appropriate altitude z. Physically, the apparent wave growth is due to the vertical energy flux being constant, in the absence of any dissipation mechanism. The real part of the propagation velocity is a constant, C_s, which is independent of density. For instance, the local kinetic energy density is to the given approximation $\frac{1}{2}M\overline{n}\widetilde{u}^2$. Since $\overline{n}$ is decreasing with altitude, we must have $\widetilde{u}$ increasing to keep the energy flux constant. This altitude increase in amplitude of the fluctuating velocity can then appear as an instability.

The present problem has a counterpart in neutral gravitational atmospheres (Hines 1960). For the plasma problem, the apparent growth will be counteracted by kinetic effects, i.e., Landau damping. For neutral atmospheres it will be by viscosity.

The present summary, as well as other related studies (Parkinson & Schindler 1969), seem to give an exact solution to a linear problem describing inhomogeneous conditions. By a smart choice of variables we seem to avoid having to Fourier transform products of functions depending on the spatial variable z. Unfortunately, this is only seemingly so. The shortcoming becomes evident if we relax the assumption of quasi-neutrality. Introducing isothermally Boltzmann distributed electrons $n_e = \overline{n}(z)\exp(e\phi/\kappa T_e)$ into Poisson's equation we have

$$\frac{\partial^2 e\phi/\kappa T_e}{\partial z^2} = \frac{e^2\overline{n}(z)}{\varepsilon_0 \kappa T_e}\left(\exp\left(\frac{e\phi}{\kappa T_e}\right) - \eta\right) \equiv \frac{1}{\lambda_{De}^2(z)}\left(\exp\left(\frac{e\phi}{\kappa T_e}\right) - \eta\right). \tag{15.28}$$

There is no need to linearize the equation here. From (15.28) we find that the local Debye length varies with z and we have $\lambda_{De}(z \to \infty) \to \infty$. This means that no matter how long wavelength λ_0 we choose for the initial perturbation, there will always be some value of $z = z_c$ where $\lambda_0 = \lambda_{De}$ and the assumption of quasi-neutrality breaks down for $z \geq z_c$. Our result is in reality local, although it may not be obvious from the outset.

- **Exercise:** The dispersion relation (15.27) can be solved for real k giving complex ω. This would typically be an initial value problem. The equation can also be solved for real ω, giving complex k. This case will correspond to a real frequency being applied at a fixed position, and the waves propagating with complex wavenumbers. Compare the two types of solutions and demonstrate that we have a cut-off frequency and a stop-band for the latter problem, but not for the former.
- **Exercise:** Consider a "composite" ionosphere in a gravitational field. with a majority ion with mass M_1 and density n_1, and a minority ion with mass M_2 and density $n_2 \ll n_1$. Determine the equilibrium density profiles, and discuss mass ratios $M_1/M_2 < 1$ and $M_1/M_2 > 1$, respectively.
- **Exercise:** Consider the ionosphere consisting of electrons and one single type of ions, but allow for $T_i \neq 0$. Determine the equilibrium density profile for this case.

Ⓢ **Exercise:** What is the effect of finite ion temperatures on the dispersion relation in a gravitational plasma atmosphere? Let the ion component be isothermal in the steady state conditions. Illustrate the difference between isothermal and adiabatic ion dynamics. Ⓢ

16

Fluid Models for Nonlinear Electrostatic Waves: Magnetized Case

In order to describe the space-time evolution of weakly nonlinear high frequency electrostatic plasma waves in magnetized plasmas we follow the analysis of Section 14.1. Again we have two limits: one where the ions are considered immobile, and one where we allow for a slow ion motion. This chapter will not consider nonlinear ion waves, only note that there are some similarities between the nonlinear upper-hybrid wave model discussed here and models for weakly nonlinear lower-hybrid waves (Shapiro et al. 1993). As for the unmagnetized plasma, only the nonlinear dynamics of plane waves is discussed here. For laboratory experiments this condition is easily realized, but in nature we will usually encounter broad spectra. Studies of also the nonlinear evolution of electron wave spectra in weakly magnetized plasmas are reported in the literature (Kasymov et al. 1985, Kono & Pécseli 2017).

16.1 Cold electrons with immobile ions

The analysis of Section 14.1.1 can be readily generalized to nonlinear electron waves in magnetized plasmas. The "expense" for the closed analytical expressions is also here the assumption of cold electrons. We have the homogeneous magnetic field $\mathbf{B} \parallel \widehat{\mathbf{z}}$. For plane waves propagating along $\mathbf{B}$, we trivially recover the unmagnetized case from Section 14.1.1. In principle we can let plane waves propagate in any given direction, but we restrict the analysis to electrostatic waves propagating $\perp \mathbf{B}$. For this case we have (Davidson & Schram 1968, Davidson 1972) the basic fluid equations for the cold electron motion as

$$\begin{aligned}
\frac{\partial}{\partial t}u_x + u_x\frac{\partial}{\partial x}u_x &= -\frac{e}{m}E - \omega_{ce}u_y \\
\frac{\partial}{\partial t}u_y + u_x\frac{\partial}{\partial x}u_y &= \omega_{ce}u_x \\
\frac{\partial}{\partial t}u_z + u_x\frac{\partial}{\partial x}u_z &= 0, \qquad (16.1)
\end{aligned}$$

with the electrostatic electric field $\mathbf{E} = E\widehat{\mathbf{x}}$, by assumption. The equations are completed by the continuity equation and Poisson's equation also here. For discussions of the assumption of electrostatic waves for the present case we refer to Section 12.2.2.

Introducing the Lagrangian variables just as in Section 14.1.1, we readily find

$$\frac{\partial^2}{\partial \tau^2}u_x = -\omega_{uh}^2 u_x,$$

with $\omega_{uh} \equiv \sqrt{\omega_{pe}^2 + \omega_{ce}^2}$ being the upper-hybrid frequency also here. We find the solutions

$$
\begin{aligned}
u_x(x_0,\tau) &= U_x(x_0)\cos\omega_{uh}\tau + \omega_{uh}X(x_0)\sin\omega_{uh}\tau \\
u_y(x_0,\tau) &= U_y(x_0) + \frac{\omega_{ce}}{\omega_{uh}}\left(U_x(x_0)\sin\omega_{uh}\tau + \omega_{uh}X(x_0)(1-\cos\omega_{uh}\tau)\right) \\
E(x_0,\tau)\frac{e}{m} &= \frac{\omega_{pe}^2}{\omega_{uh}}U_x(x_0)\sin\omega_{uh}\tau - \omega_{pe}^2 X(x_0)\cos\omega_{uh}\tau \\
&\quad -(\omega_{ce}^2 X(x_0) + \omega_{ce}U_y(x_0)).
\end{aligned} \tag{16.2}
$$

We also find $u_z(x_0,\tau) = U_z(x_0)$, trivially. As far as initial conditions are concerned, we find that the condition on the electric field is coupled to $U_y(x_0)$, i.e., we have $E(x_0,0) = -(m/e)(\omega_{uh}^2 X(x_0) - \omega_{ce}U_y(x_0)$. Poisson's equation is a relationship that holds for all times, in particular also $t=0$, giving $dE(x_0,0)/dx = -(e/\varepsilon_0)(n-n_0) = (m/e)\left(\omega_{ce}dU_y(x_0)/dx_0 - \omega_{uh}^2 dX(x_0)/dx_0\right)$, or

$$
\frac{n_e(x_0,0)}{n_0} - 1 = \frac{\omega_{uh}^2}{\omega_{pe}^2}\frac{dX(x_0)}{dx_0} - \frac{\omega_{ce}}{\omega_{pe}^2}\frac{dU_y(x_0)}{dx_0}.
$$

At least formally, the one-dimensional problem is solved. Again we find that the fundamental frequency of the oscillations is here ω_{uh}, irrespective of the wave amplitude (Davidson & Schram 1968, Davidson 1972), in the sense that the initial conditions re-occur at intervals of $2\pi/\omega_{uh}$. Although the magnetic field changes details in the wave motions, the results are basically the same as for the unmagnetized one-dimensional wave propagation analyzed in Section 14.1.1. Important modifications for the weakly nonlinear electron waves are found by allowing for mobile ions, as in Section 14.1.2.

- **Exercise:** Prove the relations (16.2).

- **Exercise:** We have two conditions: $n \geq 0$ at all times and n finite. Demonstrate that the latter condition coincides with the requirement for the transformation from x_0 to x being unique (i.e., meaning that a position (x,t) can be reached from only one $(x_0,0)$ position). Demonstrate that the requirements listed imply the conditions

$$
\frac{n(x_0,0)}{n_0} - \frac{\omega_{ce}}{\omega_{pe}^2}\frac{dU_y(x_0)}{dx_0} > \frac{\omega_{pe}^2 - \omega_{ce}^2}{2\omega_{pe}^2},
$$

and

$$
\frac{1}{\omega_{pe}}\left|\frac{dU_x(x_0)}{dx_0}\right| < \sqrt{2\left(\frac{n(x_0,0)}{n_0} - \frac{\omega_{ce}}{\omega_{pe}^2}\frac{dU_y(x_0)}{dx_0}\right) - \frac{\omega_{pe}^2 - \omega_{ce}^2}{\omega_{pe}^2}}.
$$

Demonstrate that these requirements are consistent with the results for the unmagnetized case when $\omega_{ce} \to 0$.

16.2 Mobile ions

Section 16.1 assumed the ions to be immobile and gave results which did not dramatically deviate from those found in the unmagnetized case in Section 14.1. The analysis will be substantially modified by allowing for a slow ion motion. The analysis here follows the ideas outlined in Section 14.1.2. Discussions with Profs. E. Mjølhus and L. Stenflo on this section were most helpful.

Basically, the moral of the analysis of the homogeneous and isotropic case was that the evolution to lowest order was described by the *linear cold plasma approximation*, which in that case is trivial, $\omega^2 = \omega_{pe}^2$, to be corrected by a thermal dispersion contribution on the same level as the nonlinear frequency shift, with the dominant part originating from the local change in plasma density induced by ponderomotive forces (Dysthe et al. 1982). We thus start out also here with the cold plasma equations for harmonically varying electric fields, $\mathbf{E} \sim \exp(-i\omega t)$, and have from Section 12.2.1 the relation (12.17) reproduced here

$$\nabla \times (\nabla \times \mathbf{E}) = \frac{\omega^2}{c^2} \underline{\varepsilon} \cdot \mathbf{E},$$

in terms of

$$\underline{\varepsilon} \equiv \begin{pmatrix} \varepsilon_\perp & i\mu & 0 \\ -i\mu & \varepsilon_\perp & 0 \\ 0 & 0 & \varepsilon_\parallel \end{pmatrix},$$

where we introduced

$$\varepsilon_\parallel \equiv 1 - X, \qquad \varepsilon_\perp \equiv 1 - \frac{X}{1-Y^2}, \qquad \text{and} \qquad \mu \equiv \frac{|Y|X}{1-Y^2},$$

with $X \equiv \omega_{pe}^2/\omega^2$ and $Y \equiv \omega_{ce}/\omega$, as before; see Section 12.2.1. We can also write

$$\underline{\varepsilon} = \varepsilon_\parallel \widehat{\mathbf{z}\mathbf{z}} + \varepsilon_\perp (\widehat{\mathbf{x}\mathbf{x}} + \widehat{\mathbf{y}\mathbf{y}}) - i\mu(\widehat{\mathbf{x}\mathbf{y}} - \widehat{\mathbf{y}\mathbf{x}}),$$

in terms of the orthogonal Cartesian unit vectors, with $\widehat{\mathbf{z}}$ being along the magnetic field, assumed homogeneous. We assume again electrostatic high frequency waves by writing $\mathbf{E} = -\nabla\phi$. By this assumption we reduce (12.17) to

$$\nabla \cdot (\underline{\varepsilon} \cdot \nabla\phi) = 0,$$

which is nothing but Poisson's equation without external charges. With a little effort we then find

$$\frac{\partial}{\partial z}\left(\varepsilon_\parallel \frac{\partial}{\partial z}\phi\right) + \nabla_\perp \cdot (\varepsilon_\perp \nabla_\perp \phi) + i\widehat{\mathbf{z}} \cdot \nabla_\perp \mu \times \nabla_\perp \phi = 0. \tag{16.3}$$

The important point now (Dysthe et al. 1982) is that within the cold plasma approximation we can have $\varepsilon_\parallel = \varepsilon_\parallel(\mathbf{r})$, $\varepsilon_\perp = \varepsilon_\perp(\mathbf{r})$, and similarly $\mu = \mu(\mathbf{r})$ accounting for spatial variations in plasma parameters, i.e., variations in density and magnetic field through variations in X and Y. The harmonic *time* variation is accounted for by the frequency ω entering X and Y.

For constant magnetic fields and plasma densities we have the simpler result

$$\varepsilon_\parallel \frac{\partial^2}{\partial z^2}\phi + \varepsilon_\perp \nabla_\perp^2 \phi = 0. \tag{16.4}$$

When $\varepsilon_\parallel$ and $\varepsilon_\perp$ have opposite sign, this equation is of the hyperbolic type (Whitham 1974). The characteristic cones are called "resonance cones," with frequency dependent opening angles given by

$$\tan^2\theta = \frac{k_\perp^2}{k_\parallel^2} = -\frac{\varepsilon_\parallel(\omega)}{\varepsilon_\perp(\omega)}.$$

The equation (16.4) is hyperbolic for certain frequency ranges determined by $\varepsilon_\parallel(\omega)$ and $\varepsilon_\perp(\omega)$. With $\mathbf{B} \neq 0$ we have $\omega \in \{0, \omega_{ce}\}$ and $\omega \in \{\omega_{pe}, \omega_{uh}\}$ for weak magnetic fields, or $\omega \in \{0, \omega_{pe}\}$ and $\omega \in \{\omega_{ce}, \omega_{uh}\}$ for strong magnetic fields, as discussed in Section 15.1. We denoted the upper-hybrid frequency ω_{uh} as before. In the homogeneous cold plasma limit, the dispersion is *directional*, in the sense that as long as we keep the direction of the wave-vector $\mathbf{k}$ constant, the wave has a constant frequency for any wave-number. The frequency changes when the direction of the wave-vector is changed. To retain the concept of weakly nonlinear narrow frequency band waves we *either* keep the direction of wave propagation constant, or alternatively consider, for instance, a high frequency wave $\omega \in \{\omega_{pe}, \omega_{uh}\}$, and assume that the magnetic field is so weak that the frequency band $\{\omega_{pe}, \omega_{uh}\}$ is narrow. The latter assumption is the one most often discussed in the literature (Dysthe et al. 1984, Dysthe et al. 1985*a*).

Except for the special cases mentioned, high frequency electrostatic waves in magnetized plasmas are generally *broad frequency band* processes, in contrast to the unmagnetized case, where $\omega \approx \omega_{pe}$ to lowest order.

We now specify the spatial plasma density variation by writing $n = n_0 + \widetilde{n}$, and introduce $\eta \equiv \widetilde{n}/n_0$, allowing the replacement of X by $X_0(1+\eta)$, where now X_0 refers to the background density n_0. Magnetic fields will be considered constant and homogeneous. We will assume that $\eta \ll 1$, and obtain from (16.3) the expression

$$\frac{\partial}{\partial z}\left(\eta\frac{\partial}{\partial z}\phi\right) + \frac{1}{1-Y_0^2}\nabla_\perp\cdot(\eta\nabla_\perp\phi) - i\frac{Y_0}{1-Y_0^2}\widehat{\mathbf{z}}\cdot\nabla_\perp\eta\times\nabla_\perp\phi$$
$$-\frac{1}{X_0}\left(\frac{\partial}{\partial z}\varepsilon_{0\parallel}\frac{\partial}{\partial z}\phi + \nabla_\perp\cdot(\varepsilon_{0\perp}\nabla_\perp\phi)\right) = 0, \tag{16.5}$$

where we can have the subscript $_0$ on ε without confusing it with the vacuum dielectric constant ε_0, since we have also included $\parallel$ and $\perp$ in the subscripts, indicating that $\varepsilon_{0\parallel}$ and $\varepsilon_{0\perp}$ are obtained by X_0 and Y_0. In obtaining (16.5) we used $\nabla\times\nabla\phi = 0$ one place.

From (16.5) it is now, at least in principle, trivial to obtain a partial differential equation for the temporal evolution by the standard replacement $\omega \to i\partial/\partial t$. All we have to do is to insert X_0 and Y_0 which contain the ω-variation, multiply the equation by X_0 and $1-Y_0^2$ and then ω a few times, and then finally replace ω by the partial time differentiation, as indicated.

In reality, we are only interested in narrow band processes. We thus assume the spectrum associated with the process to be narrow banded around a characteristic natural oscillation frequency ω_0, just as in Chapter 14. The assumption of a narrow band process is readily incorporated by the replacement $\omega \to \omega_0 + i\partial/\partial t$, and ignoring the second derivative $\partial^2/\partial t^2$. This gives

$$2\frac{i}{\omega_0}\frac{\partial}{\partial t}\left(\frac{\partial^2}{\partial z^2}\phi + \frac{1}{(1-Y_0^2)^2}\nabla_\perp^2\phi\right)$$
$$+\frac{\partial}{\partial z}\left(\eta\frac{\partial}{\partial z}\phi\right) + \frac{1}{1-Y_0^2}\nabla_\perp\cdot(\eta\nabla_\perp\phi) - i\frac{Y_0}{1-Y_0^2}\widehat{\mathbf{z}}\cdot\nabla_\perp\eta\times\nabla_\perp\phi$$
$$-\frac{1}{X_0}\left(\frac{\partial}{\partial z}\varepsilon_{0\parallel}\frac{\partial}{\partial z}\phi + \nabla_\perp\cdot(\varepsilon_{0\perp}\nabla_\perp\phi)\right) = 0, \tag{16.6}$$

where now the subscript $_0$ on Y and X also implies that ω_0 has been inserted.

Until now we have ignored wave dispersion due to thermal effects. The lowest order thermal corrections can formally be included by writing

$$\varepsilon \approx \varepsilon_c + \frac{1}{2}\mathbf{kk}{:}\nabla_k\nabla_k\varepsilon\bigg|_{k=0}, \tag{16.7}$$

where the subscript $|_{k=0}$ indicates that the ∇_k-derivative should be evaluated at $k=0$, and the subscript $_c$ indicates the cold plasma dielectric tensor.

By converting $\mathbf{k}$ to the spatial differential operator we obtain

$$\varepsilon \approx \varepsilon_c - \frac{1}{2}\nabla\nabla{:}\nabla_k\nabla_k\varepsilon\bigg|_{k=0}. \tag{16.8}$$

Making this replacement we have

$$\frac{1}{X_0}\left(\frac{\partial}{\partial z}\varepsilon_{0\|}\frac{\partial}{\partial z}\phi+\nabla_\perp\cdot(\varepsilon_{0\perp}\nabla_\perp\phi)\right)\to$$
$$\frac{1}{X_0}\left(\frac{\partial}{\partial z}\varepsilon_{0\|}\frac{\partial}{\partial z}\phi+\nabla_\perp\cdot(\varepsilon_{0\perp}\nabla_\perp\phi)\right)-\frac{1}{X_0}\nabla\cdot(\nabla\nabla:\nabla_k\nabla_k\varepsilon|_{k=0})\cdot\nabla\phi.$$

After some algebra (Dysthe et al. 1985*a*), we find

$$-\frac{1}{X_0}\nabla\cdot(\nabla\nabla:\nabla_k\nabla_k\varepsilon|_{k=0})\cdot\nabla\phi=$$
$$\frac{u_{The}^2}{\omega_0^2}\left(\frac{1}{(1-Y_0^2)(1-4Y_0^2)}\nabla_\perp^4+\frac{\partial^4}{\partial z^4}+\frac{2-Y_0^2+Y_0^4/3}{(1-Y_0^2)^3}\nabla_\perp^2\frac{\partial^2}{\partial z^2}\right)\phi,$$

in terms of the electron thermal velocity, u_{The}. By this, we have obtained a closed equation which relates the high frequency potential ϕ to the local plasma density variations. How the density variation $\eta\equiv\tilde{n}/n_0$ is obtained from ϕ will be discussed later on, but we have seen the basic principles already in Chapter 14. More detailed discussions were given by Dysthe et al. (1978), Dysthe et al. (1984) and Dysthe et al. (1985*a*). In the limit of a vanishing magnetic field, i.e., $Y_0\to 0$, we find the limit of homogeneous and isotropic thermal plasmas

$$-\frac{1}{X_0}\nabla\cdot(\nabla\nabla:\nabla_k\nabla_k\varepsilon|_{k=0})\cdot\nabla\phi=\frac{u_{The}^2}{\omega_0^2}\left(\nabla_\perp^4+\frac{\partial^4}{\partial z^4}+2\nabla_\perp^2\frac{\partial^2}{\partial z^2}\right)\phi=\frac{u_{The}^2}{\omega_0^2}\nabla^4\phi,$$

as expected.

Like lower-hybrid waves (Hsu & Kuehl 1983, Shapiro et al. 1993), also upper-hybrid waves have an increasing electromagnetic component as the characteristic scale length is increased (Dysthe et al. 1985*a*, Dysthe et al. 1982). These effects will in general imply that possible stationary solutions will be "leaking" due to a small amplitude long wavelength electromagnetic wave component that can escape.

16.3 Simplified special cases

The foregoing analysis was kept in rather general terms, and the ensuing analytical model ended up being quite complicated. It is appropriate to consider a few simplified cases, which allow a description based on considerably simpler models. We present results for the two simplest cases, propagation along or perpendicular to **B**. The arbitrary oblique direction was discussed by Dysthe et al. (1985*a*).

16.3.1 B-parallel propagation

For the primary direction of wave propagation being along the magnetic field we have $X_0=1$, i.e., $\omega_0=\omega_{pe}$, and find

$$i\frac{2}{\omega_{pe}}\frac{\partial^2}{\partial z^2}\frac{\partial}{\partial t}\phi-\frac{Y_0^2}{1-Y_0^2}\nabla_\perp^2\phi+3\frac{u_{The}^2}{\omega_{pe}}\frac{\partial^4}{\partial z^4}\phi=\frac{\partial}{\partial z}\left(\eta\frac{\partial\phi}{\partial z}\right). \tag{16.9}$$

Note the change in sign of the second term for $Y_0^2<1$ and $Y_0^2>1$. The result (16.9) is closely related to the one discussed by Krasnosel'skikh and Sotnikov (1977). In terms of the electrostatic electric field component $E_z \equiv -\partial\phi/\partial z$, we can write for weak magnetic fields

$$\frac{\partial^2}{\partial z^2}\left(i\frac{\partial}{\partial t}E_z + \frac{3}{2}\omega_{pe}\lambda_{De}^2\frac{\partial^2}{\partial z^2}E_z\right) - \frac{\omega_{uh}^2}{2\omega_{pe}}\nabla_\perp^2 E_z = \frac{\partial^2}{\partial z^2}(\eta E_z). \tag{16.10}$$

16.3.2 B-perpendicular propagation

For the primary waves propagating in the direction perpendicular to **B**, we have $\omega = \omega_{uh}$, the upper-hybrid frequency. At the upper hybrid resonance we have $Y_0^2 < 1$. We obtain for this case

$$i\frac{2}{\omega_{uh}}\frac{1}{1-Y_0^2}\nabla_\perp^2\frac{\partial}{\partial t}\phi + Y_0^2\frac{\partial^2}{\partial z^2}\phi + 3\frac{u_{The}^2}{\omega_{uh}^2}\frac{1}{1-4Y_0^2}\nabla_\perp^4\phi =$$
$$\nabla_\perp\cdot(\eta\nabla_\perp\phi) - iY_0\widehat{\mathbf{z}}\cdot(\nabla_\perp\eta\times\nabla_\perp\phi). \tag{16.11}$$

Not much is known about the solutions for this equation.

16.4 Models for the low frequency response

The model equations for the (complex) amplitude of the high frequency electric field contain the bulk plasma density as a parameter. Various models can be suggested for a closure of the equations, i.e., relating η to ϕ, but we emphasize that the basic equations are applicable also for the case where a steady state density variation η is prescribed.

16.4.1 Quasi static response

The simplest relation between the slowly varying plasma density and the high frequency electric field is obtained by the quasi static assumption relating $\overline{n}/n_0 \equiv \eta$ and $|\nabla\phi|^2$ as for the unmagnetized case. This will be an appropriate approximation when the time evolution is very slow.

16.4.2 Low frequencies, $\omega \ll \Omega_{ci}$

For very low frequencies, $\omega \ll \Omega_{ci}$, we can assume that the ions are moving along the magnetic field lines as "pearls on a string." In that limit a one-dimensional version of (14.28) will suffice.

$$\frac{\partial^2}{\partial t^2}\eta - C_s^2\frac{\partial^2}{\partial z^2}\eta = \frac{\varepsilon_0}{4Mn_0}\frac{\partial^2}{\partial z^2}|\nabla\phi|^2. \tag{16.12}$$

The previously mentioned quasi static response is readily obtained from (16.12) in the limit where $\partial^2\eta/\partial t^2$ is negligible.

16.4.3 Ion cyclotron waves

When the frequencies become comparable to or larger than the ion cyclotron frequency, the simple one-dimensional model (16.12) is no longer applicable. It is relatively straightforward, although it takes some time, to obtain the appropriate equation (Dysthe & Pécseli 1978, Dysthe et al. 1978)

$$\left(\left(\frac{\partial^2}{\partial t^2} - C_s^2\nabla^2 + \Omega_{ci}^2\right)\frac{\partial^2}{\partial t^2} - C_s^2\Omega_{ci}^2\frac{\partial^2}{\partial z^2}\right)\eta$$
$$= \frac{\varepsilon_0}{4Mn_0}\left(\nabla^2\frac{\partial^2}{\partial t^2} + \Omega_{ci}^2\frac{\partial^2}{\partial z^2}\right)|\nabla\phi|^2. \tag{16.13}$$

This equation contains both magnetized ion acoustic waves and electrostatic ion cyclotron waves by the linear dispersion relation $(\omega^2 - k^2 C_s^2)(1 - \Omega_{ci}^2/\omega^2) = k_\perp^2 C_s^2$, and the presence of both branches gives rise to a "banded" structure in a parameter space spanned by KC_s/Ω_{ci}^2 and $(\mathbf{u}_g \cdot \widehat{\mathbf{e}}/C_s)^2$ with $\widehat{\mathbf{e}} \equiv \mathbf{K}/K$ and $\mathbf{u}_g$ being the group velocity vector of the electron waves (Dysthe & Pécseli 1978, Dysthe et al. 1978). For a given wavevector $\mathbf{K}$, one of the low frequency wave branches (electrostatic ion cyclotron or ion acoustic) can be driven above and some below resonance: this gives rise to the intricate stability diagram for the modulationally unstable waves. The case where both branches are driven either above or below resonance will resemble the unmagnetized case.

16.4.4 Lower-hybrid response

The quasi static response of the ion dynamics represents a rather extreme limit, and more generally applicable dynamic equations should be sought after. If the relevant low frequency response is characterized by frequencies well below the ion cyclotron frequency, some evident simplifications can be introduced. A more general analysis includes also ion cyclotron waves as mentioned before. This analysis can be formulated in a limit where electron inertia is ignored. We present here an equation which takes into account also the lower-hybrid frequency, where both electron as well as ion inertia is important for the low frequency density response (Dysthe et al. 1984). For very high plasma densities with $\Omega_{ci} \ll \Omega_{pi}$, where we can assume quasi neutral plasma dynamics, see Section 15.3, the lower hybrid frequency is $\Omega_{lh} \approx \sqrt{\Omega_{ci}\omega_{ce}}$. The effects of plasma density variations induced by nonlinearities are negligible here. For density variations to have an effect we need the plasma frequency to enter explicitly in the expression in Ω_{lh}, see Section 15.3, and the assumption of quasi neutrality has to be relaxed for the model of the low frequency dynamics. In order to retain the full lower-hybrid dynamics, we therefore include Poisson's equation explicitly in the analysis, and consequently the electron and ion dynamics are distinguished. The density in the model equations for the *high frequency dynamics* is therefore explicitly identified as the *electron* density. The equation accounting for the variations in the slowly varying normalized electron density $\overline{n}_e/n_0 \equiv \eta_e$ as induced by weakly nonlinear high frequency waves becomes

$$\begin{aligned}
&\left(\frac{\partial^4}{\partial t^4}\nabla^2 + \Omega_{lh}^2\left(\nabla^2 + \frac{M}{m}\frac{\partial^2}{\partial z^2}\right)\frac{\partial^2}{\partial t^2}\right)\eta_e \\
&-\frac{\omega_{pe}^2}{\omega_{pe}^2+\omega_{ce}^2} C_s^2\left(\omega_{ce}^2\frac{\partial^2}{\partial z^2} + \frac{\partial^2}{\partial t^2}\nabla^2\right)\nabla^2\eta_e \\
&-\frac{1}{\omega_{pe}^2+\omega_{ce}^2}\left(u_{Thi}^2\omega_{ce}^2\nabla^2 + u_{The}^2\left(\omega_{ce}^2\frac{\partial^2}{\partial z^2} + \frac{\partial^2}{\partial t^2}\nabla^2\right)\right)\frac{\partial^2}{\partial t^2}\nabla^2\eta_e = \\
&\frac{1}{\omega_{pe}^2+\omega_{ce}^2}\left(\frac{\partial^2}{\partial t^2} + \Omega_{pi}^2\right)\nabla^2\left(\frac{\partial^2}{\partial t^2}\nabla\cdot\mathbf{I} + \omega_{ce}\widehat{\mathbf{b}}\cdot\left(\nabla\times\frac{\partial}{\partial t}\mathbf{I}\right) + \omega_{ce}^2\frac{\partial}{\partial z}I_z\right). \qquad (16.14)
\end{aligned}$$

We introduced the abbreviations $C_s^2 \equiv u_{Thi}^2 + u_{The}^2 m/M$, and $\Omega_{lh}^2 \equiv \Omega_{pi}^2/(1+\omega_{pe}^2/\omega_{ce}^2)$, ignoring terms containing products of the ion and electron thermal velocities like $u_{Thi}^2 u_{The}^2$. We introduced the vector $\mathbf{I} \equiv \overline{\widetilde{\mathbf{u}}\cdot\nabla\widetilde{\mathbf{u}}}$, with the fluctuating electron velocity being

$$\widetilde{\mathbf{u}} = i\frac{e}{m}\frac{\omega}{\omega^2-\omega_{ce}^2}\left(\nabla\widetilde{\phi} + i\frac{\omega_{ce}}{\omega}\widehat{\mathbf{b}}\times\nabla\widetilde{\phi} - \frac{\omega_{ce}^2}{\omega^2}\widehat{\mathbf{b}}\frac{\partial}{\partial z}\widetilde{\phi}\right),$$

with $\widehat{\mathbf{b}} \equiv \mathbf{B}/B \parallel \widehat{\mathbf{z}}$. For a more complete description of the ponderomotive forces, see Washimi and Karpman (1976), Kono et al. (1981) or Statham and ter Haar (1983). When including electrostatic lower hybrid and ion cyclotron waves in the low frequency response, care should be taken concerning the ordering of parameters (Kono & Pécseli 2017).

The nonlinear developments of upper-hybrid waves have been studied numerically (Lin & Lin 1981) and experimentally (Cho et al. 1982) but, at least in the opinion of the present author, not nearly as much as Langmuir waves in unmagnetized plasmas, and a lot of interesting work remains here. The problem can have relevance also for the nonlinear evolution of lower hybrid waves observed in the Earth's ionosphere.

17

Linear Drift Waves

A magnetized inhomogeneous plasma can sustain a low frequency electrostatic wave type called "drift-waves" because they propagate in the direction of the electron diamagnetic drift. The driving mechanism is the pressure gradient perpendicular to the externally imposed magnetic field. The plasma can reach an energetically favorable state by smoothing out the gradients in pressure, and there is thus free energy available for driving the waves linearly unstable. Alternatively it can be argued that the source of free energy is the gradient drift current associated with the pressure gradient.

Historically, the waves were identified as low frequency fluctuations (well below the ion cyclotron frequency), located near the edge of the plasma column in linear plasma devices (Hendel et al. 1968, Rowberg & Wong 1970, Rogers & Chen 1970, Politzer 1971, Timofeev & Shvilkin 1976). Extensive experimental studies were carried out in Q-machines. Detailed investigations of the stability of drift waves were made by, for instance, Chen (1964), Chen (1965*a*), Chen (1966), and Schlitt and Hendel (1972), for collisional drift waves in particular. The following summary is in debt to some unpublished lecture notes by Chen (1970).

Since density gradients are unavoidable in all laboratory experiments involving magnetically confined hot plasmas, the drift wave instability is often called a "universal" instability since it is associated with such gradients.

It is interesting that the theoretical description of electrostatic drift waves, linear as well as nonlinear, to a very large extent is similar to that describing, for instance, Rossby waves in stratified media in rotating systems (Brekhovskikh & Goncharov 1985, Nycander 1994). The analogy is rooted in the observation that the Lorentz force and the Coriolis force have identical analytical forms, differing only in the physical meaning of the variables.

17.1 Drift wave basics

The basic features of electrostatic drift waves can be described by a simple model. The expense for this simplicity is that the result that the waves are linearly stable. This is in contradiction with observations where the waves are found to be spontaneously excited due to a linear plasma instability.

17.1.1 A simple reference model ⓚ

The geometry of the problem is shown in Fig. 17.1, giving the directions of the magnetic field and the density gradient. The magnetic field will be assumed homogeneous in all that follows. Guided by the experimental observations, we consider low frequency electrostatic waves well below the ion cyclotron frequency Ω_{ci}

$$\omega \ll \Omega_c \,, \tag{17.1}$$

and assume that the waves are propagating in a direction almost perpendicular to the magnetic field so the **B**-parallel component of the wave-vector k_z is small in such a way that

$$u_{ti} \ll \omega/k_z \ll u_{te} \,, \tag{17.2}$$

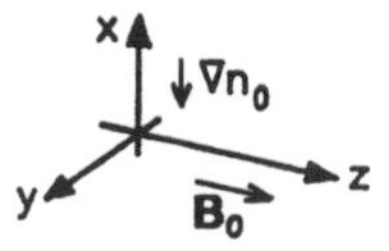

FIGURE 17.1: *Geometry of the problem with a plasma density gradient perpendicular to a homogeneous magnetic field.*

in terms of the ion and electron thermal velocities u_{ti} and u_{te}, respectively. The relation (17.2) is a rather mild restriction for many practical situations. Consistently with (17.1) and (17.2) the ion fluid velocity in the direction $\perp \mathbf{B}$ is approximated by the $\mathbf{E}\times\mathbf{B}/B^2$-velocity, while a **B**-parallel ion velocity component is ignored. The ion continuity equation becomes

$$\frac{\partial}{\partial t}n_i - \nabla\cdot\left[\frac{\nabla\phi\times\mathbf{B}}{B^2}n_i\right] = 0, \tag{17.3}$$

where the electrostatic potential was introduced, i.e., $\mathbf{E} = -\nabla\phi$. The electron momentum equation gives for the **B**-parallel velocity component

$$n_e m\left(\frac{\partial}{\partial t}u_{e\|} + \mathbf{u}_e\cdot\nabla u_{e\|}\right) = -\kappa T_e\frac{\partial}{\partial z}n_e + en_e\frac{\partial}{\partial z}\phi, \tag{17.4}$$

where the electron motion is assumed to be isothermal in view of (17.2). The electron inertia effects, i.e., the left-hand side of (17.4), can be neglected, since they are $(\omega/k_z)/u_{te}$ times smaller than the terms on the right-hand side, giving

$$\frac{\partial}{\partial z}\ln n_e = \frac{e}{\kappa T_e}\frac{\partial}{\partial z}\phi. \tag{17.5}$$

Note that we exclude waves propagating in the direction strictly perpendicular to the magnetic field lines by the inequality (17.2).

Now, strictly speaking, we cannot take the logarithm of a dimensional quantity, i.e., the density n_e in (17.5). It has to be normalized somehow, and as long as the normalization parameter is independent of z, the relation (17.5) can easily be made correct. We use the x-dependent unperturbed density, which by integration of (17.5), gives the isothermal Boltzmann relation

$$n_e = n_0(x)e^{e\phi/\kappa T_e}. \tag{17.6}$$

The reason for choosing this particular normalization in (17.5) is that it gives $n_e = n_0$ when the plasma is unperturbed, $\phi = 0$. A spatially and temporally varying electrostatic potential implies that $n_e(\mathbf{r},t) = n_0(x) + \widetilde{n}(\mathbf{r},t)$. For small perturbations, we can linearize (17.6) to give $\widetilde{n}(\mathbf{r},t)/n_0(x) \approx e\phi(\mathbf{r},t)/\kappa T_e$.

Physically, (17.6) implies that the electrons can flow along the magnetic field lines from wave crest to wave trough in order to establish an isothermal Boltzmann equilibrium in agreement with the inequality (17.2). The process is quasi static in the sense that the electrons can be assumed to adjust instantaneously when seen on the slow time scale (17.1) characterizing the ion dynamics. The coefficient $n_0(x)$ in (17.6) can be understood as an imposed boundary condition at a position where $\phi = 0$. This is a natural interpretation in connection with, e.g., Q-machine experiments where the plasma is produced by surface ionization on a hot plate with good electrical conductivity, but an inhomogeneous radial temperature, implying a corresponding variation in the plasma density.

Finally, Poisson's equation becomes redundant by the assumption of quasi-neutrality (a times called *the plasma approximation* (Chen 2016), see also Section 11.2), $n_e \approx n_i \equiv n$, an approximation which can be readily justified for perturbations with scale length much larger than the electron Debye length, $\lambda_{De} \equiv (\varepsilon_0 \kappa T_e/e^2 n_e)^{1/2}$. In this limit only the bulk plasma density n will enter the analysis. Implicitly it is assumed that also $\omega \ll \omega_{pi}$, the ion plasma frequency. Equations (17.3) and (17.6) are then readily combined to

$$\frac{\partial}{\partial t}\phi - \left[\frac{\kappa T_e}{eBn_0(x)}\frac{\partial}{\partial x}n_0(x)\right]\frac{\partial}{\partial y}\phi = 0, \tag{17.7}$$

or in vector notation

$$\frac{\partial}{\partial t}\phi - \frac{\kappa T_e}{e}\frac{\nabla\phi \times \mathbf{B}}{B^2}\cdot\frac{\nabla n_0(x)}{n_0(x)} = 0. \tag{17.8}$$

The ion motion is here incompressible, since $\nabla\cdot(\nabla\phi\times\mathbf{B}) = 0$. Nevertheless, the fluctuations are associated with variations in plasma density, since the ions move in and out along a density gradient, which is here taken in the x-direction. Equation (17.7) shows that all perturbations propagate without change in shape in the y-direction $\perp \nabla n_0$ and $\mathbf{B}$, with the electron diamagnetic drift velocity

$$u_{De} = -\frac{\kappa T_e}{eB}\frac{1}{n_0(x)}\frac{dn_0(x)}{dx}, \tag{17.9}$$

see also Section 11.1.1. The perturbations in one x-position do not affect the perturbations at other x-positions (Manheimer 1977) since no derivatives $\partial\phi/\partial x$ appear, even though $\phi = \phi(x,y,z,t)$. With the initial assumptions we implicitly assumed the z-variation to be slow so that no $\partial\phi/\partial z$ terms appear, although the assumption of $k_z \neq 0$ is essential for allowing the electrons to reach an isothermal Boltzmann equilibrium. Note that the result (17.7) is linear in ϕ, although no linearization assumptions were actually made!

The plasma approximation essential for (17.7). This is a restriction, but there is no evidence for wavelengths $\sim \lambda_{De}$ to be of particular importance for drift wave models. By relaxing the assumption of quasi neutrality we find small corrections to the dispersion relation for short wavelengths.

It might be surprising that the electron diamagnetic drift appears: this drift (and the diamagnetic current derived from it) is a phenomenon associated with conditions described in a plane perpendicular to **B**. We have nowhere considered this limiting case, on the contrary, see the discussion of (17.2). From basic dimensional reasoning (see Appendix A) we can, however, argue that if we need a characteristic velocity, then (17.9) is the only combination available given the characteristic quantities κT_e, B, e and n_0'/n_0.

By introducing an "effective" ion Larmor radius $a_i \equiv C_s/\Omega_{ci}$, with $C_s^2 \equiv \kappa T_e/M$, and a "scale length" for the density gradient, $\mathcal{L} \equiv n_0/|dn_0/dx|$, we can interpret the magnitude of the electron diamagnetic drift as the ion sound speed weighted by the ratio of a microscopic and macroscopic length scale, $|u_{De}| = C_s(a_i/\mathcal{L})$. Unfortunately, the notation "effective ion Larmor radius" for a_i is established in the literature. However, with $T_i = 0$ no ions have a gyro-radius of this magnitude; it is merely a relevant parameter describing a length scale for the plasma dynamics.

Although the result (17.8) has obvious shortcomings in many respects, the basic arguments can, on the other hand, not be *entirely* wrong, and consequently we expect (17.8) to be the "backbone" of any more detailed mathematical description of electrostatic drift waves, be it linear or nonlinear.

17.1.2 Physical description Ⓚ

It might be advantageous to present the results in the foregoing section in the form of a physical picture, following (Chen 1965*c*). Thus Fig. 17.2 illustrates the basic elements of the derivation. A solid line gives equi-density contour of the perturbation and because of (17.6) the equi-potential lines are in phase with this. This will remain correct even without linearizing (17.6). At position ②

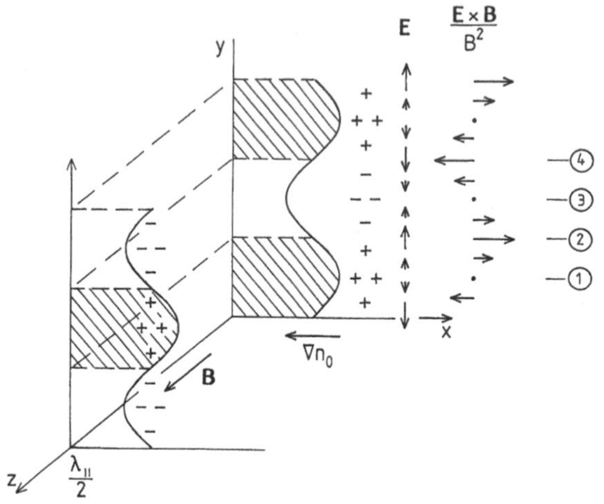

FIGURE 17.2: *Basic elements of drift wave properties. The solid line indicates an equi-density contour, while electric fields and* $\mathbf{E}\times\mathbf{B}$*-ion velocities are indicated by small arrows. Charge densities are indicated by* $+$ *and* $-$ *symbols. Regions with enhanced density are shaded for clarity. For the model used for this figure we find the drift waves to be marginally stable, in contradiction with experimental observations.*

in Fig. 17.2 the ion gyro-center velocity is maximum while the density perturbation, n, is vanishing here. A little later, particles from the denser region of the initial distribution have been brought in and n becomes positive at position ②. Similarly, at position ④ the density will decrease and the perturbation moves upward in Fig. 17.2 in the same direction as the diamagnetic drift. To find the magnitude of the phase velocity we proceed as when deriving (17.7) to find (17.9). The average value of the ion cross-field velocity in the region of density increase (shaded region) is zero, just as in the region of density depletion; the waves are marginally stable, and there is no net plasma transport along the density gradient.

One important modification from this simplified picture comes from charge separation effects we did not take into account. First, since E_y is time varying, ion inertia will cause the ions to drift in the y-direction as well as the x-direction. Second, since the ions have a much larger Larmor radius than the electrons, and since E_y varies in space, the ions will see an average E_y smaller than what the electrons see. The $\mathbf{E}\times\mathbf{B}$ drift will be smaller for ions, and there will be a charge separation because this drift is along ∇n_0. This finite Larmor radius (FLR) effect will be covered in a later section. These two small effects both tend to cause a charge build up and hence an additional electric field E_y. The resulting equation will now contain also derivatives of ϕ with respect to x. However, if k_z is large enough, the flow of electrons along B will cancel the charges building up so as to satisfy the Boltzmann relation (17.6), and the waves remain linearly stable.

17.1.3 Limitations of the electrostatic assumption

The use of a fluid model (rather than a kinetic model) imposes some self-evident limitations on the results. However, also the basic assumption of electrostatic waves limits the parameter range of the validity of the results. This can be understood heuristically with reference to the inequality (17.2). If the Alfvén velocity, $U_A \equiv B/\sqrt{nM\mu_0} \equiv c\Omega_{ci}/\omega_{pi}$ where Ω_{pi} is the the ion plasma frequency and c the speed of light, happens to be comparable to or smaller than the electron thermal velocity u_{te}, the

B-parallel component of the drift waves can resonate with an Alfvén wave, and perturbations of the ambient magnetic field become important. Recalling the definition $\beta = n\kappa T/(B^2/2\mu_0)$ we rewrite the β-value of the plasma as $\beta = 2C_s^2/U_A^2$, where $C_s \equiv \sqrt{\kappa T/M}$ is the ion sound velocity. If $U_A \leq u_{te}$ we find $\beta \geq m/M$, apart from a trivial numerical factor. The electrostatic assumption thus holds for very low β-plasmas. For $\beta \geq m/M$ we have a modification of the analysis in terms of *drift Alfvén waves*; see, for instance, Kadomtsev (1965), Weiland (1992, 2000).

17.1.4 Spatially varying magnetic fields

It was important for the derivation of (17.8) that the ion $\mathbf{E} \times \mathbf{B}/B^2$-motion was incompressible. In general, this requires **B** to be a constant vector, which again implies that the β-value of the plasma must be low. If $\mathbf{B} = \mathbf{B}(\mathbf{r})$, then the ion motion is no longer incompressible and the analysis needs nontrivial modifications. In this case

$$\nabla \cdot \frac{\nabla\phi \times \mathbf{B}}{B^2} \approx \frac{2}{B^3}(\nabla\phi \times \mathbf{B}) \cdot \nabla B \neq 0.$$

We assumed that $\beta \ll 1$ so that the magnetic field can be approximated with the value it would have in the absence of plasma, implying $\nabla \times \mathbf{B} = 0$ for time stationary conditions with no currents imposed in the volume considered. Similarly we find that the diamagnetic drift is also compressible

$$\nabla \cdot \frac{\nabla p_e \times \mathbf{B}}{B^2} \approx \frac{2}{B^3}(\nabla p_e \times \mathbf{B}) \cdot \nabla B \neq 0.$$

At first sight it might seem surprising that the diamagnetic flow is compressible when $\mathbf{B} = \mathbf{B}(\mathbf{r})$, but we should recall that with inhomogeneous magnetic fields, we have true particle drifts, as discussed in Section 7.1.6, while the standard diamagnetic drift for $\mathbf{B} =$ constant has no gyro-center motions associated.

Rather general magnetic field variations can be considered (Vranješ & Jovanović 1995, Vranješ et al. 1998), when relaxing the implied assumption of $\nabla \times \mathbf{B}(\mathbf{r}) \approx 0$ also for low-β plasmas. Now also the model for $\nabla n_0(\mathbf{r})$ becomes more complicated, since we will in general assume that the equi-density contours follow magnetic flux surfaces. For steady state it will in general be required that the spatial variation of the magnetic field is the result of field aligned currents and the diamagnetic currents in plasma. This constrain is not trivial for the choice of acceptable spatial magnetic field variations.

Ⓢ **Exercise:** Take the ion continuity equation $\partial n/\partial t + \nabla \cdot (\mathbf{u}n) = 0$, and assume Boltzmann distributed electrons as in (17.6). Assume that to a good approximation we have the ion velocity $\mathbf{u} = -\nabla\phi \times \mathbf{B}/B^2$, but now with $\mathbf{B} = \mathbf{B}(\mathbf{r})$. Derive the local differential equation for ϕ. Obtain the dispersion relation for this case. Ⓢ

17.2 Simplified linear theory with cold ions: the role of ion inertia

The result of the foregoing section concerning the locality of the drift wave fluctuations seemed somewhat peculiar. This is in a certain respect a consequence of ignoring finite ion inertia as will be demonstrated in the following. Along with ion inertia we retain also a **B**-parallel ion motion. These modifications of the foregoing analysis will imply that the fluctuations become nonlocal, i.e., the dynamic equation contains derivatives of all spatial variables, in general. This generalization of the analysis is achieved at the expense of an explicit linearization of the basic equations.

The generalization of the simple results given before assumes that the basic equations are the ion continuity equation

$$\frac{\partial n}{\partial t} + \nabla \cdot (n\mathbf{u}) = 0, \tag{17.10}$$

and the momentum equation for the ions with $T_i = 0$ thereby ignoring the ion pressure term

$$M\left(\frac{\partial}{\partial t}\mathbf{u} + \mathbf{u}\cdot\nabla\mathbf{u}\right) = e(-\nabla\phi + \mathbf{u}\times\mathbf{B}). \tag{17.11}$$

This equation is rewritten in a more suitable form by taking the cross product with $\mathbf{B}$

$$\mathbf{u}_\perp = -\frac{\nabla\phi\times\mathbf{B}}{B^2} - \frac{M}{eB^2}\left[\frac{\partial}{\partial t}(\mathbf{u}\times\mathbf{B}) + \mathbf{u}\cdot\nabla(\mathbf{u}\times\mathbf{B})\right], \tag{17.12}$$

where it was explicitly used that $\mathbf{B}$ is a constant vector. The term in angular brackets is sometimes called the inertial drift and the polarization drift is derived from it. The $\mathbf{B}$-parallel component of (17.11) remains

$$M\left(\frac{\partial}{\partial t}u_\parallel + \mathbf{u}\cdot\nabla u_\parallel\right) = -e\frac{\partial}{\partial z}\phi. \tag{17.13}$$

The set of equations is completed with (17.6) for Boltzmann distributed electrons. With the assumption of quasi-neutrality $n_e \approx n_i \equiv n$ there is no distinction between ion and electron density. The ion thermal motion will, for the time being, be ignored and an ion pressure term is omitted from (17.11). According to (17.11), inertial effects are determining the $\mathbf{B}$-parallel ion velocity component, $u_\parallel$. It is logically possible to assume that the $\mathbf{B}$-perpendicular ion fluid velocity is still given by the $\mathbf{u}_{\mathbf{E}\times\mathbf{B}} = \mathbf{E}\times\mathbf{B}/B^2$-velocity. For generality we retain the effects of ion inertia also in the $\mathbf{B}$-perpendicular direction, at least to lowest order by including the corresponding lowest order correction to $\mathbf{U}_{\mathbf{E}\times\mathbf{B}}$, namely, the ion polarization drift, see also Section 7.1.5. This is achieved by iterating once the $\mathbf{B}$-perpendicular part of (17.12), to give

$$\mathbf{u}_\perp = -\frac{\nabla\phi\times\mathbf{B}}{B^2} - \frac{M}{eB^2}\left(\frac{\partial}{\partial t} - \frac{1}{B^2}\nabla\phi\times\mathbf{B}\cdot\nabla\right)\nabla_\perp\phi. \tag{17.14}$$

For consistency with the iteration scheme, both terms in the brackets must be small, i.e., $\nabla_\perp\phi$ must be slowly varying in time as well as space. The subscript $\perp$ refers to the direction perpendicular to $\mathbf{B}$. It is implicit in the derivation that ion polarization effects are to be considered as small corrections to the $\mathbf{U}_{\mathbf{E}\times\mathbf{B}}$-drifts. With the second term in (17.14) retained, the ion motion perpendicular to the magnetic field is no longer incompressible, i.e., $\nabla\cdot\mathbf{u}_\perp \neq 0$. Inserting (17.6) with the assumption of quasi-neutrality into (17.10), the plasma density can be eliminated to give

$$\frac{\partial}{\partial t}\phi + \frac{\kappa T_e}{e}\nabla\cdot\mathbf{u} + \mathbf{u}\cdot\nabla\phi + \frac{\kappa T_e}{en_0(x)}\frac{dn_0(x)}{dx}\mathbf{u}\cdot\widehat{\mathbf{x}} = 0, \tag{17.15}$$

with $\widehat{\mathbf{x}}$ being a unit vector in the x-direction, and $\mathbf{u} = \{\mathbf{u}_\perp, u_\parallel\}$ is the ion fluid velocity vector.

- **Exercise:** Take a model where Poisson's equation is retained and the ion velocity is identified by the $\mathbf{E}\times\mathbf{B}/B^2$-velocity, ignoring any ion motion along $\mathbf{B}$. Derive a nonlinear dynamic equation for the electrostatic potential by use of the ion continuity equation and the assumption of isothermal Boltzmann distribution for the electrons. Discuss the resulting equation, and estimate the relative magnitude of the various terms for different scale lengths, measured in units of the Debye length.

17.2.1 Dispersion relation

The set of nonlinear differential equations (17.13), (17.14) and (17.15) can be solved in particular cases only. In order to obtain a dispersion relation for drift waves, the equations are linearized, with stationary solutions being $\phi_0 = 0$ and $\mathbf{u}_0 = 0$. The linear equations are

$$M\frac{\partial}{\partial t}\widetilde{u}_{\parallel} = -e\frac{\partial}{\partial z}\widetilde{\phi}, \tag{17.16}$$

$$\widetilde{\mathbf{u}}_{\perp} = -\frac{\nabla_{\perp}\widetilde{\phi}\times\mathbf{B}}{B^2} - \frac{M}{eB^2}\frac{\partial}{\partial t}\nabla_{\perp}\widetilde{\phi}, \tag{17.17}$$

and

$$\frac{\partial}{\partial t}\widetilde{\phi} + \frac{\kappa T_e}{e}\left(\nabla_{\perp}\cdot\widetilde{\mathbf{u}}_{\perp} + \frac{\partial}{\partial z}\widetilde{u}_{\parallel}\right) + \frac{\kappa T_e}{en_0(x)}\frac{dn_0(x)}{dx}\widetilde{\mathbf{u}}_{\perp}\cdot\widehat{\mathbf{x}} = 0. \tag{17.18}$$

These equations can be combined to give

$$\frac{\partial^2}{\partial t^2}\left(\widetilde{\phi} - C_s^2\frac{1}{\Omega_{ci}^2}\nabla_{\perp}^2\widetilde{\phi}\right) - C_s^2\frac{\partial^2}{\partial z^2}\widetilde{\phi} - \frac{C_s^2}{\Omega_{ci}}\frac{n_0'(x)}{n_0(x)}\left(\frac{\partial^2}{\partial y\partial t}\widetilde{\phi} + \frac{1}{\Omega_{ci}}\frac{\partial^3}{\partial x\partial t^2}\widetilde{\phi}\right) = 0, \tag{17.19}$$

where we introduced the abbreviation $n_0'(x)$ for $dn_0(x)/dx$. The equation (17.19) is solved for

$$\widetilde{\phi}(x,y,z,t) = \phi_1(x)\,e^{i(k_y y + k_z z - \omega t)}. \tag{17.20}$$

The result is

$$\frac{\omega^2 C_s^2}{\Omega_{ci}^2}\left(\frac{\partial^2}{\partial x^2}\phi_1 + \frac{n_0'(x)}{n_0(x)}\frac{\partial}{\partial x}\phi_1\right) - \left(\omega^2 + \frac{\omega^2 C_s^2}{\Omega_{ci}^2}k_y^2 - C_s^2 k_z^2 + \frac{\omega C_s^2}{\Omega_{ci}}\frac{n_0'(x)}{n_0(x)}k_y\right)\phi_1 = 0. \tag{17.21}$$

Equation (17.21) is a second order differential equation for $\phi(x)$ with variable coefficients and must be solved together with some boundary conditions, for instance taking $\phi = 0$ at the boundaries $x = R_1$ and $x = R_2$. Other terms would have appeared if we had used cylindrical rather than rectilinear coordinates. The eigenfunctions of ϕ give the variation of wave amplitude with x, and the eigenvalues ω give the frequency. The frequencies are real; in the basic equations used so far there are no mechanisms which make drift waves unstable in a linearized description.

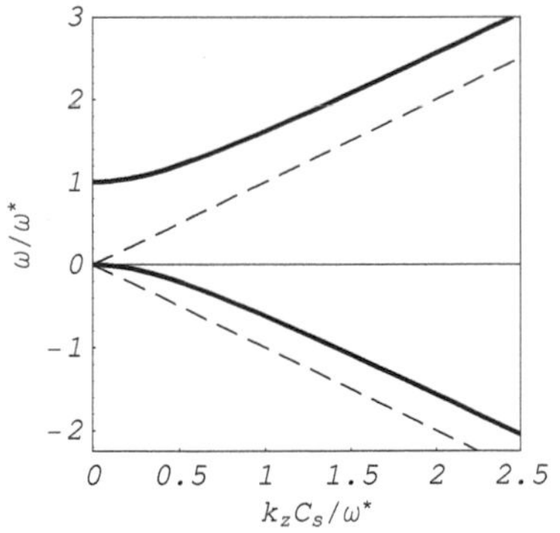

FIGURE 17.3: *Dispersion relation for electrostatic drift waves with the assumption of cold ions with $\omega^* = k_y u_{De}$ being the drift frequency. The dashed line gives the sound wave dispersion relation. To simplify the presentation, we ignored ion polarization terms, i.e., assumed $a_i = 0$.*

If n_0 is some function of x, obtained from an experiment, for instance, it is clear that the drift wave equation, of which (17.21) is a simple example, must be solved numerically for a general case. However, there are several ways to get an analytic dispersion relation. First, it is sometimes possible by a suitable choice of the function $n_0(x)$ to reduce the wave equation to a standard form, so that the solution $\phi(x)$ can be written in terms of Whittaker's functions or Hermite polynomials, etc. Second, the use of a "quadratic form" allows one to estimate $\omega(k)$ by using a trial function $\phi(x)$. Third, by a suitable transformation the wave equation can be put into the form

$$W'' + Q(x)W = 0, \tag{17.22}$$

which can be solved by the WKB approximation. The latter is, in principle, valid only if $\partial\phi/\partial x \gg \partial\phi/\partial y$. Since, in practice, the opposite is usually true, it is a better approximation to neglect $\partial\phi/\partial x$ altogether. This amounts to simply using the equation for the turning points of (17.22). In the same spirit one can neglect the derivatives of ϕ in (17.21). What remains gives an algebraic equation of $\omega(k)$, which can be interpreted as the "local" dispersion relation, i.e., local at the position x. We consequently here assume $dn_0(x)/dx/n_0(x) = d\ln n_0(x)/dx$ to be constant. This assumption is somewhat restrictive, of course, but it is not unphysical. It can be realized by an exponential density profile, $n_0(x) = N\exp(-x/L)$, where L is a scale length. The resulting dispersion relation is readily obtained from (17.21) as

$$\omega^2(1 + a_i^2 k_y^2) - \omega k_y u_{De} - C_s^2 k_z^2 = 0, \tag{17.23}$$

where $C_s = \sqrt{\kappa T_e/M}$ is the ion sound speed, u_{De} was defined in (17.9) and apart from a factor $\sqrt{2}$ the quantity $a_i = C_s/\Omega_{ci}$ can be interpreted as an effective ion Larmor radius obtained as if the ions had the temperature of the electrons. With the present exponential density profile we have $u_{De} = C_s^2/(\Omega_{ci}L)$, independent of the spatial coordinate x. The y-component of the wave-vector is perpendicular to the density gradient as well as the magnetic field. The second term in the parenthesis in (17.23) originates from the ion polarization, i.e., ion inertia, and the last term from **B**-parallel ion motion, which is again inertial. (Inertial terms are here easily recognized by explicitly containing the ion mass M.) For large k_z the dispersion relation approaches the one for ion acoustic waves. One branch, propagating in the direction of the electron diamagnetic drift, has $|\omega/k_z| > C_s$ and is called "accelerated sound;" the one propagating in the opposite direction with $|\omega/k_z| < C_s$ is the "decelerated sound" branch. In the limit of $k_z = 0$ and $a_i = 0$ the dispersion relation for the fast branch becomes identical to that obtained from (17.7). The slow branch propagates in the direction opposite to the electron diamagnetic drift. (Since ion temperature was ignored in the derivation, we have no ion diamagnetic drift.) It is customary to introduce the ion drift frequency $\omega^* = k_y u_{De}$, and the dispersion relation can then, with $a_i = 0$, be visualized in terms of the two variables ω/ω^* and $k_z C_s/\omega^*$, as shown in Fig. 17.3. Retaining ion polarization, we obtain a small dispersion of drift waves, most important in the limit of $k_z \to 0$ where $\omega = k_y u_{De}/(1 + a_i^2 k_y^2) \approx k_y u_{De}(1 - a_i^2 k_y^2)$.

17.2.2 Physical description

The results in the foregoing section can be presented in the form of a physical picture, following Chen (1965*c*). Thus Fig. 17.4 illustrates the basic elements of the derivation. A solid line gives equi-density contour of the perturbation as in Fig. 17.2 and because of (17.6) the equi-potential lines are in phase with this also here. As indicated, the ion fluid velocities deviate in general from the $\mathbf{E}\times\mathbf{B}$ velocities because of the finite ion inertia. If the electric fields were steady, an ion would drift with the local $\mathbf{E}\times\mathbf{B}$ velocity. Since, however, E_y is fluctuating the ions will experience an acceleration in the x-direction. The ion inertia opposes this acceleration and this can be expressed as an equivalent or effective gravitational acceleration $\mathbf{g} = -\partial\mathbf{u}/\partial t$. This in turn induces a drift $\mathbf{u} = (M/eB^2)\mathbf{g}\times\mathbf{B}$, giving $u_y = (M/eB)\partial u_x/\partial t = (M/eB^2)\partial E_y/\partial t \to -i(\omega/\Omega_{ci})E_y/B$. At position ② in Fig. 17.4 the electric field E_y is maximum and hence $\partial E_y/\partial t = 0$ and u_y then vanishes also. At a position ①, on the other hand, u_x vanishes but not u_y and the density perturbation differs from the case where ion inertia is neglected. Note that also in this case the average value of the ion cross-field velocity in the region of density increase (shaded region) is zero, just as it is in the region of density depletion.

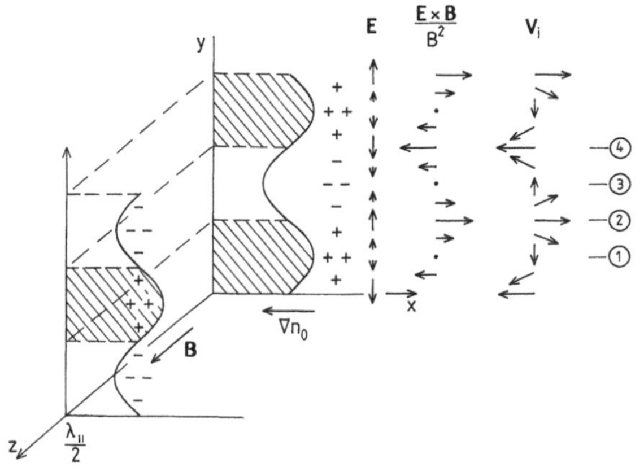

FIGURE 17.4: *Basic elements of drift wave properties with ion inertia included. The solid line indicates an equi-density contour, while electric fields and ion velocities are indicated by small arrows. Charge densities are indicated by + and − symbols. Regions with enhanced density are shaded for clarity.*

17.2.3 The electron velocity

Note that the *electron* fluid velocity was nowhere considered in the discussions of the simplified linear theory. The z-component (17.4) of the electron moment equation gives the Boltzmann distributed electrons. To the present accuracy, the **B**-perpendicular component is

$$-\kappa T_e \nabla_\perp n + en\nabla_\perp \phi - en\mathbf{u}_{e\perp} \times \mathbf{B} \approx 0, \tag{17.24}$$

in the limit of negligible electron inertia. To obtain an unperturbed solution, we assume that there is no electric field in steady state and obtain the usual gradient or diamagnetic drift $\mathbf{u}_{e\perp 0} = (\kappa T_e/e n_0 B^2)\nabla n_0 \times \mathbf{B}$.

Considering fluctuating quantities, we find by insertion of Boltzmann distributed electrons in the linearized version of (17.24) that this approximation gives identically $\widetilde{\mathbf{u}}_{e\perp} = 0$. Although this result may appear strange, it has a simple physical interpretation; the fluctuating electron fluid velocity in the **B**-perpendicular direction has two contributions, the electron $\widetilde{\mathbf{E}} \times \mathbf{B}$-drift and the gradient diamagnetic drift associated with the density variation of the wave. With Boltzmann distributed electrons, these two contributions happen to cancel exactly. The two velocities are, however, of different nature; the diamagnetic drift is, as already mentioned, not a particle drift in contrast to the $\widetilde{\mathbf{E}} \times \mathbf{B}$-drift. With $\widetilde{\mathbf{u}}_{e\perp} = 0$ we have from the electron equation of continuity that $\widetilde{u}_\parallel = (\omega/k_\parallel)e\phi/\kappa T_e$.

The cancellation which implies $\widetilde{\mathbf{u}}_{e\perp} = 0$ in the summary before is not general: if we assume an anisotropic electron pressure tensor with elements $p_\perp = \kappa T_{e\perp} n$ and $p_\parallel = \kappa T_{e\parallel} n$, we have that the **B**-parallel component of the momentum equation (17.4) gives the Boltzmann distribution $n = n_0(x)\exp(e\phi/\kappa T_{e\parallel})$ as in (17.6), while the general form of (17.24) with $e\phi = \kappa T_{e\parallel}\ln(n/n_0)$ gives

$$\begin{aligned}
\mathbf{u}_{e\perp} &= \frac{\kappa T_{e\perp}}{enB^2}\nabla_\perp n \times \mathbf{B} - \frac{\kappa T_{e\parallel}}{eB^2}\frac{n_0(x)}{n}\nabla_\perp \frac{n}{n_0(x)} \times \mathbf{B} \\
&= \frac{\kappa T_{e\perp}}{enB^2}\nabla_\perp n \times \mathbf{B} - \frac{\kappa T_{e\parallel}}{eB^2}\frac{1}{n}\left(\nabla_\perp n - \frac{n}{n_0(x)}\nabla_\perp n_0(x)\right) \times \mathbf{B} \\
&= \frac{\kappa(T_{e\perp} - T_{e\parallel})}{eB^2 n}\nabla_\perp n \times \mathbf{B} + \frac{\kappa T_{e\parallel}}{eB^2 n_0(x)}\nabla_\perp n_0(x) \times \mathbf{B},
\end{aligned} \tag{17.25}$$

where $n = n_0(x) + \widetilde{n}$. The unperturbed diamagnetic velocity obtained for $\widetilde{n} = 0$ is the same as before, but this time derived from $T_{e\perp}$. The perturbed velocity is now

$$\widetilde{\mathbf{u}}_{e\perp} = \frac{\kappa\left(T_{e\perp} - T_{e\parallel}\right)}{eB^2} \nabla_\perp \frac{\widetilde{n}}{n_0(x)} \times \mathbf{B}, \tag{17.26}$$

where $\widetilde{\mathbf{u}}_{e\perp}$ can take either sign, depending on the ratio $T_{e\perp}/T_{e\parallel}$.

For collisional plasmas we expect $T_{e\perp} = T_{e\parallel}$, and the discussion here is irrelevant, but for drift waves in collisionless plasmas we might have $T_{e\perp} \neq T_{e\parallel}$.

17.2.4 Divergence-free currents

Taking the difference between the electron and ion continuity equations in the quasi-neutral limit, $n_e \approx n_i \equiv n$, we find that the fluctuating plasma current is divergence free, $\nabla \cdot \mathbf{j} = 0$. Taking first the simple case where the ion velocity along $\mathbf{B}$ is ignored and the perpendicular part is $-\nabla\phi \times \mathbf{B}/B^2$, we find the ion current contribution to the divergence of the linearized plasma current to be $-e\nabla_\perp \cdot (n_0(x)\nabla\phi \times \mathbf{B}/B^2) = -e(n_0'(x)/B)\partial\phi/\partial y$. As mentioned before when discussing equation (17.24), the electrons will not contribute to the $\mathbf{B}$-perpendicular part, since this part of their fluid velocity vanishes to the present approximation. The component of the electron velocity *along* $\mathbf{B}$ can be obtained from the linearized electron continuity equation $\partial\widetilde{n}/\partial t + n_0(x)\partial\widetilde{u}_{ez}/\partial z = 0$, giving $\widetilde{u}_{ez} = (\omega/k_z)\widetilde{n}/n_0(x)$. (The fact that the electron velocity does not enter the electron force balance does *not* imply that this velocity vanishes.) To the present approximation, only the electrons contribute to $\partial j_z/\partial z$. You can, if you like, consider j_z as the *return current* driven by $\mathbf{j}_\perp$. The foregoing argument gives $\partial j_z/\partial z = -i\omega e\widetilde{n}$ by the electron continuity equation. With Boltzmann distributed electrons, $\widetilde{n}/n_0 = e\phi/\kappa T_e$, we readily find that the criterion for $\nabla_\perp \cdot \mathbf{j}_\perp = -\partial j_z/\partial z$ becomes $\omega = \omega^*$. When the ion polarization term is included, the additional contribution to the $\mathbf{B}$-perpendicular component of net linearized current becomes $-(Mn_0(x)/B^2)\partial\nabla_\perp\phi/\partial t$. The requirement of divergence-free fluctuating plasma currents then implies a shift in frequency, i.e., a slight deviation from ω^*. As also evident from Fig. 17.4, the ion polarization drift gives rise to accumulation of charge, inducing a (small) $\mathbf{B}$-parallel additional plasma current contributing to j_z.

The modifications introduced by anisotropic electron temperatures giving the result (17.26) do not effect the conclusions in this section, since $\nabla \cdot \widetilde{\mathbf{u}}_{e\perp} = 0$ also in that case.

17.3 Drift wave instability

The basic analysis of electrostatic drift waves in Section 17.1 demonstrated that it is relatively easy to derive a dispersion relation for these waves. The analysis gave, however, that these waves are stable, implying that they will not be observed unless they are excited by some external means. Also, it was found that the dispersion relation had *two* branches, a fast one and a slow one. It requires a much more elaborate analysis to demonstrate that one of these wave branches is unstable, while the other one will be damped under standard conditions. This detailed analysis will also give rise to some modifications of the real part of the dispersion relation, making the simplified results of illustrative value only. It turns out that the imaginary part of ω originates from a phase difference between fluctuations in density and potential. As long as (17.6) obtains, ω will be real and equal to ku_{De}. We can model the effect of a "phase shift" by first postulating, for the sake of illustration, a phase change between ϕ and n as

$$\frac{n}{n_0} = (1 - i\Delta)\frac{e\phi}{\kappa T_e}. \tag{17.27}$$

From the simplest description of the drift wave dynamics we have the relation $n/n_0 = (\omega^*/\omega)e\phi/\kappa T_e$, which here ultimately implies $\omega \approx \omega^*(1+i\Delta)$. The plasma will be unstable for $\Delta > 0$ and stable for $\Delta < 0$. The purpose of a detailed study will therefore be to account for the physical mechanisms which give rise to the phase difference between density and electrostatic potential.

One of the basic assumptions made in the simple drift wave model is the hypothesis of the electrons being Boltzmann distributed at all times. For this case the plasma was found to remain stable, so a study of *unstable* systems has to address this assumption first. The physical mechanism that allows the model $n/n_0 \approx e\phi/\kappa T_e$ is the high mobility of electrons along magnetic field lines, flowing from wave crest to wave trough. In order to have an instability, k_z must be so small that electrons *cannot* flow unimpeded from wave crest to trough along the magnetic field **B**. Then (17.6) will be violated, and it will be possible to have E_y and E_x different from what is rigorously prescribed by (17.6). There are several mechanisms which can affect the **B**-parallel mobility of the electrons:

a) Electron-neutral collisions. This effect is dominant in weakly ionized gases, such as in laboratory experimental conditions in a positive column, Q-machines with inlet of neutral gas background, or in the ionospheric E-region and the lower parts of the F-region.

b) Electron-ion collisions. This effect is the finite resistivity of a fully ionized plasma, and is the mechanism we shall consider in detail.

c) Landau damping. This effect has to do with a distortion of the electron distribution function due to the interaction of the wave of electrons with $u_z \approx \omega/k_z$. This is the mechanism responsible for drift instabilities in collisionless plasmas.

d) Inductance. This effect is the impedance to the electron flow provided by the magnetic field the flow creates. This effect is important only at high $\beta \equiv n\kappa T/(B^2/2\mu_0)$, when $\omega/k_z \approx U_A$, the Alfvén speed.

e) Electron inertia. This effect might be expected to be the dominant mechanism in the absence of collisions, but it turns out that the effect has the wrong phase and does not cause an instability except in higher order.

These mechanisms can be considered to "trigger" the instability by allowing (17.6) to be violated, so that charge separations can build up (Chen 1970). Since b) is the dominant mechanism in, for instance, Q-machine experiments and in the lower part of the ionosphere, we shall derive the dispersion relation for that case. It turns out that the results will apply qualitatively for the case a) as well.

17.4 Resistive drift waves with $T_i = 0$

The simplest description of the drift wave instability can be found from the fluid equations when the resistivity ξ is finite and T_i is zero. In this case, there are no finite Larmor radius effects since $r_L = 0$, and finite resistivity together with ion inertia causes the instability. The geometry of the problem is still the one illustrated in Fig. 17.1.

17.4.1 Basic assumptions

As already mentioned, the following analysis will rely on a fluid model. This is an approximate description, valid only for collisional plasmas, although some of the results will remain qualitatively correct also for collision*less* plasmas. A number of additional simplifying assumptions are listed in the following.

a) Cold ions $T_i \approx 0$, and warm electrons $T_e > 0$

b) $\xi \neq 0$ (finite resistivity caused by electron-ion collisions with frequency ν_{ei})

c) $\nabla \times \mathbf{E} = 0$, $\mathbf{B} =$ const. (low-β, electrostatic assumption)

d) $m/M \simeq 0$ (electron inertia is ignored)

e) $n_i = n_e \equiv n$ (quasi-neutrality, or "the plasma approximation" valid for $k\lambda_D \ll 1$ and $\omega \ll \Omega_{pi}$, the ion plasma frequency, see Section 11.2)

f) $u_{ti} \ll \omega/k_z \ll u_{te}$ in terms of ion and electron thermal velocities, noting that assumption a) is consistent by implying $u_{ti} \approx 0$

g) $\nabla p_e = \kappa T_e \nabla n$ (isothermal electrons, argued by the high electron thermal conductivity)

17.4.2 Basic equations

The basic equations for the analysis can be listed as follows.

$$Mn\left(\frac{\partial \mathbf{u}_i}{\partial t} + \mathbf{u}_i \cdot \nabla \mathbf{u}_i\right) = en\left(-\nabla\phi + \mathbf{u}_i \times \mathbf{B}\right) - n^2 e^2 \xi (\mathbf{u}_i - \mathbf{u}_e) \tag{17.28}$$

and

$$0 = -en(-\nabla\phi + \mathbf{u}_e \times \mathbf{B}) - \kappa T_e \nabla n - n^2 e^2 \xi (\mathbf{u}_e - \mathbf{u}_i) \tag{17.29}$$

are the momentum equations for ions and electrons with simplifications listed before. In retaining the electron collision term on the right-hand side of (17.29) we implicitly assume $\omega \ll \nu_{ei}$. The resistivity ξ is related to the electron-ion collision frequency ν_{ei}, i.e., the electron-ion momentum relaxation time

$$\xi = m\nu_{ei}/ne^2 \tag{17.30}$$

where

$$\nu_{ei} = \frac{ne^4}{16\pi m^2 \varepsilon_0^2}\left(\frac{m}{\kappa T_e}\right)^{3/2} \ln(\Lambda) \tag{17.31}$$

with $\ln(\Lambda)$ being the *Spitzer logarithm*, where $\Lambda = 12\pi n \lambda_D^3$. The expression (17.31) assumes singly charged ions. The resistive drags on the electrons will be important mostly for the z-component of their motion which for $\xi = 0$ mediates the balance between electron pressure and $-e\mathbf{E}$-forces giving the Boltzmann equilibrium. In the following, the resistivity will be retained only for the $\mathbf{B}$-parallel component of the electron momentum equation (17.29).

We also have the continuity equations for the ions and electrons

$$\frac{\partial n}{\partial t} + \nabla \cdot (n\mathbf{u}_i) = 0, \tag{17.32}$$

$$\frac{\partial n}{\partial t} + \nabla \cdot (n\mathbf{u}_e) = 0. \tag{17.33}$$

With the quasi-neutrality assumption there is no distinction between ion and electron densities in these equations so that (17.32)-(17.33) together give $\nabla \cdot (n(\mathbf{u}_i - \mathbf{u}_e)) = 0$, i.e., $\nabla \cdot \mathbf{j} = 0$.

The assumption of isothermal fluctuations is not crucial for the analysis since Boltzmann's constant occurs only in the ∇p_e term; an adiabatic equation of state will give the same result as these equations if κ is replaced by $\frac{5}{3}\kappa$ in the final answer.

The full set of nonlinear equations (17.28)–(17.33) cannot be solved analytically. The dispersion relation for the fluctuations can, however, be obtained by a standard linearization procedure and the stability of the waves subsequently analyzed.

17.4.3 Equilibrium

We assume first $u_{0z} = 0$ and $n_0 = n_0(x)$ in the unperturbed state. Since electron-ion collisions do not give rise to any charge separation, i.e., there are no ambipolar electric fields. The assumption of $E_0 = 0$ is then justified also, see Sections 8.8.1 and 11.3.1. Setting $\partial/\partial t = 0$, we obtain

$$Mn_0(\mathbf{u}_{i0} \cdot \nabla \mathbf{u}_{i0}) = en_0 \mathbf{u}_{i0} \times \mathbf{B} - n_0^2 e^2 \xi (\mathbf{u}_{i0} - \mathbf{u}_{e0}) \tag{17.34}$$

$$0 = -en_0 \mathbf{u}_{e0} \times \mathbf{B} - \kappa T_e \nabla n_0 - n_0^2 e^2 \xi (\mathbf{u}_{e0} - \mathbf{u}_{i0}) \,. \tag{17.35}$$

Neglecting $\mathbf{u}_{i0} \cdot \nabla \mathbf{u}_{i0}$ for the time being, adding (17.35) to (17.34), and taking the cross product with $\mathbf{B}$, we obtain

$$\mathbf{J}_0 = \kappa T_e \frac{\mathbf{B} \times \nabla n_0}{B^2} \,, \tag{17.36}$$

where $\mathbf{J}_0 \equiv en_0(\mathbf{u}_{i0} - \mathbf{u}_{e0})$ is the diamagnetic current density. The y-components of (17.34) and (17.35) yield $(\mathbf{u}_{i0} \times \mathbf{B})_y = (\mathbf{u}_{e0} \times \mathbf{B})_y = \xi j_0$. Using (17.36), we find

$$u_{i0x} = u_{e0x} = -\frac{\xi \kappa T_e}{B^2} n_0'(x) \,. \tag{17.37}$$

This is the classical diffusion velocity due to electron-ion Coulomb collisions (being in the radial direction, for cylindrical geometry). There is no current associated with this classical diffusion since electrons and ions move with the same velocity in the same direction. The assumption of $n_0(x)$ being time-stationary is thus in variance with the result: the diffusion flattens out the density profile for increasing times. Since this diffusion is assumed to be a slow process compared to the wave frequencies we are considering, we shall however ignore u_{0x}. In effect we are assuming that $\partial n_0/\partial t \ll \omega^* n_0$, where ω^* is the characteristic drift frequency, and consider $n_0(x)$ constant in time. In the following we are thus linearizing the equations around something which is not really an *exact* equilibrium solution for the initial equations (this is a trick often met in similar investigations). The procedure will be justified a posteriori by demonstrating that the drift waves can have characteristic time scales which are much shorter than the time it takes the density gradient to decay by classical diffusion alone.

The term $\mathbf{u}_{i0} \cdot \nabla \mathbf{u}_{i0}$ automatically vanishes to the desired accuracy when we set $u_{i0x} = 0$, since u_{i0y} can depend on the coordinate x only because of the assumed homogeneity in the y-direction; see Fig. 17.1. The initial assumption for obtaining (17.36) is fulfilled to the desired accuracy. The x-components of (17.34) and (17.35) give

$$u_{i0y} = 0 \,, \qquad u_{e0y} = -\frac{\kappa T}{eB} \frac{n_0'(x)}{n_0} \equiv u_{De} \,. \tag{17.38}$$

In equilibrium, there is only a diamagnetic drift of the electrons in the direction $\perp \mathbf{B}$ and $\perp \nabla n$. For a cylindrical magnetized plasma, this would correspond to the azimuthal direction. It should be emphasized here that the analytical expression results from our assumption of uniform electron temperatures which applies for many relevant physical conditions. More generally it should be recalled that the driving mechanisms for drift wave instabilities are gradients in *pressure*, $\nabla(n_0(x)T_0(x))$, so the steady state equations will generally contain also terms with $T_0'(x)$; see also Fig. 11.1.

17.5 Perturbation

Classical diffusion across magnetic field lines is ignored by keeping ξ only in the z-component of the electron equation of motion. Using a propagating plane wave $e^{i(k_y y + k_z z - \omega t)}$ as a test function (corresponding to Fourier transforming the expressions) we find

$$\left\{\widetilde{\mathbf{u}}(x,y,z,t),\widetilde{\phi}(x,y,z,t),\widetilde{n}(x,y,z,t)\right\} = \left\{\mathbf{u}_1(x),\phi_1(x),\eta(x)n_0(x)\right\} e^{i(k_y y+k_z z-\omega t)}, \qquad (17.39)$$

so that the relative density variation is $\eta = (\widetilde{n}/n_0)e^{-i(k_y y+k_z z-\omega t)}$ at any position x. This combination may not appear entirely obvious at first, but it turns out to be most convenient.

Linearizing the x-component of the electron momentum equation (17.29), we have

$$0 = en_0(-\phi' + u_{ey}B) + \kappa T_e(n_0\eta' + \eta n_0') + en_0\eta u_{0e}B, \qquad (17.40)$$

again with a "prime" denoting differentiation with respect to the spatial variable x.

The last two terms in (17.40) cancel by virtue of (17.38). Defining $\chi \equiv e\phi/\kappa T_e$ we then have

$$u_{ey} = \frac{\kappa T_e}{eB}(\chi - \eta)'. \qquad (17.41)$$

Similarly, the y and z components of (17.29) give

$$iu_{ex} = k_y\frac{\kappa T_e}{eB}(\chi - \eta) \qquad (17.42)$$

$$u_{ez} = ik_z\frac{\kappa T_e}{eB}\omega_{ce}\tau_{ei}(\chi - \eta), \qquad (17.43)$$

where $\omega_{ce}\tau_{ei} = eB/(m\nu_{ei}) = B/(n_0e\xi)$ is an important dimensionless quantity characterizing the resistivity. The velocity components u_{ex} and u_{ey} vanish for Boltzmann distributed electrons, $\chi = \eta$, as expected. At first sight it might be surprising to see the electron cyclotron frequency ω_{ce} appear: this quantity explicitly contains the electron mass m, and we have in the beginning assumed the electrons to be essentially massless particles! Note, however, that m appears explicitly in the plasma conductivity ξ, and ω_{ce} is introduced by "juggling around" with ξ.

The z-parallel electron dynamics is retained not only because of the large electron thermal velocity for finite electron temperatures; what is important here is that even a minute **B**-parallel electric field component will give the light electrons a large fluid velocity. The ions, on the other hand, will in comparison hardly move at all along **B** due to their large inertia. The **B**-perpendicular dynamics is quite different; here the ions move in the lowest approximation with the $\mathbf{E}\times\mathbf{B}$-velocity, independent of mass.

Linearizing and Fourier transforming the electron continuity equation (17.33), we have

$$(\omega - k_y u_{De})n_1 + in_0(u_{ex}' + ik_y u_{ey} + ik_z u_{ez}) + iu_{ex}n_0' = 0. \qquad (17.44)$$

Substituting for u_{ex}, u_{ey} and u_{ez} from (17.41)–(17.43), we find that the x-derivatives cancel out (i.e., the ones referring to the coordinate parallel with ∇n_0). This is a fortunate feature of the fluid equations for massless particles. Finally we have an equation which does not depend on the "shape" of the perturbation in the x-direction:

$$(\omega - k_y u_{De})\eta - \left(k_y u_{De} + ik_z^2\frac{\kappa T_e}{eB}\Omega_{ci}\tau_{ei}\right)(\chi - \eta) = 0. \qquad (17.45)$$

To simplify the notation we now introduce the following normalizations in agreement with most of the literature on drift waves.

$$\begin{aligned}
\omega^* &= k_y u_{De} = -k_y\frac{\kappa T_e}{eB}\frac{n_0'}{n_0} \equiv -k_y\frac{\kappa T_e}{M\Omega_{ci}}\frac{n_0'}{n_0}\\
\sigma_\parallel &= \frac{k_z^2}{k_y^2}(\omega_{ce}\tau_{ei})\Omega_{ci}, \quad a_i^2 = \frac{\kappa T_e}{M\Omega_{ci}^2}\\
b &= k_y^2 a_i^2, \quad b\sigma_\parallel = k_z^2\frac{\kappa T_e}{eB}(\omega_{ce}\tau_{ei}),
\end{aligned} \qquad (17.46)$$

where $\omega_{ce}\tau_{ei}$ is an important dimensionless quantity characterizing the resistivity. The quantity $a_i\sqrt{2}$ is, as before, the ion Larmor radius obtained by using the electron temperature, and b is a common parameter in small Larmor radius expansions. With this notation (17.45) becomes

$$\eta = \chi \frac{\omega^* + ib\sigma_\parallel}{\omega + ib\sigma_\parallel}. \tag{17.47}$$

This relation between the density and potential perturbations is a particularly useful one because it is derived solely on the basis of the electron equations and it must be obeyed regardless of the complications for the ion dynamics to be introduced in the following sections. (One exception, which is immaterial in the present context, is the case where there is no density gradient, $\omega^* = 0$, and infinite resistivity, $b\sigma_\parallel = 0$. In this case the relation between η and χ is undetermined because of the idealizations in the basic assumptions.) The electrostatic potential χ determined through (17.47) has no explicit dependence on the wave-number component $\mathbf{k}_\perp$, i.e., it enters only through the dispersion relation $\omega = \omega(\mathbf{k})$ to be derived in the following. In the limit $\sigma_\parallel \to \infty$ the relation (17.47) becomes $\eta = \chi$, which is simply equivalent to the linearized version of the Boltzmann relation (17.6). Finite $b\sigma_\parallel$ introduces the phase shift between η and χ which is necessary for instability.

Linearizing (17.28) and setting $\mathbf{u}_{i0} = 0$, we have for the ions

$$-i\omega \mathbf{u}_i = \frac{e}{M}(-\nabla\phi + \mathbf{u}_i \times \mathbf{B}). \tag{17.48}$$

The momentum exchange between electrons and ions in the z-direction is omitted in (17.48), since it will be immaterial due to the assumptions given in the following where the ions are taken to move predominantly in the $\perp$ **B**-direction. As a consequence the collision term is retained only in the electron dynamics which imply that the momentum exchange balance is not exactly fulfilled. This assumption would seemingly also imply that the analysis is similar to the one we would obtain by starting out with electron-neutral collisions and retaining a collisionless description for the ions. For most of the analysis it is indeed so, but one nontrivial difference is associated with the different diffusion rate for electrons and ions when collisions with neutrals is dominating; as a result, a dc electric filed builds up, in such a way that the net flux of electrons and ions becomes the same, see Section 11.3.2. Such a dc electric field is omitted in the present analysis.

For low frequencies $\omega \ll \Omega_{ci} \equiv eB/M$, the iterated solution of (17.48) containing the ion polarization drifts becomes

$$iu_{ix} = \frac{k_y\phi}{B} - \frac{\omega}{\Omega_{ci}}\frac{\phi'}{B}, \qquad u_{iy} = \frac{\phi'}{B} - \frac{\omega}{\Omega_{ci}}\frac{k_y\phi}{B}, \qquad u_{iz} = \frac{ek_z}{M\omega}\phi \approx 0, \tag{17.49}$$

where a prime $'$ denotes $\partial/\partial x$ here and in the following. In the last equation of (17.49) we have indicated that we shall consider k_z's so small that u_{iz} can be neglected; this amounts to neglecting the transition from drift waves to ion acoustic waves at large k_z/k_y. The first term in u_{ix} and u_{iy} is simply the $\mathbf{E}\times\mathbf{B}$ drift, and the second term is the ion inertia effect mentioned earlier, where only the lowest order correction is retained, just as in Section 7.1.5.

Our last equation is the linearized form of the ion continuity equation (17.32). After a Fourier transform (as before) we find

$$-i\omega n_1 + n_0(u'_{ix} + ik_y u_{iy}) + u_{ix}n'_0 = 0. \tag{17.50}$$

Substituting for u_i from (17.49) and for n_1 from (17.47), we obtain the dispersion equation

$$\phi''(x) + \frac{n'_0(x)}{n_0(x)}\phi'(x) - \left(k_y^2 + \frac{\Omega_{ci}}{\omega}\frac{n'_0(x)}{n_0(x)}k_y + \frac{\omega^* + ib\sigma_\parallel}{\omega + ib\sigma_\parallel}\frac{\Omega_{ci}^2}{C_s^2}\right)\phi(x) = 0. \tag{17.51}$$

The equation (17.51) reproduces (17.21) for $k_z \approx 0$ in the limit $\sigma_\parallel \to \infty$, as it should.

As a final test we can estimate the relative magnitude between the ion fluid velocity *along* and *perpendicular* to the magnetic field for the low frequency case relevant here, for uniform plasma densities and cold ions. We take as an estimate $\mathbf{u}_\perp \approx \mathbf{E} \times \mathbf{B}/B^2$, while $u_z \approx i(e/M)E_z/\omega$. We find

$$\frac{|\mathbf{u}_\perp|}{u_z} \approx \frac{E_\perp}{E_z} \frac{\omega}{\Omega_{ci}}.$$

It is justified to ignore the **B**-parallel ion motion, provided $E_z/E_\perp = k_z/k_\perp \ll \omega/\Omega_{ci} \ll 1$, i.e., for electrostatic waves propagating almost perpendicular to **B**. We note that in the foregoing estimate we cannot simply replace M by m to get the corresponding result for the electrons, since for that case the electron pressure is essential for the **B**-parallel motion.

17.5.1 Dispersion relation

Equation (17.51) is a complex second order differential equation for $\phi(x)$ with variable coefficients and must be solved together with some boundary conditions – typically $\phi = 0$ at $x = 0$ and $x = R$. Other terms would have appeared had we used cylindrical rather than rectilinear coordinates. The solution of such an equation is the principal problem in drift wave theory. The eigenfunctions of ϕ give the variation of wave amplitude with x, and the complex eigenvalues of ω give the frequency and growth rate. We readily note one important consequence of introducing ion inertia: the eigenfunction $\phi(x)$ is now an explicit function of x. One of the peculiarities of the simple model (17.7) was that the potential variations at one x-position were independent of what happened elsewhere.

If n_0 is some a priori known function of x, obtained from an experiment, for instance, the equation (17.51) can in general only be solved numerically. However, there are several ways to get an approximate analytic dispersion relation just as in the case of (17.21). It is sometimes possible also for complex ω to reduce the wave equation to a standard form by a suitable choice of the function $n_0(x)$, so that the solution $\phi(x)$ can be written in terms of Whittaker's functions or Hermite polynomials, etc. Alternatively, the use of a "quadratic form" allows one to estimate $\omega(k)$ by using a trial function $\phi(x)$. Third, by a suitable transformation the equation can be put into a form which can again be solved by the WKB approximation. The latter approach is, in principle, valid only if $\partial\phi/\partial x \gg \partial\phi/\partial y$. Since, in practice, the opposite is usually true, it is also here a better approximation to neglect $\partial/\partial x$ altogether and simply use the equation for the turning points. In this same spirit one can neglect the x-derivatives of ϕ in (17.51). What remains gives an algebraic equation of $\omega(k)$, which is called the "local" dispersion relation, i.e., local at, and in the vicinity of, the position x in $n_0(x)$. This is the simplest approximation and is almost universally employed also for the linearly unstable case considered here (Chen 1970).

If we neglect ϕ'' and ϕ' in (17.51) and multiply through by $a_i^2\omega(\omega + ib\sigma_\parallel)$, we obtain the desired dispersion relation for $T_i = 0$

$$\omega^2 + i\sigma_\parallel(\omega(1+b) - \omega^*) = 0. \tag{17.52}$$

In the limit of large $\sigma_\parallel/\omega^*$ and small b it is clear that $\omega \approx \omega^*$. Solving for $\omega - \omega^*$, we obtain an expression for the growth rate in this limit

$$\omega - \omega^* = -b\omega^* + i\frac{\omega^{*2}}{\sigma_\parallel}, \tag{17.53}$$

where we have substituted ω^* for ω on the right-hand side. Evidently, we have $\Re\{\omega\} \approx \omega^*(1-b)$ and $\Im\{\omega\} \approx \omega^{*2}/\sigma_\parallel$. Note that the growth rate, $\Im\{\omega\}$, is proportional to the resistivity. The other root of (17.52) is heavily damped in this limit and is unimportant. In the older literature drift waves would be called "over-stable", since both the real and imaginary parts of the frequency are nonzero. The term "instability" was then reserved for exponentially and aperiodically growing modes. This distinction has disappeared from the modern literature on plasma physics.

The wave-number variation of the real and imaginary parts of the unstable root of (17.52) is shown in the diagram in Fig. 17.5 for $b = 0$. Due to the fortunate combination of parameters, it is possible to present the dispersion relation by a single curve, although the frequency ω is a function of two variables, k_z and k_y. In reality k_y is often determined by cylindrical geometry selecting some mode numbers, and typically we have $\lambda_y \sim 2\pi R$, implying $k_y \sim 1/R$, with R being the characteristic radius of the plasma column. Often we understand k_y as a fixed wave-number, selected by the geometry of the experimental set-up, and assume k_z to be the variable wave-number component. The range of k_z-variations is in general limited by the finite length of a plasma column. Here the boundary conditions at the ends, sheath conditions in particular, are important. Inclusion of the u_{zi} terms would have made $\Re\{\omega\}$ approach the line C_s (acoustic velocity) at large k_z/k_y. The results are valid in the limit where $\omega/k_z \ll V_A$ and $\omega/k_z \ll u_{te}$ in terms of the Alfvén velocity, V_A, and the electron thermal velocity u_{te}. For the low-β plasmas considered here, the Alfvén velocity is much larger than the acoustic velocity.

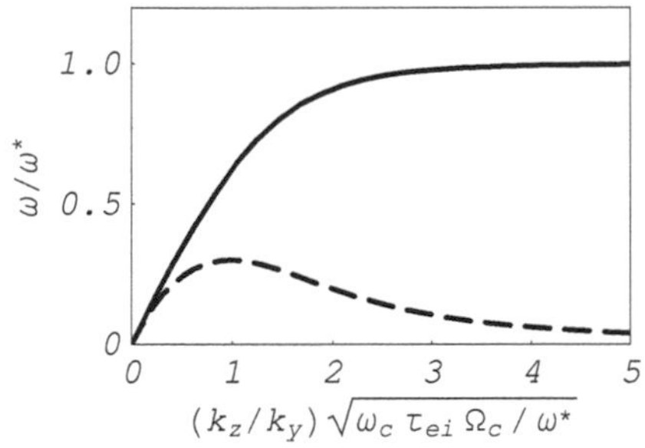

FIGURE 17.5: *Dispersion relation for resistive electrostatic drift waves with the assumption of cold ions and* $b = 0$, *with* $\omega^* = k_y u_{De}$ *being the drift frequency and* $\sigma_\parallel = (k_z/k_y)^2 \omega_{ce}\tau_{ei}\Omega_{ci}$. *The solid line gives the real part and the dashed line the imaginary part of* ω/ω^*. *The linear growth rate is maximum for* $k_y/k_z \approx \sqrt{\omega_{ce}\tau_{ei}\Omega_{ci}/\omega^*}$.

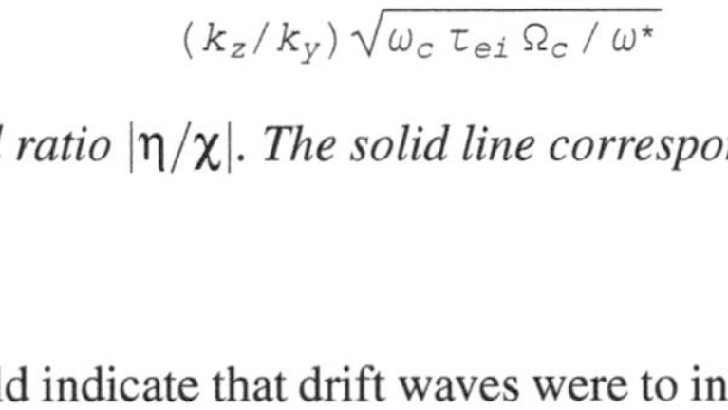

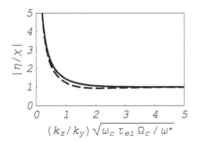

FIGURE 17.6: *The normalized ratio* $|\eta/\chi|$. *The solid line corresponds to* $b = 0$ *and the dashed line to* $b = 0.25$.

Inspection of Fig. 17.5 could indicate that drift waves were to increase in amplitude within a few periods of the fundamental frequency ω^*. This would be correct only if we could easily realize the condition for maximum growth rate, $k_y/k_z \approx \sqrt{\omega_{ce}\tau_{ei}\Omega_{ci}/\omega^*}$. In reality, the limitations of k_z due to the finite length of experiments is a nontrivial restriction. Usually we need the asymptotic variation $\Im\{\omega/\omega^*\} \approx \omega^*/\sigma_\parallel = \omega^*(k_y/k_z)^2/(\omega_{ce}\tau_{ei}\Omega_{ci})$ where $\Re\{\omega\} \approx \omega^*$. The growth rate of resistive drift waves is often found to be moderate or small.

In the limit of vanishing resistivity ($\sigma_\parallel \to \infty$ in (17.52)) we obtain $\omega = \omega^*/(1+b)$, i.e., the drift waves are marginally stable as in Section 17.1. In other words, if the electron flow along magnetic

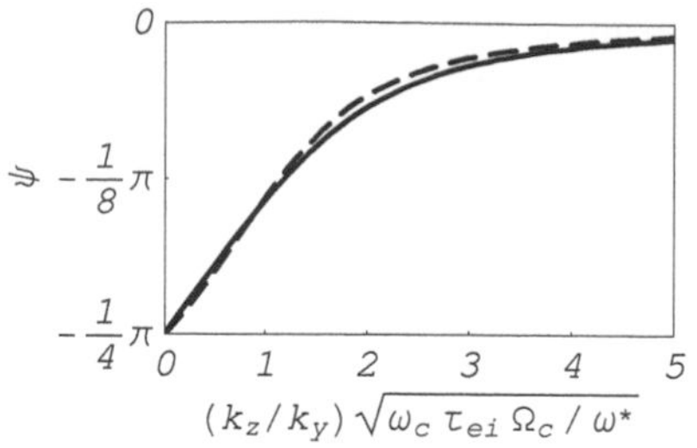

FIGURE 17.7: *The phase angle, ψ, between η and χ. The solid line corresponds to* $b = 0$ *and the dashed line to* $b = 0.25$.

field lines is unimpeded so the electrons can maintain an isothermal Boltzmann equilibrium, then the waves remain stable, irrespective of the inclusion of ion inertia in the equations for the ion dynamics, in agreement with the results of Section 17.2.1. The electron collisional motion along **B** has two effects: 1) it provides the phase difference between plasma density and potential necessary for instability, 2) it slows down the electron motion along **B** which is a stabilizing effect (Hendel et al. 1968).

17.5.2 Amplitude and phase relations

By inserting ω from (17.53) into (17.47) we find that $|\chi| < |\eta|$ in general, i.e., the potential drop between wave top and wave crest is decreased, for a given density perturbation, when the electron flow along magnetic field lines is inhibited by collisions. This was, after all, to be expected. In Fig. 17.6 we illustrate the variation of the ratio $|\eta/\chi|$ and in Fig. 17.7 the corresponding phase angle Ψ between η and χ as a function of $\sqrt{\sigma_{\parallel}/\omega^*} = (k_z/k_y)\sqrt{\omega_{ce}\Omega_{ci}\tau_{ei}/\omega^*}$. We have

$$\Psi \equiv \text{ArcTan}\left(\Im\left\{\frac{\omega^* + ib\sigma_{\parallel}}{\omega + ib\sigma_{\parallel}}\right\} \Big/ \Re\left\{\frac{\omega^* + ib\sigma_{\parallel}}{\omega + ib\sigma_{\parallel}}\right\}\right).$$

For short **B**-parallel wavelengths we have $|\eta| \sim |\chi|$ and $\Psi \sim 0$, and the electrons are almost in local Boltzmann equilibrium. As $\lambda_{\parallel}$ is increased, the phase angle decreases toward $-\pi/4$ and $|\eta|$ exceeds $|\chi|$; when the electron motion is impeded by collisions it takes a larger perturbation in density, as compared to the collisionless case, to build up a potential variation.

17.5.3 Physical description

The results from the foregoing section can be presented in the form of a physical picture, following (Chen 1965*c*). Thus Fig. 17.8 illustrates the basic elements of the derivation. A solid line gives an equi-density contour of the perturbation as in Fig. 17.2. The ion velocities deviate as in Fig. 17.4 also here from the $\mathbf{E} \times \mathbf{B}$ velocities because of the finite ion inertia terms. At position ② in Fig. 17.8 the electric field E_y is maximum and hence $\partial E_y/\partial t = 0$ and u_y then vanishes also. At a position ① on the other hand, u_x vanishes but u_y is maximum. In the vicinity of position ② ions are brought in from the equi-density contour of the perturbation as in Fig. 17.2. The ion velocities deviate as in Fig. 17.4 also here from the $\mathbf{E} \times \mathbf{B}$ velocities because of the finite ion inertia. At position ② in Fig. 17.8 the electric field E_y is maximum and hence $\partial E_y/\partial t = 0$ and u_y then vanishes also. At a position ①, on the other hand, u_x vanishes but u_y is maximum. In the vicinity of position ② ions are brought in from the main distribution by u_x while u_y, which points away from ②, acts to deplete ions. Conversely, in the neighborhood of position ④ ions are depleted by being brought back into the main distribution by a velocity u_x while the converging velocity component u_y acts to increase the local ion density. Because of finite conductivity, (17.6) is no longer accurate and the equi-potential lines are no longer in phase with the equi-density contours. The electrons have to short-circuit the ion charges by flowing along magnetic field lines, but because of the collisional drag this requires an excess electric field, over the normally existing field which balances the electron pressure gradient along **B**. As the wave propagates in the direction perpendicular to **B** as well as ∇n_0, a perturbation in ion density is propagating also *along* **B**. Because of their reduced mobility the electrons can not follow this motion and lag slightly behind, giving rise to a charge separation. The net result is a downward shift in the potential distribution, as indicated in Fig. 17.8. In the (x,y)-plane, the potential curves are shifted in the negative y-direction. This shift causes a component of u_x to appear in phase with the density perturbation n. Thus the average u_x is positive in the region where n is positive and vice versa. A positive u_x causes an increase in n where it is large already (shaded regions) due to the presence of the gradient ∇n_0, while a negative u_x causes a further decrease in density at positions where n is negative; consequently the perturbation will grow in time. This example gives the first hint for the importance of drift waves for anomalous transport of plasma across magnetic field lines. A full study of that problem under stationary conditions (in a statistical sense) will require an analysis of the saturated turbulent stage of the linear instability. If electron inertia rather than collisions was limiting the electron flow along **B**, the oscillations would remain neutrally stable.

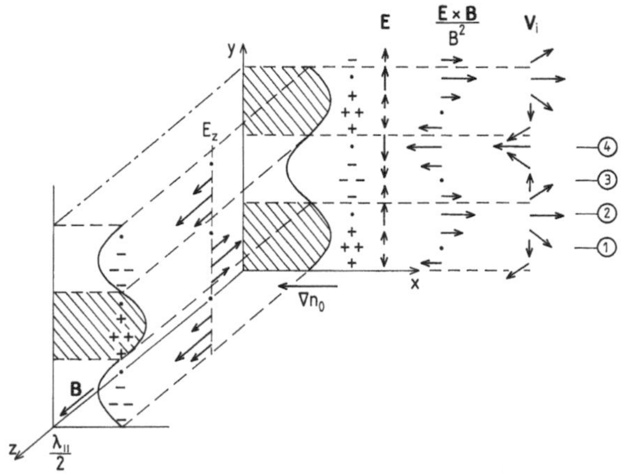

FIGURE 17.8: *Basic elements of drift wave properties for unstable conditions with ion inertia and finite resistivity included. The solid line indicates an equi-density contour, while electric fields and ion velocities are indicated by small arrows. Charge densities are indicated by + and − symbols. Regions with enhanced density are shaded for clarity. The ions have an elliptical orbit with the major axis along the density gradient, see also Section 7.1.5. When the polarization term due to ion inertia is ignored, this ellipse collapses to a straight line parallel to the density gradient.*

As the **B**-parallel wavelength is increased, it takes longer for the electrons to transport excess charge away from the (x,y)-plane. Hence the frequency decreases. Finally, at $k_z = 0$ the wave ceases to exist. This limit corresponds to a static perturbation, i.e., an arbitrary two-dimensional charge distribution can be generated by charging-up flux tubes. Since classical diffusion was ignored, these charges do not annihilate, and the perturbations remain indefinitely in the present description. This $\omega = 0$ limit for $k_z = 0$ is in this sense not physically correct; the missing small imaginary part of ω is associated with the basic assumptions concerning the zero order, or unperturbed state. The singularity in Fig. 17.6 found when $\omega \to 0$ for $k_z \to 0$ is therefore also unphysical.

The results obtained in this section are unphysical in one respect; they predict linearly unstable waves for arbitrarily large wave-numbers k_y, i.e., arbitrarily small **B**-perpendicular wavelengths. Even with the assumption of vanishing ion temperature, the basic equation breaks down when the scales become comparable to the electron Larmor radius. In reality it is more likely to be the finite *ion*-Larmor radius which sets the limit for applicability of the foregoing analysis. In the following sections, these finite ion Larmor radius effects will be studied in detail.

17.5.4 Model consistency

As already mentioned, we are making a stability analysis of a steady state plasma gradient, which is not *really* a steady state, since the collisional diffusion is eroding inhomogeneities in the plasma due to classical diffusion mediated by the electron-ion collisions. These very same collisions give rise to the finite plasma conductivity, which is responsible for the resistive drift instability. For the model to be consistent we require that the characteristic time for the drift wave dynamics is much shorter than the time it takes for the density gradient to be "flattened out". We consider first the latter process. We have the diffusion equation for the **B**-perpendicular dynamics to be $\partial n/\partial t = \nabla_\perp \cdot D\nabla_\perp n$, with D being the diffusion coefficient. In analogy with the continuity equation we can define a diffusion flux as $\mathbf{u}_D n = -D\nabla_\perp n$, implying a diffusion velocity $\mathbf{u}_D$, which we have obtained already in (17.37). Since ions and electrons diffuse "together" by electron-ion collisions, there are no charge separations associated with this process, and the diffusion velocity is the same for both species. Introducing a characteristic length scale as $\mathcal{L} \equiv n/|\nabla_\perp n|$, we obtain a characteristic time for the variation of the density gradient as $\tau_0 \equiv \mathcal{L}/|\mathbf{u}_D|$. The density scale length is a constant only for an exponential profile, but we here take a representative value, just as in the foregoing analysis. Using (17.37) we have $|u_D| = \kappa T_e \nu_{ei}/(\omega_{ce}^2 m\mathcal{L})$, giving $\tau_0 = \omega_{ce}^2 m\mathcal{L}^2/(\kappa T_e \nu_{ei})$. If we take the characteristic time for the drift waves to be $1/\omega^*$, we require $\tau_0 \gg 1/\omega^*$ for the quasi stationary density gradient model to be applicable. By simple manipulations we obtain the estimate

$$\frac{\omega_{ce}}{\nu_{ei}} \gg \frac{1}{k_y \mathcal{L}}. \tag{17.54}$$

One might have argued that the characteristic time scale is *not* the drift wave frequency as such, but the linear growth rate: after all, we will be able to actually observe a drift wave only after a time $> 1/\Im\{\omega\}$. In (17.53) we found $\Im\{\omega\} \approx \omega^{*2}/\sigma_\parallel$, and it is readily shown that $\tau_0 \gg 1/\Im\{\omega\}$ implies

$$\frac{1}{k_y a_i}\left(\frac{k_z}{k_y}\right)^2 \ll 1, \tag{17.55}$$

independent of ν and $\mathcal{L}$. We note that we generally have $k_z \ll k_y$ for drift waves, but, on the other hand, we also have $k_y a_i \ll 1$.

Implicit in the foregoing discussion we assumed that the appropriate growth rate for the drift instability was a fairly small one, obtained for $(k_z/k_y)\sqrt{\omega_{ce}\Omega_{ci}/\omega^*\nu_{ei}} \gg 1$, where $\omega \approx \omega^*$; see Fig. 17.5. By some simple manipulations, we can, however, write the normalized variable on the horizontal axes of Figs. 17.5, 17.6 and 17.7 as

$$\frac{k_z}{k_y}\sqrt{\frac{\omega_{ce}\Omega_{ci}}{\omega^*\nu_{ei}}} = \frac{k_z \ell_c}{k_y a_i}\sqrt{\frac{\nu_{ei}}{\omega^*}}, \tag{17.56}$$

where we introduced the collisional mean free path $\ell_c \equiv \sqrt{\kappa T_e/m}/\nu_{ei}$ for the electrons colliding with ions. We require in general $k_y a_i \ll 1$ as well as $k_z \ell_c \ll 1$ for the stability analysis of Section 17.5.1 to be appropriate. It is then justified to assume $k_z \ell_c / k_y a_i \sim 1$. For relevant plasma condition $\nu_{ei} \gg \omega^*$, and (17.56) is then a large number. The approximations giving (17.54) and (17.55) are therefore justified. One *might* argue that we can always obtain a growth rate $\Im\{\omega\} \approx \omega^*/4$ (see Fig. 17.5) by taking k_z small enough. In reality, this will correspond to wavelengths much larger than the length of the device, and this might be difficult to argue, even when the sheath conditions at the ends are taken into account (Chen 1965*a*, Chen 1979). The analysis becomes significantly complicated if the end conditions are taken into account; when a particle is lost to the end-plate, its energy and momentum are lost from the plasma as well. When a new particle is emitted to replace it from the surface (i.e., a hot plate in a Q-machine (Motley 1975)) this particle has to be "synchronized" with the wave, a process which again costs energy. The complexity of these processes is often jointly called *end-plate damping*. It *is* feasible to maintain a steady state plasma gradient (after all, we find one in the ionosphere, for instance), but this has to be done by sources and sinks, and these will effectively act in the same way as the end-plate damping. The important conclusion here is that a truly steady state plasma gradient can only be maintained by some mechanisms which inevitably influence the growth rate of the instability, and they should therefore be included in a detailed analysis.

- **Exercise:** Carry out the calculations giving the inequalities (17.54) and (17.55).

Ⓢ **Exercise:** Consider a Q-machine plasma with density $n_0 \approx 10^{16}$ m^{-3} (which is a rather high density for a Q-machine), a homogeneous magnetic field $B = 0.1$ T, temperature 0.2 eV and cesium ions. Assume that the plasma column of length 2 m and diameter 4 cm is terminated at both ends by a hot plate (i.e., double-ended conditions), which can be considered as conducting in the sense that the waves have nodes there. Assume the radial scale length of the density gradients to be $\ell \approx 1$ cm. Estimate the lowest order, fundamental, frequency for the drift waves. Calculate the collisional electron mean free path, the electron-ion collision frequency and estimate the maximum growth rate of the resistive drift wave instabilities. Do the waves have time to grow to a significant amplitude within a typical life time of ions in the device? Note that for Q-machines, the electron temperatures are generally only slightly larger than the ion temperatures, so that the stability analysis of Section 17.5.1 can only be taken as an approximation here. Discuss the change in conditions if one of the end plates is replaced by a cold negatively biased metal plate. Ⓢ

Ⓢ **Exercise:** Show that the fluctuating $\mathbf{E}\times\mathbf{B}/B^2$-velocity for electrostatic drift waves with Boltzmann distributed electrons, $n(\mathbf{r},t) = n_0(\mathbf{r})\exp(e\phi(\mathbf{r},t)/\kappa T_e)$, can not give rise to any net plasma loss in a homogeneously magnetized plasma with an arbitrary cross section of the plasma column. Demonstrate that this conclusion remains correct for any local relation of the form $n(\mathbf{r},t) = F(\phi(\mathbf{r},t))$. Comment on the contribution from the polarization drifts. Ⓢ

17.6 Resistive drift waves with $T_i > 0$

The foregoing analysis assumes explicitly that the ion temperature was vanishing, $T_i \approx 0$. We now relax this restriction.

17.6.1 Basic equations

The modifications of the results of the foregoing sections due to finite ion temperatures are relatively trivial as long as the electron-ion temperature ratio is large. When the two temperatures are comparable, the analysis becomes more complicated and new effects are found, concerning the drift wave stability in particular. For $T_i > 0$, we shall replace (17.28) with

$$Mn\left(\frac{\partial \mathbf{u}}{\partial t} + \mathbf{u}\cdot\nabla\mathbf{u}\right) = en(-\nabla\phi + \mathbf{u}\times\mathbf{B}) - \nabla p - \nabla\cdot\underline{\Pi} + Mn\mathbf{g}. \tag{17.57}$$

The subscript i has been suppressed on the variables $\mathbf{u}$ and n. We have neglected classical diffusion and split the divergence of the ion stress tensor into an isotropic part $\nabla p = \gamma_i \kappa T_i \nabla n$ and an anisotropic part $\nabla\cdot\Pi$. In the following we take $\gamma_i = 1$ for simplicity. The tensor $\underline{\Pi}$ is called the magnetic viscosity tensor (Braginskiĭ 1965, Hazeltine & Waelbroeck 1998), and consists of a part connected with ion-ion collisions and a part which remains in the limit $\nu_{ii} \to 0$. This collisionless viscosity is simply the FLR effect mentioned earlier. The ions have a modified $\mathbf{E}\times\mathbf{B}$ drift in a nonuniform electric field. In addition, FLR effects act to "smear-out" the ion charge distribution, i.e., the ions are to be considered as being distributed along their gyro-circle rather than associating their position with the gyro-center. It should be emphasized here that the viscosity tensor in the form given here is correct only to lowest order in an expansion of $(kr_L)^2$, where k is a wave-number characterizing the perturbation and r_L is the ion Larmor radius. A study retaining FLR effects to all orders in $(kr_L)^2$ is based on plasma kinetic theory, and is not considered here.

We have also included in (17.57) the term $Mn\mathbf{g}$ due to a gravitational field $\mathbf{g}$, because it does not complicate the analysis to do so. The gravitational force can also represent the centrifugal force of a rotating plasma column, but in that case g is no longer a constant, but varies with radius.

Deferring collisional viscosity to the next section, we may write the relevant components of $\underline{\Pi}$ in the collisionless case as follows, correct to first order in $k^2r_L^2$, as already mentioned. We have

$$\Pi_{yy} = -\Pi_{xx} = \frac{1}{2}\frac{n\kappa T_i}{\Omega_{ci}}\left(\frac{\partial u_y}{\partial x}+\frac{\partial u_x}{\partial y}\right), \quad \text{and} \quad \Pi_{xy} = \Pi_{yx} = \frac{1}{2}\frac{n\kappa T_i}{\Omega_{ci}}\left(\frac{\partial u_x}{\partial x}-\frac{\partial u_y}{\partial y}\right). \tag{17.58}$$

We shall neglect u_{zi} as in the foregoing sections. Since $\omega_{ce} \gg \Omega_{ci}$, we see right away that viscosity is unimportant for the electron dynamics in the present case, where all relevant spatial scales are much larger than the electron Larmor radius.

After some simple algebra, the contribution in (17.57) from the collisionless viscosity tensor with constant ion temperature, T_i, can be given a more compact formulation as

$$\nabla\cdot\underline{\Pi} = \frac{n\kappa T_i}{\Omega_{ci}}\left(\widehat{\mathbf{z}}\times\nabla^2\mathbf{u} - (\nabla u_y - \widehat{\mathbf{z}}\times\nabla u_x)\frac{\partial \ln n}{\partial x} - (\nabla u_x + \widehat{\mathbf{z}}\times\nabla u_y)\frac{\partial \ln n}{\partial y}\right), \tag{17.59}$$

with $\widehat{\mathbf{z}}$ being a unit vector in the z-direction. When we linearize the equations, we will assume $n = n_0(x)$, and the last term in (17.59) vanishes.

The physics behind collisionless viscosity can be explained by a simple example: assume we have an initially unperturbed plasma density with a simple ion fluid velocity field in the form $\mathbf{u} = \{u_x(x,t), 0, 0\}$, where we have z along the magnetic field lines. This flow is compressible, $\nabla\cdot\mathbf{u} \neq 0$. For a short initial time interval the ion continuity equation gives $\partial \ln n/\partial t = -\partial u_x/\partial x$ or $\partial^2 \ln n/\partial t\partial x = -\partial^2 u_x/\partial x^2$, implying that a density gradient builds up with ∇n in the x-direction. This density variation has an associated ion diamagnetic drift $u_{Di} = (\kappa T_i/eB)d\ln n(x)/dx$, which can be re-written to give $\partial u_{Di}/\partial t = (\kappa T_i/eB)\partial^2 \ln n/\partial t\partial x = -(\kappa T_i/eB)\partial^2 u_x/\partial x^2$ with $\mathbf{u}_{Di}$ perpendicular to the x-direction. The contribution of this term to $Mn\partial\mathbf{u}/\partial t$ in (17.57) becomes $(n\kappa T_i/\Omega_{ci})\partial^2 u_x/\partial x^2$, which is precisely the first term in (17.59) for the given geometry. The last two terms in (17.59) contribute when the gradients in the plasma density have increased to significant levels. We thus find that due to the collisionless ion viscosity an inhomogeneous velocity field in, say, the x-direction induces an ion drift velocity in the perpendicular direction. In some respect this is reminiscent of momentum transfer due to viscous effects as discussed in Section 2.6.1, but that effect is mediated by irreversible collisional processes, so it may be somewhat misleading to generally introduce the word "viscosity" in relation to the $\underline{\Pi}$-tensor as used in the present section.

17.6.2 Equilibrium

First we consider the equilibrium solution of (17.57). There is now an ion drift u_{i0} given by

$$\mathbf{u}_{i0} = (u_{Di} - U_G)\widehat{\mathbf{y}}, \tag{17.60}$$

where

$$u_{Di} = \frac{\kappa T_i}{eB}\frac{n_0'}{n_0} \quad \text{and} \quad U_G \equiv \frac{g}{\Omega_{ci}}.$$

We assumed $n_0'/n_0 =$ constant and $\mathbf{g} = g\widehat{\mathbf{x}} =$ constant as well, so that $u_{i0} =$ constant and $\mathbf{u}_0 \cdot \nabla \mathbf{u}_0 = 0$ and $\underline{\Pi}_0 = 0$. There will be no opportunity to confuse the electron and ion dc drifts, so we ignore the subscript $_i$ on the unperturbed ion drift velocity.

Equation (17.60) is the only place where the effect of gravity will appear; that is the reason for being cavalier in introducing gravitational forces in (17.57). In addition, we also have an electron drift u_{e0}. With the small electron mass, the effect of gravity on the electron dynamics can be ignored and (17.47) remains unchanged.

- **Comment on centrifugal forces:** It is tempting to let the gravitational acceleration g be due to a curvature in B, and set it equal to $\kappa T/MR_c$, where R_c is the radius of curvature and $\sqrt{\kappa T/M}$ is a characteristic velocity. This is, however, only formally correct. The bending of the field lines can only be achieved by introducing a gradient in the magnetic field and the ∇B-drift must then be taken into account also, giving modifications of the equilibrium and the subsequent analysis (Chen 2016). To substantiate this argument we assume that we succeeded in bending the magnetic field lines to have a curvature R in some region of space. Since we deal with low β-plasmas we require $\nabla \times \mathbf{B} = 0$ as in vacuum, i.e., the presence of the plasma does not influence the magnetic field in any appreciable way. In cylindrical coordinates we then find that $\nabla \times \mathbf{B}$ only has a z-component, since the assumed magnetic field, $\mathbf{B}$, only has a θ-component. We then have $(\nabla \times \mathbf{B})_z = (1/r)\partial_r(rB_\theta) = 0$, which implies $B_\theta \sim 1/r$; a curvature of the magnetic field is accompanied by a variation in intensity of the same field along the radius of curvature. Consequently, the curvature drifts will necessarily be accompanied by ∇B-drifts.

  The curvature and gradient-B drifts can in principle be included in the analysis as well, but are not considered here. An important consequence of the curved magnetic field lines is that the diamagnetic drift associated with the density gradient is no longer divergence free. This will have importance for the so-called ballooning modes which involve also magnetic field perturbations. Similarly, we note that the electrostatic $\mathbf{E} \times \mathbf{B}/B^2$-drift is no longer incompressible in the case where the magnetic field varies in space.

17.6.3 Perturbation

The linear part of (17.57) is

$$M\left(\frac{\partial \widetilde{\mathbf{u}}}{\partial t} + \mathbf{u}_0 \cdot \nabla \widetilde{\mathbf{u}}\right) = e(-\nabla \widetilde{\phi} + \widetilde{\mathbf{u}} \times \mathbf{B}) - \kappa T_i \nabla \eta - \frac{1}{n_0} \nabla \cdot \widetilde{\underline{\Pi}}, \tag{17.61}$$

where we used $\nabla n/n = \nabla \ln n = \nabla \ln(n_0 + \widetilde{n}) \approx \nabla n_0/n_0 + \nabla \eta$. We shall make the local approximation from the outset and neglect x-derivatives of $\mathbf{u}$, η and ϕ. In evaluating the tensor $\widetilde{\underline{\Pi}}$, there will appear terms $n_0 \nabla \widetilde{\mathbf{u}}$ and $\widetilde{n} \nabla \mathbf{u}_0$; the latter vanishes because we have taken u_0 =constant.

Again we consider one Fourier component, or make the equivalent "plane wave approximation". Only waves propagating along $\widehat{\mathbf{y}}$, in the direction perpendicular to the magnetic field as well as the density gradient, will be studied, i.e., $k_z \approx 0$ and $k_x = 0$ also. To ease the notation we do not include the subscript $_1$ anymore: with the experience gained from foregoing sections, hopefully this is no longer needed. With these assumptions, (17.58) yields

$$\begin{aligned} \Pi_{yx} &= -\frac{1}{2} i k_y \frac{n_0 \kappa T_i}{\Omega_{ci}} u_y, & \Pi'_{yx} &= -\frac{1}{2} i k_y \frac{\kappa T_i}{\Omega_{ci}} n'_0 u_y, \\ \Pi_{xx} = -\Pi_{yy} &= -\frac{1}{2} i k_y \frac{n_0 \kappa T_i}{\Omega_{ci}} u_x, & \Pi'_{xx} &= -\frac{1}{2} i k_y \frac{\kappa T_i}{\Omega_{ci}} n'_0 u_x, \end{aligned} \tag{17.62}$$

with a prime again denoting differentiation with respect to x. To simplify the notation we define

$$\omega_i \equiv k_y u_{Di} = k_y \frac{\kappa T_i}{M \Omega_{ci}} \frac{n'_0}{n_0} \quad \text{and} \quad r_L^2 \equiv \frac{2\kappa T_i}{M \Omega_{ci}^2}, \quad \theta \equiv \frac{T_i}{T_e}, \tag{17.63}$$

so that

$$\omega_i = -\theta\omega^* \quad \text{and} \quad b = k_y^2 \frac{r_L^2}{2\theta}.$$

The x and y components of (17.61) then give

$$\left(\omega - k_y u_0 + \frac{1}{2}\omega_i\right) u_x = i\Omega_{ci}\left(1 - \frac{1}{2}\theta b\right) u_y$$

$$\left(\omega - k_y u_0 + \frac{1}{2}\omega_i\right) u_y = k_y \frac{\kappa T_e}{M}(\chi + \theta\eta) - i\Omega_{ci}\left(1 - \frac{1}{2}\theta b\right) u_x. \tag{17.64}$$

The waves are electrostatic and $\mathbf{E}$ is along $\mathbf{k} \parallel \widehat{\mathbf{y}}$, but the ions have an oscillating velocity component also along $\widehat{\mathbf{x}}$. The equations (17.64) refer to the *ion* dynamics, and the electron temperature appears solely because of the normalizations used for χ and θ. The equations for u_x and u_y may be solved by writing $\alpha = \omega - k_y u_0 + \frac{1}{2}\omega_i$, $\beta = 1 - \frac{1}{2}\theta b$ and eliminating u_y and u_x, respectively. We then have

$$(\alpha^2 - \beta^2\Omega_{ci}^2)u_x = i\beta\Omega_{ci}k_y \frac{\kappa T_e}{M}(\chi + \theta\eta)$$

$$(\alpha^2 - \beta^2\Omega_{ci}^2)u_y = \alpha k_y \frac{\kappa T_e}{M}(\chi + \theta\eta). \tag{17.65}$$

When the effect of gravity is small, the term α^2 can be dropped for most relevant cases when $T_i \le T_e$, because it is of relative order b^2. This can be seen by writing

$$\left|\frac{\alpha}{\beta\Omega_{ci}}\right| \approx \frac{k_y u_0}{\Omega_{ci}} = k_y \frac{\kappa T_i}{M\Omega_{ci}^2}\frac{1}{R} = \frac{k_y^2 r_L^2}{2k_y R} \lesssim b \quad \text{for} \quad k_y R > 1, \quad \theta \lesssim 1,$$

with R being a typical length scale for the density gradient. With this simplification, (17.65) becomes

$$u_x = -i\frac{k_y(\chi + \theta\eta)}{\Omega_{ci}\left(1 - \frac{1}{2}\theta b\right)}\frac{\kappa T_e}{M}, \tag{17.66}$$

$$u_y = -\frac{\left(\omega - k_y u_0 + \frac{1}{2}\omega_i\right) k_y(\chi + \theta\eta)}{\Omega_{ci}^2\left(1 - \frac{1}{2}\theta b\right)^2}\frac{\kappa T_e}{M}. \tag{17.67}$$

The linearized ion equation of continuity is now written

$$(\omega - k_y u_0)\eta - k_y u_y + i u_x \frac{n_0'}{n_0} = 0. \tag{17.68}$$

Inserting (17.66) and (17.67) into (17.68), we find a fortunate cancellation of terms. This might look almost like a coincidence, but is in reality due to a subtle difference between fluid drifts and real particle drifts. The final result is

$$(\omega - k_y u_0 - \theta\omega^*)\eta + (b(\omega - k_y u_0) - \omega^*)\chi = 0, \tag{17.69}$$

by omitting second order terms in b.

A dispersion relation is obtained by eliminating η or χ, using (17.47). It is relatively easy to retain the α^2-term in (17.65), but the resulting dispersion relation will be considerably complicated, which makes the effort hardly worthwhile, other limitations considered.

17.6.3.1 Comments on the cancellation of terms in the viscosity tensor

The cancellation of the terms referred to before can be argued in more detail, to make the physics more evident. It is readily shown that the collisionless viscosity term gives a contribution $\mathbf{u}_\pi \equiv$

$(1/n_0eB)\widehat{\mathbf{z}} \times \nabla \cdot \underline{\Pi}$ to the ion fluid velocity. This term is, however, canceled by the polarization drift when inserted into the $\nabla \cdot (n\mathbf{u})$-term in the continuity equation (Weiland 1992). To substantiate this statement, we first calculate $\nabla \cdot (n_0(x)\mathbf{u}_\pi)$ and find, using (17.59), including only terms which are linear in $n_0'(x)/n_0(x)$

$$\begin{aligned}\nabla \cdot (n_0 \mathbf{u}_\pi) &= \mathbf{u}_\pi \cdot \nabla n_0 + n_0 \nabla \cdot \mathbf{u}_\pi \\ &= -\frac{1}{4} r_L^2 \left(\nabla n_0 \cdot \nabla_\perp^2 \mathbf{u} + n_0 \nabla_\perp^2 \nabla \cdot \mathbf{u} + n_0 (n_0'/n_0) \nabla^2 u_x \right). \end{aligned} \tag{17.70}$$

Assuming $\nabla \cdot \mathbf{u} \approx 0$, we find $\nabla \cdot (n_0 \mathbf{u}_\pi) \approx -\frac{1}{2} r_L^2 \nabla n_0 \cdot \nabla^2 \mathbf{u}$.

We then consider the polarization drift $\mathbf{u}_p \equiv (\widehat{\mathbf{z}} \times \partial \mathbf{u}/\partial t + \mathbf{u} \cdot \nabla(\widehat{\mathbf{z}} \times \mathbf{u}))/\Omega_{ci}$, which contains two terms. By linearization we calculate the contribution from the last term in $\mathbf{u}_p$ to $\nabla \cdot (n\mathbf{u})$, obtaining

$$\frac{n_0}{\Omega_{ci}} (\mathbf{u}_{Di} \cdot \nabla) \nabla \cdot (\widehat{\mathbf{z}} \times \mathbf{U}) = -\frac{1}{2} r_L^2 (\widehat{\mathbf{z}} \times \nabla n_0) \cdot \nabla \left(\frac{\partial u_x}{\partial y} - \frac{\partial u_y}{\partial x} \right), \tag{17.71}$$

where the zero order velocity $\mathbf{u}_{Di} = \frac{1}{2} r_L^2 \Omega_{ci} (\widehat{\mathbf{z}} \times \nabla n_0)/n_0$ was again assumed constant. Combining the contributions to $\nabla \cdot (n\mathbf{u})$ from $\mathbf{u}_\pi$ and $\mathbf{u}_p$ we find after simple manipulations the result $\frac{1}{2} r_L^2 \nabla n_0 \cdot \nabla(\nabla \cdot \mathbf{u}) \approx 0$ again with the assumption of incompressibility; the two contributions cancel to the desired accuracy. Physically, this result is a consequence of diamagnetic drifts being non-convective, since the gyrating ions do not move on average. Note, however, that the cancellation is not *exact* if we include terms of higher order than those considered here, and it relies in particular on the assumption that the fluctuations in plasma density are incompressible to lowest order, $\nabla \cdot \mathbf{u} \approx 0$.

Consequently, to the accuracy argued here the only contribution we find is from the time derivative part of the polarization drift, i.e.,

$$\nabla \cdot (n(\mathbf{u}_\pi + \mathbf{u}_p)) \approx \nabla \cdot \left(\frac{n_0}{\Omega_{ci}} \frac{\partial}{\partial t} (\widehat{\mathbf{z}} \times \mathbf{u}) \right),$$

to the desired accuracy. This contains two parts:

$$-\frac{1}{2} n_0 r_L^2 \frac{\partial}{\partial t} \nabla^2 \frac{e\phi}{\kappa T_i}, \qquad \text{and} \qquad -\frac{1}{2} n_0 r_L^2 \frac{\partial}{\partial t} \nabla^2 \frac{n}{n_0},$$

from the $\mathbf{E} \times \mathbf{B}$-drift and from perturbations in the ion diamagnetic drift, respectively. We have again assumed that ∇n_0 is small, and ignored higher order terms.

17.6.4 Dispersion relation

The electron equations are unchanged, so we use (17.47) to eliminate η. We also use (17.60) for u_0. With a little algebra, (17.69) then yields the following dispersion relation:

$$\omega(\omega - \omega_i) + k_y U_G \left(\omega^* b^{-1} + \omega \right) + i\sigma_\parallel \left(\omega - \omega^* + k_y U_G + b(\omega - \omega_i + k_y U_G) \right) = 0. \tag{17.72}$$

The last term bk_yU_G in (17.72) can be ignored for wavelengths much longer than the ion Larmor radius. The dispersion relation can readily be analyzed in various limits, as shown in the following.

17.6.4.1 Pure drift wave

For $U_G = 0$, we have the following dispersion relation for resistive drift waves:

$$\omega(\omega - \omega_i) + i\sigma_\parallel \left(\omega - \omega^* + b(\omega - \omega_i) \right) = 0, \tag{17.73}$$

where the last b-containing term is usually small, as mentioned. In the limit of $b \to 0$ and $\theta \to 0$, the previous result is recovered. In the limit of $\sigma_{\parallel}/\omega^* \gg 1$, we have

$$\omega - \omega^* \approx i\frac{\omega^*(\omega^* - \omega_i)}{\sigma_{\parallel}} = i\frac{\omega^{*2}(1+\theta)}{\sigma_{\parallel}}. \tag{17.74}$$

Compared to (17.53), the growth rate is increased by the factor $1+\theta$, representing the increase in total plasma pressure. This is reasonable from an energetic viewpoint (Chen 1970). Kinematically, the growth rate is increased through the FLR effect in the ion drifts. The ion diamagnetic drift is opposite in direction to the electron diamagnetic drift, so the net result of a finite ion temperature is an increase in the diamagnetic current which acts as the energy source for the instability.

With the approximation of neglecting the term containing b in (17.73), the result is a simple modification of the dispersion relation shown in Fig. 17.5. In particular the combination of independent variables used in that figure is retained.

In the limit of infinite conductivity along magnetic field lines, $\sigma_{\parallel} \to \infty$, we find

$$\omega = \frac{\omega^*(1-\theta b)}{1+b}. \tag{17.75}$$

In the limit where the electrons are isothermally Boltzmann distributed, the finite ion Larmor radius effects thus give rise to an additional dispersion of the waves.

17.6.4.2 Resistive-g mode

If we retain the $k_y U_G$ terms in (17.72) and take the same limit $\sigma_{\parallel}/\omega^* \gg 1$, noting that $k_y b^{-1}\omega^* = -\Omega_{ci} n_0'/n_0$, we find

$$\omega \approx \omega^* - k_y U_G + i\frac{\omega^*(\omega^* - \omega_i) - g n_0'/n_0}{\sigma_{\parallel}}. \tag{17.76}$$

Depending on your point of view, (17.76) then shows either the increase in drift wave growth rate when $g n_0' < 0$, or the decrease in gravitational instability growth rate (see next subsection) due to finite k_z entering through $\sigma_{\parallel}$. Note that g has to be inserted with its sign included. This *gravitational instability* is often called the *Rayleigh-Taylor instability*, and it was first discussed for neutral fluids with a vertical density gradient in a constant gravitational field. The plasma equivalent is often termed the *Kruskal-Schwarzschild instability*. In Fig. 17.9 we give an illustration that can be used for a simple physical explanation of the gravitational or Rayleigh-Taylor instability. The heavy particles (ions) are influenced by gravity, the light electrons only negligibly. The ions will be drifting due to the $\mathbf{g} \times \mathbf{B}$-drift while electrons are nearly stationary. The resulting polarization charges (shown with $+$ and $-$ signs in Fig. 17.9) give rise to electric fields $\mathbf{E}$ that give $\mathbf{E} \times \mathbf{B}$-drifts of the bulk plasma as shown. We find that the relative phases of the density perturbation and the induced electric field are such that the perturbation on the top surface in Fig. 17.9) is stable, while it is unstable on the bottom surface. Within the simple model illustrated here the growth is aperiodic. In classical plasma literature this would be termed unstable, while conditions which are both oscillating and increasing in amplitude would be overstable.

The Richtmyer-Meshkov instability occurs when an interface between fluids of differing density is impulsively accelerated, e.g., by the passage of a shock wave. This instability can be considered the impulsive acceleration limit of the Rayleigh-Taylor instability, so the results obtained here can be illustrative for both.

- **Example:** The ionosphere in the sunlit part of the ionosphere represents a balance between ionization and recombination, with a viable model given by the Chapman ionosphere; see Section 6.3. There is a large scale vertical density gradient at the upper region (above the F-maximum) where the plasma density is decreasing due to a decrease in neutral density, while at

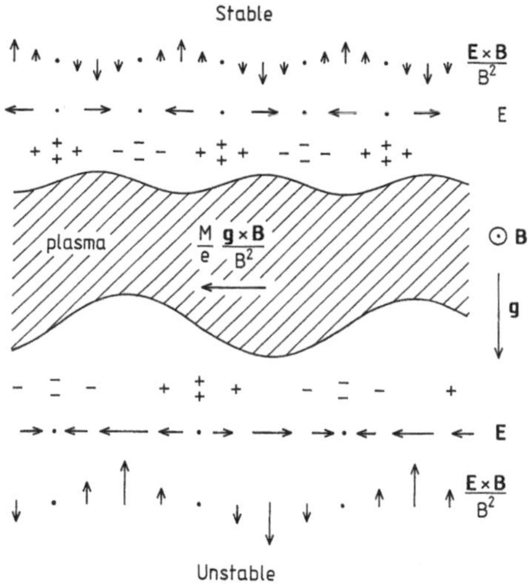

FIGURE 17.9: *Geometry for a simple physical description of the gravitational or Rayleigh-Taylor instability; see also the discussion by Rosenbluth and Longmire (1957) or Chen (2016). The figure implicitly assumes* $k_z = 0$.

the lower parts the density decreases vertically due to a decrease in the ionizing UV-radiation. The ionospheric plasma layer is supported against gravity by collisions with the neutral component, while we at the same time have continuous ionization and recombination processes. The magnetic field has a minor role here. When a part of the ionosphere enters the shadow region behind the Earth, the ionizing radiation vanishes quite suddenly, while recombination continues. This process is fastest at the bottom of the ionosphere where the density is largest. In the equatorial regions we have a transient time period, where now a plasma layer is supported against gravity mostly by the magnetic field, with a configuration that is unstable with respect to the gravitational instability. A detailed analysis has to take into account the neutral component and the collisions and recombinations.

Ⓢ **Exercise:** Consider the case where a uniform cold collisionless plasma is supported against gravity by a horizontal magnetic field (Rosenbluth & Longmire 1957). Assume that there is a sharp boundary between plasma and vacuum as in Fig. 17.9. In the case where this boundary is unperturbed, it is readily demonstrated that the proposed conditions form an equilibrium. In particular, we have a current driven by gravity, $\mathbf{J}_0 = ne(\mathbf{u}_i - \mathbf{u}_e)$, with $\mathbf{u}_i = -(Mg/eB)\widehat{\mathbf{y}}$ and $\mathbf{u}_e = (mg/eB)\widehat{\mathbf{y}} \approx 0$. The equilibrium is, however, not stable. Initially, we let the boundary be slightly perturbed by a spatially harmonic displacement as $\Delta x = a\sin(ky)$ with $a \ll \lambda = 2\pi/k$. Demonstrate that the plasma is indeed unstable and give an expression for the time evolution of the amplitude of the perturbation, $a = a(t)$. Ⓢ

Ⓢ **Exercise:** Can a magnetic field support a "blob" of plasma (or rather a plasma filled magnetic flux-tube) against a gravitational field (Chandrasekhar 1960)? See Fig. 17.10 for an illustration of a cross section of such a magnetic flux-tube. Write the expression for the polarization of the plasma due to gravity and then the electric field resulting from this polarization. It gives a simplification to use the dielectric function (8.48) here, but you can do without it. Give an expression for the acceleration of the plasma blob: is it different from zero?

Consider now an infinitely extended magnetized plasma, where a magnetic flux tube is partly depleted of plasma, corresponding to a reversed version Fig. 17.10 where the surroundings will

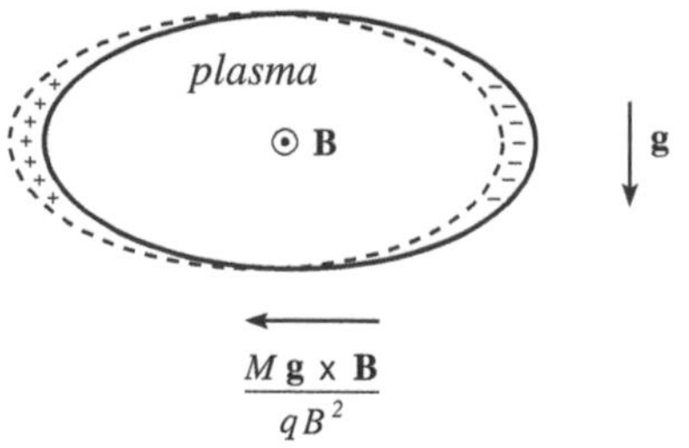

FIGURE 17.10: *Illustration of a "blob" of plasma placed in a horizontal magnetic field* **B**. *The figure should be interpreted as a cross section of a plasma-filled magnetic flux-tube. The gravitational acceleration* **g** *acts in the vertical direction. The flow direction of positive charges, $q > 0$, is given by an arrow. The magnetized plasma becomes polarized by the gravitational field as indicated.*

represent plasma and what is termed "plasma" on Fig. 17.10 will be evacuated. How are the polarization charges distributed in this case? Give a qualitative discussion of what happens. ⓢ

- **Exercise:** Let $k_z \neq 0$ and assume the electrons to be Boltzmann distributed at all times. Discuss liner drift waves for the case with a sharp interface between the plasma and vacuum as in Fig. 17.9.

17.6.5 FLR stabilization of flute modes

The *flute mode* at $k_z = 0$ in the dispersion relation in Fig. 17.5 can be made unstable by gravity. When $k_z = 0$ we have $\sigma_{\parallel} = 0$ and (17.72) becomes

$$\omega^2 + (k_y U_G - \omega_i)\,\omega - g\frac{n_0'}{n_0} = 0. \tag{17.77}$$

In the absence of the middle term, one obtains the usual growth rate $\gamma = (-g n_0'/n_0)^{1/2}$ for what is known as the interchange, Rayleigh-Taylor, or gravitational flute instability resulting when $g n_0'/n_0 < 0$. In the limit of vanishing ion temperature, the instability is simple to describe phenomenological; see, for instance, Fig. 17.9. Electrons with their small mass are hardly affected by gravity, while the ions drift with a $\mathbf{U}_G = M\mathbf{g}\times\mathbf{B}/eB^2$-velocity to set up polarization electric fields in an initial perturbation. These in turn enhance (instability) or diminish (stability) the initial perturbation, depending on the sign of $\mathbf{g}\cdot\nabla n_0$.

The $k_y U_G$ term in (17.77) arises from finite ion inertia, and the ω_i term from finite ion Larmor radius. Together, these terms lower the growth rate and reduce it to zero for sufficiently large k_y. This is known as FLR stabilization. The stabilization originates from the "smearing out" of the ion charge distribution due to the finite Larmor radii, which counteracts the charge separation caused by the ion gravity drift; see Fig. 17.9. The full dispersion relation is readily obtained as

$$\omega = \frac{1}{2}(\omega_i - k_y U_G) \pm \frac{1}{2}\sqrt{(\omega_i - k_y U_G)^2 + 4g\frac{n_0'}{n_0}},$$

with the condition for instability being $g n_0'/n_0 < -\frac{1}{4}(\omega_i - k_y U_G)^2$. The dispersion relation is illustrated in Fig. 17.11 for $4g n_0'/n_0\omega_i = -1.5$ and -0.5, as well as 0.5 and 1.5. The most unstable perturbation has $k_y = \omega_i/U_G$, and has vanishing phase velocity.

We can again consider the relation (17.47) between normalized density η and normalized electrostatic potential χ. In the limit $\sigma_{\parallel} \to 0$ we find now a finite ratio $|\eta/\chi|$ and the relative phase angle is also changed from the zero ion temperature case; see Fig. 17.7. The finite ion temperature modifies ω entering (17.47), even in the limit of vanishing gravitational drifts.

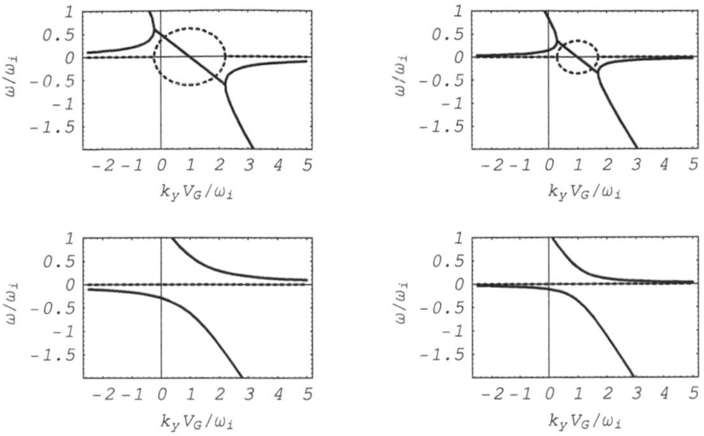

FIGURE 17.11: *Dispersion relations in normalized units, for the Rayleigh-Taylor instability (17.77), shown with* $4gn_0'/n_0\omega_i = -1.5$ *and* -0.5*, respectively, for instability in the top figures. The real parts of the frequency are given by solid lines, the imaginary parts with dashed lines. The lower figures are for stable conditions, with* $4gn_0'/n_0\omega_i = 1.5$ *and* 0.5*, to the left and right sides, respectively.*

17.7 Drift waves with ion viscosity

Equation (17.73) predicts that drift waves are always unstable regardless of k_y and k_z, as long as they are both finite. Experimentally, it was found that drift waves could be stabilized by lowering B, and hence increasing $k_y^2 r_L^2$. To explain this, we must include the collision terms in the ion viscosity tensor $\underline{\Pi}$. If we use the full tensor and take the limit $\Omega_{ci}^2\tau_{ii}^2 \gg 1$ with τ_{ii} being the ion-ion collision time (Braginskií 1965), we find that for motions perpendicular to $\mathbf{B}$ the collisional part of $\nabla_\perp \cdot \underline{\Pi}$ can be written as

$$\nabla_\perp \cdot \underline{\Pi}_c = -\mu_\perp \nabla_\perp^2 \mathbf{u}, \quad \text{where} \quad \mu_\perp \equiv \frac{n\kappa T_i}{4\Omega_{ci}^2 \tau_{ii}}, \quad \text{with} \quad \tau_{ii} = \frac{32\pi M^2 \varepsilon_0^2}{ne^4 \ln(\Lambda)} \left(\frac{\kappa T_i}{M}\right)^{3/2}. \tag{17.78}$$

Here $\ln(\Lambda)$ is the usual Coulomb logarithm. To account for ion-ion collisions, we merely add a term $\mu_{\perp 0}\nabla_\perp^2 \mathbf{u}$ to the right-hand side of (17.61). When (17.61) has been divided by Mn_0, this term becomes

$$-k_y^2 \frac{\mu_{\perp 0}}{Mn_0} = -\frac{1}{4}k_y^2 \frac{\theta a_i^2}{\tau_{ii}} = -\frac{1}{4}\theta b \nu_{ii} \equiv -\sigma_\perp, \tag{17.79}$$

where $\nu_{ii} = 1/\tau_{ii}$ is the ion-ion collision frequency. The solution of (17.61) which replaces (17.66) is now

$$u_x \frac{n_0'}{n_0} = i\omega^* \frac{(\chi + \theta\eta)}{1 - \frac{1}{2}\theta b} \quad \text{(unchanged)}, \tag{17.80}$$

and for (17.67)

$$u_y = -k_y a_i^2 \left(\omega - k_y u_0 + \frac{1}{2}\omega_i + i\sigma_\perp\right) \frac{(\chi + \theta\eta)}{\left(1 - \frac{1}{2}\theta b\right)^2}. \tag{17.81}$$

We assumed $|\omega + i\sigma_\perp|^2 \ll \Omega_{ci}^2$. The $\sigma_\perp$ term represents an ion flux in the y-direction, which can short circuit the fluctuating electric field in that direction if b is large enough; when this happens, the wave becomes damped.

Some qualitative features of the stabilization of drift waves by ion viscosity can be anticipated from the outset. Thus, inspection of the linear dispersion relation shown in Fig. 17.5 indicates that the growth rate is small for very large and very small values of the parameter $\sigma_{\parallel} = (k_z/k_y)^2\omega_{ce}\tau_{ei}\Omega_{ci}$. With ion viscosity being a small effect, we expect stabilization to appear primarily where the growth rate is small, i.e., at very large or very small **B**-parallel wavelengths.

We omitted the analysis of the Rayleigh-Taylor instability in this section so that $U_G = 0$ here.

17.7.1 Dispersion relation

Equation (17.81) is now inserted into the ion equation of continuity (17.68), and η is eliminated by virtue of the electron equation (17.47). Straightforward algebra yields the local dispersion relation

$$(\omega + i\sigma_{\perp})(\omega - \omega_i) + i\sigma_{\parallel}\left(\omega - \omega^* + b(\omega - \omega_i) + ib\sigma_{\perp}(1+\theta)\right) = 0. \tag{17.82}$$

The calculations can be simplified by noting that the contributions from the terms $Mn_0\mathbf{u}_0 \cdot \nabla\mathbf{u}$ and the collisionless part of $\nabla \cdot \underline{\Pi}_1$ in (17.61) happen to cancel to the order desired. While this is not exactly true, since the term $b(\omega - \omega_i)$ in (17.82) would be replaced by $b(\omega + \theta\omega^*)$, there is no practical difference for $\omega \approx \omega^*$.

Equation (17.82) can be simplified in the long-$\lambda_{\parallel}$ and short-$\lambda_{\parallel}$ limits, as discussed in the following. Consider first $\omega \leq O(\omega^*)$ and $\sigma_{\perp} \leq O(\omega^*)$, and assume $b\sigma_{\parallel} \ll \omega^*$. The two terms with b in (17.82) will be neglected compared to the first term in the equation. The dispersion relation then becomes

$$(\omega + i\sigma_{\perp})(\omega - \omega_i) + i\sigma_{\parallel}(\omega - \omega^*) = 0, \tag{17.83}$$

which clearly reduces to (17.73) for $\nu_{ii} \to 0$. We can obtain an explicit expression for the growth rate in the $\sigma_{\parallel} \gg \omega^*$, or short-$\lambda_{\parallel}$, limit. This is possible provided b can satisfy the inequality $\sigma_{\parallel} \gg \omega^* \gg b\sigma_{\parallel}$. Solving (17.83) for $\omega - \omega^*$ and letting as before $\omega \approx \omega^*$ on the right-hand side, we obtain

$$\omega - \omega^* = -\frac{\sigma_{\perp}}{\sigma_{\parallel}}(\omega^* - \omega_i) + \frac{i\omega^*(\omega^* - \omega_i)}{\sigma_{\parallel}}. \tag{17.84}$$

In this limit, the growth rate is not affected by $\sigma_{\perp}$, but the real part of $\omega - \omega^*$ shows that the assumption $\omega \approx \omega^*$ that we made is valid only for $\sigma_{\perp} \ll \sigma_{\parallel}$, which need not always be satisfied. For more general solutions we consider individual cases (Chen 1970).

17.7.2 Long-$\lambda_{\parallel}$ limit

To obtain the condition for viscous stabilization, equation (17.83) is rewritten as follows:

$$\omega^2 - (\omega_i - i\sigma)\omega - i\omega^*(\sigma_{\parallel} - \theta\sigma_{\perp}) = 0, \tag{17.85}$$

where $\sigma \equiv \sigma_{\parallel} + \sigma_{\perp}$ and $\omega_i = -\theta\omega^*$ was used. If the quantity $\Delta\sigma \equiv \sigma_{\parallel} - \theta\sigma_{\perp}$ vanishes, we have a real solution $\omega = 0$ and a damped solution $\omega = \omega_i - i\sigma$ propagating in the ion diamagnetic drift direction. To examine the stability threshold, we should therefore consider $\omega \approx 0$. Assuming $\omega \ll \omega^*$ and neglecting the ω^2 term, we obtain

$$\omega = \frac{i\omega^*\Delta\sigma}{i\sigma - \omega_i}. \tag{17.86}$$

For $\sigma \ll |\omega_i|$, this simplifies to

$$\omega = \frac{\omega^*}{\omega_i^2}\sigma\Delta\sigma + \frac{i\Delta\sigma}{\theta}. \tag{17.87}$$

For $\sigma \gg |\omega_i|$, we obtain instead

$$\omega = \omega^*\frac{\Delta\sigma}{\sigma} + \frac{i\theta\omega^{*2}}{\sigma}\frac{\Delta\sigma}{\sigma}. \tag{17.88}$$

In either case, $\sigma_{\|} = \theta\sigma_{\perp}$ is a necessary and sufficient condition for marginal stability. If $\sigma_{\|}$ is held constant by keeping the ratio $(k_z/k_y)^2$ constant and $\sigma_{\perp}$ decreased from this value by increasing k_y, then both $\Im(\omega)$ and $\Re(\omega)$ will increase from 0. Physically, this means that the increased ion conductivity across magnetic field lines short circuits wave crests and wave troughs. When, on the other hand, $\sigma_{\perp}$ is *in*creased from the threshold value $\Delta\sigma = 0$, we find a slightly damped solution which propagates in the direction *opposite* to that of the electron diamagnetic drift, i.e., in the direction of the *ion* diamagnetic drift.

Alternatively, we can again start at marginal stability, but now keep $\sigma_{\perp}$ with a constant k_y and increase $\sigma_{\|}$ by turning the wave vector by increasing $(k_z/k_y)^2$, and find that the waves become unstable. This part can be understood by considering the dispersion relation in Fig. 17.5 in the long $\lambda_{\|}$-limit, where it is seen that an increase in $\sigma_{\|}$ gives a corresponding increase in growth rate, which in the present case compensates the damping effect of ion viscosity. Similarly, we can explain also the stabilization by *de*crease in $\sigma_{\|}$ for fixed $\sigma_{\perp}$.

- **Exercise:** Prove that a solution of (17.83) for $\sigma_{\|} \gg \omega^* \gg b\sigma_{\|}$, found by straightforward expansion, is

$$\omega = \omega^* \frac{\Delta\sigma}{\sigma} + \frac{i(1+\theta)\omega^{*2}}{\sigma} \frac{\sigma_{\|}\Delta\sigma}{\theta\sigma^2}, \tag{17.89}$$

which reduces to (17.84), (17.87) and (17.88) in the proper limits.

It may be interesting to note that the mode described by (17.85) remains in the limit where the plasma inhomogeneity vanishes, i.e., when $\omega^* \to 0$ and $\omega_i \to 0$. For this case we find $\omega = -i\sigma$. This is a purely damped mode. If, in particular, we let $k_z \to 0$, we have

$$\omega = -i\sigma_{\perp} = -\frac{i}{4\tau_{ii}}\theta k_y^2 a_i^2 . \tag{17.90}$$

This limiting case if often called the "convective cell" limit (Shukla et al. 1984, Sugai et al. 1983). This mode is, as we see, damped by ion viscosity. Experience has shown that it can be spontaneously excited, and that it is often very important for anomalous plasma transport across magnetic field lines (Cheng & Okuda 1977).

17.7.3 Amplitude and phase relations

Since $\omega = 0$ at threshold, the phase shift relation (17.47) reduces to

$$\eta = \chi \frac{\omega^* + ib\sigma_{\|}}{ib\sigma_{\|}} \approx \frac{\omega^*}{ib\sigma_{\|}} \chi . \tag{17.91}$$

In the long-$\lambda_{\|}$ limit, n_1 and ϕ_1 are thus 90° out of phase at threshold, and $|\eta| \gg |\chi|$. In the zero viscosity case, n_1 and ϕ_1 were in phase when $\Im(\omega) \to 0$, which occurred for $\sigma_{\|} \to \infty$; see Fig. 17.7.

More generally, we can consider the relation (17.47) between normalized density η and normalized electrostatic potential χ. The finite ion viscosity modifies ω entering (17.47). In the limit $\sigma_{\|} \to 0$, the "flute-mode" limit, we find now a finite ratio $|\eta/\chi|$ and the relative phase angle is also changed from the zero ion temperature case; see Fig. 17.7. Solving (17.82) for the case where $\sigma_{\|} = 0$, we find the solutions $\omega = \omega_i$ and $\omega = -i\sigma_{\perp}$. Using (17.47) we find $\eta/\chi = \omega^*/\omega_i = -T_e/T_i \equiv -1/\theta$, and $\eta/\chi = i\omega^*/\sigma_{\perp}$, respectively. The latter result implicitly contains k_y through $\sigma_{\perp}$ and ω^*. For large k_y, i.e., wavelengths shorter than the scale length for the density gradient, we find $|\chi| \gg |\eta|$, which is a property often found to characterize flute-modes.

Often we see flute-modes, with $k_{\|} = 0$ being distinguished from drift waves with $k_{\|} \neq 0$. Since both limiting cases belong to the same linear dispersion relation, this distinction seems somewhat artificial, but is nonetheless widespread.

17.7.4 Short-$\lambda_\parallel$ limit

If we keep the b terms in (17.82), we obtain another stabilization threshold for short-$\lambda_\parallel$. For large $\sigma_\parallel/\omega^*$, we would expect $\omega \approx \omega^*$; see, for instance, the dispersion relation in Fig. 17.5. Consequently we let $\omega = (1+\varepsilon)\,\omega^*$ and expand in the small parameter ε to obtain

$$-\varepsilon\omega^* = (1+\theta)\,\frac{\omega^{*2} - b\sigma_\parallel\sigma_\perp + i(\sigma_\perp + b\sigma_\parallel)\omega^*}{i\sigma_\parallel(1+b) + (2+\theta)\omega^* + i\sigma_\perp}. \tag{17.92}$$

For $\sigma_\parallel \gg \omega^*$, $\sigma_\parallel \gg \sigma_\perp$, this reduces, after some algebra, to

$$\omega - \omega^* = \frac{1+\theta}{\sigma_\parallel}\left[-\left(\sigma_\perp + b\sigma_\parallel\right)\omega^* + i\left(\omega^{*2} - b\sigma_\parallel\sigma_\perp\right)\right]. \tag{17.93}$$

There is therefore a stability threshold at $b\sigma_\parallel\sigma_\perp = \omega^{*2}$.

17.7.4.1 Long-$\lambda_\parallel$ and short-$\lambda_\parallel$ stabilization points

We have both long-$\lambda_\parallel$ and short-$\lambda_\parallel$ stabilization points. The long-$\lambda_\parallel$ stabilization criterion, $\theta\sigma_\perp > \sigma_\parallel$, with the definitions (17.46) and (17.79), works out to be

$$\frac{1}{4}\,\frac{T_i}{T_e}\,\frac{k_y^4}{k_z^2}\,\frac{\kappa T_i}{M\Omega_{ci}^2}\,\frac{m}{M\Omega_{ci}^2}\,\frac{\nu_{ii}}{\nu_{ei}} > 1. \tag{17.94}$$

or

$$\frac{B}{k_y} < \sqrt{\lambda_\parallel} \times \text{const.} \tag{17.95}$$

for given fixed plasma parameters. In (17.94) we introduced some "cosmetic" manipulations with the mass ratio in order to have only the ion cyclotron frequency appearing. For given $\lambda_\parallel$, the drift waves are stabilized at sufficiently small B/k_y.

The short-$\lambda_\parallel$ stabilization criterion, $b\sigma_\parallel\sigma_\perp > \omega^{*2}$, is, on the other hand, found to be

$$\frac{1}{4}b^2\,\frac{T_i^2}{T_e^2}\,\frac{k_z^2}{k_y^2}\,\omega_{ce}\Omega_{ci}\,\frac{\nu_{ii}}{\nu_{ei}} > \omega^{*2}. \tag{17.96}$$

The ratio ν_{ii}/ν_{ei} depends on the mass and temperature ratios but not on density (except for a weak dependence through the Spitzer logarithm).

17.7.5 An apparent paradox

The foregoing analysis has demonstrated that ion-ion collisions can stabilize drift waves. This stabilization can be understood as a short circuiting of the electric field from wave crest to wave trough. This observation is, however, apparently in conflict with the well-known result that ion-ion (or electron-electron) collisions do not give rise to any plasma conductivity or plasma transport across magnetic field lines, see Section 8.8.1. The basic statement is that for two particles of the same kind, the center-of-mass is the same *after* the collision as it was before. It is important to emphasize that the derivation of this standard result is based on the assumption of a vanishing or a spatially constant electric field. In the present case the electric field of the wave is harmonically varying in space, and ion-ion collisions *will* in this case give rise to a net current across magnetic fields in response to this inhomogeneous electric field. See also the discussion in Section 8.8.1 and the exercise given there.

17.8 Experimental observations of low frequency electrostatic drift waves

Electrostatic drift waves have been extensively studied in Q-machine plasmas (Politzer 1971, Motley 1975), with one of the most detailed, now classical, investigations carried out by Hendel et al. (1968) at relatively large plasma densities, $10^{17} - 10^{18}$ m^{-3}, so that ion viscosity damped most of the large azimuthal mode-numbers. Some of their results obtained detecting density fluctuations with Langmuir probes are summarized in Fig. 17.12. In particular, we note that the $m = 2,3,4$ modes appear one at a time with only a slight overlap, as the magnetic field is increased. It was found that the natural frequency of oscillations was close to $\frac{1}{2}\omega^*$, which is what we expect to be the linearly most unstable wave frequency; see Fig. 17.5. To obtain this result it was necessary to compensate for the azimuthal rotation of the entire plasma column, which is induced by a radial electric field component, inherent in the set-up. The amplitude of the relative density fluctuations, $\tilde{n}/n_0$, was found to maximize at the position with the largest density gradient, as expected.

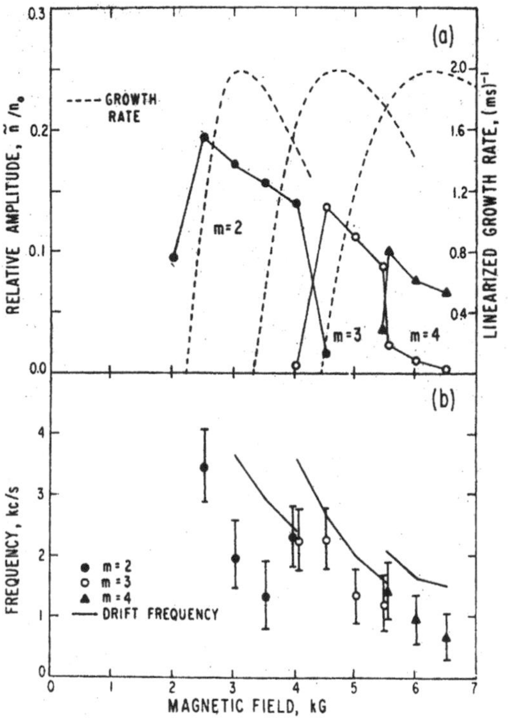

FIGURE 17.12: *Experimental results of drift wave investigations reprinted with permission from Hendel et al. (1968), copyright 1968 by the American Physical Society. The experiments were carried out in a Q-machine plasma at large densities. Figure a) shows relative mode amplitudes for varying magnetic fields, with dashed lines giving the linearized growth rate of the instability for reference. Figure b) shows the frequency of the oscillations. Here, solid lines give the analytical drift wave frequency ω^* for reference, as obtained for the given conditions.*

Considering the uncertainties involved in the measurements, the agreement with analytical results can be considered as good. We bear in mind that the rotation of the plasma column gives rise to a slight increase in the azimuthal current by the centrifugal force acting predominantly on the ions. These small corrections were unaccounted for in the theoretical results. The analysis has been generalized by Chu et al. (1969) to include the effects of a uniform rotation.

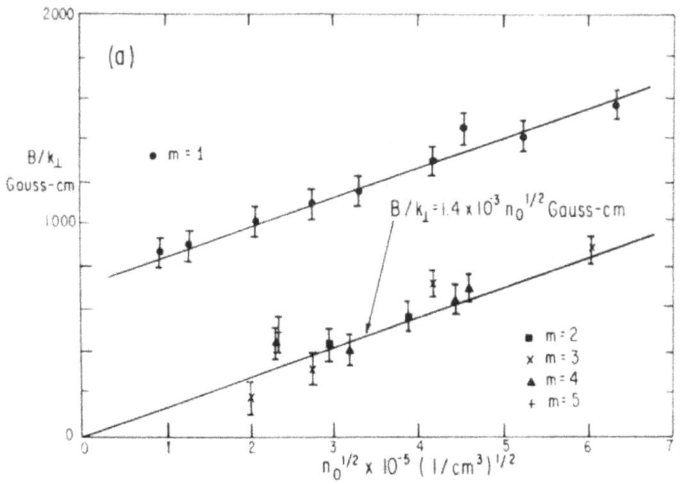

FIGURE 17.13: *Experimental studies of the parameter variation of the expression for marginal stability of drift waves, as given by (17.95). Reprinted with permission from Hendel et al. (1968), copyright 1968 by the American Physical Society. The investigations were carried out in a Q-machine plasma at large densities.*

The theoretical expression for marginal stability of drift waves (17.95) indicates a $\sqrt{n}$ scaling of B/k_y. Experimental results (Hendel et al. 1968) shown in Fig. 17.13 indicate that this predicted variation is in excellent agreement with observations.

By introducing both Cs and K ions in the system, Hendel et al. (1968) were able to change the average mass, $\overline{M}$, by a factor of more than 3 (from 39 to 132 in atomic mass units), and found results indicating a scaling $\overline{M}^{3/8}$ of $B/k_\perp$, again in good agreement with (17.94)–(17.95). The parameter scaling is in near excellent agreement with theory, but a numerical agreement concerning the threshold is obtained only by scaling down the magnetic field by a factor of 1.5, or, if you like, the threshold value of ν_{ii} is $(1.5)^4 \approx 5$ times what it should be for a given magnetic field.

Other experiments were carried out by Schlitt and Hendel (1972) in a modified Q-machine, where the hot plates were movable along the axis of the device. The length of the device, L, and thereby the **B**-parallel wavelength, could then be varied. The range of variation available was $L =$ 15–135 cm. The two hot plates act as ideal conductors, ensuring the presence of wave-nodes at these positions, thus determining the maximum parallel wavelengths. Curves for marginal stability in a plane spanned by the parameters $B/k_\perp$ and $\lambda_\parallel$ are illustrated in Fig. 17.14 for two parameter sets. The figure demonstrates three of the stability characteristics of collisional drift waves:

1): Low field stability – the plasma is stable for all parallel wavelengths for $B/k_\perp < 0.35$.

2): Short wavelength stability – as $\lambda_\parallel$ is decreased, the critical value of $B/k_\perp$ increases until a critical parallel wavelength (here $\lambda_\parallel = 60$ cm) is reached. For shorter $\lambda_\parallel$ the plasma is stable for all $B/k_\perp$.

3): High field stability – the four observations of critical values for $B/k_\perp > 0.8$ are points where the $m = 1$ mode is stabilized by **B**-parallel ion motion for large $B/k_\perp$.

Schlitt and Hendel (1972) also investigated the growth rate of the linearly unstable drift waves. The waves could be at least partially suppressed by applying a short pulse (duration 20 μs) with an amplitude of -25 V to one of the end plates (the precise mechanism for the observed stabilization seems not to be understood, nor studied in any detail). After the "switch-off" of the pulse, the drift waves were found to grow from the noise level, and their growth rate could be determined. In the *stable* regime, waves could be excited by application of a −5 V pulse, and their subsequent damping investigated. Also these results were in good agreement with theory. Also Rogers and Chen (1970) investigated the linear growth rate of the drift instabilities, using a feed-back stabilization technique which could be switched on and off. When the stability condition was changed from stable to unstable, the growth of the instability could be directly investigated. These results for the growth rates of the waves were also in good agreement with analytical results. All in all, it can be concluded that the linear theory of resistive drift waves seems to be on firm ground when compared with experimental results. It is therefore reasonable to continue the investigations by also studying the nonlinearly saturated state of the instability.

- **Exercise:** Consider a uniformly rotating magnetized cold plasma column. Assume that the rotation is imposed by an externally imposed radially varying parabolic electrostatic potential (Pécseli 1982). Assume the electron mass to be negligible. What is the angular electron rotation velocity? What is the corresponding velocity for the cold ion component, where you are assumed to include the centrifugal force? Given a radially varying plasma density, what is the azimuthal current?

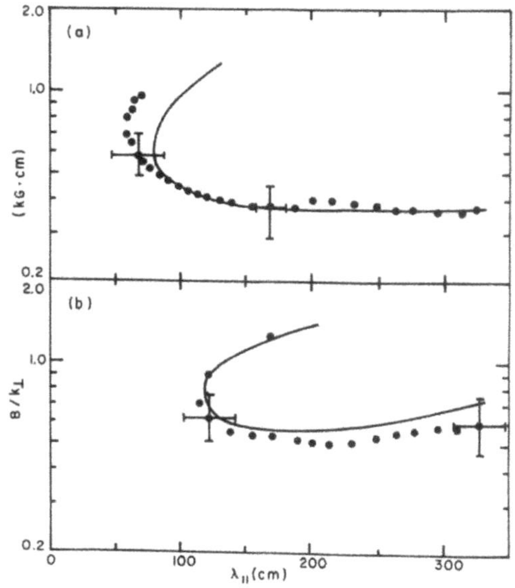

FIGURE 17.14: *Experimental results of drift wave investigations. Reprinted with permission from Schlitt and Hendel (1972), copyright 1972 by the American Institute of Physics. The experiments were carried out in a Q-machine plasma at large densities. Solid lines give theoretical curves for marginal stability with the magnetic field scaled by a factor of 1.5. The stable regions are toward the edges of the figures. In a) we have* $n = 2.5 \times 10^{10}$ cm^{-3}, *in b)* $n = 2 \times 10^{11}$ cm^{-3}.

17.9 Drift waves at larger frequencies

When the plasma density gradient becomes steep, the electron and ion drift velocities can become of the same order of magnitude as the ion acoustic sound speed. The drift wave frequency increases and eventually the assumption $\omega \ll \Omega_{ci}$ will break down. The dispersion relation will have additional roots corresponding to drift-cyclotron waves, which can be seen as ion cyclotron waves, see Section 15.2, modified by the density gradient. The propagating in the electron diamagnetic drift direction can become unstable for some finite k_z provided $k_y < n_0'/n_0$. The waves are excited when the drift wave phase velocity is comparable to the phase velocity of the ion cyclotron waves, requiring the stated large electron diamagnetic drift velocities. This means that $\Omega_{ci}/k_y \approx \omega^*/k_y$ or $\Omega_{ci} \approx \omega^* = -k_y a_i^2 \Omega_{ci} n_0'/n_0$ giving $k_y a_i^2 n_0'/n_0 \approx 1$. The small Larmor radius expansion implied in the derivation breaks down here and an accurate analysis has to be based on a kinetic theory (Mikhailovskii & Timofeev 1963).

We have one more ion wave as a possible candidate for coupling to electrostatic drift waves, the lower hybrid wave (Davidson et al. 1977), which is also a wavetype involving ion motion. Lower hybrid drift waves have importance in ionospheric environments (Shukla & Mamun 2002*b*) and they have been observed by instrumented space crafts (Treumann et al. 1991). When the β-value of the plasma is increased, we can have ω^*/k_z to be comparable to the Alfvén speed. In this limit, the drift waves can couple to Alfvén waves and the oscillations acquire an electromagnetic component (Kadomtsev 1965). The validity of the quasi-neutral approximation (i.e., $\omega^* \ll \Omega_{pi}$, see Section 11.2) needs to be checked when modeling these conditions.

17.10 Velocity shear driven instabilities

Discussions of velocity shear instabilities do not strictly speaking belong in a section on *universal* instabilities. It is perfectly feasible to imagine a magnetized inhomogeneous plasma without velocity shears. Reality is, however, something else. In a Q-machine, for instance, any radial density variation of the plasma column will be accompanied by a corresponding variation of the plasma potential (see Section 9.1), giving rise to a radial dc-electric field, $\mathbf{E}_\perp$. The result is in general a sheared azimuthal $\mathbf{E}_\perp \times \mathbf{B}/B^2$-drift, which can be unstable (Kent et al. 1969, Huld et al. 1991). Electrostatic drift wave instabilities often have modest growth rates, and a contribution from velocity shear instabilities can give rise to misleading conclusions if they are not taken properly into account. Here we will be concerned with velocity shears in the direction $\perp \mathbf{B}_0$, but note that also the magnetic field aligned velocities can have a shear, which can also give rise to unstable waves (Koepke & Reynolds 2007, Koepke 2008).

17.10.1 Velocity shear instabilities: flute modes

Here we consider slowly varying electrostatic structures in a homogeneous magnetic field, with the assumption that the scale lengths along $\mathbf{B}$ are infinite. The physical arguments for assuming Boltzmann distributed electrons are then not valid. Instead we assume the electron and ion species being coupled through Poisson's equation. Subtraction of the electron and ion continuity equations and use of (11.13) gives

$$\left(\frac{\partial}{\partial t} - \frac{1}{B^2}\nabla_\perp \phi \times \mathbf{B} \cdot \nabla_\perp\right) \nabla^2 \phi = 0, \tag{17.97}$$

which uniquely determines the evolution of the electrostatic plasma potential when the initial condition is given. The bulk plasma velocity is here $\mathbf{U} = -\nabla_\perp \phi \times \mathbf{B}/B^2$, and the vorticity deduced from it is $|\nabla_\perp \times \mathbf{U}| = |\nabla_\perp^2 \phi|$, which, by Poisson's equation shows that the charge density uniquely determines the vorticity in the present simplified two-dimensional model. The equation (17.97) is inherently nonlinear; upon linearization the result is trivially $\partial\phi/\partial t = 0$.

Consider here a basic flute-type model (17.97), and assume that, for some reason, we have an inhomogeneous background potential $\phi_0(x)$ in the system. This implies that a consistent electric field is of the form $\mathbf{E}_0 = -\nabla\phi_0(x) = \{E_x(x), 0\}$, which in turn imposes a sheared plasma velocity $\mathbf{U}_0 = \{0, E_x(x)/B\}$ in the direction transverse to the magnetic field lines. We note that in this form $\phi_0(x)$ is an *exact* solution of (17.97). We now make a perturbation analysis, and introduce $\phi = \phi_0(x) + \phi_1$. Ignoring second order terms, we find by trivial manipulations the equation

$$\frac{\partial}{\partial t}\nabla^2\phi_1 + \frac{E_x(x)}{B}\widehat{\mathbf{y}}\cdot\nabla\nabla^2\phi_1 + \widehat{\mathbf{z}}\times\nabla\phi_1\cdot\left(\frac{\partial^2}{\partial x^2}\frac{E_x(x)}{B}\widehat{\mathbf{x}}\right) = 0 \tag{17.98}$$

or

$$\frac{\partial}{\partial t}\nabla^2\phi_1 + U_0\frac{\partial}{\partial y}\nabla^2\phi_1 - U_0''\frac{\partial}{\partial y}\phi_1 = 0, \tag{17.99}$$

where $U_0'' \equiv \partial^2 U_0(x)/\partial x^2$ is the gradient of the zero order vorticity. The equation (17.99) is inhomogeneous only with respect to the x-variable, so we are free to Fourier transform it with respect to y as well time t, i.e., write $\phi_1 = \phi(x)\exp(-i(\omega t - k_y y))$. By insertion, we find the equation

$$\frac{d^2}{dx^2}\phi(x) - k_y^2\phi(x) - \frac{k_y U_0''(x)}{k_y U_0(x) - \omega}\phi(x) = 0. \tag{17.100}$$

For general profiles, $U_0''(x)$, the equation (17.100) can be solved only by numerical methods (Horton et al. 1987). We consider a special case with $U_0(x) = V\tanh(x/a)$, and find here a simplified solution by analyzing the inner and outer regions, respectively.

The outer solution is obtained by taking large values of $|x|$, where $U_0''(x) \approx 0$. Physically acceptable solutions for the electrostatic potential must be finite for $|x| \to \infty$, recalling that the potential ϕ_1 is the deviation from the unperturbed potential $\phi_0(x)$, associated with the sheared $U_0(x)$ velocity. In this case it is easy to obtain $\phi(x) = C\exp(k_y x)$ for $x < 0$ and $\phi(x) = D\exp(-k_y x)$ for $x > 0$. For the inner solution, we can take $U_0(x) \approx Vx/a$, and have $U_0''(x) \approx 0$. Consequently, also in this case it is easy to solve (17.100). The general solution is $\phi(x) = A\exp(k_y x) + B\exp(-k_y x)$. The full solution is then obtained by fitting the inner and outer solutions at $x = a$ and $x = -a$. In effect we hereby approximate the true variation $U_0(x)$ with $-U_0$ for $x < -a$, U_0 for $x > a$, and $U_0 x/a$ for $-a < x < a$. We argue on physical grounds that the potential ϕ must be a continuous function at $x = \pm a$, otherwise we would experience infinite electric fields. At the discontinuities we *can* have discontinuous electric fields, but this we have to live with. We then have

$$\begin{aligned} x = a \quad &: \quad De^{-k_y a} = Ae^{k_y a} + Be^{-k_y a} \\ x = -a \quad &: \quad Ce^{-k_y a} = Ae^{-k_y a} + Be^{k_y a}. \end{aligned} \tag{17.101}$$

From (17.100) we can obtain two boundary conditions at $x = a$ and $x = -a$. We thus integrate (17.100) from $x = -a - \varepsilon$ to $x = -a + \varepsilon$, and then let $\varepsilon \to 0$. In this limit only the derivative terms contribute, and we find the *jump conditions* (Chandrasekhar 1961, Baumjohann & Treuman 1997, Marshall 2001) to be

$$\frac{d}{dx}\phi\bigg|_{-a^-}^{-a^+} - \frac{k_y U_0'(x)}{k_y U_0(x) - \omega}\phi\bigg|_{-a^-}^{-a^+} = 0, \tag{17.102}$$

where $x = -a \pm \varepsilon \to -a^\pm$ for $\varepsilon \to 0$. In obtaining (17.102) we used that $U_0''(x)$ approximates a $\delta(x+a)$-function, which contributes no matter how small ε is, while all other terms, containing $U_0'(x)$ for instance, contribute with terms becoming negligible when $\varepsilon \to 0$.

The condition (17.102) gives that $d\phi/dx$ is discontinuous at $x = -a$ and

$$k_y(Ae^{-k_y a} - Be^{k_y a}) - k_y Ce^{-k_y a} - \frac{k_y U_0/a}{-k_y U_0 - \omega}(Ae^{-k_y a} + Be^{k_y a}) = 0.$$

Using (17.101), we find

$$B\left(2-\frac{k_yU_0/a}{k_yU_0+\omega}\right)=\frac{k_yU_0/a}{k_yU_0+\omega}e^{-2k_ya}A.$$

By an similar procedure at $x=a$ we find

$$A\left(2-\frac{k_yU_0/a}{k_yU_0-\omega}\right)=\frac{k_yU_0/a}{k_yU_0-\omega}e^{-2k_ya}B.$$

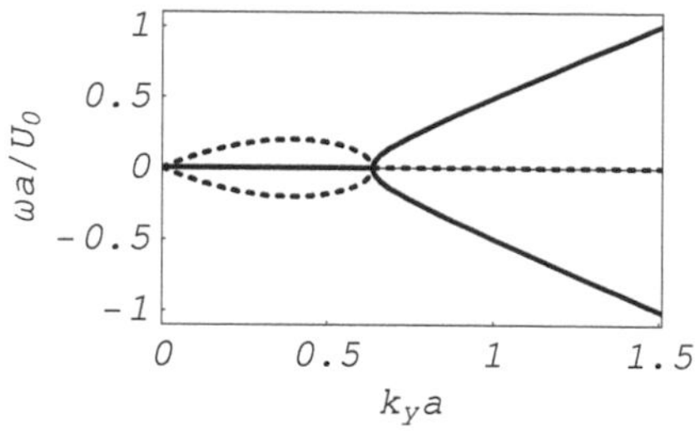

FIGURE 17.15: *Dispersion relation for Kelvin-Helmholtz, or shear flow, instability. The solid line gives the real part, the dashed line the imaginary part of* ω.

Simple manipulations finally give the dispersion relation (Horton et al. 1987)

$$\omega^2=\left(\frac{U_0}{2a}\right)^2\left[(1-2k_ya)^2-e^{-4k_ya}\right]. \tag{17.103}$$

We have two purely imaginary solutions for $k_ya<k_ca\approx 0.628$, where one corresponds to an aperiodic instability. For $k_y>k_c$ we find two propagating waves with the dispersion relation

$$\omega=\pm\left(\frac{U_0}{2a}\right)\sqrt{(1-2k_ya)^2-e^{-4k_ya}}. \tag{17.104}$$

See also Fig. 17.15 for details.

The maximum growth rate for the instability is obtained for $k_ya=\frac{1}{2}$, with the value $\omega=i\frac{1}{2}(U_0/a)/\exp(1)$, where we write $\exp(1)$ to avoid a possible confusion with the electron charge e. For this particular case we find $B=iA$, so the corresponding eigenfunction is $\phi_m=\cosh(x/2a)+i\sinh(x/2a)$ for $-a<x<a$, while $\phi_m=1+i\exp(1)$ for $x<-a$, and $\phi_m=\exp(1)+i$ for $x>a$. The potential variation associated with this simple antisymmetric eigensolution is shown in Fig. 17.16. The mode amplitude is here $[2\cosh^2(x/2a)-1]^{1/2}$, while the phase is $\theta(x)=\arctan[\tanh(x/2a)]$. Note that if the width of the shear layer $a\to\infty$ in such a way that U_0/a const., we have a constant value for the maximum growth rate, but the most unstable wave-number $k_y\to 0$, and also $k_c\to 0$. In this sense we can argue that for an infinitely wide shear, there is no finite wavelength instability, indicating that a necessary condition for instability is the *Rayleigh condition*, that the gradient of vorticity $U_0''=0$ for some x-value in the profile. The theorem is closely related to the one known as *Fjørtoft's theorem* (Marshall 2001). If we let the transition region shrink in order to let the velocity shear become a step function, i.e., let $a\to 0$ while keeping U_0 constant, we find that the growth rate of the most unstable perturbation increases indefinitely.

The potential variation in Fig. 17.16 is interesting by resembling two parallel alternating vortex chains. Since local shears will inevitably appear in any realizations of two-dimensional flows with large amplitude perturbations, this observation may give a hint for the seemingly universal presence of coupled vortex-like perturbations in two-dimensional turbulence. Numerical simulations for this problem, performed by Horton et al. (1987), indicates that this conjecture is justified.

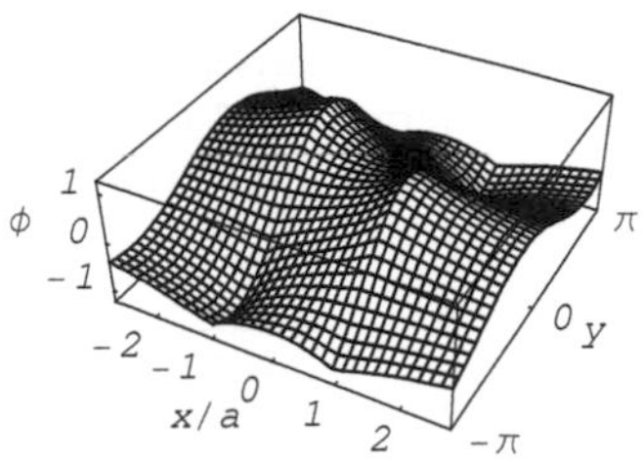

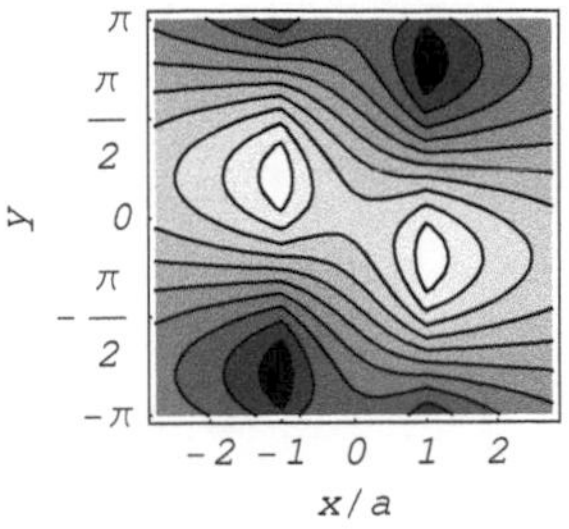

FIGURE 17.16: *Potential variation associated with the most unstable eigenfunction for the shear flow. The electrostatic potential is shown as a surface to the left, as contours to the right. The velocity shear is confined to the interval* $-1 < x/a < 1$. *The system is periodic in the y-direction.*

- **Example:** Mathematically, the relation (17.100) is a special case of the Taylor-Goldstein equation (Taylor 1931, Goldstein 1931), derived for neutral fluids in a constant gravitational field

$$\frac{d^2}{dx^2}w(x) - k_y^2 w(x) - \frac{U_0''(x)}{U_0(x) - \omega/k_y}w(x) + \frac{N^2}{(U_0(x) - \omega/k_y)^2}w(x) = 0, \tag{17.105}$$

where w is the complex amplitude of the vertical velocity. Equation (17.105) allows also for a change in fluid density through the buoyancy frequency $N \equiv \sqrt{-(g/\rho)d\rho/dx}$, with g being the (constant) gravitational acceleration. The standard Rayleigh-Taylor stability problem is obtained for vanishing flow velocity, $U = 0$, with $N \neq 0$. Instabilities of neutral fluids as described by (17.105) have been studied in great detail, theoretically as well as experimentally, by e.g., Lawrence et al. (1991).

- **Exercise:** Determine the eigenfunction associated with the marginally stable mode, where $k_y = k_c$; see also Fig. 17.15.

- **Exercise:** Determine, in terms of a, the wave-number k_c associated with the marginally stable mode, $\omega = 0$, for the case $U_0(x) = V\tanh(x/a)$, without making use of the approximate velocity profile used for obtaining (17.103).

17.10.2 Velocity shear instabilities with electron shielding

In Section 17.10 we discussed shear instabilities in the "flute limit", where the perturbations are assumed to be strictly magnetic field aligned. If we, as before, allow for a long **B**-parallel wavelength so that electrons can flow along magnetic field lines to be in an electrostatic Boltzmann equilibrium, we modify (17.100) to become

$$\frac{\partial}{\partial t}\left(\nabla^2\phi_1 - \frac{1}{\lambda_D^2}\phi_1\right) + \frac{E_x(x)}{B}\widehat{\mathbf{y}}\cdot\nabla\nabla^2\phi_1 + \widehat{\mathbf{z}}\times\nabla\phi_1\cdot\left(\frac{\partial^2}{\partial x^2}\frac{E_x(x)}{B}\widehat{\mathbf{x}}\right) = 0, \tag{17.106}$$

or

$$\frac{\partial}{\partial t}\left(\nabla^2\phi_1 - \frac{1}{\lambda_D^2}\phi_1\right) + U_0\frac{\partial}{\partial y}\nabla^2\phi_1 - U_0''\frac{\partial}{\partial y}\phi_1 = 0. \tag{17.107}$$

We find, after a Fourier transform with respect to t and y, the equation

$$\frac{d^2}{dx^2}\phi(x) - k_y^2\phi(x) - \frac{k_y U_0''(x) - \omega/\lambda_D^2}{k_y U_0(x) - \omega}\phi(x) = 0. \tag{17.108}$$

Unfortunately, this is not at all as easy to solve as (17.100), although the differences between the two basic equations appear to be modest, at first sight. The problem arises mainly when discussing the intermediate region where $U_0'' = 0$. If $\lambda_D \to \infty$, the last term in (17.100) and the equation is relatively simple to deal with, as we have seen. The outer solutions of (17.108), where $U =$ const., are easily obtained just as before as

$$\phi(x) = C\cos\left(\frac{x\sqrt{\omega - k_y^3 V\lambda_D^2 + k_y^2\lambda_D^2\omega}}{\lambda_D\sqrt{k_yV-\omega}}\right) + D\sin\left(\frac{x\sqrt{\omega - k_y^3 V\lambda_D^2 + k_y^2\lambda_D^2\omega}}{\lambda_D\sqrt{k_yV-\omega}}\right).$$

With λ_D finite, we have a problematic explicit x-dependence remaining through $U(x) = Vx/a$ in the denominator. The inner solution is for this case

$$\phi(x) = A\exp(-k_y x + a\omega/V)\,U\left(-\frac{\omega a}{2k_y^2 V\lambda_D^2}\,;0;\,2k_y x - 2a\omega/V\right)$$
$$+B\exp(-k_y x + a\omega/V)\,L\left(\frac{\omega a}{2k_y^2 V\lambda_D^2}\,;-1;\,2k_y x - 2a\omega/V\right),$$

in terms of hypergeometric functions $U(a;b;c) \equiv (1/\Gamma(a))\int_0^\infty \exp(-zt)t^{(a-1)}(1+t)^{(b-a-1)}dt$, and Laguerre polynomia $L(n;a;x)$ that satisfy the differential equation $xy'' + (a+1-x)y' + ny = 0$. The "jump-condition" remains here the same as (17.102) since $\phi(x)$ is continuous also here.

Instabilities driven by velocity shear are important in most magnetized plasmas in nature as well as in the laboratory. Often we find several instabilities contributing at the same time, and it can be difficult to sort out their respective contributions.

Comment: The analysis of electrostatic drift waves is particularly burdened by shorthand symbols for various quantities. It may be a good idea to have a list handy when reading the text. The first edition of this book has such a list on page 385.

18

Weakly Nonlinear Electrostatic Drift Waves

The analysis in the foregoing sections was based on a linearized analysis. In particular the basic equation (17.7) of Chapter 17.1 turned out to become linear, all by itself so to speak. Two essential approximations were made to derive that equation: the assumption of quasi-neutrality (the "plasma approximation") and omission of polarization drifts. In the following we elaborate in more detail the consequences of relaxing the latter of the two approximations including the nonlinear corrections.

18.1 Hasegawa-Mima equation

The relaxation of the quasi-neutrality condition implies some formal modifications of the basic equations, which, however, are unlikely to be of physical importance in comparison to the ion polarization drift, which is expected to be more important for most physically realistic conditions. The consequences of including ion polarization drifts are more important and will be studied in the following, but we retain the quasi-neutral approximation. It might be appropriate to recall here that the assumption of quasi-neutrality does not impose any conditions on $\nabla^2\phi$, see Section 11.2. First we consider the limiting case where the electrons are assumed to be in isothermal Boltzmann equilibrium (17.6). This limit was demonstrated to be linearly stable. Retaining only the **B**-perpendicular ion velocity component, the nonlinear equations (17.6), (17.10) and (17.14) are readily combined to give

$$\frac{\partial}{\partial t}\phi - \frac{C_s^2}{\Omega_c^2}\nabla_\perp\cdot\left(\frac{\partial}{\partial t} - \frac{1}{B^2}\nabla_\perp\phi\times\mathbf{B}\cdot\nabla_\perp\right)\nabla_\perp\phi = \tag{18.1}$$

$$\left(\frac{\nabla_\perp\phi\times\mathbf{B}}{B^2} - \overbrace{\frac{M}{eB^2}\left(\frac{\partial}{\partial t} - \frac{1}{B^2}\nabla_\perp\phi\times\mathbf{B}\cdot\nabla_\perp\right)\nabla_\perp\phi}^{\text{contributes to higher order}}\right)\cdot\left(\nabla_\perp\phi + \frac{\kappa T_e}{e}\frac{\nabla_\perp n_0}{n_0}\right).$$

A factor $\exp(e\phi/\kappa T_e)$ enters all terms and is canceled. The ordering implied in (18.1) is, however, not consistent with the basic assumption made by including the polarization drift in (17.14), where the second term is considered a small correction to the $\mathbf{E}\times\mathbf{B}$-ion drift. Then it is not consistent to retain the marked term in (18.1), since it contributes to higher order. We ignore also that part in the term which contains n_0'/n_0 by assuming that the density gradient is gentle. Physically this means that the scale-length of the background density variation is much larger than the effective Larmor radius, a_i, see Section 17.1. This latter assumption is not logically necessary and can be relaxed by retaining the scalar product of the polarization drift and ∇n_0. Using $\nabla_\perp\phi\times\mathbf{B}\cdot\nabla_\perp\phi = 0$ and $\nabla_\perp\cdot(\nabla_\perp\phi\times\mathbf{B}\cdot\nabla_\perp)\nabla_\perp\phi = (\nabla_\perp\phi\times\mathbf{B}\cdot\nabla_\perp)\nabla_\perp^2\phi$, we finally obtain the equation

$$\frac{\partial}{\partial t}\phi - \frac{\kappa T_e}{e}\frac{\nabla_\perp\phi\times\mathbf{B}}{B^2}\cdot\frac{\nabla_\perp n_0}{n_0} - \frac{C_s^2}{\Omega_c^2}\left(\frac{\partial}{\partial t} - \frac{1}{B^2}\nabla_\perp\phi\times\mathbf{B}\cdot\nabla_\perp\right)\nabla_\perp^2\phi = 0. \tag{18.2}$$

The first two terms are recognized as the simple result (17.7) obtained in Chapter 17.1. The third term in (18.2) is the correction from the polarization drift. We see the operator $(\partial/\partial t - B^{-2}\nabla_\perp\phi \times \mathbf{B}\cdot\nabla_\perp)$ pop up so often that it seems natural to assign it a particular importance for describing two-dimensional flows. The ordering implied in the derivation of (18.2) can be made explicit as

$$\frac{1}{\Omega_c}\frac{\partial}{\partial t} \sim \frac{\Omega}{\Omega_c} \sim a_i|\nabla \ln n_0| \sim \frac{1}{k_\| u_{Te}}\frac{\partial}{\partial t} \sim \varepsilon,$$

where $\Omega \equiv \nabla \times \mathbf{u}$ is here the ion vorticity, a_i is the ion Larmor radius derived from the electron temperature and ε is a small quantity. The restriction on $k_\| u_{T_e}$, in terms of the electron thermal velocity u_{T_e}, justifies the two-dimensional description.

Equation (18.2) is often called the Hasegawa-Mima equation after the two scientists who derived it in their analysis of weakly nonlinear drift waves (Hasegawa & Mima 1978). A similar equation, first derived by Charney (1948), applies for Rossby waves in the atmosphere and in the oceans.

Introducing dimensionless variables by normalizing length with a_i, time with $1/\Omega_c$ and potential with $\kappa T_e/e$, we have

$$\frac{\partial}{\partial t}(\nabla_\perp^2\phi - \phi) + (\hat{\mathbf{z}} \times \nabla_\perp\phi)\cdot\nabla_\perp(\nabla_\perp^2\phi) - \beta\frac{\partial}{\partial y}\phi = 0, \tag{18.3}$$

where $\beta = -d\ln n_0/dx$ was introduced as a measure of the electron diamagnetic drift. It is assumed that β can be considered as constant, at least locally. The notation is eased by introducing the Poisson brackets,

$$\{f;g\} \equiv \frac{\partial f}{\partial x}\frac{\partial g}{\partial y} - \frac{\partial f}{\partial y}\frac{\partial g}{\partial x} \equiv \hat{\mathbf{z}}\cdot(\nabla_\perp f \times \nabla_\perp g).$$

Note that $\{f;f\} = 0$, $\{f;g\} = -\{g;f\}$, $\{f+h;g\} = \{f;g\} + \{h;g\}$ and $\{f\cdot h;g\} = h\{f;g\} + f\{h;g\}$. These relations will be useful later on. A compact formulation of (18.3) can be obtained as

$$\frac{\partial}{\partial t}(\nabla_\perp^2\phi - \phi) - \beta\frac{\partial}{\partial y}\phi + \{\phi;\nabla_\perp^2\phi\} = 0, \tag{18.4}$$

or

$$\frac{\partial}{\partial t}q + \{\phi;q\} = 0, \tag{18.5}$$

with $q = \phi - \nabla_\perp^2\phi - \beta x$. We added the term βx in the time differential (the first term) in (18.4). As an alternative to (18.5) we can write

$$\left(\frac{\partial}{\partial t} + \mathbf{u}\cdot\nabla_\perp\right)q = 0, \tag{18.6}$$

with $\mathbf{u} \equiv -\nabla_\perp\phi \times \hat{\mathbf{z}}$. Equation (18.6) has the form of a continuity equation for q in incompressible flows. The equation (18.2) or (18.3) has a number of interesting properties which will be discussed in the following.

In the limit where $C_s/\Omega_c \to \infty$ and the density gradient vanishes, $\nabla_\perp n_0 = 0$, the Hasegawa-Mima equation formally reduces to the two-dimensional Euler equation

$$\left(\frac{\partial}{\partial t} + \mathbf{u}\cdot\nabla_\perp\right)\rho = 0, \tag{18.7}$$

with $\mathbf{u} \equiv -\nabla_\perp\phi \times \widehat{\mathbf{z}}$, where $\rho \equiv \nabla_\perp^2\phi = \widehat{\mathbf{z}}\cdot\nabla_\perp \times \mathbf{u}$ takes the role of vorticity in this two-dimensional model. Equation (18.7) has the form of a continuity equation for ρ, to be compared with (18.6). Seen from this point of view, it is expected that properties of the Hasegawa-Mima equation have many similarities with the equations characterizing two-dimensional turbulence in incompressible fluids. Being a generalization of the fluid vorticity, $q = \phi - \nabla_\perp^2\phi - \beta x$ is often called *potential vorticity*.

18.1.1 Linearized Hasegawa-Mima equation

In its linearized version, the Hasegawa-Mima equation becomes

$$\frac{\partial}{\partial t}(\nabla_\perp^2 \phi - \phi) - \beta \frac{\partial}{\partial y}\phi = 0. \tag{18.8}$$

The linear dispersion relation is easily obtained as

$$\omega = \frac{\beta k_y}{1+k_\perp^2} \equiv \frac{\omega^*}{1+k_\perp^2}, \tag{18.9}$$

with $k_\perp^2 \equiv k_x^2 + k_y^2$. The phase velocity is restricted to a finite range with $\omega/k_y \in [0, \beta]$ for $k_\perp^2 \in [0, \infty]$.

18.1.2 Conservation laws

Considering (18.6), it is readily concluded that q is constant along the characteristics of this equation. The same conclusion will hold for any function $F(q)$. Making explicit use of the incompressibility $\nabla \cdot \mathbf{u} = \nabla \cdot \nabla\phi \times \mathbf{B}/B^2 = 0$, it is easily shown that any integral (Casimir functional) $\iint_{-\infty}^{\infty} F(q)dxdy$ will be a constant of motion. Also $\iint_{-\infty}^{\infty} q^n dxdy$ is a constant of motion for all n. The notation in terms of Poisson brackets is particularly useful for obtaining conservation laws when the relations $\iint_{-\infty}^{\infty} f\{h;g\}dxdy = \iint_{-\infty}^{\infty} g\{f;h\}dxdy = \iint_{-\infty}^{\infty} h\{g;f\}dxdy$ are used, with the proper assumptions on the properties of the functions at $|\mathbf{r}| \to \infty$.

- **Exercise:** Using the incompressibility condition $\nabla \cdot \mathbf{u} = 0$ together with $\partial q(\mathbf{r},t)/\partial t + \mathbf{u} \cdot \nabla q(\mathbf{r},t) = 0$, prove that $d \iint_{-\infty}^{\infty} F(q)\, dxdy/dt = 0$ for any differentiable $F(q(\mathbf{r},t))$.

In particular we have a quadratic constant of motion, the energy integral given as $E \equiv \frac{1}{2}\iint_{-\infty}^{\infty} (\phi^2 + (\nabla\phi)^2)$
$dxdy$. This can be shown most simply by use of the form (18.5), which after multiplication with ϕ and integration gives

$$\iint_{-\infty}^{\infty} \left(\phi \frac{\partial}{\partial t} q + \phi\{\phi;q\}\right) dxdy = \iint_{-\infty}^{\infty} \phi \frac{\partial}{\partial t}(\nabla^2\phi - \phi)\, dxdy = 0. \tag{18.10}$$

It was used that $\iint_{-\infty}^{\infty} f\{f;g\}dxdy = 0$ for all g with the proper asymptotic relations. The term βx vanishes by the time differentiation. We note the identity $\phi\partial\nabla^2\phi/\partial t = \nabla \cdot (\phi\partial\nabla\phi/\partial t) - \frac{1}{2}\partial(\nabla\phi)^2/\partial t$. Simple manipulations of (18.10), using Green's theorem gives $\int_{-\infty}^{\infty} \nabla \cdot (\phi\partial\nabla\phi/\partial t)dxdy = \oint \phi\partial\nabla\phi/\partial t \cdot d\mathbf{n} = 0$ with $\mathbf{n}$ being a unit normal vector to the curve in the x,y-plane bounding the entire system. We took the boundary of the system to be at infinity with $\phi \to 0$ there. It is thus readily demonstrated that

$$\frac{d}{dt}\iint_{-\infty}^{\infty} \frac{1}{2}(\phi^2 + (\nabla\phi)^2)\, dxdy = 0,$$

stating the conservation of E. The two terms can be interpreted as an electrostatic potential energy ($n\phi \sim \phi^2$) and a kinetic energy ($n_0 u^2 \sim (\nabla\phi)^2$).

In a Fourier representation this conservation law takes the form

$$\frac{d}{dt}\sum_{\mathbf{k}} (1+k^2)\, |\phi_{\mathbf{k}}|^2 = 0, \tag{18.11}$$

by use of Parseval's theorem $\iint_{-\infty}^{\infty} |f|^2 dxdy = \sum_{\mathbf{k}} |f_{\mathbf{k}}|^2$.

Another quadratic constant of motion, namely the generalized enstrophy $W = \frac{1}{2}\iint_{-\infty}^{\infty} ((\nabla\phi)^2 + (\nabla^2\phi)^2)dxdy$, can be interpreted as being similar to the enstrophy defined for two-dimensional incompressible flows. The conservation of W is demonstrated by using (18.4), which after multiplication with $\nabla^2\phi$ and integration gives

$$\int\int_{-\infty}^{\infty}\nabla^2\phi\left(\frac{\partial}{\partial t}(\nabla^2\phi-\phi)-\beta\frac{\partial}{\partial y}\phi\right)dxdy=0.$$

Using

$$(\nabla^2\phi)\frac{\partial}{\partial t}\phi=\nabla\cdot\left(\nabla\phi\frac{\partial}{\partial t}\phi\right)-\nabla\phi\cdot\frac{\partial}{\partial t}\nabla\phi=\nabla\cdot\left(\nabla\phi\frac{\partial}{\partial t}\phi\right)-\frac{1}{2}\frac{\partial}{\partial t}(\nabla\phi)^2,$$

and

$$(\nabla^2\phi)\frac{\partial}{\partial y}\phi=\nabla\cdot\left(\nabla\phi(\nabla\phi\cdot\hat{\mathbf{y}})\right)-\frac{1}{2}\frac{\partial}{\partial y}(\nabla\phi)^2=\nabla\cdot\left(\nabla\phi(\nabla\phi\cdot\hat{\mathbf{y}})-\frac{1}{2}(\nabla\phi)^2\hat{\mathbf{y}}\right),$$

we obtain by use of Green's theorem and some mild restrictions on ϕ at $|\mathbf{r}|\to\infty$

$$\frac{d}{dt}\int\int_{-\infty}^{\infty}\frac{1}{2}\left((\nabla\phi)^2+(\nabla^2\phi)^2\right)dxdy=0,$$

demonstrating the conservation of generalized enstrophy, W, as defined before. For two-dimensional homogeneous and incompressible flows the enstrophy, i.e., the space-integral of the squared vorticity $\int\int_{-\infty}^{\infty}(\nabla^2\phi)^2dxdy$, is conserved. In addition to such a term, the generalized enstrophy contains a term $(\nabla\phi)^2$, which earlier was identified with the rotation of the $\nabla\phi\times\mathbf{B}$-velocity. In Fourier representation the conservation law for generalized enstrophy takes the form

$$\frac{d}{dt}\sum_{\mathbf{k}}(1+k^2)\,k^2|\phi_{\mathbf{k}}|^2=0, \tag{18.12}$$

obtained again by use of Parseval's theorem. Combination of the conservation laws of E and W gives trivially that also $\int\int_{-\infty}^{\infty}(\phi^2-(\nabla^2\phi)^2)dxdy$ is conserved by the Hasegawa-Mima equation.

It might be expected that the multitude of conserved quantities implies strong restrictions of the evolution of an initial perturbation as it is described by the Hasegawa-Mima equation. This is, however, not necessarily so, since for the continuum system there are infinitely many degrees of freedom. The constraints are of more importance when the number of degrees of freedom is restricted in, for instance, numerical solutions of the equation, when a spectral code with a finite number of Fourier modes is used. It will be found that usually only a few of the conservation laws survive this truncation in wave-numbers, conservation of E and W being two such examples. These conservations are "robust".

It should be noted that in special cases, in particular in simulations truncated at only a few Fourier modes, the number of conserved qualities can be larger (Hald 1976). The time evolutions of the systems will consequently have stronger constraints. Energy and enstrophy, as defined before, will remain conserved for all these systems, irrespective of the number of Fourier components they may contain.

18.1.3 Wave interactions

The Hasegawa-Mima equation describes weakly nonlinear electrostatic fluctuations, in particular also wave interactions. For this analysis, it is convenient to write the equations in a spectral representation by introducing

$$\phi(\mathbf{r},t)=\sum_k\phi_{\mathbf{k}}(t)e^{i\mathbf{k}\cdot\mathbf{r}}, \tag{18.13}$$

where $\phi_{\mathbf{k}}(t)=\phi^*_{-\mathbf{k}}(t)$ since $\phi(\mathbf{r},t)$ is a real-valued function. If (18.13) is inserted into the *linearized* version of (18.3), the coefficients are obtained as $\phi_{\mathbf{k}}(t)=a\exp(-i\omega_{\mathbf{k}}t)$ where $\omega_{\mathbf{k}}=\beta k_y/(1+k^2)$ is the linear dispersion relation for (18.3) and a is a small amplitude. Note that the linear properties of (18.3) are trivial in the case where the density gradient can be ignored, i.e., $\beta = 0$. In this case the result is $\nabla_\perp^2 \phi = \phi$, giving the linearized version of Debye shielding. The presence of a gradient, or rather the electron diamagnetic drift, gives rise to the linear dynamics of the waves. The nonlinear terms induce an additional time variation of the wave amplitudes $\phi_{\mathbf{k}}$, to be obtained by writing the Fourier transform of the nonlinear terms in (18.3) as a convolution sum. In this representation we have (18.3) in the form

$$(1+k^2)\frac{d}{dt}\phi_{\mathbf{k}}(t) + i\beta k_y \phi_{\mathbf{k}}(t) = \sum_{\mathbf{k}=\mathbf{k}'+\mathbf{k}''} k''^2 (\hat{\mathbf{z}} \times \mathbf{k}') \cdot \mathbf{k}''\, \phi_{\mathbf{k}'}(t)\phi_{\mathbf{k}''}(t). \tag{18.14}$$

By simple manipulations this equation is rewritten as

$$\frac{d}{dt}\phi_{\mathbf{k}}(t) + i\omega_{\mathbf{k}}\phi_{\mathbf{k}}(t) = \sum_{\mathbf{k}=\mathbf{k}'+\mathbf{k}''} \Lambda^{-\mathbf{k}}_{\mathbf{k}',\mathbf{k}''}(k''^2 - k'^2)\, \phi_{\mathbf{k}'}(t)\phi_{\mathbf{k}''}(t), \tag{18.15}$$

with the coupling coefficient being

$$\Lambda^{-\mathbf{k}}_{\mathbf{k}',\mathbf{k}''} = \frac{1}{2}\frac{(\mathbf{k}' \times \mathbf{k}'') \cdot \hat{\mathbf{z}}}{(1+k^2)},$$

and the wave-number matching condition $\mathbf{k}' + \mathbf{k}'' - \mathbf{k} = 0$. Physically, the relation (18.15) demonstrates the interaction of three wave-number components labeled by $\mathbf{k}$, $\mathbf{k}'$ and $\mathbf{k}''$, where a matching condition is fulfilled. The strength of the interaction is measured by the coupling coefficient $\Lambda^{-\mathbf{k}}_{\mathbf{k}',\mathbf{k}''}$. The coupling is large for "fat" triangles composed of the three vectors $\mathbf{k}'$, $\mathbf{k}''$ and $\mathbf{k}$, while $\Lambda^{-\mathbf{k}}_{\mathbf{k}',\mathbf{k}''}$ is small for "slim" triangles, and vanishes when the three vectors are on a straight line.

The sum in (18.15) is over all possible triplets in the actual perturbation. In general a large number, N, of different terms interact simultaneously, assuming a numerical simulation with a finite number of Fourier modes. We may, however, imagine that during a small time interval $\Delta t/N$ only one term acts, however with the N-fold intensity with all other terms "switched-off". During the next small time interval $\Delta t/N$, another term acts and so on, until all terms on the right-hand side of (18.15) have been activated (Knorr et al. 1990). This procedure is known in numerical methods as "the method of weak approximation" or as "the method of fractional steps" or simply "splitting". It can be proven that the approximation converges to the true solution if the time interval Δt is reduced. Seen in this light, the basic nonlinearity is adequately described by considering one triplet only. This interaction is described by

$$\frac{d}{dt}\phi_{\mathbf{k}}(t) + i\omega_{\mathbf{k}}\phi_{\mathbf{k}}(t) = \Lambda^{-\mathbf{k}}_{\mathbf{p},\mathbf{q}}(p^2 - q^2)\phi_{\mathbf{p}}(t)\phi_{\mathbf{q}}(t). \tag{18.16}$$

By permutation of indexes, i.e., $p \to q$, $q \to -k$ and $-k \to p$ we find

$$\frac{d}{dt}\phi_{\mathbf{p}}(t) + i\omega_{\mathbf{p}}\phi_{\mathbf{p}}(t) = \Lambda^{\mathbf{p}}_{\mathbf{q},-\mathbf{k}}(q^2 - k^2)\phi^*_{\mathbf{q}}(t)\phi_{\mathbf{k}}(t), \tag{18.17}$$

where we used the reality condition $\phi_{-\mathbf{p}} = \phi^*_{\mathbf{p}}$ and $\omega_{-\mathbf{p}} = -\omega_{\mathbf{p}}$ and recalling that for the present case $\Lambda^{-\mathbf{k}}_{\mathbf{p},\mathbf{q}}$ is real. Similarly, by $p \to -k$, $q \to p$ and $-k \to q$, we find

$$\frac{d}{dt}\phi_{\mathbf{q}}(t) + i\omega_{\mathbf{q}}\phi_{\mathbf{q}}(t) = \Lambda^{\mathbf{q}}_{-\mathbf{k},\mathbf{p}}(k^2 - p^2)\phi_{\mathbf{k}}(t)\phi^*_{\mathbf{p}}(t), \tag{18.18}$$

where $\mathbf{p} + \mathbf{q} - \mathbf{k} = 0$. Note that no explicit assumption on frequency matching has been imposed, i.e., we can have $\omega_{\mathbf{p}} + \omega_{\mathbf{q}} - \omega_{\mathbf{k}} \neq 0$ in general. The equations (18.16)–(18.18) can be solved exactly. This solution will not be obtained here, where only a qualitative analysis is attempted.

The conservation laws from the foregoing section imply

$$(1+k^2)|\phi_{\mathbf{k}}(t)|^2+(1+p^2)|\phi_{\mathbf{p}}(t)|^2+(1+q^2)|\phi_{\mathbf{q}}(t)|^2=\text{const.} \tag{18.19}$$

and

$$(1+k^2)k^2|\phi_{\mathbf{k}}(t)|^2+(1+p^2)p^2|\phi_{\mathbf{p}}(t)|^2+(1+q^2)q^2|\phi_{\mathbf{q}}(t)|^2=\text{const.} \tag{18.20}$$

- **Exercise:** Prove the relation

$$\Lambda^{-\mathbf{k}}_{\mathbf{p},\mathbf{q}}(p^2-q^2)+\Lambda^{\mathbf{p}}_{\mathbf{q},-\mathbf{k}}(q^2-k^2)+\Lambda^{\mathbf{q}}_{-\mathbf{k},\mathbf{p}}(k^2-p^2)=0$$

for the coupling coefficients.

- **Exercise:** Prove the relations (18.19) and (18.20) directly from (18.16)–(18.18).

Introducing the energy of a mode k as $(1+k^2)|\phi_{\mathbf{k}}(t)|^2 \equiv w_{\mathbf{k}}(t)$, and similarly for **p** and **q**, we have from (18.19) and (18.20)

$$w_{\mathbf{k}}(t)+w_{\mathbf{p}}(t)+w_{\mathbf{q}}(t)\equiv 2W=\text{const.} \tag{18.21}$$

and

$$k^2w_{\mathbf{k}}(t)+p^2w_{\mathbf{p}}(t)+q^2w_{\mathbf{q}}(t)\equiv 2E=\text{const.} \tag{18.22}$$

Introducing a Cartesian coordinate system, with axes in directions $w_{\mathbf{k}}$, $w_{\mathbf{p}}$ and $w_{\mathbf{q}}$, the relations (18.21)–(18.22) represent two intersecting planes. Any change in energy must satisfy $\delta w_{\mathbf{k}}+\delta w_{\mathbf{p}}+\delta w_{\mathbf{q}}=0$ and $k^2\delta w_{\mathbf{k}}+p^2\delta w_{\mathbf{p}}+q^2\delta w_{\mathbf{q}}=0$, or $\delta\mathbf{w}\cdot\mathbf{e}=0$ and $\delta\mathbf{w}\cdot\mathbf{s}=0$, where $\delta\mathbf{w}=(\delta w_{\mathbf{k}},\delta w_{\mathbf{p}},\delta w_{\mathbf{q}})$, $\mathbf{e}=(1,1,1)$ and $\mathbf{s}=(k^2,p^2,q^2)$ in the coordinate system defined before. A system point can only move in the direction $\mathbf{e}\times\mathbf{s}$, so the changes become explicitly

$$\begin{aligned}
\delta w_{\mathbf{k}} &= (q^2-p^2)\frac{d}{dt}f(t)\delta t\\
\delta w_{\mathbf{p}} &= (k^2-q^2)\frac{d}{dt}f(t)\delta t\\
\delta w_{\mathbf{q}} &= (p^2-k^2)\frac{d}{dt}f(t)\delta t.
\end{aligned} \tag{18.23}$$

The results (18.23) are easily verified by insertion into (18.21)–(18.22). The function $df(t)/dt$ with initial condition $df(t)/dy|_{t=0}=0$ can be determined by (18.16)–(18.18). We do not need this explicit result here, but rather follow Fjørtoft (1953) in considering the ratios

$$\frac{\delta w_{\mathbf{k}}}{\delta w_{\mathbf{p}}} = \frac{(q^2-p^2)}{(k^2-q^2)} \tag{18.24}$$

$$\frac{\delta w_{\mathbf{p}}}{\delta w_{\mathbf{q}}} = \frac{(k^2-q^2)}{(p^2-k^2)} \tag{18.25}$$

$$\frac{\delta w_{\mathbf{q}}}{\delta w_{\mathbf{k}}} = \frac{(p^2-k^2)}{(q^2-p^2)}. \tag{18.26}$$

Assuming for instance that $p<k<q$, we have, for example, $\delta w_{\mathbf{k}}/\delta w_{\mathbf{p}}<0$ and $\delta w_{\mathbf{q}}/\delta w_{\mathbf{k}}<0$, while $\delta w_{\mathbf{p}}/\delta w_{\mathbf{q}}>0$. From this result it is readily seen that no single of the three wave-number components can represent a source or a sink for *both* two remaining components, unless it is represented by a wave-number (i.e., inverse scale size) intermediate between the wave-numbers (or inverse scales) of the two other components (Fjørtoft 1953). Note that we can express the ratios in (18.24)–(18.26) solely in terms of p/k and q/k.

By the cos-relation for a triangle it is readily shown that we have $(k^2-q^2)/(p^2-k^2)>1$ when $k^2>2pq\cos(q,p)$, where (q,p) here denotes the angle between the sides q and p in the triangle spanned by $\mathbf{p}$, $\mathbf{q}$ and $\mathbf{k}$. When this condition is fulfilled, the numerical value of the change in energy will be largest for the component with the smallest wave-number p, and smallest for the one with the largest wave-number q. This is actually a remarkable result: in contrast to incompressible fluid turbulence in three spatial dimensions, the energy will not cascade mainly from large scales toward small scales. The conservation of energy as well as generalized enstrophy implies that if some energy is cascaded toward small scales in a triplet interaction *more* have to be cascaded toward larger scales (i.e., smaller wave numbers) for the given wave-vectors. This is the *inverse cascade*, being a characteristic of many two-dimensional flows, not only drift waves. For such systems, in case an initial perturbation is concentrated on small scales, i.e., large wave-numbers, eventually a significant part of the energy will be spreading to the long wavelength regime (Fjørtoft 1953, Merilees & Warn 1975). Other, more intuitive arguments for the inverse cascade can be found as well (Pedlosky 1987, Pécseli 2016).

A byproduct of the inverse cascade turns out to be disturbing in numerical simulations. In actual simulations it is difficult to avoid introducing an error at small scales, where the spatial resolution of the numerical code is limited. This deviation between the numerical solution and the "true" solution would usually be of little concern in other types of problems. This deviation can, however, in the analysis of, for instance, drift waves, be seen as an (unwanted) contribution to the spectral energy distribution, and the associated error will consequently spread to the long wavelength regime as well, due to the inverse cascade!

18.1.4 Coherent three wave interactions

The analysis in the foregoing section did not explicitly assume wave-vector and frequency matching for the three interacting wave components. In this section we carry out a more restrictive, yet illustrative, analysis of three interacting drift waves with restrictions imposed on the frequencies of the waves. Assume that initially a large amplitude wave $A_0\exp[-i(\omega_0 t-\mathbf{k}_0\cdot\mathbf{r})]$ with A_0 real is present, where ω_0 and $\mathbf{k}_0$ satisfy the dispersion relation. This wave is a solution to the model equations (18.16) provided the two other wave amplitudes in the nonlinear term on the right-hand side vanish (Weiland & Wilhelmsson 1977). We now assume that this solution to the equations is perturbed by two small amplitude waves $A_1(t)\exp[-i(\omega_1 t-\mathbf{k}_1\cdot\mathbf{r})]$ and $A_2(t)\exp[-i(\omega_2 t-\mathbf{k}_2\cdot\mathbf{r})]$. Assuming $A_1\ll A_0$ and $A_2\ll A_0$, we linearize the basic equations to obtain

$$\frac{d}{dt}A_1(t) = \Lambda^{\mathbf{k}_1}_{\mathbf{k}_2,-\mathbf{k}_0}(k_2^2-k_0^2)A_2^*(t)A_0e^{i\theta t}, \qquad (18.27)$$

$$\frac{d}{dt}A_2(t) = \Lambda^{\mathbf{k}_2}_{-\mathbf{k}_0,\mathbf{k}_1}(k_0^2-k_1^2)A_0A_1^*(t)e^{i\theta t}, \qquad (18.28)$$

where for the time being we allow for a non-vanishing frequency mismatch $\theta=\omega_1+\omega_2-\omega_0$, but assume exact wave-number matching $\mathbf{k}_1+\mathbf{k}_2=\mathbf{k}_0$. Solving (18.27) for $A_2^*(t)$ and inserting this into the complex conjugate (18.28) we obtain an equation for $A_1(t)$ as

$$\frac{d^2}{dt^2}A_1(t)-i\theta\frac{d}{dt}A_1(t)=\Lambda^{\mathbf{k}_1}_{\mathbf{k}_2,-\mathbf{k}_0}(k_2^2-k_0^2)\Lambda^{\mathbf{k}_2}_{-\mathbf{k}_0,\mathbf{k}_1}(k_0^2-k_1^2)A_0^2A_1. \qquad (18.29)$$

Assuming $A_1\sim\exp(-i\Omega t)$, we find

$$\Omega^2+\theta\Omega=-\Lambda^{\mathbf{k}_1}_{\mathbf{k}_2,-\mathbf{k}_0}(k_2^2-k_0^2)\Lambda^{\mathbf{k}_2}_{-\mathbf{k}_0,\mathbf{k}_1}(k_0^2-k_1^2)A_0^2 \qquad (18.30)$$

or

$$\Omega=-\frac{1}{2}\theta\pm\frac{1}{2}\left(\theta^2-4\Lambda^{\mathbf{k}_1}_{\mathbf{k}_2,-\mathbf{k}_0}(k_2^2-k_0^2)\Lambda^{\mathbf{k}_2}_{-\mathbf{k}_0,\mathbf{k}_1}(k_0^2-k_1^2)A_0^2\right)^{1/2}. \qquad (18.31)$$

The perturbation will be unstable ($\Im\{\Omega\} > 0$) if

$$4\Lambda^{\mathbf{k}_1}_{\mathbf{k}_2,-\mathbf{k}_0}(k_2^2-k_0^2)\Lambda^{\mathbf{k}_2}_{-\mathbf{k}_0,\mathbf{k}_1}(k_0^2-k_1^2)A_0^2 > \theta^2. \tag{18.32}$$

The strongest instability (largest growth rate) occurs for exact frequency matching where $\theta = 0$. This requirement imposes nontrivial restrictions on the wave-numbers of the perturbations, since the assumed perfect wave-number matching has to be retained together with the dispersion relation $\omega_j = \omega(\mathbf{k}_j)$, for all three wave-number components $\mathbf{k}_j$, with $j = 0, 1, 2$. For small amplitudes A_0 we need $\theta \approx 0$ to have instability. For enhanced amplitudes we can have instability for finite bandwidths.

The stability condition (18.32) can be discussed in more detail by considering the coupling coefficients appearing in the expression. We first note that the absolute values of all the vector products entering the $\Lambda^{-\mathbf{k}}_{\mathbf{k}',\mathbf{k}''}$-s are equal, i.e., $\mathbf{k}_1 \times \mathbf{k}_0 = \mathbf{k}_0 \times \mathbf{k}_2$, due to the assumed wave-number matching, irrespective of the magnitude of the individual wave-numbers. The sign of the coupling coefficients is then determined by the product $(k_2^2-k_0^2)(k_0^2-k_1^2)$. In particular $k_1 < k_0 < k_2$ implies instability for sufficiently large A_0^2, confirming the previous result that under certain conditions, energy can cascade from intermediate wave-numbers toward larger and smaller wave-numbers. On the other hand, the large amplitude wave in a wave-triplet cannot act as a "donor" of wave energy for two waves *both* being associated with larger (or smaller) scales.

18.1.5 Stationary solutions

Assume that a stationary solution to the Hasegawa-Mima equation exists in some frame of reference moving with the velocity u along the y-axis (Makino et al. 1981). In this case $\phi = \phi(y-ut, x)$ and we can write $\partial\phi/\partial t = -u\partial\phi/\partial y'$ with the coordinates x and $y' = y - ut$ and obtain from (18.5)

$$-u\frac{\partial}{\partial y'}q + \{\phi; q\} = 0, \tag{18.33}$$

or

$$\{\phi - ux; q\} = 0, \tag{18.34}$$

with $q = \phi - \nabla_\perp^2\phi - \beta x$ as before, and $\{-ux; q\} = -u\partial q/\partial y'$. Since q contains ϕ explicitly, both elements of the Poisson bracket in (18.34) contain ϕ. By isolating ϕ, the relation can be rewritten in a more convenient way as

$$\{\phi + ux; \nabla^2\phi + (u-\beta)x\} = 0. \tag{18.35}$$

A class of solutions to (18.35) can be found as $\nabla^2\phi + (u-\beta)x = G(\phi + ux)$ where G is an arbitrary function. In particular a simple linear relationship can be assumed, giving

$$\nabla_\perp^2\phi + (u-\beta)x = C(\phi + ux), \tag{18.36}$$

where C is a multiplicative constant. The solutions of this equation in terms of Bessel functions are in principle simple, but these solutions are generally not spatially localized, as would be required from physically acceptable pulse-like solutions. One way to get around this problem was suggested by, e.g., Larichev and Reznik (1976), who considered two spatial regions separated by a circle with radius R, i.e., an inner and an outer part of the x, y'-plane, where the two solutions were matched on the boundary; see also a detailed discussion by Flierl et al. (1980) of the related problem for perturbations of a thin fluid layer on a rotating planet. To demonstrate the procedure, new coordinates r and θ are introduced, with $x = r\sin\theta$, $y = r\cos\theta$ and $\phi = F(r)\sin\theta$. Choosing $C = -\kappa^2$ for the inner region, $r < R$, the relation (18.36) is then rewritten as

$$r^2\frac{d^2}{dr^2}F + r\frac{d}{dr}F + [(\kappa r)^2 - 1]F = [\beta - (1+\kappa^2)u]r^3. \tag{18.37}$$

By choosing $C = \rho^2$ for the outer region, $r > R$, we similarly find

$$r^2\frac{d^2}{dr^2}F + r\frac{d}{dr}F - [(\rho r)^2 + 1]F = [\beta - (1-\rho^2)u]r^3, \tag{18.38}$$

with R being a constant radius.

We now recall the relations

$$\left(r^2\frac{d^2}{dr^2} + r\frac{d}{dr} + (r^2 - 1)\right)\left\{\begin{array}{c} J_1(r) \\ Y_1(r) \end{array}\right\} = 0, \tag{18.39}$$

and

$$\left(r^2\frac{d^2}{dr^2} + r\frac{d}{dr} - (r^2 + 1)\right)\left\{\begin{array}{c} I_1(r) \\ K_1(r) \end{array}\right\} = 0. \tag{18.40}$$

As well known, the solutions have the asymptotic properties

$$J_1(r \to \infty) \to \sqrt{\frac{2}{\pi r}}\cos(r - \frac{3\pi}{4}), \qquad \text{and} \qquad Y_1(r \to \infty) \to \sqrt{\frac{2}{\pi r}}\sin(r - \frac{3\pi}{4}),$$

while

$$I_1(r \to \infty) \to \frac{1}{\sqrt{2\pi r}}e^r, \qquad \text{and} \qquad K_1(r \to \infty) \to \sqrt{\frac{\pi}{2r}}e^{-r}.$$

For the *outer* solution we discard J_1 and Y_1 because they decay too slowly for compact, finite energy pulses, and I_1 because it increases indefinitely. The only acceptable outer solution of the ones listed here is therefore K_1. This solution (modified Bessel function of second kind) can be achieved by choosing

$$\rho^2 = (u - \beta)/u, \tag{18.41}$$

to remove the inhomogeneous term containing r^3 in (18.38). For the *outer* solution we have

$$\phi(r,\theta) = -\frac{u-\beta}{\rho^2}R\frac{K_1(\rho r)}{K_1(\rho R)}\sin\theta \tag{18.42}$$

for $r > R$.

On the other hand, for $r \to 0$ we have

$$J_1(r \to 0) \to \frac{r}{2}, \qquad \text{and} \qquad Y_1(r \to 0) \to -\frac{2}{\pi r},$$

while

$$I_1(r \to 0) \to \frac{r}{2}, \qquad \text{and} \qquad K_1(r \to 0) \to \frac{1}{r}.$$

The requirement of finite *inner* solutions discards Y_1 and K_1. Since I_1 increases monotonically for all r, it is not possible to match it smoothly with a continuous derivative to the outer K_1 solution, and it is therefore discarded as well. The acceptable J_1-solution (Bessel function of the first kind) is obtained by choosing κ real in (18.37) (this solution was anticipated by taking the negative sign for the appropriate constant).

The potential ϕ and its second derivative $d^2\phi/dr^2$ are required to be continuous on $r = R$. This condition implies for the inner solution of ϕ

$$\phi(r,\theta) = \frac{u-\beta}{\kappa^2}\left(R\frac{J_1(\kappa r)}{J_1(\kappa R)} - r\left(\frac{\kappa^2}{\rho^2} + 1\right)\right)\sin\theta \tag{18.43}$$

for $r < R$, where the term proportional to r originates from the inhomogeneous solution of (18.37), using $u = \beta/(1-\rho^2)$.

The value of κ is determined by requiring also $d\phi/dr$ to be continuous at $r = R$. We use the relation

$$\frac{sJ_2(s)}{J_1(s)} = 2s^2 {\sum_n^{\infty}}' \frac{1}{\gamma_{1,n}^2 - s^2} \tag{18.44}$$

where $\gamma_{1,n}^2$ is an increasing sequence of zeroes of the function $J_1(r)$ except for the zero at $r = 0$, as indicated by the prime in (18.44) and obtain (Petviashvili & Pokhotelov 1986)

$$\sum_{n=1}^{\infty}{}' \frac{1}{(\kappa R)^2 - \gamma_{1,n}^2} = \frac{1}{2\rho R}\left(\frac{1}{\rho R} - \frac{K_1'(\rho R)}{K_1(\rho R)}\right). \tag{18.45}$$

Derivatives of order higher than two will in general be discontinuous. It can be argued that since the basic equation is accurate only to the approximation where first and second order derivatives are retained, there is no justification in requiring higher order derivatives to be continuous.

In Fig. 18.1 we show the variation of the left- and right-hand sides of (18.45) with solid and dashed lines. We have solutions of (18.45) where the two lines cross. The lowest value is found numerically as $\kappa R \simeq 4.38\,\rho R$, relating κ and ρ.

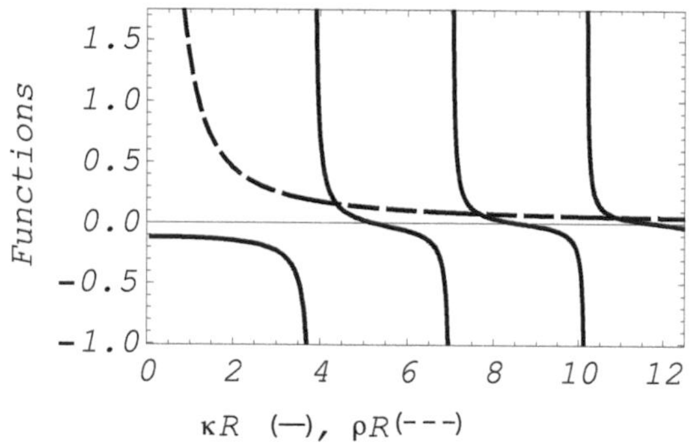

FIGURE 18.1: *Illustration of solutions of (18.45). The solid line shows the left-hand side, where the horizontal axis is in units of* κR *and the dashed line gives the right-hand side, with the axis in units of* ρR*. Note the singularities at the zeroes of the* $J_1(r)$*-function.*

We can make some simplifications at small values of κR and ρR, where the solutions are close to the singularities. As an approximation, valid in the vicinity of the poles $\kappa R \approx \gamma_{1,n}$, we have

$$2\kappa R \sum_{n=1}^{\infty}{}' \frac{1}{(\kappa R)^2 - \gamma_{1,n}^2} \approx \frac{1}{\kappa R - \gamma_{1,n}}.$$

The properties of the solutions (18.42)–(18.43) are determined by two parameters, radius R and velocity u. The amplitude, for instance, is determined when these two parameters are given. Alternatively, the amplitude can be prescribed to impose conditions on R and u. There is large freedom in selecting, for instance, the number of oscillations in the inner J_1 solution before it is matched to the outer K_1 solution, and a variety of modon solutions can be constructed. When the smallest value of the solutions for κ in (18.45) is chosen, the extremum point for the modon will correspond to the first extremum of $J_1(r)$ at $r \approx 1.8$.

An important observation concerns the *velocity* of the modon. According to (18.41) we require either $u > \beta$ or $u < 0$. Implicitly, these requirements imply that the modon velocity has to be *outside* the range of the phase velocity $\{0, \beta\}$ determined by the *linearized* equations.

In the present discussion only the possibility of constructing an exact stationary solution to the basic equation was discussed, without reference to its stability and to how such solutions actually develop. Numerical solutions of the Hasegawa-Mima equation have demonstrated that isolated modons are seemingly stable and also that structures strongly reminiscent of them develop spontaneously from rather general initial perturbations. With several such modons present in the flow, we can argue that the number of degrees of freedom is reduced in the sense that all you need is to account for the positions and basic properties of these structures, while the Fourier components need for describing the individually are phase-locked and need not be accounted for. With enhanced energy input a turbulent plasma state can be obtained where many Fourier components interact and many degrees of freedom are excited (Diamond et al. 2010, Pécseli 2016). An understanding of this state is important for describing turbulent plasma transport, in particular. The simple Hasegawa-Mima model is, however, not sufficient for modeling plasma turbulence, and some improvements are needed.

18.2 Hasegawa-Wakatani equations

A self-evident restriction of the Hasegawa-Mima equation is that it describes linearly *stable* waves or fluctuations, i.e., any perturbation has to be imposed initially. This feature is violating the basic property of drift waves as described in the foregoing chapter where it was demonstrated that drift waves are linearly *un*stable. The Hasegawa-Mima equations can, however, readily be generalized to include this feature. Following Kadomtsev (1965), we first consider the electron dynamics and write the electron continuity equation as $\partial n/\partial t + \nabla_\perp \cdot (n\mathbf{u}_e) + \partial_z(nu_\parallel) = 0$. Retaining, as in the linear analysis, collisional resistivity only in the **B**-parallel electron dynamics, we obtain first the **B**-perpendicular electron velocity as $\mathbf{u}_\perp = -\nabla_\perp \phi \times \mathbf{B}/B^2 + (T_e/e)\nabla_\perp \ln n \times \mathbf{B}/B^2$, giving $\nabla_\perp \cdot \mathbf{u}_\perp = 0$. The last term in $\mathbf{u}_\perp$ does not contribute when inserted into the electron continuity equation. Consequently, we find

$$\frac{\partial}{\partial t}\ln n + \frac{\hat{\mathbf{z}} \times \nabla_\perp \phi}{B} \cdot \nabla_\perp \ln n = \frac{1}{en}\frac{\partial}{\partial z}J_z, \tag{18.46}$$

where J_z is the z-component of the electron current, $\mathbf{J} = -en\mathbf{u}_e$. Since the ion dynamics in the z-direction are immaterial, J_z is also the *total* **B**-aligned current. An expression for J_z is obtained from the z-component of the electron momentum equation (17.29), giving

$$en\frac{\partial}{\partial z}\phi - \kappa T_e \frac{\partial}{\partial z}n + \xi en J_z = 0, \tag{18.47}$$

with the plasma resistivity included as in (17.30). From (18.47) we obtain

$$J_z = \frac{\kappa T_e}{e\xi}\frac{\partial}{\partial z}\left(\ln n - \frac{e\phi}{\kappa T_e}\right)$$

and insert it into (18.46) to obtain an equation relating density and potential fluctuations as obtained from the electron dynamics.

In the description of the ion dynamics it is customary to retain an ion viscosity term in the momentum equation with $\mu_\perp$ being the coefficient of viscosity. The equation which contains the ion polarization current to lowest order then becomes

$$\mathbf{u}_\perp = -\frac{\nabla_\perp \phi \times \mathbf{B}}{B^2} - \frac{M}{eB^2}\left(\frac{\partial}{\partial t} - \frac{1}{B^2}\nabla_\perp \phi \times \mathbf{B} \cdot \nabla_\perp\right)\nabla_\perp \phi + \frac{\mu_\perp}{neB^2}\nabla_\perp^2 \mathbf{u}_\perp \times \mathbf{B}, \tag{18.48}$$

where it was implicitly assumed that the time derivative of $\nabla_\perp \phi$ is of the same order as $(\mathbf{u}_\perp \cdot \nabla_\perp)\nabla_\perp \phi$, i.e., there are no fine-scale spatial variations in potential. Inserting (18.48) into the ion continuity equation of the form $\partial \ln n/\partial t + \nabla\cdot\mathbf{u} + \mathbf{u}\cdot\nabla \ln n = 0$ we find with the same approximations as in deriving (18.2)

$$\frac{\partial}{\partial t}\left(\ln n - \frac{1}{B\Omega_c}\nabla_\perp^2\phi\right) + \frac{\hat{\mathbf{z}}\times\nabla_\perp\phi}{B}\cdot\nabla_\perp\left(\ln n - \frac{1}{B\Omega_c}\nabla_\perp^2\phi\right) = -\frac{\mu_\perp}{neB\Omega_c}\nabla_\perp^4\phi. \tag{18.49}$$

When evaluating $(\mu_\perp/B^2)\nabla_\perp\cdot\nabla_\perp^2\mathbf{u}_\perp\times\mathbf{B}$ we used $\nabla_\perp\times\mathbf{U}\approx\hat{\mathbf{z}}\nabla_\perp^2\phi/B$ in an approximation for small $\mu_\perp$, i.e., the higher order contributions to $\mathbf{U}_\perp$ will be multiplied by a small quantity $\mu_\perp$, and can therefore be ignored. Quasi-neutrality was implicitly assumed here. Approximating $\ln n \approx \ln n_0 + n_1/n_0$ as before, we obtain an equation based solely on the ion dynamics

$$\left(\frac{\partial}{\partial t} + \frac{\hat{\mathbf{z}}\times\nabla_\perp\phi}{B}\cdot\nabla_\perp\right)\left(\frac{n_1}{n_0} - \frac{1}{B\Omega_c}\nabla_\perp^2\phi\right) + \frac{n_0'}{n_0}\frac{\partial}{\partial y}\phi = -\frac{\mu_\perp}{neB\Omega_c}\nabla_\perp^4\phi, \tag{18.50}$$

where it was explicitly used that the density gradient is "gentle" as discussed before.

Using the normalizations $\eta \equiv n/n_0$, $\chi \equiv e\phi/\kappa T_e$, $t\Omega_c$, r/a_i and $\mu/(a_i^2\Omega_c)$, z/ℓ_c, with $a_i^2 = \kappa T_e/M\Omega_c^2$ and ℓ_c being the collisional mean free path, we finally obtain the set of Hasegawa-Wakatani equations (Hasegawa & Wakatani 1983)

$$\left(\frac{\partial}{\partial t} + \hat{\mathbf{z}}\times\nabla_\perp\chi\cdot\nabla_\perp\right)\eta + \beta\frac{\partial}{\partial y}\chi = -C\frac{\partial^2}{\partial z^2}(\chi-\eta), \tag{18.51}$$

$$\left(\frac{\partial}{\partial t} + \hat{\mathbf{z}}\times\nabla_\perp\chi\cdot\nabla_\perp\right)\nabla_\perp^2\chi = -C\frac{\partial^2}{\partial z^2}(\chi-\eta) + \mu'\nabla_\perp^4\chi, \tag{18.52}$$

with $C \equiv \kappa T_e/(n_0 e^2\Omega_c\xi\ell_c^2) = \kappa T_e/(\Omega_c \nu_{ei} m\ell_c^2)$ where we introduced the resistivity ξ as given before in (17.30), and used $\hat{\mathbf{z}}\times\nabla_\perp\chi\cdot\nabla_\perp \ln n_0(x) = -(n_0'/n_0)\partial\chi/\partial y \equiv \beta\partial\chi/\partial y$, with $\beta = $ const. We see that it is natural to use different length scalings in the directions parallel and perpendicular to $\mathbf{B}$. In the equation (18.52) originating from the description of the ion dynamics we used (18.46) and then (18.47) to eliminate the terms containing $\ln n$ in (18.49). The derivation of equation (18.51) is based solely on the assumed electron dynamics, irrespective of the assumptions concerning the ions.

If the ion viscosity vanishes and the conductivity is infinite, $C\to\infty$, implying $\eta\to\chi$, the Hasegawa-Mima equation is recovered by subtracting (18.52) from (18.51). Note that while $C\to\infty$ gives $\eta\to\chi$, we cannot say anything about $C\partial^2(\chi-\eta)/\partial z^2$ in this limit, so the subtraction mentioned *is* necessary! In the opposite limit where $C\to 0$, the equation (18.52) reduces to the two-dimensional Euler equation and $\nabla_\perp^2\chi$ is advected by the $\hat{\mathbf{z}}\times\nabla_\perp\chi$-flow. We can then, at least formally, imagine the variation of χ obtained from (18.52) to be inserted into (18.51), thus acting as a "drive" for η.

Frequently, the equation (18.51) is completed by a loss term in the form $\mu_\eta\nabla_\perp^2\eta$, which makes it appear similar to the $\mu'\nabla_\perp^2(\nabla_\perp^2\chi)$-term in (18.52). The introduction of such a term is, however, purely ad hoc, and actually in variance with the basic assumptions, as discussed also in Section 10.3. For computational purposes (Camargo et al. 1995) with a modest number of spectral modes, it is often advantageous to have an exponent larger than 4 in the viscosity term, in order to have a well defined spectral subrange where the nonlinearity is important but where the influence of viscosity is negligible. For such cases a *hyper*-viscosity term of, e.g., the form $(-1)^{p+1}\mu'\nabla_\perp^{2p}\nabla_\perp^2\chi$ can be introduced, with an even exponent $2p$ for integer p.

18.2.1 Linearized Hasegawa-Wakatani equations

The set of nonlinear coupled Hasegawa-Wakatani equations can be studied in the linearized limit. In the linear form, the equations become

$$\frac{\partial}{\partial t}\eta+\beta\frac{\partial}{\partial y}\chi = -C\frac{\partial^2}{\partial z^2}(\chi-\eta) \tag{18.53}$$

$$\frac{\partial}{\partial t}\nabla_\perp^2\chi = -C\frac{\partial^2}{\partial z^2}(\chi-\eta)+\mu'\nabla_\perp^4\chi \tag{18.54}$$

By Fourier transform, it is readily shown that (18.53) reproduces (17.47). Assuming plane wave solutions, we obtain the dispersion relation

$$\omega^2+i\omega\left(Ck_z^2\left(1+\frac{1}{k_\perp^2}\right)+\mu'k_\perp^2\right)-Ck_z^2\left(i\frac{\omega^*}{k_\perp^2}+\mu'k_\perp^2\right)=0, \tag{18.55}$$

where the drift frequency $\omega^*=\beta k_y$ was again introduced. In the limit of infinite conductivity we have $C\to\infty$, and the solution is simply

$$\omega=\frac{\omega^*}{1+k_\perp^2}-i\frac{\mu'k_\perp^4}{1+k_\perp^2}, \tag{18.56}$$

reproducing the standard result for drift waves, i.e., (17.23) with $k_z=0$, where in addition a damping due to the ion viscosity is obtained. Ignoring ion viscosity, on the other hand, we find

$$\omega^2+i\omega Ck_z^2\left(1+\frac{1}{k_\perp^2}\right)-iCk_z^2\frac{\omega^*}{k_\perp^2}=0, \tag{18.57}$$

which is simply the result (17.52) for the collisional drift wave instability. The plasma becomes stable for $C=0$ as well as for $\omega^*=0$, as expected. If we in imagination (and at variance with the laws of physics!) set the plasma resistivity $\xi<0$ in C, we find that electrostatic drift waves become *stable* in the present model, while other previously stable modes now become unstable. With $\xi<0$, electrons will be diffusing *opposite* the gradient in the direction of *increasing* plasma density (see also the discussion in Section 17.3), and the drift waves become stabilized.

The Hasegawa-Wakatani equations form a closed set including driving in terms of the linear instability as described here and dissipation in terms of ion viscosity.

18.2.2 Conservation laws for the Hasegawa-Wakatani equations

As demonstrated before, the equations (18.52) and (18.51) contain the Hasegawa-Mima equation in certain limits. As we have seen, this equation has the basic conserved integrals, energy $E\equiv\frac{1}{2}\int\!\!\int\left(\phi^2+(\nabla\phi)^2\right)dxdy$ and enstrophy $W\equiv\frac{1}{2}\int\!\!\int\left((\nabla\phi)^2+(\nabla^2\phi)^2\right)dxdy$. With integration over the x,y-plane $\perp\mathbf{B}$, the two quantities can here be generalized as $E\equiv\frac{1}{2}\int\!\!\int\left(\eta^2+(\nabla_\perp\chi)^2\right)dxdy$ and $W\equiv\frac{1}{2}\int\!\!\int(\eta-\nabla_\perp^2\chi)^2dxdy$. It is customary to interpret the term η^2 in E as a potential energy density, and $\nabla_\perp^2\chi$ as a kinetic energy density to the given accuracy. This latter interpretation is indeed tempting, but should be considered with care; as we have seen, to lowest order the $\mathbf{E}\times\mathbf{B}$-motion is inertialess, and in that case the notion of kinetic energy does not really make sense.

Since the Hasegawa-Wakatani equations describe a linearly unstable situation with possibility for growing waves, the generalized energy and enstrophy is, however, no longer conserved by (18.52) and (18.51), and it is interesting to study their evolution in time. After some calculations, given by, e.g., Camargo et al. (1995), the following results can be obtained

$$\frac{d}{dt}E = \Gamma_\eta-\Gamma_\chi-\mathcal{D}_E, \tag{18.58}$$

$$\frac{d}{dt}W = \Gamma_\eta-\mathcal{D}_W, \tag{18.59}$$

with

$$\Gamma_\eta \equiv -\beta \int \eta \frac{\partial}{\partial y}\chi\, dxdy, \tag{18.60}$$

$$\Gamma_\chi \equiv C \int \left(\frac{\partial}{\partial z}(\eta-\chi)\right)^2 dxdy, \tag{18.61}$$

$$\mathcal{D}_E \equiv -\int \left(\eta \mathcal{D}_\eta - \chi \mathcal{D}_\chi\right) dxdy, \tag{18.62}$$

$$\mathcal{D}_W \equiv -\int (\eta - \nabla_\perp^2 \chi)(\mathcal{D}_\eta - \mathcal{D}_\chi) dxdy. \tag{18.63}$$

We select one **B**-parallel mode by taking $\partial/\partial z \to ik_z$. The damping terms were abbreviated as $\mathcal{D}_\eta = \mu' \nabla_\perp^4 \chi$ and $\mathcal{D}_\chi = \mu_\eta \nabla_\perp^2 \eta$, but other (even) damping exponents can be inserted as models, e.g., $\mathcal{D}_\chi = \mu_\eta \nabla_\perp^{2p} \chi$, depending on the sort of problem being considered, as discussed later.

To prove, for instance, (18.58) we write

$$\frac{d}{dt}E = \frac{1}{2}\int \frac{\partial}{\partial t}\left(\eta^2 + (\nabla_\perp \chi)^2\right) dxdy$$

and integrate by parts to obtain

$$\frac{d}{dt}E = \int \left(\eta \frac{\partial}{\partial t}\eta - \chi \frac{\partial}{\partial t}\nabla_\perp^2 \chi\right) dxdy.$$

The two partial time derivatives are now inserted from the Hasegawa-Wakatani equations to give

$$\begin{aligned} \frac{d}{dt}E = \int -\eta &\left((\hat{\mathbf{z}} \times \nabla_\perp \chi \cdot \nabla_\perp)\eta + \beta \frac{\partial}{\partial y}\chi + C\frac{\partial^2}{\partial z^2}(\chi - \eta) - \mu_\eta \nabla_\perp^2 \eta \right) \\ -\chi &\left((\hat{\mathbf{z}} \times \nabla_\perp \chi \cdot \nabla_\perp)\nabla_\perp^2 \chi - C\frac{\partial^2}{\partial z^2}(\chi - \eta) + \mu' \nabla_\perp^4 \chi \right) dxdy. \end{aligned} \tag{18.64}$$

The terms containing $\hat{\mathbf{z}} \times \nabla_\perp \chi \cdot \nabla_\perp$ do not contribute to the integral. The two linear coupling terms, those containing C, are combined to give

$$\frac{d}{dt}E = \int \left(-\beta \eta \frac{\partial}{\partial y}\chi + C(\chi - \eta)\frac{\partial^2}{\partial z^2}(\chi - \eta) - \mu_\eta \eta \nabla_\perp^2 \eta + \mu' \chi \nabla_\perp^4 \chi\right) dxdy. \tag{18.65}$$

By partial integration of the second term we readily find

$$\frac{d}{dt}E = \int \left(-\beta \eta \frac{\partial}{\partial y}\chi - C\left(\frac{\partial}{\partial z}(\chi - \eta)\right)^2 - \mu_\eta \eta \nabla_\perp^2 \eta + \mu' \chi \nabla_\perp^4 \chi\right) dxdy. \tag{18.66}$$

which is the desired result (18.58). The relation (18.59) can be proven in a similar way.

The relations (18.62) and (18.63) can be simplified somewhat in the case where $\mu' = \mu_\eta$, and the same powers, n, are used for both $\nabla_\perp^n$-operators, but there is no a priori reason for this. As an illustration we consider $\mathcal{D}_W$ in (18.63) for the general case which is almost always used, with spatial operators $\nabla_\perp^{2p}$ and $\mu' = \mu_\eta = \nu$ giving

$$\begin{aligned} \mathcal{D}_W &= -\nu \int (\eta - \nabla_\perp^2 \chi)\,(\mathcal{D}_\eta - \mathcal{D}_\chi) dxdy \\ &\equiv -\nu \int (-1)^{p+1} (\eta - \nabla_\perp^2 \chi)\, \nabla_\perp^{2p} \,(\eta - \nabla_\perp^2 \chi)\, dxdy \\ &= \nu \int \left(\nabla_\perp^p \,(\eta - \nabla_\perp^2 \chi)\right)^2 dxdy. \end{aligned} \tag{18.67}$$

Similar calculations for $\mathcal{D}_E$ give

$$\mathcal{D}_E = \nu \int \left((\nabla_\perp^p \eta)^2 + (\nabla_\perp^{p+1} \chi)^2 \right) dxdy. \tag{18.68}$$

The quantities (18.60)–(18.63) constitute sources and sinks of wave energy for the nonlinear evolution of initial conditions. Only Γ_η can act as a source; the other quantities are sinks. The density gradient β enters as a coefficient to Γ_η, which represents the rate at which free energy is extracted from the gradient; see also the discussion of plasma flux in Fig. 17.8.

Writing the x-component of the normalized velocity as $u_x = -\partial\chi/\partial y$, we can interpret $\Gamma_\eta = \beta \int n u_x dxdy$ as the net plasma flux along the density gradient in the x-direction. The term Γ_χ represents the rate with which energy is dissipated. These two quantities account for the collisional drift wave kinetics in the non viscous limit. It is important that they are active only to the extent that the electron dynamics deviates from isothermal, i.e., if $\eta = \chi$, then both Γ_χ and Γ_η vanish, in agreement with the foregoing results based on the linearized equations. The two sinks, $\mathcal{D}_\eta$ and $\mathcal{D}_\chi$, constitute viscous dissipation. Physically, these arise through collisional diffusion of electrons across magnetic field lines and due to ion viscosity, respectively.

Note that Γ_χ vanishes for flute-modes, where there is no variation in density or potential along the magnetic field lines. Numerical simulations by, e.g., Cheng and Okuda (1977) indicated that, in particular, the anomalous transport is strongly influenced by these modes, which can be excited by nonlinear coupling from drift modes with $k_z \neq 0$ to flute- or convective cell-modes with $k_z = 0$.

The most obvious damping mechanism seems to be missing in the Hasegawa-Wakatani equations. This would be smoothing out of the density gradients by anomalous diffusion caused by the waves to reduce the linear growth rate, but we have $n_0 = n_0(x)$, independent of t. The term $\hat{\mathbf{z}} \times \nabla_\perp \chi \cdot \nabla_\perp \eta$ might contain a modification of the β in the gradient term $\beta \partial\chi/\partial y$, provided the plasma density η develops a time-stationary component with a nontrivial variation along x, but being essentially constant along the y-direction. This observation is, however, of little avail as far as stabilization of the waves is concerned; if we introduce $\eta \equiv \overline{\eta}(x,t) + \tilde{\eta}(x,y,t)$ in (18.60) we find $\Gamma_\eta = -\beta \int \tilde{\eta}\, (\partial\chi/\partial y)\, dxdy$, independent of any $\overline{\eta}(x,t)$.

The Hasegawa-Wakatani model has been widely applied for analyzing weakly nonlinear drift waves. It has received some support also from experimental observations (Donnel et al. 2018).

18.2.3 Comments on the Hasegawa-Wakatani equations

In deriving the Hasegawa-Wakatani equations only the ion viscosity term contains remnants of a finite ion temperature. This is not consistent since, for instance, the collisionless part of the ion viscosity tensor is absent from the analysis, although it enters to the same order as collisional viscosity. With ion pressure and ion viscosity included, the expression for the ion velocity replacing (18.48) becomes

$$\begin{aligned}\mathbf{u}_\perp &= -\frac{\nabla_\perp \phi \times \mathbf{B}}{B^2} - \frac{M}{eB^2}\left(\frac{\partial}{\partial t} + \mathbf{u}\cdot\nabla_\perp\right)\mathbf{u}_\perp \times \mathbf{B} + \frac{\mu_\perp}{neB^2}\nabla_\perp^2 \mathbf{u}_\perp \times \mathbf{B} \\ &\quad - \frac{\kappa T_i}{neB^2}\nabla_\perp n \times \mathbf{B} - \frac{1}{neB^2}(\nabla_\perp \cdot \underline{\Pi}) \times \mathbf{B},\end{aligned} \tag{18.69}$$

which is inserted into the ion continuity equation. It was demonstrated previously in Section 17.6.4 that here the second term in the polarization drift cancels the contribution from the collisionless part of the viscosity tensor. The only remaining new term was shown to be $\partial\nabla^2\eta/\partial t$, while $\partial\nabla^2\chi/\partial t$ was included already. Consequently we find that with the present approximations, equation (18.52) is modified to

$$\left(\frac{\partial}{\partial t} + \hat{\mathbf{z}} \times \nabla_\perp \chi \cdot \nabla_\perp\right)\nabla_\perp^2 \chi + \theta \frac{\partial}{\partial t}\nabla_\perp^2 \eta = -C\frac{\partial^2}{\partial z^2}(\chi - \eta) + \mu' \nabla_\perp^4 \chi, \tag{18.70}$$

with $\theta = T_i/T_e$, while (18.51) remains unchanged. Strictly speaking, also the term containing the collisional viscosity is modified, but since this term is often modeled ad hoc there is little point in elaborating it in detail. Also, it was implicitly assumed that θ is small, i.e., terms containing products of θ, χ and η are ignored. These can easily be retained, but the resulting equations become so complicated that analytical approaches are prohibitive.

In the limit of Boltzmann distributed electrons, the linearized dispersion relation based on (18.51) and (18.70) does not reproduce (17.75) due to differences in the basic assumptions. The dispersion relation with finite electron resistivity, where $\chi \neq \eta$, is slightly more complicated than (18.57) or (18.55).

$$\omega^2 + i\omega\left(Ck_z^2\left(1+\theta+\frac{1}{k_\perp^2}\right)+\mu' k_\perp^2 - i\theta\omega^*\right) - Ck_z^2\left(\frac{i\omega^*}{k_\perp^2}+\mu' k_\perp^2\right) = 0. \tag{18.71}$$

The modification introduced by θ in the parenthesis with C as a coefficient is of minor importance, but the growth rate of the linear instability will be nontrivially modified by the changes in the other term.

The modified result (18.70) can be re-expressed in a more natural form by noting that the new term on the left side, accounting for the effects of a small but nonzero ion temperature, enters as a small correction. To lowest order we have $\partial\eta/\partial t \approx -\beta\partial\chi/\partial y$ from (18.53). By the ordering implied in the Hasegawa-Wakatani as well as the Hasegawa-Mima equations, we have here $\beta \sim \nabla_\perp \eta$ (see Section 18.1), and we have one more term contributing from the ion viscosity tensor. We can rewrite (18.70) to the desired accuracy as

$$\left(\frac{\partial}{\partial t}+\hat{\mathbf{z}}\times\nabla_\perp\chi\cdot\nabla_\perp\right)\nabla_\perp^2\chi - \theta\left(\nabla_\perp\eta\times\nabla_\perp\nabla_\perp^2\chi\cdot\hat{\mathbf{z}}+\beta\frac{\partial}{\partial y}\nabla_\perp^2\chi\right) =$$
$$-C\frac{\partial^2}{\partial z^2}(\chi-\eta)+\mu'\nabla_\perp^4\chi. \tag{18.72}$$

In this form the equations (18.51) and (18.72) incidentally reproduce the dispersion relation (17.75) if the electrons are assumed to be Boltzmann distributed. In the limit of finite electron resistivity $\chi \neq \eta$, with the modified model (18.72), the linear dispersion relation becomes

$$\omega^2 + i\omega\left(Ck_z^2\left(1+\frac{1}{k_\perp^2}\right)+\mu' k_\perp^2 - i\theta\omega^*\right)$$
$$-Ck_z^2\left(i\omega^*\left(\frac{1}{k_\perp^2}-\theta\right)+\mu' k_\perp^2\right) = 0. \tag{18.73}$$

In the flute limit, $k_z \to 0$, we have $\omega = 0$ but also weakly damped waves, $\omega = -\theta\omega^* - i\mu' k_\perp^2$ propagating backwards in the direction of the ion diamagnetic drift. These latter wave-modes depend critically on the ion temperature. The modifications introduced in (18.70) are expected to be significant for describing the coupling between drift- and flute-type modes. The splitting of the double root at $\omega = 0$ induced by $\theta \neq 0$ will be conspicuous, for instance, in numerical simulations for large times $t > 1/(\theta\omega^*)$, manifested physically by the "beating" between the two wave-modes.

In the limit of $C \to \infty$ we recover $\eta = \chi$ for Boltzmann distributed electrons. By subtracting (18.51) from (18.72) in this limit, a modified Hasegawa-Mima equation is obtained in the form

$$\left(\frac{\partial}{\partial t}+(\hat{\mathbf{z}}\times\nabla_\perp\phi)\cdot\nabla_\perp\right)((1+\theta)\nabla_\perp^2\phi-\phi) - \beta\frac{\partial}{\partial y}(\theta\nabla_\perp^2\phi+\phi) = 0$$

expressed in terms of normalized potential ϕ, where dissipation by ion viscosity is ignored. The modifications due to the finite (but small) ion temperature will be important when $\theta\nabla_\perp^2\phi \geq \phi$. It is easily demonstrated that also the modified Hasegawa-Mima equation has the basic conserved integrals, energy $E \equiv \frac{1}{2}\int\int(\phi^2+(1+\theta)(\nabla_\perp\phi)^2)\,dxdy$ and enstrophy $W \equiv \frac{1}{2}\int\int((\nabla_\perp\phi)^2+(1+\theta)(\nabla_\perp^2\phi)^2)\,dxdy$.

19

Kinetic Plasma Theory

In a fluid description of plasmas, as well as other media, we use space and time as independent variables in analyzing local densities and bulk velocities, for instance. The actual *distribution* of particle velocities in a small volume element is assumed a priori to be a Maxwellian locally. At a time t and in the vicinity of a spatial position $\mathbf{r}$, the plasma is thus characterized basically by a density $n(\mathbf{r},t)$, a bulk, or average, velocity $\mathbf{u}(\mathbf{r},t)$, and a temperature $T(\mathbf{r},t)$. We do not go into more detail here, but the basic assumption remains, namely, that as independent variables $\mathbf{r}$ and t will suffice. It is by using the concept of temperature (possibly space and time varying) that we in effect assume the velocity distribution to be a local Maxwellian. The theoretical basis for this is essentially the central limit theorem stating that the distribution of a sum of a large number of statistically independent variables is approximated by such a distribution. If the velocity of a particle at a certain time is the result of very many previous statistically independent collisions, then it can be argued that the velocity distribution is a Maxwellian to a good approximation.

Due to the long range of the Coulomb interaction, a charged particle can interact with many other particles simultaneously. This can be true even if we allow for Debye shielding; for sub-thermal velocities any charged particle in a plasma is interacting only with other charged particles within its local Debye sphere. For plasmas of interest, where the plasma parameter N_p is large, there are many charged particles in this Debye sphere. The interaction with these is *simultaneous* and not a result of many independent events since all these particles are also interacting at the same time, so the central limit theorem is no longer applicable, at least not in its traditional form.

19.1 The Vlasov equation

If we want to relax the inherent assumptions on an a priori known functional form of a velocity distribution function, the only way out is to find an equation describing the space, time and now also the *velocity* variation of such a distribution function! Now, a velocity distribution function is in general a statistical distribution function; if we repeat the experiment with unchanged macroscopic conditions, but with new conditions on a microscopic level, we will find a new distribution function, in general. The obvious remedy (or at least a temptation) is to consider a distribution function as a sum of δ-functions, with arguments at the position and velocity of each particle present in the plasma, and indeed such a model has been used as the starting point for the analysis by, for instance, Klimontovich (1967). Such a ragged, irregular function is not very appealing, and we will here bluntly assume that we will be interested only in spatial resolution on a Debye scale or larger. In this limit, the *exact* positions and velocities of the individual particles are of minor concern, and we can allow the particles to be "smeared out", so to speak, around the positions $(\mathbf{r},\mathbf{u})$ at time t. In the limit where $N_p \to \infty$, we can assume that the error becomes negligible when we do so. This implies that we can consider the particle distribution function to be a *deterministic* function, and in addition also one which will in general be smooth and differentiable, although we will keep the freedom to include also, for instance, δ-functions as limiting cases.

Recall the continuity equation (2.1), written as an equation for a flow with velocity $\mathbf{u}(\mathbf{r},t) \equiv \{u_x(\mathbf{r},t), u_y(\mathbf{r},t), u_z(\mathbf{r},t)\}$ and density $n(\mathbf{r},t)$ in configuration space spanned by the variables $\{x,y,z\}$. This continuity equation states that in order to have a density variation at a given position we must have a flow. We can generalize this idea to phase space spanned by the variables $\{x,y,z,u_x,u_y,u_z\}$. In order to have a *phase space density* variation, we must have either a velocity, just as in the standard continuity equation, or an *acceleration*. To describe the variation of the *phase space* density, $f(\mathbf{r},\mathbf{u},t)$ we introduce a sort of "hyper-velocity" $\mathcal{U} \equiv \{u_x,u_y,u_z,a_x,a_y,a_z\}$, where a_j for $j = x,y,z$ are the spatial components of the local *acceleration*. Although not written here explicitly, we *might* have that the acceleration includes velocity somehow, as discussed in more detail later on. We can now write a generalized continuity equation in phase space as

$$\frac{\partial f}{\partial t} + \nabla_{\{r,u\}} \cdot (f\,\mathcal{U}) = 0, \tag{19.1}$$

where we introduced a derivative operator

$$\nabla_{\{r,u\}} \equiv \left\{ \frac{\partial}{\partial x}, \frac{\partial}{\partial y}, \frac{\partial}{\partial z}, \frac{\partial}{\partial u_x}, \frac{\partial}{\partial u_y}, \frac{\partial}{\partial u_z} \right\}.$$

Physically, (19.1) implies that by selecting, at $t = 0$, a small region $dxdydz\,du_x du_y du_z$ in phase space at the vicinity of $(\mathbf{r}_0,\mathbf{u}_0)$, we find that the distribution function f is changing in time if 1) we have a spatial gradient in f at the position $\mathbf{r}_0$, or 2) we have a gradient with respect to the velocity at $(\mathbf{r}_0,\mathbf{u}_0)$ and the plasma volume is subject to an acceleration $\mathbf{a}$, or any combination of these two cases. In case 1) the change at $\mathbf{r}_0,\mathbf{u}_0$ is due to the local gradient moving with velocity $\mathbf{u}_0$ to give a temporal change $-\mathbf{u}_0 \cdot \nabla f$ very much like in the case of the classical continuity equation. Case 2) gives a temporal change $-\mathbf{a} \cdot \nabla_{\mathbf{u}} f$.

Stating that $\mathbf{r}$ and $\mathbf{u}$ are independent variables, we imply that any derivative of velocity with respect to position vanishes, e.g., $\partial u_i/\partial r_j = 0$ for $i \neq j$. If the acceleration is independent of velocity, the generalizations of this result to the other components of $\mathcal{U}$ are obvious. Considering, for instance, acceleration of electrons with charge $-e$ by an electric field, $\mathbf{a} = -(e/m)\mathbf{E}$, we find that this is trivially satisfied and $\partial a_i/\partial u_j = 0$ for all $\{i,j\}$. If we include the velocity dependent Lorentz term, $-e\mathbf{u} \times \mathbf{B}$, it is easily demonstrated that, due to the properties of the cross product, the j-th component of the acceleration depends on velocity component $i \neq j$, and we have $\partial a_i/\partial u_j = 0$ generally. We can then write (19.1) corresponding to incompressible motion in phase space as

$$\frac{\partial f}{\partial t} + \mathcal{U} \cdot \nabla_{\{r,u\}} f = 0. \tag{19.2}$$

Writing out the components of $\mathcal{U}$, we obtain the basic equation describing the dynamics of collisionless plasmas in the limit of large plasma parameters N_p, namely, the *Vlasov equation.*

$$\frac{\partial f}{\partial t} + \mathbf{u} \cdot \nabla f + \frac{\mathbf{K}}{m} \cdot \nabla_{\mathbf{u}} f = 0, \tag{19.3}$$

where $f = f(\mathbf{r},\mathbf{u},t)$ is the distribution of plasma particles with mass m, moving under the action of a force field $\mathbf{K} = \mathbf{K}(\mathbf{r},t)$. By $\nabla_{\mathbf{u}}$ we now understand the vector operator $\{\partial/\partial u_x, \partial/\partial u_y, \partial/\partial u_z\}$. The origin of $\mathbf{K}$ makes no difference for the discussion here: it can be prescribed or determined self-consistently through Maxwell's equations.

Defining a vector with components $\{x,y,z,u_x,u_y,u_z\}$, we find that in the phase space spanned by $\mathbf{r}$ and $\mathbf{u}$, the Vlasov equation (19.3), without any collision term, expresses a continuity equation for an incompressible phase space flow defining the evolution of f. If we follow a volume element in phase space, it will be distorted both with respect to spatial coordinates and velocity coordinates, but its phase space volume remains constant.

It may be appropriate to emphasize that a limit exists where the distribution function $f(\mathbf{r},\mathbf{u},t)$ entering the Vlasov equation (19.3) is *not* to be interpreted as a probability density! Usually, we associate distribution functions with a statement of the probability for finding a particle in a spatial interval $\{x, x+dx; y, y+dy; z, z+dz\}$ in a small time interval $\{t, t+dt\}$. This is not so here! By the distribution f, with a space-time variation described by the Vlasov equation, we will consider the *actual* distribution of particles. For a *true* distribution, this will be a very ragged function, zero most places, and finite at a particle position (or a δ function, if we assume point-like particles). To have a "well behaved" continuous function, we implicitly imply some "coarse graining" of this actual distribution function: we are interested only in scales so large that we can ignore the fine structures in $f(\mathbf{r}, \mathbf{u}, t)$. This is just like looking at a drop of ink in a glass of water: we see it as a blue continuum, and it has to be enlarged quite a lot to see the individual particles, which are giving the blue color. The validity of a continuum description depends on the length scales we consider. A reasonable length scale in most studies of plasma sciences is the Debye length or a not too small fraction thereof. Formally, a case might be constructed where the Vlasov equation becomes exact: the Vlasov limit. Here we, in imagination, subdivide each particle, electron and ion, into smaller pieces and spread them around a little. This is done in such a way that the individual charges q_j and masses $m_j \to 0$, so that $\lim_{n_j\to\infty} n_j q_j =$ const. and $\lim_{n_j\to\infty} n_j m_j =$ const. This means that we can have the ratio $e_i/m_i =$ const. during the transformation. Since we are not in other ways changing the particles, the velocities remain the same, implying that the appropriate temperatures determined by $\frac{1}{2}m_j\langle u^2\rangle = \frac{3}{2}\kappa T$ will vanish in the limit, but $n_j\kappa T$ is constant during the transition. It is readily demonstrated that the plasma frequencies ω_p and Debye lengths λ_D remain unchanged during this subdivision. Consequently, the plasma parameter $N_p \equiv n\lambda_D^3 \to \infty$, and the plasma can be seen as a continuum at any length scale in this Vlasov limit, where now also the plasma collision frequency vanishes together with the collisional resistivity; see Section 8.8.

Ⓢ **Exercise:** In Section 9.1 we discussed the equilibrium condition for the collisionless plasma in a Q-machine with some suitably defined end conditions. Postulate that we have a steady state electrostatic potential $\phi(x)$ representing the dc-potential along the Q-machine axis. Demonstrate that we have a solution of the form

$$f_e(x,u) = n_e(x)\sqrt{\frac{m}{2\pi\kappa T}}\exp\left(-\frac{1}{2}\frac{mu^2}{\kappa T}\right)$$

for the electron Vlasov equation. Give the relation between $n_e(x)$ as well as $n_i(x)$ and $\phi(x)$. Write the differential equation (without solution) that determines the self-consistent electrostatic potential. Ⓢ

19.1.1 Collisions

If the plasma is *not* collisionless, we will experience that particles are scattered out of a small phase space volume within a short time interval; this is, after all, what we consider a collision. A particle then suddenly changes its velocity, although it remains in the same narrow *spatial* volume element. If we consider a small phase space element $d\mathbf{r}d\mathbf{u}$ placed at $(\mathbf{r}_0, \mathbf{u}_0)$ we find that a particle is there just before the collision, to disappear right after the collision, to appear in some other phase space element. The spatial position in the interval $\mathbf{r}_0, \mathbf{r}_0 + d\mathbf{r}$ does not change, but its velocity does! Note that this will be true also for completely elastic collisions, where the particle changes its direction of motion, but not its energy. In these cases we have sources or sinks in phase space! Collisional effects can be retained by introducing a formal collision operator $\partial f/\partial t\,|_{col}$ on the right-hand side of (19.3). The actual form of this operator depends on the nature of the collision model, but its effect is considered to be small in the sense that the collisions are not sufficiently frequent to render the velocity distribution a local Maxwellian within time scales of interest. For the time being, we leave the collision operator in a symbolic form.

For most relevant cases, we will be dealing with electric and magnetic forces, i.e., $\mathbf{K} = q(\mathbf{E} + \mathbf{u}\times\mathbf{B})$ in terms of the Lorentz force, with q being the particle charge. For some cases relevant, for instance, for astrophysical plasmas, the expression for **K** has to be completed by a gravitational force. The electric and magnetic fields have to be determined by Maxwell's equations, where the self-consistent charge densities $\zeta \equiv q\int_{-\infty}^{\infty} f(\mathbf{r},\mathbf{u},t)d\mathbf{u}$ and current density contributions $\mathbf{J} \equiv q\int_{-\infty}^{\infty} \mathbf{u}f(\mathbf{r},\mathbf{u},t)d\mathbf{u}$ for the given species are obtained self-consistently from the Vlasov equation; this is a highly complicated nonlinear problem.

- **Exercise:** Demonstrate that the Vlasov equation without a collision term is *time reversible*, also for the case where the full Lorentz force is included in **K**. Note that collision operator $\partial f/\partial t\,|_{col}$ is written symbolically, so in spite of the unfortunate (but standard) notation it is *not* supposed to change sign by time reversal. The results concerning time reversibility explicitly refer to the collisionless case.

The collision operator formally written here as $\partial f/\partial t\,|_{col}$ is in general a very complicated expression, to write as well as to derive. Simple approximations can be used in some cases.

- We often see the simple expression

$$\left.\frac{\partial f}{\partial t}\right|_{col} = -\nu\left(f(\mathbf{r},\mathbf{u},t) - f_0(\mathbf{u})\right), \tag{19.4}$$

  where f_0 is some prescribed distribution function, often taken to be a Maxwellian, and ν is a collision frequency, where possibly $\nu = \nu(u)$. The model (19.4) is often called the BGK model after Bhatnagar et al. (1954). Unfortunately, this abbreviation is associated also with an entirely different phenomenon; see Section 22.1!

- The basic problem with the BGK collision model (19.4) is that at any spatial position, the distribution function relaxes not only to a prescribed distribution function but also to a prescribed density, $n_0 \equiv \int f_0(\mathbf{u})d\mathbf{u}$. This shortcoming can be remedied by the generalization

$$\left.\frac{\partial f}{\partial t}\right|_{col} = -\nu\left(f(\mathbf{r},\mathbf{u},t) - f_0(\mathbf{u})\frac{n(\mathbf{r},t)}{n_0}\right), \tag{19.5}$$

  where $n(\mathbf{r},t) = \int f(\mathbf{r},\mathbf{u},t)d\mathbf{u}$. The improved BGK model (19.5) is representative for *charge exchange collisions*, for instance, in particular if we let the collision frequency be velocity dependent, $\nu = \nu(u)$.

- A general, but complicated, collision model can be proposed by modeling the collision process by a Fokker-Planck equation

$$\left.\frac{\partial f}{\partial t}\right|_{col} = -\nu\nabla_{\mathbf{u}}\cdot\left((\mathbf{u}-\mathbf{U})f(\mathbf{r},\mathbf{u},t) + \frac{\kappa T}{m}\nabla_{\mathbf{u}}f(\mathbf{r},\mathbf{u},t)\right), \tag{19.6}$$

  where $\mathbf{U}(\mathbf{r},t) = \int \mathbf{u}f(\mathbf{r},\mathbf{u},t)d\mathbf{u}$ accounts for a local average drift velocity and $\nabla_{\mathbf{u}}$ is the velocity derivative operator discussed before in relation to (19.3). The expression (19.6) ensures that the steady state solution is a Maxwellian with temperature T.

We note that all three collision processes discussed here are *Markov processes*, i.e., the evolution depends on the present stage, and not on how this has been reached. The Fokker-Planck equation, in particular, is derived for such processes (Clemmow & Dougherty 1969, Pécseli 2000).

19.2 Relation between kinetic and fluid models

Since a fluid description of a plasma in a number of cases reproduces many basic wave properties, the existence of wave modes in particular, it seems intuitively reasonable to expect some sort of relation between descriptions based on a kinetic and a fluid model. Often we see the terms microscopic and macroscopic models; a kinetic model deals with *microscopic* details of velocity distributions in a fluid volume element, while a fluid model considers its macroscopic or bulk properties. The velocity distribution $f(\mathbf{r},\mathbf{u},t)$ contains all the information available, while the reduced, or bulk plasma properties are contained in plasma density $n(\mathbf{r},t) = \int_{-\infty}^{\infty} f(\mathbf{r},\mathbf{u},t)d\mathbf{u}$, mean particle flux $n(\mathbf{r},t)\mathbf{U}(\mathbf{r},t) = \int_{-\infty}^{\infty} \mathbf{u}f(\mathbf{r},\mathbf{u},t)d\mathbf{u}$, etc., using the notation $\int_{-\infty}^{\infty} d\mathbf{u}$ for $\int_{-\infty}^{\infty}\int_{-\infty}^{\infty}\int_{-\infty}^{\infty} du_x du_y du_z$. This reduced information can be obtained from the Vlasov equation by deriving its moments, as illustrated in the following. We use here capital letters for bulk plasma velocities to be distinguished from the lower case $\mathbf{u}$ which is a free variable in the Vlasov equation. To distinguish the independent vector variable $\mathbf{u}$ from the average flow velocity, we introduced the capital letter notation, $\mathbf{U}$, for the latter quantity. The fluid theory only recognizes time t and position $\mathbf{r}$ as independent variables.

First we multiply each term in (19.3) by a function of velocity $\mathsf{Q}(\mathbf{u})$, leaving for the time being the actual form of Q unspecified. We can allow Q to be a vector or tensor function, although the latter generalization will not be needed here. For later reference we introduce the notation for the averaging of a quantity as

$$\langle q(\mathbf{u})\rangle \equiv \int_{-\infty}^{\infty} q(\mathbf{u}) f(\mathbf{r},\mathbf{u},t) d\mathbf{u}\, \frac{1}{n(\mathbf{r},t)},$$

where $n(\mathbf{r},t)$ is included for normalizing the distribution with its local average value to guarantee the identity $\langle 1\rangle = 1$. The notation is directly related to a standard statistical averaging in terms of probability densities. From (19.3) we have

$$\int_{-\infty}^{\infty} \mathsf{Q}(\mathbf{u}) \left(\frac{\partial f}{\partial t} + \mathbf{u}\cdot\nabla f + \frac{\mathbf{K}}{m}\cdot\nabla_{\mathbf{u}} f\right) d\mathbf{u} = \int_{-\infty}^{\infty} \mathsf{Q}(\mathbf{u}) \left.\frac{\partial f}{\partial t}\right|_{col} d\mathbf{u}. \tag{19.7}$$

For generality, collisional effects are retained by the collision operator $\partial f/\partial t\,|_{col}$ on the right-hand side of (19.7). We assume that all integrals exist: for most (but not all) cases we can expect that the velocity distribution falls off exponentially for $|u| \to \infty$, so this assumption of convergence is not particularly restrictive.

We now consider each term in (19.7) individually. The first term gives trivially

$$\int_{-\infty}^{\infty} \mathsf{Q}(\mathbf{u}) \frac{\partial f}{\partial t} d\mathbf{u} = \frac{\partial}{\partial t}\left(n\langle \mathsf{Q}\rangle\right).$$

Since Q is independent of $\mathbf{r}$, we find after simple manipulations of the second term

$$\int_{-\infty}^{\infty} \mathsf{Q}(\mathbf{u})\mathbf{u}\cdot\nabla f d\mathbf{u} = \nabla\cdot\left(n\langle \mathbf{u}\mathsf{Q}\rangle\right).$$

The third term is a bit more complicated, in particular if $\mathbf{K}$ is velocity dependent as in the Lorentz force. This term is unspecified for the moment. We allow Q to be a scalar as well as a vector and use $\nabla_{\mathbf{u}}\cdot(f\mathbf{K}\mathsf{Q}) = (\nabla_{\mathbf{u}} f)\cdot\mathbf{K}\mathsf{Q} + f\nabla_{\mathbf{u}}\cdot(\mathbf{K}\mathsf{Q}) = \mathsf{Q}\mathbf{K}\cdot\nabla_{\mathbf{u}} f + f\nabla_{\mathbf{u}}\cdot(\mathbf{K}\mathsf{Q})$. Imposing the rather mild restriction that $(\mathsf{Q}\mathbf{K}f) \to 0$ for $|\mathbf{u}| \to \infty$ we find

$$\begin{aligned}\frac{1}{m}\int_{-\infty}^{\infty} \mathsf{Q}(\mathbf{u})\mathbf{K}\cdot\nabla_{\mathbf{u}} f d\mathbf{u} &= -\frac{1}{m}\int_{-\infty}^{\infty} \left(\nabla_{\mathbf{u}}\cdot\mathbf{K}\mathsf{Q}(\mathbf{u})\right) f d\mathbf{u} \\ &= -\frac{1}{m} n\left\langle \nabla_{\mathbf{u}}\cdot(\mathbf{K}\mathsf{Q}(\mathbf{u}))\right\rangle.\end{aligned} \tag{19.8}$$

Consequently we have deduced the following equation from the Vlasov equation

$$\frac{\partial}{\partial t}\left(n\langle \mathsf{Q}\rangle\right) + \nabla\cdot\left(n\langle \mathbf{u}\mathsf{Q}\rangle\right) - \frac{n}{m}\left\langle \nabla_{\mathbf{u}}\cdot(\mathbf{K}\mathsf{Q}(\mathbf{u}))\right\rangle = \int_{-\infty}^{\infty} \mathsf{Q}(\mathbf{u}) \left.\frac{\partial f}{\partial t}\right|_{col} d\mathbf{u}. \tag{19.9}$$

The various terms are functions of **r** and t only; the independent variable **u** has vanished by the $\int d\mathbf{u}$ integrations.

We now let $\mathsf{Q} = 1$ and find from (19.9)

$$\frac{\partial}{\partial t}n + \nabla \cdot (n\mathbf{U}) - \frac{n}{m}\langle \nabla_{\mathbf{u}} \cdot \mathbf{K} \rangle = 0, \tag{19.10}$$

making use of the observation that collisions cannot change the local plasma density. At a collision the particle changes its velocity, but not its position. For forces which are independent of velocity, gravity or electric forces for instance, the last term in (19.10) is trivially vanishing. For the Lorentz force $\mathbf{u} \times \mathbf{B}$ we find that the force in a certain direction does not depend on the velocity component in that particular direction, so also for this case we recover the continuity equation

$$\frac{\partial}{\partial t}n + \nabla \cdot (n\mathbf{U}) = 0. \tag{19.11}$$

This is a scalar equation, giving the evolution of $n = n(\mathbf{r}, t)$ if the velocity $\mathbf{U} = \mathbf{U}(\mathbf{r}, t)$ is given. This relation seems so obvious, so it might be reasonable to add a few thoughts: in fluid mechanics we are used to identifying particles with fluid elements. Because the collisions are so frequent there, we might assume the same particles occupy the same fluid element unless it is too small, i.e., particles are assumed to escape from a small volume by diffusion, which is a slow process. We have, on the other hand, argued that plasmas in most relevant conditions can be considered to be collisionless! Electrons and ions pass through a small reference volume, so to speak, without any collisions at all. The physics of collisionless plasmas and classical fluids are different on the microscopical level, but the continuity equation applies in either case.

To find an equation for $\mathbf{U} = \mathbf{U}(\mathbf{r}, t)$ we let $\mathsf{Q} = \mathbf{u}$ in (19.9) and find

$$\frac{\partial}{\partial t}(n\mathbf{U}) + \nabla \cdot (n\langle \mathbf{uu} \rangle) - \frac{n}{m}\langle \nabla_{\mathbf{u}} \cdot (\mathbf{Ku}) \rangle = \frac{\mathbf{P}}{m}, \tag{19.12}$$

with

$$\frac{\mathbf{P}}{m} \equiv \int_{-\infty}^{\infty} \mathbf{u} \frac{\partial f}{\partial t}\bigg|_{col} d\mathbf{u}.$$

Be careful; $\mathbf{Ku} \neq \mathbf{uK}$ is a dyad product.

The equation (19.12) with (19.11) gives n and **u** provided we have an equation for $\langle \mathbf{uu} \rangle$. Evidently we could now introduce the tensorial function $\underline{\mathsf{Q}}(\mathbf{u}) = \mathbf{uu}$ and repeat the whole procedure, but it is also clear that this will not help much; we will find an equation containing higher order terms in **u**, which will need more equations, etc. In a way this could be expected from the outset; the Vlasov equation has three independent variables, time as a scalar and two vector variables, position and velocity. The hierarchy of equations, where (19.11) and (19.12) are the first ones, reflects the information contained in the infinity of values for the velocity variable. The basis for the fluid model is a physical argument for truncating the hierarchy of equations.

To simplify (19.12) we introduce a new vector as $\mathbf{w} = \mathbf{u} - \langle \mathbf{u} \rangle \equiv \mathbf{u} - \mathbf{U}$, where $\langle \mathbf{w} \rangle = 0$. Evidently we have the dyad products

$$\langle \mathbf{uu} \rangle = \mathbf{UU} + \langle \mathbf{Uw} \rangle + \langle \mathbf{wU} \rangle + \langle \mathbf{ww} \rangle.$$

The term $\langle \mathbf{wU} \rangle$ vanishes since $\langle \mathbf{wU} \rangle = \langle \mathbf{w} \rangle \mathbf{U} = 0$ and similarly for $\langle \mathbf{Uw} \rangle$. Now equation (19.12) becomes

$$\frac{\partial}{\partial t}(n\mathbf{U}) + \nabla \cdot (n\mathbf{UU}) + \nabla \cdot (n\langle \mathbf{ww} \rangle) - \frac{n}{m}\langle \nabla_{\mathbf{u}} \cdot (\mathbf{Ku}) \rangle = \frac{\mathbf{P}}{m}. \tag{19.13}$$

Assume now, as a simplifying postulate, that the distribution function f is isotropic when expressed in terms of the variable **w**, i.e., it depends on $|\mathbf{w}|$ only. First calculate a tensor com-

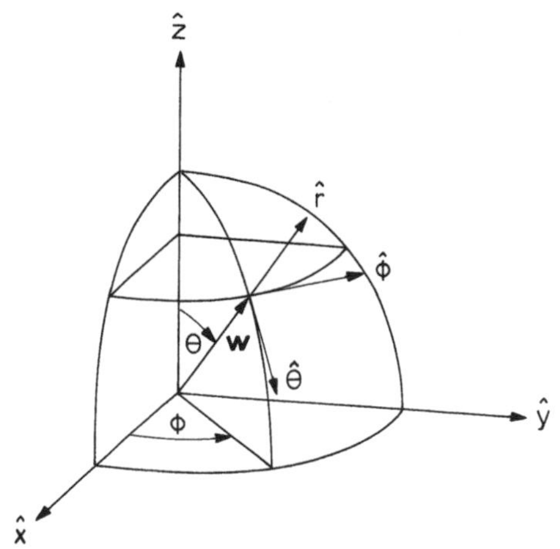

FIGURE 19.1: *The velocity components of* **w** *in spherical coordinates.*

ponent $\langle w_i w_j \rangle$ with $i \neq j$ using spherical coordinates w, θ and ϕ. We have $w_i = w\cos\theta$ and $w_j = w\sin\theta\,\cos\phi$; see Fig. 19.1.

$$\begin{aligned} \langle w_i w_j \rangle &= \int f(\mathbf{r},|\mathbf{w}|,t) w\cos\theta\, w\sin\theta\,\cos\phi\, d\mathbf{w} \\ &= \int_0^\infty \int_0^\pi \int_0^{2\pi} f(\mathbf{r},|\mathbf{w}|,t) w^2 \cos\theta\,\sin\theta\,\cos\phi\, w^2 \sin\theta\, dw\, d\theta\, d\phi, \end{aligned} \tag{19.14}$$

with $d\mathbf{w} = w^2\sin\theta dw d\theta d\phi$. The result vanishes by the ϕ integral. If, however, $i = j$ we find

$$\begin{aligned} \langle w_i w_i \rangle &= \int f(\mathbf{r},|\mathbf{w}|,t) w^2 \cos^2\theta\, d\mathbf{w} \\ &= \int_0^\infty \int_0^\pi \int_0^{2\pi} f(\mathbf{r},|\mathbf{w}|,t) w^4 \cos^2\theta\,\sin\theta\, dw\, d\theta\, d\phi \\ &= \int_0^\infty w^4 f(\mathbf{r},|\mathbf{w}|,t) \frac{4\pi}{3} dw \equiv \Upsilon(\mathbf{r},t). \end{aligned} \tag{19.15}$$

Since we are mainly interested in electric and magnetic interactions, we identify $\mathbf{K}$ with the Lorentz force where $\partial K_j/\partial u_j = 0$ and find with $\nabla_\mathbf{u}\mathbf{u} = \underline{1}$ being the unit tensor

$$\left\langle \nabla_\mathbf{u} \cdot (\mathbf{K}\mathbf{u}) \right\rangle = \left\langle (\mathbf{K}\cdot\nabla_\mathbf{u})\mathbf{u} \right\rangle + \left\langle (\nabla_\mathbf{u}\cdot\mathbf{K})\mathbf{u} \right\rangle = \langle \mathbf{K} \rangle. \tag{19.16}$$

Using these results and (19.11), a term $\mathbf{U}\nabla\cdot(n\mathbf{U})$ cancels, and we rewrite (19.13) as

$$nm\left(\frac{\partial}{\partial t}\mathbf{U} + \mathbf{U}\cdot\nabla\mathbf{U}\right) + \nabla\cdot(nm\Upsilon(\mathbf{r},t)) - nq(\mathbf{E} + \mathbf{U}\times\mathbf{B}) = \mathbf{P}, \tag{19.17}$$

where we explicitly introduced the Lorentz force for $\mathbf{K}$, implying $\langle \mathbf{K} \rangle = q(\mathbf{E} + \langle \mathbf{u} \rangle \times \mathbf{B}) \equiv q(\mathbf{E} + \mathbf{U}\times\mathbf{B})$. Recalling that $\partial\mathbf{U}/\partial t + \mathbf{U}\cdot\nabla\mathbf{U} \equiv D\mathbf{U}/Dt$ is the convective time derivative, we readily identify (19.17) as Newton's second law; the first term is the acceleration of a mass element nm, while the other terms can be interpreted as forces, the second term, in particular, as the gradient of a pressure $nm\Upsilon(\mathbf{r},t)$. In a collisionless plasma we have no a priory guarantee for a local Maxwellian distribution, but if we *assume* the velocity distribution to be such a Maxwellian, we find $\Upsilon = \kappa T/m$.

With this assumption the standard form for the pressure in ideal gases is recovered, where the temperature can be space-time varying.

The term **P** is a friction caused by collisions between particles. The "eigen-collisions", i.e., those among like-type particles, give a vanishing contribution since the individual species cannot lose net momentum by internal collisions: for such cases, the momentum lost by, e.g., one electron is gained by another. Consequently, **P** represents collisions between *different* types of particles only.

When different types of particles, electrons and various ion species interact, we can write an equation of the form (19.17) for each species. The system of equations is now closed; the hierarchy of equations is *truncated*. The expense for this was a postulate concerning the explicit form of the velocity distribution function. The postulate can be relaxed somewhat by allowing for anisotropic distributions, but clearly we cannot expect the truncated set of equations to be adequate when the assumed form for f is inapplicable. For describing Landau damping, fine details in the velocity distributions are important, and the set of equations derived here will fail to be sufficient. Only in the case where the velocity distribution function is so narrow that the deviation $\mathbf{w}$ between $\mathbf{U}$ and $\mathbf{u}$ can be ignored do we have a simple one to one correspondence between the fluid and a Vlasov model, since in this limit the tensor $\langle \mathbf{ww} \rangle$ in (19.13) can be ignored, and we are left with a closed set of equations for n and $\mathbf{U}$. This will be a "cold plasma" approximation.

- **Exercise:** It may be an instructive exercise (and a time saving one at that, at least compared to the full three-dimensional derivation given in Section 19.2) to analyze a simple Vlasov equation in one spatial and one velocity coordinate, to obtain the relation to fluid equations for that case.

19.3 Water-bag models

One particularly simple spatially one dimensional velocity distribution function deserves particular attention, in spite of being physically unrealistic: we consider distribution functions with the property of being either a constant or zero, as illustrated in Fig. 19.2. Such a model is often called a "water-bag" model, because it resembles a flexible sac filled with water (the water is assumed to retain a constant density). Albeit a bit special, this is a distribution function nonetheless, and its space-time evolution can be described by the Vlasov equation. There is no a priori requirement for the water bag to be singly connected in phase space.

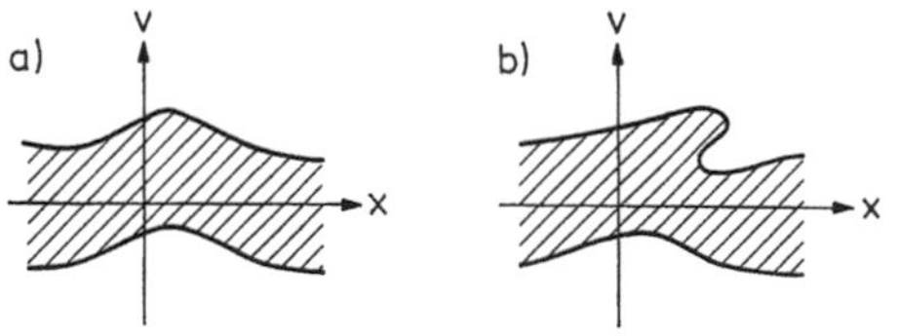

FIGURE 19.2: *Simple illustration of a phase space for the "water-bag" model for a distribution function. Shaded areas represent a constant value for the distribution, bounded by space-time varying contours. Figure b), in particular, illustrates a case with nonzero heat flux.*

Since the Vlasov equation describes incompressible motion in phase space, we need not be concerned with what is *inside* the contours of the water bag, since we know it *has* to be the initial constant phase-space density. *Outside* the contour, we have for the same reason trivially a vanishing phase-space density. The water-bag model is fully resolved if we are able to describe and solve for the dynamics of the contour forming the boundary for the water bag, or *bags* in case there are many (Roberts & Berk 1967, Berk et al. 1970). The initial outer contour, as constituted by the particles positioned on the contour, remains to be the outer contour at all later times, because of the incompressible motion in phase space. If we imagined that for some reason a small portion of the contour moved *into* the bulk of the water bag, some particles had to cross this contour to form the replacement for the new outer contour, and this cannot happen. Water bags are excellent for modeling collisionless plasma dynamics in one spatial dimension, but rather clumsy if we want to include collisions, and in practice you will never see water bags applied for collisional plasma models. It should also be emphasized that the water-bag model basically applies to one spatial and one velocity dimension: this will become more apparent later on, when we derive a dispersion relation for kinetic plasma wave models.

In a one-dimensional electrostatic model, for instance, the boundaries of the water bags for the electron distribution follow the equation

$$\frac{\partial}{\partial t}U_j + U_j\frac{\partial}{\partial x}U_j = -\frac{e}{m}E, \tag{19.18}$$

where $U_j = U_j(x,t)$ for the j-th contour. It should be emphasized that (19.18) is an equation for the *boundaries* and *not* an equation of motion for the particles constituting these boundaries! The position x is an independent variable, and we might have stationary solutions (as demonstrated later on) where $U_j = U_j(x)$, meaning that the particles at the boundary at position x move (or "stream") with velocity U_j, but the contour as such is fixed in time.

For the simple case with only two boundaries, $j = 1$, 2, the electric field is the obtained by Poisson's equation in the form

$$\frac{\partial}{\partial x}E(x,t) = -\frac{e}{\varepsilon_0}\Big(A\left(U_1(x,t) - U_2(x,t)\right) - n_0\Big), \tag{19.19}$$

where n_0 is the constant ion background density, and A is the constant value of the electron distribution function within (in this case) the two boundaries U_1 and U_2. The contours are numbered so that the electron plasma density becomes positive in (19.19). Overall charge neutrality implies $\int_{-\infty}^{\infty} n_0 dx = A\int_{-\infty}^{\infty}(U_1(x,t) - U_2(x,t))dx$ at all times, where the integration runs over all space. Generalizations to many contours are trivial.

One particularly interesting property of a single water-bag model is that it reproduces the results from fluid theory in cases where the *heat flow* is vanishing (Davidson 1972). The proof is left to the reader as an exercise. In Fig. 19.2 we illustrate the distinction between cases without and with heat flow (Davidson 1972).

- **Exercise:** What is the plasma density, the average velocity and the effective temperature of a water-bag velocity distribution function with constant level A and confined in velocity space by the boundaries $U_1(x,t)$ and $U_2(x,t)$?

19.4 Drift kinetic equation

In a strongly magnetized plasma we can adopt a sort of "hybrid" model, when discussing low frequency modes of oscillations. Here we can let the guiding center represent the particle position, and separate the motion in the **B**-parallel and **B**-perpendicular directions by writing a simplified version of the drift kinetic equation as

$$\frac{\partial}{\partial t}f + \nabla\cdot(\mathbf{u}_\perp f) + u_\parallel\frac{\partial}{\partial r_\parallel}f + \frac{\partial}{\partial u_\parallel}\left(\frac{K_\parallel}{m}f\right) = 0, \tag{19.20}$$

with

$$\mathbf{u}_\perp = \frac{\mathbf{K}\times\mathbf{B}}{qB^2} + \frac{m}{q^2B^2}\left(\frac{\partial}{\partial t}\mathbf{K}_\perp + u_\parallel \frac{\partial}{\partial r_\parallel}\mathbf{K}_\perp\right), \tag{19.21}$$

where $f = f(\mathbf{r}, u_\parallel, \mu, t)$ is the distribution of plasma particles with mass m, and magnetic moment μ, moving under the action of a force field $\mathbf{K} = \mathbf{K}(\mathbf{r}, t)$. In general we have $\mathbf{K} = q\mathbf{E}$, but also a gravitational force can contribute to $\mathbf{K}$. Since $d\mu/dt = 0$ (see Section 7.1.8), a term containing the magnetic moment μ does not appear explicitly in (19.20). For electric forces entering (19.21), we recognize the $\mathbf{E}\times\mathbf{B}/B^2$-drift as well as the polarization drift as experienced by a particle moving in a space-time varying electric field, in the direction perpendicular to the magnetic field. Even for time-stationary electric fields, the moving particle will experience a time variation $\partial\mathbf{E}_\perp/\partial t = -u_\parallel \partial\mathbf{E}_\perp/\partial r_\parallel$ of the electric field.

If we let the magnetic field be infinite, the drift kinetic equation reproduces a one-dimensional plasma model, as expected. In this limit, the charged particles move like "pearls on a string", with their physical positions being indistinguishable from the position of their gyro-center. The electric field can still have a component perpendicular to the magnetic field.

20

Kinetic Description of Electron Plasma Waves

Historically, electron plasma waves, or Langmuir waves, were those studied first by the kinetic plasma model. It might not be these waves which are the most interesting, or the most relevant, for plasmas in nature, but we take these as a reference model for discussing some of the basic problems distinguishing a fluid and a kinetic description. We have seen that a dispersion relation obtained from the linearized version of the basic fluid equations was a very useful device for understanding the plasma dynamics, so we place an emphasis in deriving a dispersion relation also for the kinetic model. For the present purpose, we assume that only high frequency fluctuations are relevant, taking for simplicity the ion component to be an immobile background of positive charge.

- **Example:** Consider first the simple one-dimensional equation

$$\frac{\partial f}{\partial t} + u\frac{\partial f}{\partial x} = 0, \tag{20.1}$$

which can be interpreted as an almost trivial limit of the Vlasov equation (19.3), with the collective interactions being "switched-off". The equation (20.1) describes a *Knudsen gas*, one where particles propagate along straight lines of orbit, without collisions; the particles are *free streaming*. The equation (20.1) is so simple that we can give an analytical solution of the time evolution of a given initial condition, while on the other hand the equation, in spite of all its simplicity, retains some elements associated with the more general Vlasov equation. We note that (20.1) is a *linear* equation. It has a trivial solution $f = F(u)$, where $\int_{-\infty}^{\infty} F(u)du = n_0$.

For a given initial condition $f(x,u,t=0) = G(x,u) + F(u)$, (20.1) has the exact solutions $f(x,u,t) = G(x-ut,u) + F(u)$ for any function G, i.e., f is constant along $x - ut =$ const., i.e., along the *characteristics* of (20.1). Take, for instance, a Gaussian distribution and a sinusoidal initial density perturbation with a constant $\eta < 1$ being a measure of the strength of the perturbation

$$G(x,u) = \sin(kx)\exp(-(u/u_0)^2)\frac{\eta}{\sqrt{\pi u_0^2}},$$

which gives

$$f(x,u,t) = \frac{\eta}{\sqrt{\pi u_0^2}} e^{-(u/u_0)^2}\sin(kx - kut) + F(u). \tag{20.2}$$

If an ideal energy analyzer could be devised and placed at an arbitrary position x_1, it would give a steady, undamped output, for all times. If, however, the analyzer has a finite resolution, the output will be damped in time. Assume, for instance, that we can only measure particle *density*, irrespective of the actual velocity of the individual particles. The resulting output from the detector is then

$$\begin{aligned}
\widetilde{n}(x,t) + n_0 &\equiv \int_{-\infty}^{\infty} f(x,u,t)du = \\
&\frac{\eta}{\sqrt{\pi u_0^2}}\int_{-\infty}^{\infty}[\sin(kx)\cos(kut) - \cos(kx)\sin(kut)]e^{-(u/u_0)^2}du + n_0 \\
&= \eta\sin(kx)e^{-(ku_0t/2)^2} + n_0,
\end{aligned} \tag{20.3}$$

where here $\widetilde{n}$ is the deviation from n_0, hence $\widetilde{n}$ can take positive as well as negative values. The result (20.3) gives an exponential damping with the *square* of time for $t \to \pm\infty$, at any position x. This means that the free streaming solution contribution to the density perturbation will vanish quite rapidly as long as the velocity distribution is like the one assumed here.

The equation (20.1) is time reversible, just like the Vlasov equation. We see that if all the information contained in the distribution function can be retained, we will be able to reconstruct the initial condition at *all* times. If, however, some of this information is lost (as it is when only density and not velocity distribution is measured) we end up with damping. In this trivial case, the damping is due to phase-mixing of freely streaming particles, but the conclusion is valid also for more complicated systems, collisionless plasma oscillations in particular.

The main lesson to be learned from the present example is that we encountered a dynamic problem without any well defined dispersion relation! Since the basic equation (20.1) is just a variant, albeit very simple, of the Vlasov equation, we might anticipate peculiar results in a discussion of that equation as well. This is after all to be expected; a dispersion relation $\omega = \omega(\mathbf{k})$ is, as we have argued before, only a way to organize the independent variables space and time. For the Vlasov equation we have *velocity* as an independent variable as well, and the concept of dispersion relation will presumably have to be re-interpreted.

- **Example:** Consider a uniform flow of freely streaming "ballistic" particles with a Maxwellian distribution with a net flow velocity U. Assume now that a flat obstacle of diameter D is placed perpendicular to the average flow velocity vector. Calculate the steady state spatial variation of the particle density in the wake behind the obstacle. Consider the problem in two as well as three spatial dimensions.

20.1 Linearized equations

Turning to the less trivial Vlasov equation (19.3), we note that in a uniform non-magnetized plasma, any distribution function $f_0(\mathbf{u})$ is a solution of (19.3), since then there are no net charges, and consequently the electrostatic field vanishes, $\mathbf{E}_0 = 0$. In thermal equilibrium in particular, a Maxwellian velocity distribution function will be a stationary solution. Perturbing such a solution slightly at $t = t_0$, we have $f = f_0 + \widetilde{f}$. Considering small amplitude fluctuations, we may linearize (19.3), i.e., ignore second order terms $\widetilde{\mathbf{E}} \cdot \nabla_{\mathbf{u}} \widetilde{f}$. To simplify the notation, we ignore from now on the $\widetilde{\ }$ on f and $\mathbf{E}$, etc. We then have the linearized version of (19.3) in the form

$$\frac{\partial f}{\partial t} + \mathbf{u} \cdot \nabla f + \frac{n_0}{m} \mathbf{K} \cdot \nabla_{\mathbf{u}} f_0(\mathbf{u}) = 0, \tag{20.4}$$

where $n_0 f_0(\mathbf{u})$ is here taken to be the unperturbed, equilibrium, electron velocity distribution, where for later convenience we use the normalization $\int_{-\infty}^{\infty} f_0(\mathbf{u}) d^3u = 1$. We might have $f_0(\mathbf{u}) = f_0(u)$, but keep the general notation here. By $\mathbf{K}$ in (20.4), we can in principle understand *any* force, but for most cases of interest we expect to have the Lorentz force $\mathbf{K} = -e\mathbf{E} - e\mathbf{u} \times \mathbf{B}$, with $-e$ being the electron charge. Gravitational forces can be important, however, for a number of important applications.

In terms of the velocity distribution function f, Maxwell's equations become

$$\nabla\cdot\mathbf{E} = -\frac{e}{m}\int_{-\infty}^{\infty} f d^3u \tag{20.5}$$

$$\nabla\times\mathbf{E} = -\frac{\partial}{\partial t}\mathbf{B} \tag{20.6}$$

$$\nabla\cdot\mathbf{B} = 0 \tag{20.7}$$

$$\nabla\times\mathbf{B} = -\mu_0 e\int_{-\infty}^{\infty}\mathbf{u} f d^3u + \frac{1}{c^2}\frac{\partial}{\partial t}\mathbf{E}. \tag{20.8}$$

It is often convenient to introduce the potentials ϕ and $\mathbf{A}$, defined through

$$\mathbf{B} = \nabla\times\mathbf{A} \qquad \text{and} \qquad \mathbf{E} = -\nabla\phi - \frac{\partial}{\partial t}\mathbf{A}.$$

These relations determine ϕ and $\mathbf{A}$, apart from a *gauge-transform* (Jackson 1960). Here, we choose the *Coulomb gauge*, where

$$\nabla\cdot\mathbf{A} = 0.$$

With this choice, we can separate rotation-free, or *longitudinal* solutions from those which are divergence-free or *transverse*.

It is a relatively trivial operation to insert the scalar and vector potentials, ϕ and $\mathbf{A}$, into the basic equations to obtain

$$\frac{\partial f}{\partial t} + \mathbf{u}\cdot\nabla f + \frac{en_0}{m}\left(\nabla\phi + \frac{\partial}{\partial t}\mathbf{A} - \mathbf{u}\times(\nabla\times\mathbf{A})\right)\cdot\nabla_{\mathbf{u}} f_0(\mathbf{u}) = 0, \tag{20.9}$$

together with

$$\nabla^2\phi = \frac{e}{\varepsilon_0}\int_{-\infty}^{\infty} f d^3u \tag{20.10}$$

$$\left(\frac{\partial^2}{\partial t^2} - c^2\nabla^2\right)\mathbf{A} = -\frac{e}{\varepsilon_0}\int_{-\infty}^{\infty}\mathbf{u} f d^3u - \frac{\partial}{\partial t}\nabla\phi, \tag{20.11}$$

with the given choice of gauge. The general problem, with both ϕ and $\mathbf{A}$ different from zero, is interesting, but usually the analysis is restricted to longitudinal oscillations where $\mathbf{A} = \text{const}$.

20.2 Kinetic dispersion relations

By restricting the analysis to electrostatic fluctuations with $\mathbf{A} = 0$, a closed set of equations is obtained by the linearized Vlasov equation and Poisson's equation.

$$\nabla\cdot\mathbf{E} = -\nabla^2\phi = -\frac{e}{\varepsilon_0}n = -\frac{e}{\varepsilon_0}\int_{-\infty}^{\infty} f(\mathbf{r},\mathbf{u},t)d^3u, \tag{20.12}$$

which relates the electron density to the electric potential, and thereby the electric field. Here $\mathbf{E}$, ϕ, n as well as f are all time varying, but related by an equation that is used for describing electrostatics, hence justifying the name "electrostatic waves". Since the role of the ion component is here solely to provide a background which ensures overall charge neutrality, the ion density n_0 does not appear when n measures the *deviation* of the local electron density from the average plasma density n_0. By the choice $\mathbf{A} = 0$, we see from (20.11) that the electron current $-e\int_{-\infty}^{\infty}\mathbf{u} f(\mathbf{r},\mathbf{u},t)d^3u$ exactly cancels Maxwell displacement current, as required for electrostatic fluctuations.

Fourier transforming (20.4) with respect to the temporal as well as the spatial variables we find

$$-i(\omega - \mathbf{u}\cdot\mathbf{k})f(\mathbf{k},\mathbf{u},\omega) + i\frac{en_0}{m}\phi(\mathbf{k},\omega)\mathbf{k}\cdot\nabla_{\mathbf{u}} f_0(\mathbf{u}) = 0. \tag{20.13}$$

We want to use this expression to find the perturbation of the electron density, $n(\mathbf{k},\omega) \equiv \int_{-\infty}^{\infty} f(\mathbf{k},\mathbf{u},\omega)d^3u$, and insert it into the Fourier transformed Poisson equation. To do this we have to divide by $(\omega - \mathbf{u}\cdot\mathbf{k})$ in (20.13) to get an explicit expression for $f(\mathbf{k},\mathbf{u},\omega)$, and then integrate with respect to velocity. This gives problems, unfortunately; we do not a priori know how to deal with the singularity at velocities where $\omega = \mathbf{u}\cdot\mathbf{k}$. One might throw in a principal value sign, but there are obviously no a priori given justifications for doing so. In a way, this is *the* basic problem in linear plasma kinetic theory, and different methods have been devised to overcome it. We take the simplest one first, and discuss some of the others later. Compared to the other methods for approaching the problem, the author (Pécseli 2000) was tempted to give this section the heading "Kinetic theory without tears".

Having seen the problems arising by the standard Fourier transform, we return to the original form of the linearized Vlasov equation (20.4). Given an initial condition, $f(\mathbf{r},\mathbf{u},t_0)$ for the problem, we integrate (20.4) along characteristics, and obtain

$$f(\mathbf{r},\mathbf{u},t) = f(\mathbf{r}-\mathbf{u}(t-t_0),\mathbf{u},t_0) + \frac{en_0}{m}\int_{t_0}^{t} \mathbf{E}(\mathbf{r}-\mathbf{u}(t-t'),t')\cdot\nabla_{\mathbf{u}} f_0(\mathbf{u})dt'. \tag{20.14}$$

We assumed from the outset that only the electrostatic case will be considered, and assumed $\mathbf{K} = -e\mathbf{E}(\mathbf{r},t)$ in (19.3). Note that the relation (20.14) is *not* just another way of writing (20.4); we have made a choice by considering the *causal* time evolution of a given initial condition, $f(\mathbf{r},\mathbf{u},t_0)$. The state at a time t depends only on the values of f and $\mathbf{E}$ at times *before* t, and not on the future! Readers unfamiliar with integration of partial differential equations can check the validity of (20.14) simply by insertion into (20.4). The initial condition $f(\mathbf{r},\mathbf{u},t_0)$ is trivially satisfied.

By integrating (20.14) with respect to velocity $\mathbf{u}$ we obtain the electron density $n(\mathbf{r},t)$, and find

$$\begin{aligned} n(\mathbf{r},t) &= \int_{-\infty}^{\infty} f(\mathbf{r}-\mathbf{u}(t-t_0),\mathbf{u},t_0)\,d^3u \\ &+ \frac{en_0}{m}\int_{-\infty}^{\infty}\int_{t_0}^{t} \mathbf{E}(\mathbf{r}-\mathbf{u}(t-t'),t')\cdot\nabla_{\mathbf{u}} f_0(\mathbf{u})dt'd^3u. \end{aligned} \tag{20.15}$$

Since the functions $e^{i\mathbf{k}\cdot\mathbf{r}}$ form a complete, orthogonal set, we can expand the spatial variation by a Fourier transform, and consider only one arbitrary spatial Fourier component denoted by $\mathbf{k}$. With the assumption of electrostatic fluctuations, $\mathbf{E} = -\nabla\phi$, we have $\mathbf{E}(\mathbf{r},t) \to -i\mathbf{k}\phi(\mathbf{k},t)e^{i\mathbf{k}\cdot\mathbf{r}}$. In this case the integration of velocity components in the directions perpendicular to $\mathbf{k}$ are trivial, and we are left with a formally one-dimensional relation for the electron density of the form

$$\begin{aligned} n(k,t)e^{ikx} &= \int_{-\infty}^{\infty} g(x-u_x(t-t_0),u_x,t_0)\,du_x \\ &- \frac{en_0}{m}ik\int_{-\infty}^{\infty}\int_{t_0}^{t} \phi(k,t')e^{ikx-u_x(t-t')}\frac{d}{du_x}F_0(u_x)dt'du_x. \end{aligned} \tag{20.16}$$

We replaced f by a new function g in (20.16), to emphasize that we integrated with respect to two of the velocity variables. The reduced distribution function, F_0, is what remains after integration with respect to u_y and u_z. For simplicity we write $F_0'(u_x) \equiv dF_0(u_x)/du_x$ in the following. The density is now inserted into Poisson's equation $-k^2\phi(\mathbf{k},t) = en(\mathbf{k},t)/\varepsilon_0$, as appropriate for the actual Fourier component, giving the result

$$\begin{aligned} k^2\phi(k,t) &= -\frac{e}{\varepsilon_0}\int_{-\infty}^{\infty} g(x-u_x(t-t_0),u_x,t_0)\,du_x \\ &+ \frac{e^2n_0}{m\varepsilon_0}ik\int_{-\infty}^{\infty}\int_{t_0}^{t} \phi(k,t')e^{-iu_x(t-t')}F_0'(u_x)dt'du_x. \end{aligned} \tag{20.17}$$

For any non-singular initial condition $g(x,u_x,t_0)$, we find that the first term on the right-hand sides of (20.15) or (20.17) vanish by "phase mixing" at large times, as in (20.3), and this contribution will be ignored in the following. In the same limit we let $t_0 \to -\infty$ without loss of generality. Note that the omission of this first term is not without consequences: if to be consistent we also ignore the first term in (20.14), we are no longer able to reproduce an initial condition by "changing the arrow of time". In these limits, we can write

$$k^2\phi(k,t) = \frac{e^2 n_0}{m\varepsilon_0} ik \int_{-\infty}^{\infty}\int_{-\infty}^{t} \phi(k,t')e^{-iu_x(t-t')}F_0'(u_x)dt'du_x. \tag{20.18}$$

This result is now trivially rewritten by introducing the new variable $\tau \equiv t - t'$, and in terms of Heaviside's step function, $H(\tau) = 1$ for $\tau > 0$ and $H(\tau) = 0$ for $\tau < 0$, giving

$$k^2\phi(k,t) = \frac{e^2 n_0}{m\varepsilon_0} ik \int_{-\infty}^{\infty} F_0'(u_x) \int_{-\infty}^{\infty} H(\tau)e^{-iku_x\tau}\phi(k,t-\tau)d\tau du_x. \tag{20.19}$$

Fourier transforming with respect to time, we note that the right-hand side of (20.19) contains a convolution of $H(\tau)e^{-iu_x\tau}$ and $\phi(k,t-\tau)$. The Fourier transform of this convolution is the product of the Fourier transforms of $H(\tau)e^{-iku_x\tau}$ and $\phi(k,\tau)$. We denote the transform of the potential by $\phi(k,\omega)$. The Fourier transform of the product $H(\tau)e^{-iu_x\tau}$ is the transform of $H(\tau)$ shifted by ku_x, giving $\pi\delta(\omega - ku_x) - iP/(\omega - ku_x)$, with P denoting that the principal value has to be taken upon integration. By insertion we find that now the transforms of the electrostatic potential cancel, and we are left with the dispersion relation in the form

$$1 - \frac{\omega_{pe}^2}{k^2} P \int_{-\infty}^{\infty} \frac{F_0'(u_x)}{u_x - \omega/k} du_x - i\pi \frac{\omega_{pe}^2}{k^2} F_0'(\omega/k) = 0. \tag{20.20}$$

This dispersion relation is *complex*, and gives the solution of ω in an implicit form, for given wavenumber k. We shall consider the solution $\omega = \omega(k)$ in a moment.

- **Example:** To obtain the Fourier transform of Heaviside's step function we may use, for instance, that the Fourier transform of the function defined as

$$G(t) = \begin{cases} A\exp(-at) & \text{for} \quad t > 0 \\ 0 & \text{for} \quad t < 0 \end{cases}$$

is the complex function

$$\check{G}(\omega) = A\frac{a - i\omega}{a^2 + \omega^2}.$$

In the limit $a \to 0$ we find that $G(t)$ approaches Heaviside's step function, and we can argue that in the same limit $\check{G}(\omega)$ will approach its Fourier transform. In particular we recall that $a/(a^2 + \omega^2) \to \pi\delta(\omega)$ for $a \to 0$. The symmetry of $\Im\{\check{G}(\omega)\}$ gives rise to the principal value sign introduced before.

To obtain a physical interpretation of (20.20) we introduce an analytical expression of the dielectric function associated with a plasma in its kinetic description. We first recall the general definition of the relative dielectric function of a dispersive medium given by (5.89) as

$$\varepsilon_0\varepsilon(k,\omega) = \varepsilon_0 - \frac{en(k,\omega)}{k^2\phi(k,\omega)},$$

where again only one Fourier component was considered. Using (20.16) and the same arguments as those giving (20.19), we obtain

$$n(k,t) = \frac{en_0}{m} ik \int_{-\infty}^{\infty} F_0'(u_x) \int_{-\infty}^{\infty} H(\tau)e^{-iku_x\tau}\phi(k,t-\tau)d\tau du_x, \tag{20.21}$$

to be inserted into (5.89). Fourier transforming the resulting expression with respect to time as we did to obtain (20.20), the relative dielectric function is obtained as

$$\varepsilon(k,\omega) = 1 - \frac{\omega_{pe}^2}{k^2} P \int_{-\infty}^{\infty} \frac{F_0'(u_x)}{u_x - \omega/k} du_x - i\pi \frac{\omega_{pe}^2}{k^2} F_0'(\omega/k), \tag{20.22}$$

for real ω and real k. The dielectric function is complex with an imaginary part, which for the case $F_0'(\omega/k) < 0$ implies dissipation: Landau damping. Physically, this implies that in order to maintain the real frequency ω, we have to supply energy to the system. Comparing (20.20) and (20.22) we again find that the dispersion relation for the electrostatic plasma waves is found by $\varepsilon(k,\omega) = 0$.

Finally we note that the dielectric function (20.22) satisfies the Kronig-Kramers relations (Yeh & Liu 1972, Pécseli 2000); see Section 5.9.2. Had we taken the principal value when integrating f from (20.13) without further ado, the final result would *not* have satisfied the Kronig-Kramers relations, and actually this *should* have aroused suspicion in the first place.

We can obtain a result like (20.22) by heuristic arguments also. Assume that we have managed to produce a set-up including a complicated array of thin wires, etc., which makes it possible to excite one selected wavelength, or rather one wavenumber **k**. We fill the experiment with a homogeneous plasma, and apply a frequency ω in order to determine experimentally the plasma dielectric properties as given by $\varepsilon(\mathbf{k},\omega)$ using the relation (20.22). We are free to let the amplitude increase slowly from zero to a certain level to have a causal response, but in this case we do *not* have a strictly harmonic signal. We can approximate the described amplitude variation as $\exp(-i\omega_R t)\exp(\omega_I t)$, where we write a complex frequency as $\omega \equiv \omega_R + i\omega_I$, with both ω_R and ω_I real. To obtain the density perturbation produced by the applied potential variation, we determine f from (20.13) and integrate with respect to velocity. This time, we do not have problems with singularities, since ω is complex. We subsequently let $\omega_I \to 0$, and recover (20.22), although, admittedly, with a somewhat uneasy feeling about the mathematical correctness of the procedure.

It seems as if inclusion of a simple damping term will make the singularity at $u = \omega/k$ disappear. This is, however, not necessarily so. We might, for instance, take the simplest collision term from Section 19.1.1, and write the linearized Vlasov equation for the present problem as

$$\frac{\partial f}{\partial t} + \mathbf{u}\cdot\nabla f + \frac{en_0}{m}\nabla\phi\cdot\nabla_{\mathbf{u}} f_0(\mathbf{u}) = -\nu f, \tag{20.23}$$

with ν being a constant collision frequency. It is, however, trivially observed that by introducing a new distribution function of the form $f_* \equiv f\exp(\nu t)$ we recover (20.4) for f_* and nothing has been gained.

- **Exercise:** Demonstrate that the dispersion relation (20.20) reproduces the standard cold plasma result, $\omega^2 = \omega_{pe}^2$, for $u_{The} \to 0$ for a Maxwellian plasma.

- **Exercise:** Apply the dispersion relation (20.20) to a Lorentzian model velocity distribution function of the form

$$f_0(u) = \frac{1}{\pi}\frac{a}{a^2+u^2},$$

  and demonstrate how you can solve the integral analytically. Would you have any objection to using this model for electron plasma waves in general? (Hint: what is the temperature of the electron gas in this model?)

- **Exercise:** Apply the dispersion relation (20.20) to the model velocity distribution function with

$$f_0(u) = \begin{cases} \dfrac{1}{a} + \dfrac{u}{a^2} & \text{for} \quad -a < u < 0 \\[2ex] \dfrac{1}{a} - \dfrac{u}{a^2} & \text{for} \quad 0 < u < a \\[2ex] 0 & \text{for} \quad |u| > a. \end{cases}$$

What is the average velocity and the temperature of the electron gas with this distribution?

20.2.1 Landau damping the easy way

Using the results of Sections 5.10.6 and 20.2, we are now in a position to derive the expression for the *linear Landau damping*. Basically, we want to find solutions to $\varepsilon(k,\omega) = 0$ for given k. In Section 5.10.6 we learned how to obtain the real and imaginary parts of the frequency from the dielectric function, given that the damping was small, $\omega_R \gg \omega_I$.

Taking the expression for the dielectric function (20.22) and using the real part as in (5.107), we find that the real part of the frequency corresponding to a given real wavenumber is determined by solution of

$$1 - \frac{\omega_{pe}^2}{k^2} P \int_{-\infty}^{\infty} \frac{F_0'(u_x)}{u_x - \omega/k} du_x = 0, \tag{20.24}$$

assuming that in a first approximation, the imaginary part of the frequency is small. This requirement implies by the imaginary part of (20.22) together with (5.108) that $F_0'(\omega_R/k) \approx 0$. Without specifying F_0 we are of course not able to determine for which phase velocities ω_R/k this criterion can be fulfilled, but we know one thing for certain: for large velocities we *must* have $F_0'(u_x) \to 0$ and $F_0(u_x) \to 0$, since the velocity distribution has to be integrable. It is therefore safe to look for a solution of (20.24) in the limit of $\omega_R/k \to \infty$. In this limit, we expect that the fine details of F_0 are immaterial, and that we might as well make the replacement $F_0(u_x) \to \delta(u_x)$. Then it is almost trivial to solve (20.24), and we find the result

$$\omega_R \approx \omega_{pe}, \tag{20.25}$$

independent of k to the present approximation, i.e., the cold plasma result. Intuitively this seems reasonable; for large phase velocities we would expect thermal effects to be $O(u_{The}/(\omega/k)) \ll 1$, and the real part of the frequency to be given by ω_{pe}, apart from small corrections. For the *imaginary part* the situation is different; here the cold plasma result gives a zero imaginary part, and with finite temperatures all we have *are* these small terms. To determine these we use the result (5.108) where now to the first approximation we have $\varepsilon(k,\omega) \approx 1 - (\omega_{pe}/\omega_R)^2$, giving $\partial\varepsilon(k,\omega_R)/\partial\omega_R \approx 2/\omega_{pe}$, which, by the imaginary part of (20.22), in turn implies

$$\omega_I(k) \approx \frac{1}{2}\pi \frac{\omega_{pe}^3}{k^2} F_0'(\omega_{pe}/k). \tag{20.26}$$

For a negative derivative of $F_0(u_x)$ at large velocities, which is the most obvious thing to expect, we thus find a damping (i.e., $\omega_I < 0$) of the Langmuir waves given by the value of this derivative taken at the relevant phase velocity, and then multiplied by some constant. Essentially, this is the Landau damping. We note that the cold plasma result is recovered in the sense that $\omega_I(k) \to 0$ in case $F_0(u) = \delta(u)$. As the phase velocity of the waves goes to infinity, $\omega_R/k \to \infty$, for $k \to 0$ the damping also disappears.

An interesting observation is, however, that there is a possibility for a phase velocity interval, where the derivative of the dispersion function is *positive*, implying that the waves can be increasing in amplitude! The plasma is then *unstable*. The phase velocity interval of these linearly unstable waves *has* to be limited, since $F_0(u_x) > 0$ has to approach zero at infinity with a negative derivative. We can, however, imagine that the plasma is penetrated by a low density electron beam with a large scatter in velocities, in such a way that the plasma frequency remains to be determined by the density of the background plasma, and that the local (positive) derivative of the distribution function remains small, to have $\omega_I \ll \omega_R$, as assumed from the outset. The range of phase velocities that corresponds to unstable modes will be finite and confined to the interval where $F_0'(u_x) > 0$.

In Fig. 20.1 we show the real and imaginary parts of the frequency for a Langmuir wave assuming that the velocity distribution of the electrons is a Maxwellian, $F_0(u) = \sqrt{m/(2\pi\kappa T_e)}$

$\exp(-mu^2/(2\kappa T_e))$, i.e., $F_0'(u) = -\sqrt{1/2\pi}(m/\kappa T_e)^{3/2} u \exp(-mu^2/(2\kappa T_e))$. The waves are damped in this case, so we show $|\omega_I|$. Note that we here have the Landau damping to be negligible for $k\lambda_D < 0.1$, i.e., $\omega_I < \omega_R 10^{-4}$. For $k\lambda_D \approx 1$, on the other hand, we find that the waves damp out within one oscillation period, so it is hardly meaningful to talk about wave-like motion for such short wavelengths.

Note that the high phase velocities having the smallest damping also have the smallest group velocities. This observation has evident consequences when relating temporal and spatial dampings by the arguments outlined in Section 5.10.2.

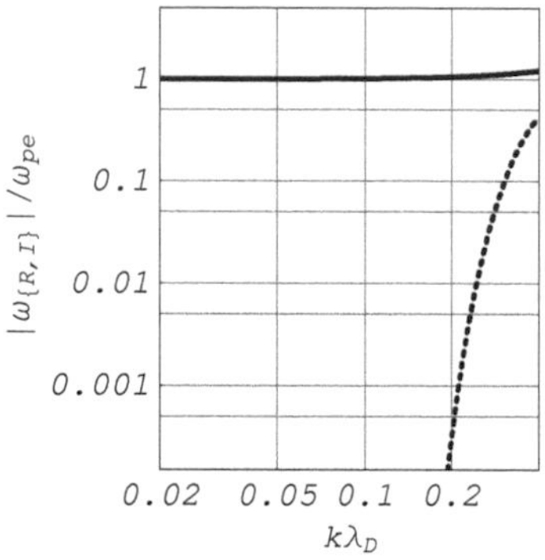

FIGURE 20.1: *Dispersion relation showing normalized real (solid line) and imaginary parts (dashed line), ω_R and $|\omega_I|$, of the frequency of a Langmuir wave as a function of normalized wavenumber assuming the velocity distribution function to be a Maxwellian. In the real part we included a small thermal correction.*

20.2.2 Physical arguments for Landau damping

We have demonstrated that in spite of its time reversibility, the Vlasov equation contains solutions which are damped in time. This is in a "certain sense" only; after all, the Vlasov equation is for the distribution function, and we observe the damped solutions by studying the fluctuations in plasma density or electrostatic potential, which contain reduced information. If we *have* the information of the entire phase space variation of the distribution function we would be able to recover the initial condition at all later times. If we only have access to the density or the potential, this is no longer possible! One can with little effort go into a more detailed discussion of the damping by use of a physical description, as summarized later. Also other physically realizable systems (albeit with rather special conditions) can be found, where time reversible processes give rise to seemingly irreversible losses (Guio & Pécseli 2016).

First of all we make clear that the electron plasma waves discussed in Section 20.2.1 are different in nature from surface waves on water, for instance. For the latter wave-type, small parcels of water perform basically a circular motion, while the waveform is propagating. (Recall how a piece of cork, for instance, is moving on the surface, as the wave passes.) Because the individual water molecules are constantly colliding, no individual particles are able to follow with the wave in classical fluids or gases. In a collisionless plasma, the situation is different. Here we have particles with distributed velocities moving along straight line orbits in the absence of the wave, or other perturbations. (In Section 20.2.1 we ignored dc magnetic fields.) *With* the wave present we can distinguish basically two types of particles; some are *resonant*, having a velocity close to the phase velocity of the waves, and the others are *non-resonant*. To be more specific, we consider those particles that are resonant in the rest frame of the wave, i.e., the frame of reference which in the laboratory frame is moving with the phase velocity where the wave potential is $\phi_0 \cos(kx)$. Assume for the moment that the wave amplitude ϕ_0 is constant. The total energy of a particle with charge q located in a plane electrostatic wave can here be written as

$$W = \frac{1}{2}Mu^2 + q\phi_0(\cos(kx) - 1),$$

with ϕ_0 being the peak potential associated with the wave. Particles with $W < 0$ are *trapped* in the sense that they have closed trajectories in phase space around the local wave minima (for positively charged particles, maxima for negative). The other particles, with $W > 0$, are *free*. We make a local parabolic approximation around the minimum, giving the equation of motion

$$M\frac{d^2}{dt^2}X = q\phi_0 k^2 X.$$

In terms of the particle position $X = X(t)$, we can define a characteristic frequency

$$\omega_B \equiv \sqrt{\left|\frac{q}{M}\phi_0\right| k^2},$$

being the *bounce frequency* for the trapped particles. The bounce *period* is $\tau_B \equiv (2\pi/k)\sqrt{M/|q\phi_0|} = \lambda\sqrt{M/|q\phi_0|}$, in terms of the wavelength λ. For the linearized problem we considered in Section 20.2.1 we assumed small amplitudes and the bounce time is therefore very long. Within this time, we will find that particles which are slower than the phase velocity of the wave will be accelerated, while those faster will be decelerated. The interaction of a particle and a wave is often illustrated with reference to a surfer on a large amplitude coastal wave (Chen 2016).

Figure 20.2 may help to explain the interaction between the wave and the particles. The figure is best understood by using a reference system moving with the wave-phase velocity ω/k. The contours labeled 1 and 2 correspond to particle phase space trajectories, where the particles are faster than the phase velocity, while they are *slower* for 5 and 6. The labels 3 and 4 give the separatrix between free and trapped orbits, i.e., orbits inside the separatrix (between contours 3 and 4) are samples of trapped particle trajectories in phase space. The top part of the figure shows the velocity distribution function, and below in a two-dimensional phase space, we find selected particle orbits. Closed elliptical orbits represent trapped particles. (Note that the physical dimensions of the two axes "velocity" and "position" in phase space are different, so the ratio of the major and minor axes in the elliptic trapped particle orbits is solely a matter of representation!) Some selected free particles are indicated, where a black filled circle indicates particles which move toward an increasing potential energy, and hence lose kinetic energy, while particles shown by open circles *gain* kinetic energy. Trapped particles are shown by asterisks. The trapped particle at B is losing kinetic energy, while the one at position A is gaining kinetic energy. We now note that the open circles at position B are deeper into the distribution (i.e., further to the left) than the filled circles at position A. Similarly, the filled circles at B are further to the right than the open circles at A. By inspection of the velocity distribution, which is here chosen to be a Maxwellian, we readily note that there are more particles gaining energy as compared to those losing energy. Since the system as such is energy conserving, we can argue that this energy must come from the wave, which is consequently being damped: this is the physical mechanism for Landau damping (Chen 2016). Note that as the amplitude of the wave is decreased, in order to make a linear theory apply, the number of trapped particles is steadily decreasing, while the number of free, or untrapped, particles increases correspondingly. The linear Landau damping does *not* solely rely on the trapped particles, contrary to many physical models that have been suggested.

Waves with phase velocities $\omega_R(k)/k \to \omega_{pe}/k \to \infty$ as $k \to 0$ will *always* be damped, but the damping reduces to zero, $\omega_I(k) \to 0$, in the limit where the phase velocity is infinite. Physically

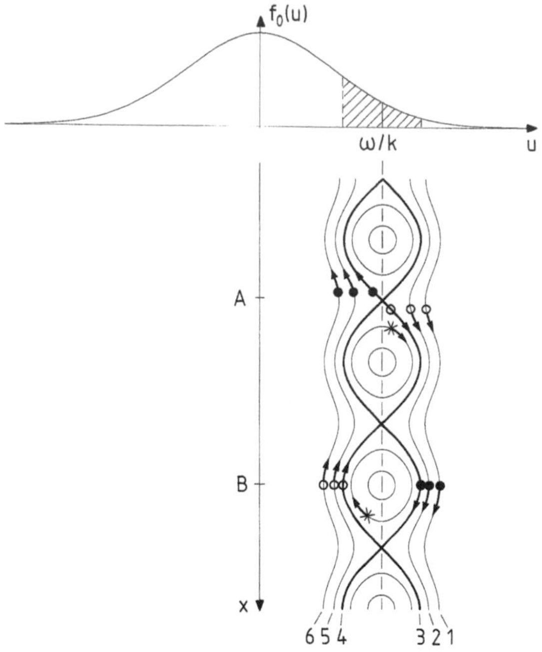

FIGURE 20.2: *Illustration for discussions of Landau damping, adapted from Chen (2016). The figure shows free and trapped particle orbits by thin lines, while the separatrix is indicated by a heavier line. The top of the figure gives the electron velocity distribution, with the shading giving the trapping region around the phase velocity, ω/k.*

this means that there are no particles for the waves to interact with, in the limit of infinite phase velocities.

Note also that the unstable waves, with $\omega_I(k) > 0$ as mentioned before, can be understood by discussing Fig. 20.2 for the case where the slope of the velocity distribution function is positive in a velocity interval containing the phase velocity $\omega_R(k)/k$.

The linear theory applies for times until the acceleration or deceleration due to the wave-particle interaction changes the particle orbits significantly from the unperturbed orbits (O'Neil 1965, Drummond 2004). For a plane wave, this will happen on a "trapping" time scale, which as we have seen is $\sim \tau_B \equiv (2\pi/k)\sqrt{M/|q\phi_0|} = \lambda\sqrt{M/|q\phi_0|}$. As an order of magnitude, we can use the time $\mathcal{L}\sqrt{M/|q\phi_0|}$, where $\mathcal{L}$ is a characteristic length scale for the potential structure. In this way we are not restricted to plane waves, but may consider localized pulses, for instance. At times larger than τ_B, the distribution function is distorted locally, and differs from the assumed $F_0(u)$.

Ⓢ **Exercise:** Let the velocity distribution function be $f_0(u)$ in a one dimensional plasma model. At a time $t = 0$, a periodic electrostatic perturbation is applied to the plasma, in such a way that it propagates as a plane wave $\phi(x) = \phi_0 \sin(kx - \omega t)$ at later times. Give an analytical expression (as much as you can) for the ratio of the number of trapped particles to the total number of particles. See also Fig. 20.2 for an illustration of the problem. Analyze the special case with f_0 being a Maxwellian. Ⓢ

- **Exercise:** For a given wavelength, determine the amplitude of a harmonic electrostatic wave so that the bounce frequency of a trapped electron equals (approximately) the electron plasma frequency.

20.2.3 Landau damping the hard way

The results obtained in Section 20.2.1 came very easily. Unfortunately, among other things, that derivation gives little insight into, for instance, the *history* of the problem (Landau 1946, Jackson 1960), and also it relies on some simplifying (although reasonable) simplifications. In this section we present a more thorough and rigorous derivation, although (fortunately!) we end up with the foregoing results in the appropriate limits.

We can try first to see in more detail what happens if we make the usual Fourier transform with respect both to the spatial and the temporal variable of the linearized Vlasov equation. With simple manipulations we find

$$-i\omega f + i\mathbf{u}\cdot\mathbf{k}f + i\frac{en_0}{m}\phi\mathbf{k}\cdot\nabla_{\mathbf{u}}f_0(\mathbf{u}) = 0, \tag{20.27}$$

We now need to find $n \equiv \int_{-\infty}^{\infty} f d^3u$ to be inserted into Poisson's equation, so we can eliminate the potential ϕ. We have

$$n = \frac{en_0}{m}\phi\int_{-\infty}^{\infty}\frac{\mathbf{k}\cdot\nabla_{\mathbf{u}}f_0(\mathbf{u})}{\omega - \mathbf{k}\cdot\mathbf{u}}d^3u. \tag{20.28}$$

Without loss of generality, we can assume $\mathbf{k}$ to be along the x-axis, and integrate with respect to u_y and u_z, giving a simple relation without scalar products

$$n = \frac{en_0}{m}\phi\int_{-\infty}^{\infty}\frac{F_0'(u)}{\omega/k - u}du, \tag{20.29}$$

where $F_0'(u) \equiv dF_0(u)/du$. Now, this relation (20.29) can be problematic: if $F_0(u)$ is differentiable at and in the vicinity of $u = \omega/k$, and $F_0'(u = \omega/k) = 0$, then everything is fine, and the integral is well behaved, but what if $F_0'(u = \omega/k) \neq 0$? Then the integral in (20.29) is singular! Vlasov was aware of this problem, and suggested that if the problem arose, then the principal value of the integral should be taken, but as we have already seen, this is just one of the infinitely many possible choices, and not necessarily the correct one. Landau (1946) pointed out that the problem could be located to the Fourier transform with respect to the temporal variable. If we instead use the Laplace transform for an initial value problem, the real transform variable ω is replaced by a complex transform variable, and then we no longer have any singularities when integrating in (20.29) with respect to real variable u. No textbook including kinetic plasma theory is complete without a summary of this pioneering work, so it is presented in the next section. The disposition assumes that the reader is familiar with Laplace transform.

20.2.3.1 Solution by Laplace transform

We perform again the spatial Fourier transform, but use the Laplace transform for the temporal variable. When transforming the temporal derivative we have to know the initial value of the perturbed distribution function, which we assume to be given as $f(x,u,t=0) = g(u)\exp ikx$, where we made the problem one dimensional from the outset. The exponential variation with the spatial variable is rather trivial and is not written out explicitly. Often, an index k is inserted as a reminder of the wavenumber, i.e., by writing $f_k(u,t)$, for instance. However, since k enters explicitly in the expressions because of the presence of a spatial derivative, there will never be any doubt that we are dealing with a spatial Fourier transform, so the index k is omitted in the following to make the notation simpler.

For the temporal Laplace transform, denoted by $\mathcal{L}t\{\}$, of a time-differentiation we have

$$\mathcal{L}t\left\{\frac{\partial f}{\partial t}\right\} \equiv \int_0^{\infty}\frac{\partial f}{\partial t}e^{-st}dt = s\mathcal{L}t\{f\} - f(u,0^+),$$

introducing the limiting value $f(u,0^+) \equiv \lim_{t \to 0} f(u,t)$ through *positive* t-values. For the present conditions we have $f(u,0^+) = g(u)$. With a little algebra, which is left to the reader, we arrive at

$$\phi(s) = -i\frac{e}{\varepsilon_0 k^2}\int_{-\infty}^{\infty}\frac{g(u)}{uk - is}du\left(1 - \frac{e^2 n_0}{\varepsilon_0 mk}\int_{-\infty}^{\infty}\frac{F_0'(u)}{uk - is}du\right)^{-1}. \tag{20.30}$$

With this result we have a singularity in the denominators $1/(uk - is)$ at $u = is/k$, but as already mentioned, we need not worry about these, as long as we integrate along the real velocity axis u. In the following we will use the abbreviation

$$D(s,k) \equiv 1 - \frac{e^2 n_0}{\varepsilon_0 mk^2}\int_{-\infty}^{\infty}\frac{F_0'(u)}{u - is/k}du, \tag{20.31}$$

for the denominator in (20.30). We recognize the electron plasma frequency ω_{pe} in the coefficient to the integral.

The inverse transform of $\phi(s)$ is given by

$$\phi(t) = \frac{1}{2\pi i}\int_{x_0 - i\infty}^{x_0 + i\infty}\phi(s)e^{st}ds, \tag{20.32}$$

where the integration path in the complex s-plane runs along a vertical line with real part $x_0 > 0$, chosen so that *all* singularities (poles) are to the left of this line. The singularities $u = is/k$ in the complex velocity plane are at $u = -\Im\{s\}/k + ix_0/k$, which is of no concern when we integrate along the real velocity axis. Note that the singularity in the u-plane is *above* the real axis for $k > 0$, and *below* for $k < 0$, since $x_0 > 0$! The present summary uses the standard form for the Laplace transform variable. An alternative form that makes the relation to a complex variable Fourier transform more clear is obtained by the replacement $s \to -i\omega$, where now ω has a positive imaginary part.

- **Example:** It might be interesting to note that (20.30) has a certain similarity with results from the theory of lumped electric circuits (Pécseli 1985). Assume that we have a mess of connected passive elements like resistors, capacitors and self-inductances. A generator (e.g., a variable battery) is applied between two arbitrary points in the circuit at time $t = 0$, and a detector (e.g., a voltmeter) between two other arbitrary points or positions. After a Laplace transform, a general expression is found for the transfer function. It has the form

  $$Y(s) = \frac{N(s)}{D(s)}X(s) \equiv H(s)X(s),$$

  where Y is the response (as, for instance, measured by the mentioned voltmeter) to the applied perturbation X induced, for instance, by the generator mentioned. The transfer function H can in general be written as the ratio between two functions N and D, which has the form of polynomia in the Laplace transform variable s, in case we deal with networks of the form mentioned. For the circuit problem, the denominator depends solely on the circuit as such, while the *numerator* depends on the circuit *as well* as the specific points where the generator is inserted (Horowitz & Hill 1989). The denominator will, just as in the case of the plasma response in (20.30), depend on the general nature of the medium, while the numerator contains information of the perturbation, as well as the specific points where the response is obtained.

With (20.30) and (20.32), we have in principle solved the problem, and have in particular managed to avoid all singularities in the velocity integrals. Unfortunately, this solution is of little or no practical use! To reach any substantial conclusions we have to introduce some simplifications into the results by manipulating the expressions in the complex plane by use of the residue theorems. For this purpose we note first that the singularities of $\phi(s)$ can be due to poles in the numerator in

(20.30), or to zeroes in the *denominator*. We are free to select problems with well behaved $g(u)$, in the sense that they give rise to singularities only at $\Re\{s\} \to -\infty$. The choice of a Maxwellian distribution for $g(u)$ has this property, for instance. Note that we cannot altogether avoid singularities of the numerator of (20.30): any complex function *must* have at least one pole and one zero (Phillips 1957), but these can be at infinite values of the argument! We will here concentrate on singularities in (20.30) originating from the zeroes in the denominator, but should like to emphasize that allowing for a free choice of $g(u)$ can give an unexpected richness in the time evolution of the potential (Hayes 1961, Hayes 1963, Pécseli 1985). Basically, the trick is to select a non-Maxwellian function $g(u)$ in such a way that all its zeroes cancel the zeroes of the denominator, and then in addition introduce singularities at the desired places. One could expect that the resulting $g(u)$ would look very weird, but this need not necessarily be so. In some cases rather peculiar modes of wave propagations have been observed in laboratory plasmas, "pseudo-waves" (Alexeff et al. 1968), which do not seem to follow dispersion relations predicted by theory. It is most likely that the excitation mechanism is producing a "peculiar" perturbation distribution function which can explain these observations (Ichikawa 1970).

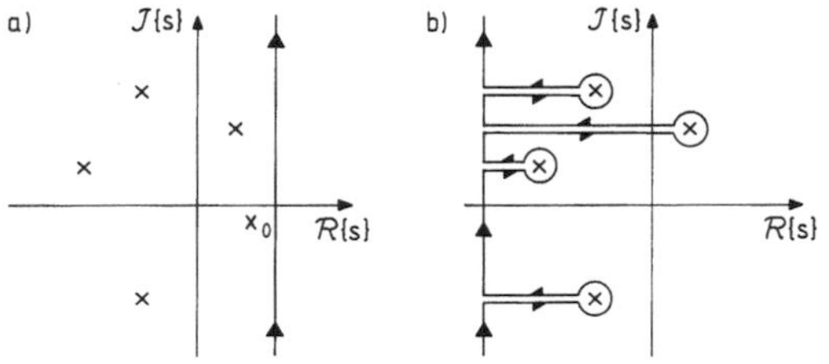

FIGURE 20.3: *Integration contours for the inverse Laplace transform in the complex s-plane. The two integration contours give the same result, but the one in b) invites an approximation that is not evident in a).*

In order to be able to make some simplifications of the result (20.32) with (20.30) we now introduce some changes in the integration path, as shown in Fig. 20.3. The vertical integration path is moved to the left: as long as we do not encounter any singularity, this is perfectly legal. Eventually, we *must* encounter such a singularity, and when we do, the path is distorted as shown in Fig. 20.3b. For stable plasmas, these small islands around the singularities will be in the left half plane; for *unstable* plasma there will in addition be one or more in the right half plane. By this procedure, the integration path is still to the left of all poles of (20.30). However, we cannot change the integration path in the s-plane without encountering problems with the integration path in the u-plane! As s is changed, also the singularity in the velocity plane is moving, as shown in Fig. 20.4. Just when we cross the imaginary axis in s, we have the singularity $u = is/k$ back at the real u-axis, where all our problems began. We therefore define the numerator as well as the denominator in (20.30) by their analytic continuations, when we move s-integration as in Fig. 20.3b. This analytic continuation is best illustrated by Fig. 20.4, showing how we deform the u-integration contour as we manipulate the s-plane, by keeping the u-integration contour *below* the singularity at all times (*above* for the case where $k < 0$). This procedure implies that we have

$$\begin{aligned}
D(s,k) &= 1 - \left(\frac{\omega_{pe}}{k}\right)^2 \int_{-\infty}^{\infty} \frac{F_0'(u)}{u - is/k} du \quad \text{for } \Re\{s\} > 0 \\
D(s,k) &= 1 - \left(\frac{\omega_{pe}}{k}\right)^2 \int_{-\infty}^{\infty} \frac{F_0'(u)}{u - is/k} du + i2\pi \left(\frac{\omega_{pe}}{k}\right)^2 F_0'\left(\frac{is}{k}\right) \quad \text{for } \Re\{s\} < 0 \\
D(s,k) &= 1 - \left(\frac{\omega_{pe}}{k}\right)^2 P\int_{-\infty}^{\infty} \frac{F_0'(u)}{u - is/k} du + i\pi \left(\frac{\omega_{pe}}{k}\right)^2 F_0'\left(\frac{is}{k}\right) \quad \text{for } \Re\{s\} = 0 \quad (20.33)
\end{aligned}$$

with real k. With these definitions we can perform the change in s-integration without problems (Nicholson 1983). The result is that the inversion (20.32) gives a number of pole contributions (from the small circles in Fig. 20.3b) and a contribution from the vertical straight integration path. The latter contribution can be made arbitrarily small by moving the integration path to the left in Fig. 20.3b, and we have

$$\phi(t) \approx \sum_j \phi(s_j) e^{s_j t}, \tag{20.34}$$

where the poles are labeled with the index j. It will be implicitly assumed in the following that we only need to be concerned with poles originating from zeroes of $D(s,k)$, although, as noted, the more general problem also involves pole contributions from the numerator in (20.30).

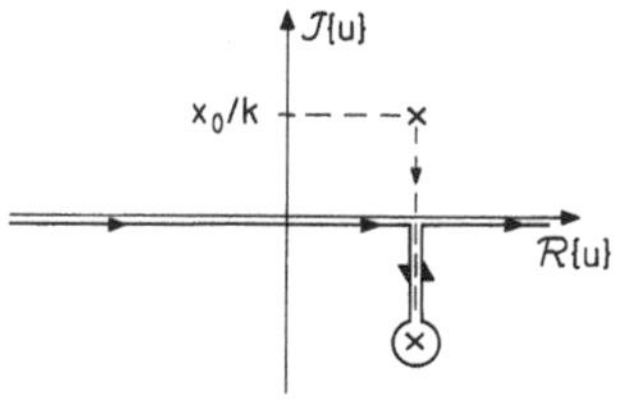

FIGURE 20.4: *Integration path in the complex u-plane. The figure assumes $k > 0$. For $k < 0$, the singularity is* below *the real u-axis, and the figure has to be mirrored with respect to the horizontal axis.*

In order to make $D(s,k)$ look more familiar, we make the change in variables $s \to -i\omega$, and look for weakly damped solutions for

$$D(\omega,k) = 1 - \left(\frac{\omega_{pe}}{k}\right)^2 \fint_{-\infty}^{\infty} \frac{F_0'(u)}{u - \omega/k} du = 0, \tag{20.35}$$

where the notation $\fint$ is an abbreviation for the velocity integration path defined by (20.33) and Fig. 20.4. It is often called the "Landau integration path" or "Landau contour". For the special case where $F_0(u)$ is a Maxwellian, we can express $D(\omega,k)$ in terms of the so-called Z-function; see a detailed discussion in Appendix C.

The poles have a real part and an imaginary part, $s_j = \Re\{s_j\} + i\Im\{s_j\}$, the imaginary part giving the oscillation frequency and the real part a damping or growth. With $\omega \to is$ we have $\omega_j = \Re\{\omega_j\} + i\Im\{\omega_j\} = -\Im\{s_j\} + i\Re\{s_j\}$. For late times, we are obviously most interested in the solution with the smallest damping rate, where we for the time being assume that the plasma is stable. Assuming a Maxwellian plasma, we will obtain an approximate expression for this least damped wave-mode, implying a search for the zero of $D(s,k)$ being closest to the real axis. It is a fair question to ask whether such weakly damped solutions exist at all. We let $k \to 0$, and presume that a finite value of ω corresponds to the limiting case $k = 0$. In this case the phase velocity diverges, $\omega/k \to \infty$, and $F_0'(\omega/k) \to 0$, so there is no problem with a singularity on the velocity axis when $k \to 0$. It is easy to demonstrate that $D(\omega,k) = 0$ *has* a solution in the limit $k \to 0$, namely $\omega^2 = \omega_{pe}^2$, which is the well known cold-plasma solution. Heuristically, we might argue that for infinite phase velocities the waves are so much faster than any particle, so they "can not see any difference" between a finite and a vanishing temperature distribution. It has sometimes been argued that the special theory of relativity provides a cut-off in the available velocities, but this argument is fundamentally in error, since it requires that the *entire* analysis is based on the linearized relativistic Vlasov equation, which is not the case here. Led by this result, we look for weakly damped solutions in the vicinity of $(\omega_{pe}, 0)$. We can make a series expansion (Jackson 1960) with a small quantity uk/ω to have

$$\begin{aligned}
⨍_{-\infty}^{\infty} \frac{F_0'(u)}{u-\omega/k} du &\approx -\frac{k}{\omega}\int_{-\infty}^{\infty} F_0'(u)\left(1+\frac{ku}{\omega}+\left(\frac{ku}{\omega}\right)^2+\ldots\right) du \\
&= \left(\frac{k}{\omega}\right)^2\left(1+3\left(\frac{ku_{te}}{\omega}\right)^2+\ldots\right),
\end{aligned}$$

where by introducing a thermal electron velocity u_{The} we implicitly assumed a Maxwellian distribution for $F_0(u)$, with $\int_{-\infty}^{\infty} uF_0(u)du = 0$. Inserting this result into (20.35), we readily obtain

$$\omega^2 \approx \omega_{pe}^2 + 3k^2 u_{The}^2. \tag{20.36}$$

This is essentially the *Bohm-Gross dispersion relation* (13.10). At first sight it looks trivially like the dispersion relation we obtained by a simple fluid theory, but an important difference is actually the numerical coefficient in front of the last term: the fluid theory gave 5/3. The difference is due to the implicit fluid assumption that the local velocity distribution is at all times a Maxwellian: the kinetic model does not make any such assumption; on the contrary, it attempts to *determine* that distribution.

The second term in (20.36) is considered as a small correction to the cold plasma result. We consider an imaginary part of ω as a correction on the same level. We introduce the notation

$$I(z) \equiv ⨍_{-\infty}^{\infty} \frac{g(u)}{u-z} du,$$

where we can use (20.33) to interpret the function for real and complex z, respectively. We can differentiate $I(z)$ to find

$$I'(z) = ⨍_{-\infty}^{\infty} \frac{g(u)}{(u-z)^2} du = ⨍_{-\infty}^{\infty} \frac{g'(u)}{u-z} du,$$

etc. Making the series expansion with $z = x + iy$ where $y \ll x$

$$I(z) \approx I(x) + iyI'(x) = \left(P\int_{-\infty}^{\infty} \frac{g(u)}{u-x} du + i\pi g(x)\right) + iy\left(P\int_{-\infty}^{\infty} \frac{g'(u)}{u-x} du + i\pi g'(x)\right),$$

we use it in (20.33) we find

$$\Im\{\omega\} \approx \frac{1}{2}\pi \frac{\omega_{pe}^3}{k^2} F_0'(\omega_{pe}/k), \tag{20.37}$$

which is nothing but the previous result (20.26), as expected.

- **Example:** In order to perform a Laplace transform one should first make certain that the solutions of the equation to be transformed are exponentially bound, otherwise the transform does not exist. Landau did not bother with such petty details but later on Backus did, in one of the more elegant theoretical papers in the field (Backus 1960). If an unstable solution was to grow faster than exponentially, it would not be found by Laplace transform, and the plasma would erroneously be found to be stable. It might be felt (incorrectly) that a linearized equation should not have solutions growing faster than exponentially, but as a counter example, the linear equation $4\partial f/\partial t + \partial^2 f/\partial x^2 = 0$ has the explosive solution $(T-t)^{-1/2}\exp(x^2/(T-t))$ for $t < T$.

  Consider again the equation (20.15) here written in one spatial dimension along the direction of the selected wave-vector for the spatial variation $\exp(-i\mathbf{k}\cdot\mathbf{r})$. After some algebra, involving integration by parts with respect to u and t, we obtain

$$n(t) = \int_{-\infty}^{\infty} f(u, t_0 = 0) e^{-ikut} du - \omega_{pe}^2 \int_{-\infty}^{\infty}\int_{t_0=0}^{t}\int_{t_0=0}^{t'} n(t_2) e^{iku(t_2-t)} F_0(u) dt' dt_2 du.$$

The reader is urged to prove this result, in particular to find the trick which introduces the double time integral. Since $F_0(u) > 0$ and $\int_{-\infty}^{\infty} F_0(u)du = 1$, it follows that

$$|n(x,t)| < A(0) + \omega_{pe}^2 \int_0^t \int_0^{t'} |n(t_2)|dt'dt_2,$$

with $A(0) \equiv \int_{-\infty}^{\infty} |f(u,0)|\, du$. Denote the double integral $G(t)$. Then, if $h(t) \equiv G''(t) - \omega_{pe}^2 G(t)$, we have $|h(t)| < \int_{-\infty}^{\infty} |f(u,0)|\, du$ and $G(t) = \int_0^t \int_0^{t_1} h(t_2) e^{\omega_{pe}(2t_1 - t_2 - t)} dt_1 dt_2$, implying $0 \le G(t) \le A(0)(\cosh(\omega_{pe}t) - 1)/\omega_{pe}^2$. With this upper bound, we readily obtain

$$|n(x,t)| < A(0)\cosh(\omega_{pe}t),$$

proving that the time evolution of the plasma density is exponentially bound for $t \to \infty$ as well as $t \to -\infty$.

In the Landau analysis, the denominator $D(s,k)$ (20.31) in the expression (20.30), or its transformed version $D(\omega,k)$, enters as a sort of mathematical entity, without any obvious physical interpretation. It is, however, readily demonstrated that it is the plasma dielectric function. Thus, in Section 5.9.3 we found a suitable definition of the relative dielectric function as (5.89)

$$\varepsilon(\omega, \mathbf{k}) = \varepsilon_0 - \frac{\rho'(\omega, \mathbf{k})}{k^2 \phi(\omega, \mathbf{k})},$$

where ρ' is the charge density variation induced self-consistently by the electric field $-ik\phi$. We have that $-en$, with n given by (20.28) with the properly defined integration contour, is the induced charge density for given ϕ. We then readily find, by inspection, that $D(\omega,k)$ is the relative dielectric function for the plasma, with the implied assumption of cold immobile ions. This observation is consistent with our previous interpretation of (20.22).

Ⓢ **Exercise:** Consider first a plane Langmuir wave with wavenumber k_0. Give the corresponding frequency ω_0. Assume that the peak-to-peak value of the electrostatic potential is $2\phi_0$. Give the bounce frequency and bounce period of an electron in a local electrostatic potential well associated with the periodic wave train. Compare this result with ω_0.

Take a numerical example for a Langmuir wave, where the plasma density is $n_0 = 10^{16}$ m^{-3}, the electron temperature $T_e = 1$ eV, and the density perturbation associated with the wave is 5 %. Let the wavelength λ_0 be 1 m. What is the ratio λ_0/λ_{De}? For how many plasma periods will you expect a linear Langmuir wave model in the Landau sense to be valid? How long is this time in seconds for the given parameters? Is the plasma really collisionless? Ⓢ

• **Exercise:** Consider a simple water-bag model (see also Section 19.3) for the distribution function $F_0(u)$, where the distribution is confined within the limits $-U_1$ and U_2. Determine $-U_1$ and U_2 so that by insertion into (20.35) it exactly reproduces the real, non-dissipative, Bohm-Gross dispersion relation (20.36). What is the average, $\langle u \rangle$, and mean square velocities, $\langle u^2 \rangle$, for this distribution?

The pioneering work by Vlasov (1945) received some attention when it was first presented, but then suffered a long time negligence (presumably due to the unfair ridicule of Landau based on that unfortunate principal value integral mentioned before), but it is now properly recognized as the foundation of the mathematical description of collisionless plasma phenomena.

20.2.4 Normal-mode solution

We can give the linearized Vlasov equation an interpretation in terms of an eigenvalue equation. First we note that for any physically acceptable problem, the spatial variation of the distribution can be decomposed by the orthogonal functions $\exp(i\mathbf{r}\cdot\mathbf{k})$ with real $\mathbf{k}$. There is no need to include complex $\mathbf{k}$ since the set of functions is complete as it is. We can therefore restrict our analysis to a spatially harmonic variation and, if desired, later construct any actual spatial variation by a superposition of these functions. We let $\widetilde{f}$ denote the distribution function corresponding to $\mathbf{k}$. For the description of electrostatic electron plasma oscillations, the linearized Vlasov equation is then trivially reduced to

$$\frac{\partial \widetilde{f}}{\partial t} + i\mathbf{u}\cdot\mathbf{k}\widetilde{f} - i\frac{e^2 n_0}{\varepsilon_0 m k^2}\mathbf{k}\cdot\nabla_{\mathbf{u}} f_0(\mathbf{u})\int_{-\infty}^{\infty}\widetilde{f}d\mathbf{u} = 0, \tag{20.38}$$

where we made use of Poisson's equation (20.12) to eliminate the electrostatic potential, and introduced again the convenient normalization $\int_{-\infty}^{\infty} f_0(\mathbf{u})d\mathbf{u} = 1$. We then integrate (20.38) with respect to the velocity components perpendicular to $\mathbf{u}$, and obtain a one-dimensional representation. Note also here that in case the unperturbed velocity distribution function is anisotropic, the resulting equation depends on the direction of $\mathbf{k}$ with respect to the chosen velocity coordinate system. To ease the notation we assume from here on that u is the velocity component along $\mathbf{k}$, and $\widetilde{f}$ is the perturbation of the velocity distribution function *after* the integration mentioned.

Looking for solutions with a temporal variation $\widetilde{f} = f\exp(-i\omega t)$ we can write the Vlasov equation in the form

$$\widehat{V}f = \omega f, \tag{20.39}$$

with the operator

$$\widehat{V} \equiv uk - \frac{\omega_{pe}^2}{k}F_0'(u)\int_{-\infty}^{\infty} du. \tag{20.40}$$

The relation (20.39) has the form of an eigenvalue equation, where f is the eigenfunction corresponding to the possibly complex eigenvalue ω. We require that the eigenfunctions are normalizable, which turns out to be an important point, although it usually is rather trivial. There is no particular reason for the normalization constant to be unity, but we might as well choose that, so $\int_{-\infty}^{\infty} f du = 1$, giving $ukf - F_0'(u)\omega_{pe}^2/k = \omega f$ from (20.39). Direct integration of (20.39) then gives

$$k\int_{-\infty}^{\infty} uf du = \omega, \tag{20.41}$$

which does not really tells us anything at this point, since evaluation of $\int_{-\infty}^{\infty} uf du$ requires f to be known, and we are not there yet. We prefer to rewrite (20.39)-(20.40) to give f explicitly. Here we then make use of a result involving Dirac's δ-function, (Dirac 1947, Pécseli 2000), stating that the solution of the relation $xg(x) = F_0(x)$ is *not* $g(x) = F_0(x)/x$ as one might think, but more generally $g(x) = F_0(x)/x + \lambda\delta(x)$, with λ arbitrary, since $x\delta(x) = 0$; see, for instance, a summary by Pécseli (2000). With this in mind, we obtain

$$f = \frac{\omega_{pe}^2}{k}F_0'(u)\frac{1}{uk-\omega} + \lambda\delta(uk-\omega) \tag{20.42}$$

from (20.40), assuming ω real. To determine λ, we now use that f is normalized. The question then arises how to integrate the singularity at $u = \omega/k$. We *choose* to take the principal value, indicating it by a P, and thereby find a corresponding λ by the normalization of f. If we had taken another prescription (of the infinitely many possibilities) for performing the integral, we would simply find another λ, and f would be normalized anyhow. The result is

$$f = \frac{\omega_{pe}^2}{k^2}F_0'(u)\frac{P}{u-\omega/k} + \left(1 - \frac{\omega_{pe}^2}{k^2}P\int_{-\infty}^{\infty}\frac{F_0'(u)}{u-\omega/k}du\right)\delta(u-\omega/k). \tag{20.43}$$

Introducing P in the first term in (20.43), we emphasize that in case we want to integrate f later on, we must take the principal value here. The result (20.43) gives an exact plane wave solution to the linearized Vlasov equation for electrostatic electron plasma oscillations for any combination of ω and k! The eigenfunctions (20.43) are usually called *van Kampen modes*, or *Case-van Kampen modes* (van Kampen 1955, Case 1959, van Kampen & Felderhof 1967). As a special case we might have $1-(\omega_{pe}/k)^2 P\int_{-\infty}^{\infty} F_0'(u)/(u-\omega/k)du = 0$, and then the δ-function does not appear in (20.43).

The important observation is that we can find a solution as (20.43) for *any* combination of ω and k; the modes do not follow any dispersion relation (van Kampen 1955). In a way this is not so surprising; a dispersion relation in classical fluid models is just a way to organize the independent variables space and time, $(\mathbf{r},t)$. As emphasized already, the Vlasov equation has in addition to these also the velocity $\mathbf{u}$ as an independent variable, and there is no a priori reason to expect any dispersion relation in the traditional sense for the present problem. The first-pole analysis by Landau summarized before was only approximate after all, giving the dominant asymptotic term.

The eigenvalues of the operator $\widehat{V}$ for a given wave-vector $\mathbf{k}$ can be divided into basically two groups: real ω and complex ω. In case ω is complex, we find that also its complex conjugate ω^* is an eigenvalue. We are in such cases dealing with an unstable plasma, and the direction of the time axis does not matter; if the eigenvalue ω gives growing waves as $t\to\infty$, then ω^* will give growing waves as $t\to-\infty$, and vice versa. An unstable plasma will remain unstable even if we turn the arrow of time!

The complex eigenvalues form a discrete set; often this set is empty and the plasma is stable. The real eigenvalues can be classified in three groups: one with $\lambda\neq 0$, one where $F_0'(\omega/k)=0$ but $\lambda\neq 0$, and one where both $F_0'(\omega/k)=0$ and $\lambda=0$. The first case is the one giving the solutions (20.43) and we have, as mentioned, a continuum of such eigenvalues for any given $\mathbf{k}$. In the second case we need not worry about the principal value sign. In the last case we have a discrete set of ω's, and again this set may be empty.

The modes (20.43) are singular by containing a δ-function, and, among other things, violating any assumption of linearization which is based on f being small. However, such an objection is rather irrelevant since the normal modes, or eigenfunctions (20.43), account for a property of the linearized Vlasov *equation*, not the physics it describes. A solution to a given physical problem should be sought in terms of a *superposition* of eigenmodes where singularities are absent in the result. However, for such a superposition we have to make certain that the set of eigenfunctions is *complete*. It is nice if they are orthogonal also, but if they were not, this could be remedied.

For completeness, we introduce the *adjoint* equation (Case 1959, Case 1978), for given k, as

$$\frac{\partial f^\dagger}{\partial t}+iukf^\dagger - i\frac{\omega_{pe}^2}{k}\int_{-\infty}^{\infty} F_0'(u)f^\dagger du = 0, \tag{20.44}$$

which for harmonic oscillations can be written in terms of an adjoint operator as $\widehat{V}^\dagger f^\dagger = \omega f^\dagger$, with

$$\widehat{V}^\dagger \equiv uk - \frac{\omega_{pe}^2}{k}\int_{-\infty}^{\infty} F_0'(u)du,$$

where we note that $F_0'(u)$ here appears *inside* the integral sign. The two operators $\widehat{V}$ and $\widehat{V}^\dagger$ are not Hermitian conjugate. Again, the eigenfunctions and eigenvalues for (20.44) can be classified into a continuous set and a discrete set, where some care has to be taken in analyzing the discrete set (Case 1978). It is readily demonstrated (Case 1959) that f and $f^\dagger$ are orthogonal, i.e., $\int_{-\infty}^{\infty} f_1 f_2^\dagger du = C_\omega \delta_{\omega_1-\omega_2}$, with a normalizing constant which for the continuous set is $C_\omega \equiv \lambda(\omega,k)$.

In order to make the notation more clear, we denote the eigenfunction (20.43) $f_k(\nu,u)$, using k as a subscript, since we now consider only *one* spatial Fourier component, and introduce a phase velocity by $\nu\equiv\omega/k$. It can be demonstrated (Case 1959) by the completeness of the set of eigenfunctions that *any* electrostatic wave solution to the linearized electron Vlasov equation can be written by a superposition of eigenmodes as

$$f_k(v) = \sum_j^N a_j f_k(v_j, u) e^{-ikv_j t} + \int_{-\infty}^{\infty} A(v) f_k(v, u) e^{-ikvt} dv, \tag{20.45}$$

where the sum is over the discrete set of N modes, where for a stable plasma $N = 0$. There may actually be some double roots, but let us not go into such details here. The coefficient in the discrete set is determined as

$$a_j = \frac{1}{C_j} \int_{-\infty}^{\infty} f_{k,j}^{\dagger} f(k, u, t = 0) du$$

and for the continuous set by the relation

$$A(v) = \frac{1}{C_k(v)} \int_{-\infty}^{\infty} f_k^{\dagger} f(k, u, t = 0) du.$$

For example, assuming the set is not empty, we have for the discrete set, with ω complex, that $f_k = (\omega_{pe}/k)^2 F_0'(u)/(u - v)$ and $f_k^{\dagger} = 1/(u - v)$, giving

$$C_j = \left(\frac{\omega_{pe}}{k}\right)^2 \int_{-\infty}^{\infty} \frac{F_0'(u)}{(u - v)^2} du.$$

Assuming that the initial condition can be written as $f(x, u, t = 0) = g(u) \exp(ikx)$ we have for the continuous set

$$A(v) = \frac{\left(\frac{\omega_{pe}}{k}\right)^2 F_0'(v) P \int \frac{g(u)}{u - v} du + g(v) \left(1 - \left(\frac{\omega_{pe}}{k}\right)^2 P \int \frac{F_0'(u)}{u - v} du\right)}{\left(1 - \left(\frac{\omega_{pe}}{k}\right)^2 P \int \frac{F_0'(u)}{u - v} du\right)^2 + \left(\pi \left(\frac{\omega_{pe}}{k}\right)^2 F_0'(v)\right)^2}.$$

It can be demonstrated (Case 1959) that the results summarized here exactly reproduce the ones obtained by Fourier-Laplace transform, as obtained by Landau (1946), as expected. In the present Case-van Kampen formalism, the damping of the continuous set is due to phase mixing of normal modes with different phase velocities, for the same given wavenumber k. The wave *growth* for unstable plasmas is of a different nature, and it might be misleading simply to consider the two phenomena as being complementary manifestations of wave-particle interaction! In case we "switch-off" the charges on all particles, and let them be freely streaming (i.e., let $\omega_{pe} \to 0$), we find $A(v) = g(v)$, with $g(u)$ being the initial perturbation of the velocity distribution function.

- **Example:** Note that the initial value problem outline here might be "inverted": we can *prescribe* $A(v)$ in (20.45) and then determine the initial perturbation, which gives the prescribed weight function (Hayes 1961, Hayes 1963, Pécseli 1985)! Assume that we *insist* on having a strictly exponential damping, with an arbitrary prescribed exponent α. This is achieved by requiring $A(v) = (1/\pi)\alpha/k/(v^2 + (\alpha/k)^2))$. With a little algebra it can be demonstrated that the perturbation in the velocity distribution function which achieves this is given by

$$g(u) = \frac{1}{2\pi} \left(\frac{\varepsilon(k, u)}{\alpha/k - iu} + \frac{\varepsilon^*(k, u)}{\alpha/k + iu} \right), \tag{20.46}$$

with $\varepsilon(k, \omega)$ as the dielectric function of the plasma, rewritten by introducing phase velocities ω/k, and making the replacement $\omega/k \to u$. When introducing $\varepsilon(k, \omega)$ as in (20.46), we cancel all zeroes in the denominator of (20.30), and then we introduce the poles that we want by the choice of denominators in (20.46). It is evident that $g(u)$ is real, as it should be. It is not positive definite, but there are no reasons for it to be so, since it is a *perturbation* of a pre-existing $f_0(u)$. The distribution $g(u)$ may not be nice, but it is normalizable, and physically acceptable. Note that the prescribed damping will in the present case be the same for $t > 0$ and for $t < 0$.

• **Comment**

Reading the present summary it seems evident that for collisionless plasmas no dispersion relation exists in the strict sense of the word, and that virtually any damping can be found, with undamped waves being a limiting case: it all seems to depend on the chosen initial condition for the velocity distribution function. The question naturally arises: how come so many experiments and numerical simulations produce results in fair agreement with the simple Landau results? The answer is most likely found in the observation that the initial value problem is not the best realization of an actual experimental condition. Most experiments are best described as an oscillating test-charge problem (the charges associated with the potential perturbation applied to the antennae) where the response function is obtained by the plasma dispersion function $\varepsilon(\mathbf{k}, \omega)$. The corresponding perturbation of the velocity distribution function is the one that develops self-consistently and not one we determine. Even for cases where we *can* create an initial condition for the velocity distribution function (we will find examples for such experimental conditions in Chapter 21), it is in practice rather limited what we can do: in reality we end up with $g(u)$ being not much different from a Maxwellian. This restriction strongly reduces the freedom of time evolutions, wave damping rates in particular.

20.3 The Penrose criterion for plasma stability

We have seen that the question of the linear stability of a plasma can be decided by determining whether the plasma kinetic dielectric functions have zeroes in the upper half of a suitably defined complex plane. To obtain a useful relation for investigating this problem, we need also some basic results concerning complex functions.

First, we can demonstrate that a distribution with only *one* local maximum will necessarily be stable. Assume that the local maximum is at a velocity $u = U$ and *assume* that the plasma is unstable. Let the generally complex variable be $\omega/k \equiv \xi \equiv \xi_1 + i\xi_2$, with both ξ_1 and ξ_2 real. Using (20.33) again with the replacement $s \to -i\omega$, we have the criterion for an unstable normal plasma mode ($\xi_2 > 0$) in the form

$$\int_{-\infty}^{\infty} \frac{F_0'(u)du}{u - \xi_1 - i\xi_2} = \left(\frac{k}{\omega_{pe}}\right)^2, \tag{20.47}$$

taken (obviously) without the pole contribution from Fig. 20.4, since we assumed an unstable mode. We rewrite the real and imaginary parts of (20.47) as

$$\int_{-\infty}^{\infty} \frac{(u - \xi_1)F_0'(u)du}{(u - \xi_1) + \xi_2^2} = \left(\frac{k}{\omega_{pe}}\right)^2,$$

and

$$\int_{-\infty}^{\infty} \frac{\xi_2 F_0'(u)du}{(u - \xi_1) + \xi_2^2} = 0.$$

Multiplying the latter expression by $(U - \xi_1)/\xi_2$ and subtracting it from the first one, we find

$$\int_{-\infty}^{\infty} \frac{(U - u)F_0'(u)du}{(u - \xi_1) + \xi_2^2} = -\left(\frac{k}{\omega_{pe}}\right)^2.$$

With U being the velocity at the maximum value of the distribution, we have $F_0' > 0$ for $u < U$, and $F_0' < 0$ for $u > U$. The left side of this expression is clearly positive, while the right-hand side is negative. Consequently, the basic assumption leading to this final expression is in error, and we have no unstable solutions. In particular, a Maxwellian velocity distribution (having only one maximum) is stable, which is nice to know! We can conclude also that a necessary (but as it turns out, not sufficient) condition for *instability* is that the electron velocity distribution has more than one local maximum.

In order to study the stability of distributions with more than one local maximum, we introduce a contour C as a semi-circle in the upper half of the complex ω-space, so that it contains the real ω-axis and closes at a half circle at infinity (Jackson 1960, Schmidt 1979). We would like to know if there are zeroes for $D(\omega,k)$ inside this contour. If there are, then the plasma is unstable in the linearized description.

We now introduce the complex function

$$G\left(\frac{\omega}{k}\right) \equiv \fint_{-\infty}^{\infty} \frac{F_0'(u)}{u-\omega/k} du,$$

which is regular in the upper half plane of its complex argument. We consider the function G as a *transformation*, which maps the upper half of the complex ω/k plane enclosed by the contour C into a closed region of the G-plane. In particular we note that the points at $|\omega| \to \infty$, for finite k, are mapped into the origin, $(0,0)$, of the complex G-plane. The boundary of the half circle containing the upper half of the complex ω-plane is then "squeezed" into the origin, while the real ω-axis is comprised by the bounding curve in the complex G-plane. It is easily shown that the upper half of the complex ω-plane is mapped into the *interior* of the bounded region in the complex G-plane. It can now be concluded that *if* the closed contour in the complex G-plane contains points in the left side (the one with positive real parts), then it is always possible to find a wavenumber k which gives a solution of $D(\omega,k)=0$ for positive imaginary parts of ω. It is obvious that for this to happen, there must be at least one upwards crossing of the real G-axis by the contour in the G-plane. To find a condition for this to occur we write $G(\xi)$ for a real argument $\xi = \omega/k$ as

$$G(\xi) = P\int_{-\infty}^{\infty} \frac{F_0'(u)du}{u-\xi} + i\pi F_0'(\xi). \tag{20.48}$$

See also (20.33). The relation (20.48) is mapping the real axis ξ onto a complex G-plane. A crossing of the real axis in this z-plane obviously requires $\Im\{G\} = \pi F_0'(\xi) = 0$ for some $\xi = \xi_0$. Furthermore, for an upward crossing, we require that, for the same ξ-value, $\Im\{G\}$ changes from negative to positive values, implying that the crossing corresponds to a *minimum* for $F_0(u)$. The requirement that the crossing takes place at the *positive* G-axis (for an instability) requires $\Re\{G\} > 0$ in (20.48). To give a criterion for this, we first rewrite the real part of (20.48) as

$$\Re\{G(\xi)\} = \lim_{\varepsilon\to 0}\left(\int_{-\infty}^{\xi-\varepsilon} \frac{F_0(u)du}{(u-\xi)^2} + \int_{\xi+\varepsilon}^{\infty} \frac{F_0(u)du}{(u-\xi)^2} + \frac{F_0(\xi-\varepsilon)}{-\varepsilon} - \frac{F_0(\xi+\varepsilon)}{\varepsilon}\right). \tag{20.49}$$

We then note that we can write

$$2\frac{F_0(\xi)}{\varepsilon} = F_0(\xi)\lim_{\varepsilon\to 0}\left(\int_{-\infty}^{\xi-\varepsilon} \frac{du}{(u-\xi)^2} + \int_{\xi+\varepsilon}^{\infty} \frac{du}{(u-\xi)^2}\right),$$

which can be used to simplify (20.49). The result is actually correct even without taking the limit $\varepsilon \to 0$. We then find

$$\begin{aligned}\Re\{G(\xi_0)\} &= P\int_{-\infty}^{\infty} \frac{F_0(u)-F_0(\xi_0)}{(u-\xi_0)^2} du \\ &+ \lim_{\varepsilon\to 0}\left(\frac{F_0(\xi_0)-F_0(\xi_0-\varepsilon)}{-\varepsilon} - \frac{F_0(\xi_0+\varepsilon)-F_0(\xi_0)}{\varepsilon}\right).\end{aligned} \tag{20.50}$$

The limiting value of the last term for $\varepsilon \to 0$ is zero. We can make a local expansion of $F_0(u)$ in the neighborhood of ξ_0 as $F_0(u) \approx F_0(\xi_0) + F_0'(\xi_0)(u-\xi_0) + \frac{1}{2}F_0''(\xi_0)(u-\xi_0)^2$. Since $F_0(\xi_0)$ is a local minimum where $F_0'(\xi_0) = 0$, we can ignore the principal value sign in the first term in (20.50), i.e., the P in front of the integral sign is redundant. The plasma is linearly unstable when $\Re\{G(\xi_0)\} > 0$. We can illustrate the properties of the function G by presenting it in a complex plane, with axes $\Re\{G(\xi)\}$ and $\Im\{G(\xi)\}$, with the phase velocity ξ being a parameter. The contours in the complex G-plane are often called *Nyquist contours*, because of the similarity of the problem with the stability analysis of a feed-back amplifier (Jackson 1960).

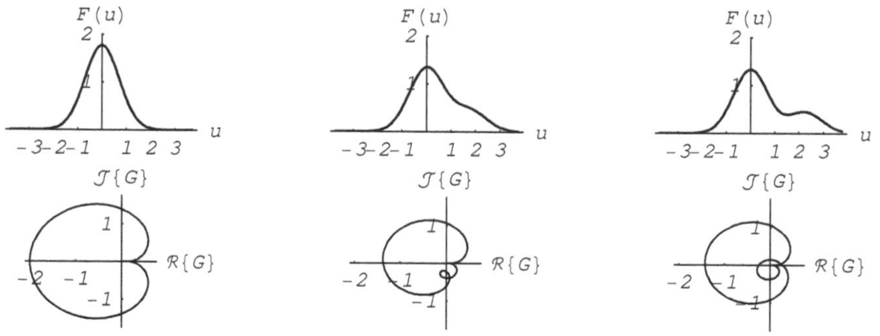

FIGURE 20.5: *Stability diagrams in the complex G-plane (bottom) and corresponding electron velocity distribution functions (top), for a 25% beam density, and beam velocities* $u_0 = 0, 1.75$ *and* 2.25, *in terms of the electron thermal velocity. The electron beam and the background electrons have the same temperature. The contours in the complex G-plane are called* Nyquist contours. *We find that the example* $u_0 = 2.25$ *corresponds to unstable conditions; the two others are stable.*

One should bear in mind that in case we have an inhomogeneous, possibly double humped electron velocity distribution, there is still a freedom in choosing the *direction* of the wave-vector, and a complete analysis should consider all directions.

The *Penrose criterion for kinetic instability of Langmuir waves*: the electron velocity distribution function has a local minimum at $u = \omega/k \equiv \xi_0$ such that

$$\int_{-\infty}^{\infty} \frac{F_0(u) - F_0(\xi_0)}{(u - \xi_0)^2} du > 0. \tag{20.51}$$

For special cases where a velocity distributions are vanishing at the local minimum, $F_0(\xi_0) = 0$, we find that the Penrose criterion is trivially satisfied since $F_0(u) \geq 0$. Such plasmas are *always* unstable. Figure 20.5 shows more general examples of stability diagrams and corresponding electron velocity distribution functions. The last figure corresponds to a linearly unstable distribution function, since the contour in the complex G-plane encircles a part of the positive $\Re\{G\}$-axis.

The nature of the instability found by the Penrose criterion is different from those met in classical continuum physics. In those cases you can find unstable conditions associated with gradients in macroscopic parameters, such as the bulk density, velocity or pressure. For the present plasma problem the bulk parameters are uniform and the source of instability is found only if the conditions are inspected in phase space. Note that the Penrose criterion only addresses the stability of the plasma as such. Details of the unstable waves, their dispersion relation in particular, has to be determined by other means.

The analysis of the present section is restrictive in the sense that it assumes electrostatic modes only, and more important an unmagnetized plasma. By imposing an external magnetic field and allowing for inhomogeneities, the stability conditions of a plasma are changed considerably, and the analysis becomes more complicated.

20.3.1 Two counter streaming cold electron beams

Consider the somewhat idealized case of an electron velocity distribution having the form of two counter streaming cold beams of equal density. For simplicity we choose the reference system such that we can write $F_0(u) = \frac{1}{2}(\delta(u-u_0)+\delta(u+u_0))$, and treat the problem in one spatial dimension. According to the Penrose criterion the plasma is unstable since it has two maxima and is zero between them. This particular case is easily solved explicitly to give the dispersion relation

$$1 = \frac{1}{2}\left(\frac{\omega_{pe}^2}{(\omega-ku_0)^2} + \frac{\omega_{pe}^2}{(\omega+ku_0)^2}\right), \tag{20.52}$$

which can be solved to give

$$\omega^2 = \frac{1}{2}\left(\omega_{pe}^2 + 2(ku_0)^2\right) \pm \frac{1}{2}\sqrt{\omega_{pe}^4 + 8(\omega_{pe}ku_0)^2}. \tag{20.53}$$

We have two types of solutions: real ω for the $+$ sign, which corresponds to propagating waves, and purely imaginary ω for the $-$ sign, corresponding to an aperiodic growth. The marginally stable mode, with $\omega = 0$, is found for $(ku_0/\omega_{pe})^2 = 1$, or in terms of the critical wavenumber $k_c = \sqrt{\omega_{pe}/u_0}$. wavenumbers $k < k_c$ give unstable solutions for the $-$ sign in (20.53). It is left as an exercise to determine the wavenumber for maximum growth, and the corresponding growth rate.

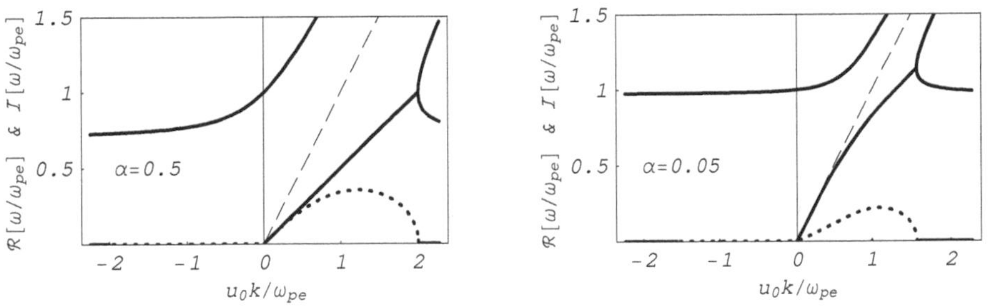

FIGURE 20.6: *Dispersion relation (real and imaginary parts of ω) for two density ratios of electron beam and background densities, α =0.5 and 0.05. The unstable modes have a counterpart with negative $\Im\{\omega\}$ that is not shown. The most unstable mode has a real part of $\omega < \omega_{pe}$ in both cases. The group velocity $u_g \equiv d\omega/dk \to \infty$ at the k-value which is marginally stable, e.g., $k = 2\omega_{pe}/u_b$ on the figure to the left. Here we have the beam propagating at a velocity u_0 while the background electrons (with the largest densities) are at rest here.*

The problem can be solved analytically for arbitrary rations for the background and beam densities, but the expressions are too lengthy to be reproduced here. We show in Fig. 20.6 the cold beam plasma dispersion relation for two ratios of the background, n_0 and beam n_b electron densities, with $\alpha \equiv n_b/n_0$. The velocity of the dense beam is taken to be zero. For two equal counter-streaming beams the symmetry gives a preferred frame of reference, but for the general case there is none. The real part of the frequency is shown with a solid line, the imaginary part with a dashed line. The growth rate of the instability is illustrated in Fig. 20.7 for varying density ratios. Note that generally we have instability for a range of frequencies $\Re\{\omega\} \in \{0, \sim\omega_{pe}\}$, where the most unstable frequency approaches ω_{pe} as $\alpha \to 0$. The case with $\alpha = 0.5$ corresponds to (20.53), albeit in a different frame of reference.

The instability does not rely on any resonant wave particle interaction and cannot properly be classified as "inverse Landau damping". Physically, we find that a tiny perturbation of the electrons in one beam perturbs the orbits of the electrons in the other beam: assume for instance, that one of the beams is *decelerated* slightly, for some reason, around $x = 0$. Then those electrons spend slightly more time in the vicinity of the origin, and consequently we have a slight surplus of negative charge. This charge then decelerates the other beam slightly, to let its electrons spend more time around $x = 0$. This increased net charge then further decelerates the first beam, and we have a feed-back leading to an instability. If we slightly *accelerate* the first beam, the entire process is repeated, but this time leading to an increase in a local *positive* charge. The present two-stream model can be considered as a limiting case for electrostatic instabilities of electron plasma waves. The other extreme limit is the one caused *solely* by a low density hot beam, which gives instability by resonant wave-particle interaction with electrons in a distribution having positive slope in the vicinity of the phase velocity of the wave. The general case can be considered as a mixture of these two limiting cases.

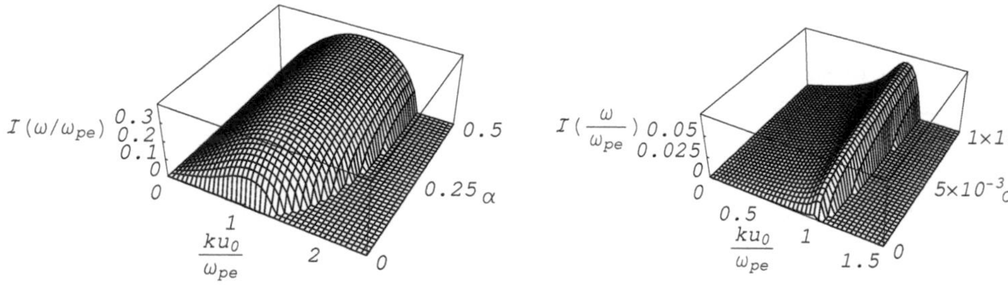

FIGURE 20.7: *The growth rate for the cold electron beam plasma instability for varying ratios of electron beam and background densities, α. The figure to the left covers the full range of density ratios, while the right-hand figure emphasizes the limit of low density beams, below 1% of the background density.*

Cold beam distributions appear as a sort of marginal case: on one hand they are accommodated by kinetic theory. If we have a distribution consisting of *one* cold beam, it can, however, also be taken as the limiting case of a fluid model for a Maxwellian, with $T_e \to 0$, while the distribution can be inserted into the plasma dielectric function as well to give a solution right away. If the distribution consists of *multiple* cold beams, it gives only a little complication in evaluating the dielectric function. A fluid model does not apply directly, but we can formally write up separate fluid-like equations for each beam, and the resulting model is readily analyzed in the same way as a simple fluid model, or rather "multi fluid" model.

Ⓢ **Exercise:** Write out the dispersion relation for the cold electron-electron two-stream instability for arbitrary ratios α of the beam and background electron densities. Demonstrate that the growth rate is given as $\Im\{\omega\} \approx (\omega_{pe}/2)(\alpha/2)^{1/3}\sqrt{3}$ for small α, where $\Im\{\omega\} \ll \Re\{\omega\}$. Ⓢ

Ⓢ **Exercise:** Consider an electron velocity distribution having the form of a "shell" in electron velocity space. If we make a "cut" along any axis, we see two δ-functions, with zero in between. Is this distribution really unstable? How about a "ring" in phase space? Ⓢ

Ⓢ **Exercise:** Demonstrate that *any* spherically symmetric velocity distribution function, $f(\mathbf{u}) = f(|\mathbf{u}|) = f(u)$, no matter how "bumpy", is stable. Ⓢ

Ⓢ **Exercise:** Consider the simple two-stream instability, where $F_0 = \frac{1}{2}\delta(u) + \frac{1}{2}\delta(u - u_0)$ as before. Take the simplest one-dimensional case. Demonstrate that there is a point on the real part of the dispersion relation where the group velocity diverges, i.e., $u_g \to \infty$. See also the discussion in Section 3.1.2 and Fig. 20.6. Ⓢ

Ⓢ **Exercise:** In Section 5.10.4 we learned that waves can have negative energy, i.e., the system has less energy *with* the wave present as compared to the case without. Simple logic should then indicate that the wave amplitude can grow if we introduce dissipation into such a system, and the wave becomes unstable. Write a set of equations for electron plasma waves with immobile ions, and a dissipative medium, here characterized by a constant conductivity σ. Let electrons with a Maxwellian velocity distribution be drifting with an average velocity that is much larger than their thermal velocity. Demonstrate that instability occurs for σ large enough, and illustrate the competition between the Landau damping and the resistivity. Consider the problem in one spatial dimension for simplicity. Ⓢ

Ⓢ **Exercise:** Let a one dimensional plasma model be composed of immobile ions of density n_0 and two Maxwellian electron components, one warm and one cold, $T_{e1} \ll T_{e2}$ with given densities so that $n_{01} + n_{02} = n_0$. Set up the kinetic linear dispersion relation for this plasma model. Demonstrate the existence of an acoustic or sound-like electron wave mode and estimate its linear Landau damping rate. Discuss the simple limiting case where $T_{e1} \to 0$, i.e. the corresponding unperturbed electron velocity distribution is $F_{01}(u) = n_{01}\delta(u)$. Discuss the case where the two electron components have a relative drift velocity (Chowdhury et al. 2017). Ⓢ

20.4 Small amplitude power theorem

Intuitively, it might be argued that the linearized Vlasov equation, or any other set of basic dynamic equations obtained by linearization of a set of nonlinear equations, cannot be used for discussions of energy and momentum, because these quantities are of second order. Therefore by construction, the linearized equations should be incapable of accounting for these, and consequently not be energy conserving. No matter how appealing this argument is, it is nonetheless incorrect, and indeed, for instance, energetics of lumped electronic circuits would be jeopardized by this argument; there are no linear circuit elements in reality. The simple relations we have all used, the standard Ohm's law, etc., are based on an implicit linearization. In reality, we are replacing the *true* circuit elements by some linearized equivalents, and assume that these models are energy conserving. It will be demonstrated that the same procedure is valid for the linearized Vlasov equation, choosing the specific case of electrostatic electron waves.

To be specific, we want to prove a theorem reminiscent of the Poynting theorem.

$$\frac{\partial}{\partial t}W + \frac{\partial}{\partial x}P + EJ = 0 \tag{20.54}$$

in terms of energy density W and energy flux densities P, where J is the plasma current density, and E the electric field, as usual.

To make life a little bit simpler, we write the linearized Vlasov equation in one spatial dimension, and consider an initial value problem, where the perturbation of the velocity distribution is written as $f(x,u,0) = g(u)\exp(ikx)$, while the unperturbed electron velocity distribution is $F_0(u)$. The ions are again considered as an immobile neutralizing background of uniform positive charge. With little loss of generality we use a frame of reference where $\int_{-\infty}^{\infty} uF_0(u)du = 0$, and take also $\int_{-\infty}^{\infty} ug(u)du = 0$.

By integration along unperturbed orbits, we have as in (20.14)

$$f(x,u,t) = g(u)e^{ik(x-ut)} + \frac{en_0}{m}e^{ik(x-ut)} \int_0^t e^{ikut'} E(t')F_0'(u)dt'. \tag{20.55}$$

We can now introduce a kinetic energy density for the particles as $W_k \equiv \frac{1}{2}m\int_{-\infty}^{\infty} u^2 f(r,u,t)du$ and the energy density associated with the electric field as $W_E \equiv \frac{1}{2}\varepsilon_0 E^2$. Since we are dealing with electrostatic fluctuations, there are no other energies present.

Consider first the "free-streaming" contribution, i.e., the first term on the right-hand side of (20.55), and indicate its contribution with a superscript f. We find

$$W_k^f = \frac{1}{2}me^{ikx}\int_{-\infty}^{\infty} u^2 g(u)e^{-ikut}du$$

and a kinetic energy flux

$$P_k^f = \frac{1}{2}me^{ikx}\int_{-\infty}^{\infty} u^3 g(u)e^{-ikut}du.$$

Obviously, we have no electric field energy density associated with the free streaming term. It is now trivially demonstrated that this free streaming kinetic energy satisfies

$$\frac{\partial}{\partial t}W_k^f + \frac{\partial}{\partial x}P_k^f = 0.$$

We therefore need not be concerned with the free streaming terms from now on.

Consider now the term accounting for the collective interaction in a linear formalism. We find two terms, one from differentiating the integrand and one from the integration limits

$$\begin{aligned}\frac{\partial}{\partial t}W_k = &-ik\frac{en_0}{2}e^{ikx}\int_{-\infty}^{\infty}\int_0^t e^{iku(t'-t)}E(t')u^3F_0'(u)dudt' \\ &+\frac{en_0}{2}e^{ikx}E(t)\int_{-\infty}^{\infty}u^2F_0'(u)du. \end{aligned} \tag{20.56}$$

The last term vanishes due to our choice of frame of reference. Similarly we find

$$\frac{\partial}{\partial x}P_k = ike^{ikx}\frac{en_0}{m}\int_{-\infty}^{\infty}\int_0^t e^{iku(t-t')}u^3E(t')F_0'(u)dudt' \tag{20.57}$$

and for the electric field energy density

$$\frac{\partial}{\partial t}W_E = \frac{1}{2}\varepsilon_0\frac{\partial}{\partial t}E^2.$$

For the assumed electrostatic fluctuations we have from Maxwell's equation (5.57)

$$\varepsilon_0\frac{\partial}{\partial t}E + J = 0, \qquad \text{giving} \qquad EJ = -E\varepsilon_0\frac{\partial}{\partial t}E = -\frac{1}{2}\varepsilon_0\frac{\partial}{\partial t}E^2.$$

Using $W = W_E + W_k$ we readily prove (20.54). Note that we have no Poynting flux for electrostatic waves, so the only energy flux density here originates from the particle kinetic energy density. The result is trivially generalized to three spatial dimensions.

Physically, (20.54) means that the second order quantities W and P follow a continuity equation where sources or sinks are $-EJ$. If we consider a spatial region without externally imposed fields or currents, we have that the only contributions to E and J are those generated by the currents and charges in the plasma. We then have $E(x,t) = ie\int_{-\infty}^{\infty} f(x,u,t)du/(k\varepsilon_0)$ and $J(x,t) = e\int_{-\infty}^{\infty} uf(x,u,t)du$. The two terms are very nearly $\pi/2$ out of phase, so upon time averaging, the average $\overline{EJ}$ is essentially vanishing. For an undamped Langmuir wave, for instance in a cold plasma, the cancellation of the two half periods would be exact. During a plasma period of oscillation, the waves have an *active* and a *reactive* phase, where the wave energy "sloshes" from field energy to particle kinetic energy due to the acceleration by the fields, and the next half period most of the energy is given back again. In the long run, the electric field energy is dissipated and, due to Landau damping, all energy is particle kinetic energy in that limit.

20.5 Experimental investigations

When first discovered, the analysis of the collisionless damping of electron plasma waves was received with great enthusiasm by the scientific community, in part because the problem provides insight into some basic problems dealing with the distinction of reversible and irreversible processes. No wonder that these wave phenomena also received great interest among experimentalists, and some very careful experiments have been carried out, by, for instance, Malmberg and Wharton (1964), who studied the damping of electron plasma waves propagating along magnetic field lines in a strongly magnetized plasma. In this case, the problem could be treated as essentially one dimensional. The finite cross section of the (long) plasma column requires parts of the theoretical analysis to be reformulated, but this turns out to be of little consequence for the damping as such.

The basic equation for the present problem can be taken to be the limiting form of the drift-kinetic equation where we let the intensity of the magnetic field go to infinity, i.e., assume $\omega_{pe} \ll \omega_{ce}$; see Section 19.4. The equation is here written in its linearized form

$$\frac{\partial}{\partial t} f + u\frac{\partial}{\partial z} f - e\frac{E_{\|}}{m}\overline{n}(r)F_0'(u) = 0, \tag{20.58}$$

where for $B \to \infty$ we have $\mathbf{u}_\perp \to 0$. In this limiting model, the electrons move as "pearls on a string", the strings being the magnetic field lines. We allow the plasma density $\overline{n}$ to vary with the radial coordinate, r, but as indicated, let the unpertubed electron velocity distribution be the same everywhere with $\int_{-\infty}^{\infty} F_0(u)du = 1$. This assumption might be a little bit optimistic, but we use that nonetheless. The electric field is assumed to be electrostatic, $E_{\|} = -\partial\phi/\partial z$. Only the **B**-parallel component of the electric field enters (24.26); the radial component appears through Poisson's equation, which has to be solved with the appropriate boundary conditions at $r = R$. We assume that the electron velocity distribution is a Maxwellian and can then introduce the plasma dispersion function; see Appendix C.

Following the analysis of Trivelpiece-Gould waves, Section 12.2.6, we obtain the relation for the electrostatic potential assumed in the form $\phi(r)\exp(-i(\omega t - kz - m\theta)$

$$\frac{1}{r}\frac{\partial}{\partial r}\left(r\frac{\partial}{\partial r}\phi\right) - \frac{m^2}{r^2}\phi - \left(\omega_{p0}^2\frac{\overline{n}(r)}{n_0}\,⨍_{-\infty}^{\infty}\frac{F_0'(u)}{u-\omega/k}du + k^2\right)\phi = 0, \tag{20.59}$$

where we took the limit $c^2 \to \infty$ (see (12.36)) and introduced a reference electron plasma frequency ω_{p0} at the position of maximum density n_0, usually in the center of the plasma column. The ratio $\overline{N} \equiv \overline{n}(r)/n_0$ is the normalized radial density variation.

To be specific, we now assume that $F_0(u)$ is a Maxwellian, so we can introduce the plasma dispersion function (see Appendix C) or the function W derived from it. We then write (20.59) as

$$\frac{1}{r}\frac{\partial}{\partial r}\left(r\frac{\partial}{\partial r}\phi\right) - \frac{m^2}{r^2}\phi + T^2\phi = 0, \tag{20.60}$$

with

$$T^2 \equiv \overline{N}(r)\,W\left(-\frac{\omega}{ku_{The}}\right)\frac{1}{\lambda_{De}^2} - k^2.$$

First we discuss the particularly simple case where the density is constant, $\overline{N}(r) = 1$. In that case, we can argue as in Section 12.2.6 that the solution fulfilling the boundary condition at $r = R$ has the form $\phi(r) = J_m(Tr)$ with $TR = p_{m,\nu}$, giving

$$W\left(-\frac{\omega}{ku_{The}}\right)\frac{1}{\lambda_{De}^2} - k^2 = \frac{p_{n,\nu}^2}{R^2}. \tag{20.61}$$

This relation can be seen as an implicit dispersion relation for the problem. In order to use the results for physically realistic laboratory conditions (Roberson 1971), we assume that a (real) frequency ω is applied to an exciter (antenna or similar) at one end of the plasma column, and that the waves are Landau damped with distance from this exciter, corresponding to complex wavenumbers $k = k_1 + ik_2$. The function W is complex with $W = W_1 + iW_2$. Writing the real and imaginary pats of the dispersion relation (20.61) we have

$$\begin{aligned} k_1^2 - k_2^2 &= \frac{1}{\lambda_{De}^2} W_1\left(-\frac{\omega}{u_{The}(k_1+ik_2)}\right) - \frac{p_{n,\nu}^2}{R^2} \\ 2k_2k_1 &= \frac{1}{\lambda_{De}^2} W_2\left(-\frac{\omega}{u_{The}(k_1+ik_2)}\right). \end{aligned}$$

With a little algebra, we rewrite these equations as

$$k_1^2 - k_2^2 = \left(\frac{W_1}{\lambda_{De}^2} - \frac{p_{n,\nu}^2}{R^2}\right)\frac{1}{1-k_2^2/k_1^2} \tag{20.62}$$

$$\frac{k_2}{k_1} = \frac{1}{2}\frac{W_2(1-k_2^2/k_1^2)}{W_1 - (p_{n,\nu}\lambda_{De}/R)^2}, \tag{20.63}$$

retaining the same argument for the W-function as before. These two relations can be solved numerically by an iterative procedure.

- **Exercise:** Find an approximate solution $k_1 = k_1(\omega)$ and $k_2 = k_2(\omega)$ for the relations (20.62)-(20.63) by using the asymptotic approximations for the W-function, see Appendix C.

We now consider the case where the plasma density is inhomogeneous in the radial direction, $\overline{N} \neq 1$. We are concerned with the lowest order mode, and consequently take the mode number $m = 0$ in (20.60). We can then write

$$\frac{\partial}{\partial r}\left(r\frac{\partial}{\partial r}\phi\right) - rk^2\phi + r\overline{N}(r)k^2\beta\phi = 0, \tag{20.64}$$

introducing

$$\beta \equiv \frac{1}{(k\lambda_{De})^2} W\left(-\frac{\omega}{u_{The}k}\right).$$

The radial plasma density variation is measurable and assumed to be known. Then the relation (20.64) has the form of a Sturm-Liouville eigenvalue problem (Davydov 1965), with β being the eigenvalue. Generally, we have infinitely many eigenvalues, which for the present case will form a *discrete* set, with possible degeneracies. By integration, we readily find from (20.64) the relation

$$\beta = \frac{\int_0^R \left(r(d\phi(r)/dr)^2 + rk^2\phi^2(r)\right)dr}{\int_0^R r\overline{N}(r)k^2\phi^2(r)dr}, \tag{20.65}$$

which gives the eigenvalue β, once the eigenfunction is determined. *If* we insert in (20.65) a function $\phi(r)$ different from an eigenfunction, it can be shown (Davydov 1965) that the right side of the relation is always larger than the lowest eigenvalue

Assume that we have found the normalized radial density variation to be well approximated by $\overline{N}(r) = \exp(-\frac{1}{2}r^2/(\sigma^2 R^2))$, where σ is a dimensionless parameter, measuring the half-width of the plasma column as fraction of the waveguide radius R. The approximation for $\overline{N}$ will often be quite reasonable for a magnetized plasma produced by a discharge, but may not be quite as good for a Q-machine plasma (see Section 9.1) where we expect a more "flat-topped" radial density distribution.

If we are able to make a "good" guess for the eigenfunction, we might be able to get an estimate β_* for the eigenvalue by use of (20.65), knowing for certain that this will be an *overestimate*, $\beta_* > \beta$. We do not have to guess "quite out of the blue", since we have reasons to expect that this eigenfunction is unlikely to deviate much from the radial Trivelpiece-Gould wave mode obtained for a *homogeneous* plasma, using $p_{0,1}$ for consistency. We have to fulfill the boundary condition at $r = R$. A suitable guess will be

$$\phi_*(r) = \exp(-hr^2/R^2) - \exp(-h),$$

where $\phi_*(r = R) = 0$, as it should. Inserting ϕ_* into (20.65) we find

$$\beta_* \approx \left(1 + \frac{(kR)^2}{2h}\right) \frac{2h + 1/(2\sigma^2)}{(kR)^2},$$

where we made the approximations $h \gg 1$ and $\exp(-1/(2\sigma^2)) \ll 1$ when carrying out the integrations. We want to minimize β_* in order to get as close as possible to the "correct" eigenvalue, for the present choice of trial function. We minimize β_* with respect to h. We obtain $d\beta_*/dh = 0$ by choosing $h = kR/\sqrt{8\sigma^2}$, giving

$$\beta_* = \left(1 + \frac{1}{\sigma k R\sqrt{2}}\right)^2,$$

where it is easily verified that this value indeed gives a minimum for β_*. Using the definition of β, we find the approximate expression for the dispersion relation

$$\frac{W\left(-\frac{\omega}{k u_{The}}\right)}{(k\lambda_{De})^2} = \left(1 + \frac{1}{\sigma k R\sqrt{2}}\right)^2.$$

By simple comparison with the previous results for a homogeneous density, we readily find that we can use those results also here if we make the replacement

$$p_{0,1} \to \sqrt{\frac{1}{2\sigma^2} + \frac{kR\sqrt{2}}{\sigma}}.$$

The important difference is that before we had a constant $p_{0,1}$, independent of k, while we now have a quantity which is a function of the wavenumber k. It is found that the present simplified results are in quite good agreement with numerical solutions of the eigenvalue relation, as long as the analytical approximation for $\overline{N}$ is adequate. The foregoing procedure estimated the *lowest* eigenvalue: if we should be interested in the others, there are methods for estimating the second lowest, if we have an estimate for the lowest one, etc., but the uncertainties rapidly become large.

The observations of Malmberg and Wharton (1964) are summarized in Figs. 20.8, where the real and the imaginary parts of the exåerimentally obtained dispersion relation of plasma waves are shown. The waves were excited by a small antenna placed centrally in the strongly magnetized plasma column. An important observation was that the spatial damping of the waves was very close to exponential. For small wave-numbers, the real part of the dispersion relation is close to the one found for Trivelpiece-Gould waves in cold plasmas; see Section 12.2.6. The dispersion relation was solved for the imaginary part of the wavenumber to give the analytical result in Fig. 20.8. It was concluded that the linear theory for Landau damping is in very good agreement with the

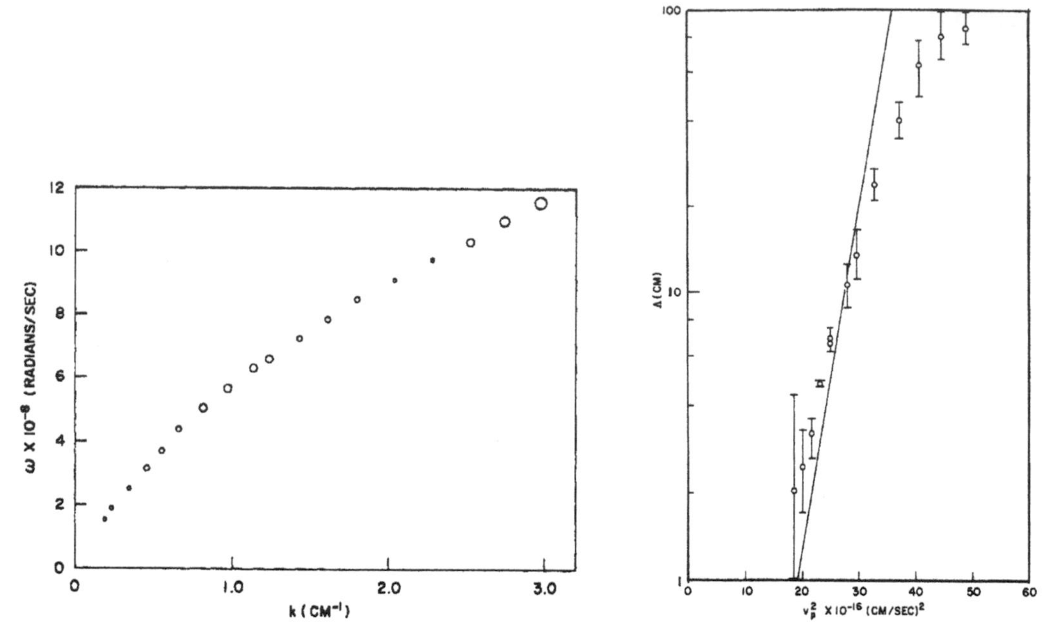

FIGURE 20.8: *Experimental results for the real part (left) and imaginary (right) parts of the dispersion relation, reproduced with permission from Phys. Rev. Lett. Malmberg and Wharton (1964). Copyright 1964 by the American Physical Society. The size of the circles for the real part gives approximately the experimental uncertainty of the measurements. The figure to the right shows the logarithm of the damping distance for varying phase velocities squared. The solid curve is the theoretical result for a Maxwellian electron velocity distribution with temperature 10.5 eV. Vertical bars indicate here the experimental uncertainty.*

observations. Very small amplitudes have to be used in order to make certain that the linear theory is applicable. Care has to be taken also to center the antenna carefully, to ensure that the assumed cylindrical symmetry implied in the analysis is really there.

- **Comment:** It interesting that only the first half of the paper by Landau (1946) is concerned with Landau damping. It is probably the most influential paper in plasma physics ever written, but in reality only the first part of it is discussed in depth in the literature! The second part concerns plasma embedded in an externally imposed spatially constant and temporally oscillating electric field with given frequency ω_0. The problem is to determine the penetration of this electric field into the plasma. In spite of the importance of the problem it has received comparatively little attention. The problem was analyzed (Landau 1946, Swanson 2003) assuming immobile ions and an ideally reflecting wall at $x = 0$, giving $f(0,u,t) = f(0,-u,t)$. The result contains a spatial damping term $\exp\left(\frac{i}{4}3\sqrt{3}\,(\omega_0 x/\omega_{pe}\lambda_{De})^{2/3}\right)$. The exponential factor containing $x^{2/3}$ rather than a simple exponential is unexpected, but the result seems to have received only limited attention.

21

Kinetic Plasma Sound Waves

The results in Section 20.2 were based on a model where the ions were considered immobile. Although it is easy to imagine cases, i.e., high frequency fluctuations, where such assumptions are justified, we would like to relax this assumption. If the full nonlinear problem is considered, this generalization implies considerable complications. However, if we restrict the analysis to small amplitude, linear, electrostatic oscillations, it is easily shown that *mobile* ions can be included by a simple generalization of the unperturbed velocity distribution in (20.13). If we replace F in (20.16) or (20.22), for instance, with $F_e(u)+(m/M)F_i(u)$, where now F_e is the unperturbed electron velocity distribution, while F_i is the ion distribution, then we recover basically the same equation as obtained from the electron Vlasov equation, with simple generalizations of the meaning of the symbols. The only problem will be that it is (obviously) not physically meaningful to insert a Maxwellian distribution for $F_e(u)+(m/M)F_i(u)$. It can be instructive for the reader to plot this effective distribution function for the case where the two constituents $F_e(u)$ and $F_i(u)$ are both Maxwellians. The procedure outlined here leads to a rather trivial repetition of the various individual steps in the mathematical analysis presented in Section 20.2, and does not really provide any new insight. We might as well go directly to the plasma dispersion relation (20.22) or the slightly more general result (20.33), and introduce the effective distribution function $F_e(u)+(m/M)F_i(u)$ there, with the result

$$\varepsilon(k,\omega) = 1 - \frac{\omega_{pe}^2}{k^2} ⨍_{-\infty}^{\infty} \frac{F_e'(u)}{u-\omega/k} du - \frac{\Omega_{pi}^2}{k^2} ⨍_{-\infty}^{\infty} \frac{F_i'(u)}{u-\omega/k} du. \tag{21.1}$$

This expression can also be understood by noting that the electron contribution to the dielectric function of the plasma is given by (20.22) and the corresponding *ion* contribution can be obtained by replacing m by M, and $F_e(u)$ by $F_i(u)$. Using (5.90), we then obtain (21.1). Evidently, (21.1) *still* contains Langmuir waves, or electron plasma waves, now including a tiny contribution from the ion motion, but the expression also allows other wave-types, characterized by low frequencies, to be considered here.

The formally correct result (21.1) for small amplitude electrostatic oscillations can be too complicated for some practical purposes, in particular when we know a priori that the electron distribution is a Maxwellian. In those cases some additional simplifications can be appreciated. We presume that the waves in question have a very low phase velocity in comparison with the electron thermal velocity: since the ion dynamics is involved this is reasonable to expect. We can therefore use the approximation

$$⨍_{-\infty}^{\infty} \frac{F_e'(u)}{u-\omega/k} du \approx ⨍_{-\infty}^{\infty} \frac{F_e'(u)}{u} du.$$

For the assumed Maxwellian distribution we have $F_e'(u) = -(mu/\kappa T_e)F_e(u)$ and we can therefore, in this limit, rewrite (21.1) as

$$\varepsilon(k,\omega) = 1 + \frac{1}{(k\lambda_{De})^2} - \frac{\Omega_{pi}^2}{k^2} ⨍_{-\infty}^{\infty} \frac{F_i'(u)}{u-\omega/k} du. \tag{21.2}$$

Again with reference to (5.90) we realize that in the presumed limit of very low phase velocities, the electron contribution to the dielectric function is in effect $1+1/(k\lambda_{De})^2$; see, for instance, (5.91). It is a simple matter to demonstrate that this is the result one will obtain by assuming that the electrons' only effect is to provide the isothermal Debye shielding of the distributed charges in the plasma.

The quasi-neutral limit is obtained by letting $(k\lambda_D)^2 \to 0$, which gives

$$\varepsilon(k,\omega) = \frac{1}{(k\lambda_{De})^2}\left(1 - \frac{\kappa T_e}{M} ⨍_{-\infty}^{\infty} \frac{F_i'(u)}{u - \omega/k} du\right). \tag{21.3}$$

In this limit the waves are dispersionless, since solutions of $\varepsilon(k,\omega) = 0$ depend on the ratio ω/k only, i.e., the phase velocity of the wave.

To save a lot of ∞'s in integration limits in the following, we simply write $\int$ when we mean $\int_{-\infty}^{\infty}$. When particular limits are needed, then these will be written explicitly.

- **Exercise:** Give the explicit expression for the plasma dielectric function in the low frequency limit, assuming cold ions and Boltzmann distributed electrons from the outset, but without assuming quasi-neutrality.

21.1 Kinetic dispersion relation for ion sound waves

Considering again electrostatic waves in a one-dimensional model, we discuss the linear dispersion relation obtained from the simple model (21.2). First we can take the simplest limit where $F_i(u) = \delta(u)$ and obtain trivially the dispersion relation from $\varepsilon(\omega,k) = 0$ as $\omega/k = \sqrt{\kappa T_e/M}/(1 + (k\lambda_{De})^2)$, consistent with the fluid result (13.27) with cold ions. In this limit the waves are not subject to any Landau damping. This is partly an artifact due to the assumption of ideally cold ions, but also the model for isothermal electrons. Even with $F_i(u) = \delta(u)$ a complete analysis would give a tiny damping due to the electrons in the velocity region close to the phase velocity of the wave. This limit requires use of the full expression (21.1), since the damping depends on details in the electron velocity distribution. For realistic electron-ion temperatures, however, this electron Landau damping turns out to be very small, and the ion damping is dominating.

It is relatively simple to include a small, finite ion temperature in the analysis. We can do this by reference to the analysis of Section 20.2.1; see also (5.108). We have $\varepsilon_1(k,\omega) \approx 1 + 1/(k\lambda_D)^2 - (\Omega_{pi}/\omega)^2$ and obtain

$$\omega_1(k) = k\sqrt{\frac{\kappa T_e/M}{1 + (k\lambda_{De})^2}} \approx k\sqrt{\frac{\kappa T_e}{M}} \tag{21.4}$$

$$\omega_2(k) = \frac{\pi}{2} k \left(\frac{\kappa T_e/M}{1 + (k\lambda_{De})^2}\right)^{3/2} F_i'\left(\frac{\omega_1}{k}\right) \tag{21.5}$$

$$\approx \frac{\pi}{2}\left(\frac{\kappa T_e}{M}\right)^{3/2} k F_i'\left(\sqrt{\frac{\kappa T_e}{M}}\right). \tag{21.6}$$

The limit of quasi-neutrality can, as before, be obtained by letting $(k\lambda_{De})^2 \to 0$. In these latter limits, we have both $\omega_1(k)/k$ and $\omega_2(k)/k$ being constants. We readily find that the phase velocity of the ion acoustic sound waves is proportional to the square root of the electron temperature at least as long as $T_i \ll T_e$. Similarly, we expect the ion Landau damping to decrease for increasing electron temperatures, at least as long as the ion velocity distribution is close to a Maxwellian. In Fig. 21.1 we show illustrative experimental results for propagating damped ion acoustic waves obtained by grid excitation in a single ended Q-machine. The results clearly demonstrate that an increase in electron temperature, with the ion temperature remaining constant, implies an increase in phase velocity and a decrease in the ion Landau damping.

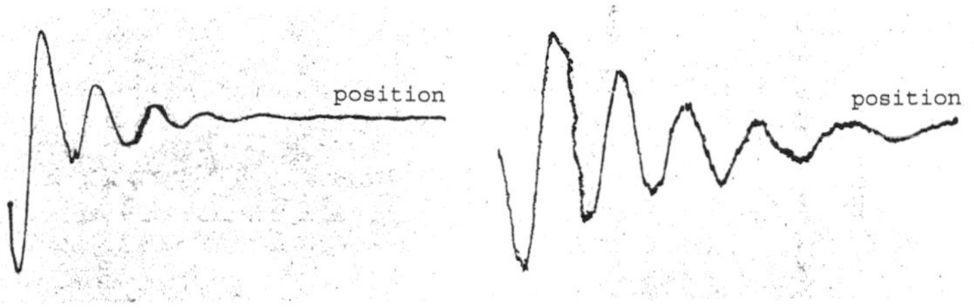

FIGURE 21.1: *Ion acoustic waves excited in a single ended Q-machine, where the electron temperature could be varied by electron cyclotron heating. The applied frequency is 10 kHz. The wave signal is shown as a function of position (increasing to the right from the exciting grid) as obtained by a lock-in amplifier, set at a fixed phase with reference to the exciter signal. The results to the left were obtained for conditions where $T_e/T_i \approx 1$, those to the right with $T_e/T_i \approx 2.5$.*

We emphasize that the foregoing results explicitly assumed that the ion Landau damping was weak, i.e., $T_i \ll T_e$. When this inequality is *not* satisfied, the approximation (5.108) is in general no longer applicable, and we have to perform a much more detailed analysis. Unfortunately, in nature as well as in many laboratory experiments (see for instance Chapter 9), we have $T_i \sim T_e$, which implies nontrivial complications in even a linear analysis.

The analysis presented here refers to the "standard" problem, where we assume an initial condition imposed on the plasma in the form of a plane wave. In a laboratory experiment, a *boundary value* problem is much more easily realized, for instance by assuming some oscillating charges imposed on the plasma in a fixed point. The fluctuations will then decay in amplitude away from the exciter, at a rate corresponding to the Landau damping. Here, a problem will arise: the plasma dispersion function has many branches also in this case (just as for Langmuir waves). In the analysis of ion acoustic waves, without restrictions on the plasma parameters, the electron/ion temperature rations in particular, the branch with the temporally least damped wave (the initial value problem) need not be the branch with the least *spatial* damping (the boundary value problem). This problem has been discussed in detail by, for instance, Gould (1964), by solving the spatial variation of the "test charge problem", where externally imposed point-charges vary harmonically with time. The first detailed laboratory observations of Landau damped ion acoustic waves were reported by Wong et al. (1964). These experiments were carried out in a magnetized Q-machine plasma (Motley 1975).

21.1.1 Unstable ion acoustic waves

We note that also (21.6) opens for the possibility of *unstable* ion sound waves when the ion distribution has a local minimum and $F_i'\left(\sqrt{\kappa T_e/M}\right) > 0$. To determine the stability of a given unperturbed ion velocity distribution function with isothermally Boltzmann distributed electrons we have to make a trivial generalization of the Penrose criterion; see Section 20.3. For Langmuir waves it was a sufficient criterion for instability that the Nyquist contour encircled some part of the positive axis: there would then always be *some* wave-numbers which would correspond to instability. The present ion acoustic problem is more restrictive. Here, the Nyquist contour has to encircle $k^2/\Omega_{pi}^2 + 1/(\Omega_{pi}\lambda_{De})^2 \geq 1/(\Omega_{pi}\lambda_{De})^2$. In the quasi-neutral approximation, the criterion reduces to encircling the point $1/(\Omega_{pi}\lambda_{De})^2 = M/\kappa T_e$. The necessary criterion remains to be that the ion distribution has a local minimum (Fried & Wong 1966).

One should bear in mind that in case we have an inhomogeneous, possibly double humped ion velocity distribution, there is still a freedom in choosing the *direction* of the wave-vector, and a complete analysis should consider all directions, just as in the case of Langmuir waves.

Particularly for linearly unstable plasmas, we might encounter restrictions in the assumption of the local Boltzmann distribution of electrons. The best example of this shortcoming is a current carrying plasma, where each of the components might very well be Maxwell distributed, but the *effective* distribution $F_e(u) + (m/M)F_i(u)$ can have a local minimum, in such a way that the appropriate Penrose criterion for instability is fulfilled. For such cases, we have to consider the full problem, which in its linearized form is given by (21.1). The Buneman instability (see Section 12.7.1) can be seen as a special limiting case for an instability where $F_e = n_0\delta(u - u_0)$ and $F_i = n_0\delta(u)$.

- **Exercise:** Consider a linearly unstable ion velocity distribution in an unmagnetized plasma. Let the model equations assume that electrons are isothermally Boltzmann distributed at all times, but do *not* make any assumption of quasi-neutrality. Following Backus (1960), demonstrate that the growth of the most unstable wave in this model will be bounded by $\exp\left(tkC_s/\sqrt{1+(k\lambda_{De})^2}\right)$, irrespective of the ion distribution. See also the discussion in Section 20.2.3.

Ⓢ **Exercise:** Write out the dispersion relation for the two cold ion-ion beam distribution for arbitrary beam/background densities and isothermal warm electrons. Determine the instability conditions for the particular case where the beam and background ion densities are equal. Comment on the difference between the analysis in one and higher spatial dimensions. Ⓢ

21.2 Basic nonlinear dynamic equation for low frequency kinetic plasma waves

The discussion of Section 21.1 was based on the dielectric function for the plasma, and indeed in many cases this will suffice, for discussions of small amplitude plane wave propagation, in particular. In order to include the ion dynamics in the full nonlinear description of kinetic plasma phenomena, the simplest but also most tedious way is to include a Vlasov equation for the ion dynamics together with the similar equation for the electrons, and let the time evolution of the two species be governed by the self-consistent electric and magnetic fields. This can develop into an extremely lengthy procedure, and it is timely to consider the possibility of simplifying approximations, similar to those in Section 21.1. One such approximation consists of assuming the the dynamics are associated with frequencies that are so low that the assumption of Boltzmann distributed, isothermal electrons can be justified, just as in (21.2). In this case a model using only one Vlasov equation, the one for the ions, can be assumed. We seek a nonlinear model which reproduces the linear results from Section 21.1.

Our investigations will be restricted to electrostatic phenomena, and the only force acting on the ions will be one originating from an electric field $\mathbf{E} = -\nabla\phi$. The fully nonlinear basic equations we can postulate from the assumption of locally isothermal Boltzmann distributed electrons are

$$\frac{\partial}{\partial t} f(\mathbf{r},\mathbf{u},t) + \mathbf{u}\cdot\nabla f(\mathbf{r},\mathbf{u},t) - \frac{e}{M}\nabla\phi(\mathbf{r},t)\cdot\nabla_u f(\mathbf{r},\mathbf{u},t) = 0, \tag{21.7}$$

for the velocity distribution $f(\mathbf{r},\mathbf{u},t)$ of singly charged ions. The electrostatic potential is derived from Poisson's equation

$$\nabla^2\phi(\mathbf{r},t) = \frac{e}{\varepsilon_0}\left(n_0 e^{e\phi(\mathbf{r},t)/\kappa T_e} - n(\mathbf{r},t)\right), \tag{21.8}$$

where the assumption of Boltzmann distributed electrons was used explicitly. Since we have made an explicit assumption concerning the electron density, there is no need to keep the subscript i on the ion density n.

For the quasi-neutral electrostatic case, with isothermally Boltzmann distributed electrons, Poisson's equation becomes redundant, and we have

$$n(\mathbf{r},t) \equiv \int_{-\infty}^{\infty} f(\mathbf{r},\mathbf{u},t)d\mathbf{u} = n_0 e^{e\phi(\mathbf{r},t)/\kappa T_e},$$

and (21.7) can be written as

$$\frac{\partial}{\partial t} f(\mathbf{r},\mathbf{u},t) + \mathbf{u}\cdot\nabla f(\mathbf{r},\mathbf{u},t) - \frac{\kappa T_e}{M}\frac{1}{n(\mathbf{r},t)}\nabla n(\mathbf{r},t)\cdot\nabla_u f(\mathbf{r},\mathbf{u},t) = 0. \tag{21.9}$$

The form (21.9) demonstrates that for small electron temperatures, the last term might be small, and the basic evolution of an initial perturbation can be well described by the particle free-streaming effects as for a Knudsen gas, as in the example in Chapter 20. Physically, this means that the electron pressure is in this case not sufficient to set up appreciable ambipolar electric fields for perturbing the ion orbits. For $T_e = 0$, the last term in (21.9) vanishes identically. When the electrons are cold, the Debye length is vanishing and all charges arising from inhomogeneous ion distributions are completely shielded. Hence, there are no electric fields to perturb the ion trajectories within this model.

The basic assumption of Boltzmann distributed electrons is appealing, but *might* be considered with some reservations for a collisionless plasma model. We can assume that the electrons have a Maxwellian distribution at infinity, and it is presumably a reasonable assumption for this case to take the electron inertia so small that electrons adjust almost instantaneously to changes in the electrostatic potential. For a monotonic variation of $\phi(x)$ (considering a one-dimensional case for simplicity) we then find that indeed $n_e = n_0 \exp(e\phi(x,t)/\kappa T_e)$. If, however, the potential has a local minimum somewhere, the low velocity part of the electron distribution in that region is "decoupled" from the rest of the distribution, and can in principle have any form, unless collisions make it a local Maxwellian also. Actually, experimental studies of Grésillon and Galison (1973) demonstrated that there were observable deviations from a local Maxwell distribution for non-monotonic potential variations. These observations should be considered in a *truly* collisionless plasma, but we bear in mind that a parameter regime exists where the ions are collisionless to a good approximation, while the electron collision time is short on the relevant time scale.

A number of linear problems associated with kinetic descriptions of plasma sound waves can be solved analytically. In this chapter we consider two, namely, radiation from a moving point charge, and the time evolution of a given initial condition, here a step-function in density. Both problems require basically the same mathematical apparatus.

21.2.1 Energy conservation

Since we have made a simplifying assumption concerning the electron dynamics, it is not readily obvious whether the system is energy conserving (although we *would* be surprised if it were not), and in particular it is not evident what expression we should interpret as an "energy". To analyze this question (Mason 1971) we multiply by $u^2/2$ and then integrate the full nonlinear ion Vlasov equation as

$$\frac{1}{2}\frac{\partial}{\partial t}\iint_{-\infty}^{\infty} u^2 f(\mathbf{r},\mathbf{u},t)d\mathbf{r}d\mathbf{u} + \frac{1}{2}\iint_{-\infty}^{\infty} \nabla\cdot\left(\mathbf{u}u^2 f(\mathbf{r},\mathbf{u},t)\right)d\mathbf{r}d\mathbf{u} =$$
$$\frac{e}{2M}\iint_{-\infty}^{\infty} u^2 \nabla\phi(\mathbf{r},t)\cdot\nabla_u f(\mathbf{r},\mathbf{u},t)d\mathbf{r}d\mathbf{u}, \tag{21.10}$$

with the short hand notation $\iint_{-\infty}^{\infty} = \int_{-\infty}^{\infty}\int_{-\infty}^{\infty}$. Using Gauss' divergence theorem on the second integral, we find that its contribution will be vanishing for periodic boundary conditions, and also if we assume infinite space with a finite support for the plasma. With the same assumption we can integrate also the term on the right-hand side to obtain

$$\begin{aligned}\frac{1}{2}\frac{\partial}{\partial t}\int\int_{-\infty}^{\infty} u^2 f(\mathbf{r},\mathbf{u},t)d\mathbf{r}d\mathbf{u} &= -\frac{e}{M}\int\int_{-\infty}^{\infty} \nabla\phi(\mathbf{r},t)\cdot\mathbf{u}f(\mathbf{r},\mathbf{u},t)d\mathbf{r}d\mathbf{u} \\ &= \frac{e}{M}\int\int_{-\infty}^{\infty} \phi(\mathbf{r},t)\nabla\cdot\mathbf{u}f(\mathbf{r},\mathbf{u},t)d\mathbf{r}d\mathbf{u}.\end{aligned}$$

We now use the ion continuity equation in the form

$$\frac{\partial}{\partial t}n+\nabla\cdot\int_{-\infty}^{\infty}\mathbf{u}f(\mathbf{r},\mathbf{u},t)d\mathbf{r}d\mathbf{u}=0$$

to obtain

$$\frac{1}{2}\frac{\partial}{\partial t}\int\int_{-\infty}^{\infty} u^2 f(\mathbf{r},\mathbf{u},t)d\mathbf{r}d\mathbf{u}=-\frac{e}{M}\int_{-\infty}^{\infty}\phi\frac{\partial}{\partial t}n\,d\mathbf{r}.$$

With the assumption of Boltzmann distributed electrons together with Poisson's equation (21.8) we can differentiate the ion density n with respect to t and obtain

$$\frac{1}{2}\frac{\partial}{\partial t}\int\int_{-\infty}^{\infty} u^2 f(\mathbf{r},\mathbf{u},t)d\mathbf{r}d\mathbf{u}-\frac{\varepsilon_0}{M}\int_{-\infty}^{\infty}\phi\frac{\partial}{\partial t}\nabla^2\phi d\mathbf{r}+\frac{e}{M}\int_{-\infty}^{\infty}\phi\frac{\partial}{\partial t}\left(n_0 e^{e\phi/\kappa T_e}\right)d\mathbf{r}=0.$$

It is important to emphasize that we take n_0 in the expression for the electron distribution $n_e = n_0\exp(e\phi/\kappa T_e)$ to be a constant here, although this assumption is not as obvious as it might appear (Pécseli et al. 1984). Generally we have n_0 independent of position, but it *might* depend on time, for instance for periodic systems.

After some straightforward algebra we find the relation

$$\frac{1}{2}M\frac{\partial}{\partial t}\int\int_{-\infty}^{\infty} u^2 f(\mathbf{r},\mathbf{u},t)d\mathbf{r}d\mathbf{u}+\frac{1}{2}\varepsilon_0\frac{\partial}{\partial t}\int_{-\infty}^{\infty}(\nabla\phi)^2 d\mathbf{r}-n_0\frac{\partial}{\partial t}\int_{-\infty}^{\infty}(\kappa T_e-e\phi)\,e^{e\phi/\kappa T_e}d\mathbf{r}=0, \quad (21.11)$$

where the first term is readily interpreted as the time derivative for a particle kinetic energy, while the second term is the energy associated with the electrostatic electric field. The last nonstandard term accounts for the energy density of the electron gas. With a suitable assumption on the time variation at $t\to-\infty$, we can integrate (21.11) with respect to time and obtain

$$\frac{1}{2}M\int\int_{-\infty}^{\infty} u^2 f(\mathbf{r},\mathbf{u},t)d\mathbf{r}d\mathbf{u}+\frac{1}{2}\varepsilon_0\int_{-\infty}^{\infty}(\nabla\phi)^2 d\mathbf{r}-n_0\int_{-\infty}^{\infty}(\kappa T_e-e\phi)\,e^{e\phi/\kappa T_e}d\mathbf{r}=\mathcal{W}, \quad (21.12)$$

where $\mathcal{W}$ is the total energy. We note that the term $n_0\int_{-\infty}^{\infty}\kappa T_e e^{e\phi/\kappa T_e}d\mathbf{r}=\kappa T_e\int_{-\infty}^{\infty}n_e d\mathbf{r}$ is the net thermal energy of the electrons. Since this energy contribution is uniquely determined by the initial conditions, we can use the "freedom" in a time variation of n_0 to make this term constant at all later times. In practice, we need quite large variations in the potential ϕ to find significant variations in this integral.

From (21.12) it is trivial to define an energy density, W. The first term contains a contribution which, in case of a Maxwellian ion distribution, also would correspond to a thermal ion energy density. The second term is the energy of the electrostatic field. The term $en_0\phi e^{e\phi/\kappa T_e}\equiv e\phi n_e$ is the potential energy density of electrons in the electrostatic potential ϕ. Fortunately, we end up with exactly the kind of expression we expected. Note that we do not need additional arguments for the electron contribution to the total energy: the electron contribution *is* already included in Poisson's equation.

It might be instructive to see the result obtained if we assume quasi-neutrality from the outset. For that case we find the relation

$$\frac{1}{2}M\frac{\partial}{\partial t}\int\int_{-\infty}^{\infty} u^2 f(\mathbf{r},\mathbf{u},t)d\mathbf{r}d\mathbf{u}-n_0\frac{\partial}{\partial t}\int_{-\infty}^{\infty}(\kappa T_e-e\phi)\,e^{e\phi/\kappa T_e}d\mathbf{r}=0, \quad (21.13)$$

replacing (21.11). The relation between density and potential is trivial in this limit

21.2.2 Energy density of an ion sound wave

We can use (21.12) or (21.13) to obtain an expression for the energy density associated with a plane sound wave. It is important to be specific here: by the wave energy, we understand the difference in total energy between a suitably chosen initial state *without* the wave and the energy of the system *with* the wave present. Here we only consider moderate amplitude waves and assume the ion temperature to be vanishing. The initial, unperturbed, state is taken to be a homogeneous plasma at rest, so that the kinetic energy vanishes and $\phi = 0$. The plasma energy density, within the model of the present chapter, is then denoted W_0 and we have the energy density associated with the perturbation expressed as

$$\Delta W \approx \frac{1}{2} M n_0 \widetilde{u}^2 + W|_{\phi=0} + \left.\frac{\partial W}{\partial \phi}\right|_{\phi=0} \phi + \frac{1}{2} \left.\frac{\partial^2 W}{\partial \phi^2}\right|_{\phi=0} \phi^2 - W_0 .$$

In terms of the perturbation in ion flow velocity $\widetilde{u}$, electric field and electrostatic potential, we find the simple result

$$\Delta W = \frac{1}{2} M n_0 \widetilde{u}^2 + \frac{1}{2} \varepsilon_0 (\nabla \phi)^2 + \frac{1}{2} n_0 \kappa T_e \left(\frac{e\phi}{\kappa T_e}\right)^2 . \tag{21.14}$$

So far the result is quite general for small amplitude perturbations. We now specify this perturbation as a plane wave, $e\phi/\kappa T_e = a\cos(\omega t - kx)$ with $a \ll 1$, restricting the problem to one spatial dimension. The frequency and wave-number have to follow the linear dispersion relation for small amplitude ion acoustic waves, which for the present case is simply $\omega^2 = k^2 C_s^2/(1 + (k\lambda_D)^2)$, with $C_s^2 \equiv \kappa T_e/M$. We have $\widetilde{u} = C_s^2 (k/\omega)(e\phi/\kappa T_e)$, giving, after some simple algebra,

$$\begin{aligned} \Delta W &= \frac{1}{2} M n_0 \left(\frac{k}{\omega}\right)^2 C_s^4 a^2 \cos^2(\omega t - kx) + \frac{1}{2} n_0 \kappa T_e (k\lambda_D)^2 a^2 \sin^2(\omega t - kx) \\ &\quad + \frac{1}{2} n_0 \kappa T_e a^2 \cos^2(\omega t - kx). \end{aligned} \tag{21.15}$$

Averaging over one period $2\pi/\omega$ gives a factor 1/2 for each of the trigonometric functions. Using the explicit form of the present dispersion relation, we ultimately find $\overline{\Delta W} = \frac{1}{2} n_0 \kappa T_e a^2 \left(1 + (k\lambda_D)^2\right)$. For the particularly simple quasi-neutral limit $(k\lambda_D)^2 \to 0$, we have $\overline{\Delta W} = \frac{1}{2} n_0 \kappa T_e a^2$.

Ⓢ **Exercise:** Consider the simple quasi-neutral limit and determine the plasma dielectric function for the present case. Demonstrate then that the general expression for the time-averaged energy density obtained in Section 5.10 is consistent with the result of the present section. Discuss also the full expression, without assuming quasi-neutrality (this is a bit more complicated). Discuss also a standing ion acoustic wave. Ⓢ

Ⓢ **Exercise:** Consider first a plane ion sound wave with wavenumber k_0. Give the corresponding frequency ω_0. Assume that the peak-to-peak value of the electrostatic potential is $2\phi_0$. Give the bounce frequency and bounce period of an ion in the local electrostatic potential well associated with the periodic wave train. Compare this result with ω_0. For how many plasma periods will you expect a linear Langmuir wave model in the Landau sense to be valid? Ⓢ

- **Exercise:** Estimate the ratio of the two last terms in (21.14) for a given characteristic length-scale L.

21.3 Sound radiation from a moving charge

In studies of wave phenomena in collisionless plasmas, the interest is traditionally concentrated on the initial value problem, typically with a plane wave given at time $t = 0$. However, *driven* plasma waves may actually be more relevant for realistic conditions, when we deal with excitation of waves by externally imposed, time-varying charges, for instance. One particular problem of this type is the excitation of waves by a charge moving through a collisionless plasma. One important lesson to be gained from this section is that it is quite possible to give *exact* solutions of the linearized Vlasov equation for some selected problems, also apart from those involving plane waves.

The starting point for this calculation is the Vlasov equation for the ions

$$\frac{\partial f}{\partial t} + \mathbf{u}\cdot\nabla f + \frac{e}{M}\mathbf{E}\cdot\nabla_{\mathbf{u}} f = 0 \tag{21.16}$$

where $f = f(\mathbf{r},\mathbf{u},t)$ is the distribution function for the ions. As an equilibrium solution to this equation we have $\mathbf{E}_0 = \mathbf{0}$ and $f_0 = f_0(\mathbf{u})$. It is an advantage for the notation to use the normalization $\int f_0(\mathbf{u})d\mathbf{u} = 1$ from now on, and take out the unperturbed equilibrium density, n_0, explicitly. We take the initial perturbation to be that of a single point charge q introduced at $t = 0$ and $\mathbf{r} = \mathbf{0}$ and immediately removed again. The disturbance can then be taken to be of the form $q\delta(\mathbf{r})\delta(t)$. If the charge q is at rest, we recover the electrostatic shielding results from Section 8.3.2. The important new effects arise due to the motion of the point charge with respect to the plasma (Wang et al. 1981).

21.3.1 Calculations in one spatial dimension

It turns out to be an advantage first to consider the one-dimensional problem, i.e., a moving slab (Guio & Pécseli 2003). We therefore use the one-dimensional version of the ion Vlasov equation, which in its linearized form is

$$\frac{\partial f}{\partial t} + u\frac{\partial f}{\partial x} - \frac{en_0}{M}\frac{\partial \phi}{\partial x}f_0'(u) = 0, \tag{21.17}$$

where we have introduced the electrostatic assumption by using $E = -\partial\phi/\partial x$.

A second relation between f and ϕ can in principle be obtained by using Poisson's equation, containing also the externally imposed charge

$$\frac{\partial^2}{\partial x^2}\phi = \frac{e(n_e - n_i)}{\varepsilon_0} - \frac{q}{\varepsilon_0}\delta(x)\delta(t). \tag{21.18}$$

Note that the dimension of q is *Coulomb* × *time*, while the dimension of density is *length*$^{-1}$, in this one-dimensional model. We shall make the assumption that the electrons are isothermally Boltzmann distributed, i.e.,

$$n_e = n_0 \exp(e\phi/\kappa T_e) \approx n_0 + n_0\frac{e\phi}{\kappa T_e}.$$

We now make a Fourier transform in spatial coordinate and a Laplace transform in time, where we use $-i\omega$ for the traditional Laplace variable s, for reasons to be evident later on. The external charge q is introduced in a uniform plasma so there are no initial conditions imposed on $f(x,u,t)$. The relations (21.17) and (21.18) then become

$$-i\omega f(k,\omega,u) + iku f(k,\omega,u) - \frac{en_0}{M}ik\phi(k,\omega)f_0(u) = 0, \tag{21.19}$$

and from Poisson's equation

$$\varepsilon_0 k^2\phi(k,\omega) = e\int_{-\infty}^{\infty} f(k,\omega,u)du - en_0\frac{e\phi(k,\omega)}{\kappa T_e} + q,$$

giving

$$(k\lambda_D)^2\frac{e\phi(k,\omega)}{\kappa T_e} = \frac{1}{n_0}\int_{-\infty}^{\infty} f(k,\omega,u)du - \frac{e\phi(k,\omega)}{\kappa T_e} + \frac{q}{en_0}, \tag{21.20}$$

where f and ϕ now denote the transformed functions. We have here used the initial condition $f(x,u,t=0)=0$ and the fact that the Laplace and Fourier transforms of a delta function is unity. Here ω is a complex number with a positive imaginary part, while k is real. By the arguments (k,ω) for the various functions, we indicate the Fourier-Laplace transformed quantities. The physical dimensions are different for the original functions and their transforms.

- **Example:** From (21.20) we can express the normalized plasma potential response to the externally applied charge q by eliminating $\int_{-\infty}^{\infty} f du$ from (21.19) to find

$$\frac{e\phi(k,\omega)}{\kappa T_e} = \frac{q/en_0}{1+(k\lambda_D)^2+\dfrac{\kappa T_e}{M}\displaystyle\int_{-\infty}^{\infty}\frac{f_0'(u)}{\omega/k-u}du}. \tag{21.21}$$

With reference to (21.2), this is what we expect, as seen by multiplying both sides of the expression by k^2 to obtain $\mathbf{k}\cdot\mathbf{k}\phi$ in Poisson's equation. If we require a finite potential response, $\phi\neq 0$, in the absence of an external charge, $q=0$, we must require the denominator to be zero. The (in general complex) solutions $\omega=\omega(k)$ will coincide for the denominator in (21.21) and the dielectric function (21.2).

We now take the quasi-neutral limit, i.e., let $(k\lambda_D)^2\to 0$, implying that the right-hand side of (21.20) is considered negligible. We can also take this limit in (21.21). Eliminating f we find

$$n_0\frac{e\phi}{\kappa T_e}-\frac{q}{e}=\int_{-\infty}^{\infty}\frac{en_0}{M}\phi\frac{f_0'(u)}{u-\omega/k}du=\frac{e\phi}{\kappa T_e}\frac{n_0\kappa T_e}{M}\int_{-\infty}^{\infty}\frac{f_0'(u)}{u-\omega/k}du. \tag{21.22}$$

To simplify the notation, we introduce $\psi\equiv n_0e\phi/\kappa T_e$ and have

$$\psi(k,\omega)=\frac{q/e}{1+\dfrac{\kappa T_e}{M}\displaystyle\int_{-\infty}^{\infty}\frac{f_0'(u)}{\omega/k-u}du}. \tag{21.23}$$

This equation can then be solved by performing the inverse Fourier and Laplace transforms. The inverse Fourier transform will be treated first. This is given by

$$\psi(x,\omega)=\frac{1}{2\pi}\int_{-\infty}^{\infty}\psi(k,\omega)e^{ikx}dk. \tag{21.24}$$

Solving this equation we follow the method proposed by Mason (1968). In the integral in the denominator we have a singularity at $u=\omega/k$. When $k<0$ the imaginary part of ω/k is negative. Thus, when we integrate along the real u-axis, the integration path is above the pole. Similarly, when $k>0$ we integrate below the pole. Therefore it is best to split the integral in (21.24) into two

$$\psi(x,\omega)=\frac{q/e}{2\pi}\left(\int_{-\infty}^{0}\psi_1(k,\omega)e^{ikx}dk+\int_0^{\infty}\psi_2(k,\omega)e^{ikx}dk\right), \tag{21.25}$$

where $\psi_{1,2}(k,\omega)$ are given by

$$\psi_{1,2}(k,\omega)=\left(1+\frac{\kappa T_e}{M}\int_{1,2}\frac{f_0'(u)}{\omega/k-u}du\right)^{-1}. \tag{21.26}$$

The two integration paths in (21.26) run above and below the pole, respectively.

If we consider the line $\omega/k=\gamma$ in the complex k-plane, as shown in Fig. 21.2, we see that in the region to the left of the line we have $\Im\{\omega/k\}<0$, i.e., the integration along the real u-axis runs above the pole. Therefore $\psi_1(k,\omega)$ is defined in this region. Similarly, $\psi_2(k,\omega)$ is defined in the

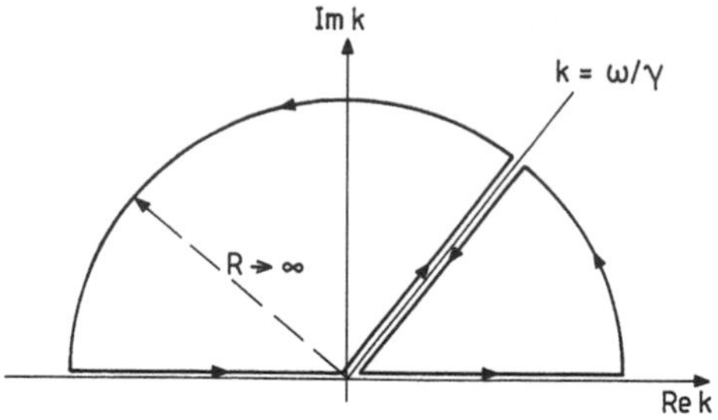

FIGURE 21.2: *Integration contours in the complex k-plane.*

region to the right of the line. Therefore we can perform an integration around a closed contour in each region. The results of these integrations will depend only on the poles of the functions in (21.26). These poles can only arise from the denominator becoming equal to zero. Now, it may be shown by a Nyquist analysis that the functions $\psi_{1,2}(k,\omega)$ are both analytical in the areas where they are defined. Thus there are no poles for the functions inside the integration contours and both contour integrals are equal to zero. In the limit $R \to \infty$, the integrals along the k-axis are then equal to the integrals along the line $k = \omega/\gamma$ with the directions shown in Fig. 21.2. Equation (21.25) will then become

$$\begin{aligned}\psi(x,\omega) &= \frac{q/e}{2\pi}\left(\int_{\infty}^{0}\psi_1(\gamma)\frac{-\omega}{\gamma^2}e^{i\omega x/\gamma}d\gamma + \int_0^{\infty}\psi_2(\gamma)\frac{-\omega}{\gamma^2}e^{i\omega x/\gamma}d\gamma\right)\\ &= \frac{q/e}{2\pi}\int_0^{\infty}\frac{\omega}{\gamma^2}\left(\psi_1(\gamma)-\psi_2(\gamma)\right)e^{i\omega x/\gamma}d\gamma. \end{aligned} \tag{21.27}$$

The inverse Laplace transform is easily performed by noting that the inverse transform of $-i\omega e^{i\omega x/u}$ is $\delta'(\frac{x}{\gamma}-t)$, where $\delta'(x)$ is the derivative of Dirac's delta function. We then get

$$\psi(x,t) = \frac{-iq/e}{2\pi}\int_0^{\infty}\frac{1}{\gamma^2}\left(\psi_1(\gamma)-\psi_2(\gamma)\right)\delta'(\tfrac{x}{\gamma}-t)d\gamma. \tag{21.28}$$

To solve this equation we make the substitution $z = x/\gamma$ whereby

$$\psi(x,t) = \frac{iq/e}{2\pi x}\int_0^{\infty}\left(\psi_1(\tfrac{x}{z})-\psi_2(\tfrac{x}{z})\right)\delta'(z-t)\mathrm{d}z.$$

After integration by parts, we obtain

$$\begin{aligned}\psi(x,t) &= \frac{-iq/e}{2\pi}\int_0^{\infty}\frac{1}{z^2}\left(\psi_1'(\tfrac{x}{z})-\psi_2'(\tfrac{x}{z})\right)\delta(x-z)dz\\ &= \frac{-iq/e}{2\pi}\left(\psi_1'(\tfrac{x}{t})-\psi_2'(\tfrac{x}{t})\right)\frac{1}{t^2}. \end{aligned} \tag{21.29}$$

Using the notation $\int_1$ and $\int_2$ as in (21.26), we make use of the fact that $\int_1 f_0'(\gamma)/(x-\gamma)d\gamma = (\int_2 f_0'(\gamma)/(x-\gamma)d\gamma)^*$ for x real, with the asterisk denoting complex conjugate. We then find with the usual integration paths

$$\psi(x,t) = \frac{q/e}{\pi t^2}h'\left(\frac{x}{t}\right), \tag{21.30}$$

where

$$
\begin{aligned}
h(\xi) &= \Im\left\{\left(1-\frac{\kappa T_e}{M}P\int_{-\infty}^{\infty}\frac{f_0'(u)}{u-\xi}du - i\pi\frac{\kappa T_e}{M}f_0'(\xi)\right)^{-1}\right\} \\
&= \pi\frac{\frac{\kappa T_e}{M}f_0'(u)}{\left(1-\frac{\kappa T_e}{M}P\int_{-\infty}^{\infty}\frac{f_0'(u)}{u-\xi}du\right)^2+\left(\pi\frac{\kappa T_e}{M}f_0'(\xi)\right)^2}.
\end{aligned}
\tag{21.31}
$$

If we assume that the unperturbed ion distribution is Maxwellian, we can express the response in terms of the plasma dispersion function (Fried & Conte 1961); see also Appendix C. We then have

$$
f_0(u)=\sqrt{\frac{M}{2\pi\kappa T_i}}e^{-\frac{Mu^2}{2\kappa T_i}} \quad \text{and} \quad f_0'(u)=-\frac{1}{\sqrt{2\pi}}\left(\frac{M}{\kappa T_i}\right)^{3/2}ue^{-\frac{Mu^2}{2\kappa T_i}}.
$$

The integral in the denominator of h can now be evaluated. By making the substitutions $y=u\sqrt{M/2\kappa T_i}=u/u_{ti}$ and $\gamma=\xi\sqrt{M/2\kappa T_i}$ with u_{ti} being the ion thermal speed we get

$$
\begin{aligned}
\int_{-\infty}^{\infty}\frac{f_0'(u)}{u-\xi}du &= \sqrt{\frac{M}{2\pi\kappa T_i}}\left(-\frac{M}{\kappa T_i}\right)\int_{-\infty}^{\infty}\frac{ue^{-\frac{Mu^2}{2\kappa T_i}}}{u-\xi}du \\
&= \frac{1}{\sqrt{\pi}}\left(-\frac{M}{\kappa T_i}\right)\int_{-\infty}^{\infty}\frac{ye^{-y^2}}{y-\gamma}dy \\
&= \frac{1}{\sqrt{\pi}}\left(-\frac{M}{\kappa T_i}\right)\left(\sqrt{\pi}+\gamma\int_{-\infty}^{\infty}\frac{e^{-y^2}}{y-\gamma}dy\right) \\
&= -\frac{M}{\kappa T_i}(1+\gamma Z(\gamma))=\frac{M}{2\kappa T_i}Z'(\gamma),
\end{aligned}
\tag{21.32}
$$

where Z is the plasma dispersion function. Then we get

$$
h(\gamma)=\frac{1}{1-\frac{\kappa T_e}{M}\frac{M}{2\kappa T_i}Z'(\gamma)}=\frac{\frac{1}{2}Q\Im\{Z'(\gamma)\}}{\left(1-\frac{1}{2}Q\Re\{Z'(\gamma)\}\right)^2+\left(\frac{1}{2}Q\Im\{Z'(\gamma)\}\right)^2},
\tag{21.33}
$$

where $Q\equiv T_e/T_i$.

Now we have the plasma response to a perturbation of the form $q\delta(x)\delta(t)$. In the following this will be denoted ψ_δ. The response to a charge moving along a path given by $x=U_0t$ is found by considering the moving charge as a continuous succession of delta functions. Analytically, this can be written as

$$
\begin{aligned}
\psi^{(1)}(x,t) &= \int_0^t\int_{-\infty}^{\infty}\delta(x'-U_0t')\psi_\delta(x-x',t-t')\mathrm{d}x'\mathrm{d}t' \\
&= \frac{-q/e}{\pi}\int_0^t\frac{1}{(t-t')^2}h'\left(\frac{x-U_0t'}{t-t'}\right)dt'.
\end{aligned}
$$

By substituting $\chi=(x-U_0t')/(t-t')$, the above integral becomes

$$
\frac{-q/e}{\pi}\int_{x/t}^{\infty}\frac{1}{x-U_0t}h'(\chi)\mathrm{d}\chi=\frac{q/e}{\pi(x-U_0t)}h\left(\frac{x}{t}\right).
\tag{21.34}
$$

The singularity at $x=U_0t$ is due to our assumption of quasi-neutrality, i.e., the results are not valid for distances shorter than the Debye length as measured from the moving particle.

Note that implicit in the analysis is a result for the perturbation of the ion velocity distribution. This can be obtained from (21.17) since we have found the space-time evolution of the potential. Considering, for instance, ψ_δ, we have

$$\frac{\partial f}{\partial t} + u\frac{\partial f}{\partial x} = \frac{qn_0}{M} f_0'(u)\, \frac{1}{\pi t^3} h''\left(\frac{x}{t}\right).$$

We can anticipate that the solution for $f(x,u,t)$ is also self similar, i.e., $f = G(u)F_u(x/t)/t^\alpha$ (Gurevich et al. 1965, Gurevich et al. 1968). We determine α by insertion in the equation and find $\alpha = 2$. We have then $f(x,u,t) = (qn_0/M)f_0'(u)F_u(x/t)/t^2$, where the function $F_u(\xi)$ is a solution of

$$(u-\xi)\frac{dF_u(\xi)}{d\xi} - F_u(\xi) = \frac{1}{\pi}h''(\xi).$$

We find solutions $F_u(\xi)$ for each choice of u, so we introduced the index u on $F_u(\xi)$. We find that $F_u(\xi)$ is singular along $u = \xi$, noting also that the equation has a homogeneous solution $a/(u-\xi)$. We have in general

$$F_u(\xi) = \frac{1}{u-\xi}\left(a + \frac{1}{\pi}\int_b^\xi h''(\gamma)d\gamma\right) = \frac{h'(\xi)}{\pi(u-\xi)},$$

with two constants a and b, where we took $a = 0$ to have only the inhomogeneous solution and b is determined by normalization of $f(x,u,t)$. The important observation here is that we have abstained from the freedom to choose the initial velocity distribution and have it developing self-consistently at application of the external charge q. Of course, the analysis could be expanded to include *both* external charges and initial conditions for the velocity distribution, but this will become rather messy!

- **Exercise:** The present analysis gives the potential response to a charge impulse, $q\delta(x)\delta(t)$, and also the response to charge q moving with constant velocity U_0, for both cases in one spatial dimension. Give the response to an oscillating charge $q\delta(x)\cos(\omega_0 t)$. There is no need to write out the result in any great detail.

21.3.2 Calculations in three spatial dimensions

The plasma response in three spatial dimensions can now be calculated in a similar way. Here we will make the postulate that the response to a perturbation $q\delta(\mathbf{r})\delta(t)$ has the following form:

$$\psi_\delta^{(3)}(r,t) = \frac{q/e}{t^n}\mathcal{D}\left(\frac{r}{t^m}\right), \tag{21.35}$$

where $\mathcal{D}$ is a function which is unspecified, for the time being.

The response in one dimension can be calculated from the postulated expression (21.35) as

$$\begin{aligned}\psi_\delta^{(1)}(x,t) &= \frac{q/e}{t^n}\int\!\!\int\!\!\int_{-\infty}^{\infty}\delta(x')\mathcal{D}\left(\frac{\sqrt{(x-x')^2+(y-y')^2+(z-z')^2}}{t^m}\right)dx'dy'dz' \\ &= \frac{q/e}{t^n}\int\!\!\int_{-\infty}^{\infty}\mathcal{D}\left(\frac{\sqrt{x^2+(y-y')^2+(z-z')^2}}{t^m}\right)dy'dz' \\ &= \frac{2\pi q/e}{t^n}\int_0^\infty \mathcal{D}\left(\frac{\sqrt{x^2+\xi^2}}{t^m}\right)\xi d\xi = \frac{\pi q/e}{t^n}\int_{x^2}^\infty \mathcal{D}\left(\frac{\sqrt{\gamma}}{t^m}\right)d\gamma,\end{aligned}$$

where first $\xi^2 = (y-y')^2 + (z-z')^2$ and later $\gamma = \xi^2 + x^2$ was introduced. This is now the one-dimensional response to a delta function in time and space. This result *must*, however, be equal to the result obtained in (21.30). We therefore have

$$\frac{\pi q/e}{t^n}\int_{x^2}^\infty \mathcal{D}\left(\frac{\sqrt{\gamma}}{t^m}\right)d\gamma = \frac{q}{\pi t^2}h'\left(\frac{x}{t}\right).$$

Differentiating this with respect to x we get

$$\frac{-2\pi q/e}{t^n} x \mathcal{D}\left(\frac{x}{t^m}\right) = \frac{q/e}{\pi t^2} h''\left(\frac{x}{t}\right)\frac{1}{t} \quad \text{giving} \quad \frac{1}{t^n}\mathcal{D}\left(\frac{x}{t^m}\right) = -\frac{1}{2\pi^2}\frac{1}{t^4}\left[\frac{t}{x}h''\left(\frac{x}{t}\right)\right]. \tag{21.36}$$

From this we see that we must have $n = 4$ and $m = 1$ and therefore

$$\psi_\delta^{(3)}(r,t) = -\frac{q/e}{2\pi^2 t^4}\left(\frac{t}{r}\right) h''\left(\frac{r}{t}\right). \tag{21.37}$$

It is implicit in the arguments that the distribution function $f_0(u)$ is isotropic, i.e., it is the same no matter what direction we use as a reference. Note the *self similarity* of this result; it depends on the ratio r/t rather than r and t separately, and is *scaled* by the factor t^{-4}.

The response to a moving charge in three dimensions can now be calculated in the same way as for the one-dimensional case. The charge is assumed to be moving in the positive z-axis with speed U_0. This yields

$$\psi(\mathbf{r},t) = -\frac{q/e}{2\pi^2}\int_0^t \frac{1}{(t-t')^4}\frac{t-t'}{\sqrt{x^2+y^2+(z-U_0t')^2}} h''\left(\frac{\sqrt{x^2+y^2+(z-U_0t')^2}}{t-t'}\right) dt'. \tag{21.38}$$

This expression can easily be evaluated numerically.

For a point charge which has been moving with constant velocity for a long time we can simplify (21.38) in the rest frame of the particle, taking the origin to be at U_0t, to give the result

$$\psi(r,z) = -\frac{q/e}{2\pi^2}\int_0^\infty \frac{1}{\gamma^4}\frac{\gamma}{\sqrt{r^2+(z-U_0\gamma)^2}} h''\left(\frac{\sqrt{r^2+(z-U_0\gamma)^2}}{\gamma}\right) d\gamma, \tag{21.39}$$

with $r^2 = x^2+y^2$, where we used that $h''(\gamma)$ decays for large values of the argument. Evidently, as expected, the disturbance induced by the moving charge is propagating with velocity U_0, having a constant spatial shape. Again we note that the solution for the electrostatic potential ϕ is singular at the particle position $(x,y,z) = (0,0,0)$ since the integral in (21.39) diverges there. This divergence is due to our assumption of quasi-neutrality.

Note that the present results are obtained without assumptions on the ratio T_i/T_e. By the assumptions of Boltzmann distributed electrons and quasi-neutrality we have $e\phi(\mathbf{r},t)/\kappa T_e = n(\mathbf{r},t)/n_0$, except at the position of the moving charge. This implies that our results apply for the relative plasma density perturbation as well.

- **Exercise:** Try to obtain the *two dimensional* expression for a moving point charge, i.e. the result corresponding to a moving thin charged wire. This does not represent any physically relevant problem (maybe with the exception of some numerical simulations), but the analysis is instructive.

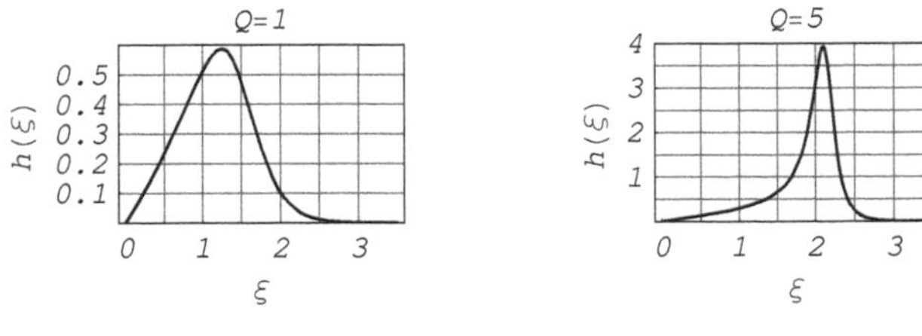

FIGURE 21.3: *The function $h(\xi)$ evaluated for two temperature ratios, $Q = 1$ and $Q = 5$.*

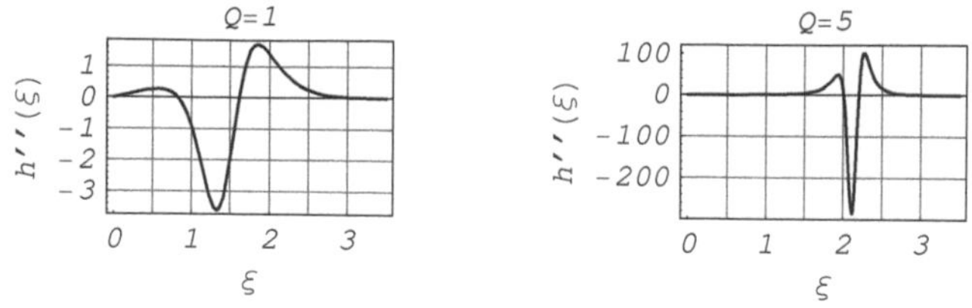

FIGURE 21.4: *The function $h''(\xi)$ evaluated for two temperature ratios, $Q = 1$ and $Q = 5$.*

21.3.2.1 Numerical results

The starting point for the calculation of the plasma response is the function $h(\xi)$. The only parameter which enters this function is the temperature ratio Q. In Fig. 21.3 the function $h(\xi)$ is shown as a function of the independent variable ξ. Figure 21.4 shows its second derivative for two different values of Q. Note how $h(\xi)$ steepens and narrows with increasing Q.

The function $h''(\xi)$ is used for calculating the plasma response to a point charge. Here we would expect the response to depend qualitatively on the speed of the injected charge and the temperature ratio. More precisely, we should see radically different behavior depending on whether the speed of the charge is below or above the ion sound speed, C_s. This is given by $C_s = \sqrt{(\gamma_i \kappa T_i + \gamma_e \kappa T_e)/M} = \sqrt{(\gamma_i + Q\gamma_e)\kappa T_i/M}$. Here, $\gamma_{i,e}$ denotes the ratio of specific heats for the ions and electrons, respectively. At relevant U_0's the electrons can be taken as isothermal while the ion response is adiabatic, i.e., $\gamma_e = 1$ and $\gamma_i = 3$. The calculated results of the response to a point charge are shown in Figs. 21.5–21.7.

We should like to emphasize one aspect of the self similarity: as already mentioned, due to the assumption of quasi-neutrality the Debye-length does not enter the problem and we have no natural length scale here. In Figs. 21.5–21.7, we use computational units for the axes. By reference to (21.39), we have that a change in units on the axes by a factor, say, β allows a change of integration variable to $\gamma' = \beta\gamma$, which ultimately results in a simple re-scaling of the response amplitude by a factor β^3. For a linear theory, as the present one, this amplitude scaling is immaterial, and only the spatial shape of the radiation pattern is relevant. In Fig. 21.5 we see the plasma response (density perturbation) to a positive point charge introduced at $t = 0$ at six different times after the injection. Here $U_0/u_{ti} = 3.5$, so the motion is supersonic. Note the "wave-fronts" propagating at an angle away from the path of the moving charge. Note also the negative perturbation which follows the positive perturbation, giving overall density conservation. The figures demonstrate the time-evolution of the polarization of the plasma as a response to the point charge that was introduced at $t = 0$.

In Fig. 21.6 is shown the response for a subsonic injection speed, $U_0 < C_s$. At subsonic speeds a semicircular perturbation propagates from origo ahead of the moving charge and the perturbation around the injected charge is more or less circular (less at increasing speeds). At supersonic speeds two almost plane "wave-fronts" appear (Guio & Pécseli 2003, Guio et al. 2008). These must travel at C_s and so must form an angle Θ with the normal to the trajectory of the charge given by $\cos\Theta = C_s/U_0$. This is analogous to the *Čerenkov radiation* known from high energy physics.

The plasma response depends strongly on the temperature ratio, Q. This is seen in Fig. 21.7. At low values of Q the perturbation is quickly damped while at high Q the ripples created by the perturbation propagate with hardly any attenuation. This is consistent with the fact that the Landau damping is strongest for low frequency acoustic plasma waves when T_e/T_i is moderate or small.

In case we want to calculate the linear plasma response to a distribution of moving charges, this can be done by simple superposition, within the present linear model.

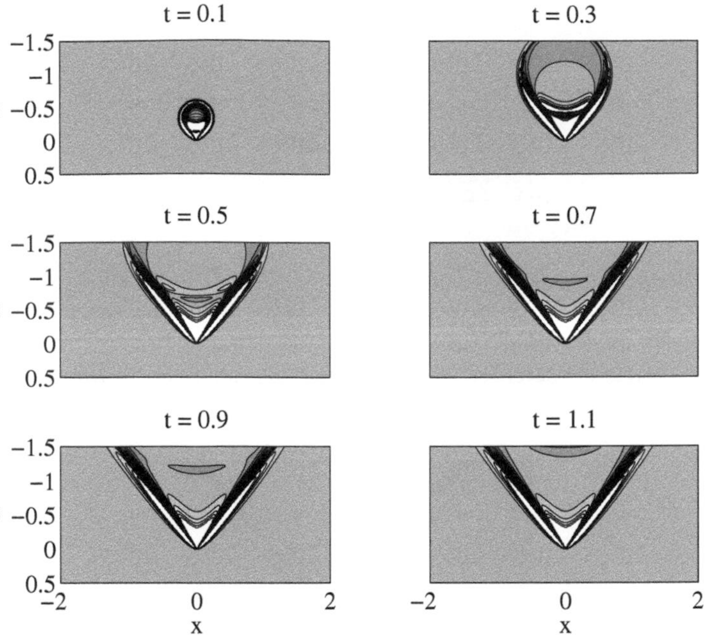

FIGURE 21.5: *The time evolution of the density response to a point charge injected at* $t = 0$. *The parameters are* $U_0/u_{ti} = 3.5$ *and* $Q = 5$. *Areas of rarefaction densities (with respect to the background) are gray or white; enhancements are black. With the assumption of quasi-neutrality and Boltzmann distributed electrons, there is no difference (apart from a constant) between plasma density and electrostatic potential. The results are shown in the rest frame of the particle. Note that the figure has been "cut off" near the origin in order to display the "wings" of the variation more clearly. The present figure is kindly provided by Dr. Patrick Guio. Related figures are found elsewhere, and here reprinted with permission from Guio and Pécseli (2003), copyright 2003, American Institute of Physics.*

21.4 A boundary value problem for wave excitation

The standard analysis refers to *initial value problems*, where a wave is excited by imposing a condition of a harmonic spatial variation at $t = 0$. In an experiment, one would prefer a *boundary value problem*, where a harmonic oscillation is imposed at a reference position, say $x = 0$, for all times. Excitation of waves in a Q-machine by a semi-transparent grid with variable bias is one such example. Unfortunately, this problem is not well posed. We consider the linearized ion Vlasov equation, here written for one spatial dimension, assuming quasi-neutrality

$$\frac{\partial f_1}{\partial t} + u\frac{\partial f_1}{\partial x} = C_s^2 f_0'(u)\frac{\partial n_1}{\partial x}, \tag{21.40}$$

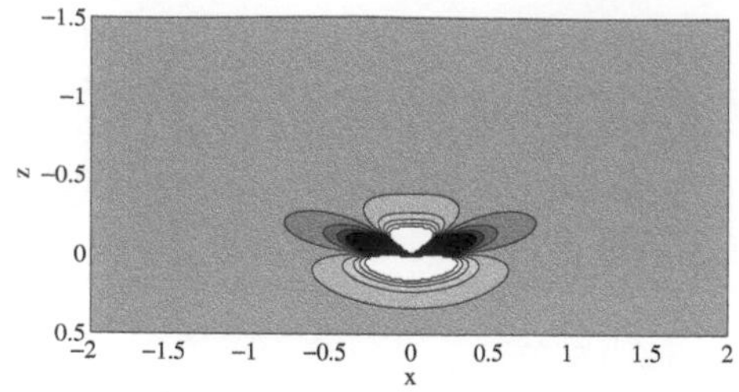

FIGURE 21.6: *Plasma response to a point charge for $Q = 5$ (implying $C_s \simeq 2.83u_{ti}$) and $U_0 = 2.0u_{ti}$. For this case the velocity of the test charge is below the sound speed. The present figure is kindly provided by Dr. Patrick Guio. Related figures are found elsewhere, and here reprinted with permission from Guio and Pécseli (2003). Copyright 2003, American Institute of Physics.*

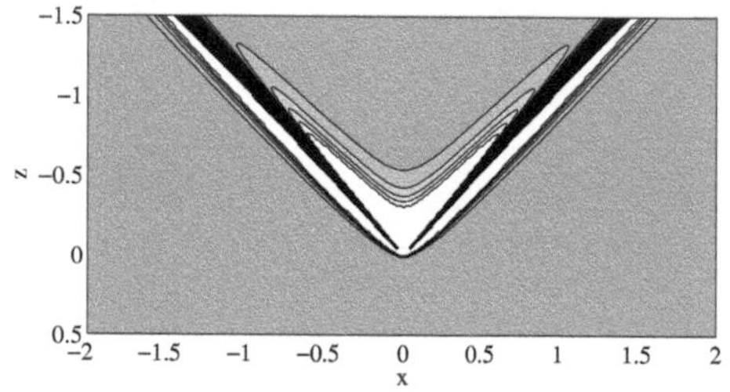

FIGURE 21.7: *Plasma response to a point charge for $Q = 7.5$. We have $U_0 = 3.5u_{ti}$. For this case the velocity of the test charge is above the sound speed. Details are otherwise as in Fig. 21.5. The present figure is kindly provided by Dr. Patrick Guio. Related figures are found elsewhere, and here reprinted with permission from Guio and Pécseli (2003). Copyright 2003, American Institute of Physics.*

with $C_s^2 \equiv \kappa T_e/M$, and $n_1(x,t) = \int_{-\infty}^{\infty} f_1 du$. For a proposed harmonic oscillation, $f_1(x=0,u,t) = g(u)\exp(-i\omega t)$, we readily obtain by integration along characteristics

$$\begin{aligned} f_1(x,u,t) &= \exp\left(-i\omega\left(t-\frac{x}{u}\right)\right)\left(g(u)+\frac{C_s^2}{u}f_0'(u)\left[\int_0^x \exp\left(-i\frac{\omega\gamma}{u}\right)\frac{\partial n_1(\gamma,t)}{\partial\gamma}d\gamma\right]_{u>0}\right. \\ &\left.+\frac{C_s^2}{u}f_0'(u)\left[\int_\infty^x \exp\left(-i\frac{\omega\gamma}{u}\right)\frac{\partial n_1(\gamma,t)}{\partial\gamma}d\gamma\right]_{u<0}\right). \qquad (21.41) \end{aligned}$$

The first integral represents particles that originate from $x = 0$ arriving at position x with the trajectories perturbed by the wave field, The second integral represents those particles which arrive from the boundary at $x \to \infty$ (and therefore have negative velocities) and also with trajectories perturbed by the wave. It could be argued that with this boundary being infinitely far away, its contribution could be ignored, since it takes the particles infinite time to reach $x = 0$. Alas, the argument is in error, since we assumed the applied signal to be an ideal harmonic oscillation, and in that case the signal is present for all times, also at $t \to -\infty$. The expression (21.41) satisfies the ion Vlasov equation (21.40) but not the proposed boundary conditions at $x = 0$ and $x = \infty$.

The problem is that the expression (21.41) is *not* consistent: we pretended to know $f_1(x=0,u,t)$, but the second term in (21.41) gives a contribution at $x = 0$ which in general is different from the one assumed. Even worse: we do not know this contribution a priori. In a single ended operation of a Q-machine, the problem is not as severe, since for a nontrivial parameter regime we might here assume that no ions have negative velocities, and $f_0'(u) = 0$ for $u < 0$, but in general we have to discard the problem as ill posed. One just *might* reformulate the problem of wave excitation by a related test charge problem, but at least in the Q-machine, this will not be a proper way to look at the mechanism for wave excitation. As already argued, in that case, it is the perturbation of the ion distribution function, and the ion density in particular, caused by grid absorption which is believed to be the significant effect.

We might, however, at least formally, get around the inherent problems in the boundary value problem formulation by choosing a slightly different point of view. One might ask: can we prescribe an *initial* value problem such that the time evolution for $t > 0$ fulfills the required condition of a harmonic oscillation at the boundary? It turns out that in terms of, for instance, the Case-van Kampen formalism, with (21.40) this problem can indeed be solved as

$$f(k,u,t) = \exp(-i\omega t)\int_{-\infty}^{\infty} C(\nu)\exp(i\omega x/\nu) f_k(\nu,u)d\nu,$$

with normal modes

$$f_k(\nu,u) = C_s^2 F_i'(u)\frac{P}{u-\nu} + \left(1 - C_s^2 P\int_{-\infty}^{\infty}\frac{F'}{u-\nu}du\right)\delta(u-\nu),$$

for a stable plasma, assuming quasi-neutrality and Boltzmann distributed electrons, as before. We defined $C_s^2 \equiv \kappa T_e/M$.

The condition on the boundary is here consistent with a distribution $g(u) = \int_{-\infty}^{\infty} C(\nu) f_k(\nu,u)d\nu$ on the boundary at $x = 0$. Note that the *initial* distribution given by $f(k,u,t=0)$ is not separable in two functions depending on position and velocity, respectively. Assuming $g(u)$ to be given, we have

$$C(u) = \frac{g_+(u)}{1 - i2\pi C_s^2 F_{i+}'} + \frac{g_-(u)}{1 + i2\pi C_s^2 F_{i-}'}$$

with

$$g_\pm(u) \equiv \frac{1}{2}g(u) \pm i\frac{1}{2\pi}P\int_{-\infty}^{\infty}\frac{g(\nu)}{u-\nu}d\nu$$

and similar definitions for $F_{i\pm}(u)$. This result formally solves our problem, but is of little practical value. From an experimental point of view, other more practical conditions must be sought.

The formulation of the boundary value problem discussed here is closely related to the *signaling problem* (Whitham 1974), which is, however, more general by being applied to nonlinear wave equations as well.

In the early days of the studies of Landau damping of ion sound waves a quite different question was discussed, at times in somewhat heated debates (the author was a PhD student at that time, and witnessed the intensity of these discussions with some surprise!): when $T_e \approx T_e$ as in Q-machines, the right-hand term in (21.40) is modest and it might be argued that it can as well be ignored, so that only the free streaming contribution discussed in the example in the beginning of Chapter 20 will remain, see also the discussion in Section 21.2. This contribution will appear as a damping, but a rather trivial one compared to the one the experiment purports to demonstrate (Hirshfield & Jacob 1968). It was argued that an unambiguous demonstration of plasma kinetic phenomena had to address the velocity distribution of the particles, which is after all the basic constituent of the plasma kinetic theory, where, in some sense, the Landau damping is just a bi-product (Andersen et al. 1971*b*). Much can be said in favor of this argument, and the following section addresses such an experiment.

21.5 A realizable initial value problem

In many cases an initial value problem can be formulated only in a formal way; it is rather difficult to imagine how to set up an initially plane wave all over space. There are of course several realizable and nontrivial examples, which are interesting for experimental tests of the results. Here we consider the one-dimensional case, where the initial perturbation consists of a step-function, so that the ion velocity distribution at $t = 0$ is $n_0 f_0(u) + \Delta n\, g(u)$ for $x < 0$ and $n_0 f_0(u)$ for $x > 0$, with $\Delta n \ll n_0$ and $\int g(u)du = 1$. The perturbation $g(u)$ of the velocity distribution function *can* differ from $f_0(u)$.

21.5.1 Introductory comments on self similar solutions

A complete understanding of the ion dynamics requires information of the velocity distribution $f(x,u,t)$ for ions with velocities u, at positions x at times t, as described by the ion Vlasov equation for ions with mass M and charge e

$$\frac{\partial}{\partial t} f + u\frac{\partial}{\partial x} f - \frac{e}{M}\frac{\partial \phi}{\partial x}\frac{\partial}{\partial u} f = 0, \tag{21.42}$$

where $E \equiv -\partial\phi/\partial x$ is the electric field and ϕ is the electrostatic potential. The ion density is $n_i = \int_{-\infty}^{\infty} f\, du$. For the relevant low-frequency dynamics we assume again the electrons to be Boltzmann distributed at all times, so that the electron dynamics are accounted for by the relation $n_e = n_0 \exp(e\phi/\kappa T_e)$ where T_e is the electron temperature.

For times exceeding the ion plasma period, $t \gg 2\pi/\Omega_{pi}$, with Ω_{pi} being the ion plasma frequency, and spatial scale-lengths exceeding the Debye length, we can assume the plasma dynamics to be quasi-neutral, $n_e \approx n_i \equiv n$. In this limit the Debye length no longer enters the problem, and we have no characteristic length in the dynamic equations. In the quasi neutral limit also the ion plasma frequency is absent, so there are no characteristic time scales present either. All we have is a velocity, the ion sound speed $C_s = \sqrt{\kappa T_e/M}$ derived from the electron temperature. For such cases, the characteristic scales are determined solely by initial or boundary conditions. For some cases the initial condition has no length scale. Heaviside's step function or the δ-function are examples of such functions. Under these conditions we can expect the plasma dynamics to evolve self similarly, in depending functionally on the ratio $\zeta \equiv x/t$ rather than on x and t independently (Gurevich et al. 1965, Gurevich et al. 1968, Denavit 1979, Singh & Schunk 1982, Nakagawa 2013), giving $\partial/\partial t \to -t^{-2}x\partial/\partial\zeta = -t^{-1}\zeta\partial/\partial\zeta$ and $\partial/\partial x \to t^{-1}\partial/\partial\zeta$, as already found in Section 21.3.1. In that problem the δ-function was introduced by a point-like moving charge.

We here assume an initial step-like density perturbation. After a little algebra and by introducing the variable $\zeta \equiv x/t$ we can reduce (21.42) to

$$(u-\zeta)\frac{\partial}{\partial\zeta} f - \frac{\kappa T_e}{M}\frac{1}{n}\frac{\partial n}{\partial\zeta}\frac{\partial}{\partial u} f = 0, \tag{21.43}$$

where we now have $f = f(\zeta,u)$ and $n = n(\zeta) \equiv \int_{-\infty}^{\infty} f(\zeta,u)du$. The collective electric field was determined by the assumption of Boltzmann distributed electrons, giving $n = n_0 \exp(e\phi/\kappa T_e)$. From (21.43) we have the implicit relation for the ion velocity distribution in the form

$$f(u,x,t) = \begin{cases} G_-(u) + C_s^2 \displaystyle\int_{-\infty}^{x/t} \frac{1}{n}\frac{\partial n}{\partial\zeta}\frac{\partial f/\partial u}{u-\zeta} d\zeta & \text{for } x/t < u, \\[2ex] G_+(u) - C_s^2 \displaystyle\int_{x/t}^{\infty} \frac{1}{n}\frac{\partial n}{\partial\zeta}\frac{\partial f/\partial u}{u-\zeta} d\zeta & \text{for } x/t > u, \end{cases} \tag{21.44}$$

where $G_-(u) \equiv f(x \to -\infty, u)$, $G_+(u) \equiv f(x \to +\infty, u)$. With the present assumptions we have $G_-(u) = (n_0 + \Delta n) f_0(u)$ and $G_+(u) = n_0 f_0(u)$, where we for simplicity introduced the unperturbed velocity distribution in the form $n_0 f_0(u)$ with the normalization $\int_{-\infty}^{\infty} f_0(u)du = 1$. This form turns out to be convenient when we linearize the equations.

21.5.2 Solution of a linearized initial value problem

In this section we shall demonstrate how a self similar solution develops naturally for a case where the initial condition has no characteristic length scale. The problem is interesting also by allowing analytical solutions of a kinetic plasma problem with collisions. It could be argued that we might then take the full result and then let the collision frequency vanish to obtain the collisionless result. The latter analysis is however much simpler than the full problem, so it is an advantage to take that separately, leaving the full problem with collisions to later.

We linearize the ion Vlasov equation written in one spatial dimension and introduce the linearized expression for the Boltzmann distributed electrons. After Laplace transform in time using s as the transformed variable (we could as well use $-i\omega$ instead as in (21.19)), with a Fourier transform in space. Assuming quasi-neutrality, the basic equations can be combined into

$$(s+iku)f_1(k,u,s) - iC_s^2 f_0'(u)k\int_{-\infty}^{\infty} f_1(k,u,s)du = f_1(k,u,t=0) \equiv \Delta n g(u)\frac{i}{k}, \tag{21.45}$$

where i/k in the initial condition on the right-hand side originates from the Fourier transform of the initial step-function. For the perturbation of the ion velocity distribution we have $\int_{-\infty}^{\infty} g(u)\,du = 1$. By (21.45) we find the result for the ion density

$$n_1(k,s) = \int_{-\infty}^{\infty} f_1(k,u,s)du = \frac{\Delta n}{k^2}\frac{\displaystyle\int_{-\infty}^{\infty}\frac{g(u)}{u-is/k}du}{\displaystyle 1-C_s^2\int_{-\infty}^{\infty}\frac{f_0(u)}{u-is/k}du}, \tag{21.46}$$

with complex s, so there are no singularities on the integration paths. The self similarity of the problem is manifested through the s/k dependence of the variables in (21.46): this is the only combination containing s. Given $n_1(k,s)$ we can find $n_1(x,t) = \int_{-\infty}^{\infty} f_1(k,u,s)du$ and then inserting the result in (21.45) to give $f_1(u,x,t)$.

We have

$$n_1(x,t) = \frac{i}{(2\pi)^2}\int_{-\infty}^{\infty}\int_{s_0-i\infty}^{s_0+i\infty} n_1(k,s)e^{ikx-+t}dsdk, \tag{21.47}$$

where s_0 is a complex constant placed to the right-hand side of all singularities in the integrand. The solution of (21.45), (21.46) and (21.47) follows the analysis of Section 21.3 also as far as the integration paths and their deformations are concerned; see also the summary given by Andersen et al. (1971*b*) and Michelsen and Pécseli (1973). We find that the space-time evolution of the problem is governed by an auxiliary function $h_0(\gamma)$ given as

$$h_0(\gamma) \equiv \Im\left\{\frac{\displaystyle P\int_{-\infty}^{\infty}\frac{g(u)}{u-\gamma}du + i\pi g(\gamma)}{\displaystyle 1-C_s^2 P\int_{-\infty}^{\infty}\frac{f_0'(u)}{u-\gamma}du - i\pi f_0'(\gamma)}\right\}. \tag{21.48}$$

We note that for the special case where $g(u) = f_0(u)$, the result (21.48) is closely related to the function h defined by (21.31).

The $h_0(\gamma)$-function is illustrated in Fig. 21.8. For simplicity we here set the two distributions equal, $f_0(u) = g(u)$, and approximated them by Maxwellians. With such models, the function $h_0(\gamma)$ can be expressed in terms of the plasma dispersion function (see Appendix C), just as $h(\gamma)$ in (21.31). Note the symmetry for the case shown here, $h_0(\gamma) = h_0(-\gamma)$.

The space-time evolution of the perturbation in ion density is obtained as

$$n_1(x,t) = \Delta n\int_{x/t}^{\infty} h_0(\gamma)d\gamma, \tag{21.49}$$

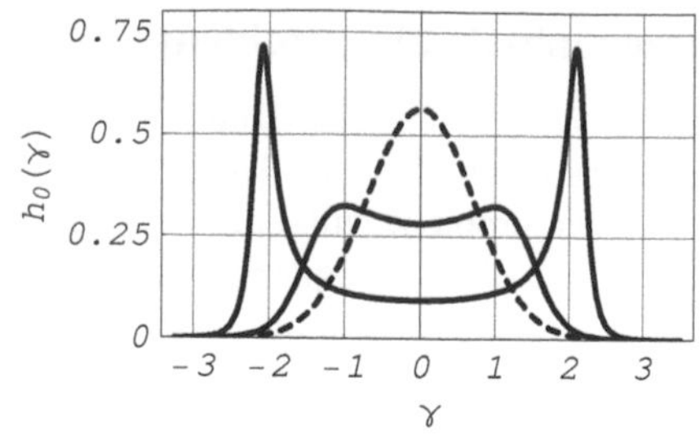

FIGURE 21.8: *Illustration of the $h_0(\gamma)$-function for the special case where $g(u) = f_0(u)$. We have results for the electron-ion temperature ratios $T_e/T_i = 0, 1$, and 5, shown with dotted, dashed, and full lines, respectively. For the present model, the value $T_e/T_i = 0$ will correspond to freely streaming ("ballistic") ions. The figure is adapted from Rekaa et al. (2013), with the permission of AIP Publishing.*

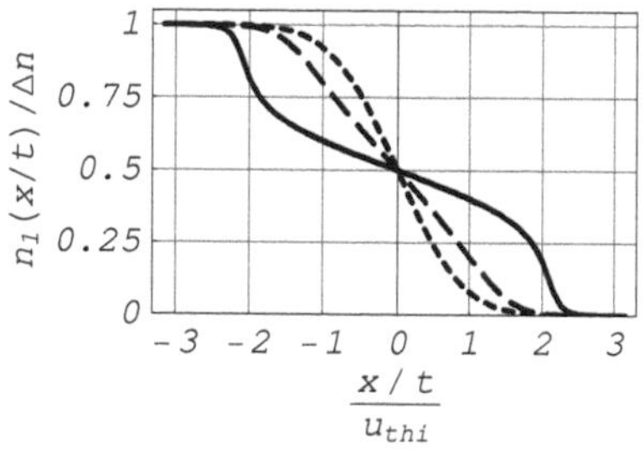

FIGURE 21.9: *Illustration of the self similar variation of the density perturbation for the temperature ratios $T_e/T_i = 0, 1$, and 5. Again, the value $T_e/T_i = 0$ will correspond to freely streaming ("ballistic") ions. See also Fig. 21.8. Note also the backward propagating rarefaction wave for $x/t < 0$. The figure is adapted from Rekaa et al. (2013), with the permission of AIP Publishing.*

with illustrations shown in Fig. 21.9. Due to the symmetry $h_0(\gamma) = h_0(-\gamma)$ we have the expansion and rarefaction waves to be anti-symmetric with respect to the point $\Delta n/2$ at $x = 0$.

The perturbation of the ion flux is is found from the ion continuity equation as

$$F_1(x,t) \equiv \int_{-\infty}^{\infty} u f_1(u,x,t)du = \Delta n \int_{x/t}^{\infty} \gamma h_0(\gamma)d\gamma. \tag{21.50}$$

The density as well as the ion flux evolves self similarly, just as the ion distribution function (Gurevich et al. 1965, Gurevich et al. 1968, Michelsen & Pécseli 1973).

With $\zeta \equiv x/t$, the linearized version of (21.43) gives

$$(u-\zeta)\frac{\partial}{\partial \zeta}f_1 - C_s^2 \frac{\partial n_1}{\partial \zeta} f_0'(u) = 0, \tag{21.51}$$

where we insert (21.49). Since we have $n_1(\zeta)$ already to be inserted in (21.51) we can find a quantity relating to the space-time varying ion kinetic energy density

$$\int_{-\infty}^{\infty} u^2 f_1(u,x,t)du = \Delta n \left(\int_{-\infty}^{\infty} u^2 g(u)du + \int_{-\infty}^{x/t} (C_s^2 - \gamma^2)h_0(\gamma)d\gamma \right), \tag{21.52}$$

found by multiplying (21.51) with u and integrating. A term containing $\int_{-\infty}^{\infty} u f_1 du$ is rewritten by use of (21.50). An integration constant is given by $\int_{-\infty}^{\infty} u^2 f_1(u,x,t)du \to \Delta n \int_{-\infty}^{\infty} u^2 g(u)du$ for $x/t \to -\infty$. Recall that $f_1(u,x,t)$ represents a deviation from $f_0(u)$ and can take negative values also.

The space-time varying local average ion velocity is given as the ratio of the ion flux and the ion density, which by use of (21.49) and (21.50) gives

$$U_1(x,t) \equiv \frac{F_1(x,t)}{n_1(x,t)} = \frac{\int_{x/t}^{\infty} \gamma h_0(\gamma)d\gamma}{\int_{x/t}^{\infty} h_0(\gamma)d\gamma}, \tag{21.53}$$

independent of Δn. For $x/t \to \infty$ we find that this velocity increases without limit (Rekaa et al. 2013), although the density of these fast particles becomes small since $n_1(x,t) = \Delta n \int_{x/t}^{\infty} h_0(\gamma)d\gamma \to 0$ here. It is interesting to note that the asymptotic limit of $U_1(\zeta \to \infty)$ is the same for all T_e/T_i, i.e. even in the absence of collective interaction (i.e. $T_e = 0$). The few particles experiencing the large initial electric fields are accelerated to high velocities, i.e. velocities exceeding the ion thermal velocity u_{The}. The present result is thus consistent with other studies (Crow et al. 1975) where it is argued that a small number of ions get accelerated to very high velocities in front of an expanding plasma pulse.

The result for the space-time variation of the perturbation of the ion velocity distribution function can be summarized as

$$f_1(u,x,t) = \begin{cases} \Delta n\, g(u) - C_s^2 \Delta n\, f_0'(u) \displaystyle\int_{-\infty}^{x/t} \frac{h_0(\gamma)}{u-\gamma} d\gamma, & \text{for } x/t < u, \\ C_s^2 \Delta n\, f_0'(u) \displaystyle\int_{x/t}^{\infty} \frac{h_0(\gamma)}{u-\gamma} d\gamma, & \text{for } x/t > u, \end{cases} \tag{21.54}$$

with $C_s^2 \equiv \kappa T_e/M$, see also (21.44) for a comparison. The distribution function $f_1(u,x,t)$ is not defined for the velocity $u = x/t$, which can here be seen as the resonant velocity. There are no singularities in the integrations in (21.54). The contribution $\Delta n g(u)$ for $x/t < u$ accounts for free streaming ions: the term that is left in case all collective interaction vanishes for $C_s \to 0$. We note again the self similarity of the result: the dependence is on the ratio x/t, and not on x and t explicitly.

Note that the present results are obtained without assumptions on the ratio T_i/T_e. They apply to Q-machine conditions as well as those met in, for instance, DP devices. For the Q-machine experiment to be discussed in Section 21.5.3, we have $T_e \approx T_i$, so the dispersive ripples at wavelengths $\lambda \sim \lambda_{De}$ are heavily Landau damped, and the restriction by the assumed quasi-neutrality is of minor importance. For larger temperature ratios, it is an advantage to retain Poisson's equation, which here introduces a characteristic length scale, λ_{De}. Also this analysis has been carried out (Mason 1970*a*, Mason 1971). In this case the self similarity of the solution is lost: self similarity is a property of problems without any characteristic length and time scales. In the quasi-neutral limit of kinetic ion-acoustic waves, we do not have either the ion plasma frequency or the Debye lengths as characteristic quantities (all we have is a velocity, C_s). It is important that the initial and/or boundary conditions must not have characteristic lengths or time scales either in order to obtain self similar solutions! Heaviside's step-function and the δ-function are examples of functions without length scales. The experiment of Michelsen and Pécseli (1973) had an initial condition with a step-like density variation.

The collective electric fields are stronger for larger temperature ratios: if we take $T_e/T_i = 1$ we see little difference as compared to the case with $T_e = 0$, indicating that dispersion by free streaming ions dominates the space-time plasma evolution for $T_e/T_i \leq 1$.

Ⓢ **Exercise:** Express the continuity equation in terms of a self similar variable $x/t \equiv \zeta$, and then demonstrate (21.50) when (21.49) is given. Ⓢ

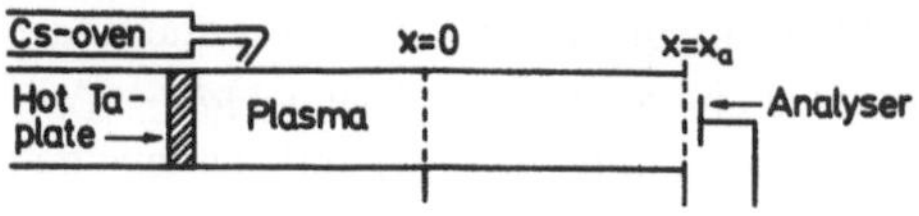

FIGURE 21.10: *Schematic diagram for the experiment of pulse propagation in a single ended Q-machine. The grid for perturbing the plasma density is placed at a reference position* $x = 0$*. The plasma column is terminated by a negatively biased metal plate, with an opening for the axially movable ion energy analyzer.*

ⓢ **Exercise:** Show an easy way to generalize the results of Section 21.5.2 to the case where we have two ion components with different masses and charges. ⓢ

ⓢ **Exercise:** Give the solution to the linear initial value problem where the ion density perturbation is $f_1(x,u,t=0) = \Delta n g(u)$ for $|x| < a/2$, and vanishes otherwise. Is the solution self similar? ⓢ

21.5.3 Experimental results

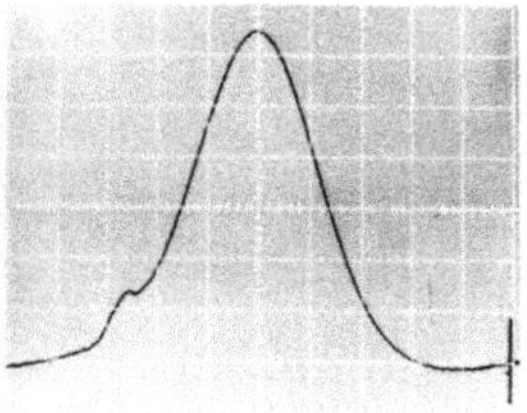

FIGURE 21.11: *Experimentally obtained ion energy velocity distribution function,* $f_0(u)$*. The horizontal scale is 0.5 V/div in energy units, kinetic + potential energy, not velocity. The zero point for ion energy is marked by a vertical line to the right. A small "bump" is the charge exchange ions, which mark the plasma potential where the ion kinetic energy is negligible, as discussed in Section 9.4. The plasma potential is here approximately* -3.9 *V.*

An experiment to test the results summarized in Section 21.5.2 was carried out in a Q-machine (Michelsen & Pécseli 1973), with the schematic diagram shown in Fig. 21.10. Details of the device, its construction, etc., are described by Motley (1975), see also Section 9.1. The present experiment is carried out in a single ended mode of operation, with a plasma column of 60 cm length and 3 cm diameter. The ion component is streaming from the hot plate toward a negatively biased terminating metal end plate. The average ion velocity (thermal and streaming velocities included) is of the order of 10^3 m/s. A semi-transparent metal grid (transparency $\approx 90\%$) is inserted perpendicular to the plasma column, between the hot plate and the end plate. The grid position is taken to be $x = 0$ for simplicity; see Fig. 21.10. Depending on the bias, this grid absorbs some of the ions. If the bias is at the plasma potential, the absorption is approximately the 10% given by the grid transparency. If the potential is negative, the grid attracts more of the ions and the absorption is larger. Note that the predominant effect of the grid is *not* described by a charge embedded in the plasma column.

If we have the grid at a negative bias, $\Phi_p - \Delta\Phi$, the plasma column has the unperturbed density $n_0 + \Delta n$ for $x < 0$, and n_0 for $x > 0$. If we at a reference time, here taken as $t = 0$, switch the grid potential to the plasma potential, Φ_p, the ion absorption is reduced, and the plasma can flow toward the end plate. The grid interacts with the ions, predominantly: the electron component follows along in a local isothermal Boltzmann equilibrium. The set-up described here can be seen as a realization of a "step-like" initial condition, with $n(x,0) = n_0 + \Delta n$ for $x < 0$ and $n(x,0) = n_0$ for $x > 0$. The strong magnetic field (approximately 0.6 T) makes the ion dynamics essentially one dimensional, at least for the present problem. The electron temperature was approximately 2200 K, as determined by the hot tantalum plate, while $T_e/T_i \approx 1.5$, and plasma densities were approximately 2×10^9 cm^{-3}.

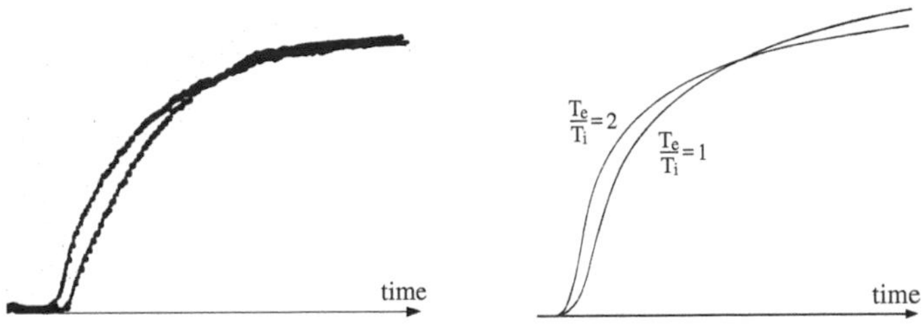

FIGURE 21.12: *Experimental and analytical results for the ion flux following an initial step-like discontinuity. The results are shown as functions of time, obtained at a fixed spatial position. Two temperature ratios, $T_e/T_i = 1$ and 2, were investigated. See also results given by Michelsen and Pécseli (1973).*

One virtue of the experiment is that the ion velocity distribution function can be obtained by direct measurement, as discussed in Section 9.4. First, the unperturbed distribution, $f_0(u)$, can be measured directly with the grid at potential Φ_p. By subtracting the measured distribution with the grid at potential $\Phi_p - \Delta\Phi$ and at Φ_p, also the distribution $g(u)$ associated with the perturbation can be determined. In principle *all* quantities needed for an analysis are thus amenable to measurements as shown by Michelsen and Pécseli (1973), and the problem is (at least in principle) free of "fitting parameters". An example for the measurement of $f_0(u)$ is shown in Fig. 21.11, where the plasma potential is marked by the charge exchange method invented by Andersen et al. (1971*a*); see also Section 9.1.

Collecting the ion flux at a downstream position, we obtain results like those shown in Fig. 21.12, where results from two electron/ion temperature ratios are shown, here as functions of t, with signals detected at a fixed spatial position. Theoretical results using (21.50) are shown as well (Michelsen & Pécseli 1973).

The time evolution of the perturbation of the ion velocity distribution is shown in Fig. 21.13, obtained with the analyzer placed at a distance of 6 cm from the grid that produces the step-like initial perturbation. The analysis was performed by use of a "wave-form eductor", an analog device with a series of condensers, which could successively be charged to a time-varying potential (this was in the "good old days", before digital electronics!). With a repetitive experiment, the noise would eventually "smooth out", although as we can see on Fig. 21.13, a small noise component remains. The vertical axes on the figures are in arbitrary units, the amplitude depending in part on how long a time the averaging was carried out. The velocities are selected by the bias on the energy analyzer with respect to the plasma potential: the resulting velocities are given in the figure caption. The main uncertainty in the velocity estimate originates from the uncertainty in the plasma potential, which enters the calculations of the ion velocity as obtained from an ion energy (kinetic + potential) with respect to ground potential. For large times, the ion distribution approaches $n_0 f_0(u) + \Delta n g(u)$. The results of Fig. 21.13 refer to the analyzer placed in one fixed position, obtaining the time evolution of the signal. The self similarity of the evolution is particularly simple to investigate with an oscilloscope, as demonstrated by Andersen et al. (1971*b*): the analyzer is moved from position x_1 to $x_2 = 2x_1$ and the time resolution of the oscilloscope is changed from Δt s/cm to $2\Delta t$ s/cm. According to the analytical expression (21.54), the picture on the oscilloscope screen should not change then. This was confirmed.

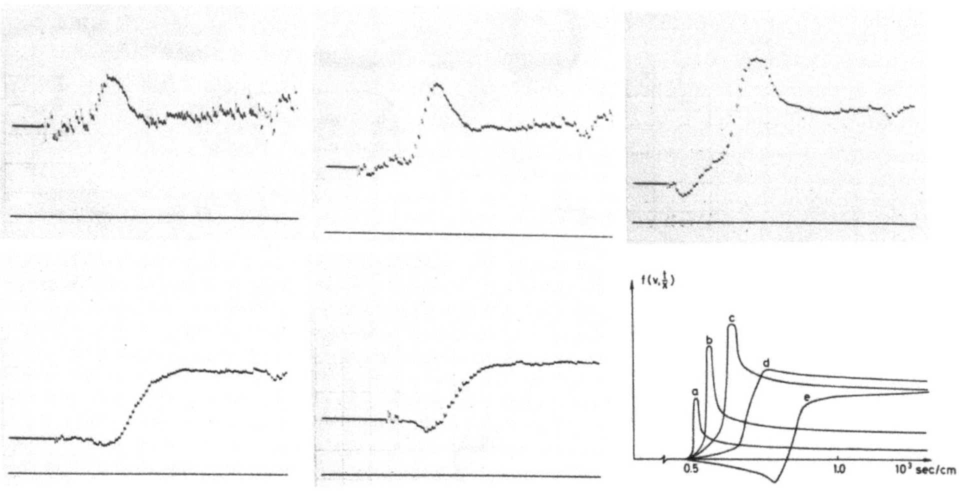

FIGURE 21.13: *Experimentally obtained time variation of the perturbed energy distribution function, measured at a fixed position. The horizontal scale has 10 μs/div. The analyzer is placed at a distance of 6 cm from the grid. The figure shows photographs of oscilloscope screens, directly reproduced. The selected velocities are a) 1930 m/s, b) 1780 m/s, c) 1590 m/s, e) 1420 m/s, and e) 1220 m/s, with an uncertainty of approximately ±2.5% for a) and b), while it is approximately ±5% for c) – e). Theoretical results are included in the last figure, with curves labeled for reference. The analysis there takes into account the finite resolution of the energy analyzer, so the singularities in (21.54) are smoothed out.*

Analytical results are also shown in Fig. 21.13. Here, the finite resolution of the ion analyzer is taken into account by integrating over velocities in an interval corresponding to the estimated resolution. The analytical results are in good agreement with the experimental observations. Since we know the temperature of the charge exchanged ions, it is possible to use the measured width of the charged exchange ion peak (see Fig. 21.11) to estimate the analyzer resolution. The distributions $f_0(u)$ and $g(u)$ are fitted with Maxwellians, allowing for average drift velocities.

The results are qualitatively simple to understand: for a monotonic density variation, the electric field is pointing in the direction *away* from the maximum density. A small fraction of the ions will thus be accelerated. This means that we will have a burst with slightly more ions with large velocities as compared to the unperturbed distribution, and slightly less ions with slow velocities. This is precisely what we find in Fig. 21.13, where we see a small "hump" of high velocity ions as the front arrives, and a depletion of the local distribution function for slow velocities.

Letting $T_e \to 0$ in the present quasi-neutral model, we recover the simple ion free streaming result discussed in Chapter 20. For the present model it is solely the electron pressure that sets up an ambipolar electric field, which subsequently sets the ions into motion. To recover the effect of ion pressure, we have to relax the quasi-neutrality approximation. Also, we note that the dispersive ripples expected for weakly dispersive ion acoustic perturbation are absent in the quasi-neutral approximation. For small temperature ratios, such as those in the present experiment, these ripples are heavily Landau damped anyhow, so this restriction is of little consequence. The results summarized in the present section received support also from numerical simulations (Rekaa et al. 2013).

Analytical results for the fully nonlinear problem are available (Gurevich et al. 1965, Gurevich et al. 1968). These studies also assume quasi-neutrality, and are implicitly restricted just as those of Michelsen and Pécseli (1973). The nonlinear effects are within the resolution of the energy analyzer with the experimental parameters relevant for Fig. 21.13.

The time evolution of an initially step-like perturbation in plasmas with large electron/ion temperature ratios was studied by Cohn and MacKenzie (1972) in a DP device; see also Section 9.2. The initial perturbation was there introduced by a collimated "flash" of light from a vacuum spark between molybdenum electrodes. The intense light ionizes a fraction of the neutral background in the device, producing a "slab" of additional plasma to the pre-existing uniform background plasma produced by the standard DP operation. Because of the large electron-ion temperature ratio, the dispersive ripples were clearly observable in this experiment. This information is lost in an analysis assuming quasi-neutrality from the outset, as in the foregoing summary. The space-time evolution of the ion distribution function was not investigated by Cohn and MacKenzie (1972).

21.6 Linearized model with ion-neutral collisions

To account for collisional interactions, the ion Vlasov equation is often modified by a simple collision term as $-\nu\,(f(u,x,t) - n_0 f_0(u))$ with normalized f_0, or its generalizations, see Section 19.1.1. In the present study we distinguish elastic ion collisions with neutrals, and charge exchange collisions. Both cases are studied numerically. For the latter processes we can obtain a closed analytical solution by introducing a physically realistic collision model in the ion Vlasov equation.

Analytical results using a general energy conserving Bhatnagar-Gross-Krook (BGK)-collision model in a kinetic description can be found in the literature (Mason 1968, Mason 1970*b*), but these results refer to discontinuities initiated by a moving piston. Here we outline analytical results directly applicable for the present initial value problem. For some special cases, the results become remarkably simple (Rekaa et al. 2013). A related problem was analyzed previously by Espedal (1971) using a Fokker-Planck collision operator, but in that case no closed analytical result could be obtained.

Since the following analysis is linear, the results for a pulse-like initial condition studied by, e.g., Andersen et al. (1971*b*) and Espedal (1971) can be found by a simple spatial differentiation. The step-like initial condition has, however, received most attention in the literature.

21.6.1 Analytical models for charge-exchange collisions

Consider a model for charge exchange collisions based on a simple form $-\nu(f_1 - n_1 f_0(u))$, see also Section 19.1.1, where $n_1 \equiv \int f_1 du$ and $\int f_0(u)du = 1$. The model is linear by construction and can be added to the right-hand side of the linearized ion Vlasov equation. It turns out to be relatively easy to retain a velocity dependence of the collision frequency $\nu = \nu(u)$ within the following analysis. The integral transforms found in the following can, however, then no longer be expressed in terms of known functions, and the practical value of the results will be limited. We therefore here take ν to be constant, and make the transformation $f_* \equiv f_1 \exp(\nu t)$, giving $n_* \equiv n_1 \exp(\nu t)$, to find

$$\frac{\partial}{\partial t} f_* + u\frac{\partial}{\partial x} f_* - \frac{T_e}{M}\frac{\partial n_*}{\partial x}\frac{\partial}{\partial u} f_0(u) - \nu f_0(u) n_* = 0. \tag{21.55}$$

A constant value for ν implies that the collisional cross sections vary as $\sigma \sim 1/u$ with varying velocity, and serves only as a solvable convenient model for charge exchange collisions. More generally, we have, however, that polarization forces ("Maxwellian molecules") give constant ν for a wide range of velocities (Trubnikov 1965, Bekefi 1966) so the model as such is not unphysical.

After a temporal Laplace transform (with complex s) and spatial Fourier transform (with real k) of (21.55) we have

$$(s + iku) f_*(k,u,s) - n_*(k,s)\left(ik\frac{T_e}{M} f_0'(u) + \nu f_0(u)\right) = f_*(k,u,t=0) \equiv i\frac{\Delta n}{k} g(u), \tag{21.56}$$

or

$$f_*(k,u,s) = i\frac{\Delta n}{k}\frac{g(u)}{s+iku} + n_*(k,s)\frac{ikC_s^2 f_0'(u) + \nu f_0(u)}{s+iku}, \tag{21.57}$$

giving

$$n_*(k,s) = \frac{i\dfrac{\Delta n}{k}\displaystyle\int_{-\infty}^{\infty}\frac{g(u)}{s+iku}du}{1 - \displaystyle\int_{-\infty}^{\infty}\frac{ikC_s^2 f_0'(u) + \nu f_0(u)}{s+iku}du}, \tag{21.58}$$

where $g(u)$ is the velocity distribution in the perturbation for $x < 0$. Integration limits are omitted for simplicity: they are $\int_{-\infty}^{\infty} du$ in all cases shown. The ensuing method of solution can be used for this general case, but a remarkable simplification results by taking $g(u) = f_0(u)$. In this case the analysis contains only one nontrivial complex function, which can then be related to the plasma dispersion function for a Maxwellian choice of $f_0(u)$. By a few elementary manipulations, we have

$$n_*(k,s) = \frac{\Delta n}{k^2}\frac{h_*(is/k)}{1+i(\nu/k)h_*(is/k)} \equiv \frac{\Delta n}{k^2}N_*(k,s), \tag{21.59}$$

where we introduced the complex function

$$h_*(is/k) \equiv \int_{-\infty}^{\infty}\frac{f_0(u)}{u-is/k}du\left(1 - C_s^2\int_{-\infty}^{\infty}\frac{f_0'(u)}{u-is/k}du\right)^{-1}, \tag{21.60}$$

closely related to (21.48). We introduce the notation $n_*^{(1)}(k,s)$ for $k < 0$ where the singularity at $is/k = u$ is below the u-axis at the velocity integration, and $n_*^{(2)}(k,s)$ for $k > 0$ where the singularity is above the u-axis. We have

$$n_*(x,s) = \frac{1}{2\pi}\int_{-\infty}^{0}\frac{\Delta n}{k^2}N_*^{(1)}(k,s)e^{ikx}dk + \frac{1}{2\pi}\int_{0}^{\infty}\frac{\Delta n}{k^2}N_*^{(2)}(k,s)e^{ikx}dk$$

and

$$n_*(x,t) = \frac{1}{i2\pi}\int_{s_0-i\infty}^{s_0+i\infty} n_*(x,s)e^{st}ds,$$

where $s_0 > 0$. The inverse Fourier transform is separated into two parts, $-\infty < k \le 0$ and $0 \le k < \infty$ as shown. We introduce the variable $\gamma \equiv is/k$ with γ real, giving $d\gamma = -is\,dk/k^2$. With this choice we have h_* to be a function of a real variable, i.e. $h_* = h_*(\gamma)$, and by (21.59) we also have $N_*(k,s) \to N_*(\gamma,s) = h_*(\gamma)/(1+(\nu/s)\gamma h_*(\gamma))$. Deforming the integration contour as shown in Fig. 21.2, we find a separation similar to the one in (21.27), here as

$$n_*(x,s) = \frac{\Delta n}{2\pi}\int_{0}^{\infty}\frac{N_*^{(1)}(\gamma,s) - N_*^{(2)}(\gamma,s)}{is}e^{-sx/\gamma}d\gamma. \tag{21.61}$$

The integrals along the circular contours in Fig. 21.2 vanish when $R \to \infty$. The integrations along the two closed contours shown (containing I and III, and II and IV, respectively) are vanishing. Inserting $\gamma = is/k$ in (21.59) we note that the denominator of $n_*(\gamma,s)$ is zero for $s = -\nu\gamma h_*(\gamma)$, recalling that in general $h_*(\gamma)$ is a complex function of real γ. The inverse Laplace transform of

$$\frac{h_*(\gamma)}{1+(\nu/s)\gamma h_*(\gamma)}\frac{e^{-sx/\gamma}}{is}$$

is $-ih_*(\gamma)\mathcal{H}(t-x/\gamma)\exp(-\gamma h_*(\gamma)\nu(t-x/\gamma))$, and gives the final result in the form

$$
\begin{aligned}
n_*(x,t) &= \frac{\Delta n}{2\pi} i \int_{x/t}^{\infty} \Big[h_*^{(2)}(\gamma)\exp\left(t\nu(\gamma - x/t)h_*^{(2)}(\gamma)\right) \\
&\qquad - h_*^{(1)}(\gamma)\exp\left(t\nu(\gamma - x/t)h_*^{(1)}(\gamma)\right)\Big]\, d\gamma \\
&= \frac{\Delta n}{\pi}\int_{x/t}^{\infty} \Big[\Im\{h_*^{(2)}(\gamma)\}\cos(\Im\{h_*^{(2)}(\gamma)\}t\nu(\gamma - x/t)) \\
&\qquad - \Re\{h_*^{(2)}(\gamma)\}\sin(\Im\{h_*^{(2)}(\gamma)\}t\nu(\gamma - x/t))\Big] \\
&\qquad \times \exp\left(-\Re\{h_*^{(2)}(\gamma)\}t\nu(\gamma - x/t)\right) d\gamma.
\end{aligned}
\tag{21.62}
$$

The two functions $h_*^{(2)}(\gamma)$ and $h_*^{(1)}(\gamma)$ are derived from h_* in (21.59) in the complex k-plane, see Fig. 21.2. The two functions are complex conjugates and $\Im\{h_*^{(2)}(\gamma)\}/\pi \equiv h_0(\gamma)$, see (21.48). By $\Re\{\}$ we understand the real part of the function in the brackets. The expression (21.62) is strongly simplified in the limit $\nu \to 0$, and we recover here the special case (21.49), where we use the definition (21.48). The self similar x/t-dependence is lost when $\nu \neq 0$. For short times, $t \ll 1/\nu$ we have solutions close to the previous self similar result, but for later times deviations develop, and the solution approaches the form characterizing diffusion equations. We find that $\Im\{h_*^{(2)}(\gamma)\} = \Im\{h_*^{(2)}(-\gamma)\}$ while $\Re\{h_*^{(2)}(\gamma)\} = -\Re\{h_*^{(2)}(-\gamma)\}$, implying that the expansion wave and the rarefaction wave are anti-symmetric as in the case without collisions, $\nu = 0$.

The result (21.62) may at first sight appear complicated. In reality it contains only elementary functions in addition to $h_0(\gamma)$ which, in turn, can here be expressed by the Z-function (Fried & Conte 1961) for Maxwellian velocity distributions or sums of such. The integral in (21.62) is readily carried out numerically, the only complication being the oscillatory integrand (Rekaa et al. 2013).

For the velocity distribution function we no longer have a closed simple expression. For the present collisional case we find by (21.55), where now $n_*(x,t)$ is known

$$
\begin{aligned}
&f_*(x,u,t)/\Delta n = g(u)\mathcal{H}(u - x/t) \\
&+ C_s^2 \frac{\partial f_0(u)}{\partial u}\int_0^t \frac{\partial n_*(x - u(t-t'),t')}{\partial x}\, dt' + \nu f_0(u)\int_0^t n_*(x - u(t-t'),t')\, dt',
\end{aligned}
\tag{21.63}
$$

where the first term (containing Heaviside's unit step function $\mathcal{H}$) corresponds to the free streaming contribution from the initial perturbation, here taken as $g(u) = f_0(u)$.

The final results are obtained as $n_1 = n_* e^{-\nu t}$ and $f_1 = f_* e^{-\nu t}$. Since the self similarity of the solution is lost for large times when $\nu \neq 0$, we no longer have any singularity at $x/t = u$: it is smoothed over by the collisions. The time-integrations in (21.63) are not easily carried out analytically due to the $\int_{x/t}^{\infty} d\gamma$-integral entering the expression for the density. For large times, the most significant contribution to $f_*(x,u,t)$ comes from the last term in (21.63). Illustrative results for the space-time evolution of the normalized density $n(x,t)/\Delta n$ are shown in Fig. 21.14. The figure shows the transition from a collisionless to a collision dominated case. The transition occurs approximately at a time $t \sim 1/\nu$ as expected. At small times, $t \ll 1/\nu$, the results are close to identical.

The result (21.63) together with (21.62) offers a theoretical result which, albeit complicated, allows for illustrating a continuous transition from a collisionless to a collision-dominated space-time evolution of the ion velocity distribution for the given step-like initial condition (Rekaa et al. 2013). The result is unique: it was made possible only by considering carefully selected conditions (e.g. quasi neutral) and corresponding plasma models.

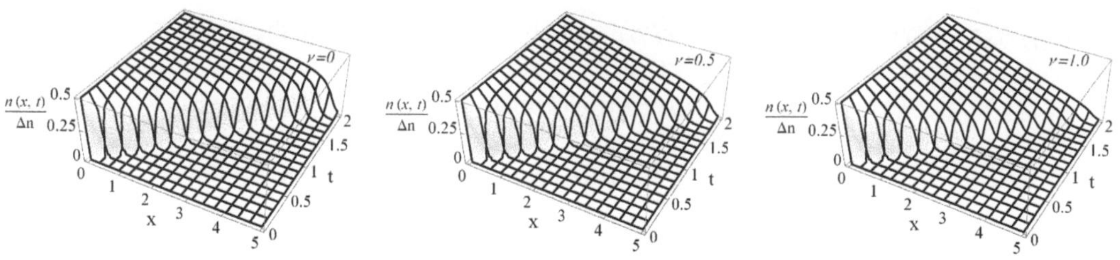

FIGURE 21.14: *Illustrative results for the space-time evolution of the normalized density $n(x,t)/\Delta n$ for the collisionless case, where the evolution is self-similar, and for two collision frequencies, $\nu = 0.5$ and 1.0. Due to the symmetry of the problem, it suffices to show the evolution for $x > 0$. We have $T_e/T_i = 7.5$.*

21.6.2 Strongly collisional regime

Using $n_1 = n_* e^{-\nu t}$ and $f_1 = f_* e^{-\nu t}$ and integrating (21.55) with respect to u we obtain the continuity equation for n_1. Multiplying the expressions with u and integrating as before, we have the momentum equation. Assuming $T_e \gg T_i$ and large ν this expression contains Fick's first law in the form $nU \approx -(C_s^2/\nu)\partial n/\partial x$ where U is the average ion velocity for the perturbation. Using this expression in the continuity equation we obtain a diffusion equation for the plasma density with diffusion coefficient $D \equiv C_s^2/\nu$. When the collisional mean free path is the shortest length scale in the system (i.e. shorter than the length-scale for the density gradient) we can approximate the dynamics by a simple diffusion equation having the relevant solution in the form

$$n_1(x,t) = \frac{\Delta n}{2}\left(1 - \text{erf}\left(\frac{x}{\sqrt{4Dt}}\right)\right), \tag{21.64}$$

in terms of the error function, $\text{erf}(x) \equiv (2/\sqrt{\pi})\int_0^x e^{-y^2}dy$. In this limit we have a dynamic length scale $\sqrt{Dt}$ characterizing the time evolution of the initial condition, and the self similar variable x/t from the analysis in Section 21.5.1 is lost. If we continue the solution illustrated in Fig. 21.14 for $t > 2$, we find a solution nearly indistinguishable from (21.64). In the diffusion limit, the ion velocity distribution remains close to $f_0(u)$, which for realistic conditions is a Maxwellian with constant temperature T_i.

Note that (21.63) with $f_1 = f_* e^{-\nu t}$ offers a general relation between the plasma density and the ion velocity distribution function as described by the ion Vlasov equation with collisional effects included. The ion velocity distribution was analyzed by Rekaa et al. (2013) for late times obtained by using the analytical form (21.64) in (21.63). The result obtained for $f_1(x,u,t)$ by this procedure is almost identical to the late time expression obtained analytically with collisional effects retained in (21.55). For large times in the collisional limit, the most significant contribution to $f_1(x,u,t)$ comes from the last term in (21.63).

22

Nonlinear Kinetic Equilibria

Nonlinear effects in the description of collisionless plasmas are concerned with two basically different types of problems. One is steady state solutions with large amplitudes, the other with dynamic wave phenomena. Some of these are to some extent trivial modifications of phenomena encountered also in fluid models. There are, however, examples of strictly kinetic nonlinear plasma phenomena, and we shall be concerned with some of these.

22.1 Equilibria

The fully nonlinear Vlasov equation can be solved at least in one spatial dimension to give a steady state electrostatic equilibrium solution without restrictions on the wave amplitudes. This observation was first made by Bernstein et al. (1957), and these solutions are hence termed "BGK-solutions". First we postulate that a frame of reference exists, where a stationary potential variation is found. The Vlasov equation written in this frame then becomes

$$u\frac{\partial}{\partial x}f_j(x,u) - \frac{q_j}{m_j}\frac{\partial}{\partial x}\phi(x)\frac{\partial}{\partial u}f_j(x,u) = 0 \tag{22.1}$$

for the electrons, and similarly for the ion component(s). Equation (22.1) is fulfilled for *any* distribution function of the form $f_j = f_j\left(\frac{1}{2}m_ju^2 + q_j\phi(x)\right)$. The index j serves as a label for electrons and various ion species.

The hitherto unspecified electrostatic potential is obtained from Poisson's equation

$$\frac{d^2}{dx^2}\phi(x) = \frac{1}{\varepsilon_0}\sum_j q_j\int_{-\infty}^{\infty} f_j\left(\frac{1}{2}m_ju^2 + q_j\phi(x)\right)du. \tag{22.2}$$

For the present one-dimensional model we have no electromagnetic effects, and ignore also steady state magnetic fields generated by possible currents in the plasma. Assume that somehow $\phi(x)$ is *given*. We shall then illustrate the remarkable observation that it is (at least formally) possible to determine distribution functions for the various plasma species which, by insertion into Poisson's equation, give precisely the prescribed potential! To do this we distinguish *free* and *trapped* particles. Free particles are those which have trajectories beginning at, say, $-\infty$ and continuing to $+\infty$. As an alternative, we have particles trapped around a potential extremum. In particular, we consider also particles beginning at, say, $+\infty$ being reflected and leaving again at $+\infty$, as trapped in a general sense! The difference between the two populations basically arises because as far as the trapped particles are concerned there must at steady state, at any spatial position, be an equal number with positive and with negative velocities. For the *free* particles we can argue that their velocity never assumes a zero value, and there are no imposed symmetries on the corresponding velocity distribution function. The exceptions are particles exactly on the *separatrix*, separating free and trapped particles in phase space. Electrons are trapped in the vicinity of potential *maxima*, positive ions around potential minima.

Assume for simplicity that we are dealing with electrons and one positively charged ion species. The index j is then replaced by e for electrons and i for ions. It is an advantage to introduce the particle energy $E_{e,i} = \frac{1}{2}m_{e,i}u^2 + q_{e,i}\phi(x)$ as a variable, and use that $du = dE_{e,i}/(m_{e,i}u) = dE_{e,i}/\sqrt{2m_{e,i}(E_{e,i} - q_{e,i}\phi(x)}$. Poisson's equation can be rewritten by distinguishing, for instance, the free and trapped electron populations as

$$\begin{aligned}\frac{d^2}{dx^2}\phi(x) &= -\frac{e}{\varepsilon_0}\int_{e\phi}^{\infty}\frac{f_i^{F,T}(E_i)}{\sqrt{2M(E_i - e\phi)}}dE_i \\ &\quad + \frac{e}{\varepsilon_0}\int_{-e\phi_{min}}^{\infty}\frac{f_e^F(E_e)}{\sqrt{2m(E_e + e\phi)}}dE_e \\ &\quad + \frac{e}{\varepsilon_0}\int_{-e\phi}^{-e\phi_{min}}\frac{f_e^T(E_e)}{\sqrt{2m(E_e + e\phi)}}dE_e, \end{aligned} \tag{22.3}$$

where ϕ_{min} is the local minimum for the electrostatic potential while M and m are the ion and electron masses, respectively. It is feasible to let the ions constitute an immobile background of positive charge and let $M \to \infty$. We assume here for simplicity that there is only one local potential minimum. In the first term we did not distinguish free and trapped ions. In the second term the distribution contains only free electrons, while the last term concerns only the *trapped* part of the electron population. The superscripts T and F are used to indicate trapped and free particle populations, respectively. For clarity, we retained the index e and i on the energy variables.

Assume now that we *prescribe* the potential $\phi(x)$, and all distributions *except* the trapped electron distribution. In this case all of the terms in (22.3) are known, with the exception of the last one. We can rewrite the equation as

$$\frac{e}{\varepsilon_0}\int_{-e\phi}^{-e\phi_{min}}\frac{f_e^T(E_e)}{\sqrt{2m(E_e + e\phi)}}dE_e = \frac{e}{\varepsilon_0}G(e\phi), \tag{22.4}$$

where now $G(e\phi)$ is a *known* function, containing all of (22.3), with the exception of the last integral.

The interesting and also important observation is now that (22.4) can be solved analytically under rather general conditions. The implication is that the resulting spatially varying distribution functions, the ones we prescribed together with the one we find from (22.4), are consistent with the prescribed potential $\phi(x)$, and thus form an exact stationary solution for the fully nonlinear coupled Vlasov-Poisson system of equations. We are not merely talking about an existence theorem here: the explicit solution of (22.4) is

$$f_e^T(E_e) = \frac{\sqrt{2m}}{\pi}\int_{e\phi_{min}}^{-E_e}\frac{dG(U)}{dU}\frac{dU}{\sqrt{-E_e - U}}, \qquad \text{for} \qquad E_e < -e\phi_{min}, \tag{22.5}$$

where for simplicity we introduced the variable $U = e\phi$. The requirement that $f_e(E_e)$ is bounded implies $G(e\phi_{min}) = 0$; see (22.4).

The analysis of Bernstein et al. (1957) is explicitly restricted to one spatial dimension, and generalizations to higher dimensionality are not at all straightforward. Some progress has, however, been made (Kato 1976, Ishibashi & Kitahara 1992). It seems, on the other hand, intuitively obvious that the results of Bernstein et al. (1957) have a good chance of being realized in a strongly magnetized plasma column, where a one-dimensional description can be justified. Again we notice, as we found for small amplitude kinetic plasma waves, that no dispersion relation exists for the finite amplitude BGK-waves discussed here.

There is one basic problem in the foregoing analysis: it is not guaranteed that the distribution function we find is non-negative, and hence it *can* be non-physical! It is hard to say how serious this problem is in reality, since many examples have been constructed with quite reasonable, and physically realistic, distribution functions. The problem can be formulated in a different way (Schamel 1986) which does not have this flaw. One can prescribe an a priori physically acceptable functional form for all the distributions containing some parameters. By varying these parameters we can then investigate what bounded potential solutions can be made consistent with these distributions. The exercise on page 143 can be interpreted as such an example.

Even in cases where a physically acceptable (i.e., non-negative) distribution function has been obtained, the stability of the resulting potential structure is not well understood, and actually surprisingly few investigations of this problem are made. Some studies were, however, conducted by, e.g., Goldman (1970), Turikov (1984) and Schamel (1986). As demonstrated in Section 22.2, experimental observations of stable BGK-equilibria have been reported in the literature.

Note that it is possible to construct BGK-equilibria also by assuming immobile ions. Since the velocity of that nonlinear structure can be anything, in principle, it can very well be of the same order of magnitude as the ion thermal velocity, in contrast to *linear* Langmuir waves. In this case, the assumption of immobile ions can no longer be maintained, and it will be found that such a BGK-solution will interact with the ion component (Saeki & Genma 1998).

The arbitrariness of constructing nonlinear solutions to the Vlasov equation can be seen as an extension of the arbitrariness we found for the linear van Kampen modes and the linear Landau damping: by manipulating the velocity distributions we could obtain almost any prescribed time evolution. We now see that this is possible also for the nonlinear collisionless plasma problem.

- **Exercise:** A distribution function need not be bounded; all we can require from a physical point of view is that all the physically relevant averages we can derive from the distribution exist. Distributions containing δ-functions are thus acceptable, in principle. What are the implications of relaxing the condition of boundedness on $f_e(E_e)$? How do you have to remedy (22.5)?

- **Exercise:** Demonstrate that we can write (22.3) in the form

$$\left(\frac{d\phi}{dx}\right)^2 + V(\phi) = \text{const.}$$

and express the pseudo-potential $V(\phi)$ in terms of ϕ, f_e and f_i for the trapped and free particles.

- **Example:** We can readily construct a BGK-equilibrium for the water-bag model; see Section 19.3. We consider the simple case, where the ions are an immobile background of uniformly distributed positive charge, and use the standard equations for the water-bag boundaries (Roberts & Berk 1967, Berk et al. 1970, Kako et al. 1971):

$$\frac{\partial}{\partial t}U_j + U_j\frac{\partial}{\partial x}U_j = \frac{e}{m}\frac{\partial}{\partial x}\phi, \tag{22.6}$$

where $U_j(x,t)$ is the velocity at a position x on the water-bag boundary referring to some unspecified frame of reference. Looking for localized stationary solutions in some appropriate frame of reference, we postulate that we can find solutions

$$\frac{\partial}{\partial x}U_j^2 = \frac{2e}{m}\frac{\partial}{\partial x}\phi$$

for all contours labeled $j = 1,2,3,\ldots$ We have the equivalent form $\frac{1}{2}mU_j^2(x) - e\phi(x) = w_j$ for electrons with charge $-e$, with w_j being an integration constant. We will assume $\phi(x) \to 0$ for $x \to \pm\infty$.

We demonstrate that such stationary solutions for $\phi(x)$ can be found for a simple model with only four boundaries: two global, with labels $j = 1$ and $j = 4$, and two local, with labels $j = 2$ and $j = 3$; see also Fig. 22.1.

With w_j being the integration constant, for the moment left unspecified, we have

$$U_j = \pm\sqrt{w_j + 2\frac{e}{m}\phi} \qquad \text{for} \qquad j = 1,2,3,4.$$

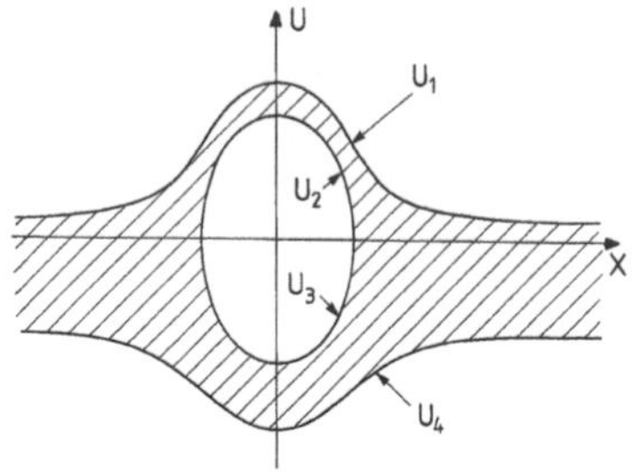

FIGURE 22.1: *Simple illustration of a phase space for the steady state "water-bag" model for a distribution function. The shaded area represents the constant value "a" for the distribution, bounded by stationary contours in the given frame of reference. The hole in the distribution gives rise to the term "phase space hole" or "phase space vortex" because of the corresponding circulation of particles in phase space (Hutchinson 2017).*

All contours are, of course, subject to the same electrostatic potential. By inspection of Fig. 22.1 we see that $U_2(x) = -U_3(x)$, since the two contours merge at the two zero crossings, and their integration constants must be the same, $w_2 = w_3$. This closed contour corresponds to trapped electrons, and subsequently we must have $\phi > 0$ there, and therefore $w_2 < 0$. The frame of reference where the phase space vortex is stationary is the one where the central part of the vortex is symmetric with respect to the x-axis. For the two other U_j's the situation is different. Since neither of these contours can go through zero, we have both w_1 and w_4 being positive. We have $U_1 \to$ const. at $x \to \pm\infty$, since we assumed $\phi \to 0$ for $x \to \pm\infty$. The average electron velocity at $x \to \pm\infty$ is $\frac{1}{2}(\sqrt{w_1} - \sqrt{w_4})$. Recall that we have chosen a frame of reference where an electrostatic structure is at rest, so we expect that in general we have a net electron flow here when the electron fluid is at rest in the laboratory frame. The expression for the separatrix (which separates free and trapped particles in phase space) is given as $\mathcal{U}(x) = \sqrt{2e\phi(x)/m}$.

From the foregoing results, we readily find the electron density as

$$
\begin{aligned}
n(x) &= a(U_1 - U_4) && \text{for} \quad 2e\phi/m + w_2 < 0 \\
n(x) &= a(U_1 - U_4 - 2U_2) && \text{for} \quad 2e\phi/m + w_2 > 0.
\end{aligned}
$$

Expressing the electron density in terms of Heaviside's step function $\mathcal{H}(\xi)$, with $\mathcal{H}(\xi < 0) = 0$ and $\mathcal{H}(\xi > 0) = 1$, we can write Poisson's equation as

$$
\begin{aligned}
\frac{d^2\phi}{dx^2} &= \frac{ea}{\varepsilon_0}\left(\sqrt{w_1 + \frac{2e}{m}\phi} + \sqrt{w_4 + \frac{2e}{m}\phi}\right. \\
&\quad \left. - 2\mathcal{H}\left(\frac{2e}{m}\phi + w_2\right)\sqrt{w_2 + \frac{2e}{m}\phi} - \left(\sqrt{w_1} + \sqrt{w_4}\right)\right), \qquad (22.7)
\end{aligned}
$$

where we used that at $x \to \pm\infty$ the electron and ion densities are the same, i.e., $n_0 = a\left(\sqrt{w_1} + \sqrt{w_4}\right)$. Multiplying both sides of (22.7) by $d\phi/dx$, we readily find after one integration

$$
\begin{aligned}
&\frac{1}{2}\left(\frac{d\phi}{dx}\right)^2 - \frac{am}{3\varepsilon_0}\left(\left(w_1 + \frac{2e}{m}\phi\right)^{3/2} + \left(w_4 + \frac{2e}{m}\phi\right)^{3/2}\right. \\
&\quad \left. - 2\mathcal{H}\left(\frac{2e}{m}\phi + w_2\right)\left(w_2 + \frac{2e}{m}\phi\right)^{3/2} - \frac{3e}{m}\left(\sqrt{w_1} + \sqrt{w_4}\right)\phi\right) = B, \qquad (22.8)
\end{aligned}
$$

where B is an integration constant; see also the discussion of the pseudo-potential in Section 4.2. Also in the present case, just as for the Korteweg-deVries equation, we can introduce a pseudo-potential in the form

$$V(\psi) \equiv w_1^{3/2} + w_4^{3/2} - (w_1 + \psi)^{3/2} - (w_4 + \psi)^{3/2} \\ +2\mathcal{H}(\psi + w_2)(w_2 + \psi)^{3/2} + \frac{3}{2}(\sqrt{w_1} + \sqrt{w_4})\psi, \qquad (22.9)$$

where we introduced the normalized potential $\psi(x) \equiv 2e\phi(x)/m$ for simplicity. We added the constant $w_1^{3/2} + w_4^{3/2}$ so that $V(\psi) = 0$ for $\psi = 0$: this is admissible, since the problem contains the integration constant B anyhow. All quantities have now physical dimensions of *velocity*^{2}. The pseudo-potential $V(\psi)$ is illustrated in Fig. 22.2 for the normalized parameters $w_1 = 1$, $w_4 = 1.5$ and $w_2 = -0.25$. In particular, we note the presence of a local minimum. Just as in Section 4.2, we note that stationary solutions for $\psi = \psi(x)$ are possible as long as the integration constant B in (22.8) takes a value between zero and the local minimum. These solutions are associated with a localized vortex-like structure or a periodic chain of such structures in phase space, as evident from Fig. 22.1. Note that we are content with an "existence theorem" for the steady state solutions; we do not give the explicit result for $\phi(x)$. The approach given here is somewhat different from the original Bernstein et al. (1957) approach. We prescribe here a functional form with some free parameters (here the w_j's), more in the spirit of Schamel (1986), and determine values for these parameters so as to give solutions of the appropriate form.

The normalized potential ψ is a continuous variable. The pseudo-potential $V(\psi)$ in Fig. 22.2 is continuous for all ψ, but it is not differentiable at $\psi = -w_2$. According to the result (22.8) this means that $d\psi/dx$ is continuous, but $d^2\psi/dx^2$ is not, meaning that the electric field associated with a phase space vortex is continuous but not always differentiable.

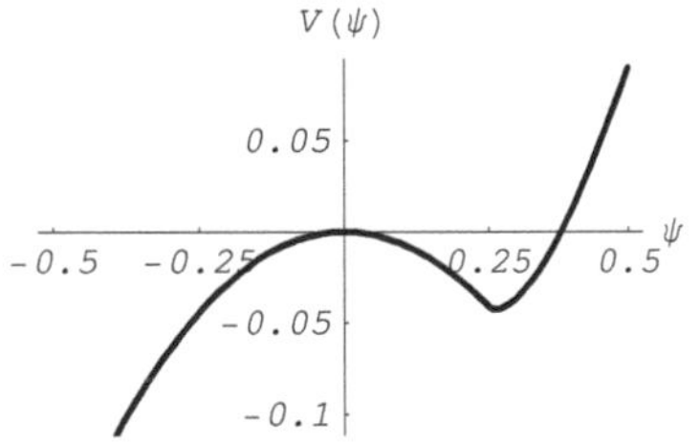

FIGURE 22.2: *Illustration of the normalized pseudo-potential $V(\psi)$ given by (22.9), for the normalized parameters $w_1 = 1$, $w_4 = 1.5$ and $w_2 = -0.25$. For the present case we find an isolated phase space vortex solution when the integration constant equals zero, i.e., $V(0)$.*

22.2 Experimental results

As we found, there are in principle infinitely many possibilities for stationary, fully nonlinear solutions to the coupled one-dimensional Vlasov-Poisson system of equations. In reality, only few are observed to occur naturally. First of all, it is not obvious, nor investigated in any detail, which of all the (infinitely many) possibilities are accessible from a realistic initial condition. Second, even if a BGK-equilibrium solution is realized, it is not at all obvious that it is stable.

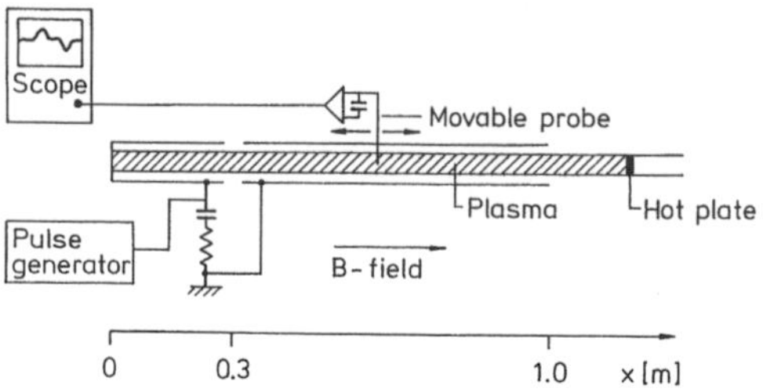

FIGURE 22.3: *Experimental set-up shown schematically (Saeki 1973). The small electric circuit at the probe illustrates a compensating network for the capacitive coupling from the probe tip to the plasma. The figure is reproduced with permission from Saeki et al. (1979), copyright 1979 by the American Physical Society.*

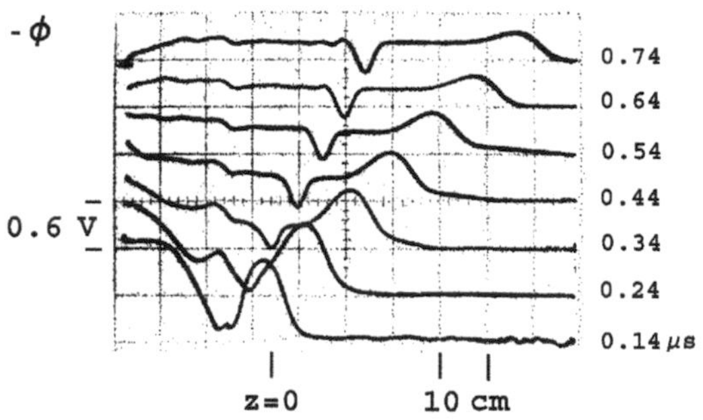

FIGURE 22.4: *Experimental observations of an electron phase space vortex excited in a strongly magnetized Q-machine plasma. The figure shows the spatial variation of the signal for selected time steps, as obtained by a sampling oscilloscope (Saeki et al. 1979).*

Here we summarize the observations of Saeki et al. (1979), obtained in a strongly magnetized ($\omega_{ce} \gg \omega_{pe}$) Q-machine plasma. Figure 22.3 shows the experimental set-up schematically (Saeki 1973, Saeki et al. 1979). The plasma column is surrounded by a metal tube, which acts like a wave-guide. The terminating part of this tube can be biased independently of the rest. By a sudden decrease in potential, electrons are accelerated toward the hot plate by the electric field in the "gap" between the two pieces of tube. The resulting electrostatic potential variation, as detected by a movable probe, is shown in Fig. 22.4 as a function of spatial position along the tube, for various times after application of the short excitation pulse. We note a fast pulse, propagating with a speed characteristic of the plasma filled wave-guide. This pulse is slowly damped by Landau damping. Behind this fast pulse we see the formation and propagation of a much slower structure which moves almost without change in shape. At first sight this is strange; the velocity of this pulse is close to the electron thermal velocity, and hence it should be heavily Landau damped. This is not so, and the numerical water-bag simulation shown in Fig. 22.5 demonstrates that the fast and slow structures appear very different when observed in phase space. The slow pulse corresponds to a phase space vortex (or "electron hole") of the type we discussed before. The analysis giving Fig. 22.1 has to be modified to account for the radial boundary conditions for the present numerical solution.

The experimental studies of Saeki et al. (1979) were extended to include excitation and interaction of *two* phase space vortices. An irreversible coalescence was observed; see example of results in

Fig. 22.6. Subsequent numerical investigations using a particle-in-cell code (Lynov et al. 1979b, Lynov et al. 1980) demonstrated that two interacting vortices *need* not always coalesce: in case their relative velocity is large, as it is in the case with head-on collisions, they may pass through each other. Investigations of the interaction between a phase space vortex and a KdV soliton showed that these two structures simply passed through each other, at least for the parameter range covered by the experimental and numerical studies. This latter result is not surprising: the phase space vortex propagates typically at particle thermal velocities, while the KdV soliton has a much larger propagation velocity, which in the experiments of Saeki et al. (1979) is determined by the radial boundary conditions. Since the relative velocity between the two structures is large, there is little time for nonlinear interaction during a collision.

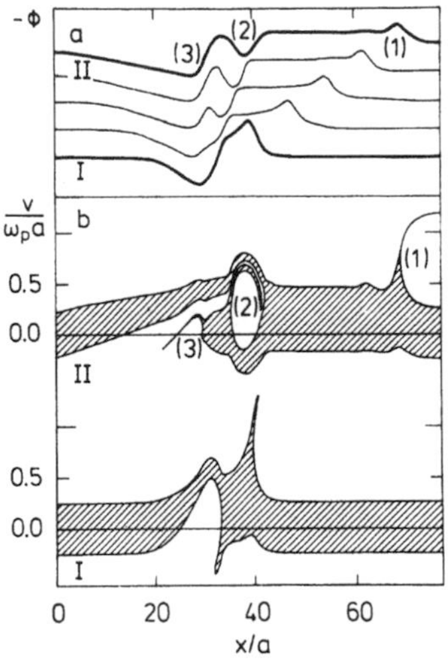

FIGURE 22.5: *Simple water-bag simulation of the experimental results shown in Fig. 22.4. The upper part of the figure shows the potential variation to be compared with Fig. 22.4; the lower part shows samples of the phase space in the simulations. By (2) we here denote the electron hole, while (1) corresponds to the soliton, which is also excited. The rarefaction wave (3) is of no interest here. The figure is reproduced with permission from Saeki et al. (1979), copyright 1979 by the American Physical Society.*

The experiments of Saeki (1973) and Saeki et al. (1979) were carried out in a strongly magnetized plasma, and can therefore be explained by a one dimensional model as the one developed by Bernstein et al. (1957). For the present problem, the boundary conditions for the electric field at the wave-guide in Fig. 22.3 also have to be taken into account. The simulations in Fig. 22.5 include these conditions.

22.2.1 Ion equilibria

The studies of Saeki et al. (1979) referred to conditions where the ion component could be considered as immobile. Later investigations (Pécseli et al. 1981, Pécseli et al. 1984) in a double-plasma device demonstrated the formation of similar structures in ion phase space. Also these results were supported by numerical simulations. Some of the first predictions of the formation of ion phase space vortices were made by Sakanaka (1972), based on numerical solutions of the full nonlinear Vlasov equation in one spatial dimension. Ion phase space vortices formed behind electrostatic shocks evolving when an ion beam was injected into a stationary background plasma. The analysis follows

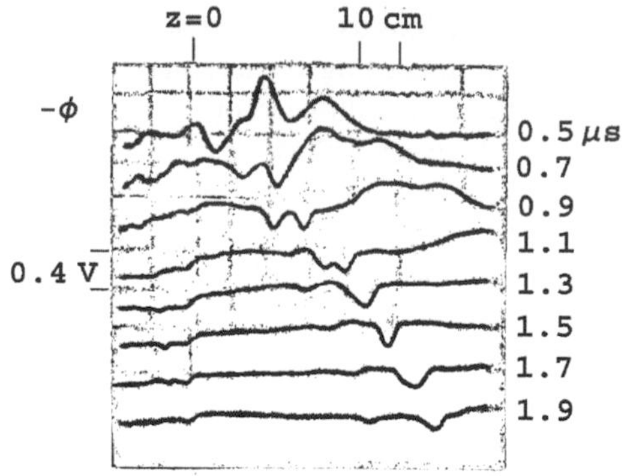

FIGURE 22.6: *Example of coalescence of two externally excited electron phase space vortices. The figure is reproduced from Guio et al. (2003) with permission.*

closely that for electron phase space vortices, where a significant simplification can be achieved by assuming the electrons to be isothermally Boltzmann distributed at all times. Numerical simulations for three-dimensional systems with externally imposed homogeneous strong magnetic fields demonstrated (Daldorff et al. 2001, Guio et al. 2003) that long-lived ion phase space vortices can be excited by ion beams or ion bursts also in dimensions higher than one.

22.2.2 Electrostatic double layers

Most of the discussion so far assumed the presence of one or more local potential minima or maxima for trapping electrons or ions. It turns out that the analysis of Section 22.1 is readily generalized to describe potential variations increasing monotonically from a low potential side to a high potential side: the so-called double layers (Block 1978, Schamel 1986, Raadu 1989). Such structures have great importance in industrial circuit breakers as well as in nature, e.g., in the Earth's ionosphere. Here we have no trapped particles, but can distinguish free or transiting particles from particles reflected at some position. Usually, a double layer is supporting a current, and the ratio of this current and the potential drop can be taken as an anomalous plasma conductivity. An ideal plasma is characterized by the small Spitzer conductivity, so an Ohmic conductivity is of no relevance here. It has been found, however, that formally also currentless double layers can be constructed (Perkins & Sun 1981, Pécseli et al. 1984). These have been observed in laboratory experiments (Charles 2007). The problem concerning stability of double layers in three-dimensional natural systems is not fully resolved, with the exception of cases with strong magnetic fields (Jovanović et al. 1982*a*).

A classification of double layers (DL) is illustrated in Fig. 22.7. For a strong DL we find electrons from the low potential side accelerated through the potential drop, to appear as a beam on the high potential side, with an average velocity determined by the potential drop (Sato et al. 1981). Slow electrons from the high potential side are reflected at some position, so the slow electron component appears as being thermal. Only very few of these electrons make it to the low potential side. For the ions the situation is similar, here with an ion beam to be observed at the low potential side. For details see the first column in Fig. 22.7. As far as the ion acceleration is concerned, it can be instructive to consult Chapter 9 for the discussion of details in the ion acceleration through the potential drop in a Q-machine.

By adjusting the plasma sources on the two sides of the double layer it is possible to make them either stationary or moving (Leung et al. 1980, Iizuka et al. 1982, Rekaa et al. 2012). In a uniform system, in a homogeneous strong magnetic field for instance, the position of a double-layer is not well defined, and its position can change appreciably by even small changes in the plasma

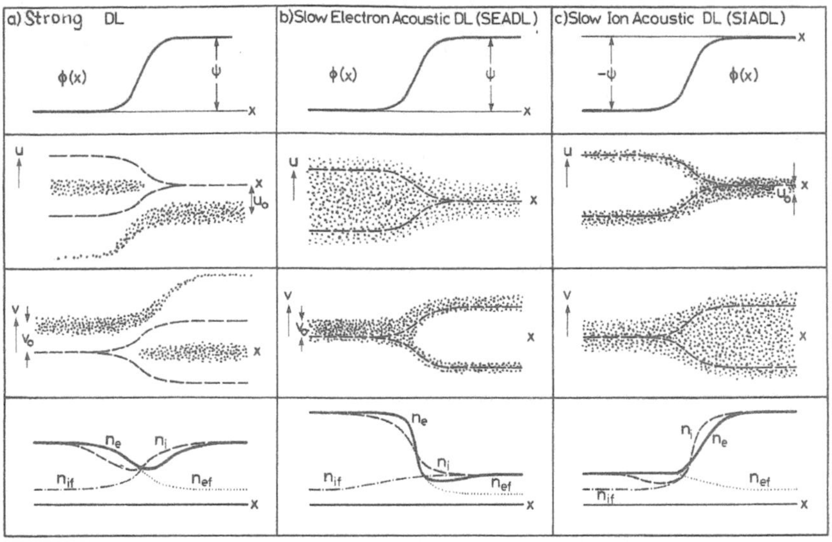

FIGURE 22.7: *Classifications of electrostatic double layers. Top frames show the spatial (one dimensional) variation of the electrostatic potential, the second frame is for the ion velocity phase space, the third frame for electron phase space, and the bottom frame shows the respective density variations of the components. The figure is reproduced from Schamel (1986) with permission from Elsevier.*

sources. An inhomogeneous magnetic field (as e.g. in the Earth's ionosphere) offers a more stable positioning of double-layers. Double-layers, both stationary and moving, have been extensively studied in laboratory plasmas (Leung et al. 1980, Sato et al. 1981).

Double layers were at first studied in spatially one dimensional systems, arguing that plasma in a strong homogeneous magnetic field can be illustrated by such models. In realistic 3 dimensional models it might be possible to assume the systems to be very long in the direction of a (not necessarily homogeneous) magnetic field, but it is unrealistic to assume the double layer to cover all space in the direction perpendicular to **B**. In the ionosphere, in particular, double layers are confined to a magnetic flux tube with some finite cross section (Jovanović et al. 1982*a*). If a satellite is approaching and then crossing the flux tube somewhere on the high potential side, while detecting the energy distribution of the electrons, it first observes a small or negligible beam energy, which is then increasing until it reaches a maximum and then decreases again as the satellite leaves the flux tube on the other side. Plotting the electron beam energy as a function of time the figure will resemble an "inverted V". Such inverted V events are taken as evidence for a double layer placed somewhere in the tube. It is not necessary for the observation that the tube is crossed through the magnetic field line with the largest potential drop. Since the double layer itself can be relatively narrow, the probability of crossing it with a space craft is small, so such an indirect evidence can be appreciated.

At a time there was some confusion and controversy concerning electrostatic double layers: care should be taken not to mistake them for being generators or batteries! A double layer (moving or stationary) represents the adjustment of a collisionless plasma to an externally imposed (and not otherwise specified) generator. This can be an ideal voltage or an ideal current generator, or something in between with a finite internal impedance. Most studies implicitly assume an ideal voltage generator, i.e., one that maintains a fixed potential difference irrespective of the load. In comparison there are only few studies including the entire physically realistic circuit (Smith 1982), although this seems to be the most proper approach. The problem here is that it is not easy to identify the proper parameters for such a realizable physical battery or generator.

Ⓢ **Exercise:** Construct a pseudo-potential for an ion phase space vortex by using a water-bag model for the ion velocity distribution function and assume that the electron component can be modeled as an isothermally Boltzmann distributed massless fluid (Pécseli et al. 1984). Demonstrate also the existence of currentless electrostatic double layers in the present model. Ⓢ

22.2.3 Mixed fluid-kinetic models

It is feasible to use models where one component is described by a set of fluid equation, the other one by kinetic dynamics. This is particularly often done in numerical simulations. An often used model takes the electron to be Boltzmann distributed at all times, while a Vlasov equation is used for the ions (Sakanaka 1972, Pécseli et al. 1984, Pécseli & Trulsen 1989). For monotonic potential variations with a source maintaining a Maxwellian velocity distribution at a distant boundary, such a model can often be justified. If, however, there are trapped electron near a local positive potential region, the basic assumption can be questioned. Experimental results for large amplitude fluctuations in collisionless plasmas (Grésillon & Galison 1973) have demonstrated significant deviations from the electron Boltzmann law. Generalizing the the results for ion acoustic solitons (see Section 14.2.2) for kinetic conditions, the agreement between analytical and experimental results was found to be somewhat improved by allowing for local deviations from the isothermal electron Boltzmann relation (Schamel 1986).

It has been suggested that so-called kappa-distributions (Marsch & Livi 1985, Hellberg & Mace 2002, Ali & Eliasson 2017, Livadiotis 2017), appearing like thermal distributions with an enhanced high velocity non-Gaussian "tail" of hot plasma are representative for conditions found in nature, the ionosphere in particular. A kappa-distribution can be seen as a generalization of a two temperature velocity distributions: this latter form is easy to model in a linear kinetic model by using two Z-functions, see Appendix C.2, but this approximation is often inadequate for high velocities. The electron velocity distribution in discharge plasmas will also often a have an enhanced contribution at high velocities. This part comes form the fast electrons in the discharge that has not yet been thermalized by collisions, for instance with the walls. Numerical simulations comparing results for kinetic ion-acoustic waves using Boltzmann distributed electrons and models with realistic kappa-distributions (Guio et al. 2003) indicate that the differences are minor for small amplitude plasma density perturbations.

23

Nonlinear Landau Damping

As discussed in Section 20.2.2, the results for linear Landau damping apply only for a limited time interval. Implicit in the linearization of the basic equations is an assumption of small wave amplitudes. This chapter studies problems where some of these assumptions are relaxed.

23.1 Nonlinear Landau damping

While linear Landau damping describes the interaction between a wave and particles which are close to the phase velocity, the nonlinear counterpart of this process affects particles with velocities in the vicinity of the wave *group* velocity. This can be a quite general phenomenon, but we consider one basic problem here, namely, the one where ions are interacting with a modulated Langmuir wave. The basic equations for the analysis consist of an equation for the weakly nonlinear electron waves, here a fluid model, and the Vlasov equation for the ions in its linearized form. Describing the electron waves by (14.21) we thus have

$$\nabla \cdot \left(\frac{\partial^2}{\partial t^2} \nabla \widetilde{\phi} - u_{The}^2 \nabla \nabla^2 \widetilde{\phi} + \omega_{pe}^2 \nabla A \right) = -\omega_{pe}^2 \nabla \cdot \left(\frac{\overline{n}}{n_0} \nabla \widetilde{\phi} \right), \tag{23.1}$$

and for the ion dynamics

$$\frac{\partial}{\partial t} f(\mathbf{r}, \mathbf{u}, t) + \mathbf{u} \cdot \nabla f(\mathbf{r}, \mathbf{u}, t) - \frac{e}{M} n_0 \nabla \overline{\phi}(\mathbf{r}, t) \cdot \nabla_u f_0(\mathbf{u}) = 0, \tag{23.2}$$

from the linearized version of (21.7), with the unperturbed ion distribution f_0 normalized to unity. The overline, denoting averaging over the rapid time scale, is introduced to distinguish the electrostatic potential of the Langmuir wave from the slowly varying potential. Kinetic effects are ignored for the electron dynamics, assuming that the Langmuir wave phase velocities ω/k are large so that linear Landau damping is negligible. We can, on the other hand, easily have the group velocity u_g close to a typical ion velocity, and these ions have the possibility for interacting with the *envelope* of the Langmuir wave. The means of interaction is the ponderomotive force exerted by the modulated Langmuir wave-train. These forces act on the electrons, creating an imbalance in the charge distribution, which gives rise to a spatially varying electric field which propagates with the group velocity. This physical mechanism is described by assuming the electrons to be isothermally Boltzmann distributed in the electrostatic field and the ponderomotive potential. We have

$$\nabla \left(\frac{1}{2} \frac{\varepsilon_0}{n_0} \overline{(\nabla \widetilde{\phi})^2} + \kappa T_e \frac{\overline{n}}{n_0} - e \overline{\phi} \right) = 0, \tag{23.3}$$

from (14.24). With the assumption of quasi-neutrality, Poisson's equation becomes redundant. These three equations describe the present problem. The analysis summarized in the following follows the exposition given by Dysthe and Pécseli (1977), but relates also to results by, for instance, Ichikawa et al. (1972), Fried and Ichikawa (1973), Weiland et al. (1978) and Shatashvili and Tsintsadze

(1982). The analysis assumes small amplitudes so that trapping of particles in the ponderomotive wave fields (see Fig. 14.2) is assumed to be unimportant.

We now introduce a plane Langmuir wave as

$$\widetilde{\phi} = \frac{1}{2}\psi e^{-i(\omega t - \mathbf{k}\cdot\mathbf{r})},$$

where the rapid oscillation is contained in the exponential factor, so that $\psi = \psi(\mathbf{r}, t)$ is slowly varying in both time and space when compared to the space and time scales $1/k$ and $1/\omega$. Note that the phase factor used by Dysthe and Pécseli (1977) has the opposite sign of the one used here.

The following analysis assumes two time scales. First, to lowest order the modulation of the Langmuir wave is propagating by the group velocity, without change of shape. It is an advantage to refer the space-time evolution of the wave to a frame moving with the group velocity, and therefore make the transformation $\tau \to t$ and $\xi \to \mathbf{r} - \mathbf{u}_g t$. The slowly varying quantities ψ, $\overline{\phi}$ and $\overline{n}$ can to lowest order be considered to be functions of $\xi = \mathbf{r} - \mathbf{u}_g t$ alone. We now introduce a small expansion parameter ε and allow for $\psi = \psi(\varepsilon\xi, \varepsilon^2\tau)$, and similarly for $\overline{\phi}$ and $\overline{n}$ to account for a slow time variation in the moving frame of reference. In the moving frame specified before, we find

$$\left(i\frac{\partial}{\partial \tau} + \frac{1}{2}\frac{du_g}{dk}\nabla_\xi^2\right)\psi = \frac{\omega_{pe}}{2}\left(\frac{\overline{n}}{n_0} - 1\right)\psi. \tag{23.4}$$

The physical interpretation of (23.4) is evident: to lowest order we have ψ constant, while higher order corrections to the time evolution are dispersion, as given by the derivative of the group velocity, and nonlinearity, as given by the right-hand side of the equation.

We now assume that the Langmuir wave is modulated by a plane wave, with wave-vector $\mathbf{K}$. The direction of $\mathbf{K}$ is in principle arbitrary with respect to the "carrier wave-vector" $\mathbf{k}$. The special case where $\mathbf{K} \perp \mathbf{k}$ is often called "filamentation" since it will appear as if the carrier wave is propagating in an environment of long parallel channels. By use of (23.2) we find

$$\overline{n}_K \equiv \int_{-\infty}^{\infty} f_K d\mathbf{u} = -n_0 \frac{e\overline{\phi}_K \Delta}{T_i},$$

where

$$\Delta \equiv \frac{T_i}{M}\left(P\int_{-\infty}^{\infty} \frac{F_0'(u)}{\widehat{\mathbf{e}}\cdot\mathbf{u}_g - u}du - i\pi F_0'(\widehat{\mathbf{e}}\cdot\mathbf{u}_g)\right), \tag{23.5}$$

in terms of the principal value of the integrand, denoted by P. We have $F_0(u) = \int_{-\infty}^{\infty} f_0 d\mathbf{u}_\perp$, with $\widehat{\mathbf{e}} \equiv \mathbf{K}/|K|$ and $\mathbf{u} = \widehat{\mathbf{e}}u + \mathbf{u}_\perp$. We integrate (14.24) and assume that the wave amplitude is vanishing at infinity to obtain

$$en_0\overline{\phi}_K = T_e\overline{n}_K + \frac{1}{4}\varepsilon_0|E|_K^2,$$

where $|E|^2 \equiv k^2|\psi|^2$. Consequently we have

$$\frac{\overline{n}_K}{n_0} = \frac{1}{4n_0T_i}\varepsilon_0|E|_K^2\frac{\Delta}{1 + (T_e/T_i)\Delta} \equiv \frac{1}{4n_0T_i}\varepsilon_0|E|_K^2(\alpha + i\beta),$$

with

$$\alpha + i\beta \equiv \frac{\Delta_1 + (T_e/T_i)(\Delta_1^2 + \Delta_1^2) + i\Delta_2}{(1 + (T_e/T_i)\Delta_1)^2 + (T_e\Delta_2/T_i)^2},$$

defining $\Delta \equiv \Delta_1 + i\Delta_2$ with both Δ_1 and Δ_2 being real. We note that for an isotropic distribution $f_0 = f(|\mathbf{u}|)$, α and β are functions of $\mathbf{K}$ only through the scalar product $\widehat{\mathbf{e}}\cdot\mathbf{u}_g$, i.e., the direction but not the magnitude of $\mathbf{K}$. For the one-dimensional case, $\mathbf{K} \parallel \mathbf{u}_g$, both α and β are consequently independent of K. We can now invert the Fourier transform to give

$$\frac{\overline{n}}{n_0} - 1 = -\alpha|\eta|^2 + \frac{\beta}{\pi}P\int_{-\infty}^{\infty}\frac{|\eta|^2}{\xi - \xi'}d\xi',$$

where we introduced the dimensionless variable $\eta \equiv k\psi\sqrt{\varepsilon_0/(4n_0T_i)}$. The mathematical origin of the convolution integral of $|\eta|^2$ and $1/(\xi-\xi')$ is the change of direction of the Landau contour for $K<0$ to $K>0$. Inserting this result into (23.4), we find in terms of the normalized variable η

$$\left(i\frac{\partial}{\partial\tau}+\frac{1}{2}\frac{du_g}{dk}\frac{\partial^2}{\partial\xi^2}\right)\eta=-\frac{\omega_{pe}}{2}\left(\alpha|\eta|^2-\frac{\beta}{\pi}P\int_{-\infty}^{\infty}\frac{|\eta|^2}{\xi-\xi'}d\xi'\right). \tag{23.6}$$

The result (23.6) is a non-linear Schrödinger equation, modified by a non-local term, which accounts for the interaction of ions with the modulated wave train. With the assumption of Boltzmann distributed electrons, we have no corresponding electron contribution, which would be small anyhow, due to the small electron mass. The electrons contribute only if their velocity distribution is strongly distorted at low velocities, having a significant derivative around velocities u_g. Approximate solutions of NLS equations with perturbation terms as found in (23.6) were studied by, for instance, Rypdal et al. (1982).

The relation (23.6) can be written as two equations if we introduce $\eta\equiv a\exp(ib)$, where a and b are both real. We have

$$\frac{\partial a}{\partial\tau} = \frac{u_g'}{2}\left(2\frac{\partial a}{\partial\xi}\frac{\partial b}{\partial\xi}+a\frac{\partial^2 b}{\partial\xi^2}\right) \tag{23.7}$$

$$\frac{\partial b}{\partial\tau} = \frac{u_g'}{2}\left(\left(\frac{\partial b}{\partial\xi}\right)^2-\frac{1}{a}\frac{\partial^2 a}{\partial\xi^2}\right)-\frac{\omega_{pe}}{2}\left(\alpha a^2-\frac{\beta}{\pi}P\int_{-\infty}^{\infty}\frac{a^2}{\xi-\xi'}d\xi'\right), \tag{23.8}$$

with $u_g'\equiv du_g/dk$. The relation (23.7) is readily given the form of a conservation equation

$$\frac{\partial a^2}{\partial\tau}-\frac{\partial}{\partial\xi}\left(u_g' a^2\frac{\partial b}{\partial\xi}\right)=0,$$

implying that $\int_{-\infty}^{\infty}a^2d\xi$ is a constant. As a consequence, the Langmuir wave energy is conserved to the present order of the analysis: all that the ions are doing is "re-shuffling" the wave energy, not dissipating it. Results like (23.6) can be obtained for a variety of wave-types, also for other conditions (Pécseli & Rasmussen 1980), where the basic physical mechanism is similar to the one discussed here, only the meaning of the symbols changes.

A simple reference solution to (23.7) and (23.8) is $a_0=$ const. and $b_0=-\frac{1}{2}\omega_{pe}\alpha a_0^2\tau$, corresponding to an unmodulated plane wave, where $-\frac{1}{2}\omega_{pe}\alpha a_0^2$ is the nonlinear, amplitude dependent, frequency shift scaling with a_0^2. Without modulation there are no contributions from the resonant ions. The nonlinear phase shift is negative, corresponding to a reduction of the local plasma density due to the combined effects of ponderomotive forces and the ambipolar electric fields. Implicit in the argument was an assumption of the plane wave being the limiting case of a very long wave-packet so that the integration constant at infinity is set to zero when we integrate (14.24).

We will now consider small perturbations of the reference solution, where the plane wave is modulated so that

$$a = a_0+\widetilde{a}\exp(-i(\Omega t-\mathbf{K}\cdot\mathbf{r})),$$
$$b = b_0+\widetilde{b}\exp(-i(\Omega t-\mathbf{K}\cdot\mathbf{r})),$$

where $\widetilde{a}$ and $\widetilde{b}$ account for the small perturbations in amplitude and phase, respectively. By insertion into the linearized versions of (23.7) and (23.8) we find

$$\Omega=\pm\sqrt{\frac{u_g'}{2}K^2\left(\frac{u_g'}{2}K^2-\omega_{pe}a_0^2(\alpha+i\beta)\right)}.$$

Clearly, for $\alpha\neq 0$ and $\beta\neq 0$, we have an unstable solution for any amplitude a_0, corresponding to Ω with a positive imaginary part for *any* wave-number of the perturbation. If we have $\beta=0$, the modulational instability will have a maximum wave-number, $K_m=\sqrt{2\omega_{pe}a_0^2\alpha/u_g'}$.

23.1.1 Fluid limit

The cold ion fluid model is found for small T_i by using $F_0 = \delta(u)$ in (23.5), which in one spatial dimension gives $\beta = 0$ and $\alpha = -(T_i/M)/(u_g^2 - C_s^2)$, where the sound speed C_s was defined before. Notice that the ion temperature was used in making η dimensionless, so there are no singularities in the limit of $T_i \to 0$. The change in sign of α for $u_g^2 < C_s^2$ to $u_g^2 > C_s^2$ is understood as the phase-shift of π when a dissipationless oscillator is driven *below* or *above* its resonance frequency. We can interpret the spatially modulated ambipolar electric field (with wave-number K) propagating with velocity u_g as a forcing field with frequency u_g/K, acting on ion sound waves with resonance frequency C_s/K. Exactly at resonance, $u_g = C_s$, we have formally the growth rate to become infinite, but the basic analysis breaks down for this case. A finite value for β limits the growth rate, and also changes the phase between the forcing and the ion acoustic response.

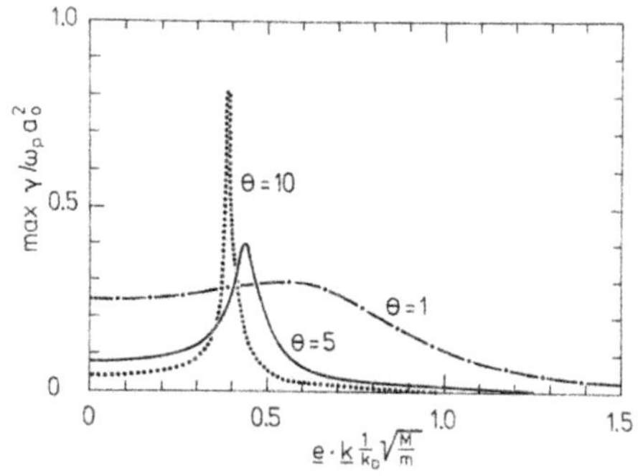

FIGURE 23.1: *Illustration of the maximal growth rate for the modulational instability for different temperature ratios in Maxwellian plasma,* $\Theta \equiv T_e/T_i$. *The figure is reproduced from Dysthe and Pécseli (1977) with the permission of the Institute of Physics Publishing.*

Of particular interest is the case where $\widehat{\mathbf{e}} \cdot \mathbf{u}_g = 0$, corresponding to a modulation in the direction transverse to the Langmuir wave field direction of propagation, i.e., to a *filamentary instability*. This will appear similarly in both fluid and kinetic models. For this particular case we have

$$\Omega = \pm\omega_{pe}\sqrt{\frac{3K^2}{2k_D^2}\left(\frac{3K^2}{2k_D^2} - a_0^2\frac{T_i}{T_i + T_e}\right)},$$

with $k_D = \sqrt{3\omega_{pe}^2/u_{The}^2} = 1/\lambda_{De}$ being the Debye wave-number where $u_{The}^2 = 3\kappa T_e/m$ is here defined with a factor of 3 and $u_g' = u_{The}^2/\omega_{pe}$. Filamentary instabilities are interesting by being absolute; see Section 3.1.7.

For the more general case, with $\widehat{\mathbf{e}} \cdot \mathbf{u}_g \neq 0$ and $T_i \neq 0$ in a kinetic model, we have that the growth rate of the modulation is always finite. We find the maximum growth rate to be $\Im\{\Omega\} = \frac{1}{2}\omega_{pe}a_0^2\sqrt{\alpha^2 + \beta^2}$, and it occurs for the modulation wave-number $K^2 = \frac{2}{3}k_{De}^2 a_0^2(\alpha^2 + \beta^2)/\alpha$. In Figs. 23.1 and 23.2 we show the normalized expression for the maximally obtainable growth rate, and also the direction for the modulational wave-number $\mathbf{K}$ which produces the maximum growth (Dysthe & Pécseli 1977).

The present analysis considered a Langmuir wave, modulated by a given wave-number K. The nonlinear Landau damping effect was due to the interaction of resonant particles and the ambipolar electric field set up by the ponderomotive forces from the wave envelope which moves with the group velocity. The analysis can readily be generalized to the case where we consider two waves

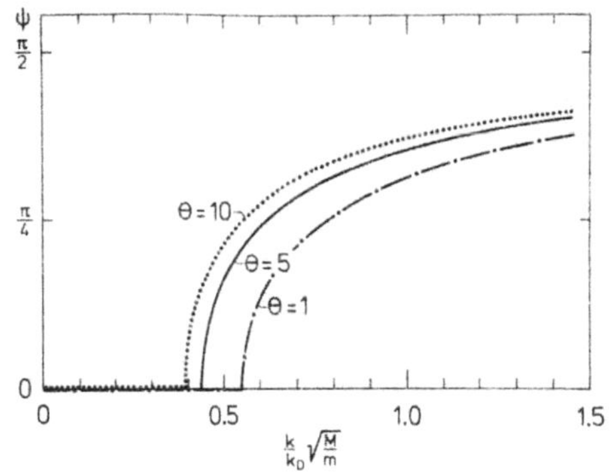

FIGURE 23.2: *Direction of the modulation wave-vector* **K** *with respect to* **k** *which gives the maximum growth rate for the modulational instability. The result is shown for different temperature ratios,* $\Theta \equiv T_e/T_i$*. The figure is reproduced from Dysthe and Pécseli (1977) with the permission of the Institute of Physics Publishing.*

(ω_1, k_1) and (ω_2, k_2) where the beating wave-field is interacting with resonant particles, i.e.,

$$\frac{\omega_1 - \omega_2}{k_1 - k_2} \approx u$$

where u is a particle velocity. When the frequencies and wave-numbers are close, the left side of this expression is close to the group velocity; see also Section 3.1.2.

The present summary considered electrostatic waves, but the basic features of the problem, the filamentation instability in particular, are also found for other wavetypes such as electromagnetic waves propagating in plasmas (Dysthe 1968). The nonlinear evolution of the filamentation can appear as a self-focusing of the wave and has received special attention. Self-focusing need not occur due to ponderomotive forces but can be caused also by modulated heating of the plasma (Dysthe et al. 1985*b*): the warm plasma expands locally and the decrease in plasma density changes the local index of refraction. These conditions can lead to instability.

When $T_e/T_i \geq M/m$ the strongly dispersive limit (Joyce et al. 1969) where $\omega \sim \Omega_{pi}$ is only weakly damped by the linear ion Landau damping, and only a weak electron Landau damping remains. In this limit also the ion waves can become modulationally unstable. This limit is discussed for instance by Berezhiani and Shatashvili (1985).

23.2 Damping of ion acoustic solitons by reflected particles

As an illustration for what happens for long times, when the linear Landau damping results cease to be valid, we consider the case where a moderate amplitude ion acoustic pulse propagates through a plasma. We assume that a second order nonlinear model suffices, and describe the weakly nonlinear sound wave by a KdV soliton model with $T_i \ll T_e$, where the potential variation of the soliton is assumed to have an amplitude ϕ_m and a width Δ. We can begin by making an important observation: linear Landau damping acts on a time scale smaller than or comparable to $\tau_B \equiv \Delta\sqrt{M/(e\phi_m)}$ see Section 20.2.2. The nonlinear effects need a relatively long time to be manifested, the soliton time (see Section 4.2), which is given as $\tau_{NL} \equiv \Delta/(3A)$ in normalized units. In physical units we have $\tau_{NL} \equiv \Delta\kappa T_e/(3C_s e\phi_m)$, giving $\tau_{NL}/\tau_B \approx \frac{1}{3}\sqrt{\kappa T_e/e\phi_m} \gg 1$, i.e., if we want to study truly nonlinear developments of a soliton, then linear Landau damping is not relevant and we have to account for the soliton-particle interaction by other means (Karpman 1979*a*).

In a first approximation we here calculate the total energy of the soliton by using (21.12) together with the results from Section 14.2.2, and subsequently use energy conservation arguments for estimating the energy exchange between particles and soliton.

First, we obtain the ion kinetic energy density as $\frac{1}{2}Mnu^2$, accurate to second order, in terms of the plasma density and the ion fluid velocity u in the rest frame of the plasma. Having the soliton moving with velocity U, we find from the ion continuity equation

$$n(u-U) = -n_0 U,$$

where we have $n = n_0$ when $u = 0$ see also Section 14.2.3. Recall that u is the ion fluid velocity in the rest frame of the plasma. For a pulse-like perturbation, we have $u \to 0$ and also $\phi \to 0$ for $x \to \pm\infty$. From the ion momentum equation we find

$$\frac{1}{2}(u-U)^2 = -\frac{e}{M}\phi + \frac{1}{2}U^2.$$

We then have

$$\frac{1}{2}Mnu^2 = MUn_0 \frac{U^2 - e\phi/M - U\sqrt{U^2 - 2e\phi/M}}{\sqrt{U^2 - 2e\phi/M}}.$$

We now recall that in terms of the soliton amplitude ϕ_m we have the soliton velocity in the rest frame of the plasma to be $U = C_s(1 + \frac{1}{3}e\phi_m/\kappa T_e)$. To second order in potential we then find the ion kinetic energy density

$$\frac{1}{2}Mnu^2 \approx \frac{1}{2}n_0\kappa T_e \left(\frac{e\phi}{\kappa T_e}\right)^2. \tag{23.9}$$

The given expressions assume cold ions, but they can be justified also for non-vanishing T_i, provided $T_e/T_i \gg 1$.

We have given the soliton shape $\phi(x) = \phi_m \operatorname{sech}^2\left(x\sqrt{e\phi_m/(6\kappa T_e \lambda_{De}^2)}\right)$, recalling that the dispersion of the ion acoustic dispersion relation to third order, k^3, is given as $\alpha = \frac{1}{2}\lambda_D^2 C_s$ for cold ions; see (13.28) and the discussion in Section 14.2.2. We now integrate the ion kinetic energy to give

$$\begin{aligned} \mathcal{W}_{kin} &= \frac{n_0\kappa T_e}{2}\int_{-\infty}^{\infty} \phi_m^2 \operatorname{sech}^4\left(x\sqrt{\frac{e\phi_m}{6\kappa T_e\lambda_{De}^2}}\right) dx \\ &= \sqrt{\frac{8}{3}}\left(\frac{e\phi_m}{\kappa T_e}\right)^{3/2} n_0\kappa T_e\lambda_D. \end{aligned}$$

Note that we integrated with respect to one spatial variable only. The soliton solution is independent of y and z, so the result is the energy per unit area in the direction perpendicular to the x-axis. The dimension of n_0 is also here *length*$^{-3}$.

The electric field energy is obtained as

$$\mathcal{W}_E = \frac{1}{2}\varepsilon_0 \int_{-\infty}^{\infty}\left(\frac{d}{dx}\phi_m \operatorname{sech}^2\left(x\sqrt{\frac{e\phi_m}{6\kappa T_e\lambda_{De}^2}}\right)\right)^2 dx = \frac{4}{15}\sqrt{\frac{2}{3}} n_0\kappa T_e\lambda_D\left(\frac{e\phi_m}{\kappa T_e}\right)^{5/2},$$

which is small compared to the ion kinetic energy for relevant amplitudes, $e\phi_m < \kappa T_e$.

To the present accuracy, the electrons give a contribution to the energy density of the soliton with a term of the same magnitude as the kinetic energy density. The total soliton energy is thus

$$\mathcal{W} = 4\sqrt{\frac{2}{3}}\left(\frac{e\phi_m}{\kappa T_e}\right)^{3/2} n_0\kappa T_e\lambda_D, \tag{23.10}$$

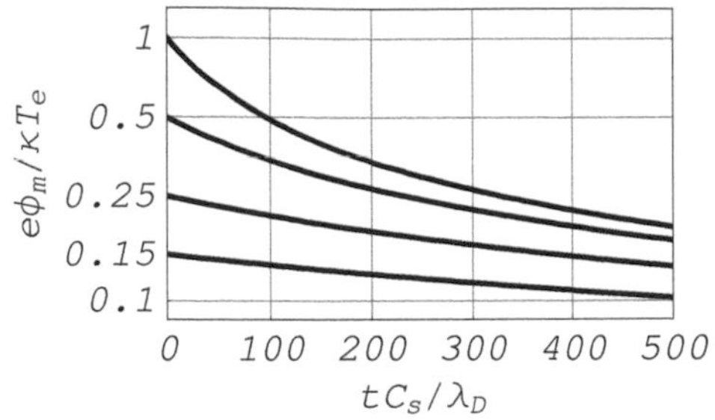

FIGURE 23.3: *Illustration of the damping of an ion acoustic soliton as described by (23.12). The figure uses normalized units, with a logarithmic vertical axis and* $C_s \equiv \sqrt{\kappa T_e/M}$*. We have* $T_e/T_i = 15$.

to the present accuracy.

Although we used expressions for cold ions in the energy estimates, we now assume that we nonetheless have a low density "tail" of resonant ions, which can exchange energy with the soliton. We also have the electrons, of course, but for practical applications, the ion contribution will dominate, unless we assume very large electron to ion temperature ratios. These ions have velocities u in the resonant region, $u \in \{u_{min}, u_{max}\} \equiv \{U_s - \sqrt{2e\phi_m/M}; U_s + \sqrt{2e\phi_m/M}\}$. We introduced the notation $U_s \equiv C_s(1 + \frac{1}{3}e\phi_m/\kappa T_e)$ for the soliton velocity in the rest frame of the plasma. The energy gain by such a resonant ion is $2MU_s(U_s - u)$, where ions can gain or loose energy, depending on the sign of $(U_s - u)$, i.e. depending on the soliton overtaking the particle, or a fast particle overtaking the soliton. The flux of these ions is $|u - C_s(1 + \frac{1}{3}e\phi_m/\kappa T_e)|n_0 f_0(u)$, where $f_0(u)$ is the ion velocity distribution function normalized to unity as before. Consequently, we can write the energy gain by resonant ions per unit time as

$$\frac{dW_{res}}{dt} = 2MU_s n_0 \int_{u_{min}}^{u_{max}} (U_s - u)|u - U_s| f_0(u) du. \tag{23.11}$$

The integration limits are $\{u_{min}, u_{max}\}$ defined before. We now equate this change in energy per time unit with the negative time derivative of the change in soliton energy obtained from (23.10). The system consisting of the soliton + the resonant particles conserves total energy. We have found the energy gained by ions accelerated by a soliton. By energy conservation we know that this energy is lost from the soliton. All soliton parameters can be expressed by the maximum soliton amplitude $\phi_m(t)$ for the KdV soliton discussed here. Since a relation between the soliton parameter and the soliton energy is known we can obtain an equation for $\phi_m(t)$. After some algebra, we find

$$\frac{d}{dt}\frac{e\phi_m}{\kappa T_e} = \frac{1}{3}\sqrt{\frac{3}{2}}\sqrt{\frac{\kappa T_e}{e\phi_m}}\frac{MU_s}{\kappa T_e \lambda_D} G(U_s, \phi_m), \tag{23.12}$$

with

$$G(U_s, \phi_m) = \int_{U_s}^{U_s + \sqrt{2e\phi_m/M}} (u - U_s)^2 f_0(u) du + \int_{U_s}^{U_s - \sqrt{2e\phi_m/M}} (u - U_s)^2 f_0(u) du,$$

recalling here that also U_s depends on ϕ_m. For Maxwellian distributions, we can express $G(U_s, \phi_m)$ in terms of error functions. Illustrative results for the damping of ion acoustic solitons are shown in Fig. 23.3 using an electron-ion temperature ratio of $T_e/T_i = 15$. Large amplitudes have a large initial damping, because the large ratio of particles being slower than or faster than the soliton. As the soliton amplitude decreases, also the damping rate is reduced. The analysis in the present section implicitly assumes that a KdV equation is applicable, and since this is correct only to second order in amplitude, we have to impose a restriction on the soliton amplitude, for instance by requiring that the nonlinear contribution to the soliton velocity $\frac{1}{3}C_s e\phi_m/\kappa T_e$ is less than $\sim 20\%$ of C_s. In Fig. 23.3 we included also too large amplitudes for the sake of illustration.

For large temperature ratios, $T_e/T_i > 25$, the number of ions interacting with the soliton becomes small, and the reflection of electrons becomes important; this contribution is neglected in Fig. 23.3, but the analysis can be carried out by the procedure used for obtaining (23.12). The assumption of Boltzmann distributed electrons can still be maintained in the basic model, since the number of resonant electrons represents only a small fraction of the electron population within the range of validity of the analysis.

By (23.12) we have in principle solved the problem. Depending on the form of $f_0(u)$ we can have damping or growth in soliton amplitude due to the interaction with resonant particles. Given $\phi_m(t)$, we can obtain the trajectory of the soliton maximum in an $\{x,t\}$-plane. We can also predict the asymptotic level of the plateau trailing the soliton (Karpman 1979*a*, Lynov 1983).

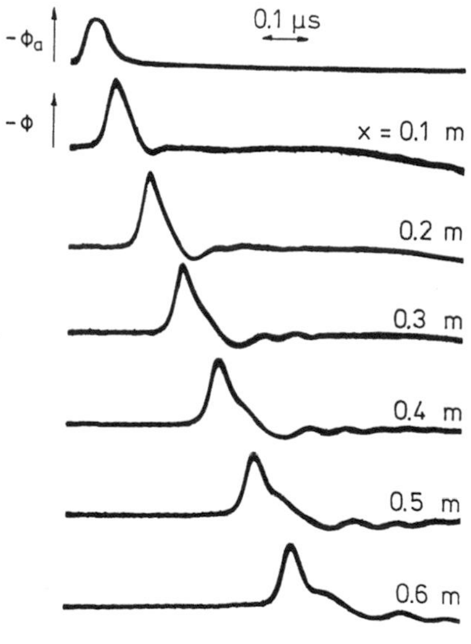

FIGURE 23.4: *Illustration of experimental results for the damping of a KdV soliton due to resonant particle interaction, here obtained for weakly dispersive Trivelpiece-Gould waves in a strongly magnetized plasma wave-guide. The figure is reprinted with permission from Karpman et al. (1980), copyright 1980, American Institute of Physics. The density of particles reflected by the soliton is here very small, and they are not noticeable on the figure, but can be seen more clearly in other results, shown by Lynov et al. (1979b). Note the evolution of a plateau and a slowly dispersing oscillating "tail" following the soliton. Trivelpiece-Gould solitons have negative polarity, so the vertical axis is inverted for the presentation here.*

The interaction of the resonant particles and the soliton represents a perturbation of the KdV equation, and therefore the general results of Section 4.2.2 are supposed to apply. In particular we anticipate the evolution of a plateau and an oscillatory tail following the soliton; see also Fig. 4.10. We need not present the specific analytical form of the perturbation term to reach this conclusion, but it was given by Karpman (1979*a*) if needed.

In order to obtain some quantitative results, we make a series expansion of $G(U_s, \phi_m)$ in (23.12), where we here let the soliton velocity be a constant $U_s \approx C_s$ since the correction varies only with

ϕ_m, which was assumed to be small anyhow. We then have

$$G(U_s,\phi_m) = 2\left(\frac{e\phi_m}{\kappa T_e}\right)^2 f_0^{(1)}(U_s)\left(1+4\sum_{n=3}^{\infty}\frac{(2n-1)(2n-2)}{(2n)!}\frac{f_0^{(2n-3)}(U_s)}{f_0^{(1)}(U_s)}\left(\frac{e\phi_0}{\kappa T_e}\right)^{n-2}\right),$$

where $f_0^{(m)}$ denotes the m-th derivative of $f_0(u)$. To lowest order, we can write the relation (23.12) as

$$\frac{d}{dt}\frac{e\phi_m}{\kappa T_e} = \frac{2}{3}\sqrt{\frac{3}{2}\frac{MU_s}{\kappa T_e \lambda_D}}\left(\frac{e\phi_m}{\kappa T_e}\right)^{3/2} f_0^{(1)}(U_s),$$

which can readily be integrated to give

$$\phi_m = \frac{\phi_0}{(1+\nu t)^2}, \tag{23.13}$$

where the damping constant is given as

$$\nu = -\frac{1}{3}\sqrt{\frac{3}{2}}\sqrt{\frac{e\phi_0}{\kappa T_e}\frac{MU_s}{\kappa T_e \lambda_D}} f_0^{(1)}(U_s).$$

For $f_0(u)$ being a Maxwellian, for instance, we have $f_0^{(1)}(U_s) < 0$, and the soliton is damped, but for other cases we can have growth which is faster than exponential, at least as long as the analysis remains applicable. If we use the parameters for a Maxwellian distribution, we find that the numerical value of ν obtained before comes quite close to the damping constant for the linear ion acoustic Landau damping. It is important, however, that the damping in (23.13) is *slower* than exponential!

Unfortunately, the elegant result (23.13) has limited applicability. This limitation can be illustrated by considering the next correction term in the series expansion in $G(U_s,\phi_m)$. In this case we have

$$\frac{d}{dt}\frac{e\phi_m}{\kappa T_e} = \frac{2}{3}\sqrt{\frac{3}{2}\frac{MU_s}{\kappa T_e \lambda_D}}\left(\frac{e\phi_m}{\kappa T_e}\right)^{3/2} f_0^{(1)}(U_s)\left(1+\frac{1}{9}\frac{f_0^{(3)}(U_s)}{f_0^{(1)}(U_s)}\frac{2e\phi_m}{M}\right).$$

For an order of magnitude estimate we can use a Maxwellian ion velocity distribution, $f_0(u) = (2\pi\sigma)^{-1/2}\exp(-u^2/2\sigma)$, with $\sigma \equiv \kappa T_i/M \ll U_s^2$. For the last correction term in the parenthesis to be small we require $(U_s^2/\sigma)(e\phi_0/\kappa T_i) \ll 5$, which is far from realistic in experimental conditions, when we at the same time require that the nonlinearities should be manifested in a reasonable time, i.e., that the soliton time should be moderate. It is most likely that (23.12) has to be solved numerically for realistic and relevant cases; see also Fig. 23.3.

A quasi-linear model for wave-particle interactions for weakly dispersive waves was formulated by Dysthe et al. (1986) using elements of the analysis summarized in the present section.

Weakly nonlinear high frequency electron plasma waves in magnetized waveguides, Trivelpiece-Gould waves, have many properties in common with ion acoustic waves (Manheimer 1969, Rasmussen 1978), and results equivalent to those summarized in the present section were demonstrated for the first time by Karpman et al. (1980) for that type of wave. Some analytical results are given in more detail by Lynov et al. (1979). The soliton damping as well as the formation of a trailing plateau and an oscillatory "tail" is illustrated in Fig. 23.4. The qualitative agreement with the analytical results of Section 4.2.2 is excellent; see also Fig. 4.10. The basic elements of the analysis outlined in the present section can be generalized also to other waveforms and their interactions with particles. A number of such studies are reported in the literature by, e.g., Kono (1986).

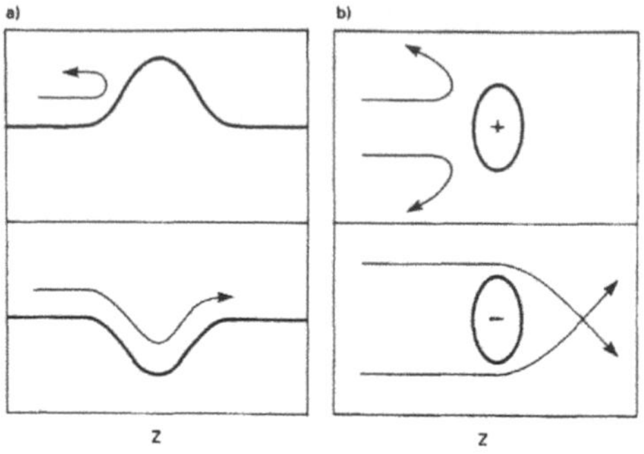

FIGURE 23.5: *Schematic illustration of the interaction of a positive charged particle and a potential enhancement and a potential depletion, respectively. In Fig. a) we show a model in one spatial dimension along the z-coordinate, while Fig. b) illustrates the same problem in 2 spatial dimensions.*

23.3 Wave-particle interaction in one and in higher dimensions

The analysis in Section 23.2 assumes dynamics in one spatial dimension. By inspection it will appear that there is a significant difference in the interaction of, say, an ion and a positive or a negative potential variation. In Fig. 23.5a) we illustrate this difference schematically. We find here that a slow ion can be reflected by a potential enhancement: this is what happens for an ion interacting with a soliton as described in Section 23.2. Should the same ion interact with a stationary potential depletion, it would pass through it and emerge on the other side with its initial velocity. If we now consider the problem in 2 or 3 spatial dimensions for localized potential structures as illustrated in Fig. 23.5b) for the 2-dimensional case, the situation becomes different. In this case we can have deflection (i.e., a momentum change) of the ion in both cases, although details of the processes will differ.

The present problem has little relevance for KdV-solitons interacting with particles since these structures are one dimensional by construction. The question will however deserve scrutiny for more general and physically realistic conditions. Strictly speaking, the insight gained by Fig. 23.5a) has relevance only along the symmetry line in Fig. 23.5b).

24

Quasi-linear Theory

The stability analysis of a collisionless plasma, electron plasma oscillations in particular, were based on a linearized model. In that description, amplitudes of unstable waves continue to grow exponentially, and the analysis eventually ceases to be valid when the assumption of linearization brakes down. It is a quite obvious question to ask what sort of mechanism eventually arrests the wave growth. Quasi-linear theory (Drummond & Pines 1964, Bernstein & Engelmann 1966, Vedenov 1968, Stix 1992, Swanson 2003) is one of the models which can answer that question, at least for some cases.

The basic equations for an electrostatic version of quasi-linear theory are written (in one spatial dimension) in a Fourier transformed version as

$$\frac{\partial f_k(u,t)}{\partial t} + iku f_k(u,t) + \frac{e}{m} ik\phi_k(t)\frac{\partial f_0(u,t)}{\partial u} = -\frac{e}{m}{\sum_q}' i(k-q)\phi_{k-q}(t)\frac{\partial f_q(u,t)}{\partial u}, \tag{24.1}$$

where the subscripts k or q refer to one particular spatial Fourier component. The prime on $\sum$ in (24.1) indicates that the $q = 0$ term is not to be included in the summation, as this term is taken separately on the left side. It gives a slight advantage, to begin with at least, to use Fourier series rather than Fourier integrals (Champeney 1973). We introduced Fourier components

$$\phi_k(t) = \frac{1}{\mathcal{L}}\int_0^{\mathcal{L}} \phi(x,t)e^{-ikx}dx, \qquad \text{and} \qquad \phi(x,t) = \sum_{n=-\infty}^{\infty} \phi_{k_n}(t)e^{ik_n x}, \tag{24.2}$$

where $k = 2n\pi/\mathcal{L}$ with $n = 0, \pm 1, \pm 2, \ldots$ as indicated with the subscript $_n$ on k in (24.2). In particular, Parseval's theorem (Champeney 1973) has the form

$$\frac{1}{\mathcal{L}}\int_0^{\mathcal{L}} \phi^2(x,t)dx = \sum_{n=-\infty}^{\infty} |\phi_{k_n}(t)|^2. \tag{24.3}$$

The physical dimensions of $\phi_k(t)$ and $\phi(x,t)$ are the same with the present definition of the Fourier transform, but you *can* have other definitions (Stix 1992), involving also factors of 2π.

We have taken $e > 0$ so the charge on an electron is $-e$. The electrostatic potential is coupled to the distribution function through Poisson's equation, $\nabla^2\phi = (e/\varepsilon_0)(n_e - n_0) \equiv (e/\varepsilon_0)(\int_{-\infty}^{\infty} f(x,u,t)du - n_0)$, which is Fourier transformed to give

$$k^2\phi_k(t) = -\frac{e}{\varepsilon_0}\int_{-\infty}^{\infty} f_k(u,t)du, \tag{24.4}$$

for $k \neq 0$. We now ignore wave-wave interactions, i.e., terms where the potential ϕ_{k-q} from one wave-number $k - q$ is mixed with the contribution from the distribution function f_q corresponding to another wave-number q, being aware that this assumption is not so obviously correct (Tsunoda et al. 1987). The summation on the right side of (24.1) is then omitted except for $k = 0$. This particular term simply corresponds to the interaction of the modes labeled q with themselves, and is therefore not a wave-wave interaction term. The Fourier component $f_k(u,t)$ corresponding to $k = 0$ is then the spatial average of the velocity distribution function. The relation (24.1) is simplified as

$$\frac{\partial f_k(u,t)}{\partial t} + iku f_k(u,t) + \frac{e}{m} ik\phi_k(t)\frac{\partial f_0(u,t)}{\partial u} = 0 \tag{24.5}$$

for $k \neq 0$, while the $k = 0$ equation is

$$\frac{\partial f_0(u,t)}{\partial t} = i\frac{e}{m}\sum_q q\phi_{-q}(t)\frac{\partial f_q(u,t)}{\partial u}. \tag{24.6}$$

The prime on the summation sign can be omitted since ϕ_0 can be assumed to vanish. It is important to note that the set of equations (24.4)–(24.6) is time reversible, just as the Vlasov equation when it is coupled with Poisson's equation for the analysis of electron plasma waves. We now solve (24.5) to give

$$f_k(u,t) = f_k(u,t_0)e^{-iku(t-t_0)} - \frac{e}{m}ik\int_{t_0}^{t} e^{-iku(t-\tau)}\phi_k(\tau)\frac{\partial f_0(u,\tau)}{\partial u}d\tau. \tag{24.7}$$

This result can easily be proven by insertion into (24.5). Hitherto, the analysis does not distinguish a stable plasma with Landau-damped waves and an unstable plasma with oscillations increasing in amplitude. We now restrict the analysis to the unstable situation and ignore the first term in (24.7) for large t. Let the lower limit of integration $t_0 \to -\infty$ and introduce the new variable $s = t - \tau$, giving

$$f_k(u,t) = -i\frac{e}{m}k\int_0^{\infty} e^{-ikus}\phi_k(t-s)\frac{\partial f_0(u,t-s)}{\partial u}ds \tag{24.8}$$

which, inserted into (24.6), gives

$$\frac{\partial f_0(u,t)}{\partial t} = \left(\frac{e}{m}\right)^2\frac{\partial}{\partial u}\int_0^{\infty}\sum_q q^2 e^{-iqus}\phi_{-q}(t)\phi_q(t-s)\frac{\partial f_0(u,t-s)}{\partial u}ds. \tag{24.9}$$

It is trivially demonstrated that $d\int_{-\infty}^{\infty} f_0(u,t)du/dt = 0$ in (24.9), i.e., the particle density is conserved.

Consistent with the assumption of a slow growth of the waves it can be assumed that $f_0(u,t)$ varies much slower than $\phi_k(t)$ for all k, since we have to lowest approximation $\phi_k(t) \sim \exp(-i\omega_{pe}t)$. This assumption will be used later on and can be verified a posteriori. Using Poisson's equation with (24.8) we find a closed expression which in principle determines the electrostatic potential

$$\phi_k(t) = i\left(\frac{e^2}{m\varepsilon_0 k}\right)\int_0^{\infty}\int_{-\infty}^{\infty} e^{-ikus}\phi_k(t-s)\frac{\partial f_0(u,t-s)}{\partial u}duds. \tag{24.10}$$

In principle (24.9) and (24.10) solve the problem by constituting a closed set of equations. The solution is, however, too complicated for practical applications, and some simplifications are necessary. Assume that the wave amplitude varies slowly on a time scale given by the frequency of oscillations. We can then assume a time evolution in a WKB-sense for one Fourier component of the potential as

$$\phi_k(t) = \phi_k(t_0)e^{\int_{t_0}^{t}[-i\omega_k+\gamma_k(\tau)]d\tau}, \tag{24.11}$$

where ω_k for a given wave-number **k** is a real constant and determined by the linear dispersion relation, which is obtained as in the standard Landau treatment of (24.4) and (24.5), i.e.,

$$\omega_k = \omega_{pe}(1+\frac{3}{2}(k\lambda_D)^2) \approx \omega_{pe}, \tag{24.12}$$

where to the desired accuracy, we take $(k\lambda_D)^2$ as a small correction to ω_{pe}. The imaginary part is obtained similarly as

$$\gamma_k(t) = \frac{\pi}{2n_0}\frac{\omega_k}{k^2}\omega_{pe}^2 f_0'(u=\omega_k/k,t) \approx \frac{\pi}{2n_0}\frac{\omega_{pe}^3}{k^2}f_0'(u=\omega_k/k,t). \tag{24.13}$$

Here, the distribution function f_0 is *not* normalized, i.e., $\int_{-\infty}^{\infty} f_0(u,t)du = n_0$, explaining the division with n_0. The results (24.12) and (24.13) can alternatively be obtained also from (24.11), as can be verified by insertion. These expressions imply that the deformation of the spatially averaged velocity distribution is so slow that we can use the actual expression for $f_0'(u,t)$ at any instant when obtaining a local dispersion relation, the imaginary part of the frequency in particular.

The relation (24.11) is equivalent to

$$\frac{\partial|\phi_k(t)|^2}{\partial t} = 2\gamma_k|\phi_k(t)|^2. \tag{24.14}$$

In relation (24.9) the quantity

$$\phi_k(t-s) = \phi_k(t_0)\exp\left(-i\omega_k(t-s-t_0) + \int_{t_0}^{t-s}\gamma_k(\tau)d\tau\right)$$

will be small for large s since $\gamma_k(\tau) \geq 0$ for linearly unstable waves, where we take $t_0 \to -\infty$ so that $s > t_0$. The implication of this observation is that the largest contribution to the integral in (24.9) comes from $s \approx 0$ so as an approximation we can use

$$\frac{\partial f_0(u,t)}{\partial t} = \left(\frac{e}{m}\right)^2 \frac{\partial}{\partial u}\int_0^\infty \sum_q q^2 e^{-iqus}\phi_{-q}(t)\phi_q(t-s)ds\frac{\partial f_0(u,t)}{\partial u}. \tag{24.15}$$

The integration with respect to s is simplified and we have as an approximation

$$\int_0^\infty e^{-iqus}\phi_q(t-s)ds \approx \frac{\phi_q(t)}{i(qu-\omega_q)+\gamma_q(t)}.$$

The expression (24.15) can then be simplified to give

$$\begin{aligned}\frac{\partial f_0(u,t)}{\partial t} &= \left(\frac{e}{m}\right)^2\frac{\partial}{\partial u}\left[\sum_q \frac{q^2|\phi_q(t)|^2}{i(qu-\omega_q)+\gamma_q(t)}\frac{\partial f_0(u,t)}{\partial u}\right]\\ &\equiv \frac{\partial}{\partial u}\left[D(u,t)\frac{\partial f_0(u,t)}{\partial u}\right],\end{aligned} \tag{24.16}$$

having the form of a diffusion equation. The diffusion coefficient $D(u,t)$ is real and positive, as seen by noting the reality conditions $|\phi_q(t)|^2 \equiv \phi_q(t)\phi_q^*(t) = \phi_{-q}^*(t)\phi_{-q}(t) = |\phi_{-q}(t)|^2$, with $\omega_q = -\omega_{-q}$ and $\gamma_q = \gamma_{-q}$, giving

$$D(u,t) = 2\left(\frac{e}{m}\right)^2\sum_{q>0}\frac{q^2|\phi_q(t)|^2\gamma_q(t)}{(qu-\omega_q)^2+\gamma_q^2(t)}. \tag{24.17}$$

Outside the resonant region, where the waves are damped, we have $|\phi_k(t)|^2 \approx 0$ and consequently $D(u,t) \approx 0$ there. The fact that $D \geq 0$ is important; a diffusion equation is only well behaved, from a physical point of view, for positive diffusion coefficients.

Up to here the formulation in terms of Fourier series was advantageous, but to extend the results it is preferable to modify the expression (24.17) somewhat by standard manipulations, which can be useful also in a general context. First we assume that the points on the q-axis are very dense, in the sense that $|\phi_q(t)|^2$ changes only little from one point to the next. Recall here that when performing the Fourier transform, the distance between two points on this q-axis is $\sim 1/\mathcal{L}$, where $\mathcal{L}$ is the length of the system, so the *density* of points on the q-axis is $\sim \mathcal{L}$. The number of points in the Δq-interval is then $\mathcal{L}\Delta q/2\pi$. For large $\mathcal{L}$ and with $\Delta q \to dq$, the sum in (24.17) can then be approximated by

$$\sum_q \to \int\frac{\mathcal{L}dq}{2\pi}, \tag{24.18}$$

to give

$$D(u,t) = 2\left(\frac{e}{m}\right)^2 \int_0^{\mathcal{L}} \frac{q^2|\phi_q(t)|^2\gamma_q(t)}{(qu-\omega_q)^2+\gamma_q^2(t)} \frac{\mathcal{L}}{2\pi} dq.$$

Recall also the limiting form for the Dirac δ-function

$$\delta(x) = \lim_{\alpha\to 0} \frac{1}{\pi} \frac{\alpha}{\alpha^2+x^2}, \tag{24.19}$$

and also $\delta(qu-\omega_q) = \delta(q-\omega_q/u)/u$.

We identify γ_q in (24.17) with α in (24.19) and write in the limit of small growth rates γ_q an expression for the diffusion coefficient in the simple form

$$D(u,t) \approx \mathcal{L}\left(\frac{e}{m}\right)^2 \frac{k^2|\phi_q(t)|^2}{u}\bigg|_{q=\omega_{pe}/u} = \mathcal{L}\left(\frac{e}{m}\right)^2 \frac{|E_{q=\omega_{pe}/u}(t)|^2}{u}, \tag{24.20}$$

where we might as well use that $\omega_q \approx \omega_{pe}$ for relevant wave-number values q.

Introducing $W_k(t) \equiv \mathcal{L}\varepsilon_0|E_k(t)|^2$, we can write the equation for the quasi-linear relaxation of the beam distribution as

$$\frac{\partial f_0(u,t)}{\partial t} = \left(\frac{e^2}{m^2\varepsilon_0}\right) \frac{\partial}{\partial u} \frac{W_k(t)}{u} \frac{\partial f_0(u,t)}{\partial u}, \tag{24.21}$$

to be used with the resonance condition $k = \omega_{pe}/u$.

At first sight it might seem surprising to find the length $\mathcal{L}$ of the system entering explicitly in (24.20), but you should recall that by (24.2) we have $\mathcal{L}$ entering implicitly also through $\phi_k(t)$ and thereby $E_k(t)$.

- **Exercise:** Consider the simple diffusion equation

$$\frac{\partial}{\partial t} n(x,t) = -D\frac{\partial^2}{\partial x^2} n(x,t),$$

with $D \geq 0$. Give the solution for an initial condition $n(x,0) = \delta(x)$. The problem illustrates why it is important to have $D(u,t) \geq 0$ in (24.17).

The diffusion of electrons in phase space can qualitatively be understood as a "random walk in velocity space". An electron is resonantly interacting with waves having phase velocity close to the electron velocity. The electron gains or loses energy, to attain a different velocity. Since, however, there will in general be Langmuir waves with all sorts of phase velocities larger than the electron background thermal velocity, the electron might always find waves to interact with, only their intensities will in general vary. The wave phases will be randomly distributed and statistically independent, and give rise to a "random walk" as those modeled by diffusion equations. Note that this interpretation relies heavily on the strongly dispersive nature of the waves, and the analogy is expected to break down if weakly dispersive waves (such as ion sound waves) are to be discussed by similar models. Quasi-linear models for weakly dispersive waves are not well developed, although some weakly nonlinear models for wave-particle interactions have been proposed (Dysthe et al. 1986).

24.1 Conservation of energy and momentum

It is amazing that in spite of all the approximations, the equations are still energy conserving. To see this we obtain the time derivative of the kinetic energy, U_{res} as

$$\begin{aligned}\frac{\partial U_{res}}{\partial t} &\equiv \frac{1}{2}m\int_{res} u^2\frac{\partial f_0(u,t)}{\partial t}du \\ &= \frac{1}{2}m\int_{res} u^2\frac{\partial}{\partial u}\left[D(u,t)\frac{\partial f_0(u,t)}{\partial u}\right]du \\ &\approx -\pi m\left(\frac{e}{m}\right)^2\sum_q q^2|\phi_q(t)|^2\int_{res}\frac{u}{q}\delta\left(u-\frac{\omega_q}{q}\right)\frac{\partial f_0(u,t)}{\partial u}du, \end{aligned} \tag{24.22}$$

where the integration is restricted to the resonant region, "*res*", as indicated, since $D(u,t)\approx 0$ outside it, as already mentioned. We used $\gamma_q/((qu-\omega_q)^2+\gamma_q^2)\approx\pi\delta(qu-\omega_q)$ for small γ_q (see (24.19)); the analysis assumed small growth rates for the instability from the outset. Now recall the expression for the growth rate γ_q given in (24.13). To the actual accuracy, we can again approximate ω_q by ω_{pe} there and find after some simple calculations

$$\frac{\partial U_{res}}{\partial t} = -\frac{\varepsilon_0}{2}\frac{\partial}{\partial t}\sum_q q^2|\phi_q(t)|^2 \tag{24.23}$$

where we used (24.10). The result (24.23) implies that the total energy density in the electric field spectrum $\frac{1}{4}\varepsilon_0\sum_q q^2|\phi_q(t)|^2$ increases at a rate of one-half of the decrease in kinetic energy of the particles. The other half goes to the "sloshing" motion of the entire particle distribution, where we recall from the discussion of the wave energy density that in the limit of small q, where the cold plasma dielectric function can be used, the energy densities in the electric field and the particles are equal. The basic physics are thus retained, but the time reversibility of the original equations is lost; we have taken the asymptotic limit too many times underway! This is not really a valid criticism of quasi-linear theory; if we have a function which can be represented by a Taylor expansion, then its value at the origin can be reconstructed from an arbitrary position. If, however, this function is replaced by its asymptote (assuming it *has* one), this information is evidently lost.

24.2 Discussions of the asymptotic stage

From (24.16) we readily obtain what is sometimes called an "H-like theorem" in the form

$$\begin{aligned}\frac{1}{2}\frac{d}{dt}\int_{-\infty}^{\infty} f_0^2 du &= \int_{-\infty}^{\infty} f_0\frac{\partial}{\partial u}D\frac{\partial}{\partial u}f_0 du \\ &\quad -\int_{-\infty}^{\infty} D\left(\frac{\partial f_0}{\partial u}\right)^2 du \le 0, \end{aligned} \tag{24.24}$$

where we once integrated by parts (Bernstein & Engelmann 1966). The discussion of the asymptotic stage of the simplified quasi-linear equations is now straightforward. From (24.24) it is evident that $\int_{-\infty}^{\infty} f_0^2 du$ must decrease to some lower positive limit. This implies that $D(u,t)\,(\partial f_0/\partial u)^2\to 0$ for all u. This requires that $\partial f_0/\partial u\to 0$ everywhere that $D\neq 0$. The the velocity distribution function asymptotically develops a plateau everywhere $W_k\neq 0$, as illustrated in Fig. 24.1.

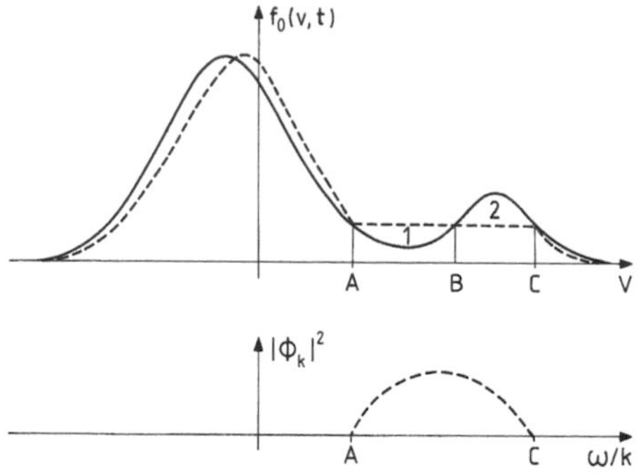

FIGURE 24.1: *The initial (solid line) and final stage (dashed line) of the velocity distribution function i quasi-linear theory, shown schematically. The lower figure shows schematically the asymptotic shape of the power spectrum of the fluctuations in electrostatic potential.*

Asymptotically we expect the wave growth to be arrested and according to (24.10) this requires $\gamma_k \to 0$ for all k, meaning $f_0'(u = \omega_k/k, t) \to 0$ according to (24.13). The plateau is formed in the distribution function by eroding the beam in such a way that the particle density remains constant. Within this simple quasi-linear theory the fluctuations reach a stationary level as $t \to \infty$ since the growth is arrested and no damping is included in the analysis. Conservation of the average particle density demonstrated before ensures that the area under the distribution function is conserved; hence the final plateau is uniquely determined by the requirement that the two areas labeled 1 and 2 on Fig. 24.1 are equal.

Note that we have at least not formally closed the set of equations, i.e., no *explicit* relation for the evolution of $W_k(t)$ has been presented in, for instance, (24.21), although it is a relatively simple matter from the derivations. We return to this question in the discussions of experimental evidence in Section 24.3. Heuristically, we can obtain a relation which has nonetheless been widely used in the literature. We have thus from (24.13) that $\gamma_k(t) \approx (\pi/2n_0)(\omega_{pe}^3/k^2) f_0'(u = \omega_k/k, t)$. Using this in (24.14) we find

$$\frac{\partial W_k(t)}{\partial t} = \frac{\pi}{n_0} \frac{\omega_{pe}^3}{k^2} W_k(t) \left. \frac{\partial f_0(u,t)}{\partial u} \right|_{u=\omega_{pe}/k}, \tag{24.25}$$

again to be interpreted with the resonance condition $u = \omega_{pe}/k$ as stated.

The relations (24.21) and (24.25) constitute a closed set of quasi-linear equations for the evolution in one spatial dimension of beam instability generated by an electron beam with a significant spread in velocities.

Ⓢ **Exercise:** Demonstrate that the equations (24.21) and (24.25) have solutions for the spectra of the form $W_k \sim k^{-\alpha}$. Determine α. Ⓢ

The simple quasi-linear model has been criticized for several reasons; first of all because the formation of the plateau is feasible only in one velocity dimension. A plateau in *three* dimensions with a finite number of particles would require the plateau level to be zero! Now, this is not really unphysical, but nevertheless it is a rather peculiar result, also being in contradiction with experimental observations, and thus the whole idea is abandoned. The reason for this shortcoming of quasi-linear theory in three dimensions is not entirely self-evident, and has been discussed some length in the literature. One bit of physics missing in the analysis presented before is associated with the fact that waves which are initially unstable with $\gamma_k \geq 0$ can later on, during the nonlinear evolution, become damped with $\gamma_k < 0$. Such a transition was ignored in the analysis. Nevertheless, quasi-linear theory played an important role in theoretical plasma physics by providing a simple description of the saturation of a basic kinetic instability, and after all it is not such a bad description of weakly unstable waves in strongly magnetized plasmas, where the particle motion along the magnetic field lines is essentially one dimensional.

The quasi-linear analysis is at times presented as a theory of plasma *turbulence*. Indeed it can be made so, but note that the foregoing summary nowhere introduced ensemble averages: in principle, at least, it is a deterministic analysis in the form given here.

24.3 Experimental results

As we have seen, the results of quasi-linear theory in one spatial dimension are relatively simple and not too difficult to obtain. To apply them to a general realistic case is, unfortunately, something quite different. One of the basic results from the analysis of this section will be a demonstration of the complexity of a realistic situation.

The results of quasi-linear theory have been compared with results from experiments carried out in a strongly magnetized plasma (Roberson et al. 1971, Roberson 1971, Llobet et al. 1985, Tsunoda et al. 1987), and good agreement with theoretical predictions was found. In their experiment, Roberson et al. (1971) and Roberson (1971) injected a narrow electron beam with circular cross-section into the plasma column from one end of the device, and the evolutions of the beam energy distribution as well as the wave spectrum were detected in the "down-stream" direction. The experiment was carried out in a strongly magnetized cylindrical plasma column, with constant radius R, i.e., as for a Trivelpiece-Gould type mode (Trivelpiece & Gould 1959); see also Section 12.2.6. The foregoing analysis referred to an infinite and uniform system solved as an initial value problem, and the analysis has to be modified in order to account for the experimental conditions.

The basic equation for the present problem can be taken to be the limiting form of the drift-kinetic equation in one spatial dimension; see Section 19.4. Following (24.1) we find for the present case

$$-i\omega f_\omega + u_z \frac{\partial}{\partial z} f_\omega - \frac{e}{m} E_\omega f_0' = \frac{e}{m} \sum_{\omega'}{}' E_{\omega-\omega'} \frac{\partial f_{\omega'}}{\partial u}, \tag{24.26}$$

where for $B \to \infty$ we can ignore $\mathbf{u}_\perp$. In this effectively one-dimensional model, the electrons move as "pearls on a string", the strings being the homogeneous magnetic field lines. Here and in the following we denote $\partial f_0/\partial u \equiv f_0'$. The prime on $\sum$ in (24.26) indicates that the $\omega' = 0$ term is not to be included in the summation. This term enters separately on the left side. We allow the plasma density to vary with the radial coordinate, r. The electric field is also here assumed electrostatic, $E_\parallel = -\partial\phi/\partial z$. Only the **B**-parallel component of the electric field enters (24.26); the radial component appears through Poisson's equation, which has to be solved with the appropriate boundary conditions at $r = R$. It is assumed that f_0 in (24.26) contains both a beam and a background component, $f_0(r, z, u) = G_0(r, u) + f_B(r, z, u)$. For conditions relevant here with small beam densities we have $\int_{-\infty}^{\infty} f_0(r, z, u)du \approx \int_{-\infty}^{\infty} G_0(r, u)du \equiv n_0(r)$.

One basic shortcoming of the previous analysis of this chapter is that it cannot directly be applied to the experimental condition, where the wave amplitude increases with *distance* from the beam-injection point (Drummond 1964, Roberson & Gentle 1971). The foregoing analysis treated an initial value problem, where a boundary value problem is appropriate here. Rather than considering an average distribution function $f_0(u,t)$, as in (24.9), we have here $f_0 = f_0(r, z, u)$. Using (24.26) for $\omega = 0$ we find

$$u \frac{\partial}{\partial z} f_0 = \frac{e}{m} \sum_{\omega}{}' E_{-\omega} \frac{\partial}{\partial u} f_\omega(r, z, u). \tag{24.27}$$

Also here the prime indicates that the $\omega = 0$ contribution should not be counted. By assuming that $E_0 = 0$, we can actually delete the prime here. Taking the z-dependence to be of the form $\exp(i\int_0^z k(z')dz')$, assuming that the wave amplitude increases slowly in the positive z-direction, we have from (24.26)

$$f_\omega(r,z,u) = \frac{e}{im}\frac{E_\omega}{(ku-\omega)}f_0',$$

and obtain (except for the u-multiplier on the left-hand side) a diffusion-like equation

$$u\frac{\partial}{\partial z}f_0 = \frac{\partial}{\partial u}D(r,z,u)\frac{\partial}{\partial u}f_0(r,z,u), \tag{24.28}$$

with a diffusion coefficient

$$D(r,z,u) \equiv \frac{e^2}{im^2}\sum_\omega \frac{|E_\omega(r,z)|^2}{ku-\omega},$$

where now the wave-number k is complex.

We assume that the waves grow in amplitude along the beam direction, as given by a complex wave-number $k \equiv k_1 + ik_2$. The diffusion coefficient is complex, with a real part being

$$D_1(r,z,u) \equiv \frac{e^2}{m^2}\sum_\omega \frac{k_2 u|E_\omega(r,z)|^2}{(k_1u-\omega)^2+(k_2u)^2}.$$

In order to find an expression for the spectrum $|E_\omega(r,z)|^2$, we assume that the electron beam is carefully centered and that its velocity V_b is larger than the maximum phase velocity of all higher order Trivelpiece-Gould modes, i.e., $V_b > \omega_{pe}R/p_{1,\nu}$ for all $\nu \geq 1$; see (12.39) in Section 12.2.6. In effect we are assuming that we only have to be concerned with the Trivelpiece-Gould mode corresponding to J_0 and its first zero.

The spatial growth of wave amplitude squared is given by

$$\frac{\partial}{\partial z}|E_\omega(r,z)|^2 = 2k_2|E_\omega(r,z)|^2. \tag{24.29}$$

To find suitable expressions for k_2 as well as k_1 for the present plasma conditions we use the general expression (5.102) for the wave energy density $\mathcal{E}$ of a wave-packet

$$\frac{\partial}{\partial t}\mathcal{E} + \mathbf{u}_g\cdot\nabla\mathcal{E} = \mathbf{J}\cdot\mathbf{E}, \tag{24.30}$$

where $\mathbf{u}_g$ is the group velocity corresponding to the the carrier wave-number of the wave-packet. (We use here $\mathcal{E}$ for the energy density, since W is in this subsection reserved for the function defined in Appendix C; see also Section 20.5.) For steady state conditions, $\partial\mathcal{E}/\partial t = 0$, and in one spatial dimension, we have

$$\mathcal{E} = \omega\varepsilon_0|E_\omega|^2\frac{\partial\varepsilon_1(\omega,k)}{\partial\omega} \approx \varepsilon_0|E_\omega|^2\frac{\omega}{(k_1\lambda_{De})^2}\frac{\partial W_1}{\partial\omega}. \tag{24.31}$$

See, for instance, (5.100) in Section 5.10.2. We introduced the real part of the function W in the dielectric function, as appropriate for a Maxwellian velocity distribution; see Appendix C.

The fluctuating current density (not to be confused with the Bessel function J) associated with a wave component (ω,k) is obtained by

$$J = -e\int_{-\infty}^{\infty} uf(r,z,u,t)du = i\frac{e^2}{m}E \fint_{-\infty}^{\infty}\frac{uf_0'(r,z,u)}{ku-\omega}du,$$

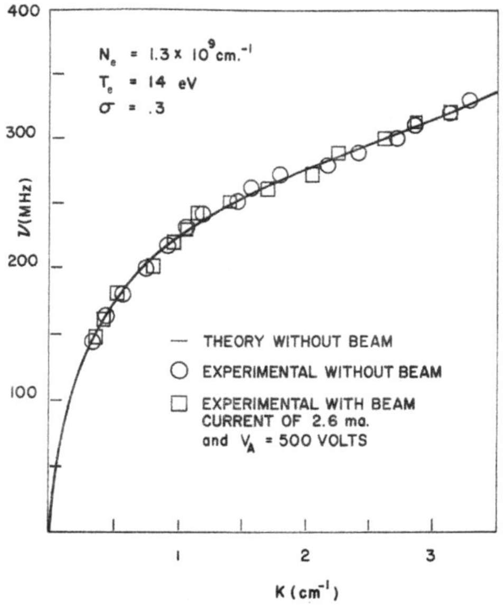

FIGURE 24.2: *The real part of the dispersion relation for the experimental test of quasi-linear theory (Roberson 1971, Roberson et al. 1971, Roberson & Gentle 1971); see also Fig. 20.8. The figure is reproduced with the permission of Prof. C. W. Roberson. A similar figure is found in Roberson et al. (1971), and permission for use has been obtained also from the American Physical Society. The radial plasma density variation was approximated by* $n(r) = n_0 \exp(-r^2/(2\sigma^2 R^2))$*, where R is the geometrical radius of the wave-guide, while* σ *is a fitting parameter, found by a least square fit to the data.*

where, in particular, the real part becomes

$$J_1 = \pi \frac{e^2}{m} E_1 \frac{\omega}{k_1^2} \frac{\partial f_0(z,r,u)}{\partial u}\bigg|_{u=\omega/k_1},$$

again assuming time-stationary conditions. We use this result for J_1 together with (24.31) in (24.30) for time-stationary conditions. After an integration over the plasma cross section we have

$$\varepsilon_0 \frac{u_g}{u_{The}^2} \frac{\partial W_1}{\partial \omega} \int_0^R r\omega_{pe}^2(r) \frac{\partial}{\partial z} |E_\omega(r,z)|^2 dr = \pi \frac{e^2}{m} \int_0^R r|E_\omega(r,z)|^2 \frac{\partial f_0(r,z,u)}{\partial u}\bigg|_{u=\omega/k_1} dr.$$

Using (24.29), we find (Book 1967)

$$k_2 = -\frac{u_{The}^2}{u_g} \frac{\pi}{\partial W_1/\partial \omega} \frac{\int_0^R r|E_\omega(r,z)|^2 f_0'(r,z,u=\omega/k_1)dr}{\int_0^R r|E_\omega(r,z)|^2 n_0(r)dr}, \tag{24.32}$$

where now we interpret $k_2 = k_2(z)$ as a "local" spatial growth rate.

We assumed that we could write the electron velocity distribution function as a sum of a beam, f_B, and a background. Usually, the beam is rather narrow in the radial direction, with a small spatial extent around $r \approx 0$. A centered beam will predominantly excite radially symmetric modes, $m = 0$ in particular. If the beam velocity is large, it preferentially excites waves with phase velocities so large that the Landau damping by the background alone can be neglected. In that case, we can write

$$k_2 \approx -\frac{u_{The}^2}{u_g} \frac{\pi}{\partial W_1/\partial \omega} |E_\omega(0,z)|^2 \frac{\int_0^R r f_B'(r,z,u=\omega/k_1)dr}{\int_0^R r|E_\omega|^2 n_0(r)dr}. \tag{24.33}$$

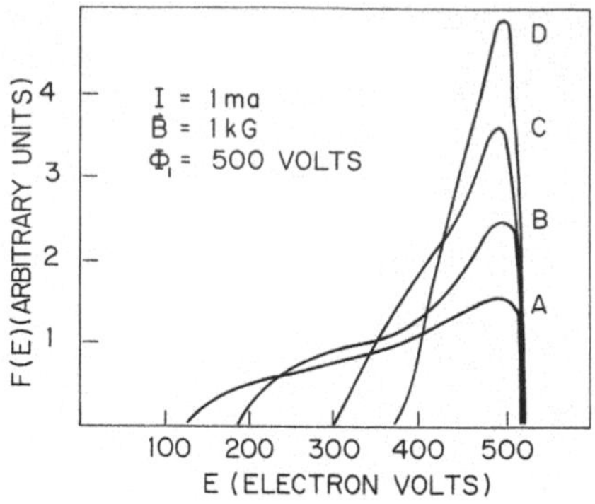

FIGURE 24.3: *Variation of the electron beam energy distribution $F(E)$ with distance from the source. Curve D is closest to the source, the distance increasing successively for C, B and A. The figure is reproduced from Roberson (1971) with the kind permission of Prof. C. W. Roberson. Experimental parameters are inserted in the figure.*

In this limit the imaginary part of the wave-number is determined by the shape of the electron beam alone.

In order to obtain an analytical expression for the saturated spectrum of the electrostatic waves in the plasma wave-guide, we combine (24.28) with the appropriate diffusion coefficient inserted, and (24.29) with (24.33). Some simplifying assumptions are necessary to reach a manageable set of closed equations. First we introduce the expansion

$$k_1(\omega) \approx k_1(\omega_0) + (\omega - \omega_0)/u_g \qquad \text{with} \qquad \omega_0 = k_1(\omega_0)u \equiv k_0 u \tag{24.34}$$

in the denominator of D entering (24.28). For Langmuir plasma waves in homogeneous isotropic plasmas, the similar expansion is trivial since we have $\omega^2 \approx \omega_{pe}^2$ to a good accuracy there. In the present case we have a significant frequency variation of the wave-number, while at the same time the group and phase velocities of the waves are substantially different, $\omega/k \neq u_g$.

We assume that the real part of the diffusion coefficient in (24.28) can be written as

$$D_1(r,z,u) \equiv \frac{e^2}{m^2}\frac{u k_2}{(1-u/u_g)^2}\sum_\omega \frac{|E_\omega(r,z)|^2}{(k_0 u - \omega)^2 + \left(\dfrac{k_2 u}{1-u/u_g}\right)^2}.$$

Assuming a long time series with duration $\mathcal{T}$, we change the summation to an integral, as in (24.18), here by the replacement $\sum_\omega \to (\mathcal{T}/2\pi)\int d\omega$. For small $\gamma_1 \equiv u/(1-u/u_g)$ we then consider $(1/\pi)k_2\gamma_1/((k_1 u - \omega)^2 + (k_2\gamma_1)^2)$ as equivalent to a δ-function, centered around $\omega_0 \approx k_0 u$ as specified also in the index of $E_\omega(r,z)$. The group velocity appears by the expansion (24.34), where also k_0 is defined. We have γ_1 small when $u_g \ll u$, which is an assumption that has to be checked a posteriori. After these manipulations we finally obtain the approximation

$$D_1(r,z,u) \approx \frac{\mathcal{T}e^2}{2m^2}\frac{|E_{\omega_0}(r,z)|^2}{1-u/u_g}.$$

Combining (24.28) and (24.29), recalling that the ratio $(\partial f_0/\partial u)/k_2$ is independent of z, we omit one $\partial/\partial z$ differentiation and have

$$\frac{\partial}{\partial z}\left(u f_0 + \frac{\mathcal{T}e^2}{4m^2}\frac{\partial}{\partial u}\frac{|E_{\omega_0}(r,z)|^2}{k_2(1-u/u_g)}\frac{\partial}{\partial u}f_0\right) = 0.$$

We now integrate this expression over the plasma column cross section, substitute the analytical result for k_2 from (24.33) and integrate from $z = 0$ to $z = \infty$. This gives

$$\frac{e^2}{m^2}\frac{\partial}{\partial u}\frac{u_g \mathcal{T}}{4\pi u_{The}^2(1-u/u_g)}\frac{\partial W_1}{\partial \omega}\int_0^R r|E_{\omega_0}(r,z\to\infty)|^2 n_0(r)dr = u\int_0^R (f_{0f}-f_{0i})rdr,$$

where we assumed that $|E_{\omega_0}(r,z=0)|^2$ is negligibly small compared to the wave field at $z \gg 0$. We introduced subscripts $_i$ for "initial" and $_f$ for "final" in f_0. We can also introduce $g_f - g_i \equiv 2\pi\int_0^R (f_{0f}-f_{0i})rdr$ to obtain

$$u_g 2\pi\int_0^R \mathcal{E}_\omega(r)\,r\,dr = \frac{2\pi m\,\omega}{\mathcal{T}k_1^2}\left(1-\frac{u}{u_g}\right)\int_{-\infty}^{u}\left(g_f(u')-g_i(u')\right)u'du' \equiv P(\omega), \tag{24.35}$$

where

$$\mathcal{E}_\omega(r) \equiv \frac{\omega\,\omega_{pe}^2(r)}{u_{The}^2 k_1^2}\frac{\partial W_1}{\partial \omega}\varepsilon_0|E_{\omega_0}(r,z\to\infty)|^2$$

is the radially varying wave energy density at frequency ω, and k_1 is the real part of the wavenumber; see also (24.31).

In (24.35) we find that the left side represents the energy flux density integrated over the plasma cross section, or the corresponding wave power per unit frequency. This is the quasi-linear result for the equilibrium power spectrum. We have the total power (i.e., energy per time unit[1]) as $\mathcal{P} = \sum_\omega P(\omega) \to (\mathcal{T}/2\pi)\int P(\omega)d\omega$. The integration is facilitated by recalling the resonance condition $u = \omega/k(\omega)$, which implies

$$\frac{du}{d\omega} = \frac{1}{k} - \frac{\omega}{k^2}\frac{dk}{d\omega} = \frac{1}{k}\left(1-\frac{u}{u_g}\right),$$

i.e., $d\omega = k\,du/(1-u/u_g)$, giving

$$\mathcal{P} = m\int_{u_1}^{u_2}\left(\int_{-\infty}^{u}\left(g_f(u')-g_i(u')\right)u'du'\right)udu = \frac{m}{2}\int_{u_1}^{u_2} u'^3\left(g_f(u')-g_i(u')\right)du',$$

where we integrated by parts, with u_1 and u_2 yet unspecified. The result here relates the wave power to a kinetic energy flux, i.e., to $u(\frac{1}{2}mu^2)$, multiplied by the appropriate velocity distribution function.

To determine the steady state power at $z \to \infty$, we find also here (just as in the initial value counterpart of the problem discussed before) that in this limit we must have $\partial|E(\omega)|^2/\partial z = 0$ as well as $\partial g_f/\partial z = 0$. This means that *either* we have $|E(\omega)|^2 = 0$ *or* $\partial g_f/\partial u = 0$ in this limit. Taking $\partial g_f/\partial u = 0$ in the interval $\{u_1, u_2\}$, and $|E(\omega = ku)|^2 = 0$ and $g_f = 2\pi\int_0^R r f_0(r,u)dr$ for all velocities outside this interval, we have the result for the saturated distribution function

$$\int_{u_1}^{u_2} u'\left(g_f(u')-g_i(u')\right)du' = 0.$$

Since we have g_f constant in this interval, we find

$$g_f(u') = \frac{1}{\int_{u_1}^{u_2} u'du'}\int_{u_1}^{u_2} u' g_i(u')du' = \frac{2}{u_2^2-u_1^2}\int_{u_1}^{u_2} u' g_i(u')du'.$$

---

[1]In Fourier transform we distinguish finite energy spectra where $2\pi\int_{-\infty}^{\infty} f(t)f^*(t)dt$ is finite, and finite power spectra where $\lim_{\mathcal{T}\to\infty}(1/2\mathcal{T})\int_{-\mathcal{T}}^{\mathcal{T}} F(t)F^*(t)dt$ exists (Champeney 1973).

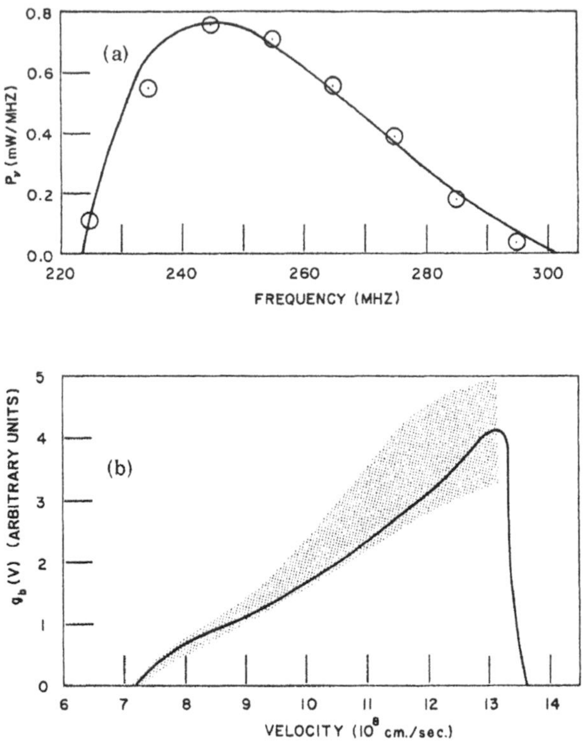

FIGURE 24.4: *Wave power spectrum in a) and inferred beam velocity distribution versus energy analyzer distribution in b). The solid line is obtained from the electronically differentiated output from the analyzer. The shading illustrates the uncertainty on the estimate (Roberson et al. 1971, Roberson & Gentle 1971). The figure is reproduced from (Roberson 1971) with the permission of Prof. C. W. Roberson. A similar figure is found in (Roberson et al. 1971), and permission for use has been obtained also from the American Physical Society. Plasma parameters are given in Fig. 24.2.*

In order to make a quantitative comparison, one has to know the linear dispersion relation, which was also determined experimentally, see Fig. 24.2. It is essential that the beam is weak, meaning that the real part of the dispersion relation is changed negligibly when the beam is introduced. This was tested explicitly, as evident from Fig. 24.2. Note that the finite radial geometry is essential for the real part of the dispersion relation, which is after all nothing but a warm-plasma counterpart of the Trivelpiece-Gould dispersion relation, see Section 12.2.6. Quasi-linear theory thus predicts the power spectrum of the waves, in particular also the total wave power, and finally also the level of the flat saturated velocity distribution also for the conditions discussed here. Experimental results for the variation of the electron beam evolution in Fig. 24.3 are in qualitative agreement with the quasi-linear analysis as summarized here. Some experimental and analytical results for the wave power spectrum and the electron beam velocity distribution are compared in more detail in Fig. 24.4. The uncertainty is there given by a shaded area. The agreement is fair, the experimental difficulties considered and also the approximations made in the analysis.

A

Dimensional Analysis

Dimensional analysis provides more than just a means for carrying out a basic check on the accuracy of some analytical results. The basic observation is that expressions like exp(1 meter) or cos(1 meter) are meaningless, while $\exp(\ell/\ell_0)$ or $\cos(\ell/\ell_0)$ with $\ell_0 = 1$ meter are perfectly meaningful.[1] If a problem contains a characteristic length scale, we expect that the solution of the problem depends in a systematic way on this scale.

Basically, it is argued that *any* physically acceptable result can be written in the form

$$f(Q_1, Q_2, \ldots, Q_n) = 0, \tag{A.1}$$

where Q_1, $Q_2, \ldots, Q_n$, etc. denotes the physical variables of the problem, with dimensions time, position, etc. Using these variables we can obtain a dimensionless form like

$$\pi \equiv Q_1^{\ell_1} \cdot Q_2^{\ell_2} \cdots Q_n^{\ell_n},$$

where the integers $\ell_1, \ell_2, \ldots, \ell_n$ are chosen to make π dimensionless.

We imagine a space spanned by the respective physical dimensions relevant for the problem. The dimension vector of any $Q = A_1^{b_1} \cdot A_2^{b_2} \cdots A_n^{b_n}$ is denoted $\mathbf{b} = (b_1, b_2, \ldots, b_n)$, with the A's denoting the relevant units such as *length, mass*, etc. We can multiply Q by any real number. All elements in the $\mathbf{b}$-vector are integer. The dimension vector of $Q_i^{\ell_i}$ is then simply $\ell_i \mathbf{b}_i$. The condition for π being dimensionless is therefore simply

$$\sum_{i=1}^{m} \ell_i \mathbf{b}_i = 0.$$

We can introduce an $m \times n$ dimensional matrix as $(\mathbf{b}_1, \mathbf{b}_2, \ldots, \mathbf{b}_m)$, where m is the number of relevant dimensions, and n is the number of variables. It is then argued that the variables can be made dimensionless, and (A.1) written as an equivalent relation

$$g(\pi_1, \pi_2, \ldots, \pi_{n-r}) = 0, \tag{A.2}$$

where now π_1, $\pi_2, \ldots, \pi_{n-r}$, etc. are all independent dimensionless variables, where r is the rank of the dimension matrix. This so called π-theorem was first proved by Buckingham (1914), but has been elaborated later by many others (Bridgman 1931, Kurth 1972, Gibbings 1982). The result is invariant with respect to a change in base units.

Written in this form, Buckingham's π-theorem can appear rather formal, maybe even of little practical use. In reality it is a very powerful theorem, but it may best be learned by applying it to some illustrative examples. We will need some basic definitions concerning matrices and sets of linear equations, which will be summarized first (Cramér 1946).

- **Example:** When A is an arbitrary matrix (not necessarily square), the determinant of any sub-matrix of A is called a *minor* of A.

[1] Logarithm functions are a bit special in this respect: while we can argue that a change in unit from $\ell_0 = 1$ meter to $\ell_0 = 1$ inch changes the functional variation of, e.g., $\exp(\ell/\ell_0)$, logarithms like $\ln(\ell/\ell_0)$ or $\log(\ell/\ell_0)$ are only *shifted* but not re-scaled.

In a square matrix $\mathsf{A} = \{A\}_{i,k}$, the cofactor $a_{i,k}$ of the element $\{A\}_{i,k}$ is the particular minor obtained by deleting the i-th row and the k-th column, multiplied by $(-1)^{i+k}$. We have the identities

$$\sum_{j=1}^{n} A_{i,j} a_{k,j} = \begin{cases} \det A & \text{for} \quad i = k \\ 0 & \text{for} \quad i \neq k \end{cases}$$

$$\sum_{j=1}^{n} A_{j,i} a_{j,k} = \begin{cases} \det A & \text{for} \quad i = k \\ 0 & \text{for} \quad i \neq k \end{cases}$$

with $\det A$ denoting the determinant of A. We also have

$$\det A = A_{\ell,\ell} a_{\ell,\ell} - \sum_{i,k=2}^{n} A_{i,\ell} A_{\ell,k} a_{\ell,\ell;i,k}$$

where $a_{\ell,\ell;i,k}$ is the cofactor of $A_{i,k}$ in the matrix $A_{\ell,\ell}$.

The *rank* of a matrix A (not necessarily square) is the greatest integer r such that A contains at least one minor of order of r, which is not zero (Cramér 1946). If all minors of A are zero, then A is a zero matrix, and in this case $r = 0$. When $\mathsf{A} = \{A\}_{m,n}$, the rank r is at most equal to the smallest of the numbers m and n.

Let the rows and columns of A be considered as vectors. If A is of rank r, it is possible to find r linearly independent rows of A, while any $r+1$ rows will be linearly dependent. The same is true for columns.

If $\mathsf{A}_1, \mathsf{A}_2, \ldots, \mathsf{A}_p$ are of ranks $r_1, r_2, \ldots, r_p$, then the rank of the sum $\mathsf{A}_1 + \mathsf{A}_2 + \cdots + \mathsf{A}_p$ is *at most* equal to the sum $r_1 + r_2 + \cdots + r_p$, while the rank of the product $\mathsf{A}_1 \cdot \mathsf{A}_2 \cdots \mathsf{A}_p$ is at most equal to the smallest of the ranks $r_1, r_2, \ldots, r_p$.

If a square matrix $\mathsf{A} = A_{n,n}$ is such that $\det A \neq 0$, then A is of rank n. Such a matrix is said to be *non-singular*, while a square matrix with $\det A = 0$ is of rank $r < n$, and is called a *singular* matrix. If an arbitrary matrix B is multiplied (pre or post) by a non-singular matrix A, then the product has the same rank as B.

If A is symmetric and of rank r, there is at least one *principal* minor of order r in A which is non-zero. Hence, in particular the rank of a diagonal matrix is equal to the number of diagonal elements which are different from zero (Cramér 1946).

- **Example:** Consider an ideal classical pendulum. We expect the expression for its frequency to depend on various physical parameters entering the problem, namely, its mass, m, and length, ℓ, and the gravitational acceleration, g_0, while we wish to describe the displacement of the pendulum by an angle, which is a function of time. We assume the maximum angular displacement to be Θ_0. We write T for *time*, L for *length* and M for *mass*, and have the dimension matrix being

	m	ℓ	Θ_0	g_0	ω
T	0	0	0	-2	-1
L	0	1	0	1	0
M	1	0	0	0	0

where we use a short hand notation $[g_0] = \{-2, 1, 0\}$ for the dimension of acceleration, with the physical dimension being $time^{-2} \times length^{1} \times mass^{0}$, and similarly for other quantities. Evidently, we have $m = 5$, $n = 3$ and the rank of the dimension matrix being $r = 2$ in this case. We have Θ_0 to be dimensionless from the beginning, so we must be able to form one more dimensionless quantity. We write from left to right in the variable on the top in the dimension matrix

$$M^{\alpha_1} \times L^{\alpha_2} \times 1 \times \left(\frac{L}{T^2}\right)^{\alpha_3} \times \left(\frac{1}{T}\right)^{\alpha_4} = 1$$

and determine the exponents α_j is such a way that the exponent of *mass*, of *time* and of *length* is each equal to zero. Evidently this requires $\alpha_1 = 0$, $\alpha_4 = -2\alpha_3$ and $\alpha_2 = -\alpha_3$, so we have $\alpha_3/\alpha_4 = -1/2$ and $\alpha_2/\alpha_4 = 1/2$, where we can choose $\alpha_4 = 2$.

Two dimensionless combinations are then Θ_0 and $\omega^2\ell/g_0$. Consequently, we can write

$$g\left(\Theta_0, \frac{\omega^2\ell}{g_0}\right) = 0,$$

where we take care not to confuse the function g with the gravitational acceleration g_0 entering the argument! We can solve this relation, at least formally, to give

$$\omega^2 = \frac{g_0}{\ell}\,\mathcal{G}(\Theta_0).$$

In view of this expression, we can reformulate the π-theorem to state that any physical quantity can be written as a dimensionally correct constant multiplied with a dimensionless function of one (or, in general, more) dimensionless quantities.

In the small amplitude limit we expect that the angular frequency of the pendulum is independent of the maximum displacement, implying $\mathcal{G}(\Theta_0) \approx$ const. there, since it does not make physical sense for the result to depend on the *sign* of Θ_0, so presumably for small Θ we can approximate $\mathcal{G} \approx \text{const.} + \alpha\Theta_0^2$, with α being some constant. This gives the (well known) result $\omega = \text{const}\sqrt{g/\ell}$. The numerical constant cannot be determined by dimensional arguments. In principle we do not know a priori *what* constant $\mathcal{G}(\Theta_0)$ approaches for small Θ.

- **Example:** Consider the Navier-Stokes equation for an incompressible fluid

$$\frac{\partial}{\partial t}\mathbf{u} + \mathbf{u}\cdot\nabla\mathbf{u} = -\frac{1}{\rho}\nabla p + \frac{\mu}{\rho}\nabla^2\mathbf{u}. \tag{A.3}$$

Assume that the geometry of the problem is properly characterized by a length scale $\mathcal{L}$ and a characteristic velocity U_0. We assume that the mass density is uniform to begin with (and remains so due to the assumed incompressibility), so that the scale length $\mathcal{L}$ solely refers to the velocity variation. We can then introduce the kinematic viscosity $\nu \equiv \mu/\rho$. The assumptions listed here apply to Reynolds' classical experiments in tubes using different cross sections and different flow velocities. The dimension matrix for this case becomes

	p	ν	ρ	u	r	t	$\mathcal{L}$	U_0
T	-2	-1	0	-1	0	1	0	-1
L	-1	2	-3	1	1	0	1	1
M	1	0	1	0	0	0	0	0

We can obtain the following dimensionless quantities $r' \equiv r/\mathcal{L}$, $u' \equiv u/U_0$, $p' \equiv p/(\rho U_0^2)$, $t' \equiv tU_0/\mathcal{L}$, and finally we can also construct *Reynolds' number* $R_e \equiv U_0\mathcal{L}/\nu$. In terms of these new variables we can rewrite the Navier-Stokes equation in the form

$$\frac{\partial}{\partial t'}\mathbf{u}' + \mathbf{u}'\cdot\nabla'\mathbf{u}' = -\nabla' p' + \frac{1}{R_e}\nabla'^2\mathbf{u}'. \tag{A.4}$$

The interesting point is now that if we have results from *one* experiment, we automatically have the results from another self-similar one, constructed in such a way that the Reynolds number is the same. Many model experiments for ship building are based on this observation. Note that the *nature* of the equations is fundamentally different when R_e is finite and when it is infinite! In the former case we have a second order differential equation, in the latter, a first order differential equation in the spatial variables.

We can understand the Reynolds' number physically by considering the ratio between the inertial forces and viscous forces, here under steady state conditions

$$\frac{\textit{inertial forces}}{\textit{viscous forces}} \sim \frac{|\rho \mathbf{u}\cdot\nabla\mathbf{u}|}{|\mu\nabla^2\mathbf{u}|} \sim \frac{\rho U^2/\mathcal{L}}{\mu U/\mathcal{L}^2} \sim \frac{\rho U\mathcal{L}}{\mu} = \frac{U\mathcal{L}}{\nu}$$

where we used $|\nabla\mathbf{u}| \sim U/\mathcal{L}$, etc. The latter equality is valid for constant mass density, using the relation between viscosity μ and kinematic viscosity ν.

Ⓢ **Exercise:** Consider a large spherical metal object in vacuum. Estimate, by dimensional analysis, the time constant τ for the initial temperature variation to even out. Given parameters are the radius R, the temperature difference ΔT between the center and the surface, the heat conductivity k and the heat capacity C per unit volume. For obtaining numerical results use the relevant numbers for iron, and let R be the radius of the Earth. Thermal expansion can be ignored. Comment on the relevance of thermal radiation. The present and following exercises were suggested by J. Højgaard Jensen in *Kvant*, the periodical of the Danish Physical Society. Ⓢ

Ⓢ **Exercise:** Placing your hand on a large cold piece of metal you will feel a cooling of your hand, indicating a heat flux from hand to metal. Use dimensional arguments to determine the initial time variation of the heat flux ("initial" meaning that your hand has not cooled down completely, so the temperature difference can be taken constant). Relevant parameters are the temperature difference ΔT between hand and metal, the heat conductivities k_h and k_m of hand and metal, respectively, and the corresponding heat capacities C_h and C_m. We have $k_h \ll k_m$. Ⓢ

Ⓢ **Exercise:** A harmonically varying signal will propagate along a telephone cable according to the expression

$$I(x,t) = I_0 e^{-\beta x}\cos\left(\omega t - \frac{\omega}{\vartheta}x\right)$$

where β (the spatial damping constant) and ϑ (the phase velocity) will depend on ω in general. For telephone cables with negligible leak-currents to the surroundings and large self-inductions it is found, however, that β and ϑ are independent of ω and depend only on the self-induction ℓ, the resistivity r, and the capacitance c of the cable, all taken per unit length (we used lower-case letters to distinguish these quantities from resistors, self-inductances and capacitors in lumped electrical circuits usually denoted by R, L and C, respectively). Determine by dimensional reasoning the variation of β and ϑ with varying ℓ, r and c. Ⓢ

A.1 Summation convention

Expressions with many summation signs can be rather cumbersome. The Einstein summation convention says that any repeated index is to be summed over, resulting in a very convenient short hand notation! Introduce a shorthand notation for the differential operator as $\nabla \equiv \{\partial/\partial x, \partial/\partial y, \partial/\partial z\} \equiv \{\partial_1, \partial_2, \partial_3\}$, or ∂_j with $j = 1,2,3$. By the Einstein convention, we can express for instance the incompressibility condition $\nabla\cdot\mathbf{u} = 0$ as $\partial_j u_j = 0$ with this elegant notation. For second rank tensors we have the double scalar product $\underline{\mathsf{A}} : \underline{\mathsf{B}} \equiv A_{ij}B_{ji} = Trace[\underline{\mathsf{A}}\cdot\underline{\mathsf{B}}]$, in particular also $\nabla\nabla : \underline{\mathrm{ab}} = \partial_j\partial_i a_i b_j$. Unfortunately, the notation used here is not uniquely adopted and we may also see $\underline{\mathsf{A}} : \underline{\mathsf{B}} \equiv A_{ij}B_{ij}$. For symmetric tensors the two definitions coincide.

- **Exercise:** Simplify $\nabla\cdot(\mathbf{u}\cdot\nabla\mathbf{u})$, see e.g. (2.24).

B

Collisional Cross Sections

B.1 Cross sections in general

Interactions between particles are at times best described in terms of *cross sections*. More generally, cross sections can be defined for almost all sorts of interactions, but as an introduction we use here a model for a beam of particles interacting with a medium consisting of a diffuse arrangement of diluted scatterers. Assume that we take a thin slab with thickness dz and area A of this scattering medium and an incident beam of, for instance, electrons with flux density $\Gamma \equiv n_e U$, where the electron velocity is U and their density is n_e. The beam direction is perpendicular to the slab in consideration. The basic interaction is first assumed to be a scattering of electrons by the particles distributed in the medium. If an electron hits a particle, it is scattered out of the beam and is assumed lost, otherwise it is collected on the other side. The slab thickness is so small that we can ignore the possibility of two particles "shadowing" each other. The entire area of scatterers seen from the direction of the electron beam is therefore $N\sigma$, where σ is the projected area of just one scatterer and N is the number of scatterers in the slab with the given thickness and area. Given the density, n_s, of scatterers, we have $N = n_s A dz$.There will be statistical variations in N, but the relative fluctuations around the average value $\langle N \rangle$ will be small if the volume is large, i.e., if A is large, so we might as well treat N as a deterministic quantity here. See Fig. B.1 for an illustration of the problem.

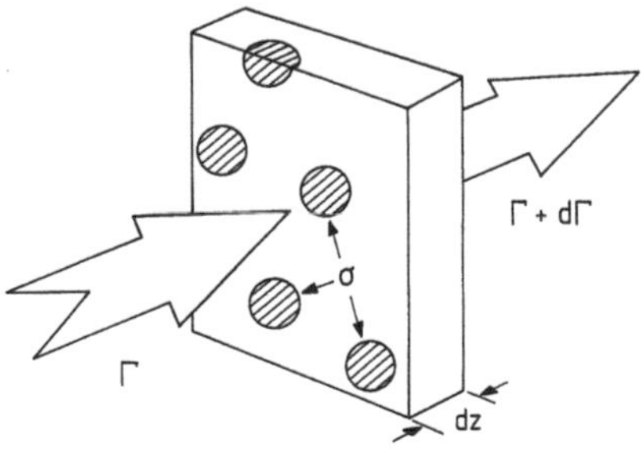

FIGURE B.1: *Illustration of the geometry for flux depletion by scattering from scatterers with cross section σ, being uniformly distributed in a statistical sense. The large arrows symbolize a spatially homogeneous flux, incident in the direction perpendicular to the surface of the reference volume shown.*

The relative change in flux is equal to the relative fraction of the entire area covered by the scatterers, in other words

$$\frac{\Gamma - (\Gamma + d\Gamma)}{\Gamma} = \frac{N\sigma}{A} = \frac{n_s \sigma A dz}{A} \tag{B.1}$$

or

$$d\Gamma = -\Gamma n_s \sigma dz, \tag{B.2}$$

where $d\Gamma$ may take either sign, i.e., we can have gain or loss when passing through the slab, depending on the physics described by σ. Provided n_s is constant, independent of the position z, we easily find the solution

$$\Gamma = \Gamma_0 e^{-\sigma n_s z} \equiv \Gamma_0 e^{-z/\ell_c} \tag{B.3}$$

where γ_0 is the incoming flux at the position $z = 0$ and we defined a characteristic length $\ell_c \equiv 1/(n_s \sigma)$ which is the distance where the incoming flux is decreased by a factor e. It is natural to interpret ℓ_c as a *mean free path* for the collisions. For a particle with velocity U, we can obtain the collision frequency as $\nu = u/\ell_c = U n_s \sigma$. Here we note that this is a dimensionally correct choice, and refer to Section 8.9 for details.

It may be timely here to add a few words on something which may appear as a technicality: we introduced ν as a collision frequency before, and by its definition, we can interpret $1/\nu$ as the average time between two "events," where we might leave the definition of what we actually mean by the interaction/scattering to later. We can define another time, namely, the one between two *encounters*, opening the possibility that two particles can meet, without affecting each other, i.e., no collision occurs. The time for this latter case is, on average, given by L/U, where L is the average distance between particles, and again U is the velocity of our reference particle. Evidently, we expect $L/U \leq \ell_c/U$. Often, we find collision frequencies for scattering processes to be related to U/L.

In case the distribution of scattering particles is inhomogeneous, we have an explicit spatial dependence of $n_s = n_s(\mathbf{r})$, and the situation becomes a bit more complicated. For the simple case where the dependence is on z only, $n_s = n_s(z)$, we find

$$\frac{d}{dz} \ln\left(\frac{\Gamma}{\Gamma_0}\right) = -\sigma n_s(z) \tag{B.4}$$

giving

$$\Gamma = \Gamma_0 e^{-\sigma \int_0^z n_s(z')dz'}, \tag{B.5}$$

where it is natural to assume σ to be constant, and take it outside the integration sign. The constant Γ_0 is determined by requiring $\Gamma = \Gamma_0$ for $z = 0$.

The concepts discussed in this appendix can be generalized considerably; we need not deal with an incoming beam of particles being scattered as they might as well be absorbed, for instance. In this case we deal with an *absorption cross section.* The basic point is that we have something coming in, and that this something is changed somewhat when it propagates through the slab of the medium it interacts with. The ideas are readily generalized to incoming radiation, etc. Note that we *can* have amplification; more may come out than what we send in! Measurements of cross section for various processes constitute a "cornerstone" in experimental physics. The reader has to consult the (vast) literature for relevant information, but we might mention some useful works concerning electron collisions by Pack and Phelps (1961), Frost and Phelps (1964) and Itikawa (1973). The mobilities of ions in a background of neutral gas are discussed by, for instance, Ellis et al. (1976).

The incoming flux is generally constituted by particles with a given velocity, and it will generally be found that the cross sections depend on the velocity, with the exception of "billiard ball"-like scattering. In general, the velocity variation can be quite complicated, but a few idealized reference cases can be mentioned:

1) The hard sphere, or "billiard ball," collisions already discussed. Here $\sigma(u) =$ const. and consequently the collision frequency varies as $\nu(u) \propto u$.

2) For interactions given by polarization forces, for so called "Maxwellian molecules" (Trubnikov 1965), we find for a wide velocity range that $\sigma \propto 1/u$, implying $\nu(u) \propto$ const.

3) For Coulomb interactions, we find that $\sigma \propto 1/u^4$, implying $\nu(u) \propto 1/u^3$.

In general we have (Bekefi 1966) that for an inverse power law dependence of the interaction force, $F \propto 1/r^s$, we find, by a dimensional analysis, the velocity dependence of the collision frequency to be $\nu(u) \propto u^{(s-5)/(s-1)}$. For instance, for Maxwellian molecules we have $s = 5$, for Coulomb interactions $s = 2$.

Note that the discussion of cross section and the collision frequencies derived from it refers to explicit processes. One of the most important applications is the question of momentum relaxation, and we note here that particles can very well collide without giving up *all* of their momentum; see, for instance, Section 8.7. As a "rule of thumb" you can assume that for isotropic collisions, where the particles leave the interaction with a velocity vector pointing in any spherical direction with equal probability, the memory for momentum loss is the same as a collision time. If it is not so, then it takes several collisions before the original momentum is lost. Momentum losses are special by involving both the magnitude and the direction of a vector. Cross sections for energy exchange and momentum exchange are generally different. We can, for instance, assume that an electron gives up all of its momentum upon collision with an ion, but it takes a number $\sim M/m$ collisions before it gives up all of its energy. Collisions between charged particles and neutrals involve many physical processes strongly depending on the energy of the particles involved. The energy needed for ionization of atoms is usually well defined, but for most laboratory (except fusion) experiments the temperatures are often too low for ionization to be important. The energy needed for excitation of neutrals is smaller, with data available in the literature. An excited atom or molecule once formed will decay to a lower level state, the ground state in particular, by emitting radiation. For collisions with molecules it is found that rotation and also vibration energies are in the range most relevant for instance in the ionosphere. Charge exchange processes can be represented as

$$A^+ + B \rightarrow A + B^+ \tag{B.6}$$

In a collision between the positive ion A^+ and the neutral atom B, an electron is transferred from B to A so that the final state consists of a neutral atom A and a positive ion B^+. The process represents an effective cooling mechanism if A^+ is an energetic ion and the B is a cold component, in which case the final state in (B.6) consists of a slow ion and a fast neutral A. The cross section of this process is generally large, in particular if A and B are the same particle species. At large relative velocities, the cross section for charge exchange falls of rapidly with velocity, but the process is significant for small energies, in Cesium plasmas in Q-machines for instance (Motley 1975).

A standard unit for cross sections is given in terms of the Bohr radius, $a_0 = 0.53 \times 10^{-10}$ m, giving $\sigma_0 \equiv \pi a_0^2 = 0.88 \times 10^{-21}$ m^2 for the corresponding geometrical cross section. The unit often used for reaction cross sections is *barn*, being equal to 10^{-28} m^2. The name, coined by scientists with a certain sense of humor, is derived from the proverbial phrase "side of a barn," i.e., something easy to hit. Values of cross sections for a given nucleus depend on the energy of the bombarding particle and the kind of reaction being considered. Boron, for example, bombarded by neutrons traveling 10^6 cm/s (or 22,500 miles per hour) has a cross section for the neutron-capture reaction of about 120 barns, whereas the cross section increases to about 1,200 barns for neutrons traveling at 10^5 cm/s. These large cross sections mean that boron is a good absorber of neutrons. The metal zirconium, on the other hand, having an absorption cross section of only 0.18 barn for the low energy neutrons that cause fission in nuclear reactors, is rather transparent to neutrons and is used to clad reactor fuel rods. The reactor fuel uranium-235 absorbs neutrons both by fission and by capture. The fission cross section is 580 barns, whereas the total cross section for neutron absorption (including the capture cross section) is 680 barns.

B.2 Mean free paths

Given the cross section of a scattering process, we can calculate the mean free path of a particle. As an aide for the discussion we take Fig. B.2, showing by the solid line in a) the orbit of a small particle undergoing collisions. The scattering particles are indicated by solid circles. Note that the illustration does not assume billiard ball collisions, i.e., the orbit does not touch the scattering particles. It is implicitly assumed that the duration of the collisional interaction is small compared to the time between collisions. In Fig. B.2b) we surround the particle trajectory by a tube having the cross section to be the scattering cross section $\sigma \approx \pi d^2$. Then in Fig. B.2c) this tube is straightened out. The volume of the tube is $\sigma \sum_j^N \ell_j$ where ℓ_j is the length of the free path between scatter number j and number $j+1$. We assume that there are N scatterings in the record available. If the density of scattering particles is n_s, then we can estimate the number of scattering particles in the tube as $\sigma n_s \sum_j^N \ell_j$. This number is varying from one realization of the process to another, but if we assume that N is large and the tube is long then the relative scatter of the number of particles in the tube is modest, and we can take this number to be deterministic. By construction, each and every one of these particles has given rise to a collision, i.e., "bendings" in the orbit of the scattered particle. The average mean free path is then the length of the total path divided by the number of such bendings

$$\ell_c = \frac{\sum_j^N \ell_j}{\sigma n_s \sum_j^N \ell_j} = \frac{1}{n_s \sigma}.$$

So far the arguments do not involve the velocities of the scattering particles or the one *being* scattered. These velocities are important for estimating the time between scatters. If the scattering particles are heavy and the scattered particle light and fast (as electrons scattered by ions, for instance), we can assume the electron energy to remain constant: the scattering process only changes the *direction* of propagation. The time between two collisions is then $\tau_j = \ell_j / u$ where u is the particle velocity, assumed constant in this time interval. If we let a light particle collide with very heavy ones, we can assume the energy to remain constant (at least for a long time) and assume that only the direction of propagation is changed. The average time between collisions is then ℓ_c / u and a collision frequency is $\nu_c = u/\ell_c$. If the collisional cross section is velocity dependent (and this is generally the case), we will have to average over the actual velocity distribution $f(u)$ to obtain the average collision frequency (Gurevich 1978).

- **Exercise:** Consider a gas consisting of identical spherical particles with diameter $2d$ and density n, with mutual collisions of the type 1 discussed in Appendix B.1. Using the results of Appendix B.2 and ignoring motion of scatterers, demonstrate that the collisional mean free path is

  $$\ell_c = \frac{1}{4\pi d^2 n}.$$

  Introducing the density as the number of particles divided by the volume, demonstrate that

  $$\ell_c = \frac{RT}{4\pi d^2 N} \frac{1}{p},$$

  where p is the pressure. You could also use $p = n\kappa T$. For constant temperature, T, we find ℓ_c to be inversely proportional to the pressure if we vary the volume.

  As posed here, the problem assumes that the particle *being scattered* is moving, while the scatterers are at rest. Consider the effect of statistically moving scatterers, using qualitative arguments to compare particles with head on and overtaking collisions.

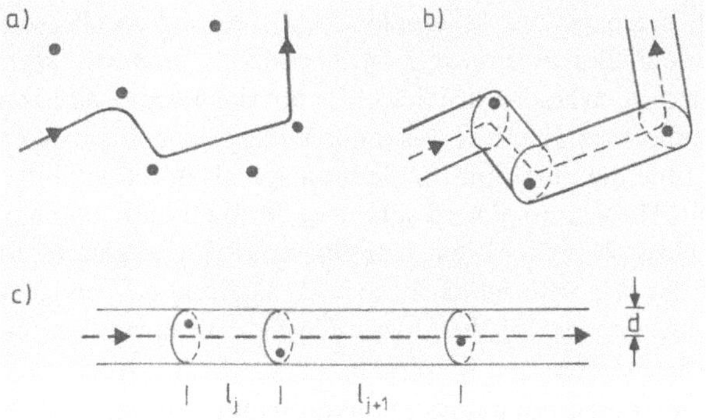

FIGURE B.2: *Illustration of orbit of a particle undergoing collisions. The scattering particles are illustrated by solid circles. The figure should be imagined as three dimensional. Figures a) and b) are in principle identical, only in b) we introduced a "tube" with the cross section being the scattering cross section πd^2, where its radius is d. In c) this tube is "straightened out", with ℓ_j being the spatial separation between collision number j and $j+1$.*

The analysis outlined here assumed that the scatterers are at rest and the only the particle *being* scattered is in motion. The results will be identical if we reverse the conditions and let the scatterers move against a fixed particle to be scattered. Only the relative motion of the particles matters. For a general analysis in a laboratory frame of reference where all particles are in motion we have to average over the relative velocities and directions of all particles. As a simple, yet useful model we can take all velocities to be the same, only the directions being random. This will give the relation $\ell_c = 3/(16\pi d^2 n)$. If we now assume that the particle velocities are given by a Maxwellian distribution function we find $\ell_c = 1/(4\sqrt{2}\pi d^2 n)$.

- **Exercise:** Assume that an incoming particle on Fig. B.1 can be scattered by *two* different types of particles, with different cross sections. Assume that both types of particles are uniformly distributed in the medium, and that the scattering processes are statistically independent. Demonstrate that the collision frequencies for the two scattering processes are *additive*, i.e., if the impurities of type 1 alone give a collision frequency ν_1 and those of type 2 give ν_2, then we have $\nu_{1+2} = \nu_1 + \nu_2$ when they are both present. Why is it important for the argument that the processes are *independent*?

B.3 A statistical model for collisions

Basically, we have to realize that collisions form a statistical process, and we have to decide on an appropriate model. The collision between a charged particle and a neutral atom depends in general on their relative velocities in a complicated manner. In order to retain the model on a tractable level it will here be assumed that the neutrals are so heavy that they might be considered almost immobile, at least compared with the moving free electron. The simplest and almost universally used model is based on the assumption that the probability of a collision in a small time interval, dt, is proportional to the length of that time interval, where ν is a constant of proportionality. Let the probability of a collision in a short time interval $\{t; t+\Delta t\}$ be

$$P(\text{one collision in } \Delta t) = \nu \Delta t, \tag{B.7}$$

with ν being a constant, independent of the velocity of the electron (Pécseli 2000). Effectively it is thus assumed that the collisional cross section is inversely proportional to the relative particle velocity. Although this is a nontrivial restriction, it is not unphysical, and is a good approximation to a number of relevant processes. In effect it is assumed that the collisional cross section is proportional to the interaction time, i.e., the time an electron spends in the vicinity of the neutral, not an unreasonable assumption. These are not hard sphere or "billiard-ball" collisions.

By assumption, the electron (or ion) leaves the neutral with a velocity component w, which is entirely independent of the one with which it arrived. There is no persistence of velocities. The constant ν is therefore independent of the "history" of the electron, i.e., on times for foregoing collisions. This is a simple example of a Markov process, which is frequently encountered in practical applications. With the present model (B.7) the probability of a free flight starting at $t=0$ and persisting at a time t can be calculated as follows:

Consider a time interval of length $t+\Delta t$ broken up into two subintervals, one of length t and one of Δt. Since the probability of a collision in Δt is independent of the prehistory, it follows that $P(0,t+\Delta t) = P(0,t)P(0,\Delta t)$, where for brevity we introduced the notation $P(0,t+\Delta t) = P(\text{no collision in the interval } t+\Delta t)$, etc. Evidently $P(0,\Delta t) = 1-\nu\Delta t$, with the probability of two or more collisions in the time interval Δt being negligible when Δt is small. Consequently $P(0,t+\Delta t) = (1-\nu\Delta t)\Delta t P(0,t)$, giving

$$\frac{P(0,t+\Delta t)-P(0,t)}{\Delta t} = -\nu P(0,t).$$

In the limit $\Delta t \to 0$ this difference equation becomes

$$\frac{dP(0,t)}{dt} = -\nu P(0,t), \tag{B.8}$$

giving the result

$$P(0,t) = e^{-\nu t}, \tag{B.9}$$

with the condition $P(0,t=0) = 1$ from (B.7), since $P(0,\Delta t) = 1-\nu\Delta t$, as before.

The average time between collisions can be found by noting that the probability of a free flight beginning at t_1 to be pertaining at t_1+t is $\exp(-\nu t)$. The probability that a free flight beginning at t_1 is terminated by a collision in the time interval $[t_1+t; t_1+t+dt]$ is then $\exp(-\nu t)\nu dt$. The average time between collisions is consequently given by

$$\int_0^\infty t\exp(-\nu t)\nu dt = \frac{1}{\nu}.$$

The average collision frequency is thus ν. The argument can be applied when "looking backwards" in time as well. Then $\exp(-\nu t)\nu dt$ is the probability that an ongoing free flight at t_1 started with a collision in a narrow time interval dt at t_1-t. Some seeming paradoxes related to this observation are summarized by, for instance, Pécseli (2000) in an appendix.

Assume now that the charged particles as well as the neutral background are embedded in a spatially uniform time-varying deterministic electric field, with intensity $E(t) = \Re\{E_0 e^{-i\omega t}\}$. The equation of motion between collisions is now $m\,dU(t)/dt = -e\Re\{E_0 e^{-i\omega t}\}$. The electron velocity between collisions is given by

$$\begin{aligned} U(t) &= U(t_1) + \frac{eE_0}{i\omega m}\Re\left\{e^{-i\omega t} - e^{-i\omega t_1}\right\} \\ &= U(t_1) + \frac{eE_0}{i\omega m}\Re\left\{\left(1-e^{-i\omega(t_1-t)}\right)e^{-i\omega t}\right\}, \end{aligned} \tag{B.10}$$

where it is assumed that the electron starts out with a velocity $U(t_1)$ after a collision at $t=t_1$. We use the capital letter U to denote the velocity of an individual electron. It is assumed implicitly that the electron starts out with a random velocity direction uniformly distributed in angle over 4π, so that the average is vanishing, $\langle U(t_1)\rangle = 0$.

The expression (B.10) remains valid until the next collision. The second term is statistically varying since the time t_1 is statistically distributed. The average value of the second term is different from zero. We can split the time interval into two parts: from t going backwards in time until $t_1 + d\Delta t$, and then a small time interval $d\Delta t$. In the first of these intervals we have no collisions, and then one collision in the last small one. The statistical distribution of $\Delta t \equiv t - t_1$ is given by the product of two probabilities, i.e., $\exp(-\nu\Delta t)\nu d\Delta t$, where Δt denotes the time since the previous collision. The average value of $U(t)$ is readily obtained as

$$\langle U(t)\rangle = -\frac{eE_0}{m}\Re\left\{\frac{e^{-i\omega t}}{\nu - i\omega}\right\}, \tag{B.11}$$

using $\int_0^\infty [1-\exp(-i\omega\tau)]\exp(-\nu\tau)\nu\, d\tau = i\omega/(\nu + i\omega)$. The result is valid also for dc conditions, where $\omega = 0$. Details of the foregoing discussions are given for instance by Boström (1964) and Pécseli (2000).

- **Exercise:** We obtained in (B.9) the probability of *no* collisions in the interval $\{0;t\}$. Demonstrate that the probability of K collisions within the interval is given by the Poisson distribution

$$P(K,t) = \frac{(\nu t)^K e^{-\nu t}}{K!}.$$

Show that the average number of collisions in a time interval $\mathcal{T}$ is

$$E(K) = \sum_{K=0}^{\infty} K\frac{(\nu\mathcal{T})^K e^{-\nu\mathcal{T}}}{K!} = \nu\mathcal{T}. \tag{B.12}$$

Evidently, this average need not be an integer number, even though K *is* integer. The result (B.12) is consistent with the average collision frequency being ν.

B.3.1 Analytical collision model

A model equation can be proposed in the form

$$\frac{d}{dt}U(t) = \frac{q}{M}E_0 e^{-i\omega t} - \nu U(t) \tag{B.13}$$

for describing collisions between neutrals and charged particles, subject to an oscillating electric field. The relation is written in the rest frame of the neutrals. We obtain trivially for this case

$$U(t) = i\frac{q}{M}E_0\frac{e^{-i\omega t}}{\omega + i\nu}.$$

The model (B.13) will be used often, also for collisional processes of other types. The physical argument in its derivation is found by noting its similarity with, for instance, a simple oscillating system subject to a restoring force and a friction. Taking, for instance, an oscillator with displacement Δx we have the equation $Md^2\Delta x/dt^2 = -K\Delta x - \nu d\Delta x/dt$, with M being the mass, $K\Delta x$ being the restoring force of a spring, and ν representing a friction with a supporting surface. Here, the friction plays the role of the collision frequency in (B.13).

By inspection it is evident that the result (B.11) is formally contained in the analytical solution of the model deterministic equation (B.13). Taking the real part, we find precisely the result for the *average* velocity obtained in (B.11). If we are not interested in the fine details in the particle velocity, we might simply identify the solution of (B.13) with the average particle velocity. The simple model

(B.13) will be used as representative for the underlying statistical process, also for more complicated space-time varying fields. It is important, however, that the model is based on a force, which can be considered constant in space, at least locally. This implies that we should use (B.13) with care in the case where the spatial field variations are on a scale comparable to, or smaller than, the collisional mean free paths.

B.3.2 Ohm's law

Consider for simplicity an unmagnetized plasma where the ions can be assumed to constitute an immobile background of positive charge density. The electron motion is (again) described by the equation

$$m\frac{d}{dt}\mathbf{U} = -e\mathbf{E} - m\nu\mathbf{U},$$

with ν representing the appropriate collision frequency. If we ignore electron inertia, i.e., the left-hand side of the relation, we find trivially

$$\mathbf{U} = -\frac{e}{m\nu}\mathbf{E} \equiv -\mu\mathbf{E}$$

where we introduce the *electron mobility* $\mu \equiv e/(m\nu)$. Inertialess models apply for slow variations or low frequencies, in this case $\omega \ll \nu$.

We now introduce the current density $\mathbf{J} = -en\mathbf{U}$, in terms of the electron density n, and have $\mathbf{J} = -en\mu\mathbf{E}$. We can introduce the conductivity as $\sigma \equiv en\mu = e^2 n/(m\nu)$ and corresponding resistivity $\xi \equiv 1/\sigma = m\nu/(e^2 n)$. If the current density is constant over a plasma cross section $\mathcal{A}$, we can introduce the net current in that cross section as $\mathbf{I} = \mathbf{J}\mathcal{A}$. If the electric field is constant in that same cross section and in addition also over a length $\mathcal{L}$, we have $\Phi = \mathcal{L}E$ in terms of a voltage difference Φ between the two end-points of $\mathcal{L}$, so that

$$I = \frac{\Phi}{R}, \qquad \text{with} \qquad R \equiv \frac{m\nu\mathcal{L}}{e^2 n\mathcal{A}}.$$

This is Ohm's law, with R being the resistance in the "circuit." With a constant R, the relation is seemingly based on a constant mobility assumption, with inertia being ignored. The electron mass enters through the collision term.

C

The Plasma Dispersion Function

Since plasmas in thermal equilibrium obviously deserve particular interest, it is found worthwhile to analyze the dielectric function for plasmas with Maxwellian velocity distributions in particular detail. The expression we so often encounter is

$$\varepsilon(k,\omega) = 1 - \left(\frac{\omega_p}{k}\right)^2 \fint_{-\infty}^{\infty} \frac{f_0'(u)}{u - \omega/k}\, du, \tag{C.1}$$

where we here want to study the special case where the velocity distribution is

$$f_0(u) = \sqrt{\frac{m}{2\pi\kappa T}}\, e^{-\frac{1}{2}mu^2/\kappa T}. \tag{C.2}$$

At this place, m can be the mass of any of the plasma constituents, and ω_p any relevant plasma frequency. We again used the symbol $\fint$ for the Landau contour of integration. By insertion we have

$$\varepsilon(k,\omega) = 1 + \frac{1}{k^2\lambda_D^2\sqrt{\pi}} \fint_{-\infty}^{\infty} \frac{xe^{-x^2}}{x - z}\, dx \equiv 1 + \frac{1}{k^2\lambda_D^2\sqrt{\pi}} I(z), \tag{C.3}$$

with

$$z \equiv \frac{\omega}{k}\sqrt{\frac{m}{2\kappa T}}.$$

By simple manipulations, the integral in (C.3) can be written as

$$I(z) = \sqrt{\pi} + z \fint_{-\infty}^{\infty} \frac{e^{-x^2}}{x - z}\, dx.$$

It is now instructive to rewrite the last term in $I(z)$ by introducing a new function

$$G(\alpha)e^{\alpha z^2} \equiv \fint_{-\infty}^{\infty} \frac{e^{-\alpha(x^2 - z^2)}}{x - z}\, dx,$$

and differentiate with respect to α to find

$$\frac{d}{d\alpha} G(\alpha)e^{\alpha z^2} = - \fint_{-\infty}^{\infty} \frac{(x^2 - z^2)e^{-\alpha(x^2 - z^2)}}{x - z}\, dx \tag{C.4}$$

$$= -2ze^{\alpha z^2} \int_0^{\infty} e^{-\alpha x^2}\, dx = -ze^{\alpha z^2}\sqrt{\frac{\pi}{\alpha}}. \tag{C.5}$$

By integration of (C.5) we have

$$G(\alpha)e^{\alpha z^2} = -2z\sqrt{\pi} \int_0^{\alpha} e^{\alpha' z^2} \frac{d\alpha'}{\sqrt{\alpha'}} + i\pi \tag{C.6}$$

$$= -2\sqrt{\pi} \int_0^{\sqrt{\alpha}} e^{(zt)^2}\, dt + i\pi \tag{C.7}$$

$$= -2\sqrt{\pi} \int_0^{z\sqrt{\alpha}} e^{t^2}\, dt + i\pi, \tag{C.8}$$

where we used $G(0) = i\pi$, recalling the integration contour implicit in $⨍$. All in all we have

$$G(\alpha) = -2\sqrt{\pi}e^{-\alpha z^2} \int_0^{z\sqrt{\alpha}} e^{t^2}\,dt + i\pi e^{-\alpha z^2}. \tag{C.9}$$

The last term in the integral $I(z)$ is now obtained by setting $\alpha = 1$, to give

$$\varepsilon(k,\omega) = 1 + \frac{1}{k^2\lambda_D^2\sqrt{\pi}}\left(1 - 2ze^{-z^2}\int_0^z e^{t^2}\,dt + iz\sqrt{\pi}e^{-z^2}\right).$$

We now introduce the so-called *plasma dispersion function* (Fried & Conte 1961, Nicholson 1983)

$$Z(z) = 2ie^{-z^2}\int_0^{iz} e^{-t^2}\,dt.$$

By differentiation of $Z(z)$ we find

$$Z'(z) = -2(1 + zZ(z)) \tag{C.10}$$

and can now write the longitudinal plasma dispersion function, for a plasma in thermal equilibrium at a temperature T, as

$$\boxed{\varepsilon(k,\omega) = 1 - \frac{1}{2(k\lambda_D)^2}Z'\left(\frac{\omega}{k}\sqrt{\frac{m}{2\kappa T}}\right),}$$

or more generally, including both the electron and the ion dynamics, as

$$\boxed{\varepsilon(k,\omega) = 1 - \frac{1}{2(k\lambda_{De})^2}Z'\left(\frac{\omega}{k}\sqrt{\frac{m}{2\kappa T_e}}\right) - \frac{1}{2(k\lambda_{Di})^2}Z'\left(\frac{\omega}{k}\sqrt{\frac{M}{2\kappa T_i}}\right),}$$

recalling the definition of the variable z. The two functions $Z(z)$ and $Z'(z)$ are shown in Figs. C.1 and C.2, respectively, as functions of a real argument. We have $\Im\{Z(0)\} = \sqrt{\pi}$. Often, a slightly different definition is used, where we introduce the function

$$\begin{aligned} W(x) &\equiv -1 + \frac{x}{\sqrt{2}}Z\left(-\frac{x}{\sqrt{2}}\right) \\ &= \frac{1}{2}Z'\left(-\frac{x}{\sqrt{2}}\right). \end{aligned} \tag{C.11}$$

By using W instead of Z', we get a slightly more elegant formulation of the plasma dielectric function as

$$\boxed{\varepsilon(k,\omega) = 1 - \frac{1}{(k\lambda_D)^2}W\left(-\frac{\omega}{k}\sqrt{\frac{m}{\kappa T}}\right).}$$

For small arguments $|x| \ll 1$ we have the expansion

$$W(x) \approx -1 + x^2 + ix\sqrt{\frac{\pi}{2}}e^{-x^2/2},$$

while for $|x| \gg 1$ we find

$$W(x) \approx \left(1 + \frac{3}{x^2}\right)\frac{1}{x} + ix\sqrt{\frac{\pi}{2}}e^{-x^2/2}.$$

Often we find that a velocity distribution function can be approximated by a sum of Maxwellians, and in such cases we can use a model containing a sum of corresponding Z-functions.

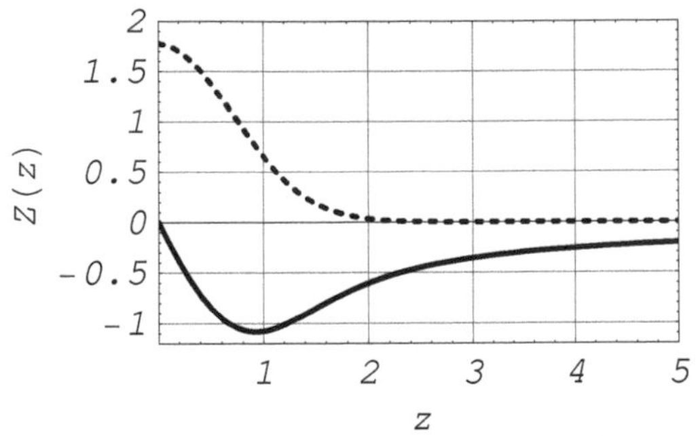

FIGURE C.1: *The plasma dispersion function, $Z(z)$, shown as a function of a real argument, z. The real part of Z is shown with a solid line, the imaginary with a dashed line. For large z we have $\Re\{Z(z)\} \approx -1/z$.*

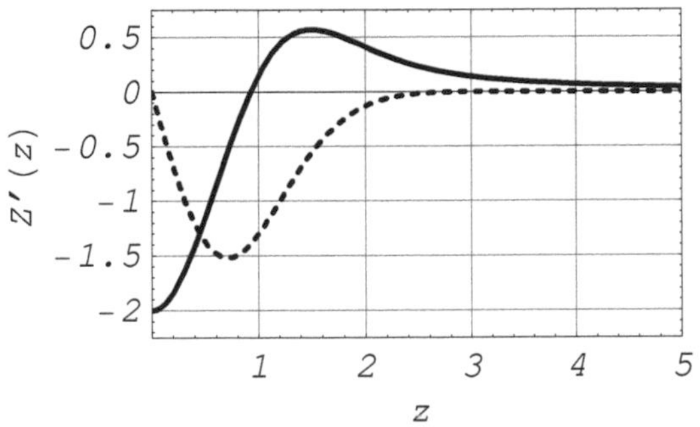

FIGURE C.2: *The derivative of the plasma dispersion function, $Z'(z)$, shown as a function of a real argument, z. The real part of Z' is shown with a solid line, the imaginary with a dashed line. For large z we have $\Re\{Z'(z)\} \approx 1/z^2$.*

C.1 Approximations for the plasma dispersion function

The function $Z'(z)$ is tabulated (Fried & Conte 1961) for real as well as complex values of z. It is, however, a great advantage to have an approximate formula which can be used on a pocket calculator, or similar. We have for real variables z that

$$\Im\{Z(z)\} = \sqrt{\pi}\exp(-z^2),$$

while for $z^2 < 6.25$ we have

$$\Re\{Z(z)\} \approx -2z\frac{((-0.03566424z^2+1.365845)z^2-4.302978)z^2+47.68008}{((z^2+6.90076)z^2+27.50551)z^2+47.67912},$$

for $6.25 < z^2 < 12.25$ we have

$$\Re\{Z(z)\} \approx -\frac{2}{z}\left(0.49916+\frac{0.251472}{-1.96079+z^2}\right),$$

for $12.25 < z^2 < 25$ we have

$$\Re\{Z(z)\} \approx -\frac{2}{z}\left(0.5002219+\frac{0.24046413}{-1.9953175+z^2}\right),$$

and for $25 < z^2$ we have

$$\Re\{Z(z)\} \approx -\frac{1}{z^3}\left(z^2+0.50003811+\frac{0.74493877}{-2.7486278+z^2}\right),$$

and yes, you better keep all those digits, in particular in case you want to use (C.10) for calculating Z', for instance. The approximation given here is very good, and to the given accuracy hardly distinguishable from the tabulated values. The problems will arise if the expression is differentiated too many times; in that case discontinuities will become noticeable at the values $z^2 = 6.25$, $z^2 = 12.25$, etc. Recall that $\Re\{Z(-z)\} = -\Re\{Z(z)\}$ and $\Im\{Z(-z)\} = \Im\{Z(z)\}$, so $z < 0$ need no special attention.

Other approximations for the plasma dispersion function can be found in the literature. Some are based on Padé approximations (Martín & González 1979, Tjulin et al. 2000, Martín et al. 1980).

C.2 Non-Maxwellian distributions

The Z-function discussed in the present Appendix is based on a Maxwellian velocity distribution. It has been suggested that so-called kappa-distributions (Marsch & Livi 1985, Ali & Eliasson 2017, Livadiotis 2017), appearing like thermal distributions with an enhanced density "tail" of hot plasma are representative for conditions found in nature, the ionosphere in particular

$$F_\kappa(u_\parallel,u_\perp) = \frac{1}{\pi^{3/2}}\frac{1}{\theta_\parallel\theta_\perp^2}\frac{\Gamma(\kappa+1)}{\kappa^{3/2}\Gamma(\kappa-1/2)}\left(1+\frac{u_\parallel^2}{\kappa\theta_\parallel^2}+\frac{u_\perp^2}{\kappa\theta_\perp^2}\right)^{-(\kappa+1)},$$

where $\theta_\parallel = [(2\kappa-3)/\kappa]^{1/2}(T_\parallel/m)^{1/2}$ and $\theta_\perp = [(2\kappa-3)/\kappa]^{1/2}(T_\perp/m)^{1/2}$, with $T_\parallel$ and $T_\perp$ being effective temperatures when $\kappa > 3/2$, here given with the Boltzmann constant included to avoid confusion with κ. The analysis giving the Z-function can be generalized for such distributions (Summers & Thorne 1992). Also mixed kappa-Maxwellian distributions have been suggested (Hellberg & Mace 2002).

D

Mathematical Theorems and Useful Relations

D.1 Gauss' theorem

Suppose that $\mathcal{V}$ is a volume with a closed surface $\mathcal{S}$ and $\mathbf{u} = \mathbf{u}(\mathbf{r})$ any vector field defined in $\mathcal{V}$ and on $\mathcal{S}$. Then, if $\mathcal{S}$ is piece-wise smooth with outward normal unit vector $\widehat{\mathbf{n}}$, and with $\mathbf{u}$ continuously differentiable

$$\iiint_{\mathcal{V}} \nabla \cdot \mathbf{u} dV = \iint_{\mathcal{S}} \mathbf{u} \cdot \widehat{\mathbf{n}} dS \tag{D.1}$$

This is Gauss' theorem, not to be confused with Gauss' law in electrostatics. We explicitly retained three and two integral signs, respectively, in order to emphasize the change in number of integration variables when we change from volume to surface integrals. In one dimension, Gauss' theorem becomes particularly simple, stating the well known expression

$$\int_a^b \frac{d}{dx} u(x)\, dx = u(b) - u(a).$$

D.2 Stokes' theorem

If a closed path $\mathcal{C}$ bounds a surface $\mathcal{S}$, then the surface integral of $\mathbf{curl}\,\mathbf{u} \equiv \nabla \times \mathbf{u}$ over $\mathcal{S}$ is equal to the line integral of $\mathbf{u}$ around the bounding curve ℓ

$$\iint_{\mathcal{S}} (\nabla \times \mathbf{u}) \cdot \widehat{\mathbf{n}} dS = \oint_{\mathcal{C}} \mathbf{u} \cdot d\ell.$$

This is Stokes' theorem.

D.3 Green's relations

By applications of Gauss' law, we can obtain Green's relations. Let, for instance, ψ and ϕ be two scalar fields. We can then apply Gauss' theorem on $\phi \nabla \psi$ to obtain

$$\iint_{\mathcal{S}} (\phi \nabla \psi) \cdot \widehat{\mathbf{n}} dS = \iiint_{\mathcal{V}} \nabla \cdot (\phi \nabla \psi)\, dV = \iiint_{\mathcal{V}} (\nabla \phi \cdot \nabla \psi + \phi \nabla^2 \psi)\, dV. \tag{D.2}$$

This is Green's first relation. It is sometimes an advantage to introduce the notation $\partial/\partial n \equiv \widehat{\mathbf{n}} \cdot \nabla$, and have

$$\iint_{\mathcal{S}} \phi \frac{\partial}{\partial n} \psi dS = \iiint_{\mathcal{V}} \left(\nabla \phi \cdot \nabla \psi + \phi \nabla^2 \psi \right) dV. \tag{D.3}$$

We can now rewrite the relation (D.3), but interchange ψ and ϕ. The two expressions can then be subtracted to give Green's second relation

$$\int\int_{\mathcal{S}}\left(\phi\frac{\partial}{\partial n}\psi-\psi\frac{\partial}{\partial n}\phi\right)dS=\int\int\int_{\mathcal{V}}(\phi\nabla^2\psi-\psi\nabla^2\phi)\,dV. \tag{D.4}$$

D.4 Solenoidal fields

A continuously differentiable *solenoidal* vector field **A** has the following equivalent characteristics (Aris 1962):

$$\begin{aligned}\nabla\cdot\mathbf{A} &= 0, \\ \int_{\mathcal{S}}\mathbf{A}\cdot\widehat{\mathbf{n}}\,dS &= 0 && \text{for any closed surface } \mathcal{S}, \\ \mathbf{A} &= \nabla\times\mathbf{a}, && \text{where } \mathbf{a} \text{ is itself solenoidal.}\end{aligned}$$

Magnetic fields represent one important example of solenoidal vector fields.

D.5 Summary

We give here a short summary of the most useful integral relations. The notation assumes a volume $\mathcal{V}$ enclosed by a surface $\mathcal{S}$, and the normal vector $\widehat{\mathbf{n}}$ pointing *out* of this surface.

$$\begin{aligned}\int_{\mathcal{V}}\nabla\phi\,dV &= \int_{\mathcal{S}}\phi\,\widehat{\mathbf{n}}\,dS \\ \int_{\mathcal{V}}\nabla\cdot\mathbf{A}\,dV &= \int_{\mathcal{S}}\mathbf{A}\cdot\widehat{\mathbf{n}}\,dS \\ \int_{\mathcal{V}}\nabla\cdot\mathsf{T}\,dV &= \int_{\mathcal{S}}\mathsf{T}\cdot\widehat{\mathbf{n}}\,dS \\ \int_{\mathcal{V}}\nabla\times\mathbf{A}\,dV &= \int_{\mathcal{S}}\widehat{\mathbf{n}}\times\mathbf{A}\,dS \\ \int_{\mathcal{V}}(\mathbf{A}\cdot\nabla\times\nabla\times\mathbf{B}-\mathbf{B}\cdot\nabla\times\nabla\times\mathbf{A})\,dV &= \int_{\mathcal{S}}(\mathbf{B}\times\nabla\times\mathbf{A}-\mathbf{A}\times\nabla\times\mathbf{B})\cdot\widehat{\mathbf{n}}\,dS\end{aligned}$$

Let now $\mathcal{S}$ be an open surface bounded by a contour $\mathcal{C}$, like a tablecloth, for instance. Let a vectorial line element on $\mathcal{C}$ be $d\ell$. We then have

$$\begin{aligned}\int_{\mathcal{S}}\widehat{\mathbf{n}}\times\nabla\phi\,dS &= \oint_{\mathcal{C}}\phi\,d\ell \\ \int_{\mathcal{S}}\widehat{\mathbf{n}}\cdot\nabla\times\mathbf{A}\,dS &= \oint_{\mathcal{C}}\mathbf{A}\cdot d\ell \\ \int_{\mathcal{S}}(\widehat{\mathbf{n}}\times\nabla)\times\mathbf{A}\,dS &= \oint_{\mathcal{C}}d\ell\times\mathbf{A} \\ \int_{\mathcal{S}}\widehat{\mathbf{n}}\cdot(\nabla\phi\times\nabla\psi)\,dS &= \oint_{\mathcal{C}}\phi\,d\psi=-\oint_{\mathcal{C}}\psi\,d\phi\end{aligned}$$

D.6 The Jacobian determinant

When transforming integrals from Cartesian to cylindrical or spherical coordinates you need to transform the volume element. Take, for instance, the Cartesian coordinates on a plane and change these to cylindrical coordinates, i.e., let $\{x,y\} \to \{r,\phi\}$ with $0 < \phi < 2\pi$, where $r = \sqrt{x^2+y^2}$ and $\phi = \arctan(y/x)$, or $x = r\cos\phi$, $y = r\sin\phi$, which gives the Jacobi determinant (or "Jacobian") as

$$J = \begin{vmatrix} \frac{\partial x}{\partial r} & \frac{\partial x}{\partial \phi} \\ \frac{\partial y}{\partial r} & \frac{\partial y}{\partial \phi} \end{vmatrix} = \begin{vmatrix} \cos\phi & -r\sin\phi \\ \sin\phi & r\cos\phi \end{vmatrix} = r.$$

We then have

$$dx\,dy \to r\,dr\,d\phi.$$

D.6.1 Cylindrical coordinates

With $\{x,y,z\} \to \{r,\phi,z\}$, where $r = \sqrt{x^2+y^2}$ and $0 < \phi < 2\pi$ with z being the axis coordinate, we have

$$dx\,dy\,dz \to r\,dr\,d\phi\,dz.$$

D.6.2 Spherical coordinates

With $\{x,y,z\} \to \{r,\theta,\phi\}$, where $r = \sqrt{x^2+y^2+z^2}$, $0 < \theta < \pi$ and $0 < \phi < 2\pi$, we have

$$dx\,dy\,dz \to r^2 \sin\theta\,dr\,d\theta\,d\phi.$$

D.7 Useful Vector Relations

A scalar is denoted by italic Roman letters, vectors by bold-face capital Roman letters and tensors by capital Greek letters. **r** is the position vector, $\underline{1}$ is the unit tensor and d is the dimensionality of the problem. We have the synonyms *div* for the operator $\nabla\cdot$ and **rot** for $\nabla\times$, but these are not used here. The rotation of a flow, or a vector field in general, can be interpreted as the rotation velocity of a small "mill" immersed in the flow, best illustrated in two spatial dimensions.

D.7.1 Some basic vector relations

If **A** is an arbitrary vector and $\widehat{\mathbf{e}}$ is a unit vector, then $\mathbf{A} = (\mathbf{A}\cdot\widehat{\mathbf{e}})\widehat{\mathbf{e}} + \widehat{\mathbf{e}} \times (\mathbf{A}\times\widehat{\mathbf{e}})$

$\mathbf{A}\times(\mathbf{B}\times\mathbf{C}) = (\mathbf{A}\cdot\mathbf{C})\mathbf{B} - (\mathbf{A}\cdot\mathbf{B})\mathbf{C}$, which implies $\mathbf{A}\times(\mathbf{B}\times\mathbf{C}) + \mathbf{B}\times(\mathbf{C}\times\mathbf{A}) + \mathbf{C}\times(\mathbf{A}\times\mathbf{B}) = 0$

$$(\mathbf{A}\times\mathbf{B})\cdot(\mathbf{C}\times\mathbf{D}) = \begin{vmatrix} \mathbf{A}\cdot\mathbf{C} & \mathbf{B}\cdot\mathbf{C} \\ \mathbf{A}\cdot\mathbf{D} & \mathbf{B}\cdot\mathbf{D} \end{vmatrix}$$

$$(\mathbf{A}\times\mathbf{B})\times(\mathbf{C}\times\mathbf{D}) = [\mathbf{C}\cdot(\mathbf{D}\times\mathbf{A})]\mathbf{B} - [\mathbf{C}\cdot(\mathbf{D}\times\mathbf{B})]\mathbf{A} = [\mathbf{A}\cdot(\mathbf{B}\times\mathbf{D})]\mathbf{C} - [\mathbf{A}\cdot(\mathbf{B}\times\mathbf{C})]\mathbf{D}$$

$$[\mathbf{A}\cdot(\mathbf{B}\times\mathbf{C})]\mathbf{D} = [\mathbf{D}\cdot(\mathbf{B}\times\mathbf{C})]\mathbf{A} + [\mathbf{A}\cdot(\mathbf{D}\times\mathbf{C})]\mathbf{B} + [\mathbf{A}\cdot(\mathbf{B}\times\mathbf{D})]\mathbf{C}$$

D.8 Differential operators

A number of useful relations involving differential operators are summarized here.

D.8.1 Some basic differential expressions

$$\nabla(ab) = a\nabla b + b\nabla a$$

$$\nabla\cdot(a\mathbf{A}) = a\nabla\cdot\mathbf{A} + \mathbf{A}\cdot\nabla a$$

$$\nabla\times(a\mathbf{A}) = a\nabla\times\mathbf{A} + (\nabla a)\times\mathbf{A}$$

$$\begin{aligned}\nabla(\mathbf{A}\cdot\mathbf{B}) &= (\nabla\mathbf{A})\cdot\mathbf{B} + \mathbf{A}\cdot\nabla\mathbf{B}\\ &= \mathbf{A}\times(\nabla\times\mathbf{B}) + \mathbf{B}\times(\nabla\times\mathbf{A}) + (\mathbf{A}\cdot\nabla)\mathbf{B} + (\mathbf{B}\cdot\nabla)\mathbf{A}\end{aligned}$$

In particular we have

$$\tfrac{1}{2}\nabla(\mathbf{A}\cdot\mathbf{A}) = \mathbf{A}\times(\nabla\times\mathbf{A}) + \mathbf{A}\cdot\nabla\mathbf{A}$$

$$\mathbf{A}\times(\nabla\times\mathbf{A}) = \tfrac{1}{2}\nabla(\mathbf{A}\cdot\mathbf{A}) - (\mathbf{A}\cdot\nabla)\mathbf{A}$$

$$\begin{aligned}\nabla^2\mathbf{A} &= \nabla(\nabla\cdot\mathbf{A}) - \nabla\times(\nabla\times\mathbf{A})\\ &\text{or } \nabla\times(\nabla\times\mathbf{A}) = \nabla(\nabla\cdot\mathbf{A}) - \nabla^2\mathbf{A}\end{aligned}$$

$$\nabla\times\nabla a = 0$$

$$\nabla\cdot\nabla\times\mathbf{A} = 0$$

$$\nabla\cdot(\mathbf{A}\times\mathbf{B}) = (\nabla\times\mathbf{A})\cdot\mathbf{B} - \mathbf{A}\cdot(\nabla\times\mathbf{B})$$

$$\nabla\cdot(\mathbf{AB}) = (\nabla\cdot\mathbf{A})\mathbf{B} + (\mathbf{A}\cdot\nabla)\mathbf{B}$$

$$\nabla\times(\mathbf{AB}) = (\nabla\times\mathbf{A})\mathbf{B} - (\mathbf{A}\times\nabla)\mathbf{B}$$

$$\nabla\times(\mathbf{A}\times\mathbf{B}) = \mathbf{A}(\nabla\cdot\mathbf{B}) - \mathbf{B}(\nabla\cdot\mathbf{A}) + (\mathbf{B}\cdot\nabla)\mathbf{A} - (\mathbf{A}\cdot\nabla)\mathbf{B}$$

$$\nabla\cdot(a\underline{\Pi}) = (\nabla a)\cdot\underline{\Pi} + a\nabla\cdot\underline{\Pi}$$

$\nabla\times\mathbf{r} = 0,$ $\nabla\mathbf{r} = \underline{1},$ $\nabla\cdot\mathbf{r} = d,$ where d is the dimensionality of the problem.

D.8.2 Differential operators in spherical geometry

Vectors are denoted by, for instance, $\mathbf{U} = \{U_r, U_\theta, U_\phi\}$, with their components in cylindrical geometry. Unit vectors are indicated by a "hat."

$$\nabla\Psi = \frac{\partial}{\partial r}\Psi\,\hat{\mathbf{r}} + \frac{1}{r}\frac{\partial}{\partial\theta}\Psi\,\hat{\theta} + \frac{1}{r\sin\theta}\frac{\partial}{\partial\phi}\Psi\,\hat{\phi}$$

$$\nabla\cdot\mathbf{U} = \frac{1}{r^2}\frac{\partial}{\partial r}(r^2U_r) + \frac{1}{r\sin\theta}\frac{\partial}{\partial\theta}(\sin\theta\,U_\theta) + \frac{1}{r\sin\theta}\frac{\partial}{\partial\phi}U_\phi$$

$$\begin{aligned}\nabla\times\mathbf{U} &= \left(\frac{1}{r\sin\theta}\frac{\partial}{\partial\theta}(\sin\theta\,U_\phi) - \frac{\partial}{\partial\phi}U_\theta\right)\hat{\mathbf{r}} + \frac{1}{r}\left(\frac{1}{\sin\theta}\frac{\partial}{\partial\phi}U_r - \frac{\partial}{\partial r}(rU_\phi)\right)\hat{\theta}\\ &\quad + \frac{1}{r}\left(\frac{\partial}{\partial r}(rU_\theta) - \frac{\partial}{\partial\theta}U_r\right)\hat{\phi}\end{aligned}$$

$$\nabla^2\Psi = \frac{1}{r^2}\frac{\partial}{\partial r}\left(r^2\frac{\partial}{\partial r}\Psi\right) + \frac{1}{r^2\sin\theta}\frac{\partial}{\partial\theta}\left(\sin\theta\frac{\partial}{\partial\theta}\Psi\right) + \frac{1}{r^2\sin^2\theta}\frac{\partial^2}{\partial\phi^2}\Psi$$

Components of the Laplacian of a vector

$$\begin{aligned}
(\nabla^2\mathbf{U})_r &= \nabla^2 U_r - \frac{2U_r}{r^2} - \frac{2}{r^2}\frac{\partial}{\partial\theta}U_\theta - \frac{2U_\theta\cot\theta}{r^2} - \frac{2}{r^2\sin\theta}\frac{\partial}{\partial\phi}U_\phi \\
(\nabla^2\mathbf{U})_\theta &= \nabla^2 U_\theta + \frac{2}{r^2}\frac{\partial}{\partial\theta}U_r - \frac{U_\theta}{r^2\sin^2\theta} - \frac{2\cos\theta}{r^2\sin^2\theta}\frac{\partial}{\partial\phi}U_\phi \\
(\nabla^2\mathbf{U})_\phi &= \nabla^2 U_\phi - \frac{U_\phi}{r^2\sin^2\theta} + \frac{2}{r^2\sin\theta}\frac{\partial}{\partial\phi}U_r + \frac{2\cos\theta}{r^2\sin^2\theta}\frac{\partial}{\partial\phi}U_\theta
\end{aligned}$$

Components for the vector $\mathbf{A}\cdot\nabla\mathbf{B}$

$$\begin{aligned}
(\mathbf{A}\cdot\nabla\mathbf{B})_r &= A_r\frac{\partial}{\partial r}B_r + \frac{A_\theta}{r}\frac{\partial}{\partial\theta}B_r + \frac{A_\phi}{r\sin\theta}\frac{\partial}{\partial\phi}B_r - \frac{A_\theta B_\theta + A_\phi B_\phi}{r} \\
(\mathbf{A}\cdot\nabla\mathbf{B})_\theta &= A_r\frac{\partial}{\partial r}B_\theta + \frac{A_\theta}{r}\frac{\partial}{\partial\theta}B_\theta + \frac{A_\phi}{r\sin\theta}\frac{\partial}{\partial\phi}B_\theta + \frac{A_\theta B_r - A_\phi B_\phi\cot\theta}{r} \\
(\mathbf{A}\cdot\nabla\mathbf{B})_\phi &= A_r\frac{\partial}{\partial r}B_\phi + \frac{A_\theta}{r}\frac{\partial}{\partial\theta}B_\phi + \frac{A_\phi}{r\sin\theta}\frac{\partial}{\partial\phi}B_\phi + \frac{A_\phi B_r + A_\phi B_\theta\cot\theta}{r}
\end{aligned}$$

Components of the divergence of a tensor

$$\begin{aligned}
(\nabla\cdot\underline{\Pi})_r &= \frac{1}{r^2}\frac{\partial}{\partial r}(r^2\Pi_{rr}) + \frac{1}{r\sin\theta}\frac{\partial}{\partial\theta}(\Pi_{r\theta}\sin\theta) + \frac{1}{r\sin\theta}\frac{\partial}{\partial\phi}\Pi_{r\phi} - \frac{\Pi_{\theta\theta}+\Pi_{\phi\phi}}{r} \\
(\nabla\cdot\underline{\Pi})_\theta &= \frac{1}{r^2}\frac{\partial}{\partial r}(r^2\Pi_{\theta r}) + \frac{1}{r\sin\theta}\frac{\partial}{\partial\theta}(\Pi_{\theta\theta}\sin\theta) + \frac{1}{r\sin\theta}\frac{\partial}{\partial\phi}\Pi_{\theta\phi} + \frac{\Pi_{r\theta}-\Pi_{\phi\phi}\cot\theta}{r} \\
(\nabla\cdot\underline{\Pi})_\phi &= \frac{1}{r^2}\frac{\partial}{\partial r}(r^2\Pi_{\phi r}) + \frac{1}{r\sin\theta}\frac{\partial}{\partial\theta}(\Pi_{\phi\theta}\sin\theta) + \frac{1}{r\sin\theta}\frac{\partial}{\partial\phi}\Pi_{\phi\phi} + \frac{\Pi_{r\phi}+\Pi_{\theta\phi}\cot\theta}{r}
\end{aligned}$$

D.8.3 Differential operators in cylindrical geometry

Vectors are denoted by, for instance, $\mathbf{U} = \{U_r, U_\theta, U_z\}$, with their components in cylindrical geometry. Unit vectors are indicated by a "hat."

$$\begin{aligned}
\nabla\Psi &= \frac{\partial}{\partial r}\Psi\,\hat{\mathbf{r}} + \frac{1}{r}\frac{\partial}{\partial\theta}\Psi\,\hat{\theta} + \frac{\partial}{\partial z}\Psi\,\hat{\mathbf{z}} \\
\nabla\cdot\mathbf{U} &= \frac{1}{r}\frac{\partial}{\partial r}(rU_r) + \frac{1}{r}\frac{\partial}{\partial\theta}U_\theta + \frac{\partial}{\partial z}U_z \\
\nabla\times\mathbf{U} &= \left(\frac{1}{r}\frac{\partial}{\partial\theta}U_z - \frac{\partial}{\partial z}U_\theta\right)\hat{\mathbf{r}} + \left(\frac{\partial}{\partial z}U_r - \frac{\partial}{\partial r}U_z\right)\hat{\theta} + \left(\frac{1}{r}\frac{\partial}{\partial r}(rU_\theta) - \frac{1}{r}\frac{\partial}{\partial\theta}U_r\right)\hat{\mathbf{z}} \\
\nabla^2\Psi &= \frac{1}{r}\frac{\partial}{\partial r}\left(r\frac{\partial}{\partial r}\Psi\right) + \frac{1}{r^2}\frac{\partial^2}{\partial\theta^2}\Psi + \frac{\partial^2}{\partial z^2}\Psi \equiv \frac{\partial^2}{\partial r^2}\Psi + \frac{1}{r}\frac{\partial}{\partial r}\Psi + \frac{1}{r^2}\frac{\partial^2}{\partial\theta^2}\Psi + \frac{\partial^2}{\partial z^2}\Psi
\end{aligned}$$

Components of the Laplacian of a vector

$$\begin{aligned}
(\nabla^2\mathbf{U})_r &= \nabla^2 U_r - \frac{U_r}{r^2} - \frac{2}{r^2}\frac{\partial}{\partial\theta}U_\theta \\
(\nabla^2\mathbf{U})_\theta &= \nabla^2 U_\theta - \frac{U_\theta}{r^2} + \frac{2}{r^2}\frac{\partial}{\partial\theta}U_r \\
(\nabla^2\mathbf{U})_z &= \nabla^2 U_z
\end{aligned}$$

Components for the vector $\mathbf{A}\cdot\nabla\mathbf{B}$

$$
\begin{aligned}
(\mathbf{A}\cdot\nabla\mathbf{B})_r &= A_r\frac{\partial}{\partial r}B_r + \frac{A_\theta}{r}\frac{\partial}{\partial\theta}B_r + A_z\frac{\partial}{\partial z}B_r - \frac{A_\theta B_\theta}{r} \\
(\mathbf{A}\cdot\nabla\mathbf{B})_\theta &= A_r\frac{\partial}{\partial r}B_\theta + \frac{A_\theta}{r}\frac{\partial}{\partial\theta}B_\theta + A_z\frac{\partial}{\partial z}B_\theta + \frac{A_\theta B_r}{r} \\
(\mathbf{A}\cdot\nabla\mathbf{B})_z &= A_r\frac{\partial}{\partial r}B_z + \frac{A_\theta}{r}\frac{\partial}{\partial\theta}B_z + A_z\frac{\partial}{\partial z}B_z
\end{aligned}
$$

Components of the divergence of a tensor

$$
\begin{aligned}
(\nabla\cdot\underline{\Pi})_r &= \frac{1}{r}\frac{\partial}{\partial r}(r\Pi_{rr}) + \frac{1}{r}\frac{\partial}{\partial\theta}\Pi_{r\theta} + \frac{\partial}{\partial z}\Pi_{rz} - \frac{\Pi_{\theta\theta}}{r} \\
(\nabla\cdot\underline{\Pi})_\theta &= \frac{1}{r}\frac{\partial}{\partial r}(r\Pi_{\theta r}) + \frac{1}{r}\frac{\partial}{\partial\theta}\Pi_{\theta\theta} + \frac{\partial}{\partial z}\Pi_{\theta z} + \frac{\Pi_{r\theta}}{r} \\
(\nabla\cdot\underline{\Pi})_\phi &= \frac{1}{r}\frac{\partial}{\partial r}(r\Pi_{zr}) + \frac{1}{r}\frac{\partial}{\partial\theta}\Pi_{z\theta} + \frac{\partial}{\partial z}\Pi_{zz}
\end{aligned}
$$

D.9 Numbers to Remember

D.9.1 Physical constants

Speed of light in a vacuum	$c = 2.9979\times10^{8}$ m/s
Gravitational constant	$G = 6.6732\times10^{-11}$ N m^2 kg^{-2}
Standard gravitational acceleration at the Earth's surface (average)	$g_n = 9.80665$ m/s^2
Permittivity of free space	$\varepsilon_0 = 8.8542\times10^{-12}$ F/m
Permeability of free space	$\mu_0 = 4\pi\times10^{-7}$ H/m
Elementary charge	$e = 1.6022\times10^{-19}$ C
Electron mass	$m = 9.1095\times10^{-31}$ kg
Proton mass	$M = 1.6726\times10^{-27}$ kg
Atomic mass unit (^{12}C$\times\frac{1}{12}$)	amu = 1.6604×10^{-27} kg
Boltzmann constant	$\kappa = 1.3807\times10^{-23}$ J/K
Planck constant	$h = 6.6256\times10^{-34}$ J s
	$\hbar \equiv h/2\pi = 1.0546\times10^{-37}$ J s
Bohr radius, $4\pi\varepsilon_0\hbar^2/e^2$ m	$a_0 = 5.2918\times10^{-11}$ m
Atomic cross section	$\pi a_0^2 = 8.7974\times10^{-21}$ m^2
Classical electron radius	$e^2/mc^2 = 2.82\times10^{-15}$ m
Avogadro number (molecules/mol)	$N_A = 6.0220\times10^{23}$
Standard pressure (760 Torr = 1 atm)	$p_0 = 1.0133\times10^{5}$ Pa
Ideal gas mol-volume	$U_{\mathrm{mol}} = 0.022414$ m^2/mol
Loschmidt number (ideal gas density at STP)	$n_L = 2.6868\times10^{25}$ m^{-3}
Approximate density of free electrons in Cu	8.5×10^{28} m^{-3}
Absolute zero temperature	$T_0 = -273.15$°C

The unit *meter* is often too large when describing length scales in matter, either in liquid or in solid state; here it may be more convenient to use Å (for Ångstrøm), where we have 1 m = 10^{10} Å.

The obvious unit for measuring temperatures is K (Kelvin), but in plasma physics one often uses electron volts, eV, to be understood as the potential ϕ that fulfills the relation $\kappa T = e\phi$. We have 1 eV $\approx 1.1604 \times 10^4$ K, or 10^3 K $\approx 86.17 \times 10^{-3}$ eV.

In MKSA or SI units we use Tesla for magnetic flux densities, but often you find it expressed in Gauss, where we have 1 G = 10^{-4} T. You might also need:

Conversion eV $\rightarrow$ J	1eV = $1.602 \cdot 10^{-19}$ J
Conversion cal $\rightarrow$ J	1 cal = 4.186 J
Conversion kWh $\rightarrow$ J	1 kWh = $3.6 \cdot 10^6$ J
Conversion erg $\rightarrow$ J	1 erg = 10^{-7} J
Conversion Btu $\rightarrow$ J	1 Btu = $1.055 \cdot 10^3$ J

and with 1 Pa $\equiv$ 1 N m^{-2}:

Atmospheric pressure $\rightarrow$ Pa	1 atm = $1.01 \cdot 10^5$ Pa
mm mercury $\rightarrow$ Pa	1 mmHg = 133.322 Pa
mm mercury $\rightarrow$ kg/m^2	1 mmHg = 13.60 kg/m^2
mm mercury $\rightarrow$ atm	1 mmHg = $1.316 \cdot 10^{-3}$ atm
atm $\rightarrow$ mmHg	1 atm = 759.88 mmHg

"Torr" is a unit of pressure based on an absolute scale, now defined as exactly 1/760 of a standard atmosphere (101.325 kPa), i.e., 1 mmHg: 1 Torr $\rightarrow 101.325/760 \approx 133.32$ Pa

The mole is defined as the amount of substance that contains as many elementary entities (e.g., atoms, molecules, ions, electrons) as there are atoms in 12 g of the isotope carbon-12 (^{12}C). Thus, by definition, one mole of pure ^{12}C has a mass of exactly 12 g. The experimentally determined value of a mole in 2006 is given as $6.02214179 \times 10^{23}$, i.e., Avogadro's number.

Often we use the notation $[*]$ indicating the physical dimension of a quantity $*$. Thus we have as an example $[\kappa] = \textit{energy temperature}^{-1}$, i.e., J K^{-1} or kg m^2 K^{-1}s^{-2}, or simply ML^{2}K^{-1}T^{-2}, with M meaning mass, L length, T time and K absolute temperature Kelvin.

D.9.2 Selected data of geophysical and astrophysical importance

- **The solar system:**

Average solar radius, $R_\odot$	6.96×10^8 m $\approx 109\ R_E$
Average Earth radius, R_E	6.37×10^6 m
Astronomical unit, *AU*	1.496×10^{11} m $\approx 215\ R_\odot$
Distance Earth to Moon	approx. 60 R_E
Orbital velocity of the Earth	approx. 30 km/s
Mass of the Sun, $M_\odot$	1.99×10^{30} kg
Mass of the Earth, M_E	5.976×10^{24} kg
Power of electromagnetic radiation from the Sun	3.86×10^{26} W
The solar constant, *S*, i.e., the average energy flux density at 1 *AU*	1370 W/m^2
The approximate magnetic moment of the Earth, $\mathcal{M}$	8×10^{22} Am^2
Earth's surface magnetic field at the magnetic poles	approx. 60×10^{-6} T

The energy radiated to the Earth's surface from the Sun is $E_S = \pi R_E^2(1-A)S$, where A is the *albedo*, being the fraction of energy radiated back into space.

Albedo, A	32%
E_S	1.1×10^{17} W
Power density of radiation absorbed by the Earth, $E_S/(4\pi R_E^2)$	230 W/m^2

The solar constant, and quantities derived from it, vary slightly with the solar cycle.

- **Resistivities:**

  For comparison with plasmas, you might need the resistivity of good conductors such as metals. We have for silver 1.6×10^{-8} Ohm m, copper 1.7×10^{-8} Ohm m, platinum 1.1×10^{-7} Ohm m, and for gold 2.4×10^{-8} Ohm m, all taken at room temperature.

D.9.3 Approximate expressions

Electron plasma frequency, assuming densities inserted in particles m^{-3}	$f_{pe} \approx 9\sqrt{n}$ Hz
	$\omega_{pe} \approx 57\sqrt{n}$ rad s^{-1}
Electron cyclotron frequency, assuming magnetic fields inserted in Tesla	$f_{ce} \approx 2.8 \times 10^{10} B$ Hz
	$\omega_{ce} \approx 17.6 \times 10^{10} B$ rad s^{-1}
Proton cyclotron frequency, assuming magnetic fields inserted in Tesla	$f_{ci} \approx 1.5 \times 10^{7} B$ Hz
	$\Omega_{ci} \approx 9.4 \times 10^{7} B$ rad s^{-1}

For order of magnitude estimates, the following approximations (accurate to within a few %) may come in handy;

$\sqrt{2\pi} \simeq 2.5$ $\qquad \pi^2 \simeq 10$ $\qquad e^3 \simeq 20$ $\qquad 2^{10} \simeq 10^3$

SI prefixes:

Number	prefix	symbol
10^{18}	exa	E
10^{15}	peta	P
10^{12}	tera	T
10^{9}	giga	G
10^{6}	mega	M
10^{3}	kilo	k
10^{2}	hecto	h
10^{1}	deca	da

Number	prefix	symbol
10^{-1}	deci	d
10^{-2}	centi	c
10^{-3}	milli	m
10^{-6}	micro	μ
10^{-9}	nano	n
10^{-12}	pico	p
10^{-15}	femto	f
10^{-18}	atto	a

There seems a little space left on this page, which might as well be used for giving tribute to a Professor Knudsen at the Danish Technical University (at that time “den Polytekniske Læreanstalt”). He wanted to give the students a “feeling” for large numbers and introduced a constant 8.1×10^{28} which is (approximately) the number of beers contained in a cubic light-year. Ever since, this has been called “Knudsen’s constant”. This is at least the way the story is passed on to me. Given a Danish beer of 0.33 l, the reader may verify that this number is more or less correct. A student immediately wanted to extend this constant by estimating a time it would take the Earth’s population to drink all this beer. Take the Earth’s population to be $\sim 10^{10}$ (which is close enough for the present purpose) and assume that each person drinks a six-pack a day. Compare the estimated time to the (assumed) lifetime of the universe, $\sim 10^{10}$ y.

Bibliography

Alexeff, I., Jones, W. D. & Lonngren, K. (1968), 'Excitation of pseudowaves in a plasma via a grid', *Phys. Rev. Lett.* **21**, 878–881.

Alfvén, H. (1939), 'On the motion of cosmic rays in interstellar space', *Phys. Rev.* **55**, 425–429.

Alfvén, H. (1950), *Cosmical Electrodynamics*, Oxford University Press, London.

Ali, S. & Eliasson, B. (2017), 'Slow test charge response in a dusty plasma with Kappa distributed electrons and ions', *Phys. Scripta* **92**, 084003.

Allen, J. E. (1992), 'Probe theory - the orbital motion approach', *Phys. Scripta* **45**, 497–503.

Allis, W. P., Buchsbaum, S. J. & Bers, A. (1963), *Waves in Anisotropic Plasmas*, MIT Press, Cambridge, Massachusetts.

Andersen, S. A., Christoffersen, G. B., Jensen, V. O., Michelsen, P. & Nielsen, P. (1971*a*), 'Determination and shaping off the ion-velocity distribution in a single-ended Q-machine', *Phys. Fluids* **14**, 728–736.

Andersen, S. A., Christoffersen, G. B., Jensen, V. O., Michelsen, P. & Nielsen, P. (1971*b*), 'Measurements of wave-particle interaction in a single ended Q-machine', *Phys. Fluids* **14**, 990–998.

Aris, R. (1962), *Vectors, Tensors, and the Basic Equations of Fluid Mechanics*, Prentice-Hall, International, London.

Backus, G. (1960), 'Linearized plasma oscillations in arbitrary electron velocity distributions', *J. Math. Phys.* **1**, 178–191.

Barston, E. M. (1964), 'Electrostatic oscillations in inhomogeneous cold plasmas', *Ann. Phys. (NY)* **29**, 282–303.

Batchelor, G. K. (1967), *An Introduction to Fluid Dynamics*, Cambridge University Press, Cambridge, UK.

Baumjohann, W. & Treuman, R. A. (1997), *Basic Space Plasma Physics*, Imperial College Press, London.

Bekefi, G. (1966), *Radiation Processes in Plasmas*, John Wiley and Sons, New York.

Bennett, W. H. (1934), 'Magnetically self-focussing streams', *Phys. Rev.* **45**, 890–897.

Berezhiani, V. I. & Shatashvili, N. L. (1985), 'Nonlinear ion plasma waves in a nonisothermal plasma', *Sov. J. Plasma Phys.* **11**, 766–769. russian original Fiz. Plazmy **11**, 1338-1343 (1985).

Berk, H. L., Nielsen, C. E. & Roberts, K. V. (1970), 'Phase space hydrodynamics of equivalent nonlinear systems: Experimental and computational observations', *Phys. Fluids* **13**, 980–995.

Bernstein, I. B. (1975), 'Geometrical optics in space- and time-varying plasmas', *Phys. Fluids* **18**, 320–324.

Bernstein, I. B. & Engelmann, F. (1966), 'Quasi-linear theory of plasma waves', *Phys. Fluids* **9**, 937–952.

Bernstein, I. B., Greene, J. M. & Kruskal, M. D. (1957), 'Exact nonlinear plasma oscillations', *Phys. Rev.* **108**, 546–550.

Bhakta, J. C. & Majumder, D. (1983), 'Effect of finite spectral width on the modulational instability of Langmuir-waves', *J. Plasma Phys.* **30**, 203–209.

Bhatnagar, P. L., Gross, E. P. & Krook, M. (1954), 'A model for collision processes in gases. I. small amplitude processes in charged and neutral one-component systems', *Phys. Rev.* **94**, 511–525.

Bishop, A. S. (1958), *Project Sherwood. The U.S. Program in Controlled Fusion*, Addison-Wesley Publishing Company, Reading, Massachussets, U.S.A.

Biskamp, D. & Parkinson, D. (1970), 'Ion acoustic shock waves', *Phys. Fluids* **13**, 2295–2299.

Blackstock, D. T. (1962), 'Propagation of plane sound waves of a finite amplitude in nondissipative fluids', *J. Acoust. Soc. Amer.* **34**, 9–30.

Blackstock, D. T. (1972), *Nonlinear acoustics (theoretical)*, American Institute of Physics Handbook, 3 edn, McGraw-Hill, chapter 3, pp. 3–183.

Bloch, S. C. (1976), 'Eighth velocity of light', *Amer. J. Phys.* **45**, 538–549.

Block, L. P. (1978), 'A double layer review', *Astrophys. Space Sci.* **55**, 59–83.

Bode, H. W. (1956), *Network Analysis and Feedback Amplifier Design*, D. van Nostrand, Princeton, New Jersey.

Book, D. L. (1967), 'Landau damping and growth of electrostatic modes with effects of spatial variation', *Phys. Fluids* **10**, 198–203.

Born, M. (1965), *Einstein's Theory of Relativity*, Dover, New York. Revised Edition prepared with the collaboration og G. Liebfriend and W. Biem.

Børve, S., Sato, H., Pécseli, H. L. & Trulsen, J. K. (2011), 'Minute-scale period oscillations of the magnetosphere', *Ann. Geophys.* **29**, 663–671.

Boström, R. (1964), 'A model for the Auroral electrojet', *J. Geophys. Res.* **69**, 4983–4999.

Boswell, R. W. (1984), 'Very efficient plasma generation by whistler waves near the lower hybrid frequency', *Plasma Phys. Contr. Fusion* **26**, 1147–1162.

Boyd, T. J. M. & Sanderson, J. J. (1969), *Plasmas Dynamics*, Thomas Nelson & Sons, London, UK.

Boyd, T. J. M. & Sanderson, J. J. (2003), *The Physics of Plasmas*, Cambridge University Press, Cambridge, UK. The authors emphasize that is is not simply a 2nd ed. of their book from 1969.

Braams, C. M. & Stott, P. E. (2002), *Nuclear Fusion. Half a Century of Magnetic Confinement Fusion Research*, Institute of Physics Publishing, Bristol, UK.

Braginskií, S. I. (1965), Transport processes in a plasma, *in* M. A. Leontovich, ed., 'Reviews of Plasma Physics', Vol. 1, Consultants Bureau, pp. 205–311.

Brekhovskikh, L. & Goncharov, V. (1985), *Mechanics of Continua and Wave Dynamics*, Springer series on Wave Phenomena, Springer Verlag, Berlin.

Bridgman, P. W. (1931), *Dimenensional Analysis*, Yale University Press, New Haven.

Brillouin, L. (1960), *Wave Propagation and Group Velocity*, Vol. 8 of *Pure and Applied Physics*, Academic Press, New York and London.

Buckingham, E. (1914), 'On physically similar systems; illustrations of the use of dimensional equations', *Phys. Rev.* **4**, 345–376.

Camargo, S. J., Biskamp, D. & Scott, B. (1995), 'Resistive drift-wave turbulence', *Phys. Plasmas* **2**, 48–62.

Cap, F. F. (1976), *Handbook on Plasma Instabilities*, Academic Press, New York.

Caruso, A. & Cavaliere, A. (1962), 'The structure of the collisionless plasma-sheath transition', *Il Nuovo Cimento (1955-1965)* **26**, 1389–1404.

Case, K. M. (1959), 'Plasma oscillations', *Ann. Phys.* **7**, 349–364.

Case, K. M. (1978), 'Plasma oscillations', *Phys. Fluids* **21**, 249–257.

Champeney, D. C. (1973), *Fourier Transforms and their Physical Applications*, Academic Press, London.

Chandrasekhar, S. (1960), *Plasma Physics*, The University of Chicago Press, Chicago. Notes compiled by S. K. Trehan after a course given by S. Chandrasekhar.

Chandrasekhar, S. (1961), *Hydrodynamic and Hydromagnetic Stability*, Oxford University Press, London.

Chang, C. T. (1991), 'Pellet-plasma interactions in tokamaks', *Phys. Reports* **206**, 143–196.

Chapman, S. & Bartels, J. (1940), *Geomagnetism*, Oxford University Press.

Charles, C. (2007), 'A review of recent laboratory double layer experiments', *Plasma Sources Sci. Technol.* **16**, R1–R25.

Charney, J. G. (1948), 'On the scale of atmospheric motions', *Geophys. Publ. Norwegian Acad. of Sciences (Geofysiske Publikasjoner)* **17**, 251–265.

Chen, F. F. (1964), 'Normal modes for electrostatic ion waves in an inhomogeneous plasma', *Phys. Fluids* **7**, 949–955.

Chen, F. F. (1965*a*), 'Excitation of drift instabilities in thermionic plasmas', *Plasma Phys. (J. Nucl. Energy Part C)* **7**, 399–417.

Chen, F. F. (1965*b*), 'Numerical computations for ion probe characteristics in a collisionless plasma', *J. Nuclear Energy. Part C, Plasma Phys., Accelerators, Thermonuclear Res.* **7**, 47–67.

Chen, F. F. (1965*c*), 'Resistive overstabilities and anomalous "diffusion"', *Phys. Fluids* **8**, 912–919.

Chen, F. F. (1965*d*), 'Saturation ion currents to Langmuir probes', *J. Applied Phys.* **36**, 675–678.

Chen, F. F. (1966), 'Microinstability and shear stabilization of a low-β, rotating, resistive plasma', *Phys. Fluids* **9**, 965–981.

Chen, F. F. (1970), Linear drift waves. UCLA Lecture notes, unpublished.

Chen, F. F. (1979), 'Axial eigenmodes for long-$\lambda_{\|}$ waves in plasmas bounded by sheaths', *Phys. Fluids* **22**, 2346–2358.

Chen, F. F. (1995), 'The "sources" of plasma physics', *IEEE Trans. Plasma Sci.* **23**, 20–47.

Chen, F. F. (2016), *Introduction to Plasma Physics and Controlled Fusion*, 3 edn, Springer, Heidelberg.

Chen, F. F. & Boswell, R. W. (1997), 'Helicons - the past decade', *IEEE Trans. Plasma Sci.* **25**, 1245–1257.

Chen, L. (1987), *Waves and Instabilities in Plasmas*, Vol. 12 of *World Scientific Lecure Notes in Physics*, World Scientific, Singapore.

Cheng, C. Z. & Okuda, H. (1977), 'Formation of convective cells, anomalous diffusion, and strong plasma turbulence due to drift instabilities', *Phys. Rev. Lett.* **38**, 708–711.

Cho, T., Yamazaki, K. & Tanaka, S. (1982), 'Observation of the upper-hybrid soliton and its recurring behavior', *J. Phys. Soc. Jap.* **51**, 988–995.

Chowdhury, S., Biswas, S., Chakrabarti, N. & Pal, R. (2017), 'Experimental observation of electron-acoustic wave propagation in laboratory plasma', *Phys. Plasmas* **24**, 062111.

Chu, T. K., Coppi, B., Hendel, H. W. & Perkins, F. W. (1969), 'Drift instabilities in a uniformly rotating plasma cylinder', *Phys. Fluids* **12**, 203–208.

Clemmow, P. C. & Dougherty, J. P. (1969), *Electrodynamics of Particles and Plasmas*, Addison-Wesley Series in Advanced Physics, Addison-Wesley, Reading, Massachusetts.

Cohn, D. B. & MacKenzie, K. R. (1972), 'Electrostatic ion-acoustic shocks produced by density steps', *Phys. Rev. Lett.* **28**, 656–658.

Cohn, D. B. & MacKenzie, K. R. (1973), 'Density-step-excited ion acoustic solitons', *Phys. Rev. Lett.* **30**, 258–261.

Cole, G. H. A. (1959), 'The pinch effect', *Science Progress (1933-)* **47**(187), 437–458.

Corum, J. F. (1980), 'Relativistic covariance and rotational electrodynamics', *J. Math. Phys.* **21**, 2360–2364.

Coulson, C. A. (1955), *Waves*, Oliver and Boyd, Edinburg.

Craik, A. D. D. (1985), *Wave Interactions and Fluid Flows*, Cambridge Monographs on Mechanics and Applied Mathematics, Cambridge University Press, Cambridge.

Cramér, H. (1946), *Mathematical Methods of Statistics*, Princeton University Press, USA.

Cramer, N. F. (2001), *The Physics of Alfvén Waves*, Wiley-VCH, Berlin.

Cranmer, S. R. (2004), 'New views of the solar wind with the Lambert W function', *American J. Phys.* **72**, 1397–1403.

Cravens, T. E. (1986), The physics of the cometary contact surface, *in* B. Battrick, E. J. Rolfe & R. Reinhard, eds, 'ESLAB Symposium on the Exploration of Halley's Comet', Vol. 250 of *ESA Special Publication*.

Crow, J. E., Auer, P. L. & Allen, J. E. (1975), 'Expansion of a plasma into a vacuum', *J. Plasma Phys.* **14**, 65–76.

Daldorff, L. K. S., Guio, P., Børve, S., Pécseli, H. L. & Trulsen, J. (2001), 'Ion phase space vortices in 3 spatial dimensions', *Europhys. Lett.* **54**, 161–167.

Dattner, A., Lehnert, B. & Lundquist, S. (1958), 'A liquid conductor model of instabilities in a pinched discharge', *J. Nuclear Energy (1954)* **7**, 293–296.

Dauxois, T. (2008), 'Fermi, Pasta, Ulam, and a mysterious lady', *Phys. Today* **61**, 55–57.

Davidson, R. C. (1972), *Methods in Nonlinear Plasma Theory*, Vol. 37 of *Pure and Applied Physics*, Academic Press, New York and London.

Davidson, R. C., Gladd, N. T., Wu, C. S. & Huba, J. D. (1977), 'Effects of finite plasma beta on the lower-hybrid-drift instability', *Phys. Fluids* **20**, 301–310.

Davidson, R. W. C. & Schram, P. P. J. M. (1968), 'Nonlinear oscillations in a cold plasma', *Nucl. Fusion* **8**, 183–195.

Davies, J. (2006), 'The Alfvén limit revisited and its relevance to laser-plasma interactions', *Laser Part. Beams* **24**, 299–310.

Davydov, A. S. (1965), *Quantum Mechanics*, Pergamon Press, Oxford.

Denavit, J. (1979), 'Collisionless plasma expansion into a vacuum', *Phys. Fluids* **22**, 1384.

Dessler, A. J. (1967), 'Solar wind and interplanetary magnetic field', *Rev. Geophys.* **5**, 1.

Diamond, P. H., Itoh, S.-I. & Itoh, K. (2010), *Modern Plasma Physics*, Vol. I. Physical Kinetics of Turbulent Plasmas, Cambridge University Press, Cambridge, UK.

Dirac, P. A. M. (1947), *The Principles of Quantum Mechanics*, 3 edn, Clarendon Press, Oxford.

Donnel, P., Morel, P., Honoré, C., Gücan, Ö., Pisarev, V., Metzger, C. & Hennequin, P. (2018), 'Drift-wave observation in a toroidal magnetized plasma and comparison with a modified Hasegawa-Wakatani model', *Phys. Plasmas* **25**, 062127.

Drazin, P. G. & Johnson, R. S. (1989), *Solitons: an Introduction*, Cambridge University Press, Cambridge, UK.

Drummond, W. E. (1964), 'Spatially growing electrostatic turbulence', *Phys. Fluids* **7**, 816–821.

Drummond, W. E. (2004), 'Landau damping', *Phys. Plasmas* **11**, 552.

Drummond, W. E. & Pines, D. (1964), 'Nonlinear plasma oscillations', *Ann. Phys.* **28**, 478–499.

Duffin, W. J. (1990), *Electricity and Magnetism*, 4 edn, McGraw-Hill, London.

Dungey, J. W. (1961), 'The steady state of the Chapman-Ferraro problem in two dimensions', *J. Geophys. Res.* **66**, 1043–1047.

Dungey, J. W. (1963), *Geophysics, the Earth's Environment*, Gordon & Breach, New York, chapter The structure of the exosphere or adventures in velocity space, p. 503.

Dysthe, K. (1966), 'Convective and absolute instability', *Nuclear Fusion* **6**, 215–222.

Dysthe, K. (1968), 'Self-trapping and self-focusing of electromagnetic waves in a plasma', *Phys. Lett. A* **27**, 59 – 60.

Dysthe, K. B., Misra, K. D. & Trulsen, J. K. (1975), 'On the linear cross-field instability problem', *J. Plasma Phys.* **13**, 249–257.

Dysthe, K. B., Mjølhus, E., Pécseli, H. L. & Stenflo, L. (1978), 'Langmuir solitons in magnetized plasmas', *Plasma Phys.* **20**, 1087–1099.

Dysthe, K. B., Mjølhus, E., Pécseli, H. L. & Stenflo, L. (1984), 'Nonlinear electrostatic wave equations for magnetized plasmas', *Plasma Phys. Contr. Fusion* **26**, 443–447.

Dysthe, K. B., Mjølhus, E., Pécseli, H. L. & Stenflo, L. (1985*a*), 'Nonlinear electrostatic wave equations for magnetized plasmas - II', *Plasma Phys. Contr. Fusion* **27**, 501–508.

Dysthe, K. B., Mjølhus, E., Pécseli, H. L. & Stenflo, L. (1985*b*), 'Thermally stimulated scattering in plasmas', *J. Appl. Phys.* **57**, 2477–2481.

Dysthe, K. B., Mjølhus, E., Pécseli, H. & Rypdal, K. (1982), 'Thermal cavitons', *Phys. Scripta* **T2**, 548–559.

Dysthe, K. B. & Pécseli, H. L. (1977), 'Non-linear Langmuir wave modulation in collisionless plasmas', *Plasma Phys.* **19**, 931–943.

Dysthe, K. B. & Pécseli, H. L. (1978), 'Non-linear Langmuir wave modulation in weakly magnetized plasmas', *Plasma Phys.* **20**, 971–989.

Dysthe, K. B., Pécseli, H. L. & Trulsen, J. (1983), 'Stochastic generation of continuous wave spectra', *Phys. Rev. Lett.* **50**, 353–356.

Dysthe, K. B., Pécseli, H. L. & Trulsen, J. (1986), 'A statistical model for soliton particle interaction in plasmas', *Phys. Scripta* **33**, 523–526.

Ellis, H. W., Pal, R. Y., McDaniel, E. W., Mason, E. A. & Viehland, L. A. (1976), 'Transport properties of gaseous ions over a wide energy range', *Atomic Data and Nuclear Data Tables* **17**, 177–210.

Elmore, W. C. & Heald, M. A. (1969), *Physics of Waves*, Dover, New York.

Eriksen, E. & Vøyenli, K. (1991), 'Quasi-electrodynamic effects in the motion of a sphere that rolls on a rotating disc', *Eur. J. Phys.* **12**, 135–141.

Eriksson, A. I., Holback, B., Dovner, P. O., Boström, R., Holmgren, G., André, M., Eliasson, L. & Kintner, P. M. (1994), 'Freja observations of correlated small-scale density depletions and enhanced lower-hybrid waves', *Geophys. Res. Lett.* **21**, 1843.

Espedal, M. S. (1971), 'The effects of ion-ion coffision on a ion-acoustic plasma pulse', *J. Plasma Phys.* **5**, 343–355.

Fermi, E. (1949), 'On the origin of cosmic radiation', *Phys. Rev.* **75**, 1169–1174.

Fermi, E., Pasta, J., Ulam, S. & Tsingou, M. (1955), Studies of non linear problems, Technical Report LA-1940, Los Alamos.

Feynman, R. P., Leighton, R. B. & Sands, M. (1963), *The Feynman Lectures on Physics*, Addison-Wesley Publishing Co., New York.

Fisher, R. K. & Gould, R. W. (1969), 'Resonance cones in the field pattern of a short antenna in an anisotropic plasma', *Phys. Rev. Lett.* **22**, 1093–1095.

Fjørtoft, R. (1953), 'On the changes in the spectral distribution of kinetic energy for twodimesional, nondivergent flow', *Tellus* **5**, 225–230.

Flierl, G. R., Larichev, V. D., McWilliams, J. C. & Reznik, G. M. (1980), 'The dynamics of baroclinic and barotropic solitary eddies', *Dynam. Atmos. Oceans* **5**, 1–41.

Fried, B. D. & Conte, S. D. (1961), *The Plasma Dispersion Function*, Academic Press, New York.

Fried, B. D. & Ichikawa, Y. (1973), 'Nonlinear Schrodinger equation for Langmuir waves', *J. Phys. Soc. Jpn.* **34**, 1073–1082.

Fried, B. D. & Wong, A. Y. (1966), 'Stability limits for longitudinal waves in ion beam-plasma interaction', *Phys. Fluids* **9**, 1084–1089.

Frost, L. S. & Phelps, A. V. (1964), 'Momentum-transfer cross sections for slow electrons in He, Ar, Kr, and Xe from transport coeffcicients', *Phys. Rev.* **136**, A1538–A1545.

Ganguli, S. B. (1996), 'The polar wind', *Rev. Geophysics* **34**, 311–348.

Gaponov, V. & Miller, M. A. (1958), 'Potential wells for charged particles in a high-frequency electromagnetic field', *Zh. Eksp. Teor. Fiz.* **34**, 242. See also (1958) Sov. Phys.-JETP, **7**, 168-169.

Gibbings, J. C. (1982), 'A logic of dimensional analysis', *J Phys. A: Math. Gen.* **15**, 1991–2002.

Godyak, V. & Sternberg, N. (2002), 'On the consistency of the collisionless sheath model', *Phys. Plasmas* **9**, 4427–4430.

Goedbloed, H. & Poedts, S. (2004), *Principles of Magnetohydrodynamics with applications to Laboratory and Astrophysical Plasmas*, Cambridge, Cambridge, UK.

Goldman, M. V. (1970), 'Theory of stability of large periodic plasma waves', *Phys. Fluids* **13**, 1281–1289.

Goldman, M. V. & Nicholson, D. R. (1978), 'Virial theory of direct Langmuir collapse', *Phys. Rev. Lett.* **41**, 406–410.

Goldstein, H. (1980), *Classical Mechanics*, 2 edn, Addison-Wesley, Reading, Massachusetts.

Goldstein, S. (1931), 'On the stability of superimposed streams of fluids of different densities', *Proc. Roy. Soc. London Ser. A* **132**, 524–548.

Goldston, R. J. & Rutherford, P. H. (1995), *Introduction to Plasma Physics*, Institute of Physics Publishing, Bristol and Philadelphia.

Good, R. H. (1999), *Classical Electromagnetism*, Saunders College Publishing, Philadelphia.

Gould, R. W. (1964), 'Excitation of ion-acoustic waves', *Phys. Rev.* **136**, A991–A997.

Grésillon, D. & Galison, P. L. (1973), 'Instantaneous electron energy distribution function in ion waves', *Phys. Fluids* **16**, 2180–2183.

Griffiths, D. J. (1999), *Introduction to Electrodynamics*, 3 edn, Prentice Hall, New Jersey, USA.

Guio, P., Børve, S., Daldorff, L. K. S., Lynov, J. P., Michelsen, P., Pécseli, H. L., Rasmussen, J. J., Saeki, K. & Trulsen, J. (2003), 'Phase space vortices in collisionless plasmas', *Nonlin. Processes Geophys.* **10**, 75–86.

Guio, P. & Forme, F. (2006), 'Zakharov simulations of Langmuir turbulence: Effects on the ion-acoustic waves in incoherent scattering', *Phys. Plasmas* **13**, 122902.

Guio, P., Miloch, W. J., Pécseli, H. L. & Trulsen, J. (2008), 'Patterns of sound radiation behind point-like charged obstacles in plasma flows', *Phys. Rev. E* **78**, 016401.

Guio, P. & Pécseli, H. L. (2003), 'Radiation of sound from a charged dust particle moving at high velocity', *Phys. Plasmas* **10**, 2667–2676.

Guio, P. & Pécseli, H. L. (2016), 'Weakly nonlinear ion waves in striated electron temperatures', *Phys. Rev. E* **93**, 043204.

Gurevich, A. L., Pariĭskaya, L. I. & Pitaevskiĭ, L. P. (1965), 'Self similar motion of rarefied plasma.', *J. Exptl. Theor. Fiz. (U.S.S.R.)* **49**, 647–654. see also (1966) Sov. Phys. JETP, **22**, 449-454.

Gurevich, A. L., Pariĭskaya, L. I. & Pitaevskiĭ, L. P. (1968), 'Self similar motion of a low-density plasma. II', *Zh. Eksp. Teor. Fiz.* **54**, 891–904. see also Sov. Phys. JETP, **27**, 476-482.

Gurevich, A. V. (1978), *Nonlinear Phenomena in the Ionosphere*, Vol. 10 of *Physics and Chemistry in Space*, Springer, New York.

Hald, O. H. (1976), 'Constants of motion in models of two-dimensional turbulence', *Phys. Fluids* **19**, 914–915.

Hama, F. R. & Nutant, J. (1961), 'Self-induced velocity on a curved vortex', *Phys. Fluids* **4**, 28–32.

Hansen, F. R., Knorr, G., Lynov, J. P., Pécseli, H. L. & Rasmussen, J. J. (1989), 'A numerical plasma simulation including finite Larmor radius effects to arbitrary order', *Plasma Phys. Contr. Fusion* **31**, 173–183.

Hanssen, A., Mjølhus, E., DuBois, D. F. & Rose, H. A. (1992), 'Numerical test of the weak turbulence approximation to ionospheric Langmuir turbulence', *J. Geophys. Res.* **97**, 12073–12091.

Hasegawa, A. (1975), *Plasma Instabilities and Nonlinear Effects*, Vol. 8 of *Physics and Chemistry in Space*, Springer-Verlag, Berlin.

Hasegawa, A. & Mima, K. (1978), 'Pseudo-three-dimensional turbulence in magnetized nonuniform plasma', *Phys. Fluids* **21**, 87–92.

Hasegawa, A. & Wakatani, M. (1983), 'Plasma edge turbulence', *Phys. Rev. Lett.* **50**, 682–686.

Hashmi, M. & van der Houven van Oordt, A. (1971), 'Production and application of a Uranium plasma in the Q-device', *in* 'Proceedings of the 3rd International Conference on Quiescent Plasmas, Elsinore September 20. - 24. 1971, Risø Report No. 250'.

Haskell, R. E. & Case, C. T. (1967), 'Transient signal propagation in losless, isotropic plasmas', *IEEE Trans. Antennas and Propagation* **AP-15**, 458–464.

Hayes, J. N. (1961), 'Damping of plasma oscillations in the linear theory', *Phys. Fluids* **4**, 1387–1392.

Hayes, J. N. (1963), 'On the non-Landau damped solutions of the linearized Vlasov equation', *Il Nuovo Cimento* **30**, 1048–1063.

Hazeltine, R. D. & Waelbroeck, F. L. (1998), *The Framework of Plasma Physics*, Perseus Books, Reading, Massachusetts.

Heald, M. A. & Wharton, C. M. (1965), *Plasma Diagnostics with Microwaves*, John Wiley & Sons, New York.

Hellberg, M. A. & Mace, R. L. (2002), 'Generalized plasma dispersion function for a plasma with a kappa-Maxwellian velocity distribution', *Phys, Plasmas* **9**, 1495–1504.

Hendel, H. W., Chu, T. K. & Politzer, P. A. (1968), 'Collisional drift waves – identification, stabilization, and enhanced plasma transport', *Phys. Fluids* **11**, 2426–2439.

Hines, C. O. (1960), 'Internal atmospheric gravity waves at ionospheric heights', *Canadian J. Phys.* **38**, 1441–1481.

Hirose, A., Ito, A., Mahajan, S. M. & Ohsaki, S. (2004), 'Relation between Hall-magnetohydrodynamics and the kinetic Alfvén wave', *Phys. Lett. A* **330**, 474–480.

Hirshfield, J. L. & Jacob, J. H. (1968), 'Free-streaming and spatial Landau damping', *Phys. Fluids* **11**(2), 411–413.

Horowitz, P. & Hill, W. (1989), *The Art of Electronics*, 2. edn, Cambridge University Press, Cambridge, UK.

Horton, W., Tajima, T. & Kamimura, T. (1987), 'Kelvin-Helmholtz instability and vortices in magnetized plasma', *Phys. Fluids* **30**, 3485–3495.

Hsu, P. & Kuehl, H. (1982), 'Focusing of lower hybrid waves by electromagnetic effects', *Nuclear Fusion* **22**, 1679–1683.

Hsu, P. & Kuehl, H. H. (1983), 'Electromagnetic effects on the self-modulation of non-linear lower hybrid waves', *Phys. Fluids* **26**, 689–699.

Huddlestone, R. H. & Leonard, S. L., eds (1965), *Plasma Diagnostic Techniques*, Academic Press, New York.

Huld, T., Nielsen, A. H., Pécseli, H. L. & Juul Rasmussen, J. (1991), 'Coherent structures in two-dimensional turbulence', *Phys. Fluids B* **3**, 1609–1625.

Hundhausen, A. J. (1972), *Coronal Expansion and Solar Wind*, Springer-Verlag, Berlin.

Hurley, J. (1961), 'Interaction of a streaming plasma with the magnetic field of a line current', *Phys. Fluids* **4**, 109–111.

Hutchinson, I. H. (1987), *Principles of Plasma Diagnostics*, Cambridge University Press, UK.

Hutchinson, I. H. (2003), 'Ion collection by a sphere in a flowing plasma: 2. non-zero Debye length', *Plasma Phys. Control. Fusion* **45**, 1477–1500.

Hutchinson, I. H. (2017), 'Electron holes in phase space: What they are and why they matter', *Phys. Plasmas* **24**, 055601.

Ichikawa, Y. H. (1970), 'Effects of bunched ion bursts on apparent nonlinear ion acoustic waves', *Phys. Fluids* **14**, 2541–2545.

Ichikawa, Y. H., Imamura, T. & Taniuti, T. (1972), 'Nonlinear wave modulation in collisionless plasmas', *J. Phys. Soc. Japan* **33**, 189–197.

Iizuka, S., Michelsen, P., Rasmussen, J. J., Schrittwieser, R., Hatakeyama, R., Saeki, K. & Sato, N. (1982), 'Dynamics of a potential barrier formed on the tail of a moving double-layer in a collisionless plasma', *Phys. Rev. Lett.* **48**, 145–148.

Ikezi, H. (1973), 'Experiments on ion-acoustic solitary waves', *Phys. Fluids* **16**, 1668–1675.

Ikezi, H., Nishikawa, K. & Mima, K. (1974), 'Self-modulation of high-frequency electric-field and formation of plasma cavities', *J. Phys. Soc. Jap.* **37**, 766–773.

Ikezi, H., Schwarzenegger, K., Simons, A. L., Ohsawa, Y. & Kamimura, T. (1978), 'Nonlinear self-modulation of ion-acoustic waves', *Phys. Fluids* **21**.

Ikezi, H., Taylor, R. J. & Baker, D. R. (1970), 'Formation and interaction of ion-acoustic solitons', *Phys. Rev. Lett.* **25**, 11–14.

Irvine, W. M. (1964), 'Electrodynamics in a rotating system of reference', *Physica* **30**, 1160–1170.

Ishibashi, N. & Kitahara, K. (1992), 'Two-dimensional Bernstein-Greene-Kruskal solution', *J. Phys. Soc. Jpn.* **61**, 2795–2804.

Itikawa, Y. (1973), 'Effective collision frequency of electrons in gases', *Phys. Fluids* **16**, 831–835.

Jackson, J. D. (1960), 'Longitudinal plasma oscillations', *J. Nucl. Energy, Part C: Plasma Phys.* **1**, 171–189.

Jackson, J. D. (1975), *Classical Electrodynamics*, 2 edn, John Wiley & Sons, New York.

Jensen, V. O. (1995), 'On stationary magnetic stresses in vacuum and magnetized plasmas', *Eur. J. Phys.* **16**, 199–200.

Johnson, E. O. & Malter, L. (1950), 'A floating double probe method for measurements in gas discharges', *Phys. Rev.* **80**, 58–68.

Jovanović, D., Lynov, J. P., Michelsen, P., Pécseli, H. L., Rasmussen, J. J. & Thomsen, K. (1982*a*), 'Three dimensional double layers in magnetized plasmas', *Geophys. Res. Lett.* **9**, 1049–1052.

Jovanović, D., Pécseli, H. L. & Thomsen, K. (1982*b*), 'Non-linear transient signal propagation in homogeneous plasma', *J. Plasma Phys.* **28**, 159–175.

Joyce, G., Lonngren, K., Alexeff, I. & Jones, W. D. (1969), 'Dispersion of ion-acoustic waves', *Phys. Fluids* **12**, 2592–2599.

Kadomtsev, B. B. (1965), *Plasma Turbulence*, Academic Press, New York.

Kadomtsev, B. B. & Karpman, V. I. (1971), 'Nonlinear waves', *Sov. Phys. Usp.* **14**, 40–60.

Kako, F. & Yajima, N. (1978), 'Interaction of ion-acoustic solitons in three-dimensional space', *J. Phys. Soc. Jpn.* **44**, 1711–1714.

Kako, F. & Yajima, N. (1982), 'Interaction of ion-acoustic solitons in two-dimensional space. II', *J. Phys. Soc. Jpn.* **51**, 311–322.

Kako, M., Taniuti, T. & Watanabe, T. (1971), 'Hole equilibria in a quasi-cold plasma', *J. Phys. Soc. Jpn.* **31**, 1820–1829.

Karpman, V. I. (1979*a*), 'The effects of the interaction between ion-sound solitons and resonance particles in a plasma', *Sov. Phys. JETP* **50**, 695–701. See also Zh. Eksp. Teor. Fiz. **77**, 1382-1395 (1979).

Karpman, V. I. (1979*b*), 'Soliton evolution in the presence of perturbation', *Phys. Scripta* **20**, 462–478.

Karpman, V. I., Lynov, J. P., Michelsen, P., Pécseli, H. L., Rasmussen, J. J. & Turikov, V. A. (1980), 'Modifications of plasma solitons by resonant particles', *Phys. Fluids* **23**, 1782–1794.

Karpman, V. I. & Shagalov, A. G. (1982), 'The ponderomotive force of a high-frequency electromagnetic field in a cold magnetized plasma', *J. Plasma Phys.* **27**, 225–238.

Kasymov, Z. Z., Näslund, E., Starodub, A. N. & Stenflo, L. (1985), 'Upper hybrid turbulence in a plasma with magnetized electrons', *Phys. Scripta* **31**, 201–204.

Kato, K. (1976), 'Two-dimensional and three-dimensional BGK waves', *J. Phys. Soc. Jpn.* **41**, 1050–1053.

Kaye, G. W. C. & Laby, T. H. (1995), *Tables of Physical and Chemical Constants*, 16 edn, Longmans, UK.

Kennedy, R. V. & Allen, J. E. (2002), 'The floating potential of spherical probes and dust grains. Part 1. Radial motion theory', *J. Plasma Phys.* **67**, 243–250.

Kent, G. I., Jen, N. C. & Chen, F. F. (1969), 'Transverse Kelvin-Helmholtz instability in a rotating plasma', *Phys. Fluids* **12**, 2140–2151.

Kentwell, G. W. & Jones, D. (1987), 'The time dependent ponderomotive force', *Phys. Rep.* **145**, 319–403.

Kim, H. C., Stenzel, L. & Wong, A. Y. (1974), 'Development of "Cavitons" and trapping of rf field', *Phys. Rev. Lett.* **33**, 886–889.

Kingsep, A. S., Chukbar, K. V. & Yan'kov, V. V. (1990), *Rev. Plasma Physics*, Vol. 16, Consultant Bureau, New York, chapter 3. Electron magnetohydrodynamics, pp. 243–291.

Kletzing, C. A., Thuecks, D. J., Skiff, F., Bounds, S. R. & Vincena, S. (2010), 'Measurements of inertial limit Alfvén wave dispersion for finite perpendicular wave number', *Phys. Rev. Lett.* **104**, 095001.

Klimontovich, Y. L. (1967), *The Statistical Theory of Non-equilibrium Processes in a Plasma*, M.I.T. Press, Cambridge, Massachusets.

Knorr, G., Hansen, F. R., Lynov, J. P., Pécseli, H. L. & Rasmussen, J. J. (1988), 'Finite Larmor radius effects to arbitrary order', *Phys. Scripta* **38**, 829–834.

Knorr, G., Lynov, J. P. & Pécseli, H. L. (1990), 'Self-organization in three-dimensional hydrodynamic turbulence', *Z. Naturforsch.* **45 a**, 1059–1073.

Knorr, G. & Merlino, R. L. (1984), 'The role of fast electrons for the confinement of plasma by magnetic cusps', *Plasma Phys. Control. Fusion* **26**, 433–442.

Koepke, M. E. (2008), 'Interrelated laboratory and space plasma experiments', *Rev. Geophys.* **46**, RG3001.

Koepke, M. E. & Reynolds, E. W. (2007), 'Simultaneous, co-located parallel-flow shear and perpendicular-flow shear in a low-temperature, ionospheric-plasma relevant laboratory plasma', *Plasma Phys. Contr. Fusion* **49**, A145–A157.

Kono, M. (1986), 'Solitons and beam reflection in an ion-beam plasma system', *Phys. Fluids* **29**, 1268–1273.

Kono, M. & Pécseli, H. L. (2016), 'Parametric decay of wide band Langmuir wave-spectra', *J. Plasma Phys.* **82**, 905820606.

Kono, M. & Pécseli, H. L. (2017), 'Stability of electron wave spectra in weakly magnetized plasmas', *J. Plasma Phys.* **83**, 905830610.

Kono, M. & Škorić, M. M. (2010), *Nonlinear Physics of Plasmas*, number 62 *in* 'Springer Series on Atomic, Optical, and Plasma Physics', Springer, Heidelberg, Germany.

Kono, M., Škorić, M. M. & ter Haar, D. (1981), 'Spontaneous excitation of magnetic-fields and collapse dynamics in a Langmuir plasma', *J. Plasma Phys.* **26**, 123–146.

Krasnosel'skikh, V. V. & Sotnikov, V. I. (1977), 'Plasma-wave collapse in a magnetized plasma', *Fiz. Plazmy* **3**, 872–879. See also Sov. J. Plasma Phys., **3**, 491-495 (1977).

Kuhn, S. (1979), 'Determination of axial steady-state potential distributions in collisionless single-ended Q-machines', *Plasma Phys.* **21**, 613–626.

Kulsrud, R. M. (2011), 'Intuitive approach to magnetic reconnection', *Phys. Plasmas* **18**, 111201.

Kuo, S. P. (2015), 'Ionospheric modifications in high frequency heating experiments', *Phys. Plasmas* **22**, 012901.

Kuper, C. G. (1968), *An Introduction to the Theory of Superconductivity*, Clarendon Press, Oxford.

Kurth, R. (1972), *Dimensional Analysis and Group Theory in Astrophysics*, Pergamon Press, Oxford, UK.

Kuznetsov, E. A. (1974), 'The collapse of electromagnetic waves in a plasma', *Zh. Eksp. Teor. Fiz.* **66**, 2037–2047. english translation Sov. Phys.-JETP, **39**, 1003-1007 (1974).

Landau, L. (1946), 'On the vibrations of the electronic plasma', *J. Phys. USSR* **10**, 25–34.

Landau, L. D. & Lifshitz, E. M. (1987), *Fluid Mechanics*, Vol. 6 of *Course of Theoretical Physics*, 2 edn, Butterworth-Heinemann, Great Britain.

Landau, L. D., Pitaevskii, L. P. & Lifshitz, E. M. (1984), *Electrodynamics of Continuous Media*, Vol. 8 of *Course of Theoretical Physics*, 2 edn, Butterworth-Heinemann, Oxford, UK.

Langmuir, I. & Kingdon, K. H. (1925), 'Thermionic effects caused by vapors of alkali metals', *Proc. Roy. Phys. Soc. Ser. A* **107**, 61–79.

Larichev, V. D. & Reznik, G. M. (1976), 'Two-dimensional solitary Rossby waves', *Doklady Akad. Nauk. SSSR* **231**, 1077–1079.

Larmor, J. (1897), 'On a dynamical theory of the electric and luminiferous medium, Part 3, relations with material media', *Phil. Trans. Roy. Soc.* **190**, 205–300.

Lawrence, G. A., Browand, F. K. & Redekopp, L. G. (1991), 'The stability of a sheared density interface', *Phys. Fluids A* **3**, 2360–2370.

Lehnert, B. (1954), 'Magneto-hydrodynamic waves in liquid Sodium', *Phys. Rev.* **94**, 815–824.

Lehnert, B. (1972), Basic features of plasma instabilities, Technical Report TRITA-EPP-72-23, Royal Institute of Technology.

Lehnert, B. & Bullard, E. C. (1955), 'An instability of laminar flow of mercury caused by an external magnetic field', *Proc. Royal Soc. London. Ser. A. Math. Phys. Sci.* **233**, 299–302.

Lehnert, B. & Sjögren, G. (1960), 'Stability of a hollow Mercury jet', *Rev. Mod. Phys.* **32**, 813–814.

Leung, P., Wong, A. Y. & Quon, B. H. (1980), 'Formation of double layers', *Phys. Fluids* **23**, 992–1004.

Lichtenberg, A. J. & Lieberman, M. A. (1983), *Regular and Stochastic Motion*, Vol. 38 of *Applied Mathematical Sciences*, Springer-Verlag, New York.

Lieberman, M. A. & Lichtenberg, A. J. (1994), *Principles of Plasma Discharges and Materials Processing*, John Wiley & Sons, Inc., New York.

Lighthill, M. J. (1960), 'Studies on magneto-hydrodynamic waves and other anisotropic wave motions', *Phil. Trans. Royal Soc. (London). Ser. A, Math. Phys. Sciences* **252**, 397–430.

Lima, F. M. S. & Arun, P. (2006), 'An accurate formula for the period of a simple pendulum oscillating beyond the small angle regime', *Am. J. Phys.* **74**, 892–895.

Limpaecher, R. & MacKenzie, K. R. (1973), 'Magnetic multipole containment of large uniform collisionless quiescent plasmas', *Rev. Scientific Instr.* **44**, 726–731.

Lin, A. T. & Lin, C.-C. (1981), 'Non-linear penetration of upper-hybrid waves induced by parametric-instabilities of a plasma in an inhomogeneous magnetic-field', *Phys. Rev. Lett.* **47**, 98–102.

Lin, C. & Zhang, X.-L. (2003), 'Exact analytical solutions of the three dimensional Debye screening in plasmas', *Phys. Scripta* **68**, 264–265.

Littlejohn, R. G. (1983), 'Variational principles of guiding centre motion', *J. Plasma Phys.* **29**, 111–125.

Livadiotis, G., ed. (2017), *Kappa Distributions, Theory and Applications in Plasmas*, Elsevier, Amsterdam, Netherlands.

Llobet, X., Bernstein, W. & Konradi, A. (1985), 'The spatial evolution of energetic electrons and plasma waves during the steady state beam plasma discharge', *J. Geophys. Res.* **90**, 5187–5196.

Lochte-Holtgreven, W., ed. (1968), *Plasma Diagnostics*, North-Holland, Amsterdam.

Lonngren, K. E. (1983), 'Soliton experiments in plasmas', *Plasma Phys. Contr. Fusion* **25**, 943–982.

Lonngren, K. & Scott, A., eds (1978), *Solitons in Action*, Academic Press, New York. Proceedings of a workshop held at Redstone Arsenal, October 26-27, 1977.

Lundquist, S. (1949), 'Experimental investigations of Magneto-Hydrodynamic waves', *Phys. Rev.* **76**, 1805–1809.

Lynov, J. P. (1983), 'Modification of Korteweg-deVries solitons in plasmas by resonant particles', *Phys. Fluids* **26**, 3262–3272.

Lynov, J. P., Michelsen, P., Pécseli, H. L. & Rasmussen, J. J. (1979), Damping of solitons by reflected particles, Technical Report Risø-M-2168, Risø National Laboratory.

Lynov, J. P., Michelsen, P., Pécseli, H. L. & Rasmussen, J. J. (1980), 'Interaction between electron holes in a strongly magnetized plasma', *Phys. Lett.* **80A**, 23–25.

Lynov, J. P., Michelsen, P., Pécseli, H. L., Rasmussen, J. J., Saeki, K. & Turikov, V. (1979b), 'Observations of solitary structures in a magnetized, plasma loaded waveguide', *Phys. Scripta* **20**, 328–335.

MacDonald, D. K. C. (1962), *Noise and Fluctuations: an Introduction*, John Wiley & Sons, New York.

Mahajan, S. M., Nikol'skaya, K. I., Shatashvili, N. L. & Yoshida, Z. (2002), 'Generation of flows in the solar atmosphere due to magnetofluid coupling', *Astrophys. J. Lett.* **576**, L161.

Makino, M., Kamimura, T. & Taniuti, T. (1981), 'Dynamics of two-dimensional solitary vortices in a low-β plasma with convective motion', *J. Phys. Soc. Japan* **50**, 980–989.

Malmberg, J. H. & Wharton, C. B. (1964), 'Collisionless damping of electrostatic plasma waves', *Phys. Rev. Lett.* **13**, 184–186.

Manheimer, W. M. (1969), 'Nonlinear development of an electron plasma wave in a cylindrical waveguide', *Phys. Fluids* **12**, 2426–2428.

Manheimer, W. M. (1977), *An Introduction to Trapped–Particle Instability in Tokamaks*, ERDA Critical Rev. Ser., National Technical Inf. Serv. Rep. TID–27157.

Märk, E. & Sato, N. (1977), 'Higher harmonics of electron plasma waves', *J. Plasma Phys.* **17**, 357–368.

Marsch, E. & Livi, S. (1985), 'Coulomb collision rates for self-similar and kappa distributions', *Phys. Fluids* **28**, 1379–1386.

Marshall, J. S. (2001), *Inviscid Incompressible Flow*, John Wiley & Sons, New York, USA.

Martín, P., Donoso, G. & Zamudio-Cristi, J. (1980), 'A modified asymptotic Padé method. Application to multipole approximation for the plasma dispersion function Z', *J. Math. Phys.* **21**, 280–285.

Martín, P. & González, M. A. (1979), 'New two-pole approximation for the plasma dispersion function Z', *Phys. Fluids* **22**, 1413–1414.

Mason, R. J. (1968), 'Electric field penetration into a plasma with a fractionally accommodating boundary', *J. Math. Phys.* **9**, 868–874.

Mason, R. J. (1970*a*), 'Structure of evolving ion-acoustic fronts in collisionless plasmas', *Phys. Fluids* **13**, 1042–1048.

Mason, R. J. (1970*b*), 'Weak shock generation according to the energy-conserving Bhatnagar-Gross-Krook kinetic equation', *Phys. Fluids* **13**, 1467–1472.

Mason, R. J. (1971), 'Computer simulation of ion-acoustic shocks. The diaphragm problem', *Phys. Fluids* **14**, 1943–1958.

Mead, G. D. (1964), 'Deformation of the geomagnetic field by the solar wind', *J. Geophys. Res.* **69**, 1181–1195.

Mead, G. D. & Beard, D. B. (1964), 'Shape of the geomagnetic field solar wind boundary', *J. Geophys. Res.* **69**, 1169–1179.

Merilees, P. E. & Warn, H. (1975), 'On energy and enstrophy exchanges in two-dimensional non-divergent flow', *J. Fluid Mech.* **69**, 625–630.

Michaelson, H. B. (1977), 'The work function of the elements and its periodicity', *J. App. Phys.* **48**, 4729–4733.

Michelsen, P. & Pécseli, H. L. (1973), 'Propagation of density perturbations in a collisionless Q-machine plasma', *Phys. Fluids* **16**, 221–225.

Michelsen, P., Pécseli, H. L., Rasmussen, J. J. & Sato, N. (1977), 'Stationary density variation produced by a standing plasma wave', *Phys. Fluids* **20**, 1094–1096.

Mikhailovkii, A. B. (1974*a*), *Theory of Plasma Instabilities*, Vol. 1: Instabilities of a Homogeneous Plasma of *Studies in Soviet Science*, Consultants Bureau, New York - London.

Mikhailovkii, A. B. (1974*b*), *Theory of Plasma Instabilities*, Vol. 2: Instabilities of an Inhomogeneous Plasma of *Studies in Soviet Science*, Consultants Bureau, New York - London.

Mikhailovskii, A. B. & Timofeev, A. V. (1963), 'Theory of cyclotron instability in a non-uniform plasma', *Sov. Phys. JETP* **17**, 621–627. See also J. Exptl. Theoret. Phys. **44**, 919-921 (1963).

Morales, G. J. & Lee, Y. C. (1977), 'Generation of density cavities and localized electric-fields in a nonuniform plasma', *Phys. Fluids* **20**, 1135–1147.

Motley, R. W. (1975), *Q Machines*, Academic Press, New York.

Munakata, K.-I. (1952), 'Some exact solutions in nonlinear oscillations', *J. Phys. Soc. Jpn.* **7**, 383–391.

Nakagawa, T. (2013), 'Ion entry into the wake behind a nonmagnetized obstacle in the solar wind: Two-dimensional particle-in-cell simulations', *J. Geophys. Res.: Space Phys.* **118**, 1–12.

Naugolnykh, K. & Ostrovsky, L. (1998), *Nonlinear Wave Processes in Acoustics*, Cambridge Texts in Applied Mathematics, Cambridge University Press, Cambridge UK.

Nayfeh, A. H. (1973), *Perturbation Methods*, John Wiley & Sons, New York.

Nelson, R. A. & Olsson, M. G. (1986), 'The pendulum - rich physics from a simple system', *Am. J. Phys.* **64**, 112–121.

Nicholson, D. R. (1983), *Introduction to Plasma Theory*, John Wiley & Sons, New York.

Nishikawa, K. (1984), 'Some aspects of plasma turbulence theory', *Prog. Theor. Phys. Suppl.* **80**, 168–179.

Nycander, J. (1994), 'Steady vortices in plasmas and geophysical flows', *Chaos* **4**, 253–267.

O'Neil, T. (1965), 'Collisionless damping of non linear plasma oscillations', *Phys. Fluids* **8**, 2255–2262.

Pack, J. L. & Phelps, A. V. (1961), 'Drift velocities of slow electrons in Helium, Neon, Argon, Hydrogen, and Nitrogen', *Phys. Rev.* **121**, 798–806.

Parker, E. N. (1965), 'Dynamical theory of the solar wind', *Space Sci. Rev.* **4**, 666.

Parkinson, D. & Schindler, K. (1969), 'Landau damping of long wavelength ion acoustic waves in a collision-free plasma with a gravity field', *J. Plasma Phys.* **3**, 13–20.

Parks, G. K. (2004), *Physics of Space Plasmas, an Introduction*, 2 edn, Westview Press, Cambridge, Massachusetts.

Pathria, R. K. (1998), *Statistical Mechanics*, 2 edn, Butterworth-Heinemann, Oxford.

Pécseli, H. L. (1982), 'Drift-wave turbulence in low-β plasmas', *Phys. Scripta* **T2/1**, 147–157.

Pécseli, H. L. (1985), 'Solitons and weakly nonlinear waves in plasmas', *IEEE Trans. Plasma Sci.* **PS-13**, 53–86.

Pécseli, H. L. (2000), *Fluctuations in Physical Systems*, Cambridge University Press, Cambridge, UK.

Pécseli, H. L. (2014), 'Modulational stability of electron plasma wave spectra', *J. Plasma Phys.* **80**, 745–769.

Pécseli, H. L. (2016), *Low frequency waves and turbulence in magnetized laboratory plasmas and in the ionosphere*, IOP Publishing, UK.

Pécseli, H. L., Armstrong, R. & Trulsen, J. (1981), 'Experimental observation of ion phase-space vortices', *Phys. Lett.* **81A**, 386–390.

Pécseli, H. L. & Engvold, O. (2000), 'Modeling of prominence threads in magnetic fields: levitation by incompressible MHD waves', *Solar Phys.* **194**, 73–86.

Pécseli, H. L., Iranpour, K., Holter, Ø., Lybekk, B., Holtet, J., Trulsen, J., Eriksson, A. & Holback, B. (1996), 'Lower hybrid wave cavities detected by the FREJA satellite', *J. Geophys. Res. Space Phys.* **101**, 5299–5316.

Pécseli, H. L. & Rasmussen, J. J. (1980), 'Nonlinear electron plasma waves in strongly magnetized plasmas', *Plasma Phys.* **22**, 421–438.

Pécseli, H. L. & Trulsen, J. (1989), 'A statistical analysis of numerically simulated plasma turbulence', *Phys. Fluids B* **1**, 1616–1636. Erratum: Phys. Fluids B **2**, 454 (1990)

Pécseli, H. L., Trulsen, J. & Armstrong, R. (1984), 'Formation of ion phase-space vortexes', *Phys. Scripta* **29**, 241–253.

Pedersen, P. O. (1927), *The Propagation of Radio Waves Along the Surface of the Earth and in the Atmosphere*, Danmarks Naturvidenskabelige Samfund, Copenhagen.

Pedlosky, J. (1987), *Geophysical Fluid Dynamics*, 2 edn, Springer-Verlag, New York.

Perkins, F. W. & Sun, Y. C. (1981), 'Double-layers without current', *Phys. Rev. Lett.* **46**, 115–118.

Petviashvili, V. I. & Pokhotelov, O. A. (1986), 'Solitary vortices in plasmas', *Sov. J. Plasma Phys.* **12**, 651. See also Fiz. Plazmy **12**, 1127-1144 (1986).

Phillips, E. G. (1957), *Functions of a Complex Variable, with Applications*, University Mathematical Texts, Oliver and Boyd, Edinburgh.

Piel, A. (2010), *Plasma Physics: An Introduction to Laboratory, Space, and Fusion Plasmas*, Springer, Heidelberg.

Piel, A. & Oelerich, G. (1985), 'Thermal structures on resonance cones in a weakly collisional plasma', *Phys. Fluids* **28**, 1366–1370.

Politzer, P. A. (1971), 'Drift instability in collisionless alkali metal plasmas', *Phys. Fluids* **14**, 2410–2425.

Porkolab, M. (1977), 'Parametric instabilities due to lower-hybrid radio frequency heating of tokamak plasmas', *Phys. Fluids* **20**, 2058–2075.

Quon, B. H., Wong, A. Y. & Ripin, B. H. (1974), 'Backscattering decay processes in electron beam-plasma interactions including ion dynamics', *Phys. Rev. Lett.* **32**, 406–409.

Raadu, M. A. (1989), 'The physics of double layers and their role in astrophysics', *Phys. Reports* **178**, 25–97

Rasmussen, J. J. (1978), 'Finite-amplitude electron-plasma waves in a cylindrical waveguide', *Plasma Phys.* **20**, 997–1010.

Rasmussen, J. J. & Rypdal, K. (1986), 'Blow-up in nonlinear Schroedinger equations-I, A general review', *Phys. Scripta* **33**, 481–497.

Reif, F. (1965), *Fundamentals of Statistical and Thermal Physics*, McGraw-Hill.

Rekaa, V. L., Pécseli, H. L. & Trulsen, J. K. (2012), 'Numerical studies of a plasma diode with external forcing', *Phys. Plasmas* **19**, 082115.

Rekaa, V. L., Pécseli, H. L. & Trulsen, J. K. (2013), 'Self-similar space-time evolution of an initial density discontinuity', *Phys. Plasmas* **20**, 072117.

Riemann, K.-U. (1991), 'The Bohm criterion and sheath formation', *J. Phys. D: Applied Phys.* **24**, 493–518.

Riemann, K.-U. (2003), 'Kinetic analysis of the collisional plasma-sheath transition', *J. Phys. D: Applied Phys.* **36**, 2811–2820.

Rietveld, M. T., Isham, B., Kohl, H., La Hoz, C. & Hagfors, T. (2000), 'Measurements of HF-enhanced plasma and ion lines at EISCAT with high-altitude resolution', *J. Geophys. Res.* **105**, 7429–7439.

Roberson, C. & Gentle, K. W. (1971), 'Experimental test of quasilinear theory of the gentle bump instability', *Phys. Fluids* **14**, 2462–2469.

Roberson, C., Gentle, K. W. & Nielsen, P. (1971), 'Experimental test of quasilinear theory', *Phys. Rev. Lett.* **26**, 226–229.

Roberson, C. W. (1971), Experimental tests of quasilinear theory, Technical report, Center for Plasma Physics and Theromonuclear Research, The University of Texas, Austin, Texas, USA. Dissertation.

Roberts, K. V. & Berk, H. L. (1967), 'Nonlinear evolution of the two-stream instability', *Phys. Rev. Lett.* **19**, 297–300.

Rogers, K. C. & Chen, F. F. (1970), 'Direct measurements of drift wave growth rates', *Phys. Fluids* **13**, 513–516.

Rohde, V., Piel, A., Thiemann, H. & Oyama, K. I. (1993), 'In situ diagnostics of ionospheric plasma with the resonance cone technique', *J. Geophys. Res.: Space Phys.* **98**, 19163–19172.

Rosenbluth, M. N. & Longmire, C. L. (1957), 'Stability of plasmas confined by magnetic fields', *Annals Phys.* **1**, 120–140.

Roth, J. R. (1995), *Industrial Plasma Engineering*, Vol. 1 Principles, Institute of Physics Publishing, Bristol and Philadelphia.

Rowberg, R. E. & Wong, A. Y. (1970), 'Collisional drift waves in the linear regime', *Phys. Fluids* **13**, 661–671.

Rudakov, L. I. & Tsytovich, V. N. (1978), 'Strong Langmuir turbulence', *Phys. Reports* **40C**, 1–73.

Rypdal, K., Lynov, J. P., Pécseli, H. L., Rasmussen, J. J. & Thomsen, K. (1982), 'Interaction of Langmuir solitons with resonant particles', *Phys. Scripta* **T2**, 534–537.

Saeki, K. (1973), 'Electron plasma wave shocks in a collisionless plasma', *J. Phys. Soc. Jpn.* **35**, 251–257.

Saeki, K. & Genma, H. (1998), 'Electron-hole disruption due to ion motion and formation of coupled electron hole and ion-acoustic soliton in a plasma', *Phys. Rev. Lett.* **80**, 1224–1227.

Saeki, K., Michelsen, P., Pécseli, H. L. & Rasmussen, J. J. (1979), 'Formation and coalescence of electron solitary holes', *Phys. Rev. Lett.* **42**, 501–504.

Sagdeev, R. Z. & Galeev, A. A. (1969), *Nonlinear Plasma Theory*, Frontiers in Physics, W. A. Benjamin Inc., Amsterdam.

Sakanaka, P. H. (1972), 'Beam-generated collisionless ion-acoustic shocks', *Phys. Fluids* **15**, 1323–1327.

Sato, N., Hatakeyama, R., Iizuka, S., Mieno, T., Saeki, K., Rasmussen, J. J. & Michelsen, P. (1981), 'Ultrastrong stationary double layers in a nondischarge magnetoplasma', *Phys. Rev. Lett.* **46**, 1330–1333.

Schamel, H. (1986), 'Electron holes, ion holes and double layers', *Phys. Reports* **140**, 161–191.

Schiff, L. I. (1939), 'A question in general relativity', *Proc. Nat. Acad. Sci.* **25**, 391–395.

Schlitt, L. G. & Hendel, H. W. (1972), 'Effects of parallel wavelength on the collisional drift instability', *Phys. Fluids* **15**, 1578–1589.

Schmidt, G. (1979), *Physics of High Temperature Plasmas*, Academic Press, New York.

Schrittwieser, R., Adamek, J., Balan, P., Hron, M., Ionita, C., Jakubka, K., Kryska, L., Martines, E., Stockel, J., Tichy, M. & Van Oost, G. (2002), 'Measurements with an emissive probe in the CASTOR tokamak', *Plasma Phys. Contr. Fusion* **44**, 567–578.

Schroeder, D. V. (2000), *An Introduction to Thermal Physics*, Addison Wesley Longman, San Francisco.

Schuck, P. W., Bonnell, J. W. & Kintner, P. M. (2003), 'A review of lower hybrid solitary structures', *IEEE Trans. Plasma Sci.* **31**, 1125–1177.

Schunk, R. W. & Nagy, A. F. (1978), 'Electron temperatures in the F region of the ionosphere: theory and observations', *Rev. Geophys. Space. Phys.* **16**, 355–399.

Sckopke, N. (1966), 'A general relation between the energy of trapped particles and the disturbance field near the Earth', *J. Geophys. Res.* **71**, 3125–3130.

Sears, F. W. & Brehme, R. W. (1968), *Introduction to the Theory of Relativity*, Addison-Wesley Publishing Co., New York.

Shapiro, V. D., Shevchenko, V. I., Solov'ev, G. I., Kalinin, V. P., Bingham, R., Sagdeev, R. Z., Ashour-Abdalla, M., Dawson, J. & Su, J. J. (1993), 'Wave collapse at the lower-hybrid resonance', *Phys. Fluids B* **5**, 3148–3162.

Shatashvili, N. L. & Tsintsadze, N. L. (1982), 'Nonlinear Landau damping phenomenon in a strongly turbulent plasma', *Phys. Scripta* **T2B**, 511–516.

Shawhan, S. D. (1979), *Magentospheric plasma waves*, Vol. III of Solar System Plasma Physics, North-Holland, chapter III.1.6.

Shivamoggi, B. K. (1988), *Introduction to Nonlinear Fluid-Plasma Waves*, Kluwer Academic Publishers, Dordrecht, Nederlands.

Shivamoggi, B. K. (2009), 'Impulse formulations of Hall magnetohydrodynamic equations', *Phys. Lett. A* **373**, 708–710.

Shukla, P. K. & Mamun, A. A. (2002*a*), *Introduction to Dusty Plasmas*, Institute of Physics Publishing, Bristol and Philadelphia.

Shukla, P. K. & Mamun, A. A. (2002*b*), 'Lower hybrid drift wave turbulence and associated electron transport coefficients and coherent structures at the magnetopause boundary layer', *J. Geophys. Res.: Space Phys.* **107**, SMP 34–1–SMP 34–8.

Shukla, P. K., Yu, M. Y., Rahman, H. U. & Spatschek, K. H. (1984), 'Nonlinear convective motion in plasmas', *Phys. Reports* **105**, 229–328.

Simon, A. (1955), 'Ambipolar diffusion in a magnetic field', *Phys. Rev.* **98**, 317–318.

Singh, N. & Schunk, R. W. (1982), 'Numerical calculations relevant to the initial expansion of the polar wind', *J. Geophys. Res.* **87**, 9154–9170.

Sivukhin, D. V. (1966), Coulomb collisions in a fully ionized plasma, *in* M. A. Leontovich, ed., 'Reviews of Plasma Physics', Vol. 4, Consultants Bureau, New York, pp. 93–241.

Sjölund, A. & Stenflo, L. (1967*a*), 'Parametric coupling between ion waves and electromagnetic waves', *App. Phys. Lett.* **10**, 201–202.

Sjölund, A. & Stenflo, L. (1967*b*), 'Parametric coupling between transverse electromagnetic and longitudinal electron waves', *Physica* **35**, 499–505.

Skiff, F., Bachet, G. & Doveil, F. (2001), 'Ion dynamics in nonlinear electrostatic structures', *Phys. Plasmas* **8**, 3139–3142.

Smith, R. A. (1982), 'A review of double layer simulations', *Phys. Scripta* **T2A**, 238–251.

Spitzer, L. (1956), *Physics of Fully Ionized Gases*, Vol. 3 of *Interscience Tracts on Physics and Astronomy*, Interscience, New York.

Statham, G. & ter Haar, D. (1983), 'Strong turbulence of a magnetized plasma. 2. the ponderomotive force', *Plasma Phys. Control. Fusion* **25**, 681–698.

Stenflo, L. (1994), 'Resonant three-wave interactions in plasmas', *Phys. Scripta* **T50**, 15–19.

Stenzel, R. L. (1999), 'Whistler waves in space and laboratory plasmas', *J. Geophys. Res.* **104**, 14379–14395.

Stern, D. (1973*a*), 'An inverse theorem about the magnetic field line velocity', *J. Geophys. Res.* **78**, 1702–1706.

Stern, D. (1973*b*), 'A study of the electric field in an open magnetospheric model', *J. Geophys. Res.* **78**, 7292–7305.

Sternberg, N. & Godyak, V. (2007), 'The Bohm plasma-sheath model and the Bohm criterion revisited', *IEEE Trans. Plasma Sci.* **35**(5), 1341–1349.

Stix, T. H. (1992), *Waves in Plasmas*, AIP, American Institute of Physics, New York.

Stratton, J. A. (1941), *Electromagnetic Theory*, McGraw-Hill Book Company.

Stringer, T. E. (1963), 'Low-frequency waves in an unbounded plasma', *J. Nucl. Energy Part C* **5**, 89–107.

Sturrock, P. A. (1994), *Plasma Physics, An Introduction to the Theory of Astrophysical, Geophysical & Laboratory Plasmas*, Cambridge University Press, Cambridge, UK.

Sugai, H., Lynov, J. P., Michelsen, P., Pécseli, H. L. & Rasmussen, J. J. (1979), 'Evolution of modulated dispersive electron waves in a plasma', *Plasma Phys.* **21**, 701–712.

Sugai, H., Pécseli, H. L., Rasmussen, J. J. & Thomsen, K. (1983), 'Evolution of externally excited convective cells in plasmas', *Phys. Fluids* **26**, 1388–1390.

Summers, D. & Thorne, R. M. (1992), 'A new tool for analyzing microinstabilities in space plasmas modeled by a generalized Lorentzian (Kappa) distribution', *J. Geophys. Res.: Space Phys.* **97**, 16827–16832.

Swanson, D. G. (2003), *Plasma Waves*, 2 edn, IoP, Institute of Physics Publishing, Bristol.

Symon, K. R. (1960), *Mechanics*, Addison-Wesley, Reading, Massachusetts.

Takahashi, K., Oishi, T., Shimomai, K., Hayashi, Y. & Nishino, S. (1998), 'Analyses of attractive forces between particles in Coulomb crystal of dusty plasmas by optical manipulations', *Phys. Rev. E* **58**, 7805–7811.

Taniuti, T. & Wei, C.-C. (1968), 'Reductive perturbation method in nonlinear wave propagation. I', *J. Phys. Soc. Jpn.* **24**, 941–946.

Taylor, G. I. (1931), 'Effect of variation in density on the stability of superposed streams of fluid', *Proc. Roy. Soc. London Ser. A* **132**, 499–523.

Taylor, R. J., MacKenzie, K. R. & Ikezi, H. (1972), 'A large double plasma device for plasma beam and wave studies', *Rev. Sci. Instr.* **43**, 1675–1678.

Thompson, W. B. (1962), *An Introduction to Plasma Physics*, Pergamon Press, Oxford, England.

Thornhill, S. G. & ter Haar, D. (1978), 'Langmuir turbulence and modulational instability', *Phys. Reports* **43**, 43–99.

Timofeev, A. V. & Shvilkin, B. N. (1976), 'Drift-dissipative instability of an inhomogeneous plasma in a magnetic field', *Sov. Phys. Usp.* **19**, 149–168.

Tisza, L. (2003), 'Remembering Eugene Wigner and pondering his legacy', *Europhys. News* **34**, 58–61.

Tjulin, A., Eriksson, A. I. & André, M. (2000), 'Physical interpretation of the Padé approximation of the plasma dispersion function', *J. Plasma Phys.* **64**(3), 287–296.

Treumann, R. A., LaBelle, J. & Pottelette, R. (1991), 'Plasma diffusion at the magnetopause: The case of lower hybrid drift waves', *J. Geophys. Res.: Space Phys.* **96**, 16009–16013.

Trivelpiece, A. W. & Gould, R. W. (1959), 'Space charge waves in cylindrical plasma colums', *J. Appl. Phys.* **30**, 1784–1793.

Trocheris, M. G. (1949), 'Electrodynamics in rotating frames of reference', *Philosophical Magazine* **40**, 1143–1154.

Trubnikov, B. A. (1965), Particle interactions in fully ionized plasmas, *in* M. A. Leontovich, ed., 'Reviews of Plasma Physics', Vol. 1, Consultants Bureau, New York, pp. 105–204.

Trulsen, J. (1971), 'Cyclotron radiation in hot magnetoplasmas', *J. Plasma Phys.* **6**, 367–400.

Trulsen, J. & Fejer, J. A. (1970), 'Radiation from a charged particle in a magnetoplasma', *J. Plasma Phys.* **4**, 825–841.

Tsunoda, S. I., Doveil, F. & Malmberg, J. H. (1987), 'Experimental test of the quasi-linear theory of the interaction between a weak warm electron-beam and a spectrum of waves', *Phys. Rev. Lett.* **58**, 1112–1115.

Turikov, V. (1984), 'Electron phase space holes as localized BGK solutions', *Phys. Scripta* **30**, 73.

Vaĭnshteĭn, L. A. (1976), 'Propagation of pulses', *Usp. Fiz. Nauk.* **118**, 339–367. See also (1976) Sov. Phys. Usp. **19**, 189-205.

Vakman, D. E. & Vaĭnshteĭn, L. A. (1977), 'Amplitude, phase, frequency - fundamental concepts of oscillation theory', *Usp. Fiz. Nauk.* **123**, 657–682. See also (1977) Sov. Phys. Usp. **20**, 1002-1016.

van Kampen, N. G. (1955), 'On the theory of stationary waves in plasmas', *Physica* **21**, 949–963.

van Kampen, N. G. & Felderhof, B. U. (1967), *Theoretical Methods in Plasma Physics*, North Holland Publishing Company, Amsterdam.

Vasyliūnas, V. V. (2001), 'Electric fields and plasma flow: what drives what?', *Geophys. Res. Lett.* **28**, 2177–2180.

Vedenov, A. A. (1968), *Theory of Turbulent Plasma*, Ilife, London.

Verdon, A. L., Cairns, I. H., Melrose, D. B. & Robinson, P. A. (2009), 'Warm electromagnetic lower hybrid wave dispersion relation', *Phys. Plasmas* **16**, 052105.

Vierinen, J., Gustavsson, B., Hysell, D. L., Sulzer, M. P., Perillat, P. & Kudeki, E. (2017), 'Radar observations of thermal plasma oscillations in the ionosphere', *Geophys. Res. Lett.* **44**, 5301–5307.

Vlasov, A. (1945), 'On the kinetic theory of an assembly of particles with collective interactions', *J. Phys. USSR* **9**, 25–40.

Vranješ, J. & Jovanović, D. (1995), 'Curvature effects on drift waves', *Phys. Scripta* **52**, 708–709.

Vranješ, J., Jovanović, D. & Shukla, P. K. (1998), 'Drift waves in plasmas with sheared flows', *Phys. Plasmas* **5**, 4300–4304.

Walker, R. J. & Russell, C. T. (1995), *Introduction to Space Physics*, Cambridge University Press, Cambridge, UK, chapter 6: Solar-wind interactions with magnetized planets, pp. 164–182.

Wang, C.-L., Joyce, G. & Nicholson, D. R. (1981), 'Debye shielding of a moving test charge in plasma', *J. Plasma Phys.* **25**(2), 225–231.

Washimi, H. & Karpman, V. I. (1976), 'Ponderomotive force of a high-frequency electromagnetic-field in a dispersive medium', *Zh. Eksp. Teor. Fiz.* **71**, 1010–1016. See also Soviet Phys. JETP, **44**, 528.

Washimi, H. & Taniuti, T. (1968), 'Propagation of ion-acoustic solitary waves of small amplitude', *Phys. Rev. Lett.* **17**, 996–998.

Watanabe, S. (1978), 'Soliton and generation of tail in nonlinear dispersive media with weak dissipation', *J. Phys. Soc. Jpn.* **45**, 276–282.

Watanabe, S., Miyakawa, M. & Yajima, N. (1979), 'Method of conservation laws for solving nonlinear Schrödinger equation', *J. Phys. Soc. Jpn.* **46**, 1653–1659.

Watanabe, S. & Yajima, N. (1984), 'K-dV soliton in inhomogeneous system', *J. Phys. Soc. Jpn.* **53**, 3325–3334.

Weber, W. J., Armstrong, R. J. & Trulsen, J. (1979), 'Ion-beam diagnostics by means of an electron-saturated Langmuir probe', *J. Appl. Phys.* **50**, 4545–4549.

Weiland, J. (1992), Low frequency modes associated with drift motions in inhomogeneous plasmas, Technical Report CTH–IEFT/PP–1992–17, Chalmers Technical University.

Weiland, J., Ichikawa, Y. H. & Wilhelmsson, H. (1978), 'A pertubation expansion for the nonlinear Schrödinger equation with applications to the influence of nonlinear Landau damping', *Phys. Scripta* **17**, 517–522.

Weiland, J. & Wilhelmsson, H. (1977), *Coherent Non-linear Interaction of Waves in Plasmas*, Vol. 88 of *International Series in Natural Philosophy*, Pergamon Press, Oxford.

Whitham, G. B. (1974), *Linear and Nonlinear Waves*, John Wiley & Sons, New York.

Wilhelmsson, H. (2000), *Fusion. A Voyage Through the Plasma Universe*, Institute of Physics Publishing, Bristol and Philadelphia.

Woan, G. (2000), *The Cambridge Handbook of Physics Formulas*, Cambridge University Press, Cambridge, UK.

Wong, A., Motley, R. W. & D'Angelo, N. (1964), 'Landau damping of ion acoustic waves in highly ionized plasmas', *Phys. Rev.* **133**, A436–A442.

Wong, A. Y. & Quon, B. H. (1975), 'Spatial collapse of beam-driven plasma waves', *Phys. Rev. Lett.* **34**, 1499–1502.

Yeh, C. & Casey, K. F. (1966), 'Reflection and transmission of electromagnetic waves by a moving dielectric slab', *Phys. Rev.* **144**, 665–669.

Yeh, K. C. & Liu, C. H. (1972), *Theory of Ionospheric Waves*, Vol. 17 of *International Geophysics Series*, Academic Press, New York and London.

Yoshida, Z. & Giga, Y. (1990), 'Remarks on spectra of operator rot', *Mathematische Zeitschr.* **204**, 235–245.

Yoshizawa, A., Itoh, S.-I. & Itoh, K. (2003), *Plasma and Fluid Turbulence. Theory and Modelling*, Plasma Physics, Institute of Physics, Bristol and Philadelphia.

Zabusky, N. J. (2005), 'Fermi-Pasta-Ulam, solitons and the fabric of nonlinear and computational science: History, synergetics, and visiometrics', *Chaos: Interdisciplinary J.. Nonlinear Sci.* **15**, 015102.

Zabusky, N. J. & Kruskal, M. D. (1965), 'Interaction of "Solitons" in a collisionless plasma and the recurrence of initial states', *Phys. Rev. Lett.* **15**, 240–243.

Zakharov, V. E. (1972), 'Collapse of Langmuir waves', *Zh. Eksp. Teor. Fiz.* **62**, 1745–1759. See also (1972) Sov. Phys.-JETP, **35**, 908-914.

Zakharov, V. E. & Kuznetsov, E. A. (1974), 'Three–dimensional solitons', *Zh. Eksp. Teor. Fiz.* **66**, 594–597. See also Sov. Phys.-JETP, **39**, 285-286.

Zakharov, V. E. & Shabat, A. B. (1972), 'Exact theory of two-dimensional self focusing and one-dimensional modulation of waves in nonlinear media.', *Sov. Phys. JETP* **34**, 62–69.

Index

Milton Keynes UK
Ingram Content Group UK Ltd.
UKHW051909111001
449569UK00027B/1409

9 781032 236421

Milton Keynes UK
Ingram Content Group UK Ltd.
UKHW031138141024
449569UK00024B/1253

9 781138 490611